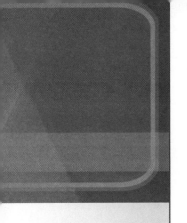

Intermediate Algebra

Intermediate Algebra

Functions and Authentic Applications

Second Edition

Jay Lehmann

Pearson Education, Inc., Upper Saddle River, New Jersey 07458

CIP Data on file at Library of Congress.

Project Manager: Mary Beckwith
Editor in Chief: Christine Hoag
Production Editor: Brittney Corrigan-McElroy
Senior Managing Editor: Linda Mihatov Behrens
Executive Managing Editor: Kathleen Schiaparelli
Vice President/Director of Production and Manufacturing: David W. Riccardi
Media Project Manager, Developmental Math: Audra J. Walsh
Assistant Managing Editor, Math Media Production: John Matthews
Assistant Manufacturing Manager/Buyer: Michael Bell
Manufacturing Manager: Trudy Pisciotti
Executive Marketing Manager: Eilish Collins Main
Marketing Project Manager: Barbara Herbst
Marketing Assistant: Annett Uebel
Development Editor: Deena Cloud
Editor in Chief, Development: Carol Trueheart
Art Director: Maureen Eide
Interior Designer/Cover Designer: John Christiana
Cover Image Specialist: Karen Sanatar
Art Editor: Thomas Benfatti
Creative Director: Carole Anson
Director of Creative Services: Paul Belfanti
Cover Photo: Roger Allyn Lee/Superstock
Art Studio: Laserwords

© 2004, 2000 by Pearson Education, Inc.
Pearson Education, Inc.
Upper Saddle River, New Jersey 07458

Printed in the United States of America
10 9 8 7 6 5 4 3

ISBN 0-13-101060-3

Pearson Education LTD.
Pearson Education Australia PTY, Limited
Pearson Education Singapore, Pte. Ltd.
Pearson Education North Asia Ltd.
Pearson Education Canada, Ltd.
Pearson Education de Mexico, S.A. de C.V.
Pearson Education–Japan
Pearson Education Malaysia, Pte. Ltd.

To Keri, who believed in this work
when I doubted and who celebrated
it with me when I remembered
—preliminary edition

And to Dylan, who is a continual
reminder that new beginnings
are always possible—first edition

And to Keri, whose committment
to spiritual growth and
fullfillment has inspired me
to do the same—second edition

Contents

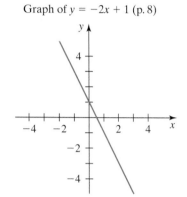

Graph of $y = -2x + 1$ (p. 8)

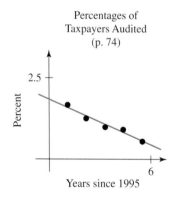

Percentages of Taxpayers Audited (p. 74)

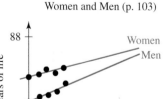

U. S. Life Expectancies of Women and Men (p. 103)

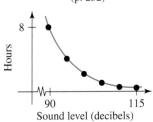

Costs of Television Ad
Slots During the
Super Bowl (p. 172)

Safe Exposure Time
(to Music at Rock Concerts)
(p. 232)

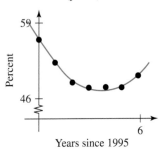

Percentages of Americans
Who Are Pro-Choice
(p. 293)

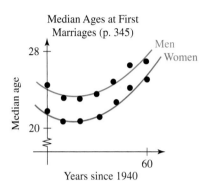

Median Ages at First Marriages (p. 345)

8 RATIONAL FUNCTIONS *357*

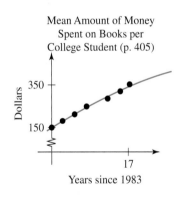

Mean Amount of Money Spent on Books per College Student (p. 405)

9 RADICAL FUNCTIONS *428*

Percentages of Births "in Spite of Contraception" (p. 470)

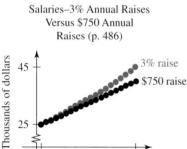

Salaries–3% Annual Raises
Versus $750 Annual
Raises (p. 486)

10 MODELING WITH SEQUENCES AND SERIES *476*

11 ADDITIONAL TOPICS *509*

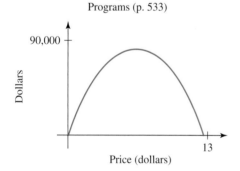

Daily Revenue from
Selling Baseball
Programs (p. 533)

A REVIEWING PREREQUISITE MATERIAL *559*

Preface

The question of common sense is always 'What is it good for?'—a question which would abolish the rose and be answered triumphantly by the cabbage. —James Russell Lowell, 1819–1891, American poet, editor.

The above quote suggests James Russell Lowell must have taken intermediate algebra. How many times have your students asked, "What is it good for?" After years of responding, "You'll find out in the next course," I began a five-year quest to develop a more satisfying and substantial response to my students' query.

Curve Fitting Approach

Although there are many ways to center an intermediate algebra course around authentic applications, I chose a curve fitting approach for several reasons. A curve fitting approach

- allows great flexibility in choosing interesting, authentic, current situations to model.
- emphasizes concepts related to functions in a natural, substantial way.
- deepens students' understanding of functions because it requires them to describe functions graphically, numerically, symbolically, and verbally.
- serves as a glue to hold together the many diverse topics of a typical intermediate algebra course.

To curve fit, students learn the following four-step modeling process:

1. Examine the data set to determine which type of model, if any, to use.
2. Find an equation for the model.
3. Verify that the model fits the data.
4. Use the model to make estimates and predictions.

This four-step process weaves together topics that are crucial to the course. Students must notice numerical patterns from data displayed in tables, recognize graphical patterns in scattergrams, find formulas of functions, graph functions, evaluate functions, and solve equations.

Curve fitting not only fosters cohesiveness within a chapter, it also creates a parallel theme for each chapter that introduces and discusses another function. This structure enhances students' abilities to observe similarities and differences between fundamental functions such as linear functions, exponential functions, and quadratic functions. Specifically, this structure has exponential functions directly follow linear functions so that students can easily observe the dual nature of these two functions (comparing the slope addition property to the base multiplier property). Additionally, logarithms can then be studied relatively early in the course, when students still have plenty of energy to learn about an unfamiliar function.

With many intermediate algebra texts, the first third of the course reviews topics typically taught in elementary algebra. For this reason, some students find it hard to stay interested in the course because they have "seen it all before." To address this issue, there is content new to most students in Sections 1.1, 1.4, and 1.6 and throughout most of Chapters 2–11. Section 1.1 sets the tone that this course will be different, interesting, alive, and relevant, inviting students' creativity into the classroom.

Response from math departments using the first edition have been enthusiastic. They feel that their students have learned a great deal from the text. Their students have found the first edition easy to read and the applications very interesting. Instructors have noticed a quantum leap forward in students' participation, morale, and success in the course.

Math departments' enthusiasm notwithstanding, there were requests from reviewers and class testers. In response, I have reorganized content and added some topics for the second edition.

NEW IN THE SECOND EDITION

Data Sets

About 90% of the data sets have been updated or replaced. Data sets were replaced because of changes in trends or because new data sets were more interesting and/or better illustrated concepts. The number of application exercises has been increased by about 15%.

New Sections

- Many reviewers wanted more discussion about polynomials of degree greater than two. To meet this need, the new Section 6.4 discusses how to factor such polynomials.
- There is a new Section 8.7 on direct variation and inverse variation. These topics are discussed within the context of curve fitting, a departure from most other texts that work with models but don't use data.
- The new Section 11.5 includes value, interest, and revenue applications. Some reviewers want these types of word problems included because they require a different form of deductive reasoning than curve fitting applications do. However, the problems in this section are distinct from those in other texts because students use the equations for authentic purposes that lend themselves to concepts that are relevant to functions.
- A brief Section A.3 on absolute value has been added.

Revised or Expanded Sections

- Section 5.1 on inverse functions has been rewritten. The section has widened its scope from the inverse of a linear function to the inverse of any invertible function. The inverse of a nonlinear invertible function is described in terms of a table and a graph. The section explains how to graph an inverse function by reflecting the graph of an invertible function across the line $y = x$, setting the stage for graphing logarithmic functions later in the chapter.
- The scope of Sections 6.2, 6.4, and 6.5 have been expanded from quadratic expressions and equations to polynomial expressions and equations.
- Section 8.4 on complex rational expressions has been expanded to include two methods to simplify these expressions.
- The cumulative review sections have been greatly expanded. Some of the chapter review sections have also been expanded.
- To offer students more technological support, the number of graphing calculator screen dumps has been increased.

Reorganization

- Natural logarithms has been moved from Chapter 11 to Section 5.6 so that all types of logarithms discussed in the text are in one chapter.
- Most reviewers do not teach sums and differences of cubes. So, this topic has been moved from Chapter 8 to the Additional Topics Chapter 11.
- Factoring by grouping has been moved from Chapter 8 to Section 6.4 to join other factoring techniques discussed in Chapter 6.

Explorations

Many instructors find the explorations to be a valuable collaborative learning tool and they requested that more explorations be added. To meet this need and also offer instructors more choices, the number of Explorations has been increased by about 50%.

Labs

Several reviewers expressed interest in having students perform research on the Internet. In response, web site addresses have been added to the Used Car Lab and the Sports Lab. The Temperature Scales Web Lab and the China and India Populations Web Lab have also been added.

CONTINUED FROM THE FIRST EDITION

Technology

To make effective use of students' time in and out of class, the text assumes that students have access to technology such as the TI-83 graphing calculator in the classroom and at home. Technology like this allows students to create scattergrams and verify the fit of a model quickly and accurately.

The text also supports instructors in holding students accountable for all aspects of the course without the aid of technology. For example, although students are encouraged to use technology to verify a model's fit to some data, students must also be able to sketch graphs by hand. Likewise, although a few exercises suggest that a student use technology to find a regression equation, the intent of most modeling exercises is that students use pencil and paper to derive a formula for a model by selecting two or three points (depending on the type of model). For such a pencil and paper exercise, a regression equation is included in the answers section because it is often difficult or impossible to anticipate which points a student will choose in order to find a reasonable equation.

After students solve symbolic problems by hand, technology allows them to reinforce their numerical and graphical intuition because they use graphing calculator tables or graphs to verify their work. Technology also empowers students to explore mathematical concepts and skills that would be too time consuming or difficult to do by hand.

Explorations

Each section of the text contains one or two explorations that support student investigation of a concept or skill. These explorations have been selected from hundreds that other instructors and I have class tested during the past five years.

An exploration can be used as a collaborative activity during class time or as part of a homework assignment. Some explorations lead students to reflect on concepts and skills of the current section. Other explorations are directed-discovery activities that introduce key concepts and skills to be discussed in the next section. All explorations are optional, so an instructor is free to assign as many explorations as the course schedule allows.

For students who enjoy learning by hands-on activities, explorations open the door for them to see the wonder and beauty of mathematics. It empowers all students to become active explorers of mathematics.

Conceptual Development

As a complement to the explorations, many examples show how to investigate concepts. They also illustrate how to perform skills of the course. This way, students learn *why* they perform skills to solve problems as well as how to solve the problems.

Homework sets have been designed to give an instructor maximum flexibility. Instructors can choose from a wide variety of exercises to balance conceptual-based, skill-based, and application-based exercises.

Key Points

Each section contains a "Key Points" feature that summarizes the definitions, properties, concepts, and skills of that section. This gives students section-by-section support in completing homework assignments. Students can also refer to this feature when preparing for quizzes and tests. A section's key points can serve as a prompt for students to reread portions of the section that address concepts that the students would like to consider further. Each chapter also has a "Key Points of This Chapter" feature.

Tips on Succeeding in This Course

Many sections close with tips that are intended to help students succeed in the course. By sprinkling these tips throughout the text, students can experiment with and assimilate new behaviors over time.

Taking It to the Lab Sections

Lab assignments have been included at the end of most chapters to increase students' understanding of concepts and the scientific method. These labs reinforce the idea that mathematics is useful. They are also an excellent avenue for more in-depth writing assignments.

Many of the labs involve curve fitting data obtained by physical experiments. Students can use data provided in the labs or collect data of their own. The "Topic of Your Choice" labs require students to choose a topic, research the topic, collect data, and analyze this data.

Two or more of these labs can be assigned at the same time to form a project assignment. Project papers include opening and summary paragraphs, typed responses in paragraph form, carefully drawn graphs and tables, and an attractive cover page. During the last three weeks of the course, each of my students uses a linear function, an exponential function, and a quadratic function to model and analyze three situations that they have chosen. This project assignment serves as an excellent review for an important portion of the course.

Margin Comments

Margin "Notes" support ideas in adjacent paragraphs. "Graphing Calculator" margin comments give support to students who are using graphing calculators.

Additional Topics Chapter

Not all topics typically taught in an intermediate algebra course can be connected with a curve fitting approach at the appropriate level. Through extensive polling of instructors across the country, I have assembled in Chapter 11 a collection of such topics that were "musts." Each section contains a "Section Quiz" feature. The union of the section quizzes can be used as a set of review exercises for Chapter 11. For instructors who wish to "cut and paste" sections from Chapter 11 into earlier chapters, these quizzes can be appended to the appropriate chapter review exercises.

Appendix A: Reviewing Prerequisite Material

Appendix A has been included to remind students of important topics typically addressed in an elementary algebra course. Examples and exercises are included in each section.

Appendix B: Using a TI-82 or TI-83 Graphing Calculator

Appendix B contains step-by-step instructions for the TI-82 and TI-83 graphing calculators. A subset of this appendix can serve as a tutorial early in the course. Additionally, when a calculator skill is first required in the text, there is a margin comment referring students to the appropriate section in Appendix B.

RESOURCES FOR INSTRUCTORS

Instructor's Resource Manual

This manual contains suggestions for course pacing and homework assignments. It also discusses how to incorporate technology and how to structure lab and project assignments.

There are section-by-section suggestions for lectures and for the explorations. The manual also contains some additional explorations to give instructors more options. The manual includes chapter tests. Because these tests contain hard-to-find data sets, they can be a useful resource even for instructors who prefer to create their own tests.

Instructor's Solutions Manual

The manual includes solutions to the even-numbered exercises in the homework sections in the text.

MathPro 5 (Instructor Version)

- Online, customizable tutorial, diagnostic, and assessment program for anytime, anywhere tutorial support.
- Diagnostic option identifies student skills, provides individual learning plan, and supports the student with tutorial reinforcement.
- Course management tracking of tutorial and testing activity.

MathPro 4 (Network Version)

- Enables instructors to create either customized or algorithmically generated practice quizzes.
- Network based reports and summaries for a class or student or for cumulative or select scores.

TestGen with Quiz Master CD-ROM Windows/Macintosh

RESOURCES FOR STUDENTS

MathPro 5 (Student Version)

- Online, customizable tutorial, diagnostic, and assessment software.
- Algorithmically generated exercises.

MathPro 4 (Student Version)

- Available on CD-ROM for stand alone use or can be networked in the school laboratory.
- Algorithmically generated exercises.

Key Topics CD Videos

Prentice Hall Tutor Center

- Obtain help for examples and exercises in *Lehmann, Intermediate Algebra: Functions and Authentic Applications, Second Edition* via toll free phone, fax, or email.
- Accessed via a registration number bundled with a new text or purchased separately with a used book. Visit http://www.prenhall.com/tutorcenter to learn more.

Student Solutions Manual

This manual contains solutions for the odd-numbered exercises in the Homework sections in the text.

Companion Website

The website contains additional resources for students and instructors.
   Companion website address
   www.prenhall.com/lehmann

Getting in Touch

I would greatly appreciate receiving your comments regarding this text. I would also like to hear about any data sets you would like to share. If you have questions regarding the text, please ask and I will respond. Finally, let me know if you or your department is interested in attending a workshop to discuss the text in greater detail.

Thank you for your interest in preserving the rose.

Jay Lehmann
MathnerdJay@aol.com

To the Student

You are about to embark on an exciting journey. In this course you will not only learn more about algebra, you will also learn how to apply algebra to describe and make predictions about authentic situations. This text contains data from hundreds of different kinds of situations. Most of the data have been collected from recent newspapers and Internet postings, so the information is current and usually of interest to the general public.

Working with true data will make mathematics more meaningful.   While working with data from true-to-life situations, you will learn the meaning of mathematical concepts. As a result, the concepts will be easier to learn, since they will be connected to familiar contexts. Second, you will see that almost any situation can be seen from a mathematical view. This will help you to understand the situation and make estimates and/or predictions.

Many of the problems you will explore in this course involve data collected during a scientific experiment or a census. The practical way to deal with such data sets is to use technology. So, a graphing calculator or computer system is required.

Applying mathematics to authentic situations is a lifelong skill.   In addition to working with data sets in this text, your instructor may assign some of the labs. Here you will collect data through experiment or research. This will give you a more complete picture of how you can use the approaches presented in this text in everyday life and possibly in your lifelong careers.

Hands-on explorations are rewarding and fun.   Learning is similar to exercising. The more often you work out, the stronger you will be. The more opportunities you take to make an intuitive leap or discover a pattern, the stronger your abilities as a critical thinker will be.

This text contains explorations that will allow you to *discover* concepts, rather than hear or read about them. Most explorations contain step-by-step instructions that lead you toward discovering new concepts.

You will also find that discovering a concept greatly improves your chances of remembering it. Since discovering a concept is exciting, it is more likely to leave a lasting impression on you. Over the years, students have remarked to me time and time again that they never dreamed that learning math could be so much fun.

Working in a team will make learning mathematics comfortable.   Discovering concepts with others is fun and rewarding. With this in mind, your instructor may have you work with a team of students to complete the explorations. As a team you will want to share your discoveries with each other. When you do, keep in mind that the student who does the explaining is getting as much, if not more, out of the exercise as the listener(s). To explain a mathematical concept, you must be able to describe it clearly and concisely, thus sharpening your understanding of the concept.

This text contains special features to support you in succeeding.   Each section contains a "Key Points" feature to help you learn, review, and retain concepts and skills addressed in the section. After reading a section, look through the Key Points and decide whether you understand the concepts and skills described. If you do not understand a Key Point, then it is time to reread the portion of the section that addresses that point. This feature can also help you prepare for quizzes and exams. However, there is no substitute for regular, careful reading of the text and your notes, and doing lots of the exercises.

Each section also contains a "Tips on Succeeding in This Course" feature. These tips are meant to inspire you to try new strategies to help you succeed in this course and future courses. Some tips may remind you of strategies that you have used successfully in the past, but have forgotten. If you browse through all of the tips early in the course you can take advantage of as many of them as you wish. Then, as you progress through the text, you'll be reminded of your favorite strategies once more.

I have also included a review of key concepts and skills from elementary algebra in Appendix A. Before the course begins or shortly after it starts, consider reading this appendix and completing the exercises. If you need more review, refer to an elementary algebra text or ask your instructor, a tutor, or a friend.

Feel free to contact me. It is my pleasure to read and respond to e-mails from students who are using my text. If you have any questions or comments about the text, feel free to contact me.

Jay Lehmann
My e-mail: MathnerdJay@aol.com
Companion website: www.prenhall.com/lehmann

ACKNOWLEDGMENTS

I am extremely grateful to my wife, Keri, who shouldered so much of our family's needs while I worked on this edition. Her excitement and deepening commitment to lead a spiritual life has inspired me to do the same, which has added liveliness and authenticity to this work. And to our preschooler, Dylan, who continues to teach me of joyous responsibility and that there is no limit to how deep love can be.

I would also like to acknowledge the following faculty at the College of San Mateo. Special thanks go to Robert Biagini-Komas for his support in giving me insightful feedback about the text through countless phone calls, e-mails, my more-than-occasional office visits from across the hall, and many entertaining conversations at Ausiello's about math, music, and life. Thanks to Cheryl Gregory and Bob Hasson for helpful e-mails sent at odd hours of the night.

I would also like to acknowledge the following people at Prentice Hall. I am very grateful to Karin Wagner, for overseeing both the preliminary and first editions and hanging in there with me as I found my footing while making significant changes from one to the other. To Mary Beckwith, for taking the time to understand my vision of this edition and thinking outside of the box for ways to come closer to that vision. Finally, to Deena Cloud, for her optimism that helped carry this project and for always keeping students' needs in the forefront of my mind.

I would also like to thank the following reviewers whose thorough, thoughtful comments helped me sculpt this text to its current form.

Gwen Autin, *Southeastern Louisiana University*
Nancy Brien, *Middle Tennessee State University*
William P. Fox, *Francis Marion University*
Kathryn M. Gundersen, *Three Rivers Community College*
Diane Mathios, *De Anza College*
Jane E. Mays, *Grand Valley State University*
Scott McDaniel, *Middle Tennessee State University*
Jason L. Miner, *Santa Barbara City College*
Ernest Palmer, *Grand Valley State University*
Jody Rooney, *Jackson Community College*
John Szeto, *Southeastern Louisiana University*
Janet E. Teeguarden, *Ivy Tech State College*
Ollie Vignes, *Southeastern Louisiana University*

I would also like to thank the reviewers of the first edition whose comments also contributed to this second edition.

Joe Berland, *Chabot College*
Laurie Burton, *Central Washington University*
Mark Clark, *Palomar College*
Joseph Ediger, *Portland State University*
Tracey Hoy, *College of Lake County*
Joyce Hunington, *Walla Walla Community College*
Kathy Kopelousos, *Lewis and Clark Community College*
Peg McPartland, *Golden Gate University*
Terrie Nichols, *Cuyamaca College*
Joyce Smith, *Chattanooga State Technical Community College*
Marguerite Smith, *Merced College*
Charles Stevens, *Skgit Valley College*
Mary Wilson, *Austin Community College*

A special thanks goes to the following class testers of the first edition who widened my perspective on how the text could be written to accommodate various styles of passionate teaching.

Ignacio Alarcon, *Santa Barbara City College*
James Edmondson, *Santa Barbara City College*
Sharareh Massooman, *Santa Barbara City College*
Ernest Palmer, *Grand Valley State University*

Linear Functions

I see a certain order in the universe and math is one way of making it visible.

—*May Sarton, As We Are Now, 1973.*

1.1 USING QUALITATIVE GRAPHS TO DESCRIBE SITUATIONS

Objectives

▸ Describe situations using qualitative graphs.

▸ Identify independent variables and dependent variables.

▸ Know the meaning of an *intercept* of a curve.

▸ Identify increasing curves and decreasing curves.

How often have you seen or heard predictions? Perhaps you have read a prediction about world population or the number of Internet users. Or maybe you have seen a news report about the estimated age of an artifact discovered at an archeological site. It is virtually impossible to find a business magazine that does not make predictions about the economy. Many of these estimates and predictions are made using mathematics.

A major objective of this text is to help you view the world in a mathematical manner. This viewpoint will allow you to recognize important patterns—patterns that will enable you to make estimates and predictions like the ones mentioned above.

To begin, in this section you will learn how to describe situations using a special type of graph called a **qualitative graph**. This is a graph that has no scaling on the axes.

Note

"Scaling" consists of the tick marks and their numbers on the axes.

Example 1 Interpreting a Graph

Since 1985, Michael Jordan has endorsed a successful line of shoes, called Air Jordan®. Let p represent the retail price of Air Jordan shoes and t represent the number of years since 1985. (For example, $t = 1$ represents the year 1986.) The prices of the shoes are described by the qualitative graph displayed in Fig. 1. What does the graph tell us?

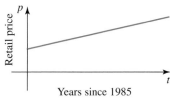

Figure 1 Retail price of Air Jordan shoes

Solution

The graph (or curve) tells us that the retail price of Air Jordans has steadily increased. The curve is said to be **linear** because it forms a straight line. ∎

Since the retail price of Air Jordans changes from year to year, we can say that the price p *depends* on the year. Due to inflation and to increasing popularity of the

shoes, the price increases over time. Note that the year does *not* depend on the price. Raising or lowering the price of the shoes has no effect on the passage of time. Time is *independent* of the price. Because p depends on t, we call p the **dependent variable**, and since time is independent of price, we call t the **independent variable**.

Note that in Fig. 1 we let the horizontal axis be the t-axis and the vertical axis be the p-axis. We usually match the horizontal axis with the independent variable and the vertical axis with the dependent variable (see Fig. 2).

Note

For a review of constants and variables, see Section A.5.

Dependent
variable

Independent
variable

Figure 2 Match the vertical axis with the dependent variable and the horizontal axis with the independent variable

Example 2   Interpreting a Graph

Let A represent the average age when men first married and let t represent the number of years since 1900. In Fig. 3, the graph describes the relationship between the variables t and A for the years between 1900 and the present. What does the graph tell us?

Solution

The graph tells us that the average age when men first married decreased each year for a while and then increased each year after that. We say that the curve sketched in Fig. 3 is **quadratic** because it has the shape of a **parabola**.   ∎

In Fig. 3, note that the curve and the A-axis intersect. The point of intersection is an A-**intercept**. Two more examples of intercepts are shown in Fig. 4.

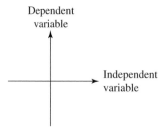

Figure 3 The average age when men first married

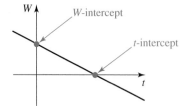

Figure 4 Intercepts of a line

In Examples 3–5, we sketch qualitative graphs that describe given situations.

Example 3   Sketching a Qualitative Graph

Let *C* represent the cost of a 30-second ad slot during the Super Bowl at *t* years since 1987. Each year, the cost has increased by more than the previous increase. Sketch a qualitative graph that describes the relationship between *C* and *t*.

Solution

Since the cost of an ad varies according to the year, the variable *C* is the dependent variable and *t* is the independent variable. We label the axes accordingly (see Fig. 5). Because ads were not free in 1987 ($t = 0$), the *C*-intercept is above the origin. Since the costs are increasing, we sketch an increasing curve. Since each increase is more than the last increase, the curve should "bend" upward.

Some **exponential curves** have shapes similar to the shape of the curve sketched in Fig. 5.   ■

Notice that the curve in Fig. 5 goes upward from left to right. We say the curve (and the graph) is **increasing**. If a curve goes downward from left to right, we say the curve (and the graph) is **decreasing** (see Figs. 6 and 7).

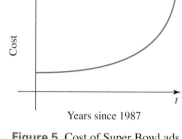

Figure 5 Cost of Super Bowl ads

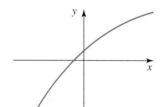

Figure 6 Increasing curve

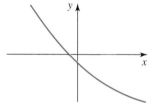

Figure 7 Decreasing curve

In this chapter and future chapters, we will discuss the curves mentioned in this section more thoroughly and use them to make predictions—linear curves in this chapter and Chapter 2, exponential curves in Chapters 4 and 5, and quadratic curves in Chapters 6 and 7. In Chapter 3 you will make predictions using two linear curves.

Example 4   Sketching a Qualitative Graph

Some hot coffee is poured into a cup at room temperature. Let *F* represent the temperature (in degrees Fahrenheit) of the coffee at *t* minutes since the coffee was poured. Sketch a qualitative graph that describes the relationship between the variables *t* and *F*.

Solution

Note that *F* depends on *t*, so we let the vertical axis be the *F*-axis and the horizontal axis be the *t*-axis (see Fig. 8). Since the coffee cools with time, the curve should be decreasing. Further, the curve should show that the drop in temperature during any minute is less than the drop in temperature in the previous minute (why?).

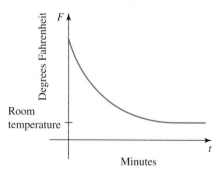

Figure 8 Temperature of a cup of coffee

The coffee's temperature will not go below room temperature, so the curve should eventually level off.   ■

Note that in each situation we have explored so far, the independent variable has represented time. Let's explore a situation where the independent variable stands for something else.

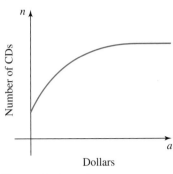

Figure 9 Spending on advertising, and CD sales

Note

| Quadrant II | Quadrant I |
| Quadrant III | Quadrant IV |

Example 5 Sketching a Qualitative Graph

Suppose that the latest Radiohead CD is about to be released. Let n represent the number of CDs that will be sold if a dollars are spent on advertising. Sketch a qualitative graph that describes the relationship between the variables a and n.

Solution

It makes sense that the number of CDs sold will be related to the amount of money spent on advertising, so n is the dependent variable and the n-axis is the vertical axis (see Fig. 9). Since a is the independent variable, the a-axis is the horizontal axis. Because n and a must both be nonnegative (why?), the qualitative curve is in *Quadrant I* (and one point of it is on the n-axis).

Even if no money is spent on advertising, some CDs will be sold. So the n-intercept should be above the origin. The more money spent on advertising, the greater the sales, so the curve should be increasing. There are only so many people, however, who would buy the CD no matter how much advertising is done, so the curve should eventually level off. ∎

EXPLORATION
Sketching a qualitative graph

A bathtub is filled with water and then the plug is pulled out. Let V represent the volume of water in the tub at t seconds after the plug is pulled out.

1. Which variable is the dependent variable? Which variable is the independent variable? Explain.

2. Sketch a qualitative graph that describes the relationship between V and t. Explain.

Taking It One Step Further

3. Carefully describe an experiment you could run to check that the shape of your curve is correct. In particular, explain how you could measure the volume of water at various times. Ask your instructor if you should run such an experiment.

EXPLORATION
Looking ahead: Connections between an equation and its graph

Graphing Calculator

See Sections B.3, B.4, and B.6.

Graphing Calculator

See Section B.5.

Graphing Calculator

See Section B.23.

In this problem you will work with the equation $y = 2x + 1$.

1. For the equation $y = 2x + 1$, substitute 1 for x and 3 for y. Notice that the new equation is a true statement. We say that the ordered pair, (1, 3), *satisfies* the equation $y = 2x + 1$.

2. Using ZDecimal, graph the equation $y = 2x + 1$.

3. Use TRACE to name the coordinates of three points that lie on the graph of $y = 2x + 1$. Do these coordinates satisfy the equation $y = 2x + 1$?

4. Pick three points that do *not* lie on the graph of $y = 2x + 1$. Do their ordered pairs satisfy the equation $y = 2x + 1$?

5. Based on your findings in steps 3 and 4, form a conjecture about which points satisfy the equation $y = 2x + 1$. The phrase "form a conjecture" means to make a guess, or form a theory. [**Hint**: Your conjecture should say something about the points that lie on or off the line.]

6. Test your conjecture. This means that you should check whether your conjecture is true for points other than the ones with which you have worked so far.

Tips on Succeeding in This Course

When responding to a question, use complete sentences, not phrases. Also, give reasons to support your response. For example, if you state that in Example 5 you believe that the advertising curve should be an increasing curve, explain *why* you think that this is true.

Key Points
OF THIS SECTION

A *qualitative graph* is a graph without scaling on the axes.

Qualitative graph

If two variables are related, the variable that depends on the other is the *dependent variable*. The other variable is the *independent variable*.

Dependent and independent variables

When sketching graphs, we match the vertical axis with the dependent variable and the horizontal axis with the independent variable.

The point(s) of intersection of a curve and an axis, say the t-axis, is called a t-*intercept*.

Intercept

If a curve goes upward from left to right, the curve is an *increasing curve*.

Increasing curve

If a curve goes downward from left to right, the curve is a *decreasing curve*.

Decreasing curve

HOMEWORK 1.1

 SSM Tutor Center   **MathPro 4** MathPro 4   **MathPro 5** MathPro 5   CDs/Videos   www

1. For parts a–d, the deer population in a forest is described during the years between 2000 and the present. Let p represent the deer population in the forest and t represent the number of years since 2000. Match each graph in Fig. 10 with each scenario.

   a. The population steadily decreased.

   b. The population steadily increased.

   c. The population remained steady.

   d. The population decreased for a while and then increased.

2. Match each graph in Fig. 11 with each scenario. Let A represent the amount of rain (in inches) that has fallen in t hours.

   a. The rain fell harder and harder.

   b. It rained softly, and then the rain stopped. After a while, it began raining hard.

   c. It rained hard, and then the rain stopped. After a while, it began raining softly.

   d. The rain fell more and more softly.

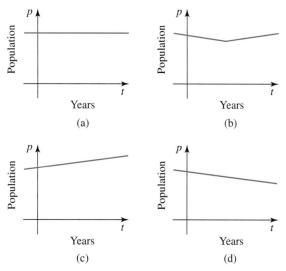

Figure 10 Exercise 1

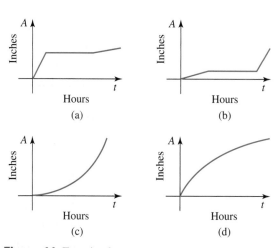

Figure 11 Exercise 2

Sketch a graph that shows the relationship between the variables defined in each of the following exercises. Justify your graph. Various "correct" graphs are possible.

3. Let *F* represent the temperature at a specific outdoor location at *t* hours after 6 A.M. on a specific day. (So *t* = 0 represents 6 A.M.)

*[**Note:** Remember to provide reasons to support your graphs.]*

4. A person left her home to go on a run. After a while, she stopped to rest. She then walked home. Let *s* represent the speed of this person at *t* minutes after she began her run.

5. An airplane flew from New York to Chicago. Let *h* represent the height of the airplane at *t* seconds after takeoff.

6. Let *h* represent the height of a tennis ball at *t* seconds after it was dropped. (Allow for bounces.)

7. The number of people undergoing laser eye surgery has increased since 1996. Let *n* represent the number of people who have undergone laser eye surgery during the year that is *t* years since 1996.

8. The percentage of smokers in the United States has steadily declined since 1965. Let *P* represent the percentage of smokers in the United States at *t* years since 1965.

9. The percentage of major firms that perform drug tests on employees and/or job applicants increased from 1987 to 1996 and decreased thereafter. Let *p* represent the percentage of firms that perform drug tests at *t* years since 1987.

10. The percentage of the U.S. population that is foreign born decreased from 1950 to 1970 and increased thereafter. Let *p* represent the percentage of the U.S. population that is foreign born at *t* years since 1950.

11. A commuter left home, drove toward her workplace, got some gas, then continued driving toward her workplace. Let *g* represent the amount of gas in the gas tank at *t* minutes after the commuter left home.

12. Let *L* represent the loudness of the sound of a foghorn that is *d* miles away from you.

13. Let *h* represent the height of a specific person at age *a* years.

14. Let *n* represent the number of species in existence and *d* represent the total amount of deforestation.

15. Let *T* represent the time it takes to complete a certain task after a person has consumed *x* ounces of beer within one hour.

16. Let *s* represent the maximum speed that a specific speedboat can travel when going downstream on a river whose current is *c* miles per hour.

17. At noon, a person began to breathe in. Let *V* represent the volume of air in this person's lungs at *t* seconds after noon.

18. Let *S* represent the speed of a car being driven on a level road when the end of the accelerator is *d* inches from the floor of the car.

19. Let *n* represent the number of times a person can lift a set of weights that weigh *w* pounds.

20. Let *p* represent the pulse rate (beats per minute) of a runner whose speed is *s* miles per hour.

21. Let *n* represent the number of people in the United States who would be willing to purchase a new Honda Civic® CX that is being sold for a price of *p* dollars.

22. Let *A* be the angle (in degrees) between the hour hand and the minute hand of a clock at *t* minutes from midnight.

23. Let *A* represent the area of a circle that has radius *r*.

24. In an experiment, some air is blown into a balloon. The balloon is then tied so that no air can enter or leave the balloon. If the balloon is squeezed, the pressure of the air inside the balloon increases. Let *P* represent the pressure of the air inside the balloon when its volume is *V* cubic inches.

25. Write a scenario to match each graph in Fig. 12. Refer to the variables *x* and *y* in your description.

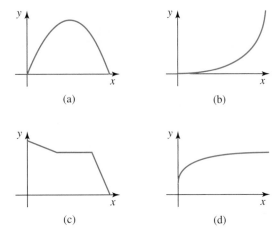

Figure 12 Exercise 25

26. Explain how to sketch a qualitative graph that describes a given situation.

1.2 SKETCHING GRAPHS OF LINEAR EQUATIONS

Objectives

▸ Know the meaning of *satisfy, solution,* and *solution set.*

▸ Know the meaning of a *graph* of an equation.

▸ Know which equations have graphs that are lines.

▸ Sketch graphs of linear equations.

▸ Find intercepts of linear equations.

▸ Sketch graphs of horizontal and vertical lines.

In Section 1.1 we worked with many types of curves. For the rest of this chapter we will focus on lines and an algebraic way to represent them.

To begin, consider the equation $y = 3x - 1$. Let's find y when $x = 2$:

$$y = 3(2) - 1$$
$$= 6 - 1$$
$$= 5$$

Note

Recall that we multiply before we subtract. See Section A.8 for a review of the order of operations.

So $y = 5$ when $x = 2$. This means that the equation $y = 3x - 1$ becomes a true statement when we substitute 2 for x and 5 for y:

$$y = 3x - 1$$
$$5 \stackrel{?}{=} 3(2) - 1$$
$$5 \stackrel{?}{=} 5$$
$$\text{true}$$

We say that the ordered pair $(2, 5)$ *satisfies* the equation $y = 3x - 1$. When using ordered pair notation such as $(2, 5)$, we list the value of x in the first position and the value of y in the second position.

DEFINITION *Solution of an equation*

We say that an ordered pair (a, b) is a **solution** of an equation of two variables if the ordered pair satisfies the equation. The **solution set** of an equation is the set of all ordered pairs that satisfy the equation.

Note

A set is a container. Much like a garbage can contains garbage, a solution set contains solutions.

For example, the ordered pair $(4, 11)$ is a solution of the equation $y = 3x - 1$, since $(4, 11)$ satisfies the equation:

$$y = 3x - 1$$
$$11 \stackrel{?}{=} 3(4) - 1$$
$$11 \stackrel{?}{=} 11$$
$$\text{true}$$

We can describe the solution set of an equation using a graph.

Example 1   Graphing an Equation

Sketch the graph of $y = -2x + 1$.

Solution

We begin by arbitrarily choosing the values 0, 1, and 2 to substitute for x:

$y = -2(0) + 1$	$y = -2(1) + 1$	$y = -2(2) + 1$
$= 0 + 1$	$= -2 + 1$	$= -4 + 1$
$= 1$	$= -1$	$= -3$
Solution: $(0, 1)$	Solution: $(1, -1)$	Solution: $(2, -3)$

Note

For a review of performing operations with signed numbers, see Section A.4.

The points $(-2, 5)$ and $(-1, 3)$ are also solutions. We organize our findings in Table 1. In Fig. 13 we plot the five solutions from Table 1. Note that a line contains the five points. It turns out that every point on the line is a solution to the equation $y = -2x + 1$.

Note

For a review of plotting points, see Section A.1.

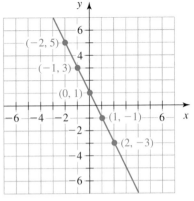

Figure 13 Graph of $y = -2x + 1$

Table 1 Solutions of $y = -2x + 1$	
x	y
-2	5
-1	3
0	1
1	-1
2	-3

It also turns out that points that do not lie on the line do not satisfy the equation. For example, the point $(3, 4)$ does not lie on the line and it does not satisfy the equation $y = -2x + 1$:

$$y = -2x + 1$$
$$4 \stackrel{?}{=} -2(3) + 1$$
$$4 \stackrel{?}{=} -6 + 1$$
$$4 \stackrel{?}{=} -5$$
$$\text{false}$$

Graphing Calculator

See Sections B.3, B.4, and B.6.

Graphing Calculator

To enter $y = -2x + 1$, press $\boxed{(-)}$ $\boxed{2}$ $\boxed{X,T,\Theta,n}$ $\boxed{+}$ $\boxed{1}$. The key $\boxed{-}$ is used for subtraction and the key $\boxed{(-)}$ is used for negative numbers as well as for taking opposites.

We can verify our graph in the ZDecimal window (see Fig. 14).

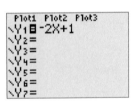

Figure 14 Use ZDecimal to verify our graph

When there is scaling on the axes, we refer to the x-axis and y-axis together as a **Cartesian coordinate system** or **coordinate system**.

Our observations made in Example 1 suggest the meaning of a graph.

Meaning of a Graph

All points that satisfy an equation form the **graph** of the equation. Points that do not satisfy an equation are not part of the graph.

In Example 1 we saw that the graph of the equation $y = -2x + 1$ is a line. Note that the equation $y = -2x + 1$ has the form $y = mx + b$ (where $m = -2$, $b = 1$).

Graphs of Equations That Can Be Put into $y = mx + b$ Form

If an equation can be put into the form

$$y = mx + b$$

where m and b are constants, then the graph of the equation is a line.

For example, the graphs of the equations

$$y = 3x - 7, \quad y = -2x, \quad \text{and} \quad y = 4$$

are lines.

Example 2 Graphing a Linear Equation

Sketch the graph of $4y - 8x + 12 = 0$.

Solution

First, we isolate y on one side of the equation.

$$
\begin{aligned}
4y - 8x + 12 &= 0 \\
4y - 8x + 12 + 8x &= 0 + 8x && \text{Add } 8x \text{ to both sides.} \\
4y + 12 &= 8x && \text{Combine like terms.} \\
4y + 12 - 12 &= 8x - 12 && \text{Subtract 12 from both sides.} \\
4y &= 8x - 12 && \text{Simplify.} \\
\frac{4y}{4} &= \frac{8x}{4} - \frac{12}{4} && \text{Divide both sides by 4.} \\
y &= 2x - 3 && \text{Reduce.}
\end{aligned}
$$

Note

For a review of solving literal equations, see Section A.10.

Since the equation $y = 2x - 3$ is of the form $y = mx + b$, we know that its graph is a line. Next, we calculate three solutions for $y = 2x - 3$ in Table 2. Although we can sketch a line using as few as two points, we plot a third point as a check. Then, we plot the points listed in Table 2 and sketch the line through them (see Fig. 15).

Note

If the third point is not in line with the other two, then we know that we have computed or plotted at least one of the points incorrectly. Plotting a fourth point usually will reveal the faulty point.

Table 2 Solutions of $y = 2x - 3$	
x	**y**
0	$2(0) - 3 = -3$
1	$2(1) - 3 = -1$
2	$2(2) - 3 = 1$

Figure 15 Graph of $y = 2x - 3$

We can view the graph using ZStandard followed by ZSquare as a partial check. This check will not reveal whether we isolated y correctly, because we entered $y = 2x - 3$ rather than the *original* equation $4y - 8x + 12 = 0$ (see Fig. 16).

Graphing Calculator

See Sections B.3, B.4, and B.6.

Figure 16 Graph of $y = 2x - 3$

Example 3 Using the Distributive Law to Help Graph a Linear Equation

Sketch the graph of $3(2y - 5) = 2x - 3 - 8x$.

Solution

First, we use the distributive law on the left side of the equation and combine like terms on the right side.

Note

For a review of the distributive law, see Section A.6.

$$3(2y - 5) = 2x - 3 - 8x$$

$$6y - 15 = -6x - 3 \qquad \text{Distributive law, combine like terms.}$$

$$6y - 15 + 15 = -6x - 3 + 15 \qquad \text{Add 15 to both sides.}$$

$$6y = -6x + 12 \qquad \text{Simplify.}$$

$$\frac{6y}{6} = \frac{-6x}{6} + \frac{12}{6} \qquad \text{Divide both sides by 6.}$$

$$y = -x + 2 \qquad \text{Reduce.}$$

Next, we calculate three solutions for $y = -x + 2$ in Table 3. Then we plot the points listed in Table 3 and sketch the line that contains them (see Fig. 17).

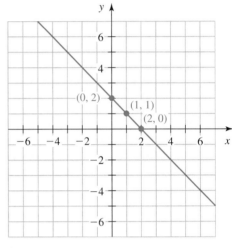

Table 3 Solutions of $y = -x + 2$	
x	y
0	$-(0) + 2 = 2$
1	$-(1) + 2 = 1$
2	$-(2) + 2 = 0$

Figure 17 Graph of $y = -x + 2$

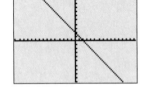

Figure 18 Graph of $y = -x + 2$

We can view the graph of $y = -x + 2$ using ZStandard followed by ZSquare as a partial check (see Fig. 18). ∎

Example 4 Graphing an Equation That Contains Fractions

Sketch the graph of $y = \dfrac{1}{2}x - 1$.

Solution

Note that $\dfrac{1}{2}$ times an even number is an integer. So, in Table 4 we use even number values for x to avoid fractional values for y. We plot the points and sketch a line that contains the points (see Fig. 19).

Table 4 Solutions of $y = \dfrac{1}{2}x - 1$	
x	y
0	$\dfrac{1}{2}(0) - 1 = -1$
2	$\dfrac{1}{2}(2) - 1 = 0$
4	$\dfrac{1}{2}(4) - 1 = 1$

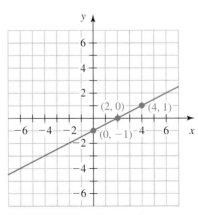

Figure 19 Graph of $y = \dfrac{1}{2}x - 1$

We can use ZDecimal to verify the graph (see Fig. 20).

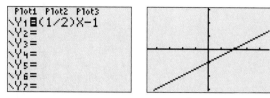

Figure 20 Verify the graph of $y = \dfrac{1}{2}x - 1$

■

Sometimes we find the *intercepts of an equation* (the intercepts of its graph) to help us sketch the graph of the equation. Since an x-intercept is on the x-axis, we know that its y-coordinate is 0 (see Fig. 21). Since a y-intercept is on the y-axis, we know that its x-coordinate is 0.

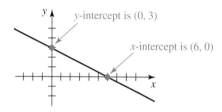

Figure 21 Intercepts of a line

Intercepts of an Equation

For an equation containing the variables x and y:

- To find the x-coordinate of each x-intercept, substitute 0 for y and solve for x.
- To find the y-coordinate of each y-intercept, substitute 0 for x and solve for y.

Example 5   Using Intercepts to Sketch a Graph

1. Find the x-intercept for the line $y = -2x + 4$.
2. Find the y-intercept for the line $y = -2x + 4$.
3. Sketch the graph of the line $y = -2x + 4$.

Note

Since the graph of the equation $y = -2x + 4$ is a line, we say "the line $y = -2x + 4$" as shorthand for "the graph of the equation $y = -2x + 4$."

Solution

1. To find the x-intercept, we substitute 0 for y and solve for x.

$$0 = -2x + 4$$
$$0 + 2x = -2x + 4 + 2x \qquad \text{Add } 2x \text{ to both sides.}$$
$$2x = 4 \qquad \text{Simplify.}$$
$$x = 2 \qquad \text{Divide both sides by 2.}$$

Note

For a review of solving linear equations in one variable, see Section A.9.

The x-intercept is $(2, 0)$.

2. To find the y-intercept, we substitute 0 for x and solve for y.

$$y = -2(0) + 4 = 4$$

The y-intercept is $(0, 4)$.

3. We list an additional solution of $y = -2x + 4$ in Table 5 and sketch the graph in Fig. 22.

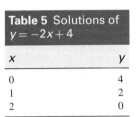

Table 5 Solutions of $y = -2x + 4$

x	y
0	4
1	2
2	0

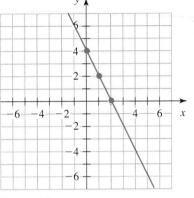

Figure 22 Graph of $y = -2x + 4$

We use ZStandard followed by ZSquare to verify our graph and the intercepts (see Fig. 23).

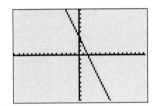

Figure 23 Verify the graph of $y = -2x + 4$

Example 6 Graphing a Vertical Line

Sketch the graph of $x = 3$.

Solution

Note that x must be 3, but y can have any value. Some solutions of $x = 3$ are listed in Table 6. We see in Fig. 24 that the graph of $x = 3$ is a vertical line.

Table 6 Solutions of $x = 3$

x	y
3	-2
3	-1
3	0
3	1
3	2

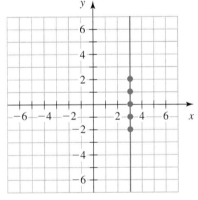

Figure 24 Graph of $x = 3$

Graphing Calculator

We cannot use a graphing calculator to graph $x = 3$. To graph an equation using a graphing calculator, the left side of the equation must be y and the right side must be a constant or an expression in terms of x.

Example 7 Graphing a Horizontal Line

Sketch the graph of $y = -5$.

Solution

Note that y must be -5, but x can have any value. Some solutions of $y = -5$ are listed in Table 7. We see in Fig. 25 that the graph of $y = -5$ is a horizontal line.

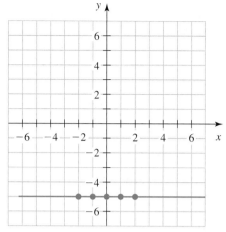

Table 7 Solutions of $y = -5$	
x	y
-2	-5
-1	-5
0	-5
1	-5
2	-5

Figure 25 Graph of $y = -5$

We can use ZStandard to verify the graph (see Fig. 26).

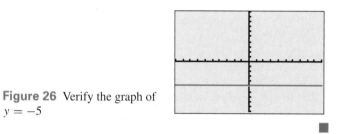

Figure 26 Verify the graph of $y = -5$

In Example 6 we saw that the graph of the equation $x = 3$ is a vertical line. In Example 7 we saw that the graph of the equation $y = -5$ is a horizontal line. These examples suggest forms of equations for vertical and horizontal lines.

Equations for Vertical and Horizontal Lines

If a and b are constants, then:

- An equation of the form $x = a$ has a vertical line as its graph (see Fig. 27).
- An equation of the form $y = b$ has a horizontal line as its graph (see Fig. 28).

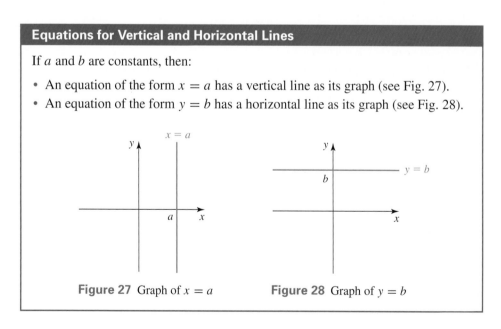

Figure 27 Graph of $x = a$ **Figure 28** Graph of $y = b$

For example, the graphs of the equations $x = 4$ and $x = -7$ are vertical lines. The graphs of the equations $y = 6$ and $y = -2$ are horizontal lines.

If the graph of an equation is a line, we call the equation a **linear equation**. Linear equations can be put into one of the forms $y = mx + b$ or $x = a$. Any equation that can be put into the form $x = a$ has a vertical line as its graph. Any equation that can be put into the form $y = mx + b$ has a nonvertical line as its graph.

EXPLORATION
Looking ahead: Graphical significance of m and b

1. In this problem, you will explore the graphical significance of the constant m for equations of the form $y = mx$. Using ZDecimal, graph the equations in the order listed and describe what you observe.

$$y = 0.3x$$
$$y = 0.7x$$
$$y = x$$
$$y = 2x$$
$$y = 3x$$

Do the same with:

$$y = -0.3x$$
$$y = -0.7x$$
$$y = -x$$
$$y = -2x$$
$$y = -3x$$

2. In this problem, you will explore the graphical significance of the constant b for equations of the form $y = 2x + b$. Using ZDecimal, graph the equations in the order listed and describe what you observe.

$$y = 2x - 2$$
$$y = 2x - 1$$
$$y = 2x$$
$$y = 2x + 1$$
$$y = 2x + 2$$

3. So far you have graphed equations of the forms $y = mx$ (where $b = 0$) and $y = 2x + b$ (where $m = 2$). Now graph more equations of the form $y = mx + b$ until you are confident that you know the graphical significance of m and b for any values m and b.

Note

To form a conjecture means to make a guess or form a theory.

4. Form a conjecture about the graph of $y = mx + b$ in each of the following situations. Test each conjecture by checking whether your conjecture is true for values of m and b other than the ones you have worked with so far.

a. m is zero
b. m is positive
c. m is negative
d. m is a large positive number
e. m is a positive number near zero

f. m is a negative number near zero
g. $m < -10$ (for example $m = -20$)
h. b is equal to 5
i. b is equal to -3
j. b is equal to 0

Tips on Succeeding in This Course

For each hour of class time, you should study for at least two hours outside of class. If you feel that your math background is weak, you may need to spend more time studying.

Key Points
OF THIS SECTION

We say that an ordered pair (a, b) is a *solution* of an equation in two variables if the ordered pair satisfies the equation. The *solution set* of an equation is the set of all ordered pairs that satisfy the equation.

Solution and solution set of an equation in two variables

All points that satisfy an equation form the *graph* of the equation. Points that do not satisfy an equation are not part of the graph.

Meaning of a graph

If an equation can be put into one of the forms $y = mx + b$ or $x = a$, where $a, b,$ and m are constants, then the graph of the equation is a line. The equation is called a *linear equation*.

Linear equation

To sketch the graph of a linear equation containing the variables x and y:
1. Create a table of ordered-pair solutions of the equation by substituting values for one variable and solving for the other variable.
2. Plot the ordered pairs from your table.
3. Sketch the line that passes through the plotted points.

Graphing a linear equation

For an equation containing the variables x and y:
- To find the x-coordinate of each x-intercept, substitute 0 for y and solve for x.
- To find the y-coordinate of each y-intercept, substitute 0 for x and solve for y.

Finding intercepts

When one variable is "missing":
- An equation of the form $x = a$ has a vertical line as its graph.
- An equation of the form $y = b$ has a horizontal line as its graph.

Vertical and horizontal lines

HOMEWORK 1.2

 SSM Tutor Center   MathPro 4   MathPro 5   CDs/Videos   www

*Use pencil and paper to graph each equation. Verify your graph by using ZStandard followed by ZSquare on your graphing calculator. [**Graphing Calculator:** See Sections B.3, B.4, and B.6. Also, recall that the key $\boxed{-}$ is used for subtraction, and the key $\boxed{(-)}$ is used for negative numbers as well as for taking opposites.]*

1. $y = 4x + 3$

2. $y = -2x + 8$

3. $y = 3x - 10$

4. $y = -x - 1$

5. $y = 2x + 2$

6. $y = -3x - 2$

7. $y = 3x - 5$

8. $y = -4x + 12$

9. $y = 2x$

10. $y = x + 1$

11. $y = -3x$

12. $y = x$

13. $9x - 3y = 0$

14. $0 = 4y - 4x$

15. $3y - 6x = 12$

16. $10x - 5y = 20$

17. $2y - 6x - 14 = -4$

18. $6y - 4x - 1 = 7y - 2x - 4$

19. $y = 3x + 4 - 2x + 3 - 5(x - 2)$

20. $2(y - 3) = 4(x + 1)$

21. $-3(y - 5) = 2(3x - 6)$

22. $-(x + 2) - y = x + 3$

23. $y = \dfrac{1}{3}x$

24. $y = -\dfrac{3}{4}x$

25. $y = -\dfrac{1}{2}x + 1$

26. $y = \dfrac{2}{3}x - 2$

27. In this exercise you will use your graphing calculator to sketch the graphs of equations of the form $y = mx + b$.
   a. Use your graphing calculator to graph $y = -4.1x + 8.7$. Is the graph a line? (Here, $m = -4.1$ and $b = 8.7$.)
   b. Use your graphing calculator to graph $y = 6$. Is the graph a line? (Here, $m = 0$ and $b = 6$.)
   c. Create and graph at least two more equations of the form $y = mx + b$. Are the graphs lines?

28. A student says that the graph of the equation $y + x^2 = 5x + x^2 + 1$ is not a line since the equation for a line does not have an x^2 term in it. What would you tell this student?

*[**Note:** When responding to questions in this text, always use sentences, not phrases.]*

29. a. Use your calculator to graph the following equations.
   i. $y = 2$
   ii. $y = -2$
   iii. $y = 5.4896$
   b. Describe the graph of $y = b$, where b is a constant.

30. a. Use pencil and paper to graph each equation.
   i. $x = 4$
   ii. $x = -5$
   iii. $x = 2.5$
   b. Describe the graph of $x = a$, where a is a constant.

Use pencil and paper to graph each equation.

31. $x = 6$ **32.** $y = 3$ **33.** $y = -4$

34. $x = -3$ **35.** $y = 0$ **36.** $x = 0$

*Use ZDecimal to graph the following equations on your graphing calculator. Use TRACE to find the coordinates of a point on the graph. Then verify that the ordered pair for that point satisfies the equation. [**Graphing Calculator**: See Sections B.6 and B.5.]*

37. $y = -3x + 1$ **38.** $2x - 3y = 6$

39. $0.83x = 4.98y - 2$

40. a. Use your graphing calculator to draw the graph of the line $y = -2.43x + 1.89$.

     b. Use ZDecimal followed by TRACE to help you create a table of ordered-pair solutions for the equation. Your table should contain at least five ordered pairs.

Find the x-intercept and y-intercept for each equation. If an intercept does not exist, say so.

41. $y = 2x + 10$ **42.** $y = -3x - 12$

43. $2x + 3y = 12$ **44.** $5x - 4y = 20$

45. $y = 3x$ **46.** $y = -2x$

47. $y = 3$ **48.** $x = -2$

49. The graph of an equation is sketched in Fig. 29. Create a table of ordered-pair solutions of this equation. Your table should contain at least five ordered pairs.

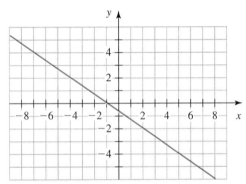

Figure 29 Exercise 49

50. For $y = mx + b$ with $m \neq 0$, find the x-intercept in terms of m and b. Find the y-intercept.

For exercises 51–58, refer to Fig. 30.

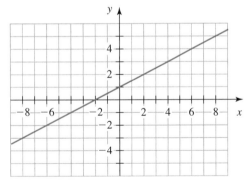

Figure 30 Exercises 51–58

51. Estimate y when $x = 4$.   **52.** Estimate y when $x = 0$.

53. Estimate y when $x = -6$.   **54.** Estimate y when $x = 3$.

55. Estimate x when $y = 4$.   **56.** Estimate x when $y = 0$.

57. Estimate x when $y = 0.5$.

58. Estimate x when $y = -1.5$.

59. A person lowers his hot-air balloon by gradually releasing air from the balloon. Let x represent the number of minutes that the person has been releasing air from the balloon, and let y represent the height of the balloon (in feet). Assume that the relationship between x and y is described by the equation $y = -200x + 800$.

     a. Sketch a graph of $y = -200x + 800$.

     b. Find the y-intercept of $y = -200x + 800$. What does it mean in terms of the situation?

     c. Find the x-intercept of $y = -200x + 800$. What does it mean in terms of the situation?

60. A person fills up her car's gas tank and then drives for a long time. Let x represent the driving time (in hours) since fueling up, and let y represent the number of gallons of gas in the car's gas tank. Assume that the relationship between x and y is described by the equation $y = -2x + 10$.

     a. Sketch a graph of $y = -2x + 10$.

     b. Find the y-intercept of $y = -2x + 10$. What does it mean in terms of the situation?

     c. Find the x-intercept. What does it mean in terms of the situation?

61. The graph of an equation is sketched in Fig. 31. Which of the points A, B, C, D, E, or F satisfy the equation?

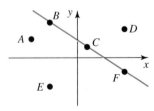

Figure 31 Exercise 61

62. The graphs of $y = ax + b$ and $y = cx + d$ are sketched in Fig. 32. For each part, decide which of the points A, B, C, D, E, or F:

     a. satisfy the equation $y = ax + b$.

     b. satisfy the equation $y = cx + d$.

     c. satisfy both equations.

d. do not satisfy either equation.

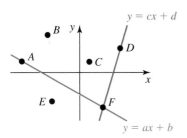

Figure 32 Exercise 62

63. Find the ordered pair(s) that satisfies both of the following equations. Explain.
$$y = 2x + 1$$
$$y = -3x + 6$$
[**Hint**: Graph the equations on the same coordinate system.]

64. The graph of the equation $y = mx + 3$ contains the point $(2, 11)$. What is the constant m?

65. The graph of the equation $y = 2x + b$ contains the point $(7, 5)$. What is the constant b?

66. Explain why the y-coordinate of an x-intercept is 0. Explain why the x-coordinate of a y-intercept is 0.

67. Describe how to sketch a graph of a linear equation. Also, describe the meaning of a graph.

1.3 SLOPE OF A LINE

Objectives

▸ Know the meaning of the *slope* of a nonvertical line.

▸ Calculate the slope of a nonvertical line.

▸ Know the relationship between slopes of parallel lines.

▸ Know the relationship between slopes of perpendicular lines.

In this section we measure the steepness of lines.

To begin, consider the sketch of two ladders leaning against a building in Fig. 33. Which ladder is steeper?

Figure 33 Ladder A and ladder B leaning against a building

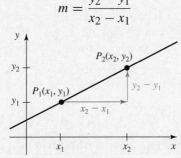

Note that ladder B is steeper, even though ladder A reaches a higher point on the building. To measure the steepness of each ladder, we must compare the *vertical* distance from the base of the building to the top of the ladder with the *horizontal* distance from the foot of the ladder to the building.

To measure the steepness of a nonvertical line, we find the *slope* of the line. We use the letter m to represent the slope.

DEFINITION *Slope of a nonvertical line*

Let $P_1(x_1, y_1)$ and $P_2(x_2, y_2)$ be two distinct points of a nonvertical line (see Fig. 34). The slope of the line is
$$m = \frac{y_2 - y_1}{x_2 - x_1}$$

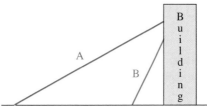

Figure 34 The slope of the line is
$$m = \frac{y_2 - y_1}{x_2 - x_1}$$

When going from P_1 to P_2, the difference $y_2 - y_1$ represents the vertical change called the **rise** and the difference $x_2 - x_1$ represents the horizontal change called the **run**.

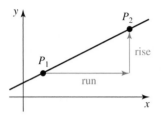

Figure 35 The rise and run between two points

> ### Slope Is Equal to the Rise over the Run
>
> $$\text{slope} = \frac{\text{rise}}{\text{run}}$$
>
> This formula means that the slope of a nonvertical line is equal to the ratio of the rise to the run (in going from one point on the line to another point on the line). See Fig. 35.

Here we list the directions (right, left, up, or down) in verbal and graphical forms associated with the signs of rises and runs:

Sign of rise or run	Direction (verbal)	Direction (graphical)
run is positive	goes to the right	
run is negative	goes to the left	
rise is positive	goes up	
rise is negative	goes down	

Example 1   Finding the Slope of a Line

Find the slope of the line that contains the points $(1, 2)$ and $(5, 4)$.

Solution

Using the formula for slope, we have

$$m = \frac{4-2}{5-1} = \frac{2}{4} = \frac{1}{2} \qquad \text{or} \qquad m = \frac{2-4}{1-5} = \frac{-2}{-4} = \frac{1}{2}$$

Or, by plotting points, we find that if the run is 4, then the rise is 2 (see Fig. 36). So, the slope is $m = \dfrac{\text{rise}}{\text{run}} = \dfrac{2}{4} = \dfrac{1}{2}$.

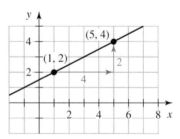

Figure 36 The slope is $\dfrac{2}{4} = \dfrac{1}{2}$

Example 2   Finding the Slope of a Line

Find the slope of the line that contains the points $(2, 3)$ and $(5, 1)$.

Solution

Review

$-\dfrac{a}{b} = \dfrac{-a}{b} = \dfrac{a}{-b}$

$$m = \frac{1-3}{5-2} = \frac{-2}{3} = -\frac{2}{3} \qquad \text{or} \qquad m = \frac{3-1}{2-5} = \frac{2}{-3} = -\frac{2}{3}$$

Or, by plotting points, we find that if the run is 3, then the rise is -2 (see Fig. 37). So, the slope is $\dfrac{-2}{3} = -\dfrac{2}{3}$. ■

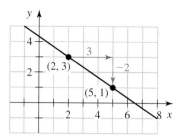

Figure 37 The slope is $\dfrac{-2}{3} = -\dfrac{2}{3}$

In Example 1 the *increasing* line has *positive* slope. In Example 2 the *decreasing* line has *negative* slope.

Consider the increasing line in Fig. 38. In going from P_1 to P_2, we see that the run and rise are both positive, so the slope of the increasing line is positive. (A positive number divided by a positive number gives a positive number.)

Now consider the decreasing line in Fig. 39. In going from P_1 to P_2, we see that the run is positive and the rise is negative, so the slope of the decreasing line is negative. (A negative number divided by a positive number gives a negative number.)

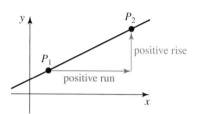

Figure 38 An increasing line has positive slope

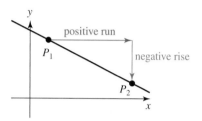

Figure 39 A decreasing line has negative slope

Slopes of Increasing and Decreasing Lines

- An increasing line has positive slope (see Fig. 38).
- A decreasing line has negative slope (see Fig. 39).

Example 3 Finding the Slope of a Line

Find the slope of the line that contains the points $(-9, -4)$ and $(12, -8)$.

Solution

$$m = \frac{-8 - (-4)}{12 - (-9)} = \frac{-8 + 4}{12 + 9} = \frac{-4}{21} = -\frac{4}{21}$$

Since the slope is negative, we know the line is decreasing. ■

Consider the two lines sketched in Fig. 40. Note that line l_2 is steeper than line l_1. In Fig. 41 we see that for line l_1, if the run is 1, the rise is 2. Here we calculate the slope of line l_1:

$$\text{Slope of line } l_1 = \frac{\text{rise}}{\text{run}} = \frac{2}{1} = 2$$

In Fig. 42 we see that for line l_2, if the run is 1, the rise is 4. Here we calculate the slope of line l_2:

$$\text{Slope of line } l_2 = \frac{\text{rise}}{\text{run}} = \frac{4}{1} = 4$$

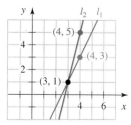

Figure 40 Two lines with different slopes

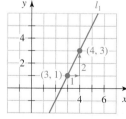

Figure 41 Line with smaller slope

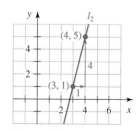

Figure 42 Line with larger slope

Note that the slope of line l_2 is greater than the slope of line l_1, which is what we would expect because line l_2 is steeper than line l_1. In general, for two nonparallel increasing lines, the steeper line will have the greater slope.

Example 4 Finding the Slope of a Horizontal Line

Find the slope of the line that contains the points $(2, 3)$ and $(6, 3)$.

Solution

We plot the points $(2, 3)$ and $(6, 3)$ and sketch the line that contains the points (see Fig. 43).

The formula for slope gives

$$m = \frac{3 - 3}{6 - 2} = \frac{0}{4} = 0$$

So, the slope of the horizontal line is zero. It makes sense that the horizontal line has slope equal to zero because such a line has "no steepness." ∎

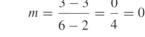

Figure 43 The horizontal line has slope equal to zero

Example 5 Investigating the Slope of a Vertical Line

Find the slope of the line that contains the points $(4, 2)$ and $(4, 5)$.

Solution

We plot the points $(4, 2)$ and $(4, 5)$ and sketch the line that contains the points (see Fig. 44).

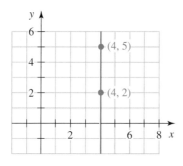

Figure 44 The vertical line has undefined slope

We use the formula for slope.

$$m = \frac{5 - 2}{4 - 4} = \frac{3}{0}$$

Since division by zero is undefined, the slope of the vertical line is *undefined*. ∎

Slopes of Horizontal and Vertical Lines

- A horizontal line has slope equal to zero (see Fig. 45).
- A vertical line has undefined slope (see Fig. 46).

Figure 45 A horizontal line has slope equal to zero

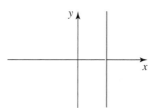

Figure 46 A vertical line has undefined slope

Example 6   Finding Slopes of Parallel Lines

Find the slopes of the parallel lines l_1 and l_2 sketched in Fig. 47.

Solution

For each line, if the run is 3, the rise is 1. So, the slope of each parallel line is

$$m = \frac{\text{rise}}{\text{run}} = \frac{1}{3}$$

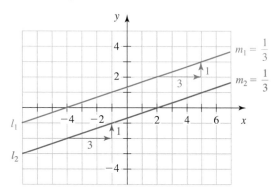

Figure 47 Two parallel lines

It makes sense that parallel nonvertical lines have equal slope, since parallel lines have the same steepness.

Slopes of Parallel Lines

If the lines l_1 and l_2 are parallel nonvertical lines, then the slopes of the lines are equal:

$$m_1 = m_2$$

Also, if two distinct lines have equal slopes, then they are parallel.

Example 7   Finding Slopes of Perpendicular Lines

Find the slope of the perpendicular lines l_1 and l_2 in Fig. 48.

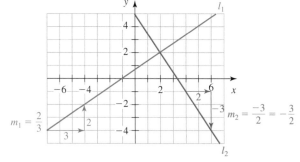

Figure 48 Two perpendicular lines

Solution

The slope of line l_1 is

$$m_1 = \frac{2}{3}$$

The slope of line l_2 is

$$m_2 = \frac{-3}{2} = -\frac{3}{2}$$

In Example 7 the slope $-\dfrac{3}{2}$ is the opposite of the reciprocal of the slope $\dfrac{2}{3}$.

Slopes of Perpendicular Lines

If the lines l_1 and l_2 are perpendicular nonvertical lines, then the slope of one line is the opposite of the reciprocal of the slope of the other line.

$$m_2 = -\frac{1}{m_1}$$

Also, if the slope of one line is the opposite of the reciprocal of another line's slope, then the lines are perpendicular.

Example 8   Finding Slopes of Parallel and Perpendicular Lines

A line l_1 has slope $\dfrac{3}{7}$.

1. If line l_2 is parallel to line l_1, find the slope of line l_2.
2. If line l_3 is perpendicular to line l_1, find the slope of line l_3.

Solution

1. The slopes of the lines l_2 and l_1 are equal, so line l_2 has slope $\dfrac{3}{7}$.

2. The slope of the line l_3 is the opposite of the reciprocal of $\dfrac{3}{7}$, or $-\dfrac{7}{3}$. ■

EXPLORATION
For a line, rise over run is constant

1. A line is sketched in Fig. 49. Plot the points $A(-2, -5)$, $B(1, 1)$, and $C(3, 5)$. (If plotted correctly, these points will lie on the line.)

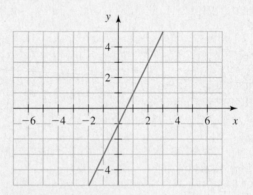

Figure 49 Use different pairs of points to calculate the slope

2. Find the slope of the line using points A and B.
3. Find the slope of the line using points B and C.
4. Find the slope of the line using points A and C.
5. Find the slope of the line using two other points of your choice.
6. What do you notice about the slopes you have calculated? Does it matter which two points on a line are used to find the slope of the line?

EXPLORATION
Looking ahead: Slope addition property

1. Complete Table 8.

Table 8 Solutions for Three Equations

$y = 2x + 1$		$y = 3x - 5$		$y = -2x + 6$	
x	y	x	y	x	y
0		0		0	
1		1		1	
2		2		2	
3		3		3	
4		4		4	

2. In Table 8, the x-coordinates increase by 1 each time. For each equation, what do you notice about the y-coordinates? Compare what you notice with the coefficient of x for each equation.

3. Form a conjecture. (This means to describe what the pattern from part 2 would be in general for any equation of the form $y = mx + b$.)

4. Then test your conjecture. (You could test your conjecture by creating an equation of the form $y = mx + b$ and checking to see if it behaves as your conjecture states.)

5. Substitute 1 for x in the equation $y = mx + b$. Then substitute 2 for x. Then substitute 3. Explain why these results suggest that your conjecture is correct.

Note

Recall that to make a conjecture means to make a guess or form a theory.

Tips on Succeeding in This Course

Some students avoid visiting their instructors during office hours because they do not want to pull their instructors away from their work. But remember, helping students during office hours *is* part of an instructor's job. Keep in mind that your instructor wants you to succeed and hopes that you take advantage of all opportunities to learn.

It is a good idea to come prepared for office visits. For example, if you are having trouble with a concept, attempt some related exercises and bring your work so that your instructor can see where you are having difficulty. If you miss a class, it is helpful to first read the material, borrow class notes, and try completing assigned exercises before visiting an instructor, so that you get the most out of the visit.

Key Points
OF THIS SECTION

Slope is a measure of the steepness of a line.

Meaning of slope

The slope m of a nonvertical line is equal to the ratio of the rise to the run. In symbols, $m = \dfrac{\text{rise}}{\text{run}} = \dfrac{y_2 - y_1}{x_2 - x_1}$, where (x_1, y_1) and (x_2, y_2) are two distinct points of the line.

Calculating slope

An increasing line has positive slope.

A decreasing line has negative slope.

Graphical significance of slope

A horizontal line has slope equal to zero.

A vertical line has undefined slope.

Slopes of parallel lines	Parallel nonvertical lines have equal slopes. Also, distinct lines with equal slopes are parallel.
Slopes of perpendicular lines	

If two nonvertical lines are perpendicular, then the slope of one line is the opposite (change the sign) of the reciprocal of the slope of the other line. Also, if the slope of one line is the opposite (change the sign) of the reciprocal of the other line, the two lines are perpendicular.

HOMEWORK 1.3

SSM
Tutor Center

MathPro 4
MathPro 4

MathPro 5
MathPro 5

CDs/Videos

www

Find the slope of the line passing through the two given points. Also state whether the line is increasing, decreasing, horizontal, or vertical.

1. $(2, 3)$ and $(5, 9)$
2. $(1, 8)$ and $(5, 4)$
3. $(-4, 10)$ and $(2, -2)$
4. $(-6, 2)$ and $(-4, 5)$
5. $(-4, -9)$ and $(-2, -1)$
6. $(-2, 8)$ and $(6, 4)$
7. $(3, -7)$ and $(8, -4)$
8. $(-6, 3)$ and $(8, -2)$
9. $(-11, -2)$ and $(-8, 1)$
10. $(-1, -6)$ and $(-2, -5)$
11. $(0, 0)$ and $(1, 1)$
12. $(0, 0)$ and $(100, 100)$
13. $(1.2, 5.4)$ and $(3.9, 2.6)$
14. $(-3.9, 2.2)$ and $(-5.1, -7.4)$
15. $(8.94, -17.94)$ and $(21.13, -2.34)$
16. $(-25.41, 82.78)$ and $(-11.26, -66.66)$
17. $(2, 6)$ and $(7, 6)$
18. $(3, 1)$ and $(3, 5)$
19. $(-9.5, 6.4)$ and $(-9.5, 2.8)$
20. $(-3.7, -5.4)$ and $(-7.7, -5.4)$
21. For each line sketched in Fig. 50, determine whether the line's slope is positive, negative, zero, or undefined.

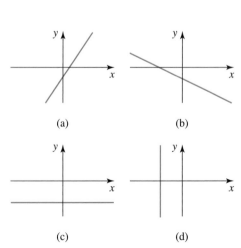

Figure 50 Exercise 21

22. Compute the slope of the line with x-intercept $(5, 0)$ and y-intercept $(0, -2)$.
23. Find the slope of the line sketched in Fig. 51.

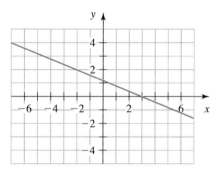

Figure 51 Exercise 23

24. Find the slope of the line sketched in Fig. 52.

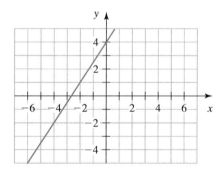

Figure 52 Exercise 24

Using the indicated slopes of lines l_1 and l_2, determine whether the lines are parallel, perpendicular, or neither. Assume that the lines are not the same.

25. $m_1 = 2, m_2 = 2$
26. $m_1 = 5, m_2 = -\dfrac{1}{5}$
27. $m_1 = 6, m_2 = -6$
28. $m_1 = 1, m_2 = -1$
29. $m_1 = \dfrac{2}{7}, m_2 = -\dfrac{7}{2}$
30. $m_1 = \dfrac{3}{5}, m_2 = \dfrac{5}{3}$
31. $m_1 = \dfrac{7}{4}, m_2 = \dfrac{4}{7}$
32. $m_1 = \dfrac{5}{8}, m_2 = -\dfrac{5}{8}$
33. $m_1 = 0, m_2$ is undefined
34. m_1 is undefined, m_2 is undefined

Sketch a line that has the indicated slope.

35. $m = \dfrac{2}{5}$
36. $m = 3$ $\left[\text{**Hint**: } 3 = \dfrac{3}{1}\right]$

37. $m = -\dfrac{4}{3}$ $\left[\textbf{Hint: } -\dfrac{4}{3} = \dfrac{-4}{3}\right]$

38. $m = -2$

39. $m = 0$

40. m is undefined

41. Sketch a line with slope 2 and another line with slope 3. Which line is steeper?

42. Sketch a line with slope 2 and another line with slope -2. Which line is steeper?

43. A person goes on a road trip. A portion of road A steadily climbs 120 feet over a horizontal distance of 4000 feet. A portion of road B steadily climbs 160 feet over a horizontal distance of 6500 feet. Which road is steeper? Explain.

44. While taking off, airplane A steadily climbs 2500 feet over a horizontal distance of 8000 feet. Airplane B steadily climbs 3100 feet over a horizontal distance of 9500 feet. Which plane is climbing more steeply? Explain.

45. Ski run A steadily declines 90 yards over a horizontal distance of 300 yards. Ski run B steadily declines 125 yards over a horizontal distance of 450 yards. Which run is steeper? Explain.

46. A ski run declines with constant steepness from the top of a mountain to a chairlift called Kangaroo. Then the run continues to decline with a different constant steepness to end at a restaurant on the mountain. The horizontal distance for the entire run is 1300 yards for a vertical decline of 415 yards. The horizontal distance from the top of the mountain to the Kangaroo chairlift is 400 yards with vertical decline of 100 yards. Find (the absolute value of) the "slopes" of each part of the run.

In each of the following exercises, sketch the graph of a line that meets the description. Also find the slope of your sketched line.

47. An increasing line that is nearly horizontal

48. A decreasing line that is nearly horizontal

49. A decreasing line that is nearly vertical

50. An increasing line that is nearly vertical

In each of the following exercises, sketch the graph of a line that meets the description.

51. The slope is a large positive number.

52. The slope is a positive number near zero.

53. The slope is a negative number near zero.

54. The slope is less than -5.

55. A line contains the points $(2, 7)$ and $(3, 10)$. Find three more points that lie on the line.

56. A line contains the points $(-6, -4)$ and $(-3, 1)$. Find three more points that lie on the line.

57. a. First, carefully sketch a graph of the equation. Next, find the slope of your sketched line using the formula $\dfrac{\text{rise}}{\text{run}}$.

　i. $y = 2x + 1$

　ii. $y = 3x - 5$

　iii. $y = -2x + 6$

b. Compare the slope of each line to the coefficient of x in the corresponding equation.

58. Recall that if $P_1(x_1, y_1)$ and $P_2(x_2, y_2)$ are two distinct points of a nonvertical line, then the slope of the line is $m = \dfrac{y_2 - y_1}{x_2 - x_1}$.

a. Use the formula $\dfrac{y_2 - y_1}{x_2 - x_1}$ to find the slope of the line that contains the points $P_1(2, 3)$ and $P_2(7, 5)$.

b. Use the formula $\dfrac{y_1 - y_2}{x_1 - x_2}$ to find the slope of the line that contains the points $P_1(2, 3)$ and $P_2(7, 5)$.

c. Compare your results from parts a and b.

d. Show that $\dfrac{y_2 - y_1}{x_2 - x_1} = \dfrac{y_1 - y_2}{x_1 - x_2}$, where $P_1(x_1, y_1)$ and $P_2(x_2, y_2)$ are two distinct points of a nonvertical line. Here is the first step:

$$\dfrac{y_2 - y_1}{x_2 - x_1} = \dfrac{y_2 - y_1}{x_2 - x_1} \cdot \dfrac{-1}{-1}$$

e. When calculating the slope of a line using two given points on a line, does it matter which point is called P_1 and which point is called P_2? Explain.

59. In this exercise you will explore the relationship between three lines that pass through the origin $(0, 0)$, where the slope of one of the lines is the reciprocal of the slope of one of the other lines.

a. Use graph paper to carefully sketch the lines that pass through the origin $(0, 0)$ with slopes 5, 1, and $\dfrac{1}{5}$.

b. Sketch the lines that pass through the origin $(0, 0)$ with slopes $\dfrac{2}{5}$, 1, and $\dfrac{5}{2}$.

c. Sketch the lines that pass through the origin $(0, 0)$ with slopes $\dfrac{3}{4}$, 1, and $\dfrac{4}{3}$.

d. What pattern do you notice from your graphs in parts a–c?

e. A line with slope m is sketched in Fig. 53. Sketch a line with slope $\dfrac{1}{m}$ that passes through the origin $(0, 0)$. Assume that the scalings on both axes are the same.

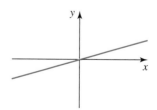

Figure 53 Exercise 59

60. a. A square has vertices at $(3, 1)$ and $(3, 7)$. How many possible positions are there for the other two vertices? Find the coordinates for each possibility.

b. A parallelogram has vertices at $(-7, -2)$, $(3, 1)$, and $(-4, 2)$. How many possible positions are there for the fourth vertex? Find the coordinates for each possibility. [**Hint**: Try drawing different line segments between the given vertices.]

61. Suppose that a line with slope $-\dfrac{2}{3}$ contains a point P. A point Q lies three units to the right and two units down from point P. A point S lies three units to the left and two units up from point P. Does the line contain the point Q? the point S? Explain.

62. Describe the meaning of the slope of a line. As part of your description, sketch various types of lines and give the slope for each line. For each sketch, explain why the slope assignment makes sense. For example, you could sketch a horizontal line, state that the slope is zero, and explain why it makes sense that the slope of a horizontal line is zero in terms of rise and run.

1.4 GRAPHICAL, NUMERICAL, AND SYMBOLIC SIGNIFICANCE OF SLOPE

I love mathematics. Mathematics is the language of nature and force. Math is directly behind most technical innovations and has also been used to solve many of the mysteries of life.
—*Tim P., student*

Objectives

▶ Know graphical, numerical, and symbolic meanings of slope.

▶ Know the meaning of m and b for an equation of the form $y = mx + b$.

▶ Use slope and y-intercept to help sketch the graph of a linear equation with defined slope.

In this section we discuss how to use the equation of a nonvertical line to find the line's slope.

Example 1 Finding the Slope of a Line

Find the slope of the line $y = 2x + 1$.

Solution

We list solutions using $x = 0, 1, 2, 3$ in Table 9 and sketch the graph of the equation in Fig. 54.

Table 9 Solutions of $y = 2x + 1$

x	y
0	1
1	3
2	5
3	7

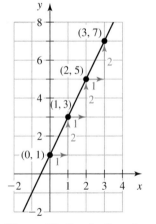

Figure 54 Graph of $y = 2x + 1$

If the run is 1, the rise is 2 (see Fig. 54). So, the slope is

$$m = \frac{\text{rise}}{\text{run}} = \frac{2}{1} = 2$$

From Example 1, we can make three observations about the slope of a nonvertical line.

1. For $y = 2x + 1$, the coefficient of x is 2, which is the slope.
2. If the run is 1, the rise is 2 (the slope). See Fig. 54.
3. As the value of x increases by 1, the value of y increases by 2 (the slope). See Table 9.

Example 2 Finding the Slope of a Line

Find the slope of the line $y = -3x + 8$.

Solution

Some solutions are listed in Table 10 and a graph is sketched in Fig. 55.

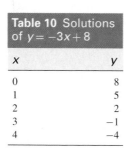

Table 10 Solutions of $y = -3x + 8$

x	y
0	8
1	5
2	2
3	−1
4	−4

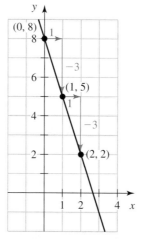

Figure 55 Graph of $y = -3x + 8$

The slope is $m = \dfrac{-3}{1} = -3$. ■

From Example 2, we can make three observations.

1. For $y = -3x + 8$, the coefficient of x is −3, which is the slope.
2. If the run is 1, the rise is −3 (the slope). See Fig. 55.
3. As the value of x increases by 1, the value of y changes by −3 (the slope). See Table 10.

Our first observations made after both Examples 1 and 2 suggest a general property about slope.

Finding Slope from a Linear Equation

For a linear equation of the form

$$y = mx + b$$

m is the slope of the line.

Example 3 Identifying Parallel or Perpendicular Lines

Decide whether the lines $y = \dfrac{5}{6}x + 3$ and $12y - 10x = 5$ are parallel, perpendicular, or neither.

Solution

For the line $y = \dfrac{5}{6}x + 3$, the slope is $\dfrac{5}{6}$. To find the slope of the line $12y - 10x = 5$, we begin by isolating y.

$$12y - 10x = 5$$

$$12y - 10x + 10x = 5 + 10x \qquad \text{Add } 10x \text{ to both sides.}$$

$$12y = 10x + 5 \qquad \text{Simplify.}$$

$$\frac{12y}{12} = \frac{10}{12}x + \frac{5}{12} \qquad \text{Divide each side by 12.}$$

$$y = \frac{5}{6}x + \frac{5}{12} \qquad \text{Reduce.}$$

For $y = \dfrac{5}{6}x + \dfrac{5}{12}$, the slope is $\dfrac{5}{6}$, the same as the slope for the line $y = \dfrac{5}{6}x + 3$. Therefore, the two distinct lines are parallel. We use ZStandard followed by ZSquare to draw the lines on the same coordinate system (see Fig. 56).

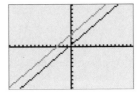

Figure 56 Graphs of the two parallel lines

Our second observations made after both Examples 1 and 2 suggest a general property of a line $y = mx + b$.

Vertical Change Property

For a line $y = mx + b$, if the run is 1, the rise is the slope m. (See Figs. 57 and 58.)

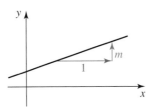

Figure 57 Vertical change property for positive slope

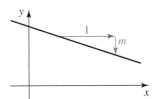

Figure 58 Vertical change property for negative slope

When sketching graphs of equations of the form $y = mx + b$, it is helpful to know the y-intercept. Substituting 0 for x in the equation $y = mx + b$ gives

$$y = m(0) + b = b$$

which shows that the y-intercept is $(0, b)$.

---

y-intercept of a Linear Equation

For an equation of the form

$$y = mx + b$$

the y-intercept is $(0, b)$.

For example, in Example 1, the line $y = 2x + 1$ has y-intercept $(0, 1)$. In Example 2, the line $y = -3x + 8$ has y-intercept $(0, 8)$.

DEFINITION *Slope-intercept form*

If an equation is of the form

$$y = mx + b$$

we say it is in **slope-intercept** form.

Example 4   Using Slope to Help Graph a Linear Equation

Sketch the graph of $y = 3x - 1$.

Solution

Note that the y-intercept is $(0, -1)$ and the slope is $3 = \dfrac{3}{1} = \dfrac{\text{rise}}{\text{run}}$. To graph:

1. Plot the y-intercept $(0, -1)$.
2. From $(0, -1)$ look 1 unit to the right and 3 units up to plot the point $(1, 2)$. See Fig. 59.
3. Sketch the line that contains these two points (see Fig. 60).

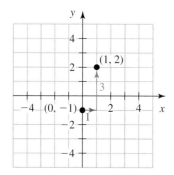

Figure 59 Plot $(0, -1)$. Then look 1 unit to the right, 3 units up, to plot $(1, 2)$

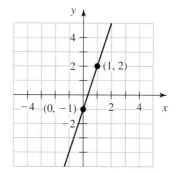

Figure 60 Sketch the line containing $(0, -1)$ and $(1, 2)$

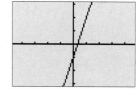

Figure 61 Use ZDecimal to verify our graph

We use ZDecimal to verify our graph (see Fig. 61).

Using Slope to Graph a Linear Equation

To sketch the graph of an equation in the form $y = mx + b$:

1. Plot the y-intercept $(0, b)$.
2. Use $m = \dfrac{\text{rise}}{\text{run}}$ to plot another point. For example, if $m = -\dfrac{2}{5} = \dfrac{-2}{5}$, then look 5 units to the right (from the y-intercept) and 2 units down to plot another point.
3. Sketch the line that passes through the two plotted points.

Example 5   Using Slope to Help Graph a Linear Equation

Sketch the graph of $2x + 3y = 6$.

Solution

First, we write the equation in slope-intercept form.

$$2x + 3y = 6$$

$$2x + 3y - 2x = 6 - 2x \qquad \text{Subtract } 2x \text{ from both sides.}$$

$$3y = -2x + 6 \qquad \text{Simplify.}$$

$$\frac{3y}{3} = \frac{-2x}{3} + \frac{6}{3} \qquad \text{Divide each side by 3.}$$

$$y = -\frac{2}{3}x + 2 \qquad \text{Reduce, } \frac{-a}{b} = -\frac{a}{b}$$

The y-intercept is $(0, 2)$ and the slope is $-\dfrac{2}{3} = \dfrac{-2}{3} = \dfrac{\text{rise}}{\text{run}}$. To graph:

1. Plot the y-intercept $(0, 2)$.

2. From $(0, 2)$ look 3 units to the right and 2 units down to plot the point $(3, 0)$. See Fig. 62.

3. Then sketch the line that contains these two points (see Fig. 63).

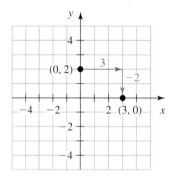

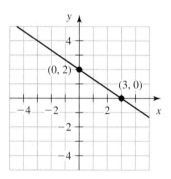

Figure 62 Plot $(0, 2)$. Then look 3 units to the right, 2 units down, to plot $(3, 0)$

Figure 63 Sketch the line containing $(0, 2)$ and $(3, 0)$

We can verify our result by checking that $(0, 2)$ and $(3, 0)$ are both solutions of $2x + 3y = 6$. ■

Our third observations made after both Examples 1 and 2 suggest a general property of a table of solutions of an equation of the form $y = mx + b$.

Slope Addition Property

For an equation of the form $y = mx + b$, if the value of the independent variable increases by 1, then the value of the dependent variable changes by the slope m.

For example, consider the equation $y = -5x + 4$. We know that as the value of x increases by 1, the value of y changes by -5.

Example 6 Identifying Possible Linear Equations

Some solutions of four equations are listed in Table 11. Which of the equations could possibly be linear?

Solution

1. Equation 1 might be linear. For the solutions in the table, as the value of x increases by 1, the value of y changes by -3.

2. Equation 2 might be linear. For the solutions in the table, as the value of x increases by 1, the value of y changes by 5.

3. Equation 3 is not linear. As the value of x increases by 1, the value of y does not change by the same number.

4. Equation 4 might be $y = 8$, which is a linear equation.

Table 11 Solutions for Four Equations

Equation 1		Equation 2		Equation 3		Equation 4	
x	y	x	y	x	y	x	y
1	23	4	12	0	3	50	8
2	20	5	17	1	6	51	8
3	17	6	22	2	12	52	8
4	14	7	27	3	24	53	8
5	11	8	32	4	48	54	8

EXPLORATION
Drawing lines with various slopes

1. On your calculator, graph a group of lines (a *family of lines*) to make a starburst like the one in Fig. 64. List the equations of your lines.

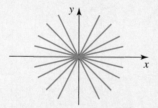

Figure 64 A starburst

2. On your calculator, graph a family of lines to make a starburst like the one in Fig. 65. List the equations of your lines.

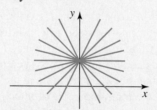

Figure 65 Another starburst

3. Summarize what you have learned about slope from this exploration, this section, and Section 1.3.

EXPLORATION
Looking ahead: Finding an equation of a line

Some solutions for four linear equations are provided in Table 12. Find an equation for each of the four lines.

Table 12 Solutions for Four Linear Equations

Equation 1		Equation 2		Equation 3		Equation 4	
x	y	x	y	x	y	x	y
0	5	3	10	2	7	1	17
1	8	4	8	4	12	3	13
2	11	5	6	6	17	5	9
3	14	6	4	8	22	7	5

Tips on Succeeding in This Course

In preparing for a quiz, test, or final exam, it is helpful to consider how the concepts that you have learned are interconnected. You can do this by building a *mind map*. A mind map for concepts relating to slope is illustrated in Fig. 66.

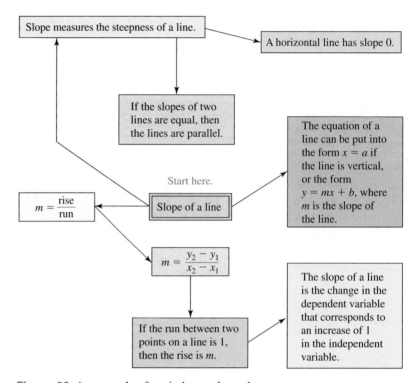

Figure 66 An example of a mind map about slope

A mind map allows you to attach entries as they occur to you, rather than being forced to list things in a more restricted way. It is also a personal creation that thus is particularly meaningful to its builder and owner. For example, some people enjoy using interesting shapes and bright colors to sketch their "thought bubbles."

To begin a mind map, put the main topic in the middle of the page, then attach concepts that relate to the main topic. Next, attach concepts that relate to each of these newly added topics, and so on, drawing lines or arrows along the way to suggest how the topics connect.

Key Points
OF THIS SECTION

Properties of an equation of the form $y = mx + b$

Properties of equations of the form $y = mx + b$:
- The graph of the equation is a line.
- The constant m is the slope of the line, a measure of the line's steepness.
- If $m > 0$, the line is increasing.
- If $m < 0$, the line is decreasing.
- If $m = 0$, the line is horizontal.
- If m is undefined, the line is vertical.
- If the value of x increases by 1, then the value of y changes by the slope m.
- If the run is 1, the rise is the slope m.
- The y-intercept of the line is $(0, b)$.

To sketch a graph of an equation in the form $y = mx + b$:

1. Plot the y-intercept $(0, b)$.

2. Use $m = \dfrac{\text{rise}}{\text{run}}$ to plot another point. For example, if

   $m = -\dfrac{3}{2} = \dfrac{-3}{2}$, then look 2 units to the right (from the y-intercept) and 3 units down to plot another point.

3. Sketch the line that passes through the two plotted points.

Graphing an equation in the form $y = mx + b$

HOMEWORK 1.4

SSM
Tutor Center

MathPro 4
MathPro 4

MathPro 5
MathPro 5

CDs/Videos

www
www

Determine the slope and the y-intercept for each linear equation. Use the slope and the y-intercept to hand-sketch the graph of the equation. Verify your graph by using ZStandard followed by ZSquare on your graphing calculator.

1. $y = 6x + 1$

2. $y = 3x + 6$

3. $y = -2x + 7$

4. $y = -3x + 5$

5. $y = \dfrac{5}{4}x - 2$

6. $y = -\dfrac{4}{3}x + 1$

7. $y = -\dfrac{3}{7}x + 2$

8. $y = \dfrac{1}{2}x - 8$

9. $y = -\dfrac{5}{3}x - 1$

10. $y = \dfrac{2}{9}x - 5$

11. $y + x = 5$

12. $4x + 3y = 12$

13. $-7x + 2y = 10$

14. $3y - 5x = 6$

15. $3(x - 2y) = 9$

16. $2(y - 3) = 8x$

17. $2x - 3y + 9 = 12$

18. $-5x - 15y + 20 = 0$

19. $7x - 2(y + 2) = 0$

20. $3(y - 4) + 1 = 2(x + 6) - 2$

21. $y = 4x$

22. $y = -7x$

23. $y = -1.5x + 3$

24. $y = 0.25x - 2$

25. $y = x$

26. $y = -x$

27. $y = 4$

28. $y = 0$

29. $y + 2 = 0$

30. $y - 3 = 0$

31. Four sets of points are described in Table 13. For each set, decide whether there is a line that passes through every point. If so, find the slope of that line. If not, decide whether there is a line that comes close to every point.

Table 13 Four Sets of Points

Set 1		Set 2		Set 3		Set 4	
x	y_1	x	y_2	x	y_3	x	y_4
1	6	1	5.9	1	3	50	90
2	106	2	5.6	11	8	51	80
3	205	3	5.3	13	13	52	70
4	305	4	5.0	20	18	53	60
5	406	5	4.7	40	23	54	50
6	505	6	4.4	90	28	55	40

32. Four sets of points are described in Table 14. For each set, decide whether there is a line that passes through every point. If so, find the slope of that line. If not, decide whether there is a line that comes close to every point.

Table 14 Four Sets of Points

Set 1		Set 2		Set 3		Set 4	
x	y_1	x	y_2	x	y_3	x	y_4
0	50	3	2	1	8	5	1
1	47	5	5	2	8	5	9
2	44	7	8	3	8	5	10
3	41	9	11	4	8	5	40
4	38	11	14	5	8	5	46
5	35	13	17	6	8	5	99

33. Some values for four linear equations are provided in Table 15. Complete the table.

Table 15 Values for Four Linear Equations

Equation 1		Equation 2		Equation 3		Equation 4	
x	y	x	y	x	y	x	y
1	12	23	69	1	-7	30	5.0
2	15	24	61	2	-2	31	4.4
3		25		3		32	
4		26		4		33	
5		27		5		34	
6		28		6		35	

34. Some values for four linear equations are provided in Table 16. Complete the table.

Table 16 Values for Four Linear Equations

Equation 1		Equation 2		Equation 3		Equation 4	
x	y	x	y	x	y	x	y
0	16	0		1	36	10	80
1		1		2		11	
2	30	2	2	3		12	
3		3		4		13	
4		4		5		14	70
5		5	14	6	16	15	

35. Graphs of four equations are shown in Fig. 67. State the signs of the constants m and b for the $y = mx + b$ form of each equation.

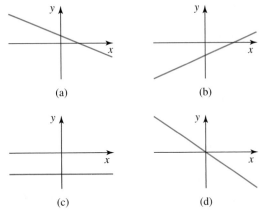

(a)   (b)

(c)   (d)

Figure 67 Exercise 35

36. On your calculator, graph a family of five parallel lines like those in Fig. 68. List the equations of your lines.

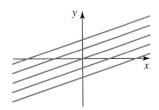

Figure 68 A family of parallel lines

Determine whether each pair of lines is parallel, perpendicular, or neither. Explain.

37. $y = 4x + 7$ and $y = 4x - 3$

38. $y = \frac{4}{9}x - 5$ and $y = -\frac{9}{4}x - 5$

39. $y = \frac{3}{8}x + 1$ and $y = \frac{8}{3}x + 4$

40. $y = \frac{2}{5}x$ and $y = \frac{2}{5}x - 3$

41. $2x + 3y = 6$ and $4x + 6y = 7$

42. $4x + y = 6$ and $x - 4y = 5$

43. $5x - 3y = 1$ and $3x + 5y = -2$

44. $8x - 4y = 1$ and $y = 8(x - 4) + 1$

45. $x = 2$ and $y = 5$ **46.** $x = -3$ and $y = 7$

47. $y = x$ and $y = -x$ **48.** $x = 0$ and $y = 0$

49. A person fills up his car's gas tank and then drives for a long time. Let x represent the driving time (in hours) since fueling up and let y represent the number of gallons of gas in the car's gas tank. Assume that the relationship between x and y is described by the equation $y = -3x + 18$.

   a. Complete Table 17.

   b. By how much is the amount of gas in the tank decreasing each hour? Compare your result to the slope of $y = -3x + 18$. Discuss your observation in terms of the slope addition property.

Table 17 Amounts of Gas in a Car's Gas Tank

Driving Time (hours) x	Amount of Gas (gallons) y
0	
1	
2	
3	
4	
5	
6	

   c. If the person is driving at about 60 mph, what is the gas mileage of the car? [**Hint**: *Gas mileage* is the number of miles that the car can travel on 1 gallon of gas.]

50. A person lowers her hot-air balloon by gradually releasing air from the balloon. Let x represent the number of minutes that the person has been releasing air from the balloon and let y represent the height of the balloon (in feet). Assume that the relationship between x and y is described by the equation $y = -400x + 2400$.

   a. Complete Table 18.

Table 18 Heights of a Balloon

Time (minutes) x	Height (feet) y
0	
1	
2	
3	
4	
5	
6	

   b. By how much is the height of the hot air balloon decreasing each minute? Compare your result to the slope of $y = -400x + 2400$. Discuss your observation in terms of the slope addition property.

51. Let y represent a person's salary (in thousands of dollars) after working x years at a company. Assume that the relationship between x and y is described by the equation $y = 2x + 26$.

   a. Complete Table 19.

Table 19 Salaries

Time at Company (years) x	Salary (thousands of dollars) y
0	
1	
2	
3	
4	

   b. By how much does the person's salary increase each year? Compare your result to the slope of $y = 2x + 26$. Discuss your observation in terms of the slope addition property.

52. Let y represent a college's enrollment (in thousands of students) and let x represent the number of years that the college has been open. Assume that the relationship between and x and y is described by the equation $y = 0.5x + 6$.

a. Complete Table 20.

Table 20 Enrollments

Number of Years College Has Been Open x	Enrollment (thousands of students) y
0	
1	
2	
3	
4	

b. By how much does the college's enrollment increase each year? Compare your result to the slope of $y = 0.5x + 6$. Discuss your observation in terms of the slope addition property.

53. a. Use your graphing calculator to graph the line $y = x$ using the ZDecimal WINDOW settings displayed in Fig. 69. [**Graphing Calculator:** See Section B.7.]

```
WINDOW
Xmin=-4.7
Xmax=4.7
Xscl=1
Ymin=-3.1
Ymax=3.1
Yscl=1
Xres=1
```

Figure 69 WINDOW for exercise 53a

b. Graph $y = x$ using the WINDOW settings displayed in Fig. 70.

```
WINDOW
Xmin=-4.7
Xmax=4.7
Xscl=1
Ymin=-1
Ymax=1
Yscl=1
Xres=1
```

Figure 70 WINDOW for exercise 53b

c. Sketch $y = x$ using the WINDOW settings displayed in Fig. 71.

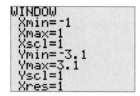

```
WINDOW
Xmin=-1
Xmax=1
Xscl=1
Ymin=-3.1
Ymax=3.1
Yscl=1
Xres=1
```

Figure 71 WINDOW for exercise 53c

d. Compare the results of parts a and b. Explain the different views of $y = x$. Do the same for parts a and c.

e. Can you make the sketch of $y = x$ appear to be a decreasing line by adjusting the WINDOW settings? Can you make the graph of $y = x$ cross the y-axis at a point other than $(0, 0)$ (the *origin*) by changing the WINDOW settings? Describe all possible appearances of the graph of $y = x$ as a result of using various WINDOW settings.

54. a. Try using your graphing calculator to graph $y = 0.0005x + 0.003$. You might observe that no graph appears to be drawn on your calculator screen. This does not mean that your calculator is broken! To locate the line, think about its slope and y-intercept. Then TRACE to confirm your suspicion. Try changing your WINDOW settings so that the graph can be clearly seen. Record a WINDOW setting that allows the graph to be seen.

b. Use your graphing calculator to graph the equation $y = -3x + 20,000$. Record a WINDOW setting that allows the graph to be seen.

c. Use your graphing calculator to sketch a graph of the equation $y = 4000x - 0.04$. Record a WINDOW setting that allows the graph to be seen.

55. A line passes through the point $(0, -3)$ with slope 2. What is an equation for the line?

56. A line passes through the point $(0, 4)$ with slope -5. What is an equation for the line?

57. A line passes through the point $(3, 8)$ and has slope 5.

a. Use pencil and paper to sketch the graph of the line.

b. Find an equation for the line. [**Hint:** Use part a to find b for $y = mx + b$].

c. Use your graphing calculator to verify that your equation is correct.

58. A line passes through the point $(-2, 7)$ and has slope -6.

a. Use pencil and paper to sketch the graph of the line.

b. Find an equation for the line.

c. Use your graphing calculator to verify that your equation is correct.

59. A line passes through the point $(2, 6)$ and has slope $-\dfrac{3}{8}$.

a. Use pencil and paper to sketch the graph of the line.

b. Estimate an equation for the line.

c. Use your graphing calculator to verify that your equation is correct.

60. a. Find the slope of each line: $y = 2$, $y = -5$, and $y = 3.72$.

b. Find the slope for any linear equation of the form $y = k$.

61. a. Find the slope for each line: $x = 3$, $x = -6$, and $x = -7.9$.

b. Find the slope for any equation of the form $x = k$.

62. Determine the slope and the y-intercept for the following linear equation. Assume that a, b, and c are constants, $b \neq 0$, and x and y are variables.

$$ax + by = c$$

63. A student says that the line $2x + 3y = 6$ has slope 2 because the coefficient of x is 2. What would you tell the student?

64. A student says that the line $y = 3x - 5$ has slope $3x$. What would you tell the student?

65. Explain why the slope addition property makes sense. Include a table of ordered pairs for a linear equation in your explanation.

66. The graph of a linear equation can be sketched by
- plotting points,
- using (sometimes) the slope and the y-intercept, and
- using (sometimes) the x-intercept and the y-intercept.

Discuss how to use each method to sketch the graph of a linear equation. For each method, describe the types of equations, if any, for which you would sketch graphs using that method.

1.5 FINDING LINEAR EQUATIONS

Objectives

▶ Find an equation of a line.

▶ Know the *point-slope form* of a linear equation.

In Section 1.4 we investigated the meaning of the slope of a nonvertical line. In this section we discuss how to use the slope of such a line to help us find an equation of the line.

Example 1   Using Slope and a Point to Find an Equation of a Line

Find an equation of the line that has slope $m = 3$ and contains the point $(2, 5)$.

Solution

Recall that the equation for a nonvertical line can be put into the form

$$y = mx + b$$

Since $m = 3$, we have

$$y = 3x + b$$

To find b, recall that any point on the graph of an equation must satisfy that equation. In particular, the point $(2, 5)$ should satisfy the equation $y = 3x + b$:

$$5 = 3(2) + b$$
$$5 = 6 + b$$
$$5 - 6 = 6 + b - 6$$
$$-1 = b$$

Now, we substitute -1 for b in $y = 3x + b$:

$$y = 3x - 1$$

We can use a graphing calculator to verify that the graph of $y = 3x - 1$ contains the point $(2, 5)$. See Fig. 72. ■

Graphing Calculator

To check that $y = 3x - 1$ contains $(2, 5)$, see Section B.5 on "TRACE."

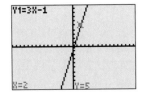

Figure 72 Check that the line contains $(2, 5)$

In Example 1 we found an equation of a line using a point and the slope of the line. We can also find an equation of a line using two points.

Example 2   Using Two Points to Find an Equation of a Line

Find an equation of the line that contains the points $(-2, 6)$ and $(3, -4)$.

Solution

First, we find the slope of the line:

$$m = \frac{-4 - 6}{3 - (-2)} = \frac{-10}{3 + 2} = \frac{-10}{5} = -2$$

So, we have $y = -2x + b$. Since the line contains the point $(3, -4)$, we substitute 3 for x and -4 for y:

$$-4 = -2(3) + b$$
$$-4 = -6 + b$$
$$-4 + 6 = -6 + b + 6$$
$$2 = b$$

So, the equation is $y = -2x + 2$. We can use a graphing calculator to check that the graph of $y = -2x + 2$ contains both $(-2, 6)$ and $(3, -4)$. See Fig. 73.

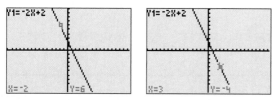

Figure 73 Check that the line contains both $(-2, 6)$ and $(3, -4)$

Finding a Linear Equation from Two Points

To find an equation of the line passing through two given points whose x-coordinates are different:

1. Use the formula $m = \dfrac{y_2 - y_1}{x_2 - x_1}$ to find the slope of the line containing the two points.

2. Substitute the value you found for m into the equation $y = mx + b$.

3. To find b, substitute the coordinates of one of the given points into your equation from step 2 and solve for b.

4. Substitute the values you found for m and b into the equation $y = mx + b$.

5. Use your graphing calculator to check that the graph of your equation contains the two given points.

In Example 3 we find an equation of a line whose slope is a fraction.

Example 3   Using Two Points to Find an Equation of a Line

Find an equation of the line passing through $(-3, -5)$ and $(2, -1)$.

Solution

First, we find the slope of the line:

$$m = \frac{-1 - (-5)}{2 - (-3)} = \frac{-1 + 5}{2 + 3} = \frac{4}{5}$$

So, we have $y = \dfrac{4}{5}x + b$. Since the line contains the point $(2, -1)$, we substitute 2 for x and -1 for y:

$$-1 = \frac{4}{5}(2) + b \qquad \text{Substitute 2 for } x \text{ and } -1 \text{ for } y.$$

$$-1 = \frac{8}{5} + b \qquad \frac{4}{5}(2) = \frac{4}{5}\left(\frac{2}{1}\right) = \frac{8}{5}$$

$$5 \cdot (-1) = 5 \cdot \frac{8}{5} + 5 \cdot b \qquad \text{Multiply both sides by 5.}$$

$$-5 = 8 + 5b \qquad 5 \cdot \frac{8}{5} = \frac{5}{1} \cdot \frac{8}{5} = \frac{8}{1} = 8$$

$$-13 = 5b \qquad \text{Subtract 8 from both sides.}$$

$$-\frac{13}{5} = b \qquad \text{Divide both sides by 5, } \frac{-13}{5} = -\frac{13}{5}$$

So, the equation is $y = \dfrac{4}{5}x - \dfrac{13}{5}$. We can use a graphing calculator to check that the graph of $y = \dfrac{4}{5}x - \dfrac{13}{5}$ contains both $(-3, -5)$ and $(2, -1)$. See Fig. 74.

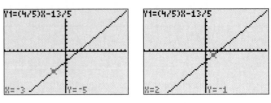

Figure 74 Check that the line contains both $(-3, -5)$ and $(2, -1)$

◼

In Example 4 we will find an equation of a line that contains a given point and is parallel to a given line. Recall that parallel nonvertical lines have equal slopes.

Example 4   Finding an Equation of a Line Parallel to a Given Line

Find an equation of a line l that contains the point $(5, 3)$ and is parallel to the line $y = 2x - 3$.

Note

Although we will use a symbolic approach to find the equation for l, a graph helps us to see how the pieces of the puzzle fit together.

Solution

We use a symbolic method to find an equation for the line l (see Fig. 75).

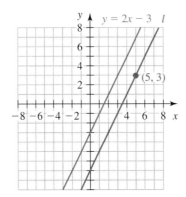

Figure 75 Find the equation of line l

For $y = 2x - 3$, the slope is 2. So, the slope of parallel line l is also 2. An equation for line l is $y = 2x + b$. To find b, we substitute the coordinates of $(5, 3)$ into the equation $y = 2x + b$.

$$3 = 2(5) + b$$
$$b = -7$$

An equation for l is $y = 2x - 7$. We use a graphing calculator to verify our equation (see Fig. 76).

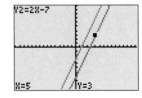

Graphing Calculator

To trace on $y = 2x - 7$, see Section B.5.

Figure 76 Check that the line contains $(5, 3)$ and is parallel to $y = 2x - 3$

◼

Recall that if two nonvertical lines are perpendicular, then the slope of one line is the opposite reciprocal of the slope of the other line.

Example 5 Finding an Equation of a Line Perpendicular to a Given Line

Find an equation of the line l that contains the point $(2, 5)$ and is perpendicular to the line $-2x + 5y = 10$.

Solution

First, we isolate y in the equation $-2x + 5y = 10$.

$$-2x + 5y = 10$$
$$-2x + 5y + 2x = 10 + 2x \qquad \text{Add } 2x \text{ to both sides.}$$
$$5y = 2x + 10 \qquad \text{Simplify.}$$
$$\frac{5y}{5} = \frac{2}{5}x + \frac{10}{5} \qquad \text{Divide both sides by 5.}$$
$$y = \frac{2}{5}x + 2 \qquad \text{Simplify.}$$

For the line $y = \frac{2}{5}x + 2$, the slope is $m = \frac{2}{5}$. The slope of the line l is the opposite reciprocal of $\frac{2}{5}$, or $-\frac{5}{2}$. An equation for l is $y = -\frac{5}{2}x + b$. To find b, we substitute the coordinates of the given point $(2, 5)$ into $y = -\frac{5}{2}x + b$.

$$5 = -\frac{5}{2}(2) + b$$
$$5 = -\frac{5}{2}\left(\frac{2}{1}\right) + b$$
$$5 = -5 + b$$
$$10 = b$$

An equation for l is $y = -\frac{5}{2}x + 10$. We use ZStandard followed by ZSquare to verify our work (see Fig. 77).

 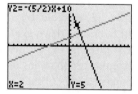

Figure 77 Check that the line contains $(2, 5)$ and is perpendicular to $-2x + 5y = 10$

Graphing Calculator

For the lines to look perpendicular, we need to use ZSquare.

In Example 6 we will use a graphical approach to find the equation of a line.

Example 6 Finding an Equation of a Line Perpendicular to a Given Line

Find an equation of the line l that contains $(4, 3)$ and is perpendicular to the line $x = 2$.

Solution

We want to find an equation for the line l that is sketched in Fig. 78. An equation for the horizontal line l is $y = 3$.

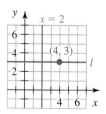

Figure 78 Find an equation for the line l

We can also find an equation of a line by another method. Suppose that a nonvertical line has slope m and contains the point (x_1, y_1). Then, if (x, y) represents a different point on the line, the slope of the line is:

$$\frac{y - y_1}{x - x_1} = m$$

Multiplying both sides of the equation by $x - x_1$ gives:

$$\frac{y - y_1}{x - x_1} \cdot (x - x_1) = m(x - x_1)$$

$$y - y_1 = m(x - x_1)$$

Note

Although we assumed that (x, y) is different from (x_1, y_1), note that (x_1, y_1) is a solution of the equation: $y_1 - y_1 = m(x_1 - x_1)$, or $0 = 0$, a true statement.

We say that this linear equation is in **point-slope form**.

Point-Slope Form

If a nonvertical line has slope m and contains the point (x_1, y_1), then an equation for the line is:

$$y - y_1 = m(x - x_1)$$

Example 7 Using Point-Slope Form to Find an Equation of a Line

A line has slope $m = 2$ and contains the point $(3, -8)$. Find an equation of the line.

Solution

Substituting $x_1 = 3$, $y_1 = -8$, and $m = 2$ into the equation $y - y_1 = m(x - x_1)$ gives:

$$
\begin{aligned}
y - (-8) &= 2(x - 3) \\
y + 8 &= 2x - 6 &\text{Distributive law} \\
y + 8 - 8 &= 2x - 6 - 8 &\text{Subtract 8 from both sides.} \\
y &= 2x - 14 &\text{Simplify.}
\end{aligned}
$$

We can use a graphing calculator to check that the graph of $y = 2x - 14$ contains the point $(3, -8)$. ■

So far in this course, we have worked with equations, graphs, and tables. In Section 1.2 we sketched a graph by going from an equation to a table and then to a graph. In Section 1.4 we sketched a graph by going directly from an equation to a graph. In exercises 69–71 at the end of this section you will find an equation of a line by going from its graph to an equation. Many times throughout the rest of the course we will use combinations of the six paths indicated in Fig. 79.

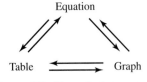

Figure 79 Six paths between equations, tables, and graphs

EXPLORATION
Deciding which points to use to find an equation of a line

1. **a.** Use the method shown in Example 2 to find an equation of the line that contains the points $(1, 2)$ and $(3, 8)$.
   b. In part a, you used one of the points $(1, 2)$ or $(3, 8)$ to find the constant b for the equation $y = 3x + b$. Now find b using the other point.
   c. Does it matter which ordered pair is used to find the constant b? Explain.

2. Now imagine any line that is not parallel to either axis. Choose four points on the line. Name the points A, B, C, and D.

   a. Use the points A and B to find an equation of the line. Write your equation in slope-intercept form $y = mx + b$.

   b. Use the points C and D to find an equation of the line. Write your equation in slope-intercept form.

   c. Are the equations you found in parts a and b the same? Explain.

Tips on Succeeding in This Course

Do you have difficulty with math? If so, do you ever tell yourself (or others) that you are not good at it? This is called *negative self-talk*. The sad thing is, the more you say this to yourself (or others), the more likely your subconscious will believe it and you *will* do poorly in math.

You can counteract years of negative self-talk by telling yourself with conviction that you are good at math.

"I am good at algebra and I continue to improve at it"

It might seem strange to state that something is true that hasn't happened yet, but it works! Such statements are called *affirmations*.

There are three guiding principles for getting the most out of saying affirmations.

1. Say affirmations that imply that the desired event has already happened or is currently happening. For example, say

   "I am good at algebra"   not   "I will be good at algebra"

2. Say affirmations in the positive. For example, say

   "I attend each class"   not   "I don't cut classes"

3. Say affirmations with conviction.

If you would like to learn more about affirmations, the book *Creative Visualization* (Bantam Books, 1985), by Shakti Gawain is an excellent resource.

Key Points
OF THIS SECTION

Finding an equation of a line that contains two given points	To find an equation of the line passing through two given points whose x-coordinates are different: 1. Use the formula $m = \dfrac{y_2 - y_1}{x_2 - x_1}$ to find the slope of the line containing the two points. 2. Substitute the value you found for m into the equation $y = mx + b$. 3. Use one of the given points to find b. For example, if one of the points is $(2, 5)$ and the slope is 3, then substitute 2 for x and 5 for y in the equation $y = 3x + b$ and solve for b. 4. Substitute the values you found for m and b into the equation $y = mx + b$. 5. Use your graphing calculator to check that the graph of your equation contains the two given points.
Point-slope form	If a nonvertical line has slope m and contains the point (x_1, y_1), then an equation for the line is $y - y_1 = m(x - x_1)$. We say that such an equation is in *point-slope form*.

HOMEWORK 1.5

 SSM Tutor Center   MathPro 4   MathPro 5   CDs/Videos   www

Find an equation of the line that has the given slope and contains the given point. Use your graphing calculator to verify that the graph of your equation passes through the given point. Also check that the sign of m agrees with whether your line is increasing or decreasing.

1. $m = 3$, $(5, 2)$
2. $m = 5$, $(-3, -1)$
3. $m = -2$, $(3, -9)$
4. $m = -4$, $(-2, -8)$
5. $m = 1.6$, $(2.1, 3.8)$
6. $m = -3.24$, $(-5.2, 1.9)$
7. $m = \dfrac{3}{5}$, $(20, 7)$
8. $m = \dfrac{2}{3}$, $(6, -1)$
9. $m = -\dfrac{1}{6}$, $(2, -3)$
10. $m = -\dfrac{1}{4}$, $(-2, 3)$
11. $m = -\dfrac{5}{2}$, $(-3, -4)$
12. $m = \dfrac{3}{4}$, $(-5, 2)$
13. $m = 0$, $(1, 2)$
14. $m = 0$, $(-3, -4)$
15. m is undefined, $(3, 7)$
16. m is undefined, $(-5, 1)$

Find an equation of the line passing through the two given points. Verify your equation using your graphing calculator.

17. $(2, 3)$ and $(4, 5)$
18. $(3, 5)$ and $(7, 1)$
19. $(-2, 6)$ and $(3, -4)$
20. $(-4, -6)$ and $(-2, 0)$
21. $(-8, -6)$ and $(-4, -14)$
22. $(-4, -1)$ and $(-2, -7)$
23. $(0, 0)$ and $(1, 1)$
24. $(0, 8)$ and $(4, 5)$
25. $(5.1, 3.9)$ and $(7.4, 2.2)$
26. $(-9.4, 7.1)$ and $(3.9, -2.3)$
27. $(3.4, 1.2)$ and $(6.8, 4.9)$
28. $(-7.13, -2.21)$ and $(-4.99, -7.78)$
29. $(2, 1)$ and $(7, 5)$
30. $(3, 2)$ and $(5, 9)$
31. $(-5, -7)$ and $(-3, -2)$
32. $(2, -1)$ and $(5, -3)$
33. $(-4, 2)$ and $(2, -5)$
34. $(-5, -3)$ and $(-2, -4)$
35. $(2, 5)$ and $(4, 5)$
36. $(9, 4)$ and $(9, 6)$
37. $(-3, -4)$ and $(-3, 6)$
38. $(4.2, -6.8)$ and $(5.3, -6.8)$

Find an equation of the line that contains the given point and is parallel to the given line. Use your graphing calculator to verify your result.

39. $y = 3x + 1$, $(4, 5)$
40. $y = 4x - 6$, $(1, 4)$
41. $y = -2x + 7$, $(-3, 8)$
42. $y = -x + 2$, $(2, -3)$
43. $y = \dfrac{1}{2}x - 3$, $(4, 1)$
44. $y = -\dfrac{2}{3}x - 1$, $(6, -3)$
45. $3x - 4y = 12$, $(3, 4)$
46. $5x + 2y = 10$, $(4, -1)$
47. $6y - x = -7$, $(-3, -2)$
48. $3y + 5x = -11$, $(-1, -4)$
49. $y = 6$, $(2, 3)$
50. $y = -4$, $(3, -1)$
51. $x = 2$, $(-5, 4)$
52. $y = -x$, $(-2, -5)$

Find an equation of the line that contains the given point and is perpendicular to the given line. Use ZStandard followed by ZSquare with your graphing calculator to verify your result.

53. $y = 2x + 5$, $(3, 8)$
54. $y = 5x - 4$, $(2, 1)$

55. $y = -3x + 7$, $(-1, 7)$

56. $y = -6x - 13$, $(-3, -2)$

57. $y = -\dfrac{2}{5}x + 3$, $(2, 7)$

58. $y = \dfrac{1}{3}x - 4$, $(1, -2)$

59. $4x - 5y = 7$, $(10, 3)$

60. $5x + 2y = -9$, $(6, -1)$

61. $-2x + y = 5$, $(-3, -1)$

62. $-3x - 4y = 12$, $(-1, 2)$

63. $x = 5$, $(2, 3)$

64. $x = -1$, $(-4, -2)$

65. $y = -3$, $(2, 8)$

66. $y = 7$, $(1, -1)$

67. Let y represent the value (in thousands of dollars) of a car when it is x years old. Some pairs of values for x and y are listed in Table 21.

Table 21 Values of a Car	
Age (years) x	Value (thousands of dollars) y
0	19
1	17
2	15
3	13
4	11

Find an equation that describes the relationship between x and y.

68. Let y represent a person's salary (in thousands of dollars) after working at a company for x years. Some pairs of values for x and y are listed in Table 22.

Table 22 Salaries	
Time at Company (years) x	Salary (thousands of dollars) y
0	25
1	28
2	31
3	34
4	37
5	40

Find an equation that describes the relationship between x and y.

69. Find an equation for the line sketched in Fig. 80. Check your equation with your graphing calculator.

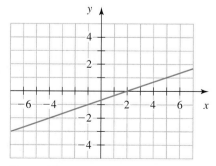

Figure 80 Exercise 69

70. Find an equation for the line sketched in Fig. 81. Check your equation with your graphing calculator.

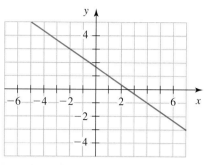

Figure 81 Exercise 70

71. Find an equation for the line sketched in Fig. 82. Check your equation with your graphing calculator.

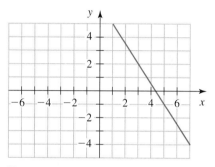

Figure 82 Exercise 71

72. Find equations of two perpendicular lines that intersect at $(3, 1)$.

73. Decide whether it is possible for a line to have the indicated number of x-intercepts. If it is possible, find an equation of such a line. If it is not possible, explain why.

a. No x-intercepts

b. Exactly one x-intercept

c. Exactly two x-intercepts

d. Infinite number of x-intercepts

74. Decide whether it is possible for a line to have the indicated number of y-intercepts. If it is possible, find an equation of such a line. If it is not possible, explain why.

a. No y-intercepts

b. Exactly one y-intercept

c. Exactly two y-intercepts

d. Infinite number of y-intercepts

75. Is there a line that contains all of the given points? If so, find an equation for it. If not, find an equation of a line that contains most of the given points.

$(-4, 15)$, $(-1, 9)$, $(3, 1)$, $(4, -1)$, $(9, -11)$

76. Is there a line that contains all of the given points? If so, find an equation for it. If not, find an equation of a line that contains most of the given points.

$(-3, 7)$, $(-1, 5)$, $(1, 1)$, $(3, -3)$, $(4, -5)$

77. Create a table of seven pairs of values for x and y that meet the given criteria.

a. Each point lies on the line $y = 3x - 6$.

b. Each point lies close to, but not on, the line $y = 3x - 6$.

c. The points do not lie close to the line $y = 3x - 6$, but all of them do lie close to another line. Also, provide an equation for the other line.

78. Suppose that a set of points all lie 0.5 unit above the line $y = -4x + 3$. Find an equation of the line that passes through the points of the set.

79. a. Find an equation of a line with slope -4.

b. Find an equation of a line with y-intercept $\left(0, \frac{3}{7}\right)$. Verify your result with your graphing calculator.

c. Find an equation of a line that contains the point $(-2, 8)$. Verify your result with your graphing calculator.

d. Determine whether there is a line that has slope -4, has y-intercept $\left(0, \frac{3}{7}\right)$, and contains the point $(-2, 8)$. Explain.

80. Find an equation of the line whose x-intercept is $(-2, 0)$ and whose y-intercept is $(0, 3)$. Verify your result with your graphing calculator.

81. Describe how to find an equation of a line that contains two given points. Also, explain how you can verify that the graph of the equation contains the two points.

1.6 FUNCTIONS

Objectives

▸ Know the meanings of *relation*, *domain*, *range*, and *function*.

▸ Identify functions using the vertical line test.

▸ Know the definition of a *linear function*.

In this chapter we have described the relationship between two variables using graphs, tables, and equations. For example, Table 23 describes a relationship between variables x and y. This relationship is also described graphically in Fig. 83.

Table 23 A Relation Described by a Table

x	y
3	2
4	1
5	3
5	4

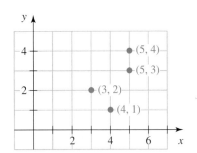

Figure 83 The same relation described by a graph

We call the set of ordered pairs listed in Table 23 a *relation*. This relation consists of four ordered pairs.

Each of the following equations describes a relation consisting of an infinite number of ordered pairs.

$$y = x + 3 \qquad y^2 = x \qquad x^2 + y^2 = 4$$

For the relation described in Table 24, we can think of the values of x as being sent to the values of y (see Fig. 84).

Table 24 Another Relation Described by a Table

x	y
1	2
2	4
2	5
3	6

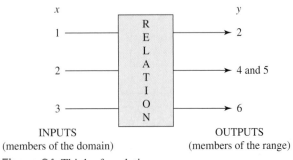

Figure 84 Think of a relation as an input-output machine

We can think of a relation as a machine, where values of x are "inputs" and values of y are "outputs." The **domain** of a relation is the set of all inputs, and the **range** of a relation is the set of all outputs.

Note that the input $x = 2$ is sent to *two* outputs $y = 4$ and $y = 5$. There is a special type of relation called a **function**, where each input is sent to exactly *one* output. So, the relation described in Table 24 is not a function.

DEFINITION *Function*

A **function** is a relation where each input gives exactly one output.

Example 1   Deciding Whether an Equation Describes a Function

Is the relation $y = x + 2$ a function? Also, find the domain and range of the relation.

Solution

Let's consider some input-output pairs (see Fig. 85).

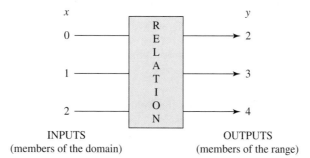

Figure 85 The "increasing by 2" relation: $y = x + 2$

INPUTS
(members of the domain)

OUTPUTS
(members of the range)

Any input gives *one* output, namely the input increased by 2, so the relation $y = x + 2$ is a function.

The domain of the relation $y = x + 2$ is the set of all real numbers, since we can add 2 to *any* real number. The range of $y = x + 2$ is also the set of real numbers, since any real number is the output of the number that is 2 units less than it. ∎

Example 2   Deciding Whether an Equation Describes a Function

Is the relation $y = \pm x$ a function?

Solution

If $x = 1$, then $y = \pm 1$. So, the input $x = 1$ yields *two* outputs $y = -1$ and $y = 1$. Therefore, the relation $y = \pm x$ is not a function. ∎

Example 3   Deciding Whether an Equation Describes a Function

Is the relation $y^2 = x$ a function?

Solution

Let's consider the input $x = 4$. We substitute 4 for x and solve for y:

$$y^2 = 4 \qquad \text{Substitute 4 for } x.$$
$$y = -2 \quad \text{or} \quad y = 2 \qquad (-2)^2 = 4, 2^2 = 4$$

Note that the input $x = 4$ yields *two* outputs $y = -2$ and $y = 2$. So the relation $y^2 = x$ is not a function. ∎

Table 25 Input-Output Pairs for a Relation

x (input)	y (output)
0	2
1	3
1	5
2	7
3	10

Example 4 Deciding Whether a Table Describes a Function

Is the relation described by Table 25 a function?

Solution

Note that the input $x = 1$ gives *two* outputs $y = 3$ and $y = 5$; so the relation is not a function. ∎

Example 5 Deciding Whether a Graph Describes a Function

Is the relation described by the graph in Fig. 86 a function?

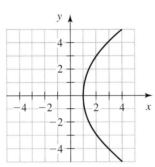

Figure 86 Graph of a relation

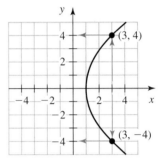

Figure 87 The input $x = 3$ gives two outputs $y = -4$ and $y = 4$

Note

We did not have to work with the input $x = 3$. We could have chosen *any* input greater than 1 and shown that it has two outputs.

Solution

See Fig. 87. Note that the input $x = 3$ gives *two* outputs $y = -4$ and $y = 4$. So the relation is not a function. ∎

Notice that the relation described in Example 5 is not a function because some vertical lines intersect the graph more than once.

Vertical Line Test

A relation is a function if each vertical line intersects the graph of the relation at no more than one point. Otherwise, the relation is not a function.

Example 6 Deciding Whether a Graph Describes a Function

Is the relation described by the circle sketched in Fig. 88 a function?

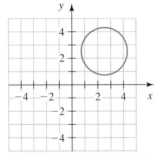

Figure 88 Graph of a circle

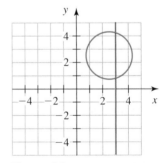

Figure 89 The graph of the circle does not describe a function

Solution

Since the vertical line sketched in Fig. 89 intersects the circle more than once, we know that the relation is not a function. ∎

Example 7　Deciding Whether an Equation Describes a Function

Is the relation $y = 2x + 1$ a function?

Solution

We begin by sketching the graph of $y = 2x + 1$ in Fig. 90. Note that each vertical line intersects the line $y = 2x + 1$ at one point. So the relation $y = 2x + 1$ is a function.

■

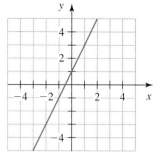

Figure 90　Graph of $y = 2x + 1$

In Example 7 we saw that the line $y = 2x + 1$ is a function. In fact, any nonvertical line is a function, since it passes the vertical line test.

DEFINITION　*Linear function*

A **linear function** is a relation whose equation can be put in the form

$$y = mx + b$$

where m and b are constants.

Note

Vertical lines do not describe functions. Why?

In this chapter we have made many observations about linear equations. Since a linear function can be described by a linear equation, these observations tell us about linear functions. Here, we summarize what we know about a linear function $y = mx + b$:

1. The graph of the function is a line.
2. The constant m is the slope of the line, a measure of the line's steepness.
3. If $m > 0$, the line is increasing.
4. If $m < 0$, the line is decreasing.
5. If $m = 0$, the line is horizontal.
6. If the value of x increases by 1, then the value of y changes by the slope m.
7. If the run is 1, the rise is the slope m.
8. The y-intercept of the line is $(0, b)$.

Finally, since a linear equation of the form $y = mx + b$ is a *function*, we know that each input gives exactly one output.

EXPLORATION
Vertical line test

1. Consider the relation described by Table 26. Is the relation a function? Explain. Now plot the points on a coordinate system. What do you notice about these points?
2. Consider the relation described by Table 27. Is the relation a function? Explain. Now plot the points on a coordinate system. What do you notice about these points?

Table 26 A Relation Described by a Table

x	y
2	1
2	5
2	7

Table 27 A Relation Described by a Table

x	y
4	2
4	3
4	6

3. Make a general conjecture about the graph of a relation that is not a function.

4. Determine whether each graph in Fig. 91 is the graph of a function.

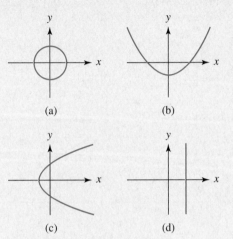

Figure 91 Which graphs describe functions?

EXPLORATION

Looking ahead: Linear modeling

The production of the Chevrolet Camaro® has decreased since 1996 (see Table 28).

Table 28	Camaro Production
Year	Camaro Production (thousands)
1996	67
1997	56
1998	48
1999	41
2000	42
2001	29

Source: *BlueOvalNews, Independent Voice of the Ford Community, www.bonforums.com/ sales/sales_mustangleader.htm. Accessed 4/7/02.*

1. Let C represent the Camaro production (in thousands). Also, let t represent the number of years since 1995. For example, $t = 1$ represents 1996, because 1996 is 1 year since 1995. So, the 1996 production of 67 thousand Camaros can be represented by $t = 1$ and $C = 67$. The information in Table 28 can be summarized with a table of values for t and C. Create such a table by filling in the missing entries in Table 29.

2. Plot the points (t, C) that you listed in Table 29 on a coordinate system like the one in Fig. 92.

3. When examining your graph, what do you notice about the arrangement of your plotted points?

4. Sketch a line that comes close to the six data points.

5. Use the line to predict when 10 thousand Camaros will be produced.

Table 29	Values of t and C for the Camaro Data
Number of Years Since 1995 t	Camaro Production (in thousands) C
1	67
2	56
3	
4	
	42
	29

6. Use the line to estimate the number of Camaros produced in 2002.

7. Find the *t*-intercept of the line. What does the *t*-intercept mean in terms of the production of Camaros? Will this prediction happen for certain? Explain.

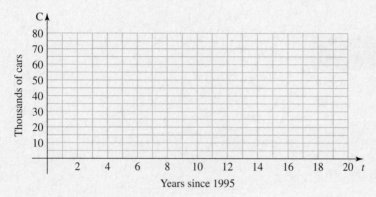

Figure 92 Plot the data points

Tips on Succeeding in This Course

When studying for an exam (or quiz), try creating your own exam to take for practice. To create your exam, select several homework exercises from each section on which you will be tested. Choose a variety of exercises that address concepts that your instructor has emphasized. Your test should include many exercises that are moderately difficult and some that are challenging. Completing such a practice test will help you reflect on important concepts and help you pin down what types of problems you need to study more.

It is a good idea to work on the practice exam for a predetermined time period. Doing so will help you get used to a timed exam, build your confidence, and lower your anxiety for the real exam.

If you are studying with another student, you can each create a test and then take each other's test. Or you can create a test together and each take it separately.

Key Points
OF THIS SECTION

A *relation* is a set of ordered pairs.	**Relation**
The *domain* of a relation is the set of all inputs. The *range* of a relation is the set of all outputs.	**Domain and range**
A *function* is a relation where each input gives exactly one output.	**Function**
How to use the vertical line test:	**Vertical line test**

- If every vertical line intersects the graph of a relation at no more than one point, then the relation is a function.
- If a vertical line intersects the graph of a relation in at least two points, then the relation is not a function.

A *linear function* is a relation that can be described by an equation of the form $y = mx + b$, where m and b are constants.	**Linear function**
The graph of a linear function is a nonvertical line.	**Graph of a linear function**

HOMEWORK 1.6

SSM
Tutor Center

MathPro 4
MathPro 4

MathPro 5
MathPro 5

CDs/Videos

www
www

1. Some (but not all) ordered pairs for four relations are listed in Table 30. Which of these relations could possibly be functions?

Table 30 Which Relations Might Be Functions? (Exercise 1)

Relation 1		Relation 2		Relation 3		Relation 4	
x	y_1	x	y_2	x	y_3	x	y_4
1	1	3	27	0	4	5	10
2	3	4	24	1	4	6	20
3	5	5	21	2	4	7	30
3	7	6	18	3	4	8	40
4	9	7	15	4	4	8	50

2. Some (but not all) ordered pairs for four relations are listed in Table 31.
   a. Which of the relations could possibly be functions?
   b. Which of the relations could possibly be linear functions?

Table 31 Which Relations Might Be Functions? (Exercise 2)

Relation 1		Relation 2		Relation 3		Relation 4	
x	y_1	x	y_2	x	y_3	x	y_4
1	3	5	27	0	50	3	11
2	4	5	24	1	45	4	13
3	5	5	21	2	40	5	17
3	6	5	18	3	35	6	25
4	7	5	15	4	30	7	40

3. For a certain relation, an input is sent to two different outputs. Could the relation possibly be a function? Explain.

4. A relation's graph contains the points $(2, 3)$ and $(5, 3)$. Is it possible that the relation is a function? Explain.

5. For a certain relation, two different inputs are sent to the same output. Could the relation possibly be a function? Explain.

6. A relation's graph contains the points $(4, 5)$ and $(4, 9)$. Is it possible that the relation is a function? Explain.

7. Determine whether each graph in Fig. 93 is the graph of a function.

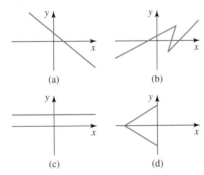

(a) (b)

(c) (d)

Figure 93 Exercise 7

8. Determine whether each graph in Fig. 94 is the graph of a function.

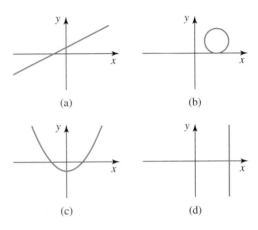

(a) (b)

(c) (d)

Figure 94 Exercise 8

Determine whether each relation is a function.

9. $y = 5x - 1$

10. $y = -3x + 8$

11. $2x - 5y = 10$

12. $4x + 3y = 24$

13. $y = 2$

14. $3y + 1 = 7$

15. $x = -3$

16. $x = 0$

17. $7x - 2y = 21 + y + x$

18. $2x + 5y - 6 = 9 + 5(y - 1)$

19. Is a nonvertical line the graph of a function? Explain.

20. Is a vertical line the graph of a function? Explain.

21. Is a circle the graph of a function? Explain.

22. Is a semicircle that is the "upper half" of a circle the graph of a function? Explain.

23. Create an equation (different than the ones in this section) that describes a function. Next, sketch a graph of the function and create a table that lists five ordered pairs of the function. Explain why your relation is a function.

24. Sketch a graph of a relation (different than the ones in this section) that is not a function. Next, create a table that lists five ordered pairs of the relation. Explain why your relation is not a function.

25. Sketch the graph of a relation where the input $x = 2$ gives exactly two outputs and the input $x = 6$ gives exactly one output. Is the relation a function? Explain.

26. Sketch the graph of a relation where the input $x = -4$ gives exactly three outputs and the input $x = 5$ gives exactly one output. Is the relation a function? Explain.

Decide whether each relation is a function. Explain.

27. $y = \sqrt{x}$ [**Hint:** Sketch a graph.]

28. $|y| = x$ [**Hint:** Substitute 2 for x, then solve for y.]

29. $y^4 = x$ [**Hint:** Substitute 16 for x, then solve for y.]

30. $y = x^4$

31. $y = x^3$

32. $y^3 = x$

33. A student tries to determine whether the relation $y = x^2$ is

a function. He finds that the inputs $x = -3$ and $x = 3$ both give the same output $y = 9$. The student concludes that the relation is not a function. Is the student's conclusion correct? Explain.

34. A student tries to determine whether the relation $|y| = |x|$ is a function. He finds that the input $x = 0$ gives exactly one output $y = 0$. The student concludes that the relation is a function. Is the student's conclusion correct? Explain.

35. Explain how you can determine whether a relation is a function.

CHAPTER SUMMARY

Key Points OF THIS CHAPTER

A *qualitative graph* is a graph without scaling on the axes.	**Qualitative graph**
When sketching graphs, we usually match the vertical axis with the dependent variable and the horizontal axis with the independent variable.	**Axes of a graph**
All points that satisfy an equation form the *graph* of the equation. Points that do not satisfy an equation are not part of the graph.	**Graph of an equation**

If a and b are constants, then: **Vertical and horizontal lines**

* For an equation of the form $x = a$, the graph is a vertical line.
* For an equation of the form $y = b$, the graph is a horizontal line.

Let (x_1, y_1) and (x_2, y_2) be two distinct points of a nonvertical line. The *slope* of   **Calculating slope**
the line is

$$m = \frac{\text{rise}}{\text{run}} = \frac{y_2 - y_1}{x_2 - x_1}$$

To sketch a graph of a function in the form $y = mx + b$:   **Graphing a function**

1. Plot the y-intercept $(0, b)$.

2. Use the fact that $m = \dfrac{\text{rise}}{\text{run}}$ to plot another point. For example, if $m = -\dfrac{3}{2} = \dfrac{-3}{2}$, then look 2 units to the right (of the y-intercept) and 3 units down to plot another point.

3. Then sketch the line that passes through your two plotted points.

To find an equation of the line passing through two given points whose x-coordinates   **Finding an equation of a line**
are different: **that contains two given points**

1. Use the formula $m = \dfrac{y_2 - y_1}{x_2 - x_1}$ to find the slope of the line containing the two points.

2. Substitute the value that you find for m into the equation $y = mx + b$.

3. Use one of the given points to find b. For example, if one of the points is $(1, 6)$ and the slope is 4, then substitute 1 for x and 6 for y into the equation $y = 4x + b$ and solve for b.

4. Substitute the values you find for m and b into the equation $y = mx + b$.

5. Use your graphing calculator to check that the graph of your equation contains the two given points.

A *relation* is a set of ordered pairs.	**Relation**
The *domain* of a relation is the set of all inputs. The *range* of a relation is the set of all outputs.	**Domain and range**
A *function* is a relation where each input gives exactly one output.	**Function**
A relation is a function if each vertical line intersects the graph of the relation at no more than one point. Otherwise, the relation is not a function.	**Vertical line test**

Properties of a function of the form $y = mx + b$

Properties of a function that can be described by an equation of the form $y = mx + b$:

- The function is called a linear function.
- The graph of the function is a line.
- The constant m is the slope of the line, a measure of the line's steepness.
- If $m > 0$, the line is increasing.
- If $m < 0$, the line is decreasing.
- If $m = 0$, the line is horizontal.
- If the value of x increases by 1, then the value of y changes by the slope m.
- If the run is 1, the rise is m.
- The y-intercept of the line is $(0, b)$.

CHAPTER 1 REVIEW EXERCISES

Sketch a qualitative graph that shows the relationship between the variables defined in each exercise. Justify your sketch.

1. Let L represent the length of a candle t minutes after it is lit.

2. Let w represent the weight of a person at age t years.

3. Let T represent the amount of time it takes to cook a marshmallow that is d inches from a campfire.

4. Let M represent the monthly car insurance payment for a car that is worth d dollars.

5. Let p represent the number of points a student scores on an exam if the student reviews the material for t hours.

6. Write a scenario to match the graph in Fig. 95. Refer to the variables x and y in your description.

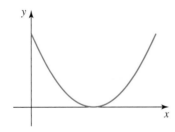

Figure 95 Exercise 6

For exercises 7–10, refer to Fig. 96.

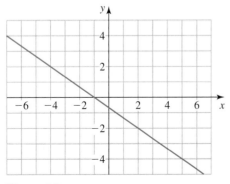

Figure 96 Exercises 7–10

7. Estimate y when $x = 5$.

8. Estimate y when $x = 0$.

9. Estimate x when $y = 2$.

10. Estimate x when $y = 0$.

Find the slope of the line passing through the two given points.

11. $(2, 7)$ and $(5, 4)$

12. $(-3, -2)$ and $(2, -5)$

13. $(-9, -7)$ and $(-1, -3)$

14. $(1.6, 2.8)$ and $(3.2, 8.9)$

15. $(-3, -5)$ and $(-2, -5)$

16. $(4, -1)$ and $(4, 3)$

Use pencil and paper to sketch the graph for each equation.

17. $y = 2x - 3$

18. $y = -3x + 10$

19. $y + 2x = 0$

20. $y = 7$

21. $3x - 2y = 12$

22. $2(x - 2y) - 1 = 7$

23. $y = -\dfrac{2}{3}x + 5$

24. $0.2y - 0.6x = 0.3$

25. $-3(y + 2) = 2x + 9$

26. $y = -0.25x + 2$

Determine whether the given lines are parallel, perpendicular, or neither.

27. $2x + 5y = 7$ and $y = \dfrac{2}{5}x + 7$

28. $x = -2$ and $y = 4$

Find an equation of the line that has the given slope and contains the given point.

29. $m = -4, (-3, 7)$

30. $m = -\dfrac{2}{3}, (5, -4)$

Find the equation of the line passing through the two given points.

31. $(3, 8)$ and $(4, 5)$

32. $(3, 1)$ and $(7, 9)$

33. $(2.8, 3.7)$ and $(6.4, 1.1)$

34. $(-3, -2)$ and $(2, 6)$

35. $(-4, 6)$ and $(2, -2)$

36. $(3, -2)$ and $(3, 5)$

37. Find an equation for each line sketched in Fig. 97.

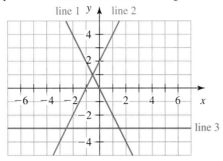

Figure 97 Exercise 37

38. Complete Table 32 so that all of the ordered pairs are contained by the same line.

Table 32 Points on a Line (Exercise 38)	
x	*y*
2	20
3	
4	
5	
6	4
7	

39. Graphs of the functions $y = 0.5x + 6$ and $y = -1.5x + 3$ are shown in Fig. 98. Find an equation for the line labeled "line 1" also sketched in Fig. 98. You may assume that line 1 and the line $y = 0.5x + 6$ are parallel.

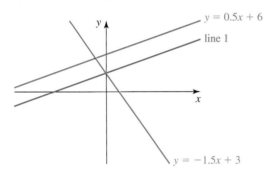

Figure 98 Exercise 39

40. Find the *x*-intercept and *y*-intercept of the function $3x - 5y = 17$.

41. The graphs of $y = ax + b$ and $y = cx + d$ are sketched in Fig. 99.

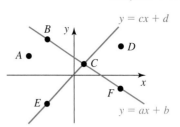

Figure 99 Exercise 41

For each part, decide which of the points A, B, C, D, E, or F

   a. satisfy the equation $y = ax + b$.

   b. satisfy the equation $y = cx + d$.

   c. satisfy both equations.

   d. do not satisfy either equation.

42. Find an equation of the line with slope $m = -\dfrac{3}{2}$ that contains the point $(-3, 8)$.

43. Find an equation of a line that contains the point $(-2, 5)$ and that is parallel to the line $y = 3x - 6$.

44. Decide whether it is possible for a line to have an infinite number of *x*-intercepts. If it is possible, find an equation for such a line. If it is not possible, explain why.

45. Some (but not all) ordered pairs for four relations are listed in Table 33. Which of these relations could possibly be functions? Explain.

Table 33 Which Relations Might Be Functions? (Exercise 45)							
Relation 1		Relation 2		Relation 3		Relation 4	
x	y_1	*x*	y_2	*x*	y_3	*x*	y_4
1	12	3	27	0	7	2	1
2	15	4	24	1	7	2	2
3	18	4	21	2	7	2	3
4	21	5	18	3	7	2	4
5	24	6	15	4	7	2	5

46. Sketch a graph of a function. Explain why your sketch is correct.

Determine whether each relation is a function.

47. $y = -4x - 7$

48. $5x - 6y = 3$

49. $x = 9$

50. $y = 1$

51. $y = x^2$

52. $y^2 = x$

CHAPTER 1 TEST

1. A student eats breakfast at home. After breakfast, she begins walking to school. After a while, she jogs. After jogging for some time, she runs the rest of the way at an even faster pace. Let *d* represent how far the student is from home at *t* minutes since she started eating breakfast. Sketch a qualitative graph that describes the relationship between *t* and *d*.

2. Write a scenario to match the graph in Fig. 100. Refer to the variables *x* and *y* in your description.

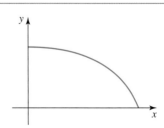

Figure 100 Exercise 2

3. Find an equation for each line sketched in Fig. 101.

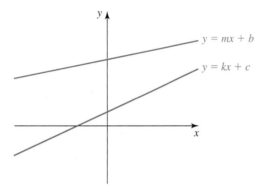

Figure 101 Exercise 3

4. Graphs of the lines $y = mx + b$ and $y = kx + c$ are sketched in Fig. 102.
   a. Which is greater, m or k? Explain.
   b. Which is greater b or c? Explain.

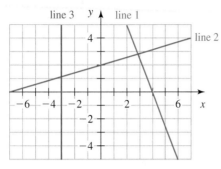

Figure 102 Exercise 4

5. Ski run A steadily declines 85 yards over a horizontal distance of 270 yards. Ski run B steadily declines 140 yards over a horizontal distance of 475 yards. Which run is steeper? Explain.

6. Use examples of linear functions and their graphs to show that slope is a measure of the steepness of a line.

7. Complete Table 34 so that all of the ordered pairs are contained by the same line.

Table 34 Points on a Line (Exercise 7)

x	y
4	
5	
6	33
7	
8	
9	45

Use pencil and paper to sketch the graph of each equation.

8. $y = -\dfrac{1}{5}x + 4$

9. $2x - 3y = 9$

10. Find the slope of the line that contains the points $(-3, 2)$ and $(5, -8)$.

11. A line contains the points $(2, 8)$ and $(5, 6)$. Find three more points that lie on the line.

12. Find an equation of the line with slope -3 that contains the point $(-2, 5)$.

13. Find an equation of the line passing through the points $(-3, 7)$ and $(2, -5)$.

14. Is there a line that contains all of the given points? If so, find an equation for it. If not, find an equation that contains most of the points.

$$(-2, 5), (0, 2), (2, -3), (3, -5), (5, -9)$$

15. Find the equation of a line that contains the point $(4, -1)$ and that is perpendicular to the line $3x - 5y = 20$.

16. Find the x-intercept and y-intercept of the function $2y + 5 = 4(x - 1) + 3$.

17. Create a table, graph, and equation that all describe the same function. Explain why your relation is a function.

18. Sketch the graph of a relation that is *not* a function. Explain.

19. Determine whether the relation described by $y = \pm\sqrt{x}$ is a function. Explain.

20. Determine whether the relation described by $y = -2x + 5$ is a function. Explain.

Modeling with Linear Functions

I have learned throughout my life as a composer chiefly through my mistakes and pursuits of false assumptions, not by my exposure to founts of wisdom and knowledge.

—*Igor Stravinsky*

2.1 USING LINES TO MODEL DATA

Objectives

▸ Know the meaning of *scattergram*, *linear model*, and *model breakdown*.
▸ Make estimates and predictions using a linear model.
▸ Find intercepts of a linear model.

In this chapter we discuss how to use linear functions to make estimates and predictions about true-to-life situations. In this section we use graphs of linear functions to make these estimates and predictions.

Example 1   Using a Graph to Describe a True-to-Life Situation

The Grand Canyon is one of the most beautiful landmarks of the United States. Yet tourists' enjoyment of the attraction is diminished somewhat by the 6000 cars per day searching for the park's 2400 parking spaces (*New York Times*, 1/28/02). The numbers of visitors for various years are listed in Table 1.

Solution

Let v represent the number (in millions) of visitors in the year that is t years since 1960. For example, $t = 10$ represents 1970, since 1970 is 10 years after 1960. Then we can describe the data with a table of values for t and v (see Table 2).

Table 1 Visitors to the Grand Canyon	
Year	Number of Visitors (millions)
1960	1.2
1970	2.3
1980	2.6
1990	3.8
2000	4.8

Source: *GRCA Annual Visitation—1915 to Present, www.nps.gov/grca/facts/yrstat.htm*

Table 2 Visitors to the Grand Canyon	
Number of Years Since 1960 t	Number of Visitors (millions) v
0	1.2
10	2.3
20	2.6
30	3.8
40	4.8

Next, we plot the (t, v) data points as the points shown in Fig. 1. It makes sense to think of v as the dependent variable, so we let the vertical axis be the v-axis. Since t is the independent variable, the horizontal axis is the t-axis. Note that we write the units "Millions of visitors" and "Years since 1960" on the appropriate axes. The ordered pair

(10, 2.3) indicates that when $t = 10$, $v = 2.3$ or in 1970 there were 2.3 million visitors. In general, we list the value of the independent variable first, followed by the value of the dependent variable.

Note

A scattergram should have scaling on both axes and labels indicating the variable names and the scale units.

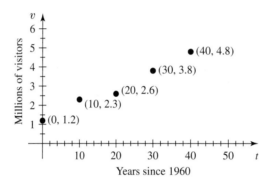

Figure 1 Scattergram for Grand Canyon data

The graph of the plotted data pairs in Fig. 1 is called a **scattergram**. Note that we can sketch a line that comes close to (or on) the data points (see Fig. 2). Since the data points all lie close to one line, we say that t and v are **approximately linearly related**.

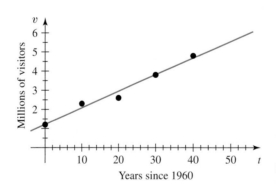

Figure 2 Linear model for Grand Canyon data

Note

The line does go exactly through the data points for 1960 and 1990.

The line in Fig. 2 provides a more meaningful visual description of the relationship between t and v than did the simple scattergram of Example 1. It should be clear, however, that this description is not exact. For example, the line does not describe exactly what happened in any of the years 1970, 1980, and 2000, as the line does not contain the data point for any of these years. However, the line does come very close to these data points, so it does suggest very good approximations for these years.

Note that the line in Fig. 2 is nonvertical, so it is the graph of a linear function. The process of choosing a linear function to represent the relationship between the number of visitors in a year and the number of years since 1960 is an example of *modeling*. In general, a *model* is a mathematical description of a true-to-life situation.

DEFINITION *Model*

A **model** is a mathematical description of a true-to-life situation.

Note

Many models such as the Grand Canyon model are useful, but they often are not *exact* mathematical descriptions of true-to-life situations.

We call the Grand Canyon linear function a *linear model*. In later chapters, we will discuss other types of models. The term "model" is being used in much the same way as it is used in "airplane model." Just as an airplane designer can use the behavior of an airplane model in a wind tunnel to predict the behavior of an actual airplane, a linear model can be used to predict what might happen in a situation in which two variables are approximately linearly related.

DEFINITION *Linear model*

A **linear model** is a linear function that describes the relationship between two quantities for a true-to-life situation.

Every linear model is a linear function. So, depending on our choice of emphasis, we can refer to the Grand Canyon model as a "model" or a "function."

However, not every linear function is a linear model. Models are used only to describe situations. Functions are used both to describe situations *and* to describe certain *mathematical* relationships between two variables. For example, if the equation $y = 2x$ is not being used to describe a situation, then it is a function, not a model.

Since the Grand Canyon data points all lie close to (or on) our linear model, it seems reasonable that data points for visitors in years between 1960 and 2000 not shown in Table 2 might also lie close to the line. Similarly, it is reasonable that data points for at least a few years before 1960 or for at least a few years after 2000 might also lie near the line.

Example 2   Using a Linear Model to Make an Estimate and a Prediction

1. Use the linear model shown in Fig. 2 to predict the number of visitors in 2010.
2. Use the linear model to estimate in what year there were 4 million Grand Canyon visitors.

Solution

1. The year 2010 corresponds to $t = 50$ because $2010 - 1960 = 50$. To estimate the number of visitors, we locate the point on the linear model where the t-coordinate is 50 and see that the v-coordinate is about 5.6 (see Fig. 3). So, according to the model, there will be 5.6 million visitors in 2010.

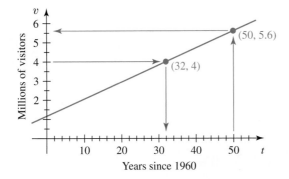

Figure 3 Using the Grand Canyon model to make an estimate and a prediction

2. To find the year of 4 million visitors, we locate the point on the linear model where the v-coordinate is 4 and see that the t-coordinate is about 32. So, *according to the linear model*, there were 4 million visitors in 1992. ◼

To help us determine whether to use a linear function to model some data, we sketch a scattergram of the data. To illustrate, we show scattergrams of data sets 1, 2, and 3 in Figs. 4, 5, and 6, respectively. It appears that a linear function would be a reasonable model for data set 1. Data set 2 would be best modeled using a function different than a linear function. Data set 3 cannot be modeled well using *any* type of function.

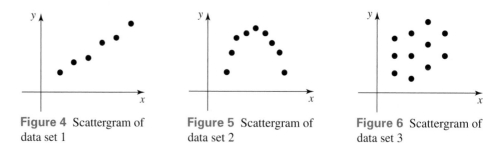

Figure 4 Scattergram of data set 1

Figure 5 Scattergram of data set 2

Figure 6 Scattergram of data set 3

If a data set is approximately linearly related, we use a scattergram of the data to guide us in sketching a reasonable linear model. But beware! It is a common student error to try to find a line that contains the greatest number of data points. However, our goal is to find a line that comes close to *all* of the data points. For example, even though model 1 in Fig. 7 does not contain any of the shown data points, it fits these data points much better than model 2, which does contain three data points.

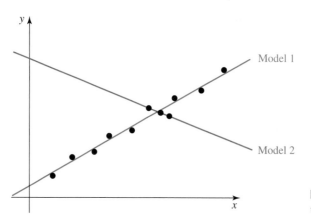

Figure 7 Comparing the fit of two models

In Example 3 we will see that even if a model describes *known* data well, the model may not give reasonable estimates and predictions for all other years.

Example 3 Intercepts of a Model, Model Breakdown

The wild Pacific Northwest salmon populations for various years are listed in Table 3.

What!
Berries again?

Table 3 Wild Pacific North-west Salmon Populations	
Year*	Population (millions)
1960	10.02
1965	10.00
1970	7.61
1975	3.15
1980	4.59
1985	3.11
1990	2.22

Source: *Golden Gate Anglers' Club*
For convenience, we assume throughout this text that, unless otherwise stated, all yearly data were collected on December 31 of the given year.

1. Let P represent the salmon population (in millions) at t years since 1950. Find a linear model that describes the relationship between t and P.

2. Find the P-intercept of the model. What does the point represent in terms of the salmon?

3. Use the model to predict when the salmon will become extinct.

Solution

1. We describe the data in terms of t and P in Table 4. Next, we sketch a scattergram (see Fig. 8). It appears that t and P are approximately linearly related, so we sketch a line that comes close to the data points. (Note that many such lines are possible.)

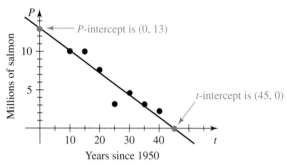

Figure 8 Intercepts of the wild Pacific Northwest salmon model

Number of Years Since 1950 t	Salmon Population (millions) P
10	10.02
15	10.00
20	7.61
25	3.15
30	4.59
35	3.11
40	2.22

Table 4 Values of t and P for the Wild Pacific Northwest Salmon Data

2. The P-intercept is $(0, 13)$, or $P = 13$ when $t = 0$ (the year 1950). According to the model, there were 13 million salmon in 1950.

3. The t-intercept is $(45, 0)$, or $P = 0$ when $t = 45$. According to the model, the salmon became extinct in 1995. Fortunately, this did not happen. A little research would show that there are wild Pacific Northwest salmon alive today, so our model gives an obviously false prediction. ■

To draw the salmon model in Fig. 8, we used a scattergram consisting of data points representing various years from 1960 to 1990. In Fig. 9 we draw the portion of the model that represents the salmon situation from 1960 to 1990 in blue and we draw the rest of the model in red. When we use the blue portion of the model to make estimates, we are performing *interpolation*. When we use the red portions of the model, we are performing *extrapolation*.

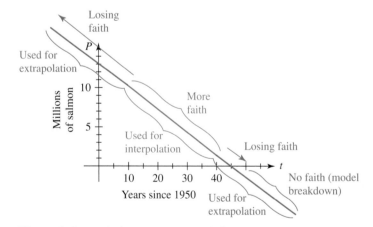

Figure 9 Interpolation versus extrapolation

Although it is possible that we could get large errors from interpolating with the salmon model, we have more faith in our results from interpolating than from extrapolating. That's because the blue portion of the model comes close to several known data points whereas we have no idea if the red portion of the model comes close to *any* data points.

When extrapolating, our faith declines more and more as we stray farther and farther from the blue portion of the salmon model. In fact, we have no faith in the portion of the model from 1995 on, because the model predicts nonpositive salmon populations for these years. We say *model breakdown* has occurred for these years.

DEFINITION *Model breakdown*

When a model yields a prediction that does not make sense or an estimate that is not a good approximation, we say that **model breakdown** has occurred.

When model breakdown occurs, it is time to modify our model or possibly rethink our modeling process. Perhaps a different model would give more reasonable predictions. We may decide that it would be helpful to gather more data to validate our choice of model.

Example 4 Modifying a Model

Table 5 Wild Pacific Northwest Salmon Populations

Year	Population (millions)
1960	10.02
1965	10.00
1970	7.61
1975	3.15
1980	4.59
1985	3.11
1990	2.22

In 2002 there were 3 million wild Pacific Northwest salmon. For each of the following scenarios, use the data for 2002 and the data in Table 5 to help sketch a model. Let P represent the wild Pacific Northwest salmon population (in millions) at t years since 1950.

1. The salmon population levels off at 10 million.
2. The salmon become extinct.

Solution

1. We represent the 3 million salmon in 2002 by the ordered pair (52, 3). We plot this data point in addition to the data points in our original scattergram and then sketch an appropriate model (see Fig 10).

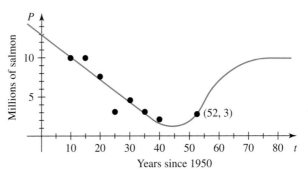

Figure 10 Population levels off at 10 million

2. See Fig. 11.

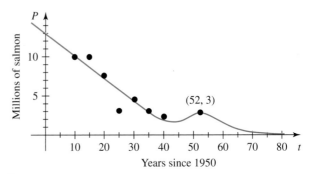

Figure 11 Population becomes extinct

EXPLORATION
How defining the independent variable affects a model

Here you will compare two salmon models that have different definitions for the independent variable t.

1. Let P represent the wild Pacific Northwest salmon population (in millions) at t years *since 1960* (see Table 6). *Carefully* sketch a scattergram of the salmon data.

2. For the scattergram in Fig. 8, the variable t is defined to be the number of years *since 1950*. Carefully sketch a line that comes close to the points in your scattergram in much the same way as the line in Fig. 8 comes close to the points in Fig. 8.

3. Find each result both by using your linear model from problem 2 and by using the linear model in Fig. 8.

    a. Find the P-intercept.

    b. Find the t-intercept.

    c. Predict the salmon population in 1988.

    d. Predict when there were 9 million salmon.

    e. Find the slope of the linear model.

4. For each part of problem 3, which of your results are the same? Explain why this makes sense in terms of the two definitions of t.

5. For each part of problem 3, which of your results are different? Explain why this makes sense in terms of the two definitions of t.

Table 6 Wild Pacific Northwest Salmon Populations

Year	Population (millions)
1960	10.02
1965	10.00
1970	7.61
1975	3.15
1980	4.59
1985	3.11
1990	2.22

EXPLORATION
Identifying types of modeling errors

Here you will explore possible causes of error for predictions based upon a linear model. You will explore these causes of error within the context of a linear model for the Pacific salmon data (see Table 7).

1. Let P represent the wild Pacific Northwest salmon population (in millions) at t years since 1950. Sketch a scattergram of the salmon data.

2. Sketch a line that comes close to the data points.

3. Use your linear model to estimate the salmon population in 1975. What is the actual number of salmon? Calculate the error in your estimate for 1975. (The error is the difference between the estimated value and the actual value.) Why is the error so great?

4. Predict the salmon population in the year 2020 using your linear model. Is this an accurate prediction? If not, why is the error so great?

5. Take another look at your graph. Is the t-axis perfectly horizontal and the P-axis perfectly vertical? Are the scalings of both axes precise? Is your line straight? How might these considerations relate to the accuracy of a prediction? Explain.

6. What are the coordinates of point S plotted in Fig. 12? Do you think you have found the correct first decimal place (tenths place) for these coordinates? How about the second decimal place?

Table 7 Wild Pacific Northwest Salmon Populations

Year	Population (millions)
1960	10.02
1965	10.00
1970	7.61
1975	3.15
1980	4.59
1985	3.11
1990	2.22

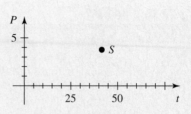

Figure 12 Problem 6

7. Problems 3–6 of this exploration suggest several possible causes of error for predictions based upon a linear model. Describe the possible causes of error.

Tips on Succeeding in This Course

It is wise to exchange phone numbers with some classmates with the understanding that if you must miss class, then you have someone to call in order to find out what you missed and what homework was assigned.

Key Points
OF THIS SECTION

Scattergram	A *scattergram* is a graph of plotted data points.
Model	A *model* is a mathematical description of a true-to-life situation.
Linear model	A *linear model* is a linear function that describes the relationship between two quantities for a true-to-life situation.
Making estimates and predictions	To make estimates and predictions using a linear model: 1. Define variables to represent the quantities under study. 2. Summarize the data using a table of values for the variables. 3. Create a scattergram by plotting the points in your table. 4. Sketch a line that comes close to the points in your scattergram. 5. Make estimates and predictions based on your sketched line.
Intercepts of a model	Suppose that a linear model is used to describe the relationship between the variables t and p. Also, assume that the line crosses both the t-axis and the p-axis. • The t-intercept is the point where the line intersects the t-axis. • The p-intercept is the point where the line intersects the p-axis.
Model breakdown	When a model gives a prediction that does not make sense or an estimate that is not a very good approximation, we say that *model breakdown* has occurred.

HOMEWORK 2.1

 SSM Tutor Center   **MathPro 4** MathPro 4   **MathPro 5** MathPro 5   CDs/Videos   www

1. In 1982 drunk drivers were responsible for 25,165 deaths in car accidents. That was 57.3% of all fatalities due to car accidents. Due to stiffer punishments, an increased number of appeals by police, and public service campaigns by organizations such as Mothers Against Drunk Driving (MADD), the percentage of fatalities in accidents due to drunk driving has gradually decreased (see Table 8).

   a. Let p represent the percentage of fatalities due to drunk driving at t years since 1980. Sketch a scattergram of the data (see Fig. 13). [**Hint**: $t = 0$ represents 1980, $t = 2$ represents 1982, and so on.]

Table 8 Percentages of Fatalities Due to Drunk Driving

Year	% Fatalities Alcohol-related
1982	57
1985	51
1988	50
1991	47
1994	40
1997	38
1999	37

Source: *National Highway Traffic Safety Administration, www.nhtsa.dot.gov. Accessed 3/12/02.*

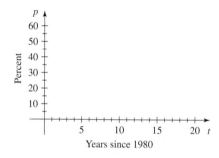

Figure 13 Fatalities due to drunk driving

b. Sketch a line that comes close to the data points.

c. Use your linear model to estimate the percentage of deaths due to drunk driving in 1992. [**Note**: When responding to exercises in this text, always use sentences, not phrases.]

d. Use your linear model to estimate when 53% of deaths were due to drunk driving.

2. The average jury awards in medical malpractice suits have increased greatly in the past decade (see Table 9). As a result, malpractice insurance rates have increased, forcing some doctors to quit practicing or to practice medicine defensively, ordering extra tests or choosing procedures that limit their risks (*New York Times*, 3/22/02).

Table 9 Average Jury Awards in Medical Malpractice Suits

Year	Average Jury Award (millions of dollars)
1994	1.2
1996	1.9
1998	3.0
1999	3.4
2000	3.5

Source: *Jury Verdict Research*

a. Let a be the average jury award (in millions of dollars) in medical malpractice suits for the year that is t years

since 1990. Sketch a scattergram of the data. [**Note**: Your scattergram should have scaling on both axes and labels indicating the variable names and the scale units.]

b. Sketch a line that comes close to the data points.

c. Use your linear model to estimate the average jury award in 1995. [**Note**: When responding to exercises in this text, always use sentences, not phrases.]

d. Use your linear model to estimate when the average jury award was $2.4 million.

3. Due to improved technology and extensive public service campaigns, the number of collisions at highway–railroad crossings per year has declined over the past two decades (see Table 10).

Table 10 Numbers of Collisions at Highway–Railroad Crossings

Year	Number of Collisions (thousands)
1988	6.6
1990	5.8
1992	5.0
1994	5.0
1996	4.3
1998	3.5
2000	3.5

Source: *Federal Highway Association*

Let n represent the number of collisions (in thousands) for the year that is t years since 1980.

a. Sketch a scattergram of the data. Extend your t-axis and n-axis so that you can make estimates about numbers of collisions before 1988 and predictions about future numbers of collisions. [**Note**: Your scattergram should have scaling on both axes and labels indicating the variable names and the scale units.]

b. Sketch a line that comes close to the data points in your scattergram.

c. Use your linear model to estimate the number of collisions in 1997. Did you perform interpolation or extrapolation to obtain your result? Explain.

d. Use your linear model to predict in which year there will be 1 thousand collisions. Did you perform interpolation or extrapolation to obtain your result? Explain.

e. Find the n-intercept of your linear model. What does it represent in terms of collisions?

f. Find the t-intercept of your linear model. What does it represent in terms of collisions? Has model breakdown occurred? Explain.

4. Repeat exercise 3, but this time let n represent the number of collisions (in thousands) for the year that is t years *since 1988*. Which of your responses for this exercise are the same as for exercise 3? Explain why it makes sense that these responses are the same. Explain why it makes sense that the other responses are different.

5. The number of Internet users in the United States steadily increased over the past decade (see Table 11).

Table 11 Number of Internet Users in the United States

Year	Number of Users (millions)
1996	39
1997	60
1998	84
1999	105
2000	122
2002	166

Source: *Jupiter MMXI*

a. Let n be the number of Internet users (in millions) in the United States at t years since 1995. Sketch a scattergram of the data.

b. Sketch a line that comes close to the data points.

c. Find the n-intercept of your linear model. What does it mean in terms of Internet users?

d. Estimate by how much the number of Internet users is increasing per year.

e. Use your linear model to predict when everyone in the United States will be an Internet user. Assume that the U.S. population is 287 million.* [**Note**: Remember, if you think model breakdown occurs, say so, say where, and explain why.]

6. The temperature at which water boils (the *boiling point*) depends on elevation; the higher you are, the lower the boiling point. At sea level, for example, water boils at 212°F, whereas at an elevation of 10,000 meters, water boils at about 151°F. The boiling points for various elevations are listed in Table 12.

Table 12 Boiling Points of Water

Elevation (in thousands of meters)	Boiling Point (°F)
0	212
1	205
2	200
5	181
10	151
20	94

Source: Thermodynamics, an Engineering Approach *by Yunus A. Cengel and Michael A. Boles*

a. Let B represent the boiling point (in degrees Fahrenheit) at an elevation of E thousand meters. Sketch a scattergram of the data.

b. Sketch a line that comes close to the data points.

c. Mount Everest, the highest mountain in the world, reaches 8850 meters at its peak. What is the boiling point of water at the peak?

d. We say that water is lukewarm if its temperature is close to body temperature (about 98.6°F). At what elevation would boiling water feel lukewarm?

e. The cooking time required to make a hardboiled egg depends on the temperature of the water. Let T represent the amount of time it takes to cook an egg in boiling water at an elevation of E thousand meters. Sketch a *qualitative* graph that describes modeling relationship between E and T.

7. In Example 1 of Section 1.1, we sketched a qualitative graph that described the relationship between prices of Air Jordan shoes and years since 1985. Now, you will use the data in Table 13 to make some predictions. (We will assume all prices quoted in this exercise are suggested retail prices.)

Table 13 Prices for Air Jordan Shoes

Name of Shoe	Year	Price per Pair (dollars)
Air Jordan I	1985	85.00
Air Jordan III	1988	99.50
Air Jordan V	1990	113.50
Air Jordan VIII	1993	135.00
Air Jordan XVII	2002	200.00

Source: *Kicksology.net > The Reviews, www.kicksology.net/ reviews/index.html, accessed 4/7/02, and* Consumer Reports.

a. Let p represent the price (in dollars) of Air Jordan shoes at t years since 1985. Create a scattergram of the data.

b. Sketch a line that comes close to the data points.

c. Use your line to estimate the price of the Air Jordan XIV shoes in 1999. The actual price was $150. By how much did the model overestimate the price?

d. Use your line to estimate the price of the Air Jordan XVI shoes in 2001. The actual price was $160. By how much did the model overestimate the price?

e. Michael Jordan was retired when both the Air Jordan XIV and Air Jordan XVI shoes were released. He returned from retirement before the Air Jordan XVII shoes were released. Give a possible explanation for why your estimates in parts c and d were overestimates.

f. On the same axes that you used to sketch your linear model in part b, now sketch a curve that might have described the price of the shoes if Michael Jordan had never returned from the retirement described in part e. [**Hint**: Your new curve will not be a linear model.]

8. Most scientists agree that Earth's average temperature has increased by about 1°F over the past 120 years. What scientists continue to debate is why this has happened and whether we should be concerned (Source: *Time*). Some scientists believe that the warming is due to carbon emissions from fossil-fuel burning (see Table 14).

Table 14 Carbon Emissions from Fossil-Fuel Burning

Year	Carbon Emissions (billions of tons)
1950	1.6
1960	2.6
1970	3.8
1980	4.9
1990	5.9

Source: *The Hard Numbers on Climate Change—a WI Press Release, www.worldwatch.org/alerts/010716.html, and A Worldwatch News Brief—World Carbon Emissions Fall, www.worldwatch.org/ alerts/990727.html. Accessed 3/18/02.*

---

*The U.S. population in 2002 was about 287 million.

Let *c* represent the carbon emissions (in billions of tons) at *t* years since 1950.

a. Sketch a scattergram of the carbon-emission data.

b. Sketch a line that comes close to the data points in your scattergram. Use your line to estimate the carbon emissions in 2000.

c. The actual amount of carbon emissions in 2000 was 6.3 billion tons. Is this amount less than, equal to, or greater than your estimate in part b? What does this comparison suggest may be starting to happen? On the same set of axes that you used to sketch your linear model in part b, now sketch a curve that describes what might be starting to happen. [**Hint**: Your new curve will not be a linear model.]

9. The values of Microsoft® stock for various times are listed in Table 15.

Table 15 Values of Microsoft Stock

Date	Number of Months Since January 1, 1998	Value of Share (dollars)
January 1, 1998	0	32
July 1, 1998	6	55
January 1, 1999	12	75
July 1, 1999	18	92
January 1, 2000	24	112

Source: *ASK Research/Stock Charting and Analysis, 139.142.147.15/index.html. Accessed 4/27/02.*

Let *v* represent the value (in dollars) of a share of Microsoft stock at *t* months since January 1, 1998.

a. Draw a scattergram for the data in Table 15.

b. Sketch a line that comes close to the data points in your scattergram. Use your line to estimate the value of a share on January 1, 2001.

c. Due to investors' sudden lack of confidence in dot-com stocks, the value of many stocks sharply declined after January 2000. The actual data for July 1, 2000 and January 1, 2001 are shown in Table 16. Create a scattergram for the data from January 1, 1998 to January 1, 2001.

Table 16 More Values of Microsoft Stock

Date	Number of Months Since January 1, 1998	Value of Share (dollars)
July 1, 2000	30	80
January 1, 2001	36	43

Source: *ASK Research/Stock Charting and Analysis, 139.142.147.15/index.html. Accessed 4/27/02.*

d. Compute the error in the prediction for January 1, 2001 that you made in part b. (The error is the difference between the estimated value and the actual value.) Explain why the error in your estimate is so large.

10. Table 17 describes the number of AIDS- and HIV-related deaths in New York City. Let *n* represent the number of AIDS- and HIV-related deaths (in thousands) at *t* years since 1980.

a. Draw a scattergram for the data in Table 17.

Table 17 AIDS and HIV Deaths in New York City, 1986–1994

Year	Number of Deaths (in thousands)
1986	2.7
1987	3.2
1988	3.8
1989	4.4
1990	4.8
1991	5.2
1992	5.7
1993	6.1
1994	7.1

Source: New York Times

b. Sketch a line that comes close to the data points in your scattergram. Use your line to estimate the number of deaths in 1997.

c. Due most likely to improved drug therapies, the number of AIDS and HIV deaths have sharply declined since 1995. The actual data from 1995 to 1997 are listed in Table 18. Create a scattergram for the data from 1986 to 1997.

Table 18 AIDS and HIV Deaths in New York City, 1995–1997

Year	Number of Deaths (in thousands)
1995	7.0
1996	5.0
1997	2.6

Source: New York Times

d. Compute the error in the prediction for 1997 that you made in part b. (The error is the difference between the estimated value and the actual value.) Explain why the error in your prediction is so large.

11. Table 19 describes the difference in the percentages of sound recording sales between pop and rap/hip-hop recordings for various years. It shows that pop recordings had greater sales in the early 1990s. However, this advantage eroded to the point where rap/hip-hop recordings had greater sales in the late 1990s.

Table 19 Percentage Differences in Sound Recordings

Year	Difference of Percentages
1990	5.2
1994	2.4
1996	0.4
1999	−0.5
2000	−1.9

Source: *Recording Industry Association of America, Inc.*

Let *D* represent the difference of the percentages at *t* years since 1990.

a. Without graphing, estimate the coordinates of the *t*-intercept for a line that comes close to the data points. What does this point represent in terms of difference in percentages of sales? If you don't see how to estimate the coordinates, create a scattergram of the data. Then estimate.

b. Without graphing, estimate the coordinates of the *D*-intercept for a line that comes close to the data points. What does this point represent in terms of difference in percentages of sales? If you don't see how to estimate the coordinates, create a scattergram of the data.

12. Have you ever noticed that it feels colder when it is a windy day? The *windchill* (or *windchill factor*) is a measure of how cold it feels as a result of being exposed to the wind. The windchill can get as low as 40 degrees below zero (Fahrenheit) in the Midwest. In 2001 the National Weather Service began using new tables and formulas for the windchill factor. The new windchill factor isn't as extreme as the old one. Table 20 provides some data on the windchill factor for various temperatures when the wind speed is 10 mph.

Table 20 Windchill Factors for a 10-mph Wind

Temperature (Fahrenheit)	Windchill (Fahrenheit)
−15	−35
−10	−28
−5	−22
5	−10
10	−4
15	3
20	9
25	15

Source: *The New Improved Wind Chill Index, www.srh.noaa.gov/ffc/html/wci.shtml. Accessed 10/22/01.*

Let *w* represent the windchill (in degrees Fahrenheit) corresponding to a temperature of *t* degrees Fahrenheit when the wind speed is 10 mph.

a. Without graphing, estimate the coordinates of the *t*-intercept for a line that comes close to the data points. What does this point represent in terms of temperature? If you don't see how to estimate the coordinates, create a scattergram of the data. Then estimate.

b. Without graphing, estimate the coordinates of the *w*-intercept for a line that comes close to the data points. What does this point represent in terms of temperature? If you don't see how to estimate the coordinates, create a scattergram of the data.

13. Describe the meaning of a linear function and a linear model. Also, is a linear function necessarily a model? Is a linear model necessarily a function?

14. When using a line to model a situation, do we usually have more faith in a result obtained by interpolating or extrapolating? Explain.

15. Which is more desirable: to find a linear model whose graph contains several, but not all, data points or to find a linear model whose graph does not contain any data points, but comes close to all data points? Include in your discussion some sketches of scattergrams and linear models.

16. Describe how to find a linear model for a situation and how to use the model to make estimates and predictions.

2.2 FINDING EQUATIONS FOR LINEAR MODELS

Objective

▸ Find an equation for a linear model.

In Section 2.1 we used graphs of linear functions to model data. In this section we use *equations* of linear functions to model data.

Example 1 Finding the Equation for a Linear Model

During the early 1990s there was great consumer demand for food products claiming to be "low fat" or "no fat." Since then, this demand has greatly declined. Table 21 shows the percentage of new products that claim to be "low fat" or "no fat" from 1996 to 2001.

Table 21 Percentage of New Products Claiming to Be "Low Fat" or "No Fat"

Year	Percent
1996	29
1997	26
1998	22
1999	17
2000	16
2001	11

Source: *Productscan Online, www.productscan.com. Accessed 5/11/02.*

Let *p* be the percentage of new food products claiming to be "low fat" or "no fat" at *t* years since 1995. Find an equation for a line that comes close to the points in the scattergram for the data.

Solution

We begin by viewing the positions of the points in the scattergram (see Fig. 14). To save time and improve accuracy of plotting points, we can use a graphing calculator to view a scattergram of the data (see Fig. 15).

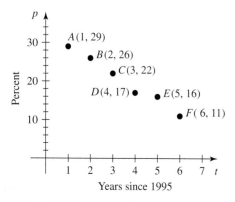

Figure 14 Scattergram for "low fat" and "no fat" food products

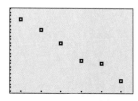

Figure 15 Graphing calculator scattergram

Graphing Calculator

See Section B.8.

Our task is to find an equation for a line that comes close to the data points. Notice that the line that contains the points $D(4, 17)$ and $E(5, 16)$ does *not* come close to the other data points. On the other hand, note that the line that passes through the points $A(1, 29)$ and $C(3, 22)$ appears to come close to the rest of the points. We will find the equation for this line.

Recall that an equation for a linear function can be put into the form $y = mx + b$, so with the variables t and p we have:

$$p = mt + b$$

We can use the points $A(1, 29)$ and $C(3, 22)$ to find the values for m and b for our chosen linear model.

First, we find the slope of this line (see Fig. 16):

$$m = \frac{22 - 29}{3 - 1} = \frac{-7}{2} = -3.5$$

Note

It is not necessary to use *data points* to find an equation. By viewing a scattergram, we can choose two non–data points that would yield a reasonable model. Section B.23 describes how to use a graphing calculator to find coordinates of points.

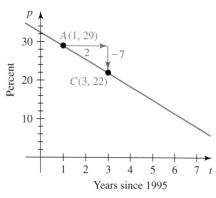

Figure 16 Points A and C of the "low-fat" scattergram

So, -3.5 can be substituted for m in the equation $p = mt + b$:

$$p = -3.5t + b$$

The constant b can be found by substituting the coordinates of the point $A(1, 29)$ into the equation $p = -3.5t + b$ and then solving for b:

$$29 = -3.5(1) + b$$
$$29 = -3.5 + b$$
$$29 + 3.5 = -3.5 + b + 3.5$$
$$32.5 = b$$

Now 32.5 can be substituted for b in the equation $p = -3.5t + b$:

$$p = -3.5t + 32.5$$

We can check the correctness of our equation by using a graphing calculator to verify that our line contains the points $(1, 29)$ and $(3, 22)$. See Fig. 17.

Graphing Calculator

See Sections B.10 and B.11.

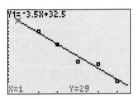 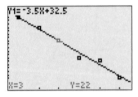

Figure 17 Checking that the model contains both $(1, 29)$ and $(3, 22)$

We benefit in many ways by viewing a scattergram of data. First, we can determine whether the data is approximately linearly related. Second, if the data is approximately linearly related, viewing a scattergram helps us choose two good points to use to find an equation for a linear model. Third, by graphing the model with the scattergram, we can assess if the model does in fact fit the data reasonably well.

In Example 1 we described the linear "low-fat" model using the linear equation $p = -3.5t + 32.5$. Depending on what aspect of the model we want to emphasize, we can refer to the model as the "low fat model," "low fat function," or "equation $p = -3.5t + 32.5$."

In Example 1 we used two points to find the model $p = -3.5t + 32.5$. There is another way to find a model. Your graphing calculator should have a built-in **linear regression** for finding an equation for a linear model. Linear regression gives the equation $p = -3.57t + 32.67$. In Fig. 18 we see that both models fit the data well.

Graphing Calculator

See Section B.16.

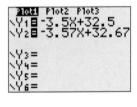 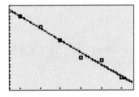

Figure 18 Comparing two "low-fat" models

Note

It appears that there is only one line since the two lines are so close together.

The linear equation found by linear regression is called a **linear regression equation** and the function described by the equation is called a **linear regression function**. The graph is called a **regression line**. (You can learn more about linear regression in a statistics course.)

Now you have two ways to find the equation of a linear model. You should understand that no matter which method is used, the objective is the same: to find an equation of a line that comes close to the data points.

Table 22 Percentages of Americans Who Smoke

Year	Percent Who Smoke
1965	42.4
1974	37.1
1979	33.5
1983	32.1
1987	28.8
1992	26.5
1995	24.7
1999	23.3

Source: *National Center for Health Statistics*

Example 2   Finding the Equation of a Linear Model

In 2001 each pack of cigarettes sold in the United States cost the country about $7.18 in medical care costs and lost productivity (Centers for Disease Control and Prevention). Due to increased public awareness of the life-threatening impact of cigarette smoking, smoking has been on the decline for the past several decades (see Table 22).

Let p represent the percentage of Americans who smoke at t years since 1900.

1. Use two well-chosen points to find an equation of a model that describes the relationship between t and p.

2. Find the linear regression model equation and line by using a graphing calculator. Compare the two models you have found.

Solution

1. We see from the scattergram in Fig. 19 that a line containing the points $A(65, 42.4)$ and $G(95, 24.7)$ comes close to the rest of the data points.

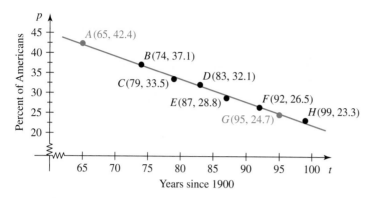

Figure 19 Smoking scattergram and linear model

Note

The zigzag lines on the axes near the origin indicate that part of the scale is missing. This is done so that a clearer view of the data points can be shown.

For an equation of the form $p = mt + b$, we first use the points $A(65, 42.4)$ and $G(95, 24.7)$ to find m.

$$m = \frac{24.7 - 42.4}{95 - 65} = -0.59$$

So, the equation has the form

$$p = -0.59t + b$$

To find b, we use the point $A(65, 42.4)$ and substitute 65 for t and 42.4 for p.

$$42.4 = -0.59(65) + b$$
$$42.4 = -38.35 + b$$
$$42.4 + 38.35 = -38.35 + b + 38.35$$
$$80.75 = b$$

So, the equation is $p = -0.59t + 80.75$.

We can use a graphing calculator to verify that the linear model contains the points $(65, 42.4)$ and $(95, 24.7)$ and comes close to the other data points (see Fig. 20).

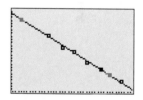

Figure 20 Verifying the smoking model

2. The regression equation is $p = -0.57t + 79.31$. The regression equation is "close" to the equation $p = -0.59t + 80.75$. Further, both models appear to fit the data well (see Fig. 21).

Figure 21 Comparing two smoking models

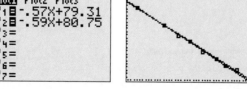

Note

It appears that there is only one line because the two lines are so close together.

Note that the slope of the smoking function $p = -0.57t + 79.31$ is -0.57, which means that the graph is a decreasing line. When the graph of a function is decreasing, we say that the function is **decreasing**. When the graph of a function is increasing, we say that the function is **increasing**.

In this section we have discussed how to find equations for linear models. In Section 2.3 we will use such equations to make estimates and predictions.

EXPLORATION
Choosing "good points" to find a model

The wild Pacific Northwest salmon data are listed in Table 23. Notice that the table now includes a first column that indicates a name for each data point. For example, point D refers to the point $(25, 3.15)$.

Table 23 Values of t and P for Wild Pacific Northwest Salmon

Name of Point	Years Since 1950 t	Population (millions) P
A	10	10.02
B	15	10.00
C	20	7.61
D	25	3.15
E	30	4.59
F	35	3.11
G	40	2.22

1. Find an equation of the line that contains points A and B.

2. Use your graphing calculator to verify that the graph of your equation passes through both points A and B. Does the line come close to the other data points?

3. If you had used the points A and F, you would have found the equation $P = -0.28t + 12.78$. Compare its graph to the graph you drew in problem 2. Explain why the graphs look so different.

4. List all pairs of points that yield equations that you think would be good linear models. (You do not have to find the equations.)

5. Several pairs of points from a scattergram yield equations that could serve as models for the data. Discuss how to choose two such data points to find an equation that comes reasonably close to all the data points.

6. It is not necessary to use data points to find an equation for a linear model. While viewing the salmon scattergram, use the arrow keys on your calculator to identify two non–data points that you feel would yield an equation of a line that is close to the data points. Find an equation of the line that contains these two points. Then, use your graphing calculator to verify that the graph of your equation comes close to the data points.

Graphing Calculator

See Section B.23.

Tips on Succeeding in This Course

Remember to use your graphing calculator to verify your work. In this section, for example, you can use your graphing calculator to check your equations. Checking your work increases your chances of catching errors and, thus, will likely improve your performance on homework assignments, quizzes, and tests.

Key Points
OF THIS SECTION

Finding an equation for a linear model

To find an equation for a linear model, given some data:

1. Create a scattergram of the data.

2. Find a line that comes close to the data points.

3. Find an equation for the line.

4. Use your graphing calculator to verify that the graph of your equation comes close to the points of the scattergram. If it doesn't, check for graphing errors and for calculation errors, or try using different points to derive your equation.

HOMEWORK 2.2

SSM
Tutor Center

MathPro 4
MathPro 4

MathPro 5
MathPro 5

CDs/Videos

www
www

Consider the scattergram of data and the graph of the model $y = mx + b$ *in the indicated figure. Sketch the graph of a linear model that better describes the data and then explain how you would adjust the values of m and b of the original model so that it would better describe the data.*

1. See Fig. 22.

Figure 22 Exercise 1

2. See Fig. 23.

Figure 23 Exercise 2

3. Find an equation of a line that comes close to the points listed in Table 24. Then use a graphing calculator to check that your line comes close to the points. [*Graphing Calculator:* See Sections B.8 and B.10.]

Table 24 Finding a Linear Model	
x	*y*
3	5
4	7
5	10
6	12
7	15

4. Find an equation for a line that comes close to the points listed in Table 25. Then use a graphing calculator to check that your line comes close to the points. [*Graphing calculator:* See Sections B.8 and B.10.]

Table 25 Finding a Linear Model	
x	*y*
1	18
4	14
5	12
7	8
10	5

5. Three students are to find a linear model for the data in Table 26. Student A uses the points (1, 5.9) and (2, 6.4), student B uses the points (3, 9.0) and (4, 11.0), and student C uses the points (5, 12.1) and (6, 15.5). Which student seems to have made the best choice of points? Explain.

Table 26 Three Students Model Data	
x	*y*
1	5.9
2	6.4
3	9.0
4	11.0
5	12.1
6	15.5
7	16.5

6. Three students are to find a linear model for the data in Table 27. Student A uses the points (3, 13.8) and (4, 10.1), student B uses the points (6, 7.8) and (7, 4.3), and student C uses the points (5, 9.1) and (8, 3.1). Which student seems to have made the best choice of points? Explain.

Table 27 Three Students Model Data	
x	*y*
3	13.8
4	10.1
5	9.1
6	7.8
7	4.3
8	3.1
9	1.1

7. Five percent of babies born in 1945 in the United States were born out of wedlock (the parents were not married). This percentage increased steadily during the past few decades (see Table 28). The increase was due to a variety of factors, including an increase in the number of unplanned births, the existence of more alternative family units, greater societal acceptance of single-parent households, as well as greater societal acceptance of a couple who chooses to have a child before getting married.

Table 28 Births out of Wedlock	
Year	Percent of Births out of Wedlock
1970	10.7
1975	14.3
1980	18.4
1985	22.0
1990	28.0
1995	32.2
2000	33.2

Source: *National Center for Health Statistics*

Let *p* represent the percentage of births out of wedlock in the United States at *t* years since 1900.

a. Use your graphing calculator to create a scattergram for the data.

b. Find an equation for a line that you think comes close to the points in the scattergram.

c. Draw your line and the scattergram in the same viewing window. Verify that the line passes through your two chosen points and that it comes close to all of the data points.

8. Repeat exercise 7, but this time let p represent the percentage of births out of wedlock in the United States at t years *since 1970*. Compare the slope of your model with the slope of the model found in exercise 7. Compare the p-intercepts. Explain why your comparisons make sense.

9. Each year thousands of people die in car accidents due to drunk drivers. Yet some people maintain that they can drink and drive safely.

Does consumption of alcohol impair one's reasoning ability and coordination? Three students decided to run an experiment to address this question.

Each trial of the experiment consisted of one student attempting to place 18 geometric-shaped blocks into matching slots in the six sides of a cube. The first try occurred after consuming no alcohol. Subsequent tries each occurred after consuming 20 milliliters (a small amount) of an alcoholic drink known as a "screwdriver" at 15-minute intervals. The other two students recorded the time it took the drinking student to fit the 18 shapes into the slots. The results of the experiment are shown in Table 29.

Table 29 Times to Complete a Task under the Influence of Alcohol

Number of Screwdrivers	Time (seconds)
0	80
1	84
2	95
3	102
4	105
5	111
6	117
7	120
8	126
9	135
10	184

Source: *Experiment conducted by Edwin P., Kevin S., and Dan S., Fall Semester, 1994*

a. Let T represent the number of seconds it took the drinking student to complete the task after n drinks. Use your graphing calculator to create a scattergram of the drinks–time data.

b. Does the data point (10, 184) fit the pattern of the rest of the points in your scattergram? What do you think happened?

c. Find an equation for a linear function that models the drinking data for the first nine drinks.

10. The American life span has been increasing over the past century. Advancements in medical services have been a significant contributor to this increase. Americans have also been armed with better information regarding nutrition, exercise, and the effects of smoking. Table 30 shows life expectancies at birth for Americans in various years.

Table 30 Life Expectancies

Year of Birth	Life Expectancy (years)
1975	72.6
1977	73.3
1980	73.7
1982	74.5
1985	74.7
1987	74.9
1990	75.4
1993	75.5
1996	76.1
2000	76.9

Source: *U.S. Bureau of the Census*

a. Let L represent the life expectancy at birth (in years) for an American born t years after 1970. Create a scattergram of the data.

b. Find an equation for a linear function that models the life expectancy data.

c. Draw your line and the scattergram in the same viewing window. Verify that the line passes through your two chosen points and that it comes close to all of the data points.

11. Killer whales are highly intelligent and social. When whales of a pod reunite after a long separation, they sometimes form a line and touch noses. The stock of killer whales in Puget Sound has declined since 1996 due to depletion of salmon, toxic pollution, and disturbances from whale-watching boats and other shipping traffic (see table 31).

Table 31 Killer Whales in Puget Sound

Year	Population
1996	97
1997	91
1998	89
1999	84
2000	82
2001	78

Source: *Killer Whale (Orcinus orca), www.biologicaldiversity.org/swcbd/ species/orca/index.html. Accessed 3/24/02.*

a. Let p represent the killer whale population in Puget Sound at t years since 1990. Use a graphing calculator to draw a scattergram of the data.

b. Find an equation for a line that you think comes close to the points in the scattergram.

c. Draw your line and the scattergram in the same viewing window. Verify that the line contains your two chosen points and that it comes close to all of the data points.

12. The percentages of colleges and universities offering distance learning (by computer) is shown in Table 32.

a. Let p represent the percentage of colleges and universities that offer distance learning at t years since 1990. Use a graphing calculator to draw a scattergram of the data.

b. Find an equation for a line that you think comes close to the points in the scattergram.

Table 32 Percentages of Colleges and Universities That Offer Distance Learning

Year	Percent
1995	33
1997	44
1998	60
1999	69
2002	85

Source: *International Data Corporation*

c. Draw your line and the scattergram in the same viewing window. Verify that the line contains your two chosen points and that it comes close to all of the data points.

13. Table 33 lists world record times for the women's 400-meter run. Let r represent the record time (in seconds) at t years since 1900.

Table 33 Women's 400-Meter Run Record Times

Year	Runner	Country	Record Time (seconds)
1957	Marlene Mathews	Australia	57.0
1959	Maria Itkina	USSR	53.4
1962	Shin Geum Dan	North Korea	51.9
1969	Nicole Duclos	France	51.7
1972	Monika Zehrt	E. Germany	51.0
1976	Irena Szewinska	Poland	49.29
1979	Marita Koch	E. Germany	48.60
1983	Jarmila Kratochvilová	Chad	47.99
1985	Marita Koch	E. Germany	47.60

Source: *WR Progression (Women), www.apulanta.fi/matti/yu/ wrprogr_Women.html#400m. Accessed 12/30/02.*

a. Use your graphing calculator to create a scattergram of the data.

b. Choose two points that lie on a line that comes close to the points in the scattergram. Use these two points to find an equation for the line.

c. Draw your line and the scattergram in the same viewing window. Verify that the line passes through your two chosen points and that it comes close to all of the data points.

14. World record times for the men's 400-meter run are listed in Table 34.

Table 34 Men's 400-Meter Run Record Times

Year	Runner	Country	Record Time (seconds)
1900	Maxie Long	USA	47.8
1916	Ted Meredith	USA	47.4
1928	Emerson Spencer	USA	47.0
1932	Bill Carr	USA	46.2
1941	Graver Klemmer	USA	46.0
1950	George Rhoden	Jamaica	45.8
1960	Carl Kaufmann	Germany	44.9
1968	Lee Evans	USA	43.86
1988	Harry Reynolds	USA	43.29
1999	Michael Johnson	USA	43.18

Source: *WR Progression (Men), www.apulanta.fi/matti/yu/ wrprogr_Men.html#400m. Accessed 12/30/02.*

Let r represent the record time (in seconds) at t years since 1900.

a. Use your graphing calculator to create a scattergram of the data.

b. Choose two points that lie on a line that comes close to the points in the scattergram. Use these two points to find an equation for the line.

c. Draw your line and the scattergram in the same viewing window. Verify that the line passes through your two chosen points and that it comes close to all of the data points.

15. In exercises 13 and 14, you found equations for the women's and men's 400-meter run record times. Equations that model the data well are:

$$r = -0.27t + 70.45 \qquad \text{women's model}$$
$$r = -0.053t + 48.08 \qquad \text{men's model}$$

where r represents the record time (in seconds) at t years since 1900.

a. Use pencil and paper to sketch graphs of both models for the years from 1900 to 2050. (If you can sort out how to use a graphing calculator to do this exercise, you may do so.)

b. Do the models predict that the women's record time will equal the men's record time? If so, what is the record time and when will the record be set?

c. Do the models predict that the women's record time will be less the men's record time? If so, in what years?

16. In exercise 8 in Section 2.1, you modeled carbon emissions using the graph of a linear model. Here, you will derive a linear equation to use as a linear model (see Table 35).

Table 35 Carbon Emissions from Fossil-Fuel Burning

Year	Emissions (billions of tons)
1950	1.6
1960	2.6
1970	3.8
1980	4.9
1990	5.9

a. Let c represent the carbon emissions (in billions of tons) at t years since 1950. Use a graphing calculator to draw a scattergram of the data.

b. Find an equation for a line that you think comes close to the points in the scattergram.

c. Draw your line and the scattergram in the same viewing window. Verify that the line passes through your two chosen points and that it comes close to all of the data points.

d. Use TRACE to estimate the amount of carbon emissions in 2000. The actual carbon emissions were 6.3 billion tons in 2000. Calculate the error in the 2000 estimate. (The error is equal to the difference between the estimated value and the actual value.)

17. The number of audits by the Internal Revenue Service (IRS) has greatly declined since 1996 due to a decrease in staff, old computers, and a law passed in 1998 defending taxpayers (see Table 36). This decline has raised concern that more people will cheat on their taxes. IRS Commissioner Charles Rossotti estimates that the decline in audits has cost the government more than $2 billion in 2001 alone.

Table 36 Percentages of Taxpayers Audited

Year	Percent
1996	1.7
1997	1.3
1998	1.0
1999	0.9
2000	0.5

Source: *Internal Revenue Service*

Congratulations !
It's a tax deduction !

a. Let p represent the percentage of taxpayers who were audited at t years since 1990. Use a graphing calculator to draw a scattergram of the data.

b. Choose two points that lie on a line that comes close to the points in the scattergram. Use these two points to find an equation for the line.

c. Draw your line and the scattergram in the same viewing window. Verify that the line contains your two chosen points and that it comes close to all of the data points.

d. Use your graphing calculator to find the regression equation for the data. [**Graphing Calculator:** See Section B.16.]

e. Use your graphing calculator to draw the scattergram and both of your models from parts b and d in the same viewing window. Discuss how well each function models the data.

18. Although the media might lead one to believe otherwise, the number of violent youth crimes has greatly declined since 1993 (see Table 37).

Table 37 Numbers of Violent Crimes Committed by Youths

Year	Number of Violent Crimes (millions)
1993	1.1
1994	1.0
1995	0.8
1996	0.8
1997	0.7
1998	0.6
1999	0.6
2000	0.5

Source: *Bureau of Justice Statistics*

a. Let n represent the number (in millions) of violent crimes committed by youths in the year that is t years since 1990. Use a graphing calculator to draw a scattergram of the data.

b. Choose two points that lie on a line that comes close to the points in the scattergram. Use these two points to find an equation for the line.

c. Draw your line and the scattergram in the same viewing window. Verify that the line contains your two chosen points and that it comes close to all of the data points.

d. Use your graphing calculator to find the regression equation for the data. [**Graphing Calculator:** See Section B.16.]

e. Use your graphing calculator to draw the scattergram and both of your models from parts b and d in the same viewing window. Discuss how well each function models the data.

19. If a student at College of San Mateo (CSM) scores at least 21 points (out of 50 points) on the placement test for intermediate algebra, that student may enroll in intermediate algebra. Using four semesters of data, the CSM mathematics department computed the percentages of intermediate algebra students who succeeded in the course (grade of A, B, or C) for various groups of scores on the placement test (see Table 38).

Table 38 Percentages of Intermediate Algebra Students Who Succeeded

Placement Score Group	Score Used to Represent Score Group	Percentage Who Succeeded in Intermediate Algebra
21–25	23	34
26–30	28	47
31–35	33	55
36–40	38	71
41–45	43	84
46–50	48	*

Source: *College of San Mateo mathematics department*
There were not enough students in this category to give useful information.

a. Let *p* represent the percentage of intermediate algebra students who succeed in the course who scored *x* points on the placement test. Use a graphing calculator to draw a scattergram of the data.

b. Find an equation for a line that you think comes close to the points in the scattergram.

c. Draw your line and the scattergram in the same viewing window. Verify that the line contains your two chosen points and that it comes close to all of the data points.

20. Explain how to find an equation for a linear model for a given situation. Also, explain how you can verify that the linear function models the situation reasonably well.

2.3 FUNCTION NOTATION AND MAKING PREDICTIONS

Mathematics is a door to the world. If you ignore it, you injure your chances at becoming successful. But if you embrace it, you hold the power to do anything you want and to be able to do it successfully. This is why I like math. —*Shaun O., student*

Objectives

▸ Use equations of linear models to make estimates and predictions.

▸ Use function notation, $f(x)$.

▸ Know the meaning of *domain* and *range* of a model.

In Section 2.2 we found equations for models. In this section we'll use such equations to make estimates and predictions. We'll also discuss how to name functions.

Example 1 Using the Equation for a Linear Model to Make Predictions

Table 39 shows the average salaries for professors at four-year public colleges and universities.

Table 39 Average Salaries for Professors

Year	Average Salary (thousands of dollars)
1975	16.6
1980	22.1
1985	31.2
1990	41.9
1995	49.1
2000	57.7

Source: *American Association of University Professors*

Let *s* represent the average salary (in thousands of dollars) at *t* years since 1900. A possible model is

$$s = 1.70t - 112.17$$

1. Verify that the model $s = 1.70t - 112.17$ represents the data well.

2. Predict the average salary in 2008.

3. Predict when the average salary will be $75,000.

Solution

1. We draw the graph of the model and the scattergram of the data in the same viewing window (see Fig. 24). It appears that the function models the data quite well.

2. The year 2008 is represented by $t = 108$, so the average salary can be found by substituting 108 for *t* in the equation $s = 1.70t - 112.17$.

$$s = 1.70(108) - 112.17$$
$$= 71.43$$

The model predicts that the average salary will be $71,430 in 2008.

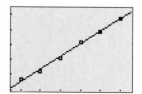

Figure 24 Check how well $s = 1.70t - 112.17$ models the data

3. The salary \$75,000 is represented by $s = 75$, so the year can be found by substituting 75 for s in the equation $s = 1.70t - 112.17$.

$$75 = 1.70t - 112.17$$
$$75 + 112.17 = 1.70t - 112.17 + 112.17$$
$$187.17 = 1.70t$$
$$1.70t = 187.17$$
$$t = \frac{187.17}{1.70}$$
$$t = 110.1$$

According to the model, the average salary will be \$75,000 in 2010. We can verify our work in both problems 2 and 3 by using a graphing calculator table (see Figs. 25 and 26).

Graphing Calculator

See Section B.15.

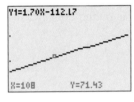

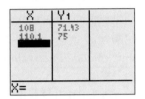

Figure 25 Putting table in "Ask" mode

Figure 26 Verify the predictions

Or, we can graphically verify our work using TRACE (see Fig. 27).

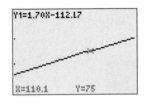

Figure 27 Verify predictions using TRACE

In Example 1, we made a prediction for the dependent variable s. We did this by substituting a value for the independent variable t in the model $s = 1.70t - 112.17$, and then we solved for s.

We also made a prediction for the independent variable t. We did this by substituting a value for the dependent variable s in the model $s = 1.70t - 112.17$, and then we solved for t.

Rather than use an equation, table, or graph each time we want to refer to a function, it would be easier to give the function a name. To see how, consider the average salary function:

$$s = 1.70t - 112.17$$

Rather than use "s" to represent the average salary, we can use "$f(t)$" (read "f of t"):

$$s = f(t)$$

Here, we have named the function "f" and can write:

$$f(t) = 1.70t - 112.17.$$

Note that the independent variable t is written within the parentheses of $f(\ )$. In general, the dependent variable can be represented by the expression formed by writing the independent variable name within the parentheses of $f(\ )$:

$$\text{dependent variable} = f(\text{independent variable})$$

For another example, consider the function $y = 2x + 1$. In this case, the dependent variable y can be represented by $f(x)$:

$$y = f(x)$$

and the function $y = 2x + 1$ can be expressed as:

$$f(x) = 2x + 1$$

Here we compare these two representations using the equations $y = 2x + 1$ and $f(x) = 2x + 1$ to find the value of y when $x = 3$:

Find y when x = 3   **Find f(x) when x = 3**

$y = 2x + 1$ $f(x) = 2x + 1$

$y = 2(3) + 1$ $f(3) = 2(3) + 1$

$y = 7$ $f(3) = 7$

Note that for both equations we found that $y = 7$ when $x = 3$. Also note that the expression $f(3)$ represents the value of y when $x = 3$.

In Example 2 we will make the same predictions as in Example 1, but now we will use $f(t)$ notation.

Example 2   Using *f(t)* Notation to Make Predictions

Make the indicated predictions using the model $f(t) = 1.70t - 112.17$, where $s = f(t)$ represents the average salary (in thousands of dollars) of professors at four-year public colleges and universities at t years since 1900.

1. Predict the average salary in 2008.

2. Predict when the average salary will be $75,000.

Solution

1. For the year 2008, we use $t = 108$. Since $f(t)$ represents s, then $f(108)$ stands for the value of s when t is 108. To find $f(108)$, we write:

$$f(t) = 1.70t - 112.17$$
$$f(108) = 1.70(108) - 112.17 \qquad \text{Substitute 108 for } t.$$
$$= 71.43$$

   According to the model, the average salary will be $71,430 in 2008, the same prediction we made in Example 1.

2. For a $75,000 average salary, we use $s = 75$. Since $f(t)$ represents s, we substitute 75 for $f(t)$ in the equation $f(t) = 1.70t - 112.17$:

$$75 = 1.70t - 112.17$$
$$75 + 112.17 = 1.70t - 112.17 + 112.17$$
$$187.17 = 1.70t$$
$$1.70t = 187.17$$
$$t = \frac{187.17}{1.70}$$
$$t = 110.1$$

So, the model predicts that the average salary will be $75,000 in the year 2010, the same prediction as in Example 1.   ∎

In Example 2 we used f to name the function

$$f(t) = 1.70t - 112.17$$

where $f(t)$ represents the average salary for professors at *public* colleges and universities. When using more than one function to model situations, naming the functions helps us to distinguish between them. For example, it turns out that the average salaries of professors at *private* colleges and universities can also be modeled well using a linear function. A good model is

$$s = 2.22t - 155.06$$

where s is the average salary (in thousands of dollars) at t years since 1900. We can keep track of this function by using g as its name:

$$g(t) = 2.22t - 155.06$$

The most common symbols used to name functions are f, g, and h.

In Example 3 we use a model to make predictions, find the intercepts of the model, and interpret the meaning of the intercepts.

Example 3   Using $g(t)$ Notation, Finding Intercepts

Table 40 Americans Who Smoke	
Year	Percent
1965	42.4
1974	37.1
1979	33.5
1983	32.1
1987	28.8
1992	26.5
1995	24.7
1999	23.3

In Example 2 in Section 2.2, we found $p = -0.59t + 80.75$, where p represents the percentage of Americans who smoke at t years since 1900 (see Table 40).

1. Write the equation for the linear function using the name g.
2. Find $g(108)$. What does the result mean in terms of Americans who smoke?
3. Find the value for t when $g(t) = 30$. What does it represent?
4. Find the p-intercept for the model. What does it represent?
5. Find the t-intercept for the model. What does it represent?

Solution

1. To use the name g, substitute $g(t)$ for p in the equation $p = -0.59t + 80.75$:

$$g(t) = -0.59t + 80.75$$

2. To find $g(108)$, we substitute 108 for t in the equation $g(t) = -0.59t + 80.75$:

$$g(108) = -0.59(108) + 80.75 = 17.03$$

So, $p = 17.03$ when $t = 108$. According to the model, 17.0% of Americans will smoke in 2008.

3. Here we substitute 30 for $g(t)$ in the equation $g(t) = -0.59t + 80.75$ and solve for t:

$$30 = -0.59t + 80.75$$
$$30 + 0.59t = -0.59t + 80.75 + 0.59t$$
$$30 + 0.59t = 80.75$$
$$30 + 0.59t - 30 = 80.75 - 30$$
$$0.59t = 50.75$$
$$t = \frac{50.75}{0.59}$$
$$t \approx 86.02$$

The model estimates that 30% of Americans smoked in 1986.

4. Since the model $g(t) = -0.59t + 80.75$ is in slope-intercept form, we see that the p-intercept is $(0, 80.75)$. So, the model estimates that 80.8% of Americans smoked in 1900. Some research would show that this estimate is too high; model breakdown has occurred.

5. To find the t-intercept, we substitute 0 for $g(t)$ and solve for t:

$$0 = -0.59t + 80.75$$
$$0 + 0.59t = -0.59t + 80.75 + 0.59t$$
$$0.59t = 80.75$$
$$t = \frac{80.75}{0.59}$$
$$t \approx 136.86$$

The t-intercept is (136.86, 0). So, the model predicts that no Americans will smoke in 2037. However, common sense suggests that this event probably won't occur.

We can verify our work in problems 2–5 by putting a graphing calculator table in "Ask" mode (see Fig. 28).   ■

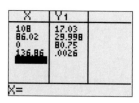

Figure 28 Verify the estimates and predictions

In Example 3 we found the intercepts of the smoking model $g(t) = -0.59t + 80.75$. The p-intercept is the point (0, 80.75), whose p-coordinate is the constant term 80.75. We found the t-intercept by substituting 0 for $g(t)$ in the equation $g(t) = -0.59t + 80.75$ and then solving for t.

Recall that a function can be viewed as a machine that transforms inputs into outputs. In Example 3, we found that the function g sends the input $t = 108$ to the output $p = 17.03$. Figure 29 illustrates this, using the $g(t)$ notation.

Figure 29 "Machine" for calculating percentage of Americans who smoke

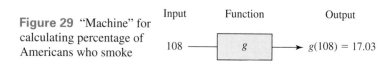

To find $g(108)$, we say we *evaluate* the function g at $t = 108$.

Example 4   Evaluating a Function

Evaluate the function $f(x) = 3x - 5$ at each given value of x.

1. $f(4)$
2. $f(-2)$
3. $f(a)$
4. $f(a + h)$

Solution

1. $f(4) = 3(4) - 5 = 7$
2. $f(-2) = 3(-2) - 5 = -11$
3. $f(a) = 3a - 5$
4. $f(a + h) = 3(a + h) - 5 = 3a + 3h - 5$ ■

Example 5   Finding Values for *x* or *f(x)*

Let $f(x) = -2x + 15$.

1. Find $f(3.82)$.
2. Find x when $f(x) = \dfrac{3}{4}$.

Solution

1. $f(3.82) = -2(3.82) + 15 = 7.36$.

2. To find x when $f(x) = \dfrac{3}{4}$, we substitute $\dfrac{3}{4}$ for $f(x)$ and solve for x.

$$\frac{3}{4} = -2x + 15 \qquad \text{Substitute } \frac{3}{4} \text{ for } f(x).$$

$$4 \cdot \frac{3}{4} = 4(-2x + 15) \qquad \text{Multiply both sides by the LCD 4.}$$

$$3 = -8x + 60 \qquad \text{Simplify, distributive law.}$$

$$8x = 57 \qquad \text{Add } 8x \text{ to both sides. Subtract 3 from both sides.}$$

$$x = \frac{57}{8} \qquad \text{Divide both sides by 8.}$$

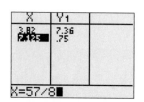

Figure 30 Verify the estimates and predictions

We can verify our work in problems 1 and 2 by putting a graphing calculator table in "Ask" mode (see Fig. 30).   ■

So far in this section, we have found inputs and outputs of a function from its equation. In Example 6, we find inputs and outputs of a function from its graph.

Example 6 Using a Graph to Find Values for x or $f(x)$

A graph of a function f is sketched in Fig. 31.

1. Estimate $f(4)$.
2. Estimate $f(0)$.
3. Find x when $f(x) = -2$.
4. Find x when $f(x) = 0$.

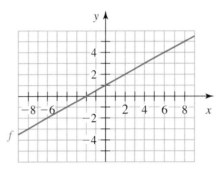

Figure 31 Problems 1–4 of Example 6

Solution

1. $f(4)$ is the value of y when $x = 4$. Since the line contains the point $(4, 3)$, we conclude that $f(4) = 3$.
2. $f(0)$ is the value of y when $x = 0$. Since the line contains the point $(0, 1)$, we conclude that $f(0) = 1$.
3. We want the value of x when $y = -2$. Since the line contains the point $(-6, -2)$, we know that $x = -6$.
4. We want the value of x when $y = 0$. Since the line contains the point $(-2, 0)$, we know that $x = -2$. ■

Recall that the domain of a function is the set of all inputs and that the range of a function is the set of all outputs. For the **domain** and **range** of a model, we consider input-output pairs only where both the input and the output make sense in terms of the situation. The domain of the model is the set of all such inputs, and the range of the model is the set of all such outputs.

Example 7 Finding the Domain and Range of a Model

A store is open from 9 A.M. to 5 P.M. Monday through Saturday each week. Let $I = f(t)$ represent an employee's weekly income (in dollars) from working t hours each week at $10 per hour.

1. Find an equation for the model f.
2. Find the domain and range of the model f.

Solution

1. The employee's weekly income (in dollars) is equal to the pay per hour times the numbers of hours worked per week:

$$f(t) = 10t$$

2. To find the domain and range of the model f, we consider input-output pairs only where both the input and the output make sense in terms of the situation. Since the store is open 8 hours a day, 6 days a week, the employee might work up to 48 hours each week. So, the domain is the set of numbers between 0 and 48 (including 48): $0 < t \le 48$.

Since the number of hours worked is between 0 and 48 hours and the pay is $10 per hour, the range is the set of numbers between 0 and 480 (including 480): $0 < f(t) \le 480$.

In Fig. 32 we illustrate the inputs 22, 35, and 48 being sent to the outputs 220, 350, and 480, respectively. We also label the part of the t-axis that represents the domain and the part of the I-axis that represents the range.

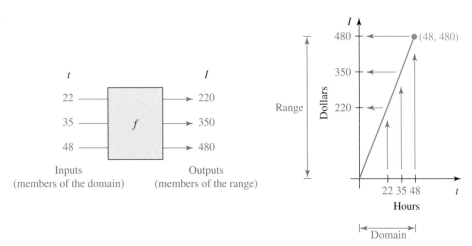

Figure 32 Domain and range of the employee income model

In this section and Section 2.2, we have discussed how to find linear models and how to use these models to make estimates and predictions. Here is a summary of this process:

Four-Step Modeling Process

1. Create a scattergram of the data to determine if there is a nonvertical line that comes close to the points. If so, choose two points (not necessarily data points) that you can use to find the equation of a linear model.

2. Find an equation for your model.

3. Verify your equation by checking that the graph of your model contains the two chosen points. If it doesn't, check for calculation errors. Also, check that the line does, indeed, come close to all the data points. If it doesn't, try using different points to derive an equation.

4. Use the equation for your model to make estimates, make predictions, and draw conclusions.

EXPLORATION
Formula for slope

1. Let $f(x) = 2x + 1$. Find each of the following and compare your three results with each other.

   a. the slope of f b. $\dfrac{f(5) - f(3)}{5 - 3}$ c. $\dfrac{f(7) - f(4)}{7 - 4}$

2. Let $g(x) = 3x + 5$. Find each of the following and compare your three results with each other.

   a. the slope of g b. $\dfrac{g(3) - g(1)}{3 - 1}$ c. $\dfrac{g(4) - g(0)}{4 - 0}$

3. Let f be a function of the form $f(x) = mx + b$. Describe $\dfrac{f(c) - f(d)}{c - d}$, where $c \neq d$. Explain.

EXPLORATION

Looking ahead: Significance of the slope and the dependent variable's intercept of a model

Time (hours) t	Distance (miles) d
0	
1	
2	
3	
4	

Table 41 Distances Traveled by an Airplane

1. A small airplane is traveling at a constant speed of 100 miles per hour. Let d represent the distance (in miles) the airplane can travel in t hours.

   a. Complete Table 41.

   b. Perform the first three steps of the modeling process.

   c. Compare the slope of your model to the speed of the airplane.

   d. What is the d-intercept? What does it mean in terms of the situation?

2. In 2000 a company is worth \$10 million. Each year, its value increases by \$2 million. Let V represent the value (in millions of dollars) of the company at t years since 2000.

   a. Complete Table 42.

   b. Perform the first three steps of the modeling process.

   c. Compare the slope of your model to the rate at which the value of the company is increasing.

   d. What is the V-intercept? What does it mean in terms of the situation?

Table 42 Values of a Company

Year Since 2000 t	Value (millions of dollars) V
0	
1	
2	
3	
4	

3. A person is in a hot-air balloon at a height of 1600 feet. The person begins gradually letting air out of the balloon, and the balloon descends at a rate of 200 feet per minute. Let H represent the height (in feet) of the balloon after t minutes of releasing air from the balloon.

   a. Complete Table 43.

   b. Perform the first three steps of the modeling process.

   c. Compare the slope of your model to the rate at which the balloon is descending.

   d. What is the H-intercept? What does it mean in terms of the situation?

Table 43 Heights of a Balloon

Time (minutes) t	Height (feet) H
0	
1	
2	
3	
4	

4. In general, what is the meaning of the slope of a model in terms of its true-to-life situation? What is the meaning of the dependent variable's intercept?

Tips on Succeeding in This Course

Do you find that you tend to make the same mistakes repeatedly throughout a math course? If so, you might find that it helps to keep a journal where you list errors you have made on assignments, quizzes, and tests. For each error you list, also include the correct solution as well as a description of the concept needed to solve the problem correctly. You can review this journal from time to time to help you avoid making these errors.

Key Points
OF THIS SECTION

When making a prediction about the dependent variable of a linear model, substitute a chosen value for the independent variable in the model, then solve for the dependent variable.

Making a prediction about the dependent variable

When making a prediction about the independent variable of a linear model, substitute a chosen value for the dependent variable in the model, then solve for the independent variable.

Making a prediction about the independent variable

Suppose that a function of the form $p = mt + b$ is used to model a situation. Also assume that the line crosses both the t-axis and the p-axis.

Intercepts of a model

- The p-intercept is $(0, b)$.
- To find the t-coordinate of the t-intercept, substitute 0 for p in the model's equation and then solve for t.

A linear function of the form $y = mx + b$ can also be described by the equation $f(x) = mx + b$; here we use $f(x)$ to represent y. In general:

$$\text{dependent variable} = f(\text{independent variable})$$

$f(x)$ notation

For the *domain* and *range* of a model, we consider input-output pairs only where both the input and the output make sense in terms of the situation. The domain of the model is the set of all such inputs, and the range of the model is the set of all such outputs.

Domain and range of a model

HOMEWORK 2.3

SSM
Tutor Center

MathPro 4
MathPro 4

MathPro 5
MathPro 5

CDs/Videos

www

*Substitute the indicated value for x and solve the resulting equation for y. [**Note:** To review solving equations, see Section A.9.]*

1. $3y + x = 17$ for $x = 2$

2. $4y - x = 15$ for $x = 5$

3. $-3y - 2x = -27$ for $x = 4$

4. $5y + 4x = 22$ for $x = -2$

5. $2y - 6x = -49$ for $x = -5$

6. $2(y - 1) + 3x = 10$ for $x = 2$

7. $-5(y + 5) = -11 + 4x$ for $x = -1$

8. $3y + 4x + 1 = 5y - 1$ for $x = -6$

9. $2.7y - 3.8x = 5.9$ for $x = 6.6$

10. $7.42x - 9.83y = 3.25$ for $x = 5.23$

Evaluate the function $f(x) = -3x + 2$ at each given value of x. Verify your result using a graphing calculator graph or table.

11. $f(1)$ **12.** $f(4)$ **13.** $f(9)$

14. $f(0)$ **15.** $f(-2)$ **16.** $f(-5)$

Evaluate the function $g(x) = 6x - 4$ at each given value of x. Verify your result using a graphing calculator graph or table.

17. $g(2.3)$ **18.** $g(-3.7)$ **19.** $g\left(\dfrac{1}{6}\right)$

20. $g\left(-\dfrac{1}{3}\right)$ **21.** $g\left(-\dfrac{2}{3}\right)$ **22.** $g\left(\dfrac{5}{6}\right)$

Evaluate the function $f(x) = -4x - 1$ at each given value of x.

23. $f(a)$ **24.** $f(2a)$ **25.** $f(a + 1)$

26. $f\left(\dfrac{a}{2}\right)$ **27.** $f(a + h)$ **28.** $f(a - h)$

For $f(x) = -3x + 7$, find the value of x that corresponds to each given value of $f(x)$. Verify your result using a graphing calculator graph or table.

29. $f(x) = 6$ **30.** $f(x) = -4$

31. $f(x) = -4.9$ **32.** $f(x) = 2.1$

For $f(x) = 2x - 5$, find the value of x that corresponds to each given value of $f(x)$. Whenever possible, verify your result using a graphing calculator graph or table.

33. $f(x) = -\dfrac{2}{3}$ **34.** $f(x) = \dfrac{3}{5}$

35. $f(x) = a$ **36.** $f(x) = a + 2$

For exercises 37–48, refer to Fig. 33.

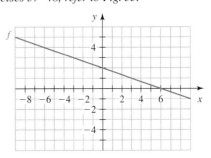

Figure 33 Exercises 37–48

37. Estimate $f(-6)$.

38. Estimate $f(0)$.

39. Estimate $f(3)$.

40. Estimate $f(4)$.

41. Estimate $f(2.5)$.

42. Estimate $f\left(-\dfrac{11}{2}\right)$.

43. Estimate x when $f(x) = 0$.

44. Estimate x when $f(x) = 1$.

45. Estimate x when $f(x) = 3$.

46. Estimate x when $f(x) = 3.5$.

47. Estimate x when $f(x) = \dfrac{1}{2}$.

48. Estimate x when $f(x) = \dfrac{5}{2}$.

For exercises 49–52, refer to Table 44.

Table 44 Values of f
(Exercises 49–52)

x	$f(x)$
0	0
1	2
2	4
3	2
4	0

49. Find $f(2)$.

50. Find $f(4)$.

51. Find x when $f(x) = 2$.

52. Find x when $f(x) = 4$.

Find the x-intercept and y-intercept of each function.

53. $f(x) = 5x - 8$

54. $f(x) = 4x + 2$

55. $f(x) = 3x$

56. $f(x) = -7x$

57. $f(x) = 5$

58. $f(x) = -2$

59. $f(x) = -2.1x - 4.2$

60. $f(x) = -3.8x - 13.5$

61. $f(x) = \dfrac{1}{2}x - 3$

62. $f(x) = -\dfrac{3}{4}x + \dfrac{1}{2}$

Solve each equation.

63. $x - 1 + 2x = 3x - 9x + 17$

64. $6(x - 2) = -15$

65. $3 - 6(2 - x) = x - (4 - x)$

66. $2.1x + 3.4 = 2.5x - 8.2$

67. $0.5(x - 3.9) + 2 = 2.12x - 7$

68. $-2(3x + 5) = -3(4x + 3)$

69. $4(x - 2) - 3(x - 1) = 2(x + 6)$

70. $\dfrac{1}{2}x + \dfrac{1}{3} = \dfrac{5}{2}$

71. $\dfrac{1}{3}x - \dfrac{1}{4} = \dfrac{2}{3}$

72. $-\dfrac{5}{6}x + \dfrac{3}{4} = \dfrac{7}{2}$

73. $-\dfrac{3}{4}x - \dfrac{5}{8} = \dfrac{1}{2}$

74. $x = x$

75. $x = x + 1$

76. $2x + 1 = 2x + 1$

77. Two students attempted to solve the equation $2x + 4 = 0$. Did one, both, or neither of the students solve the equation correctly? Explain.

Student 1	**Student 2**
$2x + 4 = 0$	$2x + 4 = 0$
$2x + 4 + 6 = 0 + 6$	$2x + 4 - 4 = 0 - 4$
$2x + 10 = 6$	$2x = -4$
$2x + 10 - 10 = 6 - 10$	$\dfrac{2x}{2} = \dfrac{-4}{2}$
$2x = -4$	$x = -2$
$2x \cdot 3 = -4 \cdot 3$	
$6x = -12$	
$\dfrac{6x}{6} = \dfrac{-12}{6}$	
$x = -2$	

78. Let $f(x) = mx + b$. Find $f(x + 1) - f(x)$ and discuss what this result means.

79. In exercise 7 of Homework 2.2, you found an equation close to $p = 0.81t - 45.86$ that models the percentage p of births out of wedlock in the United States at t years since 1900 (see Table 45).

Table 45 Births out of Wedlock

Year	Percent
1970	10.7
1975	14.3
1980	18.4
1985	22.0
1990	28.0
1995	32.2
2000	33.2

a. Rewrite the equation $p = 0.81t - 45.86$ using $f(t)$ notation.

b. Find $f(108)$. What does your result mean in terms of births out of wedlock?

c. Find the value of t so that $f(t) = 43$. What does your result mean in terms of births out of wedlock?

d. According to the model, in what year will all births be out of wedlock?

e. Estimate the percentage of births out of wedlock for 1997. The actual percentage is 32.4. What is the error in your estimate? (The error is the difference between the estimated value and the actual value.)

80. In exercise 8 of Homework 2.2, you found an equation close to $p = 0.81t + 10.59$ that models the percent p of births out of wedlock in the United States at t years *since 1970*. Repeat exercise 79, but this time use the model $p = 0.81t + 10.59$. In what instances are your answers the same? When are they different?

81. In exercise 9 of Homework 2.2, you found an equation close to $T = 5.85n + 81.18$, where T represents the number of seconds it took a drinking student to complete a task after n drinks. (Recall that the task consisted of one student attempting to place 18 shapes into matching slots in the six sides of a cube.) See Table 46.

Table 46 Times to Complete a Task

Number of Screwdrivers	Time (seconds)
0	80
1	84
2	95
3	102
4	105
5	111
6	117
7	120
8	126
9	135
10	184

a. Rewrite the equation $T = 5.85n + 81.18$ using $f(n)$ notation.

b. What is the T-intercept of the linear model? What does the T-intercept represent in terms of the experiment?

c. Is the model an increasing function or a decreasing function? What does this mean in terms of the experiment?

d. Use a graphing calculator table to find $f(0)$, $f(1)$, $f(2)$, ..., and $f(10)$. Which of these estimates has the least error? The most? Explain.

82. In exercise 10 of Homework 2.2, you found an equation close to $L = 0.16t + 72.19$, where L represents the life expectancy at birth (in years) for an American born t years after 1970 (see Table 47).

Table 47 Life Expectancies

Year of Birth	Life Expectancy (years)
1975	72.6
1977	73.3
1980	73.7
1982	74.5
1985	74.7
1987	74.9
1990	75.4
1993	75.5
1996	76.1
2000	76.9

a. Use the linear model to predict the life expectancy for an American born in 2008. Verify graphically that your prediction fits your model.

b. Use the linear model to predict the birth year in which life expectancy for an American will be 80 years. Verify your answer graphically.

c. Use the linear model to estimate what your (or someone else's) life expectancy was at birth. Given the fact that you have made it to your current age, do you think that your life expectancy is now less than, the same as, or more than it was at your time of birth?

d. Find the t-intercept of the linear model. What does the t-intercept represent in terms of life expectancies? [**Note:** Remember, if you think model breakdown occurs, say so, say where, and explain why.]

e. Use pencil and paper to sketch a *qualitative* graph that you feel describes human life expectancy at birth during the existence of the human race on Earth. [**Hint**: Your curve should be nonlinear.]

83. The number of commercial airline enplanements (number of passengers boarding aircraft) on domestic fights increased steadily during the 1990s (see Table 48). If a passenger boards two successive airplanes on each half of a round trip, that would count as four enplanements.

Table 48 Numbers of Commercial Airline Enplanements on Domestic Flights

Year	Number of Enplanements (millions)
1991	452
1993	487
1995	547
1997	599
1999	635
2000	666

Source: *Historical Air Traffic Statistics, www.bts.gov/oai/indicators/airtraffic/annual/ 1981-2001.html. Accessed 4/27/02.*

a. Let $f(t)$ represent the number of commercial airline enplanements on domestic flights (in millions) for the year that is t years since 1990. Use your graphing calculator to draw a scattergram of the data.

b. Find an equation for f. Does your model fit the data well?

c. Use your model f to estimate the number of enplanements in 2001. The actual number of enplanements in 2001 was 622 million. What is the error in your estimate? (the error is the difference between the estimated value and the actual value.)

d. The number of enplanements in 2001 was low due to the terrorist acts on aviation on September 11, 2001. Airports were closed for several days and after they reopened many people were reluctant to fly. Estimate the amount of money lost to airlines in 2001 due to the terrorist attacks by making the following assumptions:

- All trips are round trips.
- The average number of enplanements for a round trip is four (two out, two back).
- The average round-trip fare is $500.

84. In 1950, 1.8 million people lived in Detroit. However, Detroit's population has been declining over the past half century, to the extent that its population slipped below the 1-million mark in the 2000 census. Because the population fell below 1 million, Detroit lost well over $100 million in federal funds and income tax.

During the 1990s, in an effort to retain these monies, Detroit restored (rather than tearing down) abandoned buildings so as to attract additional residents. It also tried to locate all of its homeless so that everyone was counted in the 2000 census. Detroit's populations for various years are listed in Table 49.

Table 49 Detroit's Populations

Year	Population
1950	1,849,568
1960	1,670,144
1970	1,514,063
1980	1,203,368
1990	1,027,974

Source: *U.S. Census Bureau*

a. Let $P = f(t)$ represent Detroit's population at t years since 1950. ["Let $P = f(t)$ represent Detroit's population" tells you that we will use P as a name for the population and f as a name for the function being used to model the situation.] Sketch a scattergram of the data.

b. Find an equation for f. Does the graph of your equation come close to the points in the scattergram?

c. Use your model f to estimate the population in 2000. Assume that your result would have been Detroit's population count but for the city's attempts to increase the count. The actual count in 2000 was 951,270. Estimate by how much Detroit increased its population count through its efforts.

d. Find the linear model g that *contains* the data points in Table 50, where $g(t)$ is Detroit's population at t years since 1950.

Table 50 Detroit's Populations

Year	Population
1990	1,027,974
2000	951,270

Source: *FAIR—Detroit Immigration Factsheet, www.fairus.org/html/ msas/042midec.htm. Accessed 3/20/02.*

e. Find $f(60)$ and $g(60)$. If Detroit continues its attempts at attracting new residents and locating all of its homeless, which of these results will likely be the better prediction for the population in 2010? Explain.

85. Although the United States and Great Britain use the Fahrenheit (°F) temperature scale, most countries use the Celsius (°C) temperature scale. The Celsius scale was developed in such a way that 0°C is the temperature at which water freezes and 100°C is the temperature at which water boils (at sea level). Table 51 shows equivalent Fahrenheit and Celsius temperatures.

Table 51 Equivalent Temperature Readings

Celsius Reading	Fahrenheit Reading
0	32
20	68
40	104
60	140
80	176
100	212

a. Let $F = f(C)$ represent the Fahrenheit reading corresponding to a Celsius reading of C degrees. ["Let $F = f(C)$ represent the Fahrenheit reading" tells you that we will use F as a name for the Fahrenheit reading and f as a name for the function being used to model the situation.] Perform the first three steps of the modeling process. [**Modeling Process:** 1. Scattergram 2. Equation 3. Verify 4. Estimate/Predict.]

b. If the temperature is 25°C, what is the Fahrenheit reading?

c. If the temperature is 40°F, what is the Celsius reading?

86. The rate that a cricket chirps depends on the temperature of the surrounding air. You can estimate the temperature of the air by counting cricket chirps! Experiments have shown that these estimates are very reliable. In fact, crickets have an edge over mercury thermometers in that their rate of chirping almost immediately zones in on the appropriate number of chirps for some temperature, whereas the rise or fall of mercury in a thermometer lags behind a change in the air temperature. On the other hand, a temperature estimate via cricket chirps is for the air near the cricket, not necessarily where you are. Some data are provided in Table 52.

Table 52 Rates of Cricket Chirping

Temperature (Fahrenheit)	Rate (number of chirps per minute)
50	43
60	86
70	129
80	172
90	215

Source: Eric Sloane's Weather Book, *by Eric Sloane*

a. Let $g(F)$ represent the number of chirps per minute a cricket makes when the temperature is F degrees Fahrenheit. Find an equation for g. Verify that the graph of your equation comes close to the points in the scattergram of the data.

b. Find $g(73)$. What does your result mean in terms of the crickets chirping?

c. Find the value for F where $g(F) = 100$. What does your result mean in terms of the crickets chirping?

d. What is the temperature(s) at a field where crickets are not chirping?

87. In 1963 there were only 417 male-female pairs of bald eagles in the United States. However, the symbol of the United States has made a major comeback during the past three decades, going from the endangered species list, to the threatened species list, to being taken off the threatened list (see Table 53).

Table 53 Numbers of Bald Eagle Male-Female Pairs

Year	Number of Bald Eagle Pairs (thousands)
1993	4.2
1994	4.5
1995	4.9
1996	5.1
1997	5.3
1998	5.7

Source: USA Today

Let $f(t)$ represent the number of male-female pairs of bald eagles (in thousands) at t years since 1990.

a. Perform the first three steps of the four-step modeling process.

b. In 1999 the bald eagle was taken off the threatened species list. Estimate the number of bald eagle pairs in that year.

c. Find $f(15)$. What does your result mean in terms of bald eagles?

d. Find t when $f(t) = 10$. What does your result mean in terms of bald eagles?

e. Estimate by how much the number of bald eagle pairs is increasing each year. Compare your estimate to the slope of f. Explain why your comparison makes sense in terms of the slope addition property.

88. In exercise 18 in Homework 2.2, you found an equation close to $n = -0.082t + 1.30$, where n represents the number (in millions) of violent crimes committed by youths in the year that is t years since 1990 (see Table 54).

Table 54 Numbers of Violent Crimes Committed by Youths

Year	Number of Violent Crimes (millions)
1993	1.1
1994	1.0
1995	0.8
1996	0.8
1997	0.7
1998	0.6
1999	0.6
2000	0.5

a. Rewrite the equation $n = -0.082t + 1.30$ using $f(t)$ notation.

b. Find $f(2)$. What does your result mean in terms of the situation?

c. Find t when $f(t) = 0.1$. What does your result mean in terms of the situation?

d. Find the n-intercept. What does your result mean in terms of the situation?

e. Find the t-intercept. What does your result mean in terms of the situation? [**Note**: Remember, if you think model breakdown occurs, say so, say where, and explain why.]

89. In exercise 19 in Homework 2.2, you found an equation close to $p = 2.48x - 23.64$, where p represents the percentage of intermediate algebra students at College of San Mateo (CSM) who succeeded in the course (grade of A, B, or C) who scored x points on the placement test (see Table 55).

Table 55 Percentages of Intermediate Algebra Students Who Succeeded

Placement Score Group	Score Used to Represent Score Group	Percent Who Succeeded in Intermediate Algebra
21–25	23	34
26–30	28	47
31–35	33	55
36–40	38	71
41–45	43	84
46–50	48	*

Source: *College of San Mateo mathematics department*
There were not enough students in this category to give useful information.

a. Rewrite the equation $p = 2.48x - 23.64$ using $f(x)$ notation.

b. Students who score below 21 points (out of 50 points) on the placement test are not allowed to enroll in intermediate algebra. Use the model f to estimate how high the cutoff score would have to be to ensure that all students succeed in the course.

c. Use the model f to estimate for which scores no students would succeed in the course.

d. If in one semester 145 students scored in the 16–20 point range on the placement test, predict how many of these students would have succeeded in the course if they had been allowed to enroll in it. Would you advise CSM to lower the placement score cutoff to 16? Explain.

e. Table 56 shows the numbers of students in various placement-score groups for one semester.

Table 56 Placement Test Scores for One Semester

Placement Score Group	Number of Students
21–25	94
26–30	44
31–35	19
36–40	12
41–45	9
46–50	4

For students who scored at least 21 points on the placement test that semester, estimate how many succeeded in the course.

90. Table 57 shows the numbers of discharges from the military because of sexual orientation.

Table 57 Discharges From the Military Because of Sexual Orientation	
Year	Number Discharged
1991	964
1992	714
1993	678
1994	600
1995	732
1996	857
1997	997
1998	1145
1999	1034
2000	1231
2001	1273

Source: New York Times, Servicemembers Legal Defense Network, www.sldn.org/templates/press/record.html?record=508. Accessed 10/28/02.

a. Let $f(t)$ represent the number of discharges because of sexual orientation in the year that is t years since 1990. Use a graphing calculator to draw a scattergram of the data. Is the data approximately linearly related? Explain.

b. In 1994, the Clinton Administration adopted a policy of "don't ask, don't tell" for gay men and lesbians in the military. The intent of this policy was to reduce the number of discharges of gay men and lesbians. Did implementing this policy have the desired impact? Explain.

c. Find a linear function f that models the data well for the years 1994–1998.

d. Does your model f underestimate or overestimate the number of discharges on the basis of sexual orientation for the years 1999, 2000, and 2001? Explain.

e. In 2001 Fort Bragg reported that 4 service members had been discharged on the basis of sexual orientation when, in fact, 24 service members had been discharged for this reason. (The value 1273 in Table 57 takes this into account.) Assuming that there were underreportings at other military bases, use your model from part c to estimate the extent of the underreporting in 2001 (including at Fort Bragg).

91. A basement is flooded with 640 cubic feet of water. It takes 4 hours to pump out the water. Let $f(t)$ represent the number of cubic feet of water that remains in the basement after t hours of pumping out water.

a. Find a linear equation for f. [**Hint:** You are given information about two points that can be used to find an equation.]

b. Use pencil and paper to sketch a graph of f. Use a graphing calculator to verify your graph.

c. What are the domain and range of the model? Explain.

92. A person buys 12 ounces of ice cream in a cup. It takes the person 5 minutes to eat all of the ice cream. Let $f(t)$ represent the number of ounces of ice cream that remain in the cup t minutes after the person begins eating the ice cream.

a. Find a linear equation for f. [**Hint:** You are given information about two points that can be used to find an equation.]

b. Use pencil and paper to sketch a graph of f. Use a graphing calculator to verify your graph.

c. What are the domain and range of the model? Explain.

93. For a function f, assume that $f(3) = 5$. Name an input and an output of f. Also, find three possible equations for f.

94. Describe the four-step modeling process in your own words.

2.4 SLOPE IS A RATE OF CHANGE

Objectives

▸ Understand why slope is a rate of change.
▸ Find linear models using the rate-of-change property of slope.

In Section 1.4 we discussed an important property of slope.

Slope Addition Property

For a linear function, if the value of the independent variable increases by 1, then the value of the dependent variable changes by the slope.

This property will shed light on the meaning of the slope of a linear model.

Example 1   Meaning of the Slope of a Linear Model

A student drives a car at 50 miles per hour. Let $d = f(t)$ represent the distance (in miles) the student can travel in t hours.

1. Find the slope of the function f. What does it represent in terms of the distance traveled?

2. Find an equation for f.

Solution

1. We begin by creating a table of values for t and d in Table 58. Since a line contains the points plotted in Fig. 34, we will model the situation using a linear function. Note from Table 58 that as t increases by 1, d increases by 50, so the slope of the linear model is 50. Also note that the slope $m = 50$ is equal to the 50 mph speed of the car.

2. Since f is a linear function, we can write $f(t) = mt + b$. Since the d-intercept is $(0, 0)$ (see Table 58 or Fig. 34) and the slope is $m = 50$, we have $f(t) = 50t + 0$, or $f(t) = 50t$.

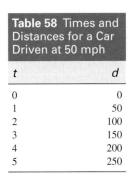

Table 58 Times and Distances for a Car Driven at 50 mph

t	d
0	0
1	50
2	100
3	150
4	200
5	250

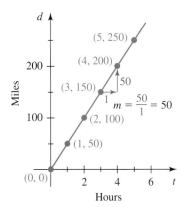

Figure 34 Car model and scattergram

In Example 1 we found that the slope of the car model is equal to 50 miles per hour, the *rate of change* of the car's position. We will explore the meaning of slope for another model in Example 2.

Example 2   Meaning of the Slope of a Linear Model

Budget® Car Rental Company charges a flat daily fee of about $40 plus $0.20 per mile for a pickup truck rental.* Let $c = g(x)$ represent the daily cost (in dollars) of renting a pickup truck that is driven x miles.

1. Find the slope of the function g. What does it represent in terms of the rental cost?

2. Find an equation for g.

Solution

1. We begin by finding values for x and c in Table 59. By the scattergram in Fig. 35, we see that we can use a linear model. As x increases by 1, c increases by 0.20, so the slope of the function g is 0.20. Note that the slope is equal to the $0.20 charge per mile.

Table 59 Daily Pickup Truck Rental Cost

x	c
1	40.20
2	40.40
3	40.60
4	40.80
5	41.00

---

*The exact flat fee was $39.95.

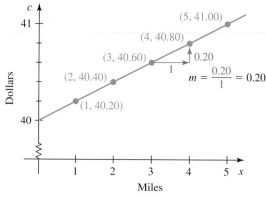

Figure 35 Car model and scattergram

Table 60 Daily Pickup Truck Rental

x	c
1	0.20(1) + 40
2	0.20(2) + 40
3	0.20(3) + 40
4	0.20(4) + 40
5	0.20(5) + 40
x	0.20x + 40

Note

Another way: Since the slope is 0.20, we know that $c = 0.20x + b$. We can find b by using the point $(5, 41)$ to substitute 5 for x and 41 for c in the equation $c = 0.20x + b$.

Graphing Calculator

See Section B.13.

2. In Table 60 we list the values of x and c, but this time the arithmetic for the c values has not been carried out, so that a pattern is more apparent. For example, if $x = 2$, then 2 miles were driven, so the cost would be 2 times $0.20 plus the flat fee of $40. Thus, c would be $0.20(2) + 40$. The pattern suggests that $c = 0.20x + 40$, or $g(x) = 0.20x + 40$.

We verify the equation by checking that the entries in the graphing calculator table in Fig. 36 equal the the entries in Table 59. We also check that the graph of our function contains the points of a scattergram for the data (see Fig. 37).

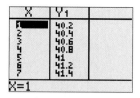

Figure 36 Use a table to verify the model

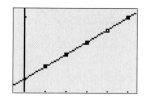

Figure 37 Use a graph to verify the model

In Examples 1 and 2, we found that the slopes of the models are 50 miles per hour and $0.20 per mile, respectively. Both of these values are examples of **rates of change**. For any linear model, slope is a rate of change.

The rate of 50 miles per hour is an example of a rate of change of distance with respect to time. The rate of $0.20 per mile is an example of a rate of change of cost with respect to distance.

In general, the slope m in the equation $y = mx + b$ is the rate of change of y (the dependent variable) with respect to x (the independent variable).

Slope is a Rate of Change Property

For a linear function, slope is the rate of change of the dependent variable with respect to the independent variable.

Example 3 Meaning of the Slope of a Linear Model

A company's profit for 2000 was $10 million. Each year, the profit increases by $3 million. Let $p = h(t)$ represent the profit (in millions of dollars) for the year that is t years after 2000.

1. Find the slope of the function h. What does it represent in terms of the company's profit?

2. Find an equation for h.

Solution

1. Since the profit increases by \$3 million per year, the slope is 3.

2. We want an equation for h of the form $p = h(t) = mt + b$. Since $p = 10$ when $t = 0$, the p-intercept is $(0, 10)$. Recall that $m = 3$, so an equation for h is $h(t) = 3t + 10$. ∎

   Notice that in each of the three situations explored in this section, a quantity changes at a constant rate. That is, the distance the car travels changes by 50 miles per hour, the pickup truck rental fee increases by \$0.20 per mile, and the company's profit increases by \$3 million per year. Also notice that each situation can be modeled *exactly* with a linear function. The following property generalizes this observation.

Constant Rate of Change Property

If the rate of change of the dependent variable with respect to the independent variable is constant, then there is a linear relationship between the variables.

EXPLORATION
Slope of a linear model for approximately linearly related data

The numbers of U.S. satellite television subscribers for various years are listed in Table 61.

Excuse me, sir, we were wondering if you could help us locate our satellite ?[1]

Table 61 Numbers of U.S. Satellite Television Subscribers

Year	Number of Subscribers (millions)
1995	4
1996	6
1997	8
1998	11
1999	13
2000	16
2001	18

Source: *Satellite Broadcasting and Communications Association*

Let $s = f(t)$ represent the number of U.S. satellite television subscribers (in millions) at t years since 1995.

1. Use a graphing calculator to draw a scattergram of the data. Does it appear that the data can be modeled well by a linear function?

2. Use the data points $(0, 4)$ and $(6, 18)$ to find a linear model.

3. What is the slope of your model? According to your model, by how much will sales increase each year?

4. Refer to Table 61 to find by how much the number of subscribers actually increased each year. List the increases in the number of subscribers from 1995 to 1996, from 1996 to 1997, and so on. How do the actual sales increases compare with the slope of your model?

5. Find the average of the increases in the number of subscribers. How does this average compare with the slope of your model?

Review

To find the average of six numbers, divide the sum of the six numbers by the number 6.

---

[1]Cartoon concept by Ross Grossman.

Table 62 Annual Fuel Consumption		
	Annual Fuel Consumption (billions of gallons)	
Year	Cars	Light Trucks
1992	65.6	40.9
1993	67.2	42.9
1994	68.1	44.1
1995	68.1	45.6
1996	69.2	47.4
1997	69.9	49.4
1998	71.7	50.5
1999	73.2	52.8

Source: *U.S. Federal Highway Administration*

Graphing Calculator

See Sections B.6 and B.18.

EXPLORATION

Looking ahead: Using a system of equations to model a situation

Fuel consumption by cars and light trucks (vans, pickups, and SUVs) is described for various years in Table 62.

The annual fuel consumptions (in billions of gallons) $C(t)$ and $L(t)$ for cars and light trucks, respectively, are modeled by the system

$$F = C(t) = 0.98t + 63.74$$
$$F = L(t) = 1.65t + 37.60$$

where t represents the number of years since 1990.

1. Find the F-intercepts of both of the functions C and L. What do these intercepts mean in terms of fuel consumption?

2. Find the slopes of both of the functions C and L. What do these slopes mean in terms of fuel consumption?

3. Your responses to problems 1 and 2 should suggest an event that will happen in the future. Describe that event.

4. Use "ZOOM OUT" and "intersect" on a graphing calculator to estimate the coordinates of the point where the graphs of C and L intersect. What does it mean in terms of fuel consumption that the graphs intersect at this point? State the year and fuel consumption for this event.

Tips on Succeeding in This Course

Do you ever feel that you understand your homework assignments, yet you perform poorly on quizzes and tests? If so, you may not be studying enough to be ready to solve problems *in a test environment*. For example, although it is a good idea to refer to your lecture notes when stumped on a homework exercise, you must continue to solve similar exercises until you can solve them *without* referring to your lecture notes (unless your instructor uses open-notebook tests). The same idea applies to getting help from someone, referring to examples in the text, looking up answers in the back of the text, or any other form of support.

Getting support to help yourself learn math is a great idea. Just make sure that you spend the last part of your study time completing exercises without such support. One way to complete your study time would be to make up a practice quiz or test for yourself to do in a given amount of time.

Key Points

OF THIS SECTION

Meaning of slope	The slope of a linear model is equal to the rate of change of the dependent variable with respect to the independent variable.
Constant rate of change	If the rate of change of the dependent variable with respect to the independent variable is constant, then the variables are linearly related.

HOMEWORK 2.4

SSM
Tutor Center

MathPro 4
MathPro 4

MathPro 5
MathPro 5

CDs/Videos

www

1. A student drives a car at 70 miles per hour. Let $f(t)$ represent the distance (in miles) that the student travels in t hours.

   a. Find the slope of f. What does it represent in terms of the distance traveled?

   b. Find an equation for f.

2. A train is moving at 40 miles per hour. Let d represent the distance (in miles) the train travels in t hours.

   a. Create a table of values for d and t and then make a scattergram.

   b. Discuss the constant rate of change property within the context of this situation.

3. In 1998, 1.7 million U.S. households paid at least some of their bills online. Each year, the number of households who do so increases by about 2.1 million. Let n represent the number of households (in millions) who pay at least some of their bills online at t years since 1998.

   a. A portion of a table of ordered pairs (t, n) is provided (see Table 63). Notice that the arithmetic for the n values has not been carried out, so that a pattern can be recognized. Copy and complete the rest of Table 63.

   Table 63 Numbers of Households That Pay their Bills Online

t	n
0	$1.7 + 0(2.1)$
1	$1.7 + 1(2.1)$
2	$1.7 + 2(2.1)$
3	$1.7 + 3(2.1)$
4	
5	
t	

   Source: *Yankee Group*

   b. Refer to the bottom row of your completed Table 63 to write an equation that describes the relationship between t and n.

   c. Use your graphing calculator to create a scattergram of the ordered pairs in your completed Table 63. Then draw a graph of your equation and the scattergram in the same viewing window to verify that you have found the correct equation. If you have not found the correct equation, find it now.

   d. What is the slope of your model? What does it represent in terms of the situation? [**Note:** Remember to include units in your description.]

4. Due to increased sensitivity to the suffering of animals, the number of animals used in laboratory experiments is being reduced. In 1984 about 200 thousand dogs were used in laboratory experiments in the United States. Since then, the number of dogs used in laboratory experiments has decreased by about 9.9 thousand dogs per year. Let d represent the number of dogs (in thousands) used in laboratory experiments in the United States at t years since 1984.

   a. A portion of a table of ordered pairs (t, d) is provided (see Table 64). Notice that the arithmetic for the d values has not been carried out, so that a pattern can be recognized. Copy and complete the rest of Table 64.

   Table 64 Dogs Used in Experiments

t	d
0	$200 - 0(9.9)$
1	$200 - 1(9.9)$
2	$200 - 2(9.9)$
3	$200 - 3(9.9)$
4	
5	
t	

   Source: Scientific American

   b. Refer to the bottom row of your completed Table 64 to write an equation that expresses d in terms of t.

   c. Use a graphing calculator table or graph to verify that you have found the correct equation.

   d. What is the slope of your model? What does it represent in terms of the situation?

5. In September 2001, AOL® offered a plan in which a customer was charged \$2.50 per hour to be online. There was also a monthly fee of \$4.95 per month. Let $f(t)$ be the total monthly charge (in dollars) if a customer is online for t hours per month.

   a. What is the slope of f? What does it represent in terms of the situation?

   b. Find an equation for f.

   c. Use a graphing calculator table or graph to verify that you have found the correct equation.

   d. AOL offered another plan that charged a flat fee of \$23.90 per month with unlimited time online. When is this plan a better deal than the other plan?

6. In 1986, major league baseball games lasted an average of 2 hours and 44 minutes. Each year, this average time has increased by approximately 1 minute. Let $g(t)$ represent the average time (in minutes) of major league baseball games in the year that is t years since 1986.

   a. What is the slope of g? What does it mean in terms of the length of baseball games?

   b. Find an equation for g.

   c. Use a graphing calculator table or graph to verify that you have found the correct equation.

   d. Predict the average time for a baseball game in 2007.

7. For Spring 2003, full-time undergraduates paid \$1224.60 for tuition at Southeastern Louisiana University. Students could rent textbooks for \$15.75 per textbook. Let $f(n)$ be the total one-semester cost (in dollars) of tuition and renting n textbooks.

   a. What is the slope of f? What does it represent in terms of the situation?

b. Find an equation for f.

c. If a full-time undergraduate paid a total of $1303.35 for tuition and renting textbooks, how many textbooks did the student rent?

8. For Spring 2003, part-time students at Centenary College paid $320 per credit for tuition and paid a part-time student fee of $10 per semester. Let $h(n)$ be the total one-semester cost (in dollars) of tuition and fee for a part-time student who is taking n credits.

a. What is the slope of h? What does it represent in terms of the situation?

b. Find an equation for h.

c. If a student's sole source of funds for college is a $3100 scholarship, how many credits worth of courses can the student take?

9. A person drives her Honda CR-V® on a road trip. At the start of the trip she fills up the 15.3-gallon tank with gasoline. During the trip, the car uses about 0.05 gallons of gas per mile. Let $g(x)$ represent the number of gallons of gasoline remaining in the tank after driving x miles.

a. What is the slope of g? What does it mean in terms of the situation?

b. Find an equation for g.

c. Find the x-intercept of g. What does it represent in terms of the situation?

d. What is the domain of g?

e. If the driver will refuel the car when 1 gallon of gasoline remains in the tank, how far can the car be driven before refueling?

10. Atmospheric pressure at sea level is 1 atmosphere (atm). Under water, pressure increases by approximately 0.0303 atm for every 1 foot increase in depth. For example, water pressure in the ocean is 1.0303 atm at a depth of 1 foot.

a. Let P represent water pressure (in atm) at a depth of d feet. Find an equation that expresses P in terms of d.

b. Use a graphing calculator table or graph to verify that you have found the correct equation.

c. What is the slope of your model? What does it represent in terms of the situation?

d. How deep must you dive for the water pressure to be twice the pressure at sea level?

11. In Example 1 in Section 2.3, we made predictions using the model $s = 1.70t - 112.17$, where s represents the average salary (in thousands of dollars) for professors at four-year public colleges and universities at t years since 1900. What is the slope of this model? What does it mean in terms of average salaries?

12. In exercise 18 in Section 2.2, you found an equation close to $n = -0.082t + 1.30$, where n represents the number (in millions) of violent crimes committed by youths in the year that is t years since 1990. What is the slope of this model? What does it mean in terms of violent crimes?

13. In exercise 13 in Section 2.2, you found an equation close to $r = -0.27t + 70.45$, where r represents the record time (in seconds) for the women's 400-meter run at t years since 1900. What is the slope of this model? What does it mean in terms of record times?

14. In Example 2 in Section 2.2, we found

$$p = -0.59t + 80.75$$

where p represents the percent of Americans who smoke at t years since 1900. What is the slope of this model? What does it mean in terms of Americans who smoke?

15. As of May 1, 2002, MCI® charged $0.99 per minute, plus a $1.50 per call surcharge, for state-to-state calls under a certain calling plan. Let c represent the cost (in dollars) of making a state-to-state call for t minutes. Are t and c linearly related? If so, what is the slope? What does the slope represent?

16. Gasoline sells for $1.85 per gallon at a certain gas station. Let T represent the total charge (in dollars) for G gallons of gas. Are G and T linearly related? If so, what is the slope? What does the slope represent?

17. The percentage of the world's population that lives in rural areas has steadily decreased for the past 50 years (see Table 65).

Table 65 Percentages of World Population Living in Rural Areas

Year	Percent
1950	70
1960	66
1970	63
1980	61
1990	57
2000	52

Source: *Food and Agricultural Organization of the United Nations*

a. Let $f(t)$ represent the percentage of world population that lives in rural areas at t years since 1950. Perform the first three steps of the four-step modeling process to find an equation for f.

b. Find the slope of f. What does it represent in this situation?

c. Use your model f to predict when no one will live in rural areas. [**Note:** Remember, if you think model breakdown occurs, say so, say where, and explain why.]

18. The number of households that have been burglarized per year has greatly decreased in the past two decades (see Table 66).

Table 66 Numbers of Households Burglarized

Year	Number of Burglaries per 1000 Households
1980	101.4
1985	75.2
1990	64.5
1995	49.3
2000	31.8

Source: *Bureau of Justice Statistics, National Crime Victimization Survey*

a. Let $f(t)$ represent the number of households that have been burglarized per 1000 households in the year that is t years since 1900. Perform the first three steps of the four-step modeling process to find an equation for f.

b. Find the slope of f. What does it represent in this situation?

c. Use your model f to predict when no homes will be burglarized.

19. Average tuitions at four-year colleges are listed in Table 67.

	Public Tuition	Private Tuition
Table 67 Average Tuitions at Four-Year Colleges		
Year	(dollars)	(dollars)
1995	2977	14,537
1996	3151	15,605
1997	3323	16,531
1998	3486	17,229
1999	3645	18,226
2000	3774	19,312

Source: *U.S. National Center for Education Statistics*

a. Let $f(t)$ represent the average tuition (in dollars) for public colleges at t years since 1990. Perform the first three steps of the four-step modeling process to find an equation for f. [**Modeling Process:** 1. Scattergram 2. Equation 3. Verify 4. Estimate/Predict.]

b. Let $g(t)$ represent the average tuition (in dollars) for private colleges at t years since 1990. Perform the first three steps of the four-step modeling process to find an equation for g.

c. What is the slope of f? The slope of g? What do these slopes tell you about tuitions?

d. A student intends to earn a bachelor's degree by attending a college for four years, starting in 2007. Compare the approximate total four-year cost of attending a public college with the approximate total four-year cost of attending a private college.

20. Prices for existing homes have gradually increased over the years (see Table 68).

Year	Median Sales Price ($1000s)
Table 68 Median Prices of Homes	
1995	110.5
1996	115.8
1997	121.8
1998	128.4
1999	133.3
2000	139.0
2001	147.5

Source: *National Association of Realtors*®

[**Note:** The median sales price is the price "in the middle."]

a. Let $p = f(t)$ represent the median price (in thousands of dollars) for homes sold during the year that is t years after 1990. Find an equation for f.

b. What is the slope of your model f? What does it mean in terms of prices?

c. Find the increase in prices from 1995 to 1996, from 1996 to 1997, and so on. Compare these increases to the slope of your model f. Explain why your observations make sense.

21. An airplane is flying at a rate of 500 miles per hour for 3 hours. It then runs into strong headwinds and travels at a rate of 400 miles per hour for 2 more hours. Let $f(t)$ represent the distance (in miles) traveled in t hours.

a. Complete a table consisting of values for t and $f(t)$. Use 0, 1, 2, 3, 4, 5 for t. You may assume that it takes no time at all for the airplane to decelerate from 500 miles per hour to 400 miles per hour.

b. Sketch a graph of the function f.

c. Discuss the assumption that it takes no time at all for the airplane to decelerate from 500 miles per hour to 400 miles per hour. Is that possible? Explain.

d. Discuss the constant rate of change property in the context of this exercise.

22. At noon, a math instructor drives her car at 50 miles per hour for 2 hours. During the next minute, she accelerates to 70 miles per hour, then travels at that speed for 3 more hours. (This means that the driver is on the road for a total of 5 hours and 1 minute.) Let d represent the distance traveled (in miles) after t hours have elapsed.

a. Complete Table 69.

Time of Day	t (in hours)	d (in miles)
Table 69 Accelerating from 50 mph to 70 mph in One Minute		
12:00	0	
1:00	1	
2:00	2	
2:01	2.017	
3:01	3.017	
4:01	4.017	
5:01	5.017	

b. Sketch a graph that describes the relationship between t and d.

For each scenario, sketch a qualitative graph that relates distance d from home to the amount of time t that has elapsed.

23. A student drives toward school at a constant rate. After a while, his favorite song plays on the radio and he drives the rest of the way at a faster constant rate.

24. A commuter is driving toward work at a constant rate. The commuter notices a police car in the right lane and slows down so that she can drive at a slower constant rate. After a while, the police car takes an exit ramp, and the commuter quickly resumes driving at the original constant rate.

25. A student drives toward school at a constant rate. After a while, he gets a flat tire, which takes some time to fix. He then continues to drive toward school, but at a faster constant rate to make up for lost time.

26. A student drives toward school at a constant rate. After a while, the student realizes that she left her graphing calculator at home. She turns around and heads home at a faster constant rate in hopes of still getting to class on time. After getting her calculator, she drives toward school at an even faster constant rate.

27. Explain what the statement "Slope is a rate of change" means. Give an example other than those given in the text.

Taking it to the lab

For each lab assignment, consult with your instructor on whether to organize your responses as a numbered list or to write them in a paragraph.

THE USED CAR LAB

In this lab you will explore the relationship between the advertised resale price of a specific make and model of a car and the age of the car. To explore this question, choose a car make and model (such as Honda Civic). Refer to used-car advertisements in a newspaper, car magazine, or web site (such as www.edmunds.com) to find the ages and prices for about 20 cars of your chosen type. You may use more than one car of a single age, but if you use more than one source of data, be careful that you are not using the *same* car twice (for example, Herb Motors' 2000 Honda Civic as listed in both *The Advertiser* and *The Herald*).

Let p represent the advertised price (in dollars) of your choice of car that is a years old. (Note that a is the age of the car, not the year in which it was made.)

Analyzing the Data

1. Include a table of data. State the source(s) of the data.
2. Using pencil and paper, sketch a scattergram of your data. If no line comes close to the data points, choose another make and model of car.
3. Find an equation whose graph comes close to the data points. Write your equation using $f(a)$ notation.
4. Sketch a graph of f on your scattergram.
5. Estimate what the advertised price should be for your choice of car if it is 10 years old.
6. Predict when your car will be worth half as much as the cheapest price listed in your table of data.
7. Find the slope of your model. What does it represent in terms of the car?
8. Find the a-intercept and the p-intercept of your model. What do these intercepts represent in terms of the car? Do you think that f models the car situation well near the intercepts? Explain.
9. For what values of a is there model breakdown for certain? Explain.
10. Sketch a qualitative graph of the relationship between your car's price and its age for all possible ages.

Taking It One Step Further

11. Let $g(a)$ represent a 10% down payment (in dollars) for your choice of car that is a years old. Find an equation for g.
12. Find $g(5)$. What does the result mean in terms of your car?
13. Compare the slope of g to the slope of f. Explain why your observation(s) makes sense in terms of your car.

THE GOLF BALL LAB

In this lab you will explore the relationship between the height of a golf ball before dropping it and its one-bounce height.*

> **Note**
>
> Your *data* is probably not a function. However, your *model* f is a function.

---

*Lab written by Jim Ryan, State Center Community College District Clovis Center, Clovis, CA.

Materials

You will need at least three people and the following items:

1. a tape measure
2. a golf ball

Recording the Data

The same person should drop the golf ball each time. A second person should always do the measuring of the height of the golf ball (from the bottom of the ball) before it is dropped. The ball should be dropped from an initial height of 12 inches. A spotter should watch to estimate the bounce height of the golf ball. Repeat this process three times, and then compute the average of the three bounce heights. Next, find average bounce heights of the golf ball for initial heights of 24 inches, 36 inches, 48 inches, 60 inches, and 72 inches. In case your instructor cannot spare the class time required to run the experiment, some data is listed in Table 70.

Analyzing the Data

1. Display your golf ball data in a table.
2. Let R represent the bounce height (in inches) of the golf ball after being dropped from an initial height of H inches. Sketch a scattergram of the golf ball data.
3. Find an equation whose graph comes close to the data. Write your equation using $f(H)$ notation.
4. Use pencil and paper to sketch a graph of f with your scattergram.
5. Use your model to estimate the bounce height for a drop height of 80 inches.
6. A golf ball is hit to a maximum height of 50 feet on a golf course. What does your model estimate the bounce height to be after one bounce? Do you think this estimate is accurate? If not, will it be an underestimate or an overestimate? Explain.
7. Find the slope of your model. What does the slope mean in terms of the golf ball? Explain.

Table 70 Drop and Bounce Heights of a Golf Ball

Drop Height (inches)	Bounce Height (inches)
12	10.0
24	20.3
36	31.0
48	44.5
60	52.0
72	64.0

Source: *Data collected by the author*

THE TEMPERATURE SCALES WEB LAB

In this lab you will work with three temperature scales: Fahrenheit, Celsius, and Kelvin. If operational, the following web sites can help you complete this lab:

Who Invented the Thermometer—Fahrenheit, Celsius, and Kelvin Scales
inventors.about.com/library/inventors/blthermometer.htm

Online Conversion—Temperature Conversion
www.onlineconversion.com/temperature.htm

About Temperature
www.unidata.ucar.edu/staff/bylynds/tmp.html#Dev

You may also have to perform searches to find additional information. To get you started with searches, parts of this lab suggest some good key words.
    Respond to the following instructions and questions.

1. Who invented the Fahrenheit, Celsius, and Kelvin temperature scales? Briefly describe each of these three inventors. [**Key words:** "history," "Fahrenheit," "Celsius," "Kelvin," and "temperature."]
2. Describe the three temperature scales. Also, discuss how they are related to each other in words.
3. For the three temperature scales, there are a total of six conversions possible. Find a web site that will perform temperature conversions for you. [**Key words:** "temperature," "conversion," "Fahrenheit," "Celsius," and "Kelvin."]
   a. Convert 47°F to the other two scales.
   b. Convert 13°C to the other two scales.
   c. Convert 35°K to the other two scales.
4. For each of the six conversions, find an equation that can be used to perform the conversion. Show how to use one equation or a pair of equations to find another of the six equations.

5. Repeat problem 3, but this time use your conversion equations.

6. Use a three-column graphing calculator table to display equivalent temperatures for the three scales. Explain how you did this and copy the table on paper.

THE WALKING STUDENT LAB

In this lab you or a classmate will walk toward a wall. While this student is walking, the lab mate will record the distance from the student to the wall at various times.

Materials

You will need at least three people and the following items:

1. a timing device
2. a tape measure
3. a Texas Instruments CBR™ unit, or a Texas Instruments CBL™ unit with a Vernier motion detector probe (in place of a timing device and tape measure). (optional)

Preparation

Have the student who will do the walking stand about 15 feet from a wall. If you aren't using the CBL or CBR unit, lay out the tape measure from the wall to the student.

Recording of Data

Have the student walk toward the wall at a steady rate. Record the distance the student is from the wall at various times. If you are not using a CBL or CBR unit, you may need several people to assist in measuring and recording the data. Try to get at least 5 data points (time and distance).

In case your instructor cannot spare the class time required to run the experiment, some data is listed in Table 71.

Analyzing the Data

1. Display your data in a table.
2. Let d represent the distance (in feet) the student was from the wall at t seconds after the student began walking. Which variable is the dependent variable?
3. Use your graphing calculator to draw a scattergram of the data.
4. Find an equation whose graph comes close to the data points.
5. Use your graphing calculator to draw a graph of your equation and the scattergram in the same viewing window. Also, sketch the graph of the equation and scattergram using pencil and paper.
6. Write your equation using $f(t)$ notation.
7. Is your line increasing or decreasing? Explain what this means in terms of the walking student.
8. Use your equation for f to estimate how far the student was from the wall at 2 seconds.
9. Use your equation for f to estimate when the student was 1 foot from the wall.
10. What is the slope of f? What does the slope represent in terms of the walking student?
11. What is the d-intercept of f? What does the d-intercept represent in terms of the walking student?
12. What is the t-intercept of f? What does the t-intercept represent in terms of the walking student?
13. Use your equation for f to estimate how far the student was from the wall after 20 seconds. Has model breakdown occurred? Explain.

Table 71 Distances between a Student and a Wall

Time (seconds)	Distance (feet)
0.4	10.55
1.6	8.52
3.3	5.62
4.1	4.19
5.4	1.91

Source: *Data collected by students in the author's class*

LINEAR LAB: TOPIC OF YOUR CHOICE

Your objective in this lab is to find a linear function that can be used to model some true-to-life situation. Your function should model a situation that has not been discussed in this text. Your first task will be to find some data. Almanacs, newspapers, magazines, and scientific journals are good resources. You may want to try searching on the Internet. Or, you can conduct an experiment. Choose something that interests you!

Analyzing the Data

1. What two variables did you explore? If you made any false starts, tell about each pair of variables that you had hoped to explore. Also, explain why you chose new variables.

2. Which variable is the dependent variable? Which variable is the independent variable?

3. Describe how you found your data. If you conducted an experiment, provide a careful description with specific details of how you went about running your experiment. If you didn't conduct an experiment, state the source of your data.

4. Include a table of your data.

5. Perform the first three steps of the four-step modeling process. (If your data are not approximately linear, find some data that are.)

6. Use pencil and paper to sketch the graph of your linear model and a scattergram of your data in the same coordinate system.

7. What is the slope of your model? What does it mean in terms of the situation you chose to model?

8. Does it make sense to you why your variables are approximately linearly related in terms of the situation you chose to model? Explain.

9. Find the intercepts of your linear model. What do they mean in terms of the situation you are modeling?

10. Make an estimate or prediction about your dependent variable based on a specific value for your independent variable. Describe what your result means in terms of the situation you are modeling.

11. Make an estimate or prediction about your independent variable based on a specific value for your dependent variable. Describe what your result means in terms of the situation you are modeling.

12. Comment on your lab experience.
    a. For example, you might address whether this lab was enjoyable, insightful, and so on.
    b. Were you surprised by any of your findings? If so, which ones?
    c. How would you improve your process for this lab if you did it again?
    d. How would you improve your process if you had more time and money?

CHAPTER SUMMARY

Key Points
OF THIS CHAPTER

A *linear model* is a linear function that describes the relationship between two quantities for a true-to-life situation. | **Linear model**

All linear models are linear functions. All functions are not linear models. Functions are not restricted to describing true-to-life situations. | **Model versus function**

When a model gives a prediction that does not make sense or an estimate that is not a very good approximation, we say that *model breakdown* has occurred. | **Model breakdown**

Four-step modeling process: | **Four-step modeling process**
1. Create a scattergram of the data.
2. Use two points to find an equation of a nonvertical line that comes close to the data points.

3. Verify your equation by checking that the graph of your model contains the two chosen points. If it doesn't, check for calculation errors. Also, check that the line does indeed come close to all the data points. If it doesn't, try using different points to derive an equation.

4. Use the linear model to make estimates and predictions.

Making a prediction about the dependent variable	When making a prediction about the dependent variable of a linear model, substitute a chosen value for the independent variable in the model, then solve for the dependent variable.
Making a prediction about the independent variable	When making a prediction about the independent variable of a linear model, substitute a chosen value for the dependent variable in the model, then solve for the independent variable.
Intercepts of a model	Suppose that a function of the form $p = mt + b$ is used to model a situation. Also assume that the line crosses both the t-axis and the p-axis. * The p-intercept is $(0, b)$. * To find the t-coordinate of the t-intercept, substitute 0 for p in the model's equation, then solve for t.
$f(x)$ notation	A linear function of the form $y = mx + b$ can also be described by the equation $f(x) = mx + b$; here we use $f(x)$ to represent y. In general: $$\text{dependent variable} = f(\text{independent variable})$$
Meaning of slope	The slope of a linear model is equal to the rate of change of the dependent variable with respect to the independent variable.
Constant rate of change	If the rate of change of the dependent variable with respect to the independent variable is constant, then there is a linear relationship between the variables.

CHAPTER 2 REVIEW EXERCISES

Solve each equation.

1. $3x - 5 = x + 8$

2. $3(x - 2) + 1 = x - 7(x + 1)$

3. $1.2(x - 3.1) = 4.8x + 7.4$ **4.** $\dfrac{2}{3}x - 1 = \dfrac{3}{4}$

Evaluate the function $f(x) = -5x + 9$ at each given value of x. Whenever possible, verify your result using a graphing calculator graph or table.

5. $f(3)$ **6.** $f(0)$ **7.** $f(-2)$

8. $f(6.89)$ **9.** $f\left(-\dfrac{2}{5}\right)$ **10.** $f(a + 1)$

For $f(x) = 2x + 3$, find the value of x that corresponds to each given value of $f(x)$. Whenever possible, verify your result using a graphing calculator graph or table.

11. $f(x) = 9$ **12.** $f(x) = 0$ **13.** $f(x) = -6$

14. $f(x) = 1.81$ **15.** $f(x) = \dfrac{2}{3}$ **16.** $f(x) = a$

For exercises 17–24, refer to Fig. 38.

17. Estimate $f(3)$.   **18.** Estimate $f(0)$.

19. Estimate $f(-1)$.   **20.** Estimate $f(-3)$.

21. Estimate x when $f(x) = 3$.

22. Estimate x when $f(x) = 0$.

23. Estimate x when $f(x) = 2$.

24. Estimate x when $f(x) = -4$.

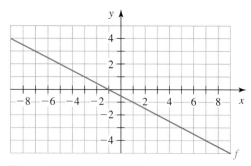

Figure 38 Exercises 17–24

For exercises 25–28, refer to Table 72.

Table 72 Values of f (Exercises 25–28)	
x	$f(x)$
0	1
1	2
2	4
3	3
4	0

25. Find $f(0)$.

26. Find $f(2)$.

27. Find x when $f(x) = 0$.

28. Find x when $f(x) = 2$.

Find the x-intercept and y-intercept of each function.

29. $f(x) = -7x + 3$

30. $f(x) = -3.1x - 5.7$

31. $f(x) = 4$

32. $f(x) = -\dfrac{4}{7}x + 2$

33. Copy the graphs of the data points and the model $y = mx + b$ in Fig. 39. Sketch the graph of a linear model that better describes the data, then explain how you would adjust the slope and the y-intercept of the original model to better describe the data.

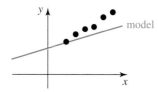

Figure 39 Exercise 33

34. A student's car has a 13-gallon gasoline tank. The car uses 1.8 gallons per hour when driven at 65 miles per hour. After filling up the gas tank, the student begins driving at 65 miles per hour and continues to drive at this speed until he runs out of gas. Let $A = f(t)$ represent the amount of gasoline (in gallons) in the gas tank at t hours after the student filled up the tank.

  a. Find an equation for f.

  b. What is the slope of f? What does it represent in terms of gasoline in the tank?

  c. What is the A-intercept of f? What does it represent in terms of gasoline in the tank?

  d. What is the t-intercept of f? What does it represent in terms of gasoline in the tank?

  e. What are the domain and range of the model? Explain.

35. The average household income in 1995 was $50,500. Each year, the income increases by about $1309. Let $g(t)$ represent the average household income (in dollars) at t years since 1995.

  a. Find the slope of g. What does it mean in terms of average income?

  b. Find an equation for g.

  c. Find $g(12)$. What does your result mean in terms of the situation?

  d. Find t where $g(t) = 70,000$. What does your result mean in terms of the situation?

36. The idea of dividing California into two states has been around since 1850, and was promoted as recently as 1993. Table 73 shows the percentages of samples of Californians who approve of splitting California into two states—North and South.

Year	Percent That Approve of the Split
1959	8.0
1965	11.0
1969	13.5
1981	17.5
1992	24.5
1993	29.0

Table 73 Splitting California into Two States

Source: *Poll conducted by the Field Institute*

Let $f(t)$ represent the percentage of Californians who at t years since 1900 approve of splitting California into two states.

  a. Find an equation for f.

  b. What will be the percentage of Californians who approve of dividing California into two states in the year 2008?

  c. Find the t-intercept of f. What does your result represent in terms of approval for splitting California into two states?

  d. Predict when a majority of Californians will be in favor of splitting California into two states. How confident are you in this prediction?

37. Although today most chief executive officers (CEOs) and presidents of large companies have at least a bachelor's degree, this was not the case in the early 1900s. In fact, the percentage of these leaders who had at least a bachelor's degree gradually increased during the 20th century (see Table 74).

Year	Percent
1900	27
1925	41
1950	63
1978	93
1985	89
1987	94

Table 74 CEOs and Presidents with at Least a Bachelor's Degree

Source: Liberal Education and the Corporation *by Michael Useem*

  a. Let $p = g(t)$ represent the percentage of CEOs and presidents of large companies who have at least a bachelor's degree at t years since 1900. Find an equation for g.

  b. Find $g(100)$. What does your result mean in terms of CEOs and presidents of large companies?

  c. Find the value for t where $g(t) = 100$. What does your result mean in terms of CEOs and presidents of large companies?

  d. For which years does model breakdown occur for certain? Explain.

  e. Sketch a qualitative graph that describes the relationship between percentage and time for the years 1776–2076.

CHAPTER 2 TEST

For exercises 1–8, refer to Fig. 40.

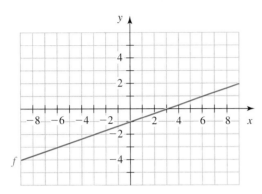

Figure 40 Exercises 1–8

1. Estimate $f(6)$.

2. Estimate $f(3)$.

3. Estimate $f(0)$.

4. Estimate $f(-5)$.

5. Estimate x when $f(x) = -3$.

6. Estimate x when $f(x) = -2$.

7. Estimate x when $f(x) = 0$.

8. Estimate x when $f(x) = 0.5$.

Let $f(x) = -4x + 7$.

9. Find $f(-3)$.

10. Find x when $f(x) = 2$.

Find the x-intercept and y-intercept for each function.

11. $f(x) = 3x - 7$

12. $g(x) = -2x$

13. $h(x) = 2.5x + 10$

14. $k(x) = \dfrac{1}{3}x - 8$

Solve each equation.

15. $2(x - 5) - 3(4x + 2) = 6x$

16. $\dfrac{3}{8}x - \dfrac{1}{4} = \dfrac{1}{2}$

17. A student drives from home toward a movie theater at a constant rate. When she arrives at the theater, she finds out that the movie she wanted to see is sold out. After staying at the theater for a few minutes, she drives toward home but at a slower constant rate. Sketch a qualitative graph that relates distance d from home to the amount of time t that has elapsed.

18. The tuition at a college was $8320 in 2000. Each year, the tuition has increased by $750. Let $D = f(t)$ represent the tuition (in dollars) for the year that is t years since 2000.

   a. Find an equation for f.

   b. What is the slope of f? What does it mean in terms of the college's tuition?

   c. What is the D-intercept? What does it mean in terms of the college's tuition?

   d. Use f to predict when the tuition will be $20,000.

19. When exercising, you should raise your pulse rate to your target pulse rate. Your target pulse rate depends on your age (see Table 75). Let $T = f(A)$ represent the target pulse rate (in beats per minute) for a person who is A years old.

Table 75 Target Pulse Rates for Exercising	
Age	Target Pulse Rate (beats per minute)
20	150
30	142
40	135
50	127
60	120
70	113

Source: Health Handbook *from Long's Drugs*

   a. Find an equation for f.

   b. Use your model to estimate the target pulse rate for someone who is 18 years old.

   c. According to the model f, at what age is a person's target pulse rate 118 beats per minute?

   d. Find the model's A-intercept. What does this mean in terms of target pulse rates? Has model breakdown occurred? Explain.

   e. Find the model's T-intercept. What does this mean in terms of target pulse rates? Has model breakdown occurred? Explain.

20. The portion of the military who are women has increased over the past three decades (see Table 76).

Table 76 Percentages of the Military Who Are Women	
Year	Percent
1965	1.2
1975	4.6
1985	9.8
1990	11.1
1997	13.6
2000	14.8

Source: *USA Today*

   a. Let $f(t)$ represent the percentage of the military who are women at t years since 1900. Find an equation for f.

   b. What is the slope of your model? What does it represent in terms of the military?

   c. Find $f(100)$. What does your result mean in terms of the military?

   d. Find the value for t where $f(t) = 100$. What does your result mean in terms of the military?

   e. For what values of t does model breakdown occur for certain? Explain.

Systems of Linear Equations

Mathematics possesses not only truth, but supreme beauty—a beauty cold and austere, like that of sculpture.
—*Lord Bertrand Russell, "A Free Man's Worship," Mysticism and Logic.*

3.1 USING GRAPHS TO SOLVE SYSTEMS

Objectives

▸ Use systems of linear equations to model data.

▸ Know the meaning of *solution* and *solution set* of a system of equations.

▸ Use a graphical approach to solve systems of linear equations.

In this chapter we discuss modeling a situation with more than one linear model.

Example 1   Using Two Models to Make a Prediction

In the United States, life expectancies for women have been longer than life expectancies for men for many years (see Table 1).

The life expectancies (in years) $W(t)$ and $M(t)$ for women and men, respectively, are modeled by the system

$$L = W(t) = 0.098t + 77.58$$
$$L = M(t) = 0.192t + 69.98$$

where t represents the number of years since 1980. Use graphs of W and M to predict when life expectancies for women and men will be equal.

Table 1 U.S. Life Expectancies of Women and Men

Year of Birth	Women (years)	Men (years)
1980	77.4	70.0
1985	78.2	71.1
1990	78.8	71.8
1995	78.9	72.5
2000	79.5	74.1

Source: *U.S. Bureau of the Census*

Solution

We begin by sketching graphs of W and M on the same coordinate system (see Fig. 1).

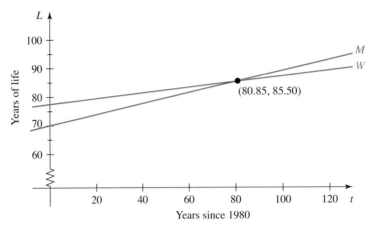

Figure 1 Life expectancy models for women and men

Note that the intersection point is approximately $(80.85, 85.50)$. So, the models predict that life expectancy for both women and men will be about 86 years in 2061. We are not very confident in this prediction, however, because it is for so far into the future.

We verify our work using "intersect" on a graphing calculator (see Figs. 2–4).

Graphing Calculator

See Sections B.7 and B.18.

Figure 2 Enter the functions

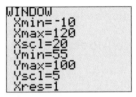

Figure 3 Set up the window

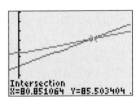

Figure 4 Find the intersection point

In Example 1 we predicted when the life expectancy of women will equal the life expectancy of men. Graphically, this event is described by the intersection point of the graphs of the two life expectancy models.

In general, if the independent variable of two models represents time, then the intersection point(s) of the graphs of these models indicates when the quantities represented by the dependent variables were or will be equal.

In this section we locate the intersection point of the graphs of two models by graphing. In Section 3.3 we discuss how to find the intersection point of two models using symbolic approaches.

In Chapter 1 we discussed the meaning of a graph.

Graph

All points that satisfy an equation form the **graph** of the equation. Points that do not satisfy an equation are not part of the graph.

Understanding graphs will help us greatly in this section.

Example 2 Finding Ordered Pairs That Satisfy Both of Two Given Equations

Find all ordered pairs that satisfy both of the equations

$$y = 2x + 1$$
$$y = -3x + 6$$

Solution

To begin, we sketch a graph of each equation on the same coordinate system (see Fig. 5).

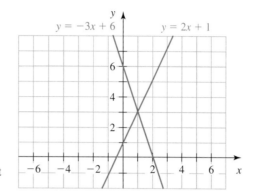

Figure 5 Intersection point is $(1, 3)$

For a point to satisfy one of the equations, it must lie on that equation's graph. For a point to satisfy *both* equations, it must lie on *both* lines. So, the only point that satisfies both equations is the intersection point $(1, 3)$.

We can verify that $(1, 3)$ satisfies both equations:

$$y = 2x + 1 \qquad\qquad y = -3x + 6$$

$$3 \stackrel{?}{=} 2(1) + 1 \qquad\qquad 3 \stackrel{?}{=} -3(1) + 6$$

$$3 \stackrel{?}{=} 3 \qquad\qquad\qquad 3 \stackrel{?}{=} 3$$

$$\text{true} \qquad\qquad\qquad \text{true} \qquad\qquad\qquad \blacksquare$$

In Example 2 we worked with the two linear equations

$$y = 2x + 1$$
$$y = -3x + 6$$

We refer to these equations as a **system of two linear equations in two variables**. It is also a **linear system**, which consists of two or more linear equations. We found that the only point that satisfies both equations is the intersection point $(1, 3)$. We call the set containing only $(1, 3)$ the **solution set** of the system.

<u>**DEFINITION** *Solution of a system*</u>

We say that an ordered pair (a, b) is a **solution** of a system of two equations in two variables if it satisfies both equations. The **solution set** of such a system is the set of all ordered pairs that satisfy both equations. We **solve** a system by finding all of its solutions.

Note

The solution of a system is a set of ordered pairs. For ease in referring to such a solution, we will often not mention "set." For example, the phrase "the solution is (a, b)" will be shorthand for "the solution set is $\{(a, b)\}$."

In general, the solution set of a system of two linear equations can be deduced by locating the intersection point(s) of the graphs of the two equations.

Example 3   Solving a System of Two Linear Equations by Graphing

Solve the system

$$y = 2x + 4$$
$$y = -x + 1$$

Solution

The graphs of the equations are sketched in Fig. 6.

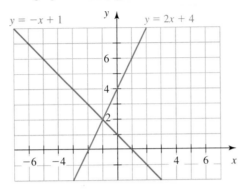

Figure 6 Intersection point is $(-1, 2)$

The solution is the intersection point $(-1, 2)$. We can verify that $(-1, 2)$ satisfies both equations:

$$y = 2x + 4$$
$$2 \overset{?}{=} 2(-1) + 4$$
$$2 \overset{?}{=} 2$$
$$\text{true}$$

$$y = -x + 1$$
$$2 \overset{?}{=} -(-1) + 1$$
$$2 \overset{?}{=} 2$$
$$\text{true}$$

We can also verify our work by using "intersect" on a graphing calculator (see Fig. 7).

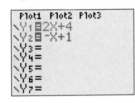
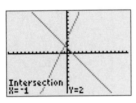

Figure 7 Verify that the intersection point is $(-1, 2)$

Example 4   Solving a System of Two Linear Equations by Graphing

Solve the system

$$\frac{1}{2}x + \frac{5}{4}y = \frac{5}{2}$$
$$y = 3x - 4$$

Solution

Note

See Sections A.8 and A.9 for a review of solving an equation with fractions.

First, we multiply both sides of the equation $\frac{1}{2}x + \frac{5}{4}y = \frac{5}{2}$ by the LCD 4.

$$\frac{1}{2}x + \frac{5}{4}y = \frac{5}{2}$$

$$4\left(\frac{1}{2}x + \frac{5}{4}y\right) = 4 \cdot \frac{5}{2} \qquad \text{Multiply both sides by the LCD 4.}$$

$$4 \cdot \frac{1}{2}x + 4 \cdot \frac{5}{4}y = 4 \cdot \frac{5}{2} \qquad \text{Distributive law}$$

$$2x + 5y = 10 \qquad \text{Simplify.}$$

$$5y = -2x + 10 \qquad \text{Subtract } 2x \text{ from both sides.}$$

$$y = -\frac{2}{5}x + 2 \qquad \text{Divide both sides by 5.}$$

Next we sketch a graph of the equations $y = -\frac{2}{5}x + 2$ and $y = 3x - 4$ (see Fig. 8).

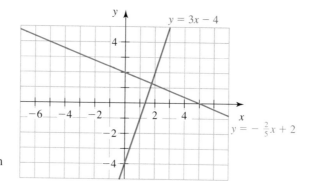

Figure 8 Approximate solution is (1.8, 1.3)

We can estimate the solution to be approximately (1.8, 1.3). However, we can get a better estimate by using "intersect" on a graphing calculator (see Fig. 9).

 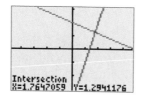

Figure 9 Using "intersect" to find the intersection point

It turns out that the pair (1.7647059, 1.2941176) found by "intersect" is accurate to seven decimal places. For ease in calculator entry, we round the coordinates to the second decimal place and check that (1.76, 1.29) approximately satisfies both equations:

$$\frac{1}{2}x + \frac{5}{4}y = \frac{5}{2} \qquad\qquad y = 3x - 4$$

$$\frac{1}{2}(1.76) + \frac{5}{4}(1.29) = \frac{5}{2} \qquad\qquad 1.29 = 3(1.76) - 4$$

$$2.4925 \approx 2.5 \qquad\qquad 1.29 \approx 1.28$$

Because (1.76, 1.29) approximately satisfies both equations, we know that (1.76, 1.29) is a good approximation of the *exact* solution, which we will learn to find in Section 3.2. ∎

Example 5   Solving an Inconsistent System

Solve the system

$$y = 2x + 1$$
$$y = 2x + 3$$

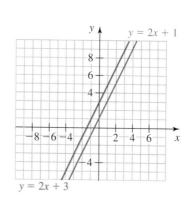

Figure 10 The solution is the empty set

Solution

Since the distinct lines have equal slopes, the lines are parallel (see Fig. 10). Parallel lines do not intersect, so there is no point that satisfies both equations. The solution set is the empty set. ∎

A linear system whose solution set is the empty set is called an **inconsistent system**.

Example 6 Solving a Dependent System

Solve the system

$$y = 2x + 1$$
$$6x - 3y = -3$$

Solution

We write the equation $6x - 3y = -3$ in slope-intercept form:

$$6x - 3y = -3$$
$$2x - y = -1 \qquad \text{Divide each side by 3.}$$
$$-y = -2x - 1 \qquad \text{Subtract } 2x \text{ from both sides.}$$
$$y = 2x + 1 \qquad \text{Multiply each side by } -1.$$

So, the graph of $6x - 3y = -3$ and the graph of $y = 2x + 1$ are the same line. The solution set of the system is the set of the infinite number of points that lie on the line $y = 2x + 1$ and on the (same) line $6x - 3y = -3$. ■

Whenever a linear system has an infinite number of solutions, we say that the system is **dependent**.

In Examples 4, 5, and 6, we have seen three types of systems:

1. A one-point-solution system. See Fig. 11.
2. An inconsistent system (empty set solution). See Fig. 12.
3. A dependent system (infinite number of solutions). See Fig. 13.

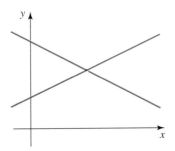

Figure 11 Graph of a system with one ordered pair solution

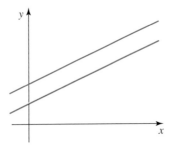

Figure 12 Graph of an inconsistent system

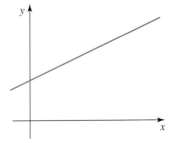

Figure 13 Graph of a dependent system

EXPLORATION
Comparing the three types of systems

1. Is the system

$$y = -2x + 3$$
$$y = -2x + 5$$

a dependent system, an inconsistent system, or a one-point-solution system? Explain.

2. Now, you will work with the system

$$y = 3x - 5$$
$$y = mx + b$$

where m and b are constants.

   a.   Find values for m and b so that the given system is inconsistent. What is the solution set of your system? Use your graphing calculator to verify your work.

   b.   Find values for m and b so that the given system is dependent. What is the solution set of your system?

   c.   Find values for m and b so that the given system is a one-point-solution system. Use "intersect" on your graphing calculator to find the one-point solution.

Graphing Calculator

See Section B.18.

3.   Now consider this general system of two linear equations.

$$y = m_1 x + b_1$$
$$y = m_2 x + b_2$$

Discuss dependent systems, inconsistent systems, and one-point-solution systems in terms of m_1, m_2, b_1, and b_2.

Tips on Succeeding in This Course

To help prepare themselves mentally and physically for competition, many exceptional athletes visualize themselves performing well at their event many times throughout their training period. For example, a runner training for the 100 meter run might imagine getting set in the starting blocks, taking off right after the gun goes off, being in front of the other runners, and so on, right up until the moment of breaking the tape at the finish line.

In an experiment to test the effectiveness of visualization, three groups of basketball players were used. The first group warmed up by shooting baskets before a game. The second group visualized shooting baskets, but did not shoot any baskets during their pre-game warm-up. The third group did not warm up nor visualize before the game. It is a fact that the visualization group not only outperformed the group that did not warm up, they also did better than the group that warmed up by shooting baskets!

You can do visualizations, too. You could visualize doing all the things you feel you need to do to succeed in this course. If you do this regularly, you will find you have better follow-through with what you intend to do. You will also feel more confident about succeeding.

Key Points

OF THIS SECTION

If $f(t)$ and $g(t)$ are models where t represents time, then the intersection point(s) of the graphs of these models indicates when the quantities represented by the dependent variables were or will be equal.

Intersection point(s) of the graphs of two models

The *solution set* of a system of two equations is the set of all ordered pairs that satisfy both equations. The solution set can be deduced by locating the intersection point(s) of the graphs of the two equations.

Solution set of a system

There are three types of systems of two linear equations in two unknowns:

Types of systems

- A *one-point-solution system* consists of two equations whose graphs intersect in one point. The solution is this point.

- A *dependent system* consists of two equations whose graphs are identical lines. The solution set is the set of points on the line.

- An *inconsistent system* consists of two equations whose graphs are parallel lines. The solution set is the empty set.

HOMEWORK 3.1

 SSM Tutor Center

 MathPro 4 MathPro 4

 MathPro 5 MathPro 5

CDs/Videos

www

Find the solution set of each system by graphing the equations by hand. If the system is inconsistent or dependent, say so. Verify any one-point solution by checking that it satisfies both equations.

1. $y = 2x + 2$
$y = -3x + 7$

2. $y = x - 5$
$y = -2x + 4$

3. $y = -\dfrac{1}{2}x + 3$
$y = \dfrac{1}{3}x + 8$

4. $y = -\dfrac{1}{4}x + 3$
$y = \dfrac{1}{2}x$

5. $y = 3(x - 1)$
$y = -2x + 7$

6. $y = x - 2$
$y = -3(x - 2)$

7. $y = 3x - 7$
$4x + 2y = 6$

8. $y = -2x + 3$
$4y - 12 = -8x$

9. $5(y - 2) = 21 - 2(x + 3)$
$y = 3(x - 1) + 8$

10. $3x + 5y = 7$
$15x = 10 - 25y$

11. $\dfrac{1}{2}x - \dfrac{1}{2}y = 1$
$\dfrac{1}{4}x + \dfrac{1}{2}y = 2$

12. $\dfrac{1}{4}x + \dfrac{1}{2}y = \dfrac{3}{2}$
$y = 2(x - 3) - 1$

Use "intersect" on a graphing calculator to solve each system with solutions rounded to the nearest hundredth. If the system is inconsistent or dependent, say so. If your solution is one ordered pair, check that it approximately satisfies both equations.
*[**Graphing Calculator:** See Section B.18.]*

13. $y = -2x + 6$
$y = 3x + 1$

14. $y = 5.437x - 2.136$
$y = -2.752x + 3.984$

15. $y = 2x - 500$
$y = -0.5x + 700$

16. $4x - 3y = 13$
$2y + 5x = 4$

17. $y = 2x - 1$
$4x - 2y = 2$

18. $y = 2x + 1$
$3y - 1 = 2(3x + 1)$

19. $y = 5x + 10$
$0.2y - x = 3$

20. $1.3y - 3.9x = 2.6$
$0.25y + 1.25x = -1.75$

21. $\dfrac{1}{2}x - \dfrac{1}{2}y = 1$
$\dfrac{1}{3}x + \dfrac{2}{3}y = 2$

22. $\dfrac{1}{3}y + x + 2 = 0$
$y = \dfrac{1}{2}x + 4$

23. The winning times for the Olympic 500-meter speed skating event have generally been decreasing over the past three decades (see Table 2).

Table 2 Olympic 500-Meter Speed Skating Times

Year	Winning Time (in seconds)	
	Women	Men
1972	43.33	39.44
1976	42.76	39.17
1980	41.78	38.03
1984	41.02	38.19
1988	39.10	36.45
1992	40.33	37.14
1994*	39.25	36.33
1998	38.21	35.59
2002	37.375**	34.615**

Source: *The Universal Almanac*
Winter games are now held in the middle of the four-year break between summer games.
**Skating times are now the average time of two heats.*

The winning times (in seconds) $W(t)$ and $M(t)$ for women and men, respectively, are modeled by the system

$$w = W(t) = -0.19t + 43.73$$
$$w = M(t) = -0.16t + 39.90$$

where t represents the number of years since 1970.

a. Use the equations for the models to estimate the winning time for women and the winning time for men in 2002. Find the errors in your estimates.

b. Compare the slope of W to the slope of M. What does your comparison tell you about the winning times for women and men?

c. Explain why your work in parts a and b suggest that there may be a time when the women's winning time will be equal to the men's winning time.

d. Use "intersect" on your graphing calculator to predict when the women's winning time will be equal to the men's winning time. Also, find that winning time.

24. Annual U.S. consumption of chicken and red meat (in pounds consumed per person) is described for various years in Table 3.

Table 3 U.S. Per-Person Consumptions of Chicken and Red Meat

Year*	Chicken (pounds)	Red Meat (pounds)
1970	27.7	131.7
1975	26.4	125.8
1980	32.7	126.4
1985	36.4	124.9
1990	42.5	112.3
1995	48.8	114.7
2000	52.9	113.5

Source: *U.S. Department of Agriculture, as reported by students Michelle F., Grace C., Tsui-Jun T., and Elizabeth V., Spring Semester, 1996*

The students' table of data has been updated to include data from 1995 and 2000.

Let $C(t)$ represent the annual consumption of chicken and $R(t)$ represent the annual consumption of red meat, both in pounds consumed per person, at t years since 1900. Consumption of each can be modeled by the system:

$$C(t) = 0.93t - 40.85$$

$$R(t) = -0.65t + 176.52$$

a. Compare $C(104)$ to $R(104)$. What does your comparison tell you about consumption of food?

b. Use "intersect" on your graphing calculator to predict when consumption of chicken will equal consumption of red meat. What will be the consumption? How confident are you in this prediction?

25. Table 4 lists women's and men's total enrollments at U.S. colleges for various years. The term *U.S. colleges* is used here to encompass community colleges, professional schools, universities, and colleges, both publicly and privately controlled.

Table 4 College Enrollments

Year	Women's Enrollment (millions)	Men's Enrollment (millions)
1980	6.0	5.4
1985	6.6	5.9
1990	7.4	6.2
1995	8.0	6.7
1998	8.6	6.9

Source: *U.S. Bureau of the Census*

Let $W(t)$ and $M(t)$ represent the enrollments (in millions) for women and men, respectively, both at t years since 1970.

a. Find equations for W and M.

b. Use "intersect" on a graphing calculator to estimate when women's and men's enrollments were approximately equal.

c. Predict the total enrollment of women and men in 2009.

26. The number of households in the United States and the number of those households with personal computers are shown for various years in Table 5.

Table 5 U.S. Households with Personal Computers

Year	Total Households (millions)	Households with PCs (millions)
1995	99.0	31.36
1996	99.6	35.39
1997	101.0	39.56
1998	102.5	43.67
1999	103.6	50.00
2000	104.8	55.49

Source: *U.S. Department of Commerce; Media Metrix*

a. Let $f(t)$ represent the total number (in millions) of U.S. households and let $g(t)$ represent the number (in millions) of U.S. households with PCs, both at t years since 1990. Find equations for f and g.

b. Use "intersect" on a graphing calculator to predict when all U.S. households will have PCs.

c. Calculate the percentages of households that had PCs in 1995, 1996, 1997, 1998, 1999, and 2000.

d. Let $p(t)$ represent the percentage of households that have PCs at t years since 1990. Use your results in part c to help you find an equation for p.

e. Use p to predict when all U.S. households will have PCs. Compare the result with your result in part b.

27. In the 2000 presidential election, the race between George W. Bush and Al Gore was extremely close—so close that there was great public concern about the inaccurate vote-counting systems, which included punch cards, lever machines, and paper ballots. Punch cards and lever systems are gradually being replaced with optical scan or other modern electronic systems (see Table 6).

Table 6 Vote-Counting Systems

Year	Percent of Registered Voters	
	Punch Card or Lever Machine	Optical Scan or Other Modern Electronic System
1990	71.7	13.5
1992	69.4	16.5
1994	63.1	25.8
1996	58.0	32.3
1998	52.9	36.4

Source: *Federal Election Commission*

a. Let $f(t)$ represent the percentage of voters using punch cards or lever machines and $g(t)$ represent the percentage of voters using optical scan or other modern electronic systems, both at t years since 1990. Find equations for f and g.

b. Use "intersect" on a graphing calculator to estimate when the percentage of voters using optical scan or other modern electronic systems will equal the percentage of voters using punch cards or lever machines. Is your result a national (even-numbered) election year? If not, in which national election year will the percentages be closest? What are those percentages?

c. There are two other ways to vote: DataVote (punch holes in ballots that haven't been perforated) and paper (voters mark their ballets, and the counting is done manually). Predict the percentage of voters that will use these other methods in the *national election* year you found in part b.

28. The graphs of $y = ax + b$ and $y = cx + d$ are sketched in Fig. 14.

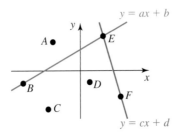

Figure 14 Exercise 28

For each part, decide which of the points A, B, C, D, E, or F

a. satisfy the equation $y = ax + b$.

b. satisfy the equation $y = cx + d$.

c. satisfy both equations.

d. do not satisfy either equation.

29. Fig. 15 shows the graphs of two linear equations. Estimate the coordinates of the solution of the system to the nearest tenth.

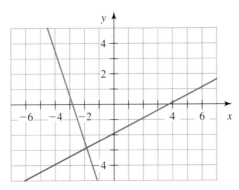

Figure 15 Exercise 29

30. Fig. 16 shows the graphs of two linear equations. Estimate the coordinates of the solution of the system to the nearest whole number. Explain. [**Hint:** Use the slope of each line.]

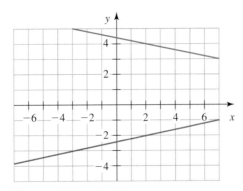

Figure 16 Exercise 30

31. Some values of a linear function f and a linear function g are listed in Table 7. Estimate the solution of a system of two equations that describe f and g.

Table 7 Values of Functions f and g (Exercise 31)

x	0	1	2	3	4	5	6	7	8
$f(x)$	30	27	24	21	18	15	12	9	6
$g(x)$	2	7	12	17	22	27	32	37	42

32. Some values of a linear function f and a linear function g are listed in Table 8. Estimate the solution of a system of two equations that describe f and g.

Table 8 Values of Functions f and g (Exercise 32)

x	0	1	2	3	4	5	6	7	8
$f(x)$	99	95	91	87	83	79	75	71	67
$g(x)$	3	5	7	9	11	13	15	17	19

For exercises 33–38, refer to Fig. 17.

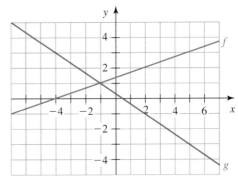

Figure 17 Exercises 33–38

33. Find $f(-4)$.

34. Find $g(-4)$.

35. Find x where $f(x) = 3$.

36. Find x where $g(x) = 3$.

37. Find x where $f(x) = g(x)$.

38. Estimate x where $g(x) = 0$.

39. Create a system of two linear equations as indicated. Verify your system graphically.

a. The solution of the system is $(2, 1)$.

b. The system is inconsistent.

c. The system is dependent.

40. Solve the system. [**Hint:** Sketch the graphs on the same coordinate system.]

$$y = 2$$
$$x = -3$$

41. Find all ordered pairs that satisfy all three equations.

$$y = x + 3$$
$$y = -2x + 9$$
$$y = 3x - 1$$

42. Find all ordered pairs that satisfy all three equations.

$$y = 2x - 5$$
$$y = 0.6x + 1$$
$$y = -1.2x + 5$$

43. Create a system of three linear equations whose solution is $(-4, 3)$. Verify your result by checking that $(-4, 3)$ satisfies all three equations. Also verify your result graphically.

44. A system of linear equations has $(-2, 3)$ and $(4, 1)$ as solutions.

   a. Find a third solution.

   b. How many solutions are there?

45. Explain why the solution(s) of a system of two linear equations is the intersection point(s) of the graphs of the two equations.

46. Describe the three types of systems of two linear equations and how to solve these systems. Also, explain how to verify your work.

3.2 USING ELIMINATION AND SUBSTITUTION TO SOLVE SYSTEMS

Objectives

▶ Use substitution to solve a system of two linear equations.

▶ Use elimination to solve a system of two linear equations.

In Section 3.1 we used graphs of linear functions to solve linear systems. In this section we use *equations* of linear functions to solve such systems.

Example 1   Solving a System Using Substitution

Solve the system

$$y = x - 1$$
$$3x + 2y = 13$$

Solution

First, we substitute $x - 1$ for y in the equation $3x + 2y = 13$:

$$3x + 2(x - 1) = 13$$

Note that by making this substitution, we now have an equation in terms of *one* variable. Next, we solve that equation for x:

$$3x + 2(x - 1) = 13$$
$$3x + 2x - 2 = 13 \qquad \text{Distributive law}$$
$$5x - 2 = 13 \qquad \text{Combine like terms.}$$
$$5x = 15$$
$$x = 3$$

> **Note**
>
> To review the distributive law and combining like terms, see Sections A.6 and A.7.

This means that the x-coordinate of the solution is 3. To find the y-coordinate, substitute 3 for x in either of the original equations and solve for y:

$$y = x - 1$$
$$y = 3 - 1$$
$$y = 2$$

So, the solution is $(3, 2)$. We can check that $(3, 2)$ satisfies both of the system's equations:

$$y = x - 1 \qquad\qquad 3x + 2y = 13$$
$$2 \overset{?}{=} 3 - 1 \qquad\qquad 3(3) + 2(2) \overset{?}{=} 13$$
$$2 \overset{?}{=} 2 \qquad\qquad 9 + 4 \overset{?}{=} 13$$
$$\text{true} \qquad\qquad\qquad \text{true}$$

Or, we can check that $(3, 2)$ is the solution by graphing the two equations and checking that $(3, 2)$ is the intersection point of the two lines (see Fig. 18). To do so on a graphing calculator we must first solve $3x + 2y = 13$ for y:

$$y = -\frac{3}{2}x + \frac{13}{2}.$$

 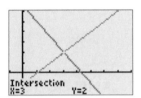

Figure 18 Verify that the intersection point is $(3, 2)$

In Example 1 we solved a system using *substitution*. In Example 2 we will solve a system using the following property.

Adding Left Sides and Right Sides of Two Equations

If $a = b$ and $c = d$, then

$$a + c = b + d$$

In words, the sum of the left sides of two equations is equal to the sum of the right sides.

For example, if we add the left sides and add the right sides of the equations $2 = 2$ and $3 = 3$, we obtain the true statement $2 + 3 = 2 + 3$.

Example 2   Solving a System Using Elimination

Solve the system

$$4x - 5y = 3$$
$$3x + 5y = 11$$

Solution

We begin by adding the left sides and adding the right sides of the two equations:

$$7x + 0 = 14 \qquad y \text{ is "eliminated."}$$

Next, we solve for x:

$$7x = 14$$
$$x = 2$$

Then, we substitute 2 for x in either of the original equations and solve for y:

$$4x - 5y = 3$$
$$4(2) - 5y = 3$$
$$8 - 5y = 3$$
$$-5y = -5$$
$$y = 1$$

The solution is $(2, 1)$. We could then check that $(2, 1)$ satisfies *both* of the original equations.

In Example 2 we solved a system by using *elimination*.

Example 3   Solving a System Using Elimination

Solve

$$3x + 2y = 18$$
$$6x - 5y = 9$$

Solution

If we add the left sides and right sides of the equations as they are now, neither of the variables would be eliminated. Therefore, we first multiply both sides of $3x + 2y = 18$ by -2, yielding the system

$$-6x - 4y = -36$$
$$6x - 5y = 9$$

Now that the coefficients of the x terms are equal in absolute value and opposite in sign, we add the left sides and add the right sides of the equations and solve for y.

$$0 - 9y = -27 \qquad x \text{ is eliminated.}$$
$$-9y = -27$$
$$y = 3$$

We substitute 3 for y in the equation $3x + 2y = 18$ and solve for x.

$$3x + 2(3) = 18$$
$$3x + 6 = 18$$
$$3x = 12$$
$$x = 4$$

The solution is $(4, 3)$. ∎

To use elimination, we must first get the coefficients of the x terms (or y terms) to be equal in absolute value and opposite in sign, before adding the left sides and adding the right sides of the equations. Note that this process is similar to finding a least common denominator.

Example 4   Solving a System Using Elimination

Solve

$$4x - 3y = -3$$
$$5x + 2y = 25$$

Solution

In order to eliminate the y terms, we perform the following multiplications:

$$4x - 3y = -3 \qquad \text{Multiply both sides by 2.}$$
$$5x + 2y = 25 \qquad \text{Multiply both sides by 3.}$$

This gives:

$$8x - 6y = -6$$
$$15x + 6y = 75$$

The coefficients of the y terms are now equal in absolute value and opposite in sign. Next, we add the left sides and add the right sides of the equations and solve for x.

$$23x + 0 = 69$$
$$23x = 69$$
$$x = 3$$

Substituting 3 for x in the equation $4x - 3y = -3$ gives:

$$4(3) - 3y = -3$$
$$12 - 3y = -3$$
$$-3y = -15$$
$$y = 5$$

The solution is (3, 5). ∎

We often use substitution to solve a system if after isolating a variable in one of the equations, there are no fractions in the equation. Otherwise, we tend to use elimination to solve the system.

The systems in Examples 1–4 have one-point solutions. What happens in either the substitution method or the elimination method when the system is inconsistent (no solution) or dependent (infinitely many solutions)?

Example 5   Using Substitution to Solve an Inconsistent System

Use substitution to solve the inconsistent system

$$y = 2x + 1$$
$$y = 2x + 3$$

Solution

We substitute $2x + 1$ for y in the equation $y = 2x + 3$ and solve for x:

$$2x + 1 = 2x + 3$$
$$1 = 3$$
$$\text{false}$$

We get a statement that is false, from which we conclude that the system has no solution (is inconsistent). ∎

Example 5 illustrates that if we solve an inconsistent system, the result is a false statement. Recall that the solution set of an inconsistent system is the empty set.

Inconsistent System of Two Equations

If the result of applying substitution or elimination is a false statement, the system is inconsistent and the solution set is the empty set.

Example 6   Using Elimination to Solve a Dependent System

Use elimination to solve the dependent system

$$y = 2x + 1$$
$$y = 2x + 1$$

Solution

Note

We can subtract the left sides and the right sides, since subtracting a number is the same as *adding* the opposite of the number.

Subtracting the left sides and subtracting the right sides of the equations gives

$$0 = 0$$

We get a statement that is true, from which we conclude that the system has infinitely many solutions (is dependent). ∎

Example 6 illustrates the idea that if we solve a dependent system, the result will be a true statement that can be put into the form $a = a$. Recall that the solution set of a dependent system of two linear equations is the set of all of the points on the (same) line.

Dependent System of Two Linear Equations

If the result of applying substitution or elimination is a true statement (one that can be put into the form $a = a$), then the system is dependent and the solution set is the set of all of the points on one line.

Example 7 Solving a Dependent System

Solve

$$y = 3x - 4$$

$$\frac{5}{2}x - \frac{5}{6}y = \frac{10}{3}$$

Solution

First, we multiply both sides of $\frac{5}{2}x - \frac{5}{6}y = \frac{10}{3}$ by the LCD 6.

$$\frac{5}{2}x - \frac{5}{6}y = \frac{10}{3}$$

$$6\left(\frac{5}{2}x - \frac{5}{6}y\right) = 6 \cdot \frac{10}{3} \qquad \text{Multiply both sides by the LCD 6.}$$

$$6 \cdot \frac{5}{2}x - 6 \cdot \frac{5}{6}y = 6 \cdot \frac{10}{3} \qquad \text{Distributive law}$$

$$15x - 5y = 20 \qquad \text{Simplify.}$$

Then we substitute $3x - 4$ for y in the equation $15x - 5y = 20$.

$$15x - 5(3x - 4) = 20$$

$$15x - 15x + 20 = 20$$

$$20 = 20$$

$$\text{true}$$

Since $20 = 20$ is a true statement (of the form $a = a$), we conclude that the system is dependent and the solution set of the system is the set of points on the line $y = 3x - 4$, or the (same) line $\frac{5}{2}x - \frac{5}{6}y = \frac{10}{3}$. ∎

In Figs. 19, 20, and 21, we summarize the three types of systems, the results of applying substitution or elimination, and the solutions.

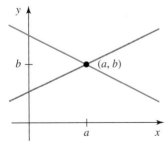

Figure 19 One-point-solution system. Result: $x = a$, $y = b$ Solution set: $\{(a, b)\}$

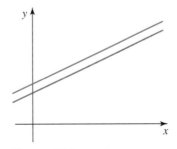

Figure 20 Inconsistent system. Result: False statement. Solution set: Empty set

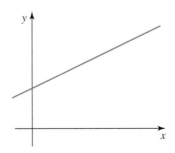

Figure 21 Dependent system. Result: True statement. Solution set: Set of points of line

EXPLORATION
Comparing techniques of solving systems

Consider the system

$$2x + y = 4$$
$$x = 5 - 2y$$

1. Use substitution to solve the system.
2. Use elimination to solve the system.
3. Use graphing to solve the system.
4. Compare your results from 1, 2, and 3.
5. Give an example of a one-point-solution system that is easiest to solve using substitution. Also give an example of such a system that is easiest to solve using elimination. Finally, give an example of such a system that is easiest to solve by graphing. Explain. Also, solve your three systems.

EXPLORATION
Using graphs to solve an equation in one variable

1. **a.** Solve

    $$y = 2x + 1$$
    $$y = 7 - x$$

    using substitution. Use "intersect" on a graphing calculator to verify your result.

    b. Solve $2x + 1 = 7 - x$. Check that your solution satisfies the equation.

    c. Compare your solution to the equation in part b to the x-coordinate of the solution to the system in part a.

2. **a.** Solve

    $$y = 3x - 5$$
    $$y = x + 3$$

    using substitution. Use "intersect" to verify your result.

    b. Solve $3x - 5 = x + 3$. Check that your solution satisfies the equation.

    c. Compare your solution to the equation in part b to the x-coordinate of the solution to the system in part a.

3. Solve $3x - 7 = 8 - 2x$ using a symbolic method. Explain how you can verify your solution by using "intersect."

4. It is difficult to find the exact solution of the equation $x^3 = 5 - x$. Find the approximate solution by using "intersect."

Graphing Calculator

To enter $y = x^3$, press $\boxed{\text{X,T,}\Theta\text{,}n}$ $\boxed{\wedge}$ 3.

Tips on Succeeding in This Course

If you finish a quiz or exam early, it pays to verify your answers with cross-checks. For example, suppose you decide by using elimination that the solution to the system

$$2x + 3y = 9$$
$$4x + 5y = 17$$

is $(3, 1)$. There are several ways to verify your answer. You could check that $(3, 1)$ satisfies both equations. You could graph each equation and check that the intersection point is $(3, 1)$. Or, you could solve the system using substitution.

Key Points

A system of equations can be solved by elimination, substitution, or graphing. All three methods give the same result.

Methods for solving a system

To use substitution, isolate a variable in one equation and then substitute for it in the other equation.

Substitution method

To use elimination, vertically align the variable terms of your equations. If needed, multiply both sides of one equation by a number (and if necessary multiply both sides of the other equation by another number) so that the coefficients of a variable are equal in absolute value and opposite in sign. Then, add the left sides and the right sides of the equations.

Elimination method

There are three types of systems of two linear equations in two unknowns: one-point-solution systems, dependent systems, and inconsistent systems. There are ways to recognize the types when solving by substitution or elimination.

Recognizing types of systems

1. For a one-point-solution system, both substitution and elimination give the coordinates of the solution point.
2. For an inconsistent system, both elimination and substitution give a false statement. When this happens, state that the solution set is the empty set.
3. For a dependent system, both substitution and elimination give a true statement that can be put into the form $a = a$. When this happens, state that the solution set is the set of points of the line.

HOMEWORK 3.2

 SSM Tutor Center

 MathPro 4

 MathPro 5

 CDs/Videos   www

Solve each system using substitution. Verify your solution graphically or by checking that it satisfies both equations in the system.

1. $y = x - 5$
$x + y = 9$

2. $4x - 2y = 12$
$x = 3y - 2$

3. $2x - 3y = -1$
$x = 4y + 7$

4. $y = 11 - 3x$
$y = 2x + 1$

5. $3x - 5y = 29$
$y = 2(x - 5)$

6. $y = 5 - 2x$
$7x - 3y - 11 = 0$

7. $3x - 2y = 18$
$x = -4y - 8$

8. $9x - 2y = -38$
$y = -5x$

9. $y = 99x$
$y = 100x$

10. $y = x$
$y = -x$

11. $y = 0.2x + 0.6$
$2y - 3x = -4$

12. $x = 2 - y$
$0.6x + 0.3y = 0.3$

13. $y = \frac{1}{2}x - 5$
$2x + 3y = -1$

14. $4x - 7y = 3$
$x = \frac{2}{3}y + 4$

Solve each system using elimination. Verify your solution graphically or by checking that it satisfies both equations in the system.

15. $4x + y = 9$
$3x + y = 5$

16. $2x - 6y = 2$
$4x + 9y = 25$

17. $3x - 2y = 7$
$-6x - 5y = 4$

18. $-3x - 5y = -22$
$4x - 7y = -39$

19. $3x + 5y = 3$
$x = 8 - 4y$

20. $3x + 2y = 3$
$9x - 8y = -33$

21. $8x - 9y = -43$
$12x + 15y = 21$

22. $y = 3x + 34$
$y = -8x - 54$

23. $0.9x + 0.4y = 1.9$
$0.3x - 0.2y = 1.3$

24. $0.2x - 0.5y = 0.2$
$0.8x + 1.5y = -6.2$

25. $3(2x - 1) + 4(y - 3) = 1$
$4(x + 5) - 2(4y + 1) = 18$

26. $2(x - 3) - 3(y + 1) = -5$
$-4(x - 2) + 5(y + 3) = 13$

27. $\frac{1}{5}x + \frac{3}{2}y = 7$
$\frac{2}{5}x - \frac{9}{2}y = -16$

28. $-\frac{1}{2}x - \frac{1}{3}y = -1$
$-\frac{3}{2}x + \frac{2}{3}y = -8$

29. $\frac{2}{3}x + \frac{1}{2}y = \frac{1}{6}$
$\frac{1}{2}x + \frac{5}{4}y = \frac{11}{4}$

30. $\frac{1}{4}x + \frac{5}{2}y = 2$
$\frac{5}{6}x - \frac{1}{3}y = -2$

Solve each system using either elimination or substitution. If the system is inconsistent or dependent, say so. Verify any one-point solution graphically or by checking that it satisfies both equations in the system.

31.
$$y = 2x + 5$$
$$6x - 3y = -3$$

32.
$$5x - y = 6$$
$$10x - 2y = 12$$

33.
$$13x + 10y = -7$$
$$17x - 15y = 47$$

34.
$$3x - 5y = 10$$
$$7x + 2y = 37$$

35.
$$4x - 5y = 3$$
$$-12x + 15y = -9$$

36.
$$2x - y + 4 = 0$$
$$7 = 8x - 4y$$

37.
$$y = 2.3x - 7$$
$$y = -0.6x + 4$$

38.
$$0.5x - 0.2y = 2$$
$$1.5x + 0.4y = 1$$

39.
$$-8x + 9y = -7$$
$$6x - 15y = -3$$

40.
$$-4x + 8y = 2$$
$$6x - 12y = -3$$

41.
$$4x - 3y = 1$$
$$-8x + 6y = -5$$

42.
$$-(x - 6) + 6(y + 1) = 58$$
$$3(x + 1) - 4(y - 2) = -15$$

43.
$$y = \frac{1}{2}x + 3$$
$$2y - x = 6$$

44.
$$y = -\frac{1}{3}x + 4$$
$$x + 3y = 8$$

45.
$$\frac{5}{6}x + \frac{1}{4}y = 3$$
$$-\frac{1}{3}x + \frac{5}{2}y = 4$$

46.
$$\frac{1}{2}x - \frac{3}{4}y = -4$$
$$\frac{2}{3}x - \frac{1}{3}y = -4$$

47. Solve the system of equations three times, once by each of the three methods: elimination, substitution, and graphing. Decide which method you prefer for this system.

$$3x + y = 11$$
$$y = -2x + 9$$

48. Consider the system

$$2x + 4y = 10$$
$$3x - 7y = 2$$

a. Solve the system by eliminating the x terms.

b. Solve the system by eliminating the y terms.

c. Compare your results in parts a and b.

49. A student decides to solve the system

$$y = 2x + 3$$
$$y = 2.01x + 1$$

by graphing the equations (see Fig. 22). The student decides that the solution is the empty set. Is the student correct? If so, explain. If not, find the correct solution.

 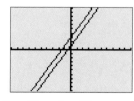

Figure 22 Graphs of two lines (Exercise 49)

50. To solve the system

$$x = 3$$
$$y = 4$$

a student adds the left sides and the right sides to get $x + y = 7$. The student thinks that the solution set is the set of points of the line $y = -x + 7$. What would you tell this student?

51. Some values of a linear function f and a linear function g are listed in Table 9. Find the solution of a system of two equations that describe the functions f and g. [**Hint**: Find equations for f and g.]

Table 9 Values of Functions f and g (Exercise 51)								
x	0	1	2	3	4	5	6	7	8
$f(x)$	3	7	11	15	19	23	27	31	35
$g(x)$	50	44	38	32	26	20	14	8	2

52. Some values of a linear function f and a linear function g are listed in Table 10. Find the solution of a system of two equations that describe the functions f and g.

Table 10 Values of Functions f and g (Exercise 52)								
x	0	1	2	3	4	5	6	7	8
$f(x)$	201	204	207	210	213	216	219	222	225
$g(x)$	6	11	16	21	26	31	36	41	46

53. Find the coordinates of the points A, B, C, D, E, and F as shown in Fig. 23. The equations of the lines l_1–l_4 are provided but no attempt has been made to sketch the lines accurately, except for showing the intersection points. Verify your results graphically.

l_1: $y = 2x + 3$
l_2: $3y + x = 30$
l_3: $y + 3x = 26$
l_4: $y = 2x - 10$

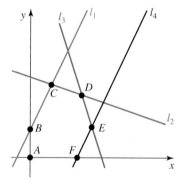

Figure 23 Exercise 53

54. Determine the constants a and b so that $(2, 3)$ is the solution of the system. Verify your result by checking that $(2, 3)$ satisfies each of your equations. Also verify your result graphically.

$$7x - 4y = a$$
$$5x + 2y = b$$

55. a. Let a, b, c, d, e, and f be constants so that the system below has a one-point solution. Solve the system.

$$ax + by = c$$
$$dx + ey = f$$

b. Use your result from part a to solve this system.

$$3x + 5y = 2$$
$$4x + 3y = 4$$

56. Explain why the solution set of a system of two linear equations is either empty, consists of exactly one solution, or consists of infinitely many solutions. Include in your explanation why such a solution set cannot have exactly two solutions.

57. Describe how to solve a system using elimination. Include in your discussion the result of solving a system by elimination if the system is a one-point-solution system, an inconsistent system, or a dependent system. Finally, describe how to solve a system using substitution.

3.3 USING SYSTEMS TO MODEL DATA

I like math because it's interesting. I feel mentally challenged whenever I see a math problem. Math is very crucial to the development of our society and it applies to virtually every aspect of our lives. Many fields require the knowledge of math, such as computer science, architecture, engineering, and so on. —*Steven W., student*

Objective

▸ Use elimination and substitution to find the intersection point of the graphs of two linear models.

In Section 3.1 we used a graphical approach to find the intersection point of the graphs of two linear models. In this section we discuss how to use elimination and substitution to find such an intersection point.

Example 1   Solving a System to Make a Prediction

In Example 1 in Section 3.1, we modeled life expectancies (in years) $W(t)$ and $M(t)$ for women and men in the United States, respectively, by the system

$$L = W(t) = 0.098t + 77.58$$
$$L = M(t) = 0.192t + 69.98$$

where t represents the number of years since 1980 (see Table 11).

Table 11 U.S. Life Expectancies of Women and Men

Year of Birth	Women (years)	Men (years)
1980	77.4	70.0
1985	78.2	71.1
1990	78.8	71.8
1995	78.9	72.5
2000	79.5	74.1

Use a symbolic method to predict when the life expectancies for women and men will be equal.

Solution

In Example 1 in Section 3.1, we found that the intersection point of the graphs of the two models represents the event when the life expectancies for women and men will be equal. We can find this intersection point using substitution (or elimination). To solve by substitution, we substitute $0.098t + 77.58$ for L in the equation $L = 0.192t + 69.98$ and solve for t.

$$0.098t + 77.58 = 0.192t + 69.98$$
$$0.098t - 0.192t = 69.98 - 77.58$$
$$-0.094t = -7.6$$
$$t \approx 80.85$$

Next, we substitute 80.85 for t in the equation $L = 0.098t + 77.58$.

$$L = 0.098(80.85) + 77.58 \approx 85.50$$

So, the approximate solution to the system is (80.85, 85.50), the same result that we found in Example 1 in Section 3.1. According to the models, the life expectancies for women and men will be equal in 2061. ■

In Example 1 we used substitution to predict when the life expectancy of women will equal the life expectancy of men. Graphically, this event is described by the intersection point of the graphs of the two life-expectancy models. In general, we locate the intersection point of the graphs of two models by graphing or by using elimination or substitution to solve a system of the models' equations.

Example 2 Solving a System to Make a Prediction

In exercises 13 and 14 in Homework 2.2, you modeled world record times for the 400-meter run (see Table 12).

Table 12 400-Meter Run Record Times

	Women		Men
Year	Record Time (seconds)	Year	Record Time (seconds)
1957	57.0	1900	47.8
1959	53.4	1916	47.4
1962	51.9	1928	47.0
1969	51.7	1932	46.2
1972	51.0	1941	46.0
1976	49.29	1950	45.8
1979	48.60	1960	44.9
1983	47.99	1968	43.86
1985	47.60	1988	43.29
		1999	43.18

The record times (in seconds) $W(t)$ and $M(t)$ for women and men, respectively, are modeled by the system

$$r = W(t) = -0.27t + 70.45$$
$$r = M(t) = -0.053t + 48.08$$

where t represents the number of years since 1900. Predict when the women's record time and the men's record time will be equal.

Solution

We solve the system

$$r = -0.27t + 70.45$$
$$r = -0.053t + 48.08$$

using substitution. We do so by substituting $-0.27t + 70.45$ for r in the equation $r = -0.053t + 48.08$.

$$-0.27t + 70.45 = -0.053t + 48.08$$
$$-0.27t + 0.053t = 48.08 - 70.45$$
$$-0.217t = -22.37$$
$$t \approx 103.09$$

Next, we substitute 103.09 for t in the equation $r = -0.27t + 70.45$ and solve for r.

$$r = -0.27(103.09) + 70.45 \approx 42.62$$

So, according to the models, the winning times for women and men will both be 42.62 seconds in 2003.

We can verify our result by using "intersect" on a graphing calculator (see Fig. 24).

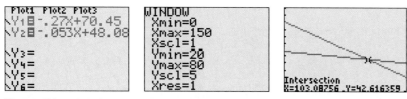

Figure 24 Verify that the intersection point is approximately $(103.09, 42.62)$

Recall that if the rate of change of the dependent variable with respect to the independent variable is constant, then there is a linear relationship between the variables. For the linear model that describes the relationship, the slope is equal to the constant rate of change. We use these ideas in Example 3.

Example 3   Solving a System to Make a Prediction

In 2002 prices for a 2001 Cadillac DeVille® were near $30,055, and prices for a 2001 Acura Integra® were near $16,565. The DeVille depreciates by $4391 each year and the Integra depreciates by $869 each year.* When will the 2001 cars have the same value?

Solution

Let $V = D(t)$ represent the value (in dollars) of a 2001 DeVille and $V = I(t)$ represent the value (in dollars) of a 2001 Integra at t years since 2002.

Since a 2001 Deville's value decreases by a *constant* $4391 each year, the function D is linear and its slope is -4391. The V-intercept is $(0, 30,055)$, since the car is worth $30,055 at year $t = 0$. So, an equation for D is:

$$V = D(t) = -4391t + 30,055$$

Similar work in finding the equation for the function I gives:

$$V = I(t) = -869t + 16,565$$

Next, we substitute $-4391t + 30,055$ for V in the equation $V = -869t + 16,565$.

$$-4391t + 30,055 = -869t + 16,565$$

Then we solve for t:

$$-4391t + 869t = 16,565 - 30,055$$
$$-3522t = -13,490$$
$$t \approx 3.83$$

We conclude that the cars will have the same value in approximately 4 years (in 2006). Next, we find the common value of the cars:

$$D(3.83) = -4391(3.83) + 30,055 \approx 13,237$$

---

*Prices quoted for four-door Cadillac DeVille sedan and four-door Acura Integra GS sedan from Edmunds® Automobile Buyer's Guide at www.edmunds.com. Accessed June 3, 2002.

So, both cars will be worth about $13,237 in 2006, according to the models. We verify our work by using "intersect" on a graphing calculator (see Fig. 25).

Figure 25 Verify that the intersection point is about (3.83, 13,237)

■

EXPLORATION
Looking ahead: Connection between a system of linear equations and a linear inequality in one variable

Recall from Example 2 that the equations

$$r = W(t) = -0.27t + 70.45 \qquad \text{Women's records}$$
$$r = M(t) = -0.053t + 48.08 \qquad \text{Men's records}$$

model, respectively, women's and men's 400-meter run record times (in seconds) at t years since 1900.

1. Use a graphing calculator to draw the graphs of W and M on the same coordinate system and find the intersection point. Then copy the graphs on a piece of paper and mark the point of intersection.

2. Now find values of t for which the models predict that the women's record time will be less than the men's record time. Which years do these values of t represent? On your graph, shade the part of the t-axis that represents these years.

3. Try a *numerical* verification using your graphing calculator. To do this, enter the functions as displayed in Fig. 26. Use the table setup displayed in Fig. 27, but replace the rectangle to the right of "TblStart=" with the value for t when the men's and women's time will be equal. Then display the tables and use the arrow keys to verify your result from problem 2. (See Section B.14 for calculator instructions.)

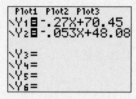

Figure 26 Entered functions

Figure 27 Table setup

4. Which of these two inequalities states that the women's record time is less than the men's record time?

$$W(t) < M(t) \qquad\qquad M(t) < W(t)$$

Note

In Section 3.4 we will discuss how to *solve* such an inequality so that we can make predictions like the one you made in problem 2 without graphing.

5. Substitute $-0.27t + 70.45$ for $W(t)$ and $-0.053t + 48.08$ for $M(t)$ in the inequality that you chose in problem 4. An inequality like this is called a **linear inequality in one variable**. This inequality becomes a true statement when any of the values for t that you found in problem 2 are substituted for t in the inequality. Explain.

Tips on Succeeding in This Course

While working on an Exploration, you should try to sort out the point of the Exploration, rather than just trying to get desired results. After completing an Exploration, it is important that you reflect on the concepts you have just learned and how these concepts fit into the framework of concepts you have learned in previous Explorations.

Key Points
OF THIS SECTION

If $f(t)$ and $g(t)$ are models where t represents time, then the intersection point(s) of the graphs of these models indicate when the quantities represented by the dependent variables were or will be equal.

Intersection point(s) of the graphs of two models

A system of linear models can be solved by using:

* graphing
* substitution
* elimination

Methods of solving a system of models

HOMEWORK 3.3

 SSM Tutor Center

 MathPro 4 MathPro 4

 MathPro 5 MathPro 5

CDs/Videos

www

1. In exercise 23 of Homework 3.1, the winning Olympic 500-meter speed skating times (in seconds) $W(t)$ and $M(t)$ for women and men, respectively, are modeled by the system

$$y = W(t) = -0.19t + 43.73$$
$$y = M(t) = -0.16t + 39.90$$

where t represents the number of years since 1970 (see Table 13). Use substitution or elimination to predict when the women's winning time will be equal to the men's winning time. Also, find that winning time.

Table 13 Olympic 500-Meter Speed Skating

Year	Winning Time (seconds) Women	Men
1972	43.33	39.44
1976	42.76	39.17
1980	41.78	38.03
1984	41.02	38.19
1988	39.10	36.45
1992	40.33	37.14
1994	39.25	36.33
1998	38.21	35.59
2002	37.375	34.615

2. In exercise 24 of Homework 3.1, the annual U.S. consumption (in pounds consumed per person) $C(t)$ and $R(t)$ of chicken and red meat, respectively, are modeled by the system

$$y = C(t) = 0.93t - 40.85$$
$$y = R(t) = -0.65t + 176.52$$

where t represents the number of years since 1900 (see Table 14).

Table 14 U.S. Per-Person Consumptions of Chicken and Red Meat

Year	Chicken (pounds)	Red Meat (pounds)
1970	27.7	131.7
1975	26.4	125.8
1980	32.7	126.4
1985	36.4	124.9
1990	42.5	112.3
1995	48.8	114.7
2000	52.9	113.5

Use substitution or elimination to predict when consumption of chicken will equal consumption of red meat. What will be the consumption? How confident are you in this prediction?

3. In exercise 25 of Homework 3.1, the enrollments (in millions) at U.S. colleges $W(t)$ and $M(t)$ for women and men, respectively, are modeled by the system

$$E = W(t) = 0.14t + 4.52$$
$$E = M(t) = 0.083t + 4.60$$

where t represents the number of years since 1970 (see Table 15). Use elimination or substitution to estimate when women's and men's enrollments were approximately equal.

Table 15 College Enrollments

| Year | Enrollment (millions) | |
	Women	Men
1980	6.0	5.4
1985	6.6	5.9
1990	7.4	6.2
1995	8.0	6.7
1998	8.6	6.9

4. In exercise 26 of Homework 3.1, the total number (in millions) of U.S. households $f(t)$ and the number (in millions) of U.S. households with PCs $g(t)$ are modeled by the system

$$f(t) = 1.21t + 92.64$$
$$g(t) = 4.82t + 6.45$$

where t represents the number of years since 1990 (see Table 16).

Table 16 U.S. Households with Personal Computers

Year	Total Households (millions)	Households with PCs (millions)
1995	99.0	31.36
1996	99.6	35.39
1997	101.0	39.56
1998	102.5	43.67
1999	103.6	50.00
2000	104.8	55.49

a. Find the slopes of f and g. What do they mean in terms of the situation?

b. Since the number of households in the United States is increasing, it is not surprising that the number of households with PCs is also increasing. Explain why your results in part a show that this is not the only reason why the number of households with PCs is increasing.

c. Predict the *percentage* of U.S. households that will have PCs in 2009.

d. Use substitution to predict when all U.S. households will have PCs.

5. In exercise 27 of Homework 3.1, the percentage of voters $f(t)$ using punch cards or lever machines and the percentage of voters $g(t)$ using optical scan or other modern electronic systems are modeled by the system

$$p = f(t) = -2.45t + 72.82$$
$$p = g(t) = 3.08t + 12.58$$

where t represents the number of years since 1990 (see Table 17).

Table 17 Vote-Counting Systems

| Year | Percent of Registered Voters | |
	Punch Card or Lever Machine	Optical Scan or Other Modern Electronic System
1990	71.7	13.5
1992	69.4	16.5
1994	63.1	25.8
1996	58.0	32.3
1998	52.9	36.4

Source: *Federal Election Commission*

Use substitution or elimination to predict when the percentage of voters using optical scan or other modern electronic systems will equal the percentage of voters using punch cards or lever machines. Is your result a national (even-numbered) election year? If not, in which national election year will the percentages be closest?

6. Student-to-teacher ratios for public and private elementary schools for various years are listed in Table 18.

Table 18 Student-to-Teacher Ratios

| Year | Student-to-Teacher Ratio | |
	Public	Private
1995	19.3	16.6
1997	18.3	16.6
1999	18.1	16.4
2001	17.9	16.3

Source: *National Center for Education Statistics*

Let $f(t)$ and $g(t)$ represent the student-to-teacher ratio for public elementary schools and private elementary schools, respectively, at t years since 1990.

a. Find equations for f and g.

b. Use substitution or elimination to predict when the student-to-teacher ratios for public and private elementary schools will be equal.

7. The *Denver Post* and the *Rocky Mountain News* are competing newspapers in Denver, Colorado. Their circulations for various years are listed in Table 19.

	Table 19 Newspaper Circulations	
Year	Denver Post Circulation (thousands)	Rocky Mountain News Circulation (thousands)
1992	255	355
1993	270	350
1994	285	345
1995	295	340

Source: *Advertising Age, www.adage.com. Accessed 3/27/02; "Auditors slap Post, Rocky," 1997-02-24, The Denver Business Journal, denver.bizjournals.com/denver/stories/1997/02/24/story2.html. Accessed 6/3/02.*

Let $D(t)$ represent the circulation (in thousands) of the *Denver Post* and $R(t)$ represent the circulation (in thousands) of the *Rocky Mountain News,* where t is the number of years since 1990. The equations for D and R are given below.

$$D(t) = 13.5t + 229$$
$$R(t) = -5t + 365$$

a. Use substitution or elimination to estimate when the two newspapers had equal circulations.

b. The newspaper with greater circulation can usually charge more for advertisements, thus increasing its revenue. The competition between the two newspapers escalated in 1997. In fact, on March 1–2, 1997, the newspapers included articles claiming that the other newspaper had made false reports of their circulation to the audit bureau. Use your result in part a to explain why it is not surprising that the competition heated up in 1997.

c. Throughout the years, both newspapers increased their circulations by giving away "bonus issues." Bonus issues are copies of the newspaper given away free to subscribers who subscribe to less than a full week. During the circulation wars from 1997–2000, both newspapers dramatically increased their distribution of bonus issues. In 2000 the combined circulation from the two newspapers was 826 thousand. Estimate the combined increase in number of bonus copies from the two newspapers due to the circulation wars. [**Hint**: Begin by finding $D(10)$ and $R(10)$.]

d. Use the models D and R to estimate the combined circulation from the two newspapers in 2001.

e. In January, 2001, the circulation wars ended when the Justice Department approved that the two newspapers could join their revenue streams under a single entity, the Denver Newspaper Agency (DNA). In 2001 the combined circulation was 638 thousand. Is your estimate in part d an underestimate or an overestimate? Give a possible explanation for this in terms of the circulation wars, DNA, and bonus copies.

8. The percentage of students studying for a master of divinity degree at theological schools for various years is listed in Table 20.

	Table 20 Students Studying for a Master of Divinity Degree	
		Percent
Year	Women	Men
1973	5	95
1983	17	83
1993	25	75
1996	28	72
1998	30	70
2000	31	69

Source: *USA Today*

a. Let $W(t)$ and $M(t)$ represent the percentage of divinity major students who are women and men, respectively, at t years since 1970. Find equations for W and M.

b. Compare the slopes of your models. What do the slopes mean in terms of the situation?

c. Use substitution or elimination to predict when the number of women studying for a master of divinity degree will equal the number of men studying for such a degree.

d. Using only the equation for W, predict when the number of women studying for a master of divinity degree will equal the number of men studying for such a degree. Compare your result with your result in part c.

9. World record times for the 1500-meter run are listed in Table 21.

Table 21 1500-Meter Run Record Times			
Women		Men	
Year	Record Time (seconds)	Year	Record Time (seconds)
1927	318	1926	231
1936	287	1941	227
1946	277	1955	220
1957	269	1980	211
1962	259	1995	207
1993	230	1998	206

Source: *WR Progression (Women), www.apulanta.fi/matti/yu/wrprogr_Women.html#1500m; WR Progression (Men), www.apulanta.fi/matti/yu/wrprogr_Men.html#1500m. Both accessed 12/30/02.*

a. Let $W(t)$ and $M(t)$ represent the record times (in seconds) for women and men, respectively, at t years since 1900. Find the regression equations for W and for M.

b. Use the regression equations to predict when the women's record time will equal the men's record time. Verify your result using "intersect" on a graphing calculator.

c. Now find the regression equation for the women's record times excluding the record set in 1927. Also, find the regression equation for the men's times excluding the record set in 1926. Use these equations to predict when the women's record time will equal the men's record time.

d. Explain why your result in part b is so different from your result in part c.

10. In 2002, the price of a 2001 Honda Accord® was about $13,205, with a depreciation of about $1124 per year.*A student had $500 in 2002 and saves $1700 each year. Assume that the student does not earn interest on his savings.

 a. Let $H(t)$ represent the value (in dollars) of a 2001 Honda Accord and $S(t)$ represent the student's total savings (in dollars), both at t years since 2002. Find equations for H and S.

 b. Predict when the student will be able to buy a 2001 Honda Accord in cash.

 c. Use a graphical method to verify your work in part b.

 d. Now assume that the student earns interest on his savings. Is your answer in part b an underestimate or an overestimate? Explain.

11. In 2002, the price of a 2001 Ford Taurus® was about $12,939, with a depreciation of about $1582 per year. The price of a 2001 Ford Escort® was about $9330, with a depreciation of about $1024 per year.†

 a. Let $V = T(t)$ represent the value (in dollars) of a 2001 Ford Taurus and $V = E(t)$ represent the value (in dollars) of a 2001 Ford Escort, both at t years since 2002. Find equations for T and E.

 b. When will the cars have the same value? What is that value?

 c. Use a graphical method to verify your work in part b.

12. Fitness USA® offers membership for a flat fee of $35 plus a monthly fee of $26. Gold's Gym® offers membership for a flat fee of $14 plus a monthly fee of $29.95, with a year's commitment.‡ Let $f(t)$ represent the total cost (in dollars) of being a member at Fitness USA and $g(t)$ represent the total cost (in dollars) of being a member at Gold's Gym, both for t months.

 a. Find equations for f and g.

 b. Use f and g to estimate when the total cost at each health club will be equal.

 c. Use a graphing calculator table or graph to verify your work in part b.

13. The tuition at college A was $6100 in 2000 and increases by $670 each year since then. The tuition at college B was $8500 in 2000 and increases by $440 each year since then.

 a. Let $A(t)$ represent the tuition (in dollars) at college A and $B(t)$ represent the tuition (in dollars) at college B at t years since 2000. Find equations for A and B.

 b. Use your models to predict when both colleges will have the same tuition. What is that tuition?

 c. Verify your result in part b graphically.

14. The enrollment at college A was 25,300 students in 2000 and declines by 600 students each year since then. The enrollment at college B was 13,200 students in 2000 and increases by 800 students each year since then.

 a. Let $A(t)$ represent the enrollment at college A and $B(t)$ represent the enrollment at college B at t years since 2000. Find equations for A and B.

 b. Use your models to predict when both colleges will have the same enrollment. What is that enrollment?

 c. Verify your result in part b graphically.

15. A person considers joining a weight-loss program. Jenny Craig® offers a promotion where members pay a start-up fee of $19 (to lose 19 pounds) and a weekly fee of $72 for Jenny Craig food. Members meet with a food counselor free of charge.

 Weight Watchers offers a promotion where members do not pay a start-up fee, but do pay a weekly fee of $17 per week to meet with a food counselor. Members purchase their own food, which usually costs the person about $60 per week.

 a. Let $J(t)$ represent the amount of money (in dollars) the person will pay for the Jenny Craig program (including food) for t weeks. Let $W(t)$ represent the amount of money (in dollars) the person will pay for the Weight Watchers program *as well as for food* for t weeks. Find equations for J and W.

 b. Use elimination or substitution to estimate how many weeks it will take for the total cost at Jenny Craig to equal the total cost at Weight Watchers (plus the cost of food). What is that total cost?

 c. Use a graphical method to verify your work in part b.

16. Company A sold $14.1 million of software products in 2000, and its sales have since increased by $1.2 million each year. Company B sold $8.7 million of software products in 2000, and its sales have since increased by $1.8 million each year.

 a. Let $A(t)$ represent sales (in millions of dollars) by company A and $B(t)$ represent sales (in millions of dollars) by company B at t years since 2000. Find equations for A and B.

 b. Use elimination or substitution to predict when sales by the two companies will be equal. What will the sales be?

 c. Use a graphing calculator table or graph to verify your result in part b.

17. In 2000 the average price of a home in a community was $214,000. The price increases by about $8000 each year. A family has $10,000 in 2000 and plans to save $300 each month. Predict when the family will be able to pay a 10% down payment on an average-priced house in the community.

18. Describe how you can find a system of linear equations to model a situation. Also, explain how you can use the system to make an estimate or prediction for the situation.

---

*Prices quoted for four-door Honda Accord DX sedan by Edmunds Automobile Buyer's Guide at www.edmunds.com. Accessed June 3, 2002.

†Prices quoted for four-door Ford Taurus LX sedan and two-door Ford Escort ZX2 by Edmunds Automobile Buyer's Guide at www.edmunds.com. Accessed June 3, 2002.

‡Both health clubs' prices depended on which program was selected.

3.4 USING LINEAR INEQUALITIES IN ONE VARIABLE TO MAKE PREDICTIONS

Objectives

▸ Know the meaning of *satisfy*, *solution*, and *solution set* as related to an inequality.
▸ Know properties of inequalities.
▸ Solve linear inequalities.
▸ Use linear inequalities to model situations.

So far in this chapter, we have estimated when one quantity will be equal to another quantity. In this section we investigate when one quantity is less than (or greater than) another quantity.

Example 1   Using Models to Compare Two Quantities

One Budget® office rents pickup trucks for $39.95 per day plus $0.19 per mile. One U-Haul® location charges $19.95 per day plus $0.49 per mile.

1. Find models that describe the one-day cost of renting a pickup truck from the companies.
2. Use graphs of your models to estimate for which mileages Budget offers the lower price.

Solution

1. Let $B(x)$ represent the one-day cost (in dollars) of driving a Budget pickup truck x miles. Let $U(x)$ represent the one-day cost (in dollars) of driving a U-Haul pickup truck x miles. Equations for B and U are:

$$C = B(x) = 0.19x + 39.95$$
$$C = U(x) = 0.49x + 19.95$$

2. First, we sketch a graph of B and U in the same coordinate system (see Fig. 28). Note that the graph of B is below the graph of U for $x > 66.67$. Since the height of a point represents a price, we see that Budget offers the lower price for mileages over approximately 66.67 miles.

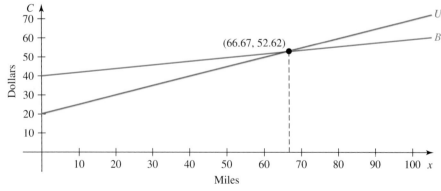

Figure 28 Budget and U-Haul models

In Example 1 we used graphing to estimate for what mileages Budget's prices are lower than U-Haul's prices. Now we turn our attention to exploring how to use *inequalities* to arrive at such an estimation.

We begin by examining properties of inequalities. What happens if we add 2 to both sides of the inequality $4 < 7$?

$$4 < 7$$
$$4+2 \overset{?}{<} 7+2$$
$$6 \overset{?}{<} 9$$

true

What happens if we add -2 to both sides of the inequality $4 < 7$?

$$4 < 7$$
$$4+(-2) \overset{?}{<} 7+(-2)$$
$$2 \overset{?}{<} 5$$

true

These examples suggest the following property.

Adding a Number to Both Sides of an Inequality

If $a < b$, then $a + c < b + c$

In words, adding a constant to both sides of an inequality *preserves the order* of the inequality.

Similar properties apply to less than or equal to, greater than, and greater than or equal to inequalities. Also, similar properties hold for subtraction, since subtracting a number is the same as adding the opposite of the number.

We can use a number line to illustrate that if $a < b$, then $a + c < b + c$. From Fig. 29 we see that if a lies to the left of b ($a < b$), then $a + c$ lies to the left of $b + c$ ($a + c < b + c$).

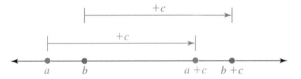

Figure 29 If a lies to the left of b, then $a + c$ lies to the left of $b + c$

What if we multiply both sides of the inequality $4 < 7$ by 2?

$$4 < 7$$
$$4(2) \overset{?}{<} 7(2)$$
$$8 \overset{?}{<} 14$$

true

Finally, what happens if we multiply both sides of $4 < 7$ by -2?

$$4 < 7$$
$$4(-2) \overset{?}{<} 7(-2)$$
$$-8 \overset{?}{<} -14$$

false

Note that when we multiply an inequality by a negative number, the order of the inequality is *not* preserved.

> **Multiplying Both Sides of an Inequality by a Number**
>
> • For a *positive* number c: If $a < b$, then $ac < bc$
> • For a *negative* number c: If $a < b$, then $ac > bc$

Similar properties apply to less than or equal to, greater than, and greater than or equal to inequalities. Also, similar rules apply for division, since dividing by a nonzero number is the same as multiplying by its reciprocal.

Consider multiplying both sides of $a < b$ by -1:

$$a < b$$
$$-1a > -1b \qquad \text{Reverse the direction of the inequality.}$$
$$-a > -b$$

So, if $a < b$, then $-a > -b$. We can use a number line to illustrate this fact. To plot the point for $-a$, we reflect the point for a across the origin of the number line (see Fig. 30).

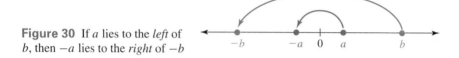

Figure 30 If a lies to the *left* of b, then $-a$ lies to the *right* of $-b$

From Fig. 30 we see that if a lies to the *left* of b ($a < b$), then $-a$ lies to the *right* of $-b$ ($-a > -b$).

We use properties of inequalities when working with *linear inequalities in one variable*. Here are some examples:

$$4x - 6 < 8 \qquad 5x \le 13 \qquad 2(x - 4) + 7 > -3 \qquad 5x \ge 2x - 6$$

We say a number **satisfies** an inequality in one variable if the inequality becomes a true statement after substituting the number for the variable. We call such a number a **solution** of the inequality.

Example 2 Finding Solutions of an Inequality

1. Does the number 2 satisfy the inequality $3x - 5 < 7$?
2. Does the number 6 satisfy the inequality $3x - 5 < 7$?
3. Find all solutions of the inequality $3x - 5 < 7$.

Solution

1. We substitute 2 for x in the inequality $3x - 5 < 7$.

$$3(2) - 5 \overset{?}{<} 7$$
$$6 - 5 \overset{?}{<} 7$$
$$1 \overset{?}{<} 7$$
$$\text{true}$$

So, 2 satisfies the inequality $3x - 5 < 7$.

2. Here we substitute 6 for x in the inequality $3x - 5 < 7$.

$$3(6) - 5 \overset{?}{<} 7$$
$$18 - 5 \overset{?}{<} 7$$
$$13 \overset{?}{<} 7$$
$$\text{false}$$

So, 6 does not satisfy the inequality $3x - 5 < 7$.

3. To find all solutions of the inequality, we use properties of inequalities.

$$3x - 5 < 7$$
$$3x - 5 + 5 < 7 + 5 \qquad \text{Add 5 to both sides.}$$
$$3x < 12$$
$$\frac{3x}{3} < \frac{12}{3} \qquad \text{Divide each side by 3.}$$
$$x < 4$$

All numbers less than 4 satisfy the inequality $3x - 5 < 7$. The solutions of the inequality are all numbers less than 4. ∎

DEFINITION *Solution of an inequality in one variable*

We say that a number is a **solution** of an inequality in one variable if it satisfies the inequality. The **solution set** of an inequality is the set of all numbers that satisfy the inequality. We **solve** an inequality by finding all of its solutions.

To solve a linear inequality in one variable, we apply properties of inequalities to isolate the variable on one side of the inequality.

Example 3   Solving a Linear Inequality

Solve the inequality $-2x \geq 10$.

Solution

We divide both sides of the inequality by -2, a negative number.

$$-2x \geq 10$$
$$\frac{-2x}{-2} \leq \frac{10}{-2} \qquad \text{Reverse the direction of the inequality.}$$
$$x \leq -5$$

Since we divided by a negative number, we reversed the direction of the inequality. The solution set is the set of all numbers less than or equal to -5. ∎

In Example 3 we found that the solution set of $-2x \geq 10$ is the set of all real numbers less than or equal to -5. We can represent these solutions graphically on a number line by shading the part of the number line that lies to the left of -5 (see Fig. 31). We draw a filled-in circle at -5 to indicate that -5 is a solution, too.

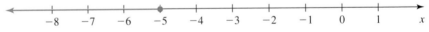

Figure 31 Graph of $x \leq -5$

If the solution set of an inequality is the set of numbers where $x < -5$, we shade the part of the number line that lies to the left of -5 and draw an *open* circle at -5 to indicate that -5 is *not* a solution (see Fig. 32).

Figure 32 Solutions of $x < -5$ graphed on a number line

More examples of inequalities with matching graphs are given in Fig. 33.
We can describe the solution set of an inequality using *interval notation*. For

example, we describe the numbers greater than 2 by $(2, \infty)$. We describe the numbers greater than or equal to 2 by $[2, \infty)$. We describe the set of real numbers by $(-\infty, \infty)$. More examples of interval notation are also shown in Fig. 33.

In Words	Inequality Notation	Graph	Interval Notation
Numbers less than 2	$x < 2$	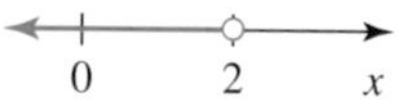	$(-\infty, 2)$
Numbers less than or equal to 2	$x \leq 2$		$(-\infty, 2]$
Numbers greater than 2	$x > 2$	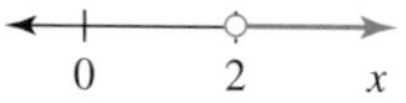	$(2, \infty)$
Numbers greater than or equal to 2	$x \geq 2$		$[2, \infty)$

Figure 33 Words, graphs, and notations for inequalities

Example 4   Solving a Linear Inequality

Solve $-3(4x - 5) - 1 \leq 17 - 6x$. Describe the solution set with inequality notation, a graph, and interval notation.

Solution

$$-3(4x - 5) - 1 \leq 17 - 6x$$

$$-12x + 15 - 1 \leq 17 - 6x \qquad \text{Distributive law}$$

$$-12x + 14 \leq 17 - 6x$$

$$-12x + 14 + 6x \leq 17 - 6x + 6x \qquad \text{Add } 6x \text{ to both sides.}$$

$$-6x + 14 \leq 17$$

$$-6x + 14 - 14 \leq 17 - 14 \qquad \text{Subtract 14 from both sides.}$$

$$-6x \leq 3$$

$$\frac{-6x}{-6} \geq \frac{3}{-6} \qquad \text{Reverse the direction of the inequality.}$$

$$x \geq -\frac{1}{2}$$

We can graph the solution set on a number line (see Fig. 34)

Figure 34 Graph of $x \geq -\dfrac{1}{2}$

or we can describe the solution set using interval notation as $\left[-\dfrac{1}{2}, \infty\right)$.

To verify our result, we check that for inputs greater than or equal to $-\dfrac{1}{2}$, the outputs of $y = -3(4x - 5) - 1$ are less than or equal to the outputs of $y = 17 - 6x$ (see Fig. 35). We do this by setting up the table so that x begins at -0.5 and increases by 1. Then, we scroll up two rows so that we can view values of x that are less than -0.5 and greater than -0.5.

Graphing Calculator

See Section B.14.

Figure 35 Verify the result

In Example 1 we used graphing to estimate for what mileages Budget offers a lower price than U-Haul. In Example 5, we solve an inequality to make the estimate.

Example 5 Using Models to Compare Two Quantities

In Example 1 we modeled the one-day pickup truck costs (in dollars) $B(x)$ and $U(x)$ at Budget and U-Haul, respectively, by the system

$$C = B(x) = 0.19x + 39.95$$

$$C = U(x) = 0.49x + 19.95$$

where x represents the number of miles driven. Use inequalities to estimate for which mileages Budget offers the lower price.

Solution

Budget offers the lower price when

$$B(x) < U(x).$$

We substitute $0.19x + 39.95$ for $B(x)$ and $0.49x + 19.95$ for $U(x)$ to get a linear inequality in one variable:

$$0.19x + 39.95 < 0.49x + 19.95$$

Then, we solve the inequality by isolating x on the left side of the inequality.

$$0.19x + 39.95 < 0.49x + 19.95$$

$$0.19x + 39.95 - 0.49x < 0.49x + 19.95 - 0.49x \qquad \text{Subtract } 0.49x.$$

$$-0.30x + 39.95 < 19.95 \qquad \text{Combine like terms.}$$

$$-0.30x + 39.95 - 39.95 < 19.95 - 39.95 \qquad \text{Subtract } 39.95.$$

$$-0.30x < -20$$

$$\frac{-0.30x}{-0.30} > \frac{-20}{-0.30} \qquad \text{Reverse the direction of the inequality.}$$

$$x > 66.\overline{6}$$

Note

Recall that $66.\overline{6}$ represents $66.666\ldots$.

Budget offers the lower price if the truck is driven over $66.\overline{6}$ miles, the same result that we found in Example 1.

To verify our result, we check that for inputs greater than $66.\overline{6}$, the outputs of $y = 0.19x + 39.95$ are less than the outputs of $y = 0.49x + 19.95$ (see Fig. 36). We do this by setting up the table so that x begins at about $66.\overline{6}$ and increases by 1. Then, we scroll up three rows so that we can view values of x that are less than $66.\overline{6}$ and greater than $66.\overline{6}$.

Figure 36 Verify the result

Note that Budget's costs in column Y_1 are less than U-Haul's costs in column Y_2 for distances above $66.\overline{6}$ miles.

EXPLORATION
Meaning of the solution set of an inequality

To start, we solve the inequality $-3x + 7 < 1$.

$$-3x + 7 < 1$$
$$-3x + 7 - 7 < 1 - 7$$
$$-3x < -6$$
$$\frac{-3x}{-3} > \frac{-6}{-3}$$
$$x > 2$$

1. Choose a number greater than 2. Check that your number satisfies the inequality $-3x + 7 < 1$.

2. Choose two more numbers greater than 2. Check that both of these numbers satisfy the inequality $-3x + 7 < 1$.

3. Choose three numbers that are *not* greater than 2. Show that each of these numbers does *not* satisfy the inequality $-3x + 7 < 1$.

4. Explain what it means when we write $x > 2$ as the last step in solving the inequality $-3x + 7 < 1$.

Tips on Succeeding in This Course

You can prepare for an exam by meeting with a study team. When forming a study team, it usually works best if you invite students who are at your level of ability in this course. This will allow everyone to make contributions and will make everyone feel more comfortable asking questions about troublesome concepts.

It usually is a good idea, also, to spend some time studying alone. This will ensure that you understand the concepts and can solve the relevant problems without help from your study team.

Key Points
OF THIS SECTION

The *solution set* of an inequality is the set of all numbers that satisfy the inequality. We *solve* an inequality when we find its solution set.

Solution set of an inequality

The solution to the inequality $2x + 1 > 11$ is $x > 5$. (Try it.) This means that all numbers greater than 5 satisfy the inequality $2x + 1 > 11$ and all numbers less than or equal to 5 do *not* satisfy the inequality $2x + 1 > 11$.

Using inequality notation to describe a solution set

The direction of an inequality is reversed when multiplying or dividing by a negative number. Using symbols, we have:

Reversing the direction of an inequality

- If $a < b$, then $ac < bc$, where c is a *positive* number.
- If $a < b$, then $ac > bc$, where c is a *negative* number.
- Similar properties hold for $a \leq b$, $a > b$, $a \geq b$, and when the operation of multiplication is replaced by division.

We can find the values of t where a quantity $f(t)$ is less than a quantity $g(t)$ by each of the following methods:

Methods of solving an inequality

- We can solve the inequality $f(t) < g(t)$ using symbols.
- We can locate the values of t where the graph of f is lower than the graph of g.
- We can use tables of input-output pairs for f and g to find values of t where the outputs of f are less than the outputs of g.

HOMEWORK 3.4

  SSM Tutor Center

MathPro 4   MathPro 4

MathPro 5 MathPro 5

CDs/Videos www

1. Use words, inequalities, and graphs to complete Fig. 37.

In Words	Inequality Notation	Graph	Interval Notation
Numbers greater than 3			
		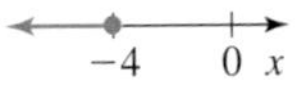	
	$x < 5$		
			$[-1, \infty)$

Figure 37 Exercise 1

2. Use words, inequalities, and graphs to complete Fig. 38.

In Words	Inequality Notation	Graph	Interval Notation
		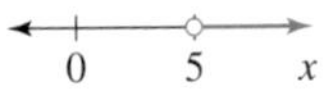	
Numbers less than or equal to -6			
			$[-\infty, 4)$
	$x \geq -3$		

Figure 38 Exercise 2

Use a symbolic method to solve each inequality. Describe the solution set with inequality notation, interval notation, and a graph. Then, use graphing calculator tables or graphs to verify your result.

3. $x + 2 \geq 5$

4. $x - 3 < 8$

5. $-x + 2 \geq 5$

6. $-7x > 21$

7. $2x + 7 < 11$

8. $15 - 4x \leq 0$

9. $9x < 4 + 5x$

10. $-3x > -15$

11. $2.1x - 7.4 \leq 10.4$

12. $6.6 - 5.2x > 7.7$

13. $2x - 3 > 7x + 22$

14. $6x - 2 \geq 4x - 14$

15. $7x - 3 \leq -2x - 21$

16. $10 - 8x \geq 2x - 6$

17. $3 - 2(x - 4) > 4x + 1$

18. $-5(2x + 4) + 1 \leq 3(x - 1)$

19. $-6.23x + 2.35 < 1.76(3 - 2.73x)$

20. $3.2 - 4.1x \leq 9.8(x + 3.5) - 5.9x$

21. $7(x + 1) - 8(x - 2) \leq 0$

22. $-3(2x + 1) > -2(x + 4)$

23. $-\dfrac{2}{3}x > 4$

24. $-\dfrac{1}{4}x \leq 2$

25. $\dfrac{1}{4}x - 2 < 3$

26. $\dfrac{5}{6}x + \dfrac{1}{2} \geq \dfrac{2}{3}$

27. $\dfrac{2}{3} - \dfrac{3}{4}x \leq \dfrac{5}{2}$

28. $\dfrac{5}{8} + \dfrac{1}{6}x > \dfrac{2}{3}$

29. $-\dfrac{1}{2}x - \dfrac{1}{6} \geq \dfrac{1}{3}$

30. $-\dfrac{3}{4}x + \dfrac{5}{2} < 1$

31. In exercise 11 of Homework 3.3, the values (in dollars) $T(t)$ and $E(t)$ for a 2001 Ford Taurus and a 2001 Ford Escort, respectively, are modeled by the system

$$V = T(t) = -1582t + 12{,}939$$
$$V = E(t) = -1024t + 9330$$

where t is the number of years since 2002. For what years will the value of the 2001 Taurus be more than the value of the 2001 Escort?

32. One Penske® office rents 15-foot trucks for a one-day fee of $39.95 plus $0.29 per mile. One Budget® location charges a one-day fee of $59.95 plus $0.19 per mile.

   a. Let $P(x)$ represent Penske's charge (in dollars) for driving x miles in one day. Let $B(x)$ represent Budget's charge (in dollars) for driving x miles in one day. Find equations for P and B.

   b. For how many miles driven is the one-day charge at Penske less than the charge at Budget?

33. One U-Haul office rents 10-foot trucks for a one-day fee of $19.95 plus $0.69 per mile. One Penske location charges a one-day fee of $29.95 plus $0.39 per mile.

   a. Let $U(x)$ represent U-Haul's charge (in dollars) and $P(x)$ represent Penske's charge (in dollars), both for driving x miles in one day. Find equations for U and P.

   b. For how many miles driven is the one-day charge at U-Haul less than the charge at Penske?

34. In exercise 6 of Homework 3.3, the student-to-teacher ratios $f(t)$ and $g(t)$ for public elementary schools and private elementary schools, respectively, are modeled by the system

$$f(t) = -0.22t + 20.16$$
$$g(t) = -0.055t + 16.92$$

where t represents the number of years since 1990 (see Table 22).

Table 22 Student-to-Teacher Ratios

	Student-to-Teacher Ratio	
Year	Public	Private
1995	19.3	16.6
1997	18.3	16.6
1999	18.1	16.4
2001	17.9	16.3

   a. Which will have a lower student-to-teacher ratio in 2008, public or private schools? By how much?

   b. Predict the years in which public schools will have a smaller student-to-teacher ratio than private schools.

   c. Predict the years in which public schools will have a larger student-to-teacher ratio than private schools.

35. In Example 1 of Section 3.1, birth-year life expectancies (in years) $W(t)$ and $M(t)$ for women and men in the United States, respectively, are modeled by the system

$$L = W(t) = 0.098t + 77.58$$
$$L = M(t) = 0.192t + 69.98$$

where t is the number of years since 1980 (see Table 23).

Table 23 U.S. Life Expectancies of Women and Men

Year of Birth	Women	Men
1980	77.4	70.0
1985	78.2	71.1
1990	78.8	71.8
1995	78.9	72.5
2000	79.5	74.1

Source: *U.S. Bureau of the Census*

a. How much longer are women likely to live than men, on average, for the birth year 2008?

b. Use a symbolic method to predict the birth years for which men will have a life expectancy longer than women. Verify your result using a graphical method.

c. A woman born in 1980 wants to choose a man to marry so that she does not outlive him.
  i. According to the linear models, should the woman marry a younger or older man? Explain your reasoning.
  ii. What are acceptable birth years for potential husbands? [**Hint**: There is a way to do this using an inequality. If you draw a blank, do this *by trial and error*.]

36. As mathematicians' work becomes more specialized, there is a greater need for them to work together so that their combined knowledge spans the topic of study (*Source: The Future of Scientific Communication*). Technology such as fax machines and the Internet have aided the process of collaboration. Table 24 shows the percentages of papers written by one or two authors for various years.

Table 24 Mathematical Papers Written by One or Two Authors

	Percent	
Year	One Author	Two Authors
1970	81	18
1975	77	20
1980	69	24
1985	66	26
1990	62	29
1995	57	32
1999	50	34

Source: *Mathematical Reviews*

Let $f(t)$ represent the percentage of papers written by one author and $g(t)$ represent the percentage of papers written by two authors.

a. Find equations for f and g.

b. In 2008, what percentage of papers will be written by one author? two authors? three or more authors?

c. Predict when more papers will be written by two authors than by one author.

37. A student has tried to solve the inequality $3x + 7 > 1$. Decide whether the student did the exercise correctly. If the student made a mistake, describe the student's error.

Student's Work

$$3x + 7 > 1$$
$$3x + 7 - 7 > 1 - 7$$
$$3x > -6$$
$$\frac{3x}{3} < \frac{-6}{3}$$
$$x < -2$$

38. A student has tried to solve the inequality $7(2 - x) \le 3x - 6$. Decide whether the student did the exercise correctly. If the student made a mistake, describe the student's error.

Student's Work

$$7(2 - x) \le 3x - 6$$
$$14 - 7x \le 3x - 6$$
$$14 - 7x - 14 \le 3x - 6 - 14$$
$$-7x \le 3x - 20$$
$$-7x - 3x \le 3x - 20 - 3x$$
$$-10x \le -20$$
$$\frac{-10x}{-10} \le \frac{-20}{-10}$$
$$x \le 2$$

39. a. List three numbers that satisfy the inequality $3(x - 2) + 1 \ge 7 - 4x$.

b. List three numbers that do not satisfy the inequality $3(x - 2) + 1 \ge 7 - 4x$.

40. Solve the inequality $2x + 1 > 2x + 1$.

41. Find values for m and c so the solution set to the inequality $mx < c$ is the set of numbers where $x > 2$.

42. Create an inequality of the form $mx + b \le ax + c$, where m, b, a, and c are constants. Then, solve your inequality.

43. A student thinks that the solution set of an inequality of the form $ax + b < cx + d$ is the set of numbers where $2 < x < 5$ (numbers between 2 and 5). Explain why this is not possible. [**Hint**: Think graphically.]

44. a. Is the following statement true? Explain.

$$\text{If } a < b, \text{ then } a - c < b - c.$$

b. Is the following statement true? Explain.

$$\text{If } a < b \text{ and } c \ne 0, \text{ then } \frac{a}{c} < \frac{b}{c}.$$

For exercises 45–48, refer to Fig. 39.

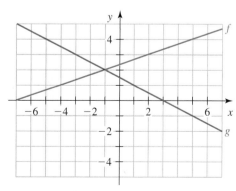

Figure 39 Exercises 45–48

Is the statement true? Explain.

45. $f(-4) > g(-4)$ **46.** $f(2) \geq g(2)$

47. $f(-1) < g(-1)$ **48.** $f(-1) \leq g(-1)$

49. Graphs of the linear functions f and g are sketched in Fig. 40. Approximate the solution set to the inequality $f(x) > g(x)$.

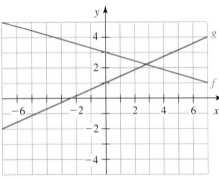

Figure 40 Exercise 49

50. Graphs of the linear functions f and g are sketched in Fig. 41. Determine which is the graph of f and which is the graph of

g if the solution set to the inequality $f(x) \leq g(x)$ is the set of numbers where $x \geq 2$.

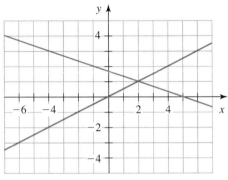

Figure 41 Exercise 50

51. Describe how to solve a linear inequality in one variable. Include a description of when you need to reverse the direction of an inequality. Also, explain why it is necessary to reverse the direction in this case. Finally, explain what you have accomplished by solving an inequality.

Taking it to the lab

For each lab assignment, consult with your instructor on whether to organize your responses as a numbered list or to write them in a paragraph.

THE SPORTS LAB

In Section 3.3 we modeled the women's and men's 400-meter run record times and predicted when the women's record time would equal the men's record time. You may have also done similar modeling for the Olympic 500-meter speed skating times and world record times for the 1500-meter run in exercises 1 and 9, respectively, in Homework 3.3. In this lab, you will research a sporting event of your choice and use linear functions to model your data for the women's and men's versions of the event. The following web sites may help you perform your research:

Track and Field Statistics
http://www.apulanta.fi/matti/yu/index_en.html
Olympics Results
http://www.ex.ac.uk/cimt/data/olympics/olymindx.htm
Index to the Olympics
http://www.hickoksports.com/history/olympix.shtml

Respond to the following instructions and questions.

1. Include tables of the women's data and the men's data you collected. State the source of your data.
2. Define any variables you used to model your data.
3. Use pencil and paper to draw sketches of scattergrams for the women's and men's data in the same coordinate system. Make sure it is clear which data points are for the women and which are for the men.
4. Find a linear function that models the women's data and a linear function that models the men's data. If you can't model your data well with linear functions, then choose another sporting event.

5. Draw graphs of your functions in the same coordinate system as your scatter-grams.

6. Compare the intercepts on the vertical axis for both of your linear functions. What does your comparison mean in terms of the sporting event?

7. Compare the slopes of your two linear functions. What does your comparison mean in terms of the sporting event?

8. Is there a point in time when your two models predict that the women's performance was or will be equal to the men's performance? If so, when did or will this happen?

9. Use your models to estimate when in the past or future the women's performance has been or will be better than the men's performance.

10. Sketch a reasonable curve on paper that describes the relationship between the performance of the men and years *for all of time*. Sketch a similar curve for the women. Sketch both curves on the same coordinate system.

THE TRUCK LAB

You are the owner of a large appliance store and are considering whether you should continue renting delivery trucks or whether you should purchase a truck.

On average, you rent a truck twice a week and drive 100 miles on each delivery day. The rental truck costs $19.99 per day plus $0.49 per mile. You must also pay 8% sales tax.

You can buy a delivery truck for $36,450, plus 8% sales tax. The truck will require $450 per year in maintenance, including sales tax. You estimate that repair costs will add about $600 in costs each year, including sales tax.

Because you must pay for gas and insurance regardless of whether you rent or buy, you decide that these costs will not influence your decision.

1. Let $R(t)$ represent the total money (in dollars) you will pay for costs related to renting a truck for t years. Let $O(t)$ represent the total money (in dollars) you will pay for costs related to owning a truck for t years. Find equations for both R and O.

2. If you choose to buy a truck, how long would you have to own the truck so that owning would be better than renting?

3. Next, take into account that if you buy a truck, you can resell it later. Use the rule of thumb that a new truck loses 20% of its value as soon as you drive it off the lot. Assume that the value of the truck will decrease by $1200 each year. How long would you have to own the truck before reselling it so that owning would be better than renting?

4. Compare your result in problem 2 with your result in problem 3. Explain why it makes sense that the results differ in this way.

5. There are two more things to consider. First, inflation will affect some of the costs. Second, if you choose to buy a truck, you will have to pay interest on a loan. Without performing calculations, discuss how these two factors would affect your answer to problem 3. Explain.

CHAPTER SUMMARY

Key Points

OF THIS CHAPTER

The *solution set* of a system of two equations is the set of all ordered pairs that satisfies both equations.	**Solution set of a system**
A system of two equations can be solved by elimination, substitution, or graphing. All three methods give the same result.	**Three methods for solving a system**

1. To use graphing, locate the intersection point(s) of the graphs of the equations.

2. To use elimination, vertically align the variable terms of your equations. If needed, multiply both sides of one equation by a number (and if necessary multiply both sides of the other equation by another number) so that the coefficients of a variable are equal in absolute value and opposite in sign. Then, add the left sides and the right sides of the equations.

3. To use substitution, isolate a variable in one equation and then substitute for it in the other equation.

Three types of systems

There are three types of systems of two linear equations in two unknowns: one-point-solution systems, dependent systems, and inconsistent systems. There are ways to recognize the types when solving by substitution or elimination.

1. For a one-point-solution system, both substitution and elimination give the coordinates of the solution point.

2. For an inconsistent system, both elimination and substitution give a false statement. When this happens, state that the solution set is the empty set.

3. For a dependent system, both substitution and elimination give a true statement that can be put into the form $a = a$. When this happens, state that the solution set is the set of points on the line.

Intersection point of two models

If $f(t)$ and $g(t)$ are models where t represents time, then the intersection point(s) of the graphs of these models indicates when the quantities represented by the dependent variables were or will be equal.

Solution set of an inequality

The *solution set* of an inequality in one variable is the set of all numbers that satisfy the inequality. We *solve* an inequality by finding its solution set.

Reversing the direction of an inequality

The direction of an inequality is reversed when multiplying or dividing by a negative number. Using symbols, we have:

- If $a < b$, then $ac < bc$, where c is a *positive* number.
- If $a < b$, then $ac > bc$, where c is a *negative* number.
- Similar properties hold for $a \leq b$, $a > b$, $a \geq b$, and when the operation of multiplication is replaced by division.

Three methods to solve an inequality

We can find the values of t where a quantity $f(t)$ is less than a quantity $g(t)$ by each of the following methods:

- We can solve the inequality $f(t) < g(t)$ using symbols.
- We can locate the values of t where the graph of f is lower than the graph of g.
- We can use tables of input-output pairs for f and g to find values of t where the outputs of f are less than the outputs of g.

CHAPTER 3 REVIEW EXERCISES

Find the solution set of the system by graphing the equations by hand.

1. $y = -\dfrac{3}{2}x + 1$

$y = \dfrac{1}{4}x - 6$

2. $3x - 5y = -1$

$y = -2(x - 4)$

Solve each system using either elimination or substitution. If the system is inconsistent or dependent, say so. Verify your solution set either graphically or by checking that one-point solutions satisfy both equations in the system.

3. $y = 2x + 5$

$y = -3x + 10$

4. $4x - 5y = -22$

$3x + 2y = -5$

5. $-2x + 3y = 7$

$-6y = -4x - 14$

6. $3x - 7y = 5$

$6x - 14y = -1$

7. $-4x - 5y = 3$

$10y = -8x - 6$

8. $y = 4.2x - 7.9$

$y = -2.8x + 1.1$

9. $y = 4.9x$

$-3.2y = x$

10. $0.4x + 0.3y = 0.4$

$-0.8x + 1.2y = 6.4$

11. $3x - 5y = 21$

$y = \frac{1}{2}x - 4$

12. $\frac{3}{5}x - \frac{2}{3}y = 4$

$-\frac{6}{5}x + \frac{8}{3}y = -4$

13. $2(3x - 4) + 3(2y - 1) = -5$

$-3(2x + 1) + 4(y + 3) = -7$

14. Solve the system of equations three times, once by each of the three methods: elimination, substitution, and graphing.

$$2x - 5y = 15$$
$$y = -2x + 9$$

15. Create a system of two linear equations as indicated. Verify your system graphically.

   a. The system is dependent.

   b. The system is inconsistent.

   c. The solution of the system is $(4, 6)$.

16. Some values of functions f and g are given in Table 25. For the system of equations that describe the functions f and g, determine two approximate solutions of the system.

Table 25 Some Values of Functions f and g (Exercise 16)

x	-4	-3	-2	-1	0	1	2	3	4
$f(x)$	2	5	8	11	14	17	20	23	26
$g(x)$	31	24	19	16	15	16	19	24	31

17. Determine the constants a and b so that $(5, 3)$ is the solution of the system. Verify your result by checking that $(5, 3)$ satisfies each of your equations. Also verify your result graphically.

$$2x + 3y = a$$
$$6x - 4y = b$$

18. Find the coordinates of the points A, B, C, D, E, and F as shown in Fig. 42. The equations of the sketched lines are provided but no attempt has been made to sketch the lines accurately except for showing the intersection points. Verify your results graphically.

$l_1: y = 3x + 4$
$l_2: 3y + 2x = 34$
$l_3: y + 4x = 28$
$l_4: y = 3x - 14$

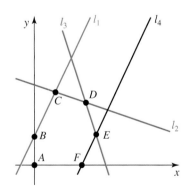

Figure 42 Exercise 18

19. Find all ordered pairs that satisfy all three equations.

$$y = -x + 2$$
$$y = -2x + 7$$
$$y = 3x - 6$$

Use a symbolic method to solve each inequality. Describe the solution set with inequality notation, interval notation, and a graph. Then, use graphing calculator tables or graphs to verify your result.

20. $-2x \le 18$

21. $3x - 8 \le 13$

22. $-8x > -3(x - 2) - 2x + 9$

23. $4.2 - 3.6x \ge 3.9(x + 2.1)$

24. $-5(2x + 3) \ge 2(3x - 4)$

25. $-\frac{2}{3}x + \frac{5}{2} < \frac{7}{3}$

26. $-\frac{3}{4}x \le \frac{5}{2}$

27. $3 - \frac{2}{5}x > x - \frac{3}{2}$

28. a. List three numbers that satisfy the inequality $7 - 2(3x + 5) < 4x + 1$.

   b. List three numbers that do not satisfy the inequality $7 - 2(3x + 5) < 4x + 1$.

29. A student has tried to solve the inequality $5 - 3x \le 11$. Decide whether the student did the exercise correctly. If the student made a mistake, describe the student's error.

Student's Work

$$5 - 3x \le 11$$
$$5 - 3x - 5 \le 11 - 5$$
$$-3x \le 6$$
$$\frac{-3x}{-3} \le \frac{6}{-3}$$
$$x \le -2$$

For exercises 30–35, refer to Fig. 43.

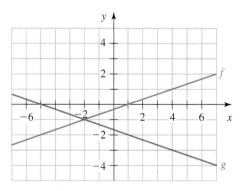

Figure 43 Exercises 30–35

30. Find $f(4)$.

31. Find $g(4)$.

32. Find x where $f(x) = 0$.

33. Find x where $g(x) = 0$.

34. Find x where $f(x) = g(x)$.

35. Find x where $f(x) > g(x)$.

36. Find values for a, b, and c so that the solution set to the inequality $ax + b > c$ is the set of numbers where $x < 5$. Use a symbolic method to verify your result.

37. One Rent A Wreck® office rents 16-foot trucks for a one-day fee of $75 plus $0.22 per mile. One U-Haul location charges a one-day fee of $29.95 plus $0.69 per mile.

   a. Let $R(x)$ represent Rent A Wreck's charge (in dollars) and $U(x)$ represent U-Haul's charge (in dollars), both for driving x miles in one day. Find equations for R and U.

   b. For how many miles driven is the one-day charge at Rent A Wreck equal to the charge at U-Haul?

   c. For how many miles driven is the one-day charge at Rent A Wreck less than the charge at U-Haul?

38. In 2000, the average price of a home in a community is $250,000, and it increases by about $9000 each year. A family has $12,000 in 2000 and plans to save $400 each month. Predict when the family will be able to pay a 10% down payment on an average-priced house in the community.

39. Petroleum consumptions by North America and the Far East for various years are shown in Table 26.

Table 26 Petroleum Consumptions

Year	Petroleum Consumption (quadrillion Btu)	
	North America	Far East
1986	38.4	22.9
1990	40.4	28.2
1995	41.5	37.3
1997	43.6	40.5
1999	45.8	41.0

Source: *Energy Information Administration*

Let $n(t)$ represent the petroleum consumption by North America and $f(t)$ represent the petroleum consumption by the Far East, both in quadrillion Btu, where t represents the number of years since 1980.

   a. Find equations for n and f.

   b. Compare the slope of n to the slope of f. What do the slopes represent in terms of the situation?

   c. Estimate when the petroleum consumption was the same in the Far East and in North America.

   d. Find values of t where $n(t) < f(t)$. What does this mean in terms of the situation?

40. As the number of job offerings in the aerospace industry continues to decrease due to the end of the Cold War, many engineers in the Los Angeles area have left the aerospace industry to work in the entertainment business, applying their expertise to amusement park rides and special effects in movies. Numbers of jobs in Los Angeles County are modeled by the system:

$$H(t) = 21.14t + 100.00$$
$$A(t) = -15.63t + 220.68$$

where $H(t)$ represents the number of motion picture and television positions (in thousands) and $A(t)$ represents the number of aerospace jobs (in thousands), both at t years since 1990 (see Table 27).

Table 27 Numbers of Jobs (in thousands)

Year	Hollywood	Aerospace
1992	143.2	200.0
1993	163.6	169.0
1994	179.6	147.7
1995	211.4	135.5
1996	225.0	138.6

Source: *New York Times*

   a. Estimate when there were the same number of jobs in Hollywood as in the aerospace industry.

   b. Predict the years in which there are more jobs in Hollywood than in the aerospace industry.

   c. In terms of employment, which of the two industries has experienced the most change? Explain.

CHAPTER 3 TEST

Solve each system using either elimination or substitution. If the system is inconsistent or dependent, say so. Verify your solution set either graphically or by checking that one-point solutions satisfy both equations in the system.

1. $\quad y = 3x - 1$
$\quad\;\; 3x - 2y = -1$

2. $2x - 5y = 3$
$\quad\;\; 6x = 15y + 9$

3. $4x - 6y = 5$
$\quad 6x - 9y = -2$

4. $\dfrac{2}{5}x - \dfrac{3}{4}y = 8$
$\quad \dfrac{3}{5}x + \dfrac{1}{4}y = 1$

5. $-4(x + 2) + 3(2y - 1) = 21$
$\quad 5(3x - 2) - (4y + 3) = -59$

6. Create a system of two linear equations that has $(5, 2)$ as its *only* solution.

7. Solve the system of equations three times, once by each of the three methods: elimination, substitution, and graphing.

$$4x - 3y = 9$$
$$y = 2x - 5$$

8. Consider the solution set to the system

$$y = 5x - 13$$
$$y = mx + b$$

where m and b are constants. If the system's solution set is the empty set, what can you say about m? About b?

Solve each inequality. Describe the solution set with inequality notation, interval notation, and a graph.

9. $2 - 10x \geq 3x + 14$

10. $3(x + 4) + 1 < 5(x - 2)$

11. $2.6(x - 3.1) > 4.7x - 5.9$

12. $-\dfrac{5}{3}x + \dfrac{1}{6} \le \dfrac{7}{4}x$

For exercises 13–16, refer to Fig. 44.

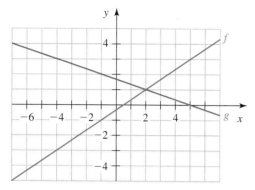

Figure 44 Exercises 13–16

13. Find $f(5)$.

14. Find x where $g(x) = 3$.

15. Find x where $f(x) = g(x)$.

16. Find x where $f(x) \le g(x)$.

17. The graphs of f and g are sketched in Fig. 45. Solve the inequality $f(x) < g(x)$.

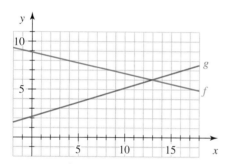

Figure 45 Exercise 17

18. a. List three values that satisfy the inequality $3x - 11 < 7 - 6x$.

b. List three values that do not satisfy the inequality $3x - 11 < 7 - 6x$.

19. Let $A(t)$ represent the enrollment (in thousands) at college A at t years since 1900. Let $B(t)$ represent the enrollment (in thousands) at college B at t years since 1900. Equations for A and B are given below:

$$A(t) = -0.12t + 31.75$$
$$B(t) = 0.08t + 9.43$$

a. Which college had a larger enrollment in 2000? By how much?

b. What are the slopes of A and B? Discuss their significance in terms of the enrollments at college A and college B.

c. When will the colleges have the same enrollment, according to the linear models?

d. For which years will college A have a larger enrollment than college B?

20. The number of two-year colleges and four-year colleges and universities for various years are listed in Table 28.

Table 28 Numbers of Two-Year and Four-Year Colleges and Universities		
Year	2-Year	4-Year
1970	891	1665
1980	1274	1957
1985	1311	2029
1990	1418	2141
1995	1462	2244
1998	1727	2343

Source: *U.S. Center for Education Statistics*

Let $f(t)$ represent the number of two-year colleges and $g(t)$ represent the number of four-year colleges and universities at t years since 1970.

a. Find equations for f and g.

b. Predict the total number of two-year and four-year colleges and universities in 2008.

c. Use f and g to predict when there will be the same number of two-year colleges as four-year colleges and universities.

d. Explain in terms of the slopes of f and g why your result in part c is so far in the future.

Cumulative Review of Chapters 1–3

1. A person places a pitcher of hot tea in the refrigerator. Let T represent the temperature of the tea t minutes after the tea was put in the refrigerator. Sketch a qualitative graph that describes the relationship between t and T.

Graph the equation.

2. $y = -\dfrac{2}{3}x + 4$

3. $5x - 3y = 15$

4. $3(x - 4) = -2(y + 5) + 4$

5. $x + 2 = 0$

6. Find the slope of the line that contains the points $(-4, 2)$ and $(3, -1)$.

7. Find an equation of a line with slope $-\dfrac{3}{5}$ that contains the point $(2, -3)$. Write your result in slope-intercept form.

8. Find an equation of a line that contains the points $(-5, -2)$ and $(-2, 3)$.

9. Find an equation of the line that contains the point $(-5, 3)$ and is perpendicular to the line $2x - 5y = 20$.

10. Find three points that lie between the lines $y = 2x + 3$ and $y = 2x + 3.1$.

For exercises 11–13, refer to Table 29, which provides some values for four linear functions.

Table 29 Values for Four Linear Functions

Equation 1		Equation 2		Equation 3		Equation 4	
x	f(x)	x	g(x)	x	h(x)	x	k(x)
0	97	4		1	23	10	−28
1		5		2		11	
2		6	4	3		12	
3	58	7		4		13	
4		8		5		14	−16
5		9	43	6	−22	15	

11. Complete Table 29.

12. Find $f(5)$.

13. Find x when $g(x) = 30$.

Let $f(x) = -\dfrac{3}{2}x + 7$.

14. Find $f(-4)$.

15. Find x when $f(x) = -4$.

16. What is the slope of f?

17. Find the x-intercept of f.

18. Find the y-intercept of f.

19. Sketch the graph of f.

For exercises 20–26, refer to Fig. 46.

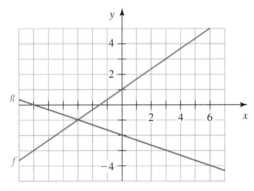

Figure 46 Exercises 20–26

20. Find $f(-6)$.

21. Find x where $g(x) = 0$.

22. Find the y-intercept of f.

23. Find the slope of g.

24. Find an equation for f.

25. Find x where $f(x) = g(x)$.

26. Find x where $f(x) \le g(x)$.

Solve the equation.

27. $-5x - 3(2x + 4) = 8 - 2x$

28. $\dfrac{7}{8} - \dfrac{1}{4}x = \dfrac{1}{2} + \dfrac{3}{8}x$

29. Determine whether the graph in Fig. 47 is the graph of a function. Explain.

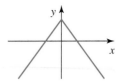

Figure 47 Exercise 29

30. Determine whether the relation $x = \sqrt{y}$ is a function.

31. Give an example of a function. Describe it using an equation, graph, and table.

Solve the system.

32. $2x + 4y = -8$
$5x - 3y = 19$

33. $3x - 7y = 14$
$\quad y = \dfrac{3}{7}x - 2$

34. Solve the system of equations three times, once by each of the three methods: elimination, substitution, and graphing.

$$2x - 5y = -4$$
$$x + 3y = 9$$

Solve. Describe the solution set with inequality notation, interval notation, and a graph.

35. $-2(4x + 5) \ge 3(x - 7) + 1$

36. $\dfrac{4}{3}x - \dfrac{1}{6} < 2 - \dfrac{5}{6}x$

37. Consider the scattergram of data and the graph of the model $y = mt + b$ in Fig. 48. Sketch the graph of a linear model that better describes the data and then explain how you would adjust the values of m and b of the original model so that it would better describe the data.

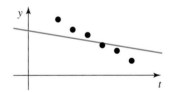

Figure 48 Exercise 37

38. a. Solve $-3x + 6 = 0$.

b. Solve $-3x + 6 < 0$. Graph the solution set on a number line.

c. Graph $y = -3x + 6$.

d. Solve the system.

$$y = -3x + 6$$
$$y = 4x - 1$$

39. Give an example of each of the following and then describe the solution set for each of your examples. [**Hint**: In some cases, it will be helpful to describe the solution set using a graph.]

a. equation in one variable

b. equation in two variables

c. system of two equations in two variables

d. an inequality in one variable

40. Greenskeepers have been mowing golf putting surfaces progressively lower over the past half century (see Table 30).

Table 30 Grass Heights on Golf Putting Surfaces

Decade	Number Used to Represent Decade	Grass Height (inches)
1950s	55	$\frac{1}{4}$
1960s	65	$\frac{7}{32}$
1970s	75	$\frac{3}{16}$
1980s	85	$\frac{5}{32}$
1990s	95	$\frac{1}{8}$

Source: *Golf Course Superintendents Association of America*

Let $h(t)$ represent the grass height (in inches) at t years since 1900.

a. Find an equation for h.

b. Find $h(105)$. What does your result mean in terms of the situation?

c. Find t when $h(t) = \frac{1}{16}$. What does your result mean in terms of the situation?

d. Find the slope of h. What does it represent in terms of the situation?

e. Find the t-intercept of h. What does it mean in terms of the situation?

41. The number of bicyclists younger than 13 hit and killed by motor vehicles was ~~446~~ children in 1975 and declines by about 12 children each year.

a. Let $n = f(t)$ represent the number of bicyclists younger than 13 that were hit and killed by motor vehicles in the year that is t years since 1975. Find an equation for f.

b. What is the slope of f? What does it mean in terms of the situation?

c. Find the n-intercept. What does it mean in terms of the situation?

d. Find the t-intercept. What does it mean in terms of the situation?

e. For which years in the future is there model breakdown for certain? Explain.

42. Company A sold 9.5 million dollars of sports equipment in 2000 and its sales have increased by 1.3 million dollars each year. Company B sold 5.2 million dollars of sports equipment in 2000 and its sales have increased by 1.8 million dollars each year.

a. Let $A(t)$ and $B(t)$ represent the sales (in millions of dollars) by company A and company B, respectively, at t years since 2000. Find equations for A and B.

b. Use pencil and paper to sketch the graphs of A and B on the same coordinate system. According to your sketch, when will sales at the companies be equal? What will those sales be?

c. Now use elimination or substitution to predict when sales at the companies will be equal. What will those sales be? Compare your results with your results in part b.

d. Solve the inequality $A(t) < B(t)$. What does your result mean in terms of the situation?

43. World record times for the 200-meter run are listed in Table 31.

Table 31 200-Meter Run Record Times

Women		Men	
Year	Record Time (seconds)	Year	Record Time (seconds)
1973	22.38	1951	20.6
1974	22.21	1963	20.3
1978	22.06	1967	20.14
1984	21.71	1979	19.72
1988	21.34	1996	19.32

Source: *WR Progression (Women), www.apulanta.fi/matti/yu/wrprogr_Women.html#200m; WR Progression (Men), www.apulanta.fi/matti/yu/wrprogr_Men.html#200m. Both accessed 12/30/02.*

a. Let $r = W(t)$ and $r = M(t)$ represent the record times (in seconds) for women and men, respectively, at t years since 1900. Find the equations for W and for M.

b. Find $W(107)$ and $M(107)$. What do your results mean in terms of the situation?

c. Compare the slope of W to the slope of M. What does your comparison tell you about the situation?

d. Explain why your results from parts b and c suggest that there may be a time when the women's record time will be equal to the men's record time.

e. Predict when the women's record time will equal the men's record time.

f. Solve $W(t) > M(t)$. What does your result mean in terms of the situation?

g. Find the t-intercepts of W and M. What do your results mean in terms of the situation?

h. Sketch qualitative graphs that describe the relationship between t and r for hundreds of years into the past and the future on the same axes.

Exponential Functions

There is not much difference between the delight a novice experiences in cracking a clever brain teaser and the delight a mathematician experiences in mastering a more advanced problem. Both look on beauty bare—that clean, sharply defined, mysterious, entrancing order that underlies all structure.

—*Martin Gardner*

4.1 PROPERTIES OF EXPONENTS

Objectives

▸ Learn the definitions for zero exponent and negative exponent.

▸ Know properties of exponents.

We begin this section by reviewing the definition of an exponent.

DEFINITION b^n

For a counting number n,

$$b^n = \underbrace{bbb \cdots b}_{n \text{ factors}}$$

we call b the **base** and n the **exponent**. We refer to b^n as "b to the nth power."

For example, $4^3 = 4 \cdot 4 \cdot 4 = 64$.

For an expression of the form $-b^n$, we compute b^n before finding the opposite. For example,

$$-2^4 = -(2^4) = -(2 \cdot 2 \cdot 2 \cdot 2) = -16$$

If we want the base to be -2, we enclose -2 in parentheses:

$$(-2)^4 = (-2)(-2)(-2)(-2) = 16$$

We can use a graphing calculator to check both computations (see Fig. 1).

Note

Recall that the *counting numbers* is the set of numbers 1, 2, 3, ... To review various sets of numbers, see Section A.2.

Graphing Calculator

To find -2^4: Press $\boxed{(-)}$ $\boxed{2}$ $\boxed{\wedge}$ $\boxed{4}$ $\boxed{\text{ENTER}}$.

```
-2^4
            -16
(-2)^4
             16
```

Figure 1 Compute -2^4 and $(-2)^4$

In this section and Section 4.2, we discuss how to use properties of exponents to simplify expressions involving exponents. There are five properties that will be useful to us in this course.

Properties of Exponents

For the following properties, m and n are counting numbers.

- $b^m b^n = b^{m+n}$
- $\dfrac{b^m}{b^n} = b^{m-n}, \quad b \neq 0$ and $m > n$
- $(bc)^n = b^n c^n$
- $\left(\dfrac{b}{c}\right)^n = \dfrac{b^n}{c^n}, \quad c \neq 0$
- $(b^m)^n = b^{mn}$

Note

We first explore the meanings of these properties for counting number exponents. Later, we explore their meanings for other numbers, too.

Example 1 Meaning of Exponential Properties

1. Explain why it makes sense that $b^2 b^3 = b^5$.
2. Explain why it makes sense that $b^m b^n = b^{m+n}$, where m and n are counting numbers.
3. Explain why it makes sense that $\left(\dfrac{b}{c}\right)^n = \dfrac{b^n}{c^n}$, where n is a counting number and $c \neq 0$.

Solution

1. By writing $b^2 b^3$ without using exponents, we see

$$b^2 b^3 = (bb)(bbb)$$
$$= bbbbb$$
$$= b^5$$

We can verify that this result is correct for various constant bases by examining graphing calculator tables for both $y = x^2 x^3$ and $y = x^5$ (see Fig. 2).

Figure 2 Comparing tables for $y = x^2 x^3$ and $y = x^5$

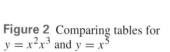

Graphing Calculator

See Section B.14.

2. We write $b^m b^n$ without using exponents.

$$b^m b^n = (\underbrace{bbb \cdots b}_{m \text{ factors}})(\underbrace{bbb \cdots b}_{n \text{ factors}}) = \underbrace{bbb \cdots b}_{m+n \text{ factors}} = b^{m+n}$$

3. We write $\left(\dfrac{b}{c}\right)^n$ without using exponents.

$$\left(\frac{b}{c}\right)^n = \underbrace{\left(\frac{b}{c}\right)\left(\frac{b}{c}\right)\left(\frac{b}{c}\right) \cdots \left(\frac{b}{c}\right)}_{n \text{ factors}} = \frac{\overbrace{bbb \cdots b}^{n \text{ factors}}}{\underbrace{ccc \cdots c}_{n \text{ factors}}} = \frac{b^n}{c^n}$$

Note

Recall that we assume that c is nonzero.

We can use properties of exponents to simplify exponential expressions. An exponential expression is simplified if:

1. There are no parentheses.
2. Each constant appears as a base as few times as possible. (For example, for $b \neq 0$, we write $b^4 + \dfrac{b^2}{b}$ as $b^4 + b$.)
3. Each numerical expression (such as 2^3 or $5 \cdot 7$) has been calculated, and each numerical fraction has been reduced.

Example 2   Simplifying Exponential Expressions

Simplify each expression.

1. $(2b^2 c^3)^5$ **2.** $(3b^3 c^4)(2b^6 c^2)$ **3.** $\dfrac{3b^7 c^6}{12b^2 c^5}$ **4.** $\left(\dfrac{24b^7 c^8}{16b^2 c^5 d^3} \right)^4$

Solution

1. $(2b^2 c^3)^5 = 2^5 (b^2)^5 (c^3)^5$ $(bc)^n = b^n c^n$

$\qquad\qquad\;\; = 32b^{10} c^{15}$ $(b^m)^n = b^{mn}$

2. $(3b^3 c^4)(2b^6 c^2) = (3 \cdot 2)(b^3 b^6)(c^4 c^2)$ Rearrange factors.

$\qquad\qquad\qquad\;\; = 6b^9 c^6$ $b^m b^n = b^{m+n}$

3. $\dfrac{3b^7 c^6}{12b^2 c^5} = \dfrac{b^{7-2} c^{6-5}}{4}$ $\dfrac{b^m}{b^n} = b^{m-n}$

$\qquad\qquad = \dfrac{b^5 c}{4}$

4. $\left(\dfrac{24b^7 c^8}{16b^2 c^5 d^3} \right)^4 = \left(\dfrac{3b^5 c^3}{2d^3} \right)^4$ $\dfrac{b^m}{b^n} = b^{m-n}$

$\qquad\qquad\qquad = \dfrac{(3b^5 c^3)^4}{(2d^3)^4}$ $\left(\dfrac{b}{c} \right)^n = \dfrac{b^n}{c^n}$

$\qquad\qquad\qquad = \dfrac{3^4 (b^5)^4 (c^3)^4}{2^4 (d^3)^4}$ $(bc)^n = b^n c^n$

$\qquad\qquad\qquad = \dfrac{81b^{20} c^{12}}{16d^{12}}$ $(b^m)^n = b^{mn}$ ■

Note

Recall that 0 is a whole number, but not a counting number.

In Example 2 the exponents are counting numbers. What would be a reasonable definition for b^0? If the property $\dfrac{b^m}{b^n} = b^{m-n}$ is to be true for $m = n$, then

$$1 = \dfrac{b^n}{b^n} = b^{n-n} = b^0, \quad b \neq 0$$

So, a reasonable definition for b^0 is 1.

Note

In exercise 57 you will see why it makes sense for 0^0 to be left undefined.

DEFINITION Zero exponent

For $b \neq 0$,

$$b^0 = 1$$

For example: $7^0 = 1$, $(-3)^0 = 1$, and $(ab)^0 = 1$, where $ab \neq 0$.

What should be the meaning of a negative integer exponent? If the property $\dfrac{b^m}{b^n} = b^{m-n}$ is to be true for $m = 0$, then

$$\frac{1}{b^n} = \frac{b^0}{b^n} = b^{0-n} = b^{-n}, \quad b \neq 0$$

So, we should define b^{-n} to be $\dfrac{1}{b^n}$.

DEFINITION *Negative integer exponent*

If $b \neq 0$ and n is a counting number, then

$$b^{-n} = \frac{1}{b^n}$$

Note

Recall that the *integers* is the set of numbers $\pm 1, \pm 2, \pm 3, \ldots$.

From the definition of a negative exponent, it follows that

$$\frac{1}{b^{-n}} = b^n$$

where $b \neq 0$ and n is a counting number. Here is the proof:

Note

A "proof" of a property is a mathematical explanation that shows that the property is true.

$$\frac{1}{b^{-n}} = 1 \div b^{-n}$$
$$= 1 \div \frac{1}{b^n}$$
$$= 1 \cdot b^n$$
$$= b^n$$

So $\dfrac{1}{b^{-n}} = b^n$, which is what we intended to show.

Simplifying an exponential expression includes writing the expression so that each exponent is positive.

Example 3   Simplifying Exponential Expressions

Simplify each expression.

1. 7^{-2} **2.** $3^{-1} + 4^{-1}$ **3.** $\dfrac{5}{b^{-3}}$

Solution

1. $7^{-2} = \dfrac{1}{7^2} = \dfrac{1}{49}$

2. $3^{-1} + 4^{-1} = \dfrac{1}{3} + \dfrac{1}{4} = \dfrac{4}{12} + \dfrac{3}{12} = \dfrac{7}{12}$

3. $\dfrac{5}{b^{-3}} = 5\left(\dfrac{1}{b^{-3}}\right) = 5b^3$ ■

It turns out that the five properties discussed at the start of this section are valid for *integer* exponents.

Properties of Exponents

For the following properties, m and n are integers, and b and c are nonzero.

- $b^m b^n = b^{m+n}$
- $\dfrac{b^m}{b^n} = b^{m-n}$
- $(bc)^n = b^n c^n$
- $\left(\dfrac{b}{c}\right)^n = \dfrac{b^n}{c^n}$
- $(b^m)^n = b^{mn}$

Example 4   Simplifying Exponential Expressions

Simplify each expression.

1. $2^{-1003} 2^{1000}$ **2.** $\dfrac{b^{-6}}{b^{-4}}$ **3.** $\dfrac{35b^{-9}c^3}{25b^{-7}c^{-5}}$ **4.** $\left(\dfrac{18b^{-4}c^7}{6b^{-3}c^2}\right)^{-4}$

Graphing Calculator

A TI-83/82 will give an "overflow" error for 2^{1000} because it is so large. It will also round 2^{-1003} to 0, although 2^{-1003} is very close to, but not equal to zero.

Solution

1.
$$2^{-1003} 2^{1000} = 2^{-1003+1000} \qquad\qquad b^m b^n = b^{m+n}$$
$$= 2^{-3}$$
$$= \frac{1}{2^3} \qquad\qquad\qquad\qquad b^{-n} = \frac{1}{b^n}$$
$$= \frac{1}{8}$$

2.
$$\frac{b^{-6}}{b^{-4}} = b^{-6-(-4)} \qquad\qquad\qquad \frac{b^m}{b^n} = b^{m-n}$$
$$= b^{-6+4}$$
$$= b^{-2}$$
$$= \frac{1}{b^2} \qquad\qquad\qquad\qquad b^{-n} = \frac{1}{b^n}$$

3.
$$\frac{35b^{-9}c^3}{25b^{-7}c^{-5}} = \frac{7b^{-9-(-7)}c^{3-(-5)}}{5} \qquad\qquad \frac{b^m}{b^n} = b^{m-n}$$
$$= \frac{7b^{-2}c^8}{5}$$
$$= \frac{7c^8}{5b^2} \qquad\qquad\qquad\qquad b^{-n} = \frac{1}{b^n}$$

4.
$$\left(\frac{18b^{-4}c^7}{6b^{-3}c^2}\right)^{-4} = (3b^{-4-(-3)}c^{7-2})^{-4} \qquad \frac{b^m}{b^n} = b^{m-n}$$
$$= (3b^{-1}c^5)^{-4}$$
$$= 3^{-4}(b^{-1})^{-4}(c^5)^{-4} \qquad\qquad (bc)^n = b^n c^n$$
$$= 3^{-4}b^4 c^{-20} \qquad\qquad\qquad (b^m)^n = b^{mn}$$
$$= \frac{b^4}{3^4 c^{20}} \qquad\qquad\qquad\qquad b^{-n} = \frac{1}{b^n}$$
$$= \frac{b^4}{81 c^{20}}$$

EXPLORATION
Properties of exponents

1. In Example 1 we showed that the statement

$$b^2b^3 = b^5$$

   makes sense by first writing the expression b^2b^3 without using exponents. For each part, show that the given statement makes sense by first writing an expression without using exponents.

   a. $(bc)^4 = b^4c^4$

   b. $\dfrac{b^7}{b^3} = b^4, \ \ b \neq 0$

   c. $(b^3)^4 = b^{12}$

2. In Example 1 we also showed that the general statements

$$b^m b^n = b^{m+n} \qquad \text{and} \qquad \left(\frac{b}{c}\right)^n = \frac{b^n}{c^n}, \ \ c \neq 0$$

   make sense for counting numbers m and n. For each part below, show that the general statement makes sense. Assume that m and n are counting numbers.

   a. $(bc)^n = b^n c^n$

   b. $\dfrac{b^m}{b^n} = b^{m-n}, \ \ $ where $m > n$ and $b \neq 0$

   c. $(b^m)^n = b^{mn}$

3. Choose values of b, c, and counting number n to show that the statement $(b + c)^n = b^n + c^n$ is false, in general.

EXPLORATION
Looking ahead: Definition of $b^{1/n}$

1. In this part of the exploration you will explore the meaning of $b^{1/2}$, where b is nonnegative.

   a. For now, do not use a calculator. Here you will explore how you should define $9^{1/2}$. You can determine a reasonable value for $9^{1/2}$ by first finding the *square* of the value:

$$(9^{1/2})^2 = 9^{\frac{1}{2}\cdot 2} = 9^1 = 9$$

   What would be a good meaning for $9^{1/2}$? [**Hint**: Can you think of a positive number whose square equals 9?]

   b. What would be a good meaning for $16^{1/2}$? For $25^{1/2}$?

   c. Now use your calculator to find $9^{1/2}$, $16^{1/2}$, and $25^{1/2}$. Is your calculator interpreting $b^{1/2}$ as you would expect?

   d. What would be a good meaning for $b^{1/2}$, where b is nonnegative?

2. In this part of the exploration you will explore the meaning of $b^{1/3}$.

   a. For now, do not use a calculator. You will explore how we should define $8^{1/3}$. We will first find the *cube* of the value:

$$(8^{1/3})^3 = 8^{\frac{1}{3}\cdot 3} = 8^1 = 8$$

   What would be a good meaning for $8^{1/3}$? Explain.

Note

Here we assume that the property $(b^m)^n = b^{mn}$ still applies.

Graphing Calculator

Instructions for $9^{1/2}$: Press 9 $\boxed{\wedge}$ $\boxed{(}$ 1 $\boxed{\div}$ 2 $\boxed{)}$.

b. What would be a good meaning for $27^{1/3}$? For $64^{1/3}$?

c. Use your calculator to find $8^{1/3}$, $27^{1/3}$, and $64^{1/3}$. Is your calculator interpreting $b^{1/3}$ as you would expect?

d. What would be a good meaning for $b^{1/3}$?

3. What would be a good meaning for $b^{1/n}$, where n is a counting number and b is nonnegative?

Note

We will discuss whether $b^{1/n}$ has meaning when b is negative in Section 4.2.

Tips on Succeeding in This Course

If you have not had passing scores on tests and quizzes during the first part of the course, it is time to honestly evaluate what the problem is, what changes should be made, and whether you can commit to making those changes.

Sometimes students must change how they study for the course. For example, Rosie was a student of mine who did poorly on exams and quizzes for the first third of the course. I was confused that she was not passing the course, since she had good attendance, was actively involved in classroom work, and was doing the homework assignments. Suddenly Rosie started getting A's on every quiz and test. I asked Rosie what had happened. She said, "I figured out that to do well in this course it was not enough practice for me to just do the exercises you assigned. So now I do a lot of extra exercises from each section."

Key Points OF THIS SECTION

Definitions for b^0 and b^{-n}

For the following definitions $b \neq 0$.

- $b^0 = 1$
- $b^{-n} = \dfrac{1}{b^n}$, where n is a counting number

Properties of exponents

For the following properties, m and n are integers and b and c are nonzero.

- $b^m b^n = b^{m+n}$
- $\dfrac{b^m}{b^n} = b^{m-n}$
- $(bc)^n = b^n c^n$
- $\left(\dfrac{b}{c}\right)^n = \dfrac{b^n}{c^n}$
- $(b^m)^n = b^{mn}$

Writing an expression without exponents

If you forget an exponential property, you can rediscover it by writing an appropriate expression without exponents. For example, if you forget that $(bc)^n = b^n c^n$, you can rediscover this property by writing $(bc)^3 = (bc)(bc)(bc) = bbbccc = b^3 c^3$.

Simplifying an expression

An exponential expression is simplified if:

1. There are no parentheses.

2. Each constant appears as a base as few times as possible. (For example, for $b \neq 0$, we write $b^2 + \dfrac{b^4}{b^3}$ as $b^2 + b$.)

3. Each numerical expression (such as 3^2 or $4 \cdot 6$) has been calculated and each numerical fraction has been reduced.

4. Each exponent is positive.

HOMEWORK 4.1

SSM
Tutor Center

MathPro 4
MathPro 4

MathPro 5
MathPro 5

CDs/Videos

www

Simplify each expression without using a calculator. Then use a calculator to verify your result. [**Note:** *To review order of operations, see Section A.8.*]

1. 2^{-1} **2.** 5^0 **3.** 3^{-2}

4. -4^2 **5.** $(-4)^2$ **6.** $(2^3)^2$

7. $2^{-1} + 3^{-1}$ **8.** $\dfrac{1}{2^{-1}} + \dfrac{1}{3^{-1}}$

Simplify each expression without using a calculator.

9. $\dfrac{7^{902}}{7^{900}}$ **10.** $4^{2003}4^{-2000}$

11. $13^{500}13^{-500}$ **12.** $(130^{-1})^{-1}$

13. $(25^3 - 411^5 + 89^2)^0$ **14.** $\dfrac{6^{200}}{2^{198}3^{199}}$

Simplify each expression.

15. $b^7 b^9$ **16.** $b^4 b^{-8}$

17. $b^5 b^{-5}$ **18.** $(4b^{-9})(5b^4)$

19. $(3b^{-7})(-2b^6)$ **20.** $(-4b^{-1}c^2)(6b^3c^{-4})$

21. $(-b^2)^4$

22. $(4b^3c^7)^2(2b^5c^4)^3$

23. $(2b^4c^{-2})^5(3b^{-3}c^{-4})^{-2}$

24. $(-5b + 3c)^0 + (7b - 4c)^0$

25. $\dfrac{b^{10}}{b^{15}}$ **26.** $\dfrac{b^{-2}}{b^2}$

27. $\dfrac{2b^{-12}}{b^{-9}}$ **28.** $\dfrac{(2b)^3}{b^7}$

Simplify each expression.

29. $\dfrac{-12b^5c^5}{14b^4c^5}$ **30.** $\dfrac{28b^{-2}c^{-3}}{4b^{-3}c^{-1}}$

31. $\dfrac{15b^{-7}c^{-3}d^8}{-45c^2b^{-6}d^8}$ **32.** $\dfrac{(2b^6c)(6b^2c^5)}{3b^4c^4}$

33. $\dfrac{(-5b^{-3}c^4)(4b^{-5}c^{-1})}{80b^2c^{17}}$ **34.** $\dfrac{(16b^{-2}c)(25b^4c^{-5})}{(15b^5c^{-1})(8b^{-7}c^{-2})}$

35. $\dfrac{(24b^3c^{-6})(49b^{-1}c^{-2})}{(28b^2c^4)(14b^{-5}c)}$ **36.** $\dfrac{(2b^{-7}c^4)^3}{5b^2c^6}$

37. $\dfrac{(3b^5c^{-2})^3}{7b^{-3}c}$ **38.** $\left(\dfrac{3b^{-2}}{2c^{-1}}\right)^{-2}$

39. $\left(\dfrac{b^5c^{-2}}{b^2c^4}\right)^2$ **40.** $\left(\dfrac{2bc^2}{5b^{-1}b^{-2}}\right)^{-1}$

41. $\left(\dfrac{5b^4c^7}{67b^{-2}c^3}\right)^0$ **42.** $\left(\dfrac{8b^{-2}c^{-7}}{12b^{-5}c^{-3}}\right)^{-3}$

43. $\left(\dfrac{7b^4c^{-5}}{14b^7c^{-2}}\right)^4$ **44.** $(42b^{-8}c^7)^{-89}(42b^{-8}c^7)^{89}$

45. $b^{-1}c^{-1}$ **46.** $b^{-1} + c^{-1}$

47. $\dfrac{1}{b^{-1}} + \dfrac{1}{c^{-1}}$ **48.** $\dfrac{1}{b^{-1}} \cdot \dfrac{1}{c^{-1}}$

Simplify each expression. Assume that n is a counting number.

49. $b^{4n}b^{3n}$ **50.** $b^{5n-1}b^{2n+4}$

51. $\dfrac{b^{7n-1}}{b^{2n+3}}$ **52.** $\dfrac{b^{3n+4}b^{n-5}}{b^{2n-3}}$

53. Two students tried to simplify an expression. Which student(s), if any, simplified the expression correctly? If any errors were made, describe them.

Student A	Student B
$(5b^2)^{-1} = -5b^{-2}$	$(5b^2)^{-1} = 5^{-1}(b^2)^{-1}$
$\phantom{(5b^2)^{-1}} = \dfrac{-5}{b^2}$	$\phantom{(5b^2)^{-1}} = 5^{-1}b^{-2}$
	$\phantom{(5b^2)^{-1}} = \dfrac{1}{5b^2}$

54. Two students tried to simplify an expression. Which student(s), if any, simplified the expression correctly? If any errors were made, describe them.

Student 1	Student 2
$\dfrac{7b^8}{b^{-3}} = 7b^{8-(-3)}$	$\dfrac{7b^8}{b^{-3}} = 7b^{8-3}$
$\phantom{\dfrac{7b^8}{b^{-3}}} = 7b^{11}$	$\phantom{\dfrac{7b^8}{b^{-3}}} = 7b^5$

55. Many students get confused with expressions such as $2^2, 2^{-1}$, $2(-1), \left(\dfrac{1}{2}\right)^2, \left(\dfrac{1}{2}\right)^{-1}, -2^2, (-2)^2,$ and $\dfrac{1}{2}$. List these numbers from least to greatest. Mention if there are any "ties."

56. Explain why 2^n is positive for any integer n. [**Hint:** Use the definition $2^{-n} = \dfrac{1}{2^n}$.]

57. In this exercise, you will explore "0^0."
   a. Simplify $5^0, 4^0, 3^0, 2^0$, and 1^0. Based on these values, what would be a reasonable value for 0^0?
   b. Simplify $0^5, 0^4, 0^3, 0^2$, and 0^1. Based on these values, what would be a reasonable value for 0^0?
   c. Explain why it might be a good idea to leave 0^0 meaningless.

58. a. Simplify $\left(\dfrac{b}{c}\right)^{-2}$.

   b. Simplify $\left(\dfrac{b}{c}\right)^{-n}$.

   c. Use your result from part b to simplify $\left(\dfrac{b}{c}\right)^{-5}$ in one step.

59. Simplify each expression.

  a. b^{-1}

  b. $(b^{-1})^{-1}$

  c. $((b^{-1})^{-1})^{-1}$

  d. $(((b^{-1})^{-1})^{-1})^{-1}$

  e. $\underbrace{((((b^{-1})^{-1})^{-1})\cdots)^{-1}}_{n\ \text{exponents}}$

60. Some students confuse the properties $b^m b^n = b^{m+n}$ and $(b^m)^n = b^{mn}$. Explain why each property makes sense and compare the properties. Give examples to illustrate your comparison.

61. Describe what it means to use exponential properties to simplify an expression. Include several examples in your description.

4.2 RATIONAL EXPONENTS

Objectives

▸ Know definitions for rational exponents.

▸ Simplify expressions that have rational exponents.

▸ Use scientific notation.

Note

Recall that a *rational number* is a number of the form $\dfrac{m}{n}$, where m is an integer and n is a counting number.

In Section 4.1 we discussed the meaning of integer exponents. In this section we explore the meaning of rational exponents.

How should we define $b^{1/n}$, where n is a counting number? If the exponential property $(b^m)^n = b^{mn}$ is to be true for $m = \dfrac{1}{2}$ and $n = 2$, then

$$\left(9^{\frac{1}{2}}\right)^2 = 9^{\frac{1}{2}\cdot 2} = 9^1 = 9$$

Since $(-3)^2 = 9$ and $3^2 = 9$, the statement suggests that a good meaning for $9^{1/2}$ is -3 or 3. We define $9^{1/2} = 3$. We call 3 the *principal second root* or **principal square root** of 9, written $\sqrt{9}$.

Also, if the property $(b^m)^n = b^{mn}$ is to be true for $m = \dfrac{1}{3}$ and $n = 3$, then

$$\left(8^{\frac{1}{3}}\right)^3 = 8^{\frac{1}{3}\cdot 3} = 8^1 = 8$$

Since $2^3 = 8$, the statement suggests that a good meaning for $8^{1/3}$ is 2. The number 2 is called the *principal third root* or **principal cube root** of 8, written $\sqrt[3]{8}$.

For $(-8)^{1/3}$, a good meaning is -2, since $(-2)^3 = -8$. We do not assign a real number value to $(-9)^{1/2}$, since no real number squared is equal to -9.

DEFINITION $b^{1/n}$

For the following definition, n is a counting number.

- If $b \geq 0$, then $b^{1/n}$ is the nonnegative number whose nth power is b.
- If $b < 0$ and n is odd, then $b^{1/n}$ is the (negative) number whose nth power is b.

$b^{1/n}$ is called the **principal nth root of b** and may be represented by $\sqrt[n]{b}$.

Example 1   Simplifying Expressions with Rational Exponents

Simplify each expression.

1. $25^{1/2}$ **2.** $64^{1/3}$ **3.** $(-64)^{1/3}$

4. $16^{1/4}$ **5.** $-16^{1/4}$ **6.** $(-16)^{1/4}$

Solution

1. $25^{1/2} = 5$, since $5^2 = 25$.

2. $64^{1/3} = 4$, since $4^3 = 64$.

3. $(-64)^{1/3} = -4$, since $(-4)^3 = -64$.

4. $16^{1/4} = 2$, since $2^4 = 16$.

5. $-16^{1/4} = -(16^{1/4}) = -2$

6. $(-16)^{1/4}$ is not a real number since the fourth power of any real number is nonnegative.

Graphing calculator checks for problems 1, 2, and 3 are shown in Fig. 3.   ■

What would be a reasonable definition for $b^{m/n}$? If the exponential properties we discussed in Section 4.1 are to hold true for rational exponents, we have:

$$8^{\frac{2}{3}} = 8^{\frac{1}{3}\cdot 2} = \left(8^{\frac{1}{3}}\right)^2 = 2^2 = 4 \quad \text{or} \quad 8^{\frac{2}{3}} = 8^{2\cdot\frac{1}{3}} = (8^2)^{\frac{1}{3}} = 64^{\frac{1}{3}} = 4$$

Likewise,

$$32^{\frac{3}{5}} = 32^{\frac{1}{5}\cdot 3} = \left(32^{\frac{1}{5}}\right)^3 = 2^3 = 8 \quad \text{or} \quad 32^{\frac{3}{5}} = 32^{3\cdot\frac{1}{5}} = (32^3)^{\frac{1}{5}} = 32{,}768^{\frac{1}{5}} = 8$$

Also,

$$32^{-\frac{3}{5}} = \frac{1}{32^{\frac{3}{5}}} = \frac{1}{8}$$

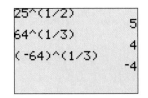

Figure 3 Checks for problems 1, 2, and 3

DEFINITION *Rational exponent*

For the following definition, m and n are counting numbers and b is any real number for which $b^{1/n}$ is a real number.

- $b^{m/n} = (b^{1/n})^m = (b^m)^{1/n}$
- $b^{-m/n} = \dfrac{1}{b^{m/n}}, \quad b \neq 0$

Example 2   Simplifying Expressions with Rational Exponents

Simplify each expression.

1. $25^{3/2}$ **2.** $(-27)^{2/3}$ **3.** $32^{-2/5}$ **4.** $(-8)^{-5/3}$

Solution

1. $25^{3/2} = (25^{1/2})^3 = 5^3 = 125$

2. $(-27)^{2/3} = ((-27)^{1/3})^2 = (-3)^2 = 9$

3. $32^{-2/5} = \dfrac{1}{32^{2/5}} = \dfrac{1}{(32^{1/5})^2} = \dfrac{1}{2^2} = \dfrac{1}{4}$

4. $(-8)^{-5/3} = \dfrac{1}{(-8)^{5/3}} = \dfrac{1}{((-8)^{1/3})^5} = \dfrac{1}{(-2)^5} = -\dfrac{1}{32}$

Graphing calculator checks for problems 1, 2, and 3 are shown in Fig. 4.   ■

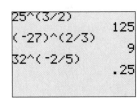

Figure 4 Checks for problems 1, 2, and 3

The exponential properties that we discussed in Section 4.1 are valid for *rational* exponents.

Properties of Exponents

For the following properties, m and n are rational numbers and b and c are any real numbers for which b^m, b^n, and c^n are real numbers.

- $b^m b^n = b^{m+n}$
- $\dfrac{b^m}{b^n} = b^{m-n}, \quad b \neq 0$
- $(bc)^n = b^n c^n$
- $\left(\dfrac{b}{c}\right)^n = \dfrac{b^n}{c^n}, \quad c \neq 0$
- $(b^m)^n = b^{mn}$

Recall these rules for adding, subtracting, and multiplying fractions:

1. $\dfrac{a}{b} \cdot \dfrac{c}{d} = \dfrac{ac}{bd}$, where b and d are nonzero numbers. For example,

$$\frac{5}{7} \cdot \frac{2}{3} = \frac{5 \cdot 2}{7 \cdot 3} = \frac{10}{21}$$

2. $\dfrac{a}{b} + \dfrac{c}{b} = \dfrac{a+c}{b}$ and $\dfrac{a}{b} - \dfrac{c}{b} = \dfrac{a-c}{b}$, where b is a nonzero number. For example,

$$\frac{5}{7} - \frac{3}{7} = \frac{5-3}{7} = \frac{2}{7}$$

3. When adding and subtracting fractions with different denominators, first find a common denominator. For example,

$$\frac{5}{6} + \frac{4}{9} = \frac{5}{6} \cdot \frac{3}{3} + \frac{4}{9} \cdot \frac{2}{2} = \frac{15}{18} + \frac{8}{18} = \frac{23}{18}$$

Example 3 Simplifying Expressions with Rational Exponents

Simplify each expression. Assume that b is positive.

1. $(4b^6)^{3/2}$ **2.** $\dfrac{b^{2/7}}{b^{-3/7}}$ **3.** $b^{2/3}b^{1/2}$ **4.** $\left(\dfrac{32b^2}{b^{12}}\right)^{2/5}$

Solution

1. $(4b^6)^{3/2} = 4^{3/2}(b^6)^{3/2}$ $(bc)^n = b^n c^n$

$\qquad = (4^{1/2})^3 b^{\frac{6}{1} \cdot \frac{3}{2}}$ $b^{m/n} = (b^{1/n})^m, \quad (b^m)^n = b^{mn}$

$\qquad = 2^3 b^{\frac{18}{2}}$

$\qquad = 8b^9$

2. $\dfrac{b^{2/7}}{b^{-3/7}} = b^{\frac{2}{7}-\left(-\frac{3}{7}\right)}$ $\dfrac{b^m}{b^n} = b^{m-n}$

$\qquad = b^{\frac{2}{7}+\frac{3}{7}}$

$\qquad = b^{5/7}$

3. $b^{2/3}b^{1/2} = b^{\frac{2}{3}+\frac{1}{2}}$ $b^m b^n = b^{m+n}$

$\qquad = b^{\frac{4}{6}+\frac{3}{6}}$ Find a common denominator.

$\qquad = b^{7/6}$

4. $\left(\dfrac{32b^2}{b^{12}}\right)^{2/5} = (32b^{2-12})^{2/5}$ $\dfrac{b^m}{b^n} = b^{m-n}$

$\qquad = (32b^{-10})^{2/5}$

$\qquad = \left(\dfrac{32}{b^{10}}\right)^{2/5}$ $b^{-n} = \dfrac{1}{b^n}$

$\qquad = \dfrac{32^{2/5}}{(b^{10})^{2/5}}$ $\left(\dfrac{b}{c}\right)^n = \dfrac{b^n}{c^n}$

$$= \frac{\left(32^{1/5}\right)^2}{b^{10 \cdot \frac{2}{5}}} \qquad b^{m/n} = \left(b^{1/n}\right)^m, \quad \left(b^m\right)^n = b^{mn}$$

$$= \frac{2^2}{b^4}$$

$$= \frac{4}{b^4} \qquad\qquad\qquad\qquad\qquad\qquad\qquad\qquad \blacksquare$$

Example 4 Simplifying an Expression with Rational Exponents

Simplify $\dfrac{\left(81b^6c^{20}\right)^{1/2}}{\left(27b^{12}c^9\right)^{2/3}}$. Assume that b and c are positive.

Solution

$$\frac{\left(81b^6c^{20}\right)^{1/2}}{\left(27b^{12}c^9\right)^{2/3}} = \frac{81^{1/2}\left(b^6\right)^{1/2}\left(c^{20}\right)^{1/2}}{27^{2/3}\left(b^{12}\right)^{2/3}\left(c^9\right)^{2/3}} \qquad (bc)^n = b^n c^n$$

$$= \frac{9b^{6 \cdot \frac{1}{2}}c^{20 \cdot \frac{1}{2}}}{\left(27^{1/3}\right)^2 b^{12 \cdot \frac{2}{3}} c^{9 \cdot \frac{2}{3}}} \qquad b^{m/n} = \left(b^{1/n}\right)^m, \quad \left(b^m\right)^n = b^{mn}$$

$$= \frac{9b^3 c^{10}}{3^2 b^8 c^6}$$

$$= \frac{9b^{-5}c^4}{9} \qquad\qquad \frac{b^m}{b^n} = b^{m-n}$$

$$= \frac{c^4}{b^5} \qquad\qquad \text{Reduce, } b^{-n} = \frac{1}{b^n} \qquad \blacksquare$$

For the remainder of this section, we will discuss how to use exponents to describe numbers in a form called *scientific notation*.

Example 5 Converting to Standard Decimal Notation

Simplify each expression.

1. 5×10^3 **2.** 5×10^{-3}

Solution

1. $5 \times 10^3 = 5 \times 1000 = 5000$

    We simplify $5 \times 10^3 = 5.0 \times 10^3$ by *multiplying* 5.0 by 10 three times and hence moving the decimal point three places to the *right*:

$$5.0 \times 10^3 = 5000.0 = 5000$$

three places to the right

2. $5 \times 10^{-3} = 5 \times \dfrac{1}{10^3} \qquad b^{-n} = \dfrac{1}{b^n}$

$$= \frac{5}{1} \times \frac{1}{1000}$$

$$= \frac{5}{1000}$$

$$= 0.005 \qquad \frac{5}{1000} \text{ is 5 thousandths.}$$

We simplify 5.0×10^{-3} by *dividing* 5.0 by 10 three times and hence moving the decimal point three places to the *left*:

$$5.0 \times 10^{-3} = 0.005$$

three places to the left ∎

In Example 5 we simplified the numbers 5×10^3 and 5×10^{-3}. Both of these numbers are in scientific notation. Here are more examples of numbers in scientific notation:

$$2.1 \times 10^{16} \qquad 7.345 \times 10^{27} \qquad -3.5 \times 10^{-15} \qquad 9.91 \times 10^{-49}$$

Note

N may equal 1 or −1, but not 10 or −10.

A number of the form $N \times 10^k$ is in **scientific notation** if N is a number between 1 and 10 (or −1 and −10) and k is an integer.

The problems in Example 5 suggest how to convert a number from scientific notation $N \times 10^k$ to standard decimal notation.

Converting from Scientific Notation to Standard Decimal Notation

To write the scientific notation $N \times 10^k$ in standard decimal notation, we move the decimal point of the number N as follows:

- If k is positive, move the decimal point k places to the right.
- If k is negative, move the decimal k places to the left.

Example 6 Converting to Standard Decimal Notation

Write each number in standard decimal notation by moving the decimal point.

1. 3.462×10^5 **2.** 7.38×10^{-4}

Solution

1. We *multiply* 3.462 by 10 five times and hence move the decimal point of 3.462 five places to the *right*:

$$3.462 \times 10^5 = 346{,}200.0$$

five places to the right

2. We *divide* 7.38 by 10 four times and hence move the decimal point of 7.38 four places to the *left*:

$$7.38 \times 10^{-4} = 0.000738$$

four places to the left ∎

Scientific notation allows us to write some extremely large or small numbers compactly.

Example 7 Converting to Scientific Notation

Write each number in scientific notation.

1. 6,257,000,000   **2.** 0.00000721

Solution

1. In scientific notation we would have

$$6.257 \times 10^k$$

If we move the decimal point of 6.257 nine places to the right, the result is 6,257,000,000. So, $k = 9$ and the scientific notation is

$$6.257 \times 10^9$$

2. In scientific notation we would have

$$7.21 \times 10^k$$

If we move the decimal point of 7.21 six places to the left, the result is 0.00000721. So, $k = -6$ and the scientific notation is

$$7.21 \times 10^{-6} \qquad \blacksquare$$

The problems in Example 7 suggest how to convert a number from standard decimal notation to scientific notation.

Converting from Standard Decimal Notation to Scientific Notation

To write a number in scientific notation, count the number of places k that the decimal point needs to be moved so that the new number N is between 1 and 10 (or -1 and -10).

- If the decimal point is moved to the left, then scientific notation for the original number is $N \times 10^k$.
- If the decimal point is moved to the right, then scientific notation for the original number is $N \times 10^{-k}$.

Example 8   Converting to Scientific Notation

Write each number in scientific notation.

1. 48,100,000   2. 0.000629

Solution

1. For 48,100,000, the decimal point needs to be moved 7 places to the left so that the new number is between 1 and 10. Therefore, the scientific notation is 4.81×10^7.

2. For 0.000629, the decimal point needs to be moved 4 places to the right so that the new number is between 1 and 10. Therefore, the scientific notation is 6.29×10^{-4}. $\qquad \blacksquare$

Calculators express numbers in scientific notation so that the numbers "fit" on the screen. To represent 6.023×10^{23}, most calculators use the notation 6.023 E 23, where E stands for exponent (of 10). Calculators represent 2.493×10^{-50} by 2.493 E -50 (see Fig. 5).

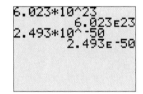

Figure 5 The numbers 6.023×10^{23} and 2.493×10^{-50}

EXPLORATION

Looking ahead: Graphical significance of a and b for y = abˣ

1. In this problem, you will explore the graphical significance of the constant b for functions of the form $y = b^x$. Use ZDecimal to graph

$$y = 1.2^x$$
$$y = 1.5^x$$
$$y = 2^x$$
$$y = 5^x$$

Graphing Calculator

To change window settings, see Section B.7.

in order and describe any patterns you observe. If you want a better view, set $Y\min = 0$.
Do the same with:
$$y = 0.3^x$$
$$y = 0.5^x$$
$$y = 0.7^x$$
$$y = 0.9^x$$

2. In this problem, you will explore the graphical significance of the constant a for functions of the form $y = a(1.1)^x$. Use ZStandard to graph
$$y = 2(1.1)^x$$
$$y = 3(1.1)^x$$
$$y = 4(1.1)^x$$
$$y = 5(1.1)^x$$

in order and describe any patterns that you observe. If you want a better view, set $Y\min = 0$.

Use ZStandard to do the same with:
$$y = -2(1.1)^x$$
$$y = -3(1.1)^x$$
$$y = -4(1.1)^x$$
$$y = -5(1.1)^x$$

If you want a better view, set $Y\max = 0$.

3. So far, you have sketched the graphs of equations of only the forms $y = b^x$ (where $a = 1$) and $y = a(1.1)^x$ (where $b = 1.1$). Graph more equations of the form $y = ab^x$ until you are confident you know the graphical significance of the constants a and b, for any possible combination of values for a and b. If you have any new insights about the graphical significance of the constants a and b, describe them.

Note

Recall that "form a conjecture" means to make a guess or form a theory.

4. Use problems 1–3 to form a conjecture about the graph of the equation $y = ab^x$ in the following situations. Test each conjecture.

    a. a is positive   **b.** a is negative
    c. $b > 1$ **d.** $0 < b < 1$
    e. $b = 1$ **f.** b is negative

5. Describe the connection between the y-intercept of $y = ab^x$ and the values of a and b.

 EXPLORATION

Looking ahead: Numerical significance of a and b for $f(x) = ab^x$

In this exploration you will investigate the nature of exponential functions of the form $f(x) = ab^x$.

1. Use your graphing calculator to create a table of ordered pairs for $f(x) = 2(3)^x$, $g(x) = 64\left(\dfrac{1}{2}\right)^x$, and a third exponential function of your choice. (See Figs. 6 and 7.) Use the following values for the x-coordinates: $0, 1, 2, \ldots, 6$.

2. **a.** What connection do you notice between the y-coordinates of each function and the base b of the function $y = ab^x$?

    b. Test your conjecture by choosing yet another exponential function, and check whether it behaves in the way you think it should.

Figure 6 Enter the three functions

Figure 7 Table setup

   c. For $f(x) = ab^x$, we have $f(0) = a$, $f(1) = ab$, $f(2) = abb$, and $f(3) = abbb$. Explain why these results suggest that your conjecture is correct.

3. a. What connection do you notice between the y-coordinates of each function and the coefficient a of the function $y = ab^x$?

   b. Test your conjecture by choosing yet another exponential function, and check whether it behaves in the way you think it should.

   c. Use pencil and paper to find $f(0)$, where $f(x) = ab^x$. Explain why your result shows that your conjecture is correct.

Tips on Succeeding in This Course

If you work an exercise by referring to a similar example in your notebook or the text, it is a good idea to try the exercise again without referring to your source of help. If you find that you need to refer to your source of help to solve the exercise a second time, consider trying the exercise a third time without help. When you successfully complete the exercise without help, take a moment to reflect on which concepts you used to work the exercise, where you had difficulty, and what the key idea was that opened the door of understanding for you.

   A similar strategy can be used in getting help from a student, instructor, or tutor.

   If this sounds like a lot of work, it is! But this work is well worth it. Although it is important to complete each assignment, it is also important to learn as much as possible while progressing through it.

Key Points

OF THIS SECTION

For the following definition, n is a counting number.

- If $b \geq 0$, then $b^{1/n}$ is the nonnegative number whose nth power is b.
- If $b < 0$ and n is odd, then $b^{1/n}$ is the (negative) number whose nth power is b.

Definition of $b^{1/n}$

For the following definition, m and n are counting numbers and b is any real number for which $b^{1/n}$ is a real number.

- $b^{m/n} = \left(b^{1/n}\right)^m = \left(b^m\right)^{1/n}$
- $b^{-m/n} = \dfrac{1}{b^{m/n}}, \quad b \neq 0$

Definition of $b^{m/n}$ and $b^{-m/n}$

For the following properties, m and n are rational numbers and b and c are any real numbers for which b^m, b^n, and c^n are real numbers.

- $b^m b^n = b^{m+n}$
- $\dfrac{b^m}{b^n} = b^{m-n}, \quad b \neq 0$
- $(bc)^n = b^n c^n$

Exponential properties

$$\bullet \ \left(\frac{b}{c}\right)^n = \frac{b^n}{c^n}, \quad c \neq 0$$

$$\bullet \ (b^m)^n = b^{mn}$$

Converting to standard decimal notation

To write the scientific notation $N \times 10^k$ in standard decimal notation, move the decimal point of the number N as follows:

- If k is positive, move the decimal point k places to the right.
- If k is negative, move the decimal point k places to the left.

Converting to scientific notation

To write a number in scientific notation, count the number of places k that the decimal point needs to be moved so that the new number N is between 1 and 10 (or -1 and -10).

- If the decimal point is moved to the left, then the scientific notation of the original number is $N \times 10^k$.
- If the decimal point is moved to the right, then the scientific notation of the original number is $N \times 10^{-k}$.

HOMEWORK 4.2

SSM
Tutor Center

MathPro 4

MathPro 5

CDs/Videos

www

Simplify each of the following without using a calculator. Then use a calculator to verify your result. [**Graphing Calculator:** Instructions for $x^{m/n}$: Press $\boxed{\land}$ $\boxed{(}$ $\boxed{m}$ $\boxed{\div}$ $\boxed{n}$ $\boxed{)}$.]

1. $16^{1/2}$
2. $27^{1/3}$
3. $1000^{1/3}$
4. $32^{1/5}$
5. $49^{1/2}$
6. $81^{1/4}$
7. $125^{1/3}$
8. $64^{1/6}$
9. $8^{4/3}$
10. $16^{3/4}$
11. $9^{3/2}$
12. $64^{2/3}$
13. $32^{2/5}$
14. $27^{4/3}$
15. $4^{5/2}$
16. $81^{3/4}$
17. $27^{-1/3}$
18. $16^{-1/4}$
19. $-36^{-1/2}$
20. $-32^{-1/5}$
21. $4^{-5/2}$
22. $9^{-3/2}$
23. $(-27)^{-4/3}$
24. $(-32)^{-3/5}$

Simplify without using a calculator. Then use a calculator to verify your result.

25. $2^{1/4}2^{3/4}$
26. $\dfrac{5^{4/3}}{5^{1/3}}$
27. $(3^{1/2})^2$
28. $3^{7/5}3^{3/5}$
29. $\dfrac{7^{1/3}}{7^{-5/3}}$
30. $(2^{2/3}5^{1/3})^3$

Simplify each expression. Assume that b and c are positive.

31. $b^{7/6}b^{5/6}$
32. $b^{1/5}b^{3/5}$
33. $b^{3/5}b^{-13/5}$
34. $b^{2/7}b^{-6/7}$
35. $(16b^8)^{1/4}$
36. $(27b^{27})^{1/3}$

37. $4(25b^8c^{14})^{-1/2}$
38. $-(8b^{-6}c^{12})^{2/3}$
39. $(b^{3/5}c^{-1/4})(b^{2/5}c^{-7/4})$
40. $(b^{-2.1}c^{7.8})(b^{5.4}c^{-10.3})$
41. $(5bcd)^{1/5}(5bcd)^{4/5}$
42. $(3b^{1/2}c^{1/3})(2b^{-3/2}c^{2/3})$
43. $(8b^6c^9)^{1/3}(16b^{20}c^4)^{1/2}$
44. $[(8b^3)^2(b^2c^8)]^{1/4}$
45. $\dfrac{b^{5.6}c^{-3.8}}{b^{2.3}c^{-1.3}}$
46. $\dfrac{b^{3/4}c^{1/2}}{b^{-1/4}c^{-1/2}}$
47. $\left(\dfrac{9b^3c^{-2}}{25b^{-5}c^2}\right)^{1/2}$
48. $\left(\dfrac{16b^{12}c^2}{2b^{-3}c^{-4}}\right)^{-1/3}$
49. $b^{3/7}b^{2/5}$
50. $16^{1/4}b^{1/4}b^{1/2}$
51. $\dfrac{b^{5/6}}{b^{1/4}}$
52. $\dfrac{b^{-2/3}}{b^{1/7}}$
53. $\left(\dfrac{8b^{2/3}}{2b^{4/5}}\right)^{3/2}$
54. $\left(\dfrac{27b^{1/3}c^{3/4}}{8b^{-2/3}c^{1/2}}\right)^{4/3}$
55. $\dfrac{(8bc^3)^{1/3}}{(81bc^3)^{1/4}}$
56. $\dfrac{(1000b^{-7}c^8)^{2/3}}{(32b^{15}c^4)^{3/5}}$

Write each number in standard decimal form.

57. 3.965×10^2
58. 8.23172×10^3
59. 2.39×10^{-1}
60. 7.46×10^{-3}
61. 5.2×10^2
62. 7.74×10^6
63. 9.113×10^{-5}
64. 7.3558×10^{-2}
65. -6.52×10^5
66. -3.006×10^{-3}
67. 9×10^5
68. 4×10^3
69. -8×10^{-2}
70. -6.1×10^0

Write each number in scientific notation.

71. 5426

72. 173,229

73. 23,587

74. 6,541,883

75. 0.00098

76. 0.08156

77. 0.0000346

78. 0.827

79. −42,215

80. −647

81. −0.008928

82. 14

83. 8.0

84. 60,000

85. −100,000

86. −0.000001

Each of the following sentences contains a number in scientific notation. Write the number in standard decimal form.

87. The U.S. population was about 2.87×10^8 in 2002.

88. The moon has a mean distance from Earth of approximately 2.389×10^5 miles.

89. The hydrogen ion concentration of human blood is about 6.3×10^{-8} moles per liter.

90. One inch is approximately 1.58×10^{-5} mile.

91. The faintest sound that humans are able to hear has an intensity of about 10^{-12} watt per square meter.

Each of the sentences in exercises 92–96 contains a number (other than a date). Write the number in scientific notation.

92. In 1989 the tanker *Exxon Valdez* spilled about 10,080,000 gallons of oil in Prince William Sound, Alaska.

93. In 1556 approximately 830,000 people died in an earthquake in Shensi, China.

94. The age of Earth is about 4,500,000,000 years.

95. The wavelength of violet light is approximately 0.00000047 meter.

96. One second is about 0.0000000317 year.

97. We can represent $\sqrt{5}$ by $5^{1/2}$. Explain.

98. To write the scientific notation $N \times 10^k$ in standard decimal notation, we move the decimal point of the number N k places to the right if k is positive. Explain.

99. List the exponent definitions and properties that are discussed in this section and Section 4.1. Also, explain how you can recognize which definition or property will help you simplify a given expression.

4.3 SKETCHING GRAPHS OF EXPONENTIAL FUNCTIONS

Objectives

▸ Know the meaning of an *exponential function*.

▸ Know the graphical significance of a and b for a function of the form $f(x) = ab^x$.

▸ Sketch the graph of an exponential function.

Now that we have discussed the meanings of exponents, we apply our understanding of exponents to sketch the graphs of *exponential functions*. Here are some examples of such functions:

$$f(x) = 2(3)^x, \qquad g(x) = -7\left(\frac{1}{2}\right)^x, \qquad h(x) = 5^x$$

Notice that for exponential functions, the variable appears as an exponent.

DEFINITION Exponential function

An **exponential function** is a function whose equation can be put into the form

$$f(x) = ab^x$$

where $a \neq 0$, $b > 0$, and $b \neq 1$. The constant b is called the **base**.

Note

Note that if $a = 0$, then $f(x) = 0 \cdot b^x = 0$; so f is linear. Also, if $b = 1$ then $f(x) = a \cdot 1^x = a$; so f is linear. You will explore why we want $b > 0$ in exercise 84.

Example 1   Graphing an Exponential Function with $b > 1$

Sketch the graph of $f(x) = 2^x$.

Solution

First, we list input-output pairs of the function f in Table 1. Note that as the value of x increases by 1, the value of y is multiplied by 2 (the base).

Next, we plot the points from Table 1 in Fig. 8.

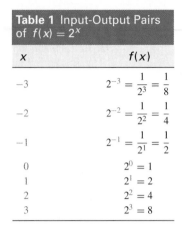

Table 1 Input-Output Pairs of $f(x) = 2^x$

x	$f(x)$
-3	$2^{-3} = \dfrac{1}{2^3} = \dfrac{1}{8}$
-2	$2^{-2} = \dfrac{1}{2^2} = \dfrac{1}{4}$
-1	$2^{-1} = \dfrac{1}{2^1} = \dfrac{1}{2}$
0	$2^0 = 1$
1	$2^1 = 2$
2	$2^2 = 4$
3	$2^3 = 8$

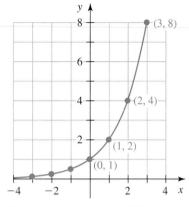

Figure 8 Graph of $y = 2^x$

Next, in Fig. 8, we sketch an increasing curve that contains the plotted points. Note how the graph shows that as the value of x increases by 1, the value of y is doubled.

We can set up a WINDOW to verify our graph (see Figs. 9–11).

Graphing Calculator

See Appendix B.7.

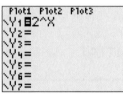

Figure 9 Enter the function

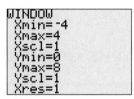

Figure 10 Set up the window

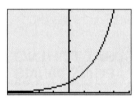

Figure 11 Graph the function

Sketching a smooth curve in Fig. 8 implies that 2^x has meaning for *any* real number exponent x. This is, indeed, true. In fact, real number exponents can be defined so that the graph of any exponential function is a smooth graph. Also, the exponential properties that we have discussed for rational exponents also apply for real number exponents. We can use a calculator to find *real number powers* of numbers.

Recall that points that lie on an equation's graph satisfy the equation, and points that do not lie on the graph do not satisfy the equation. The graph of an exponential function is called an **exponential curve**.

Example 2 Graphing an Exponential Function with $0 < b < 1$

Sketch the graph of $g(x) = 4\left(\dfrac{1}{2}\right)^x$.

Solution

Input-output pairs for g are listed in Table 2. For example,

$$g(-1) = 4\left(\frac{1}{2}\right)^{-1} = 4\left(\frac{1}{2^{-1}}\right) = 4(2^1) = 8.$$

So, $(-1, 8)$ is an input-output pair. Note that as the value of x increases by 1, the value of y is multiplied by $\dfrac{1}{2}$.

We plot the found points in Fig. 12.

Then, in Fig. 12, we sketch a decreasing exponential curve that contains the plotted points. Note how the graph shows that as the value of x increases by 1, the value of y is halved.

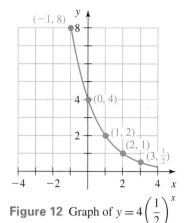

Figure 12 Graph of $y = 4\left(\dfrac{1}{2}\right)^x$

Table 2 Input-Output Pairs of $g(x) = 4\left(\frac{1}{2}\right)^x$

x	$g(x)$
-1	8
0	4
1	2
2	1
3	$\dfrac{1}{2}$

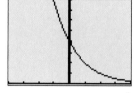

Figure 13 Graph of $y = 4\left(\dfrac{1}{2}\right)^x$

We can use a graphing calculator to verify our graph (see Fig. 13).

Examples 1 and 2 suggest the *base multiplier property* and the *increasing or decreasing property*.

Base Multiplier Property

For an exponential function of the form $y = ab^x$, if the value of the independent variable increases by 1, the value of the dependent variable is multiplied by b.

We have seen two examples of this property in Examples 1 and 2. Here are two more examples of the base multiplier property.

1. For the function $f(x) = 2(3)^x$, as the value of x increases by 1, the value of y is multiplied by 3.

2. For the function $f(x) = 5\left(\dfrac{3}{4}\right)^x$, as the value of x increases by 1, the value of y is multiplied by $\dfrac{3}{4}$.

To prove the base multiplier property for the exponential function $f(x) = ab^x$, we compare outputs for the inputs k and $k + 1$, which are 1 apart:

$$f(k) = ab^k \qquad \begin{aligned} f(k + 1) &= ab^{k+1} \\ &= ab^k b^1 \\ &= f(k)b \end{aligned}$$

Since $f(k + 1) = f(k)b$, we conclude that if the value of the independent variable increases by 1, the value of the dependent variable is multiplied by b, which is what we set out to show.

For the increasing or decreasing property, we note in Example 1 that the base b is greater than 1 and the graph is increasing. In Example 2, we note that the positive base is less than 1 and the graph is decreasing. For $f(x) = ab^x$ with $a > 0$, in general, we have the property that each multiplication by a base greater than 1 gives a larger value for y, whereas each multiplication by a positive base less than 1 gives a smaller value for y.

Increasing or Decreasing Property

Let $f(x) = ab^x$, where $a > 0$.

- If $b > 1$, then the function f is increasing. We say the function **grows exponentially** (see Fig. 14).

(Continued)

- If $0 < b < 1$, then the function f is decreasing. We say the function **decays exponentially** (see Fig. 15).

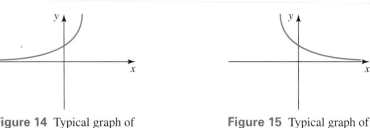

Figure 14 Typical graph of $f(x) = ab^x$, where $a > 0$ and $b > 1$

Figure 15 Typical graph of $f(x) = ab^x$, where $a > 0$ and $0 < b < 1$

When sketching the graph of an exponential function, it is helpful to first plot the y-intercept. Substituting 0 for x in the general equation $y = ab^x$ gives

$$y = ab^0 = a(1) = a.$$

So, the y-intercept is $(0, a)$.

y-intercept of an Exponential Function

For an exponential function of the form

$$y = ab^x,$$

the y-intercept is $(0, a)$.

For the function $y = 5(8)^x$, the y-intercept is $(0, 5)$. For the function $y = 4\left(\frac{1}{7}\right)^x$, the y-intercept is $(0, 4)$.

Example 3   Intercepts and Graph of an Exponential Function

Let $f(x) = 6\left(\frac{1}{2}\right)^x$.

1. Find the y-intercept of f.
2. Find the x-intercepts of f.
3. Sketch the graph of f.

Solution

1. Since $f(x) = 6\left(\frac{1}{2}\right)^x$ is of the form $f(x) = ab^x$, we know that the y-intercept is $(0, a)$, or $(0, 6)$.

2. By the base multiplier property, we know that as the value of x increases by 1, the value of y is multiplied by $\frac{1}{2}$ (see Table 3).

   When we halve a number, it becomes smaller. But no number of halvings will give a result that is zero. So, as x grows large, y will become extremely close to, but never equal, 0. Likewise, the graph of f gets arbitrarily close to, but never reaches, the x-axis (see Fig. 16). In this case, we call the x-axis a *horizontal asymptote*. We conclude that the function f does not have any x-intercepts.

3. We plot the points in Table 3 and sketch a decreasing exponential curve that contains the five points (see Fig. 16). If we had not already found a table of points, we could have plotted the y-intercept and plotted additional points by increasing the value of x by 1 and going half as high for the value of y each time.

Table 3 Input-Output Pairs of $f(x) = 6\left(\frac{1}{2}\right)^x$

x	$f(x)$
0	6
1	3
2	$\frac{3}{2}$
3	$\frac{3}{4}$
4	$\frac{3}{8}$

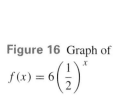

Figure 16 Graph of
$$f(x) = 6\left(\frac{1}{2}\right)^x$$

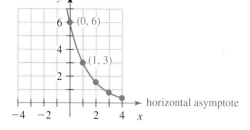

As a check, we note that according to the increasing or decreasing property, the function f is decreasing since the base $\frac{1}{2}$ is between 0 and 1. For a more thorough check, we can use a graphing calculator to verify our graph (see Fig. 17). ■

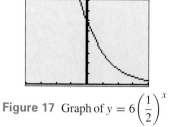

Figure 17 Graph of $y = 6\left(\frac{1}{2}\right)^x$

For each exponential function, the x-axis is a horizontal asymptote (see Fig. 18).

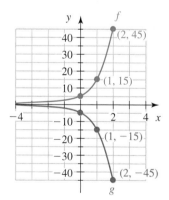

Figure 18 For each exponential function, the x-axis is a horizontal asymptote

Example 4 Graphs of Functions of the Form $y = ab^x$ and $y = -ab^x$

Sketch and compare the graphs of $f(x) = 5(3)^x$ and $g(x) = -5(3)^x$.

Solution

Input-output pairs for f and g are listed in Table 4 and plotted in Fig. 19.

In Table 4 we see that for each value of x, the outputs of g are the opposites of the outputs of f. Because of this, the graph of g is the "mirror reflection" of the graph of f with the mirror placed along the x-axis. We can find the graph of g by *reflecting* the graph of f *across the x-axis*.

Table 4 Input-Output Pairs of $f(x) = 5(3)^x$ and $g(x) = -5(3)^x$

x	$f(x)$	$g(x)$
0	5	−5
1	15	−15
2	45	−45
3	135	−135
4	405	−405

Figure 19 Graphs of $f(x) = 5(3)^x$ and $g(x) = -5(3)^x$

■

Reflection Property

The graphs of $f(x) = -ab^x$ and $g(x) = ab^x$ are reflections of each other across the x-axis.

We illustrate the reflection property as well as summarize four types of exponential curves in Figs. 20 and 21. Note that for all exponential functions, the x-axis is a horizontal asymptote.

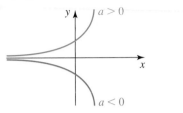

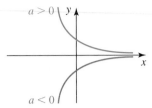

Figure 20 Typical graphs of $f(x) = ab^x$, $b > 1$

Figure 21 Typical graphs of $f(x) = ab^x$, $0 < b < 1$

Recall that for $b > 0$, b^x has meaning for any real number exponent x. So, the domain of any exponential function $f(x) = ab^x$ is the set of real numbers.

Further, Figs. 20 and 21 show that $f(x) = ab^x$ has positive outputs if $a > 0$ and negative outputs if $a < 0$. Therefore, the range of $f(x) = ab^x$ is the set of all positive real numbers if $a > 0$ and the range is the set of all negative real numbers if $a < 0$.

In Example 5 we use the graph of an exponential function f to find input or output values of f.

Example 5   Finding Values of a Function from Its Graph

The graph of f is sketched in Fig. 22.

1. Find $f(2)$.   **2.** Find x when $f(x) = 1$.

3. Find x when $f(x) = 0$.

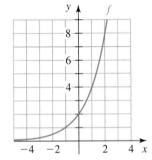

Figure 22 Graph of f

Solution

1. Since the point $(2, 8)$ lies on the graph, we conclude that $f(2) = 8$.

2. Since the point $(-1, 1)$ is the only point on the graph with $f(x) = 1$, we conclude that $x = -1$.

3. Although it may *appear* that the graph of f intersects the x-axis, recall that the graph of an exponential function gets close to, but never reaches the x-axis. So, there is no value of x where $f(x) = 0$.   ∎

EXPLORATION
Drawing families of exponential curves

For each problem, use your graphing calculator to graph a family of curves.

1. List the equations of a family of exponential curves like the ones shown in Fig. 23.

2. List the equations of a family of exponential curves like the ones shown in Fig. 24. All of these curves pass through the point $(0, 2)$.

3. Summarize what you have learned from this exploration and this section about the coefficient a and the base b for functions of the form $f(x) = ab^x$.

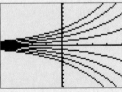

Figure 23 A family of exponential curves

Figure 24 A family of exponential curves passing through $(0, 2)$

EXPLORATION
Looking ahead: Using trial and error to find a model

In exercise 5 of Homework 2.1, you found that the number of Internet users in the United States can be modeled by a linear function. Here, you will model the number of Internet users *worldwide* (see Table 5).

Let $n = f(t)$ represent the number of users (in millions) at t years since 1995.

1. Use a graphing calculator to draw a scattergram of the data. Would it be better to model the data using a linear or exponential function? Explain.

2. Imagine an exponential function $f(t) = ab^t$ whose graph comes close to the data points in your scattergram. What is the n-intercept? What does this tell you about the value of a or b? Explain.

3. Guess a reasonable value for b for your function $f(t) = ab^t$. [**Hint**: The base multiplier property may help you guess.]

4. Substitute your values for a and b from problems 2 and 3 into the equation $f(t) = ab^t$.

5. Graph f and the scattergram in the same viewing window to see how well your model fits the data.

6. Now find better values of a and b through trial and error. When you are satisfied with your values for a and b, write the equation for f that you have found.

Table 5 Numbers of Internet Users Worldwide

Year	Number of Users (millions)
1995	34
1996	54
1997	90
1998	149
1999	230
2000	311
2001	457

Source: *ITU, www.itu.int/osg/dsg/ speeches/2001/16belgradel.pdf. Accessed 3/02/02.*

Tips on Succeeding in This Course

After writing a response to a question, read your response to make sure that it says what you intended it to say. Reading your response aloud, if you are in a place where you feel comfortable doing this, can help.

Key Points

OF THIS SECTION

An *exponential function* is a function whose equation can be put into the form $f(x) = ab^x$, where $a \neq 0$, $b > 0$, and $b \neq 1$. The constant b is called the *base*.

Exponential function

For the exponential function $f(x) = ab^x$:

Properties of an exponential function

- If the value of x increases by 1, then the value of y is multiplied by the base b.
- The y-intercept is $(0, a)$.
- The domain is the set of all real numbers.
- If $a > 0$, then the graph of f is above the x-axis and the range is the set of all positive real numbers.
- If $a < 0$, then the graph of f is below the x-axis and the range is the set of all negative real numbers.
- If $a > 0$ and $b > 1$, then f is an increasing function.
- If $a > 0$ and $0 < b < 1$, then f is a decreasing function.
- The x-axis is a horizontal asymptote.

The graphs of $f(x) = -ab^x$ and $g(x) = ab^x$ are mirror reflections of each other across the x-axis.

Reflection property

To sketch a graph of $y = ab^x$:

Graphing an exponential function

- Plot the y-intercept.

- Plot other points found by substituting values for x in the equation $y = ab^x$ and solving for y.
- Use your knowledge of the shape of an exponential curve to sketch a curve that passes through the plotted points.

Meaning of a graph

A point that lies on the graph of an equation satisfies the equation. A point that does not lie on the graph of an equation does not satisfy the equation.

HOMEWORK 4.3

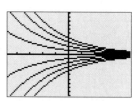

SSM
Tutor Center

MathPro 4
MathPro 4

MathPro 5
MathPro 5

CDs/Videos

www

Sketch the graph of the given function. Then use your graphing calculator to verify your graph.

1. $y = 3^x$

2. $y = 4^x$

3. $y = 10^x$

4. $y = 5^x$

5. $y = 3(2)^x$

6. $y = 2(3)^x$

7. $y = 6(3)^x$

8. $y = 3(5)^x$

9. $y = 15\left(\dfrac{1}{3}\right)^x$

10. $y = 20\left(\dfrac{1}{4}\right)^x$

11. $y = 12\left(\dfrac{1}{2}\right)^x$

12. $y = \left(\dfrac{2}{3}\right)^x$

Sketch the graph of both functions on the same coordinate system. Then use your graphing calculator to verify your graphs.

13. $f(x) = 2^x$, $g(x) = -2^x$

14. $f(x) = 3^x$, $g(x) = -3^x$

15. $f(x) = 4(3)^x$, $g(x) = -4(3)^x$

16. $f(x) = 2(10)^x$, $g(x) = -2(10)^x$

17. $f(x) = 8\left(\dfrac{1}{2}\right)^x$, $g(x) = -8\left(\dfrac{1}{2}\right)^x$

18. $f(x) = 6\left(\dfrac{1}{3}\right)^x$, $g(x) = -6\left(\dfrac{1}{3}\right)^x$

19. Four functions of the form $y = ab^x$ have been sketched in Fig. 25. Describe the constants a and b for each function.

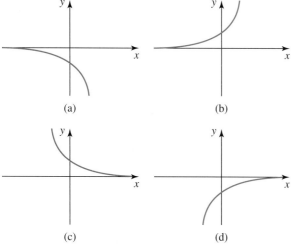

(a)

(b)

(c)

(d)

Figure 25 Exercise 19

20. Functions $f(x) = ab^x$ and $g(x) = cd^x$ are sketched in Fig. 26.

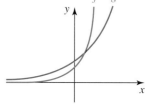

Figure 26 Graphs of $f(x) = ab^x$ and $g(x) = cd^x$

a. Which coefficient is greater, a or c? Explain.

b. Which base is greater, b or d? Explain.

21. Use your graphing calculator to graph a family of exponential curves similar to the family graphed in Fig. 27. List the equations of your family.

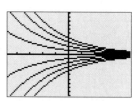

Figure 27 A family of exponential curves

22. Use your graphing calculator to graph a family of exponential curves similar to the family graphed in Fig. 28. All of these curves pass through the point $(0, -2)$. List the equations of your family.

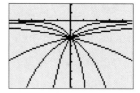

Figure 28 A family of exponential curves passing through $(0, -2)$

23. Use your graphing calculator to draw a graph similar to the one in Fig. 29. Use an equation of the form $f(x) = ab^x$, where a and b are constants that you specify. What is the equation that works?

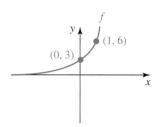

Figure 29 Exercise 23

24. Use your graphing calculator to draw a graph similar to the one in Fig. 30. Use an equation of the form $g(x) = ab^x$, where a and b are constants that you specify. What is the equation that works? Use trial and error.

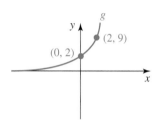

Figure 30 Exercise 24

25. Find equations of exponential functions that could correspond to the graphs sketched in Fig. 31.

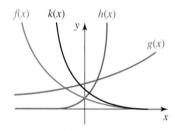

Figure 31 Exercise 25

26. a. Simplify and compare both expressions.

   i. $4(3)^2$ and 12^2

   ii. $4^2 \cdot 3^2$ and 12^2

b. Build a graphing calculator table that shows the same input values for both functions. Explain in terms of order of operations or exponential properties why the tables are the same or different.

   i. $f(x) = 4(3)^x$ and $g(x) = 12^x$

   ii. $f(x) = 4^x \cdot 3^x$ and $g(x) = 12^x$

27. Some input-output pairs of the functions f, g, h, and k are provided in Table 6. For each function, determine whether the given values suggest that the function is linear, exponential, or neither.

Table 6 Identifying Functions (Exercise 27)

x	f(x)	g(x)	h(x)	k(x)
0	13	4	48	5
1	9	12	24	55
2	5	36	12	555
3	1	108	6	5555
4	−3	324	3	55555

28. Input-output pairs for four exponential functions are listed in Table 7. Complete the table.

Table 7 Complete the Table (Exercise 28)

x	f(x)	g(x)	h(x)	k(x)
0	3	64	2	100
1	6	32	6	10
2	12	16		
3	24			
4				

29. Input-output pairs for four exponential functions are listed in Table 8. Complete the table.

Table 8 Complete the Table (Exercise 29)

x	f(x)	g(x)	h(x)	k(x)
0	5			
1		80	54	
2	20			
3		20		192
4			2	768

30. Input-output pairs for four exponential functions are listed in Table 9. Complete the table.

Table 9 Complete the Table (Exercise 30)

x	f(x)	g(x)	h(x)	k(x)
0			3	400
1		3		
2	25			
3		147		
4	1		30000	25

For exercises 31–38, refer to Fig. 32.

31. Find $f(-3)$.

32. Find $f(-1)$.

33. Find $f(0)$.

34. Find $f(1)$.

35. Find x when $f(x) = 4$.

36. Find x when $f(x) = 2$.

37. Find x when $f(x) = 1$.

38. Find x when $f(x) = -2$.

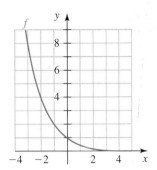

Figure 32 Graph of
f—exercises 31–38

For exercises 39–46, refer to Table 10.

Table 10 Some Values of an Exponential Function *f*

x	f(x)
0	3
1	6
2	12
3	24
4	48
5	96
6	192

39. Find $f(3)$.

40. Find $f(6)$.

41. Find $f(5)$.

42. Find $f(0)$.

43. Find x when $f(x) = 3$.

44. Find x when $f(x) = 6$.

45. Find x when $f(x) = 24$.

46. Find x when $f(x) = 96$.

47. The most expensive slot for television advertising occurs during the Super Bowl. For the 2002 Super Bowl, a 30-second slot cost $1.9 million, which works out to about $63,333 per second! The prices of most products increase over time due to inflation. What's remarkable is that the costs of 30-second ad slots have increased greatly even when adjusted for inflation (see Table 11).

Table 11 Costs of Television Ad Slots During the Super Bowl

Super Bowl	Year	Cost* for 30 Seconds (millions of dollars)
V	1971	0.3
X	1976	0.3
XV	1981	0.6
XX	1986	0.8
XXV	1991	1.0
XXX	1996	1.2
XXXIV	2000	2.0
XXXV	2001	2.3

Source: *NFL Research*

*Adjusted for inflation

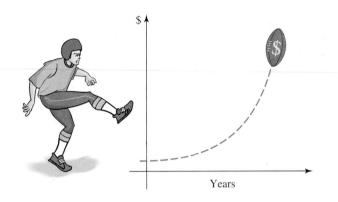

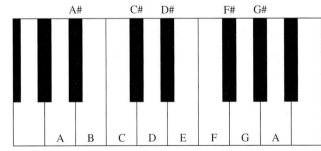

a. A linear model for the cost (in millions of dollars) of a 30-second ad slot is $L(t) = 0.062t - 0.66$ and an *exponential model* is $E(t) = 0.127(1.07)^t$, where *t* is the number of years since 1960. Draw the graph of *L*, the graph of *E*, and the scattergram of the data in the same viewing window. Describe how close each graph comes to the points in the scattergram. Also, ZOOM OUT and decide which model appears to make better estimates for the years before 1971.

b. Use *L* and *E* to estimate the cost of a 30-second ad slot during Super Bowl I in 1967. The actual cost (adjusted for inflation) was $200,000. Which function does a better job of modeling the situation for 1967?

c. In Super Bowl XXXVI in 2002, the cost of a 30-second ad slot was $1.9 million, the first decline in cost ever. The decline was likely due to the effects of a poor economy, including far fewer dot-com companies bidding for the game's 60 30-second ad slots. Use *E* to help estimate the total loss in revenue from ad slots during Super Bowl XXXVI due to the poor economy. [**Hint**: First find $E(42)$.]

48. If you place your hand on a piano and play a note, you will feel the piano vibrate. The number of vibrations per second (hertz) of a note is called its **frequency**. If you strike the keys on a piano from left to right, the frequencies of the notes increase. We use some of the letters of the alphabet, sometimes in conjunction with the "sharp" symbol ♯, to refer to these notes (see Fig. 33). The frequencies for 13 notes in a row are listed in Table 12.

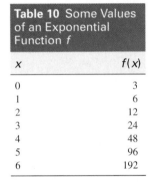

Figure 33 Names of notes on a piano

a. Let $f(n)$ represent the frequency (in hertz) of the note that is *n* notes above the note A (the one with frequency 220 hertz). Draw a scattergram of the data listed in Table 12. Will a linear function or an exponential function best model the data?

Table 12 Frequencies of Notes on a Piano

Note	Number of Notes above A	Frequency (in hertz)
A	0	220.0
A♯	1	233.1
B	2	246.9
C	3	261.6
C♯	4	277.2
D	5	293.7
D♯	6	311.1
E	7	329.6
F	8	349.2
F♯	9	370.0
G	10	392.0
G♯	11	415.3
A	12	440.0

Source: Math and Music *by Garland and Kahn*

b. Draw the graph of the function $f(n) = 220(2)^{n/12}$ and the scattergram in the same viewing window. Does the graph of f come close to the data points? [**Graphing Calculator:** Instructions for $220(2)^{x/12}$: Press 220 ⎣ (⎦ ⎣ 2 ⎦ ⎣) ⎦ ⎣ ∧ ⎦ ⎣ (⎦ ⎣ X, T,Θ,n ⎦ ⎣ ÷ ⎦ ⎣ 12 ⎦ ⎣) ⎦ .]

c. Estimate the frequency of the note D that is 17 notes above the note A (the one with frequency 220 hertz).

d. Use TRACE to find which note has a frequency of 523.25 hertz.

e. Use a graphing calculator table to find $f(0)$, $f(12)$, $f(24)$, $f(36)$, and $f(48)$. What pattern do you notice? Describe this pattern in terms of the situation.

Find the x- and y-intercepts for each function.

49. $y = 7^x$

50. $y = 8(4)^x$

51. $y = 3\left(\dfrac{1}{5}\right)^x$

52. $y = -9\left(\dfrac{2}{3}\right)^x$

53. In this exercise you will compare the function $f(x) = 100(2)^x$ and the function $g(x) = 5(3)^x$.

a. Find the y-intercept of each function.

b. What does the base multiplier property tell you about each function?

c. Based on your comments in parts a and b, which function's outputs will eventually be much greater than the other's outputs? Explain.

d. Use a graphing calculator table to verify your comments to parts a–c. To do this, enter the functions and set up a table as indicated in Fig. 34 and Fig. 35, respectively.

Figure 34 Enter the functions

Figure 35 Set up the table

54. What are the x- and y-intercepts of a function of the form $y = ab^x$, where $b > 0$?

Let $f(x) = 16^x$. Find each output and simplify without using a calculator. Then use a calculator to verify your result.

55. Find $f\left(\dfrac{1}{2}\right)$.

56. Find $f\left(-\dfrac{1}{4}\right)$.

57. Find $f\left(\dfrac{3}{2}\right)$.

58. Find $f\left(-\dfrac{3}{4}\right)$.

59. Find $f(0)$.

60. Find $f(-1)$.

Let $f(x) = 2^x + 3^x$.

61. Find $f(1)$.

62. Find $f(2)$.

63. Find $f(0)$.

64. Find $f(-1)$.

Let $f(x) = 3^x$.

65. Find x when $f(x) = 3$.

66. Find x when $f(x) = 9$.

67. Find x when $f(x) = 1$.

68. Find x when $f(x) = \dfrac{1}{3}$.

69. Without using a calculator, complete Table 13 with values for the function $f(x) = 2^x$. Then use your graphing calculator to verify your results.

Table 13 Values of the Function $f(x) = 2^x$

x	f(x)	x	f(x)
−3		1	
−2		2	
−1		3	
0		4	

70. Without using a calculator, complete Table 14 with values for the function $f(x) = 64^x$. Then use a calculator to verify your results.

Table 14 Values of the Function $f(x) = 64^x$

x	f(x)	x	f(x)
−1		0	
$-\dfrac{5}{6}$		$\dfrac{1}{6}$	
$-\dfrac{2}{3}$		$\dfrac{1}{3}$	
$-\dfrac{1}{2}$		$\dfrac{1}{2}$	
$-\dfrac{1}{3}$		$\dfrac{2}{3}$	
$-\dfrac{1}{6}$		$\dfrac{5}{6}$	
0		1	

*Use graphing calculator tables to compare each pair of functions f and g. What do you observe? Use exponential properties to show why this is so. [**Graphing Calculator:** For 2^{3x}: Press 2 ⎣ ∧ ⎦ ⎣ (⎦ 3 ⎣ X, T, Θ,n ⎦ ⎣) ⎦. Recall that if an exponent involves an operation, you must use parentheses.]*

71. $f(x) = 2^{3x}$, $g(x) = 8^x$

72. $f(x) = 2^{-x}$, $g(x) = \left(\dfrac{1}{2}\right)^x$

73. $f(x) = 2^{x+3}$, $g(x) = 8(2)^x$

74. $f(x) = 3^x 3^x$, $g(x) = 3^{2x}$

75. $f(x) = \dfrac{6^x}{3^x}$, $g(x) = 2^x$

76. $f(x) = 2^0$, $g(x) = 3^0$

77. $f(x) = \dfrac{3^{2x}}{3^x}$, $g(x) = 3^x$

78. $f(x) = 2^x 3^x$, $g(x) = 6^x$

79. $f(x) = x^{1/2}$, $g(x) = \sqrt{x}$ [**Graphing Calculator:** For $\sqrt{x}$: Press $\boxed{\text{2nd}}\ \boxed{x^2}\ \boxed{\text{X,T,}\Theta\text{,}n}\ \boxed{)}$.]

80. $f(x) = 5^{x/3}$, $g(x) = (5^{1/3})^x$

81. $f(x) = 2^x$, $g(x) = 8^{x/3}$

82. $f(x) = 25^{x/2} \cdot 5^x$, $g(x) = 25^x$

83. A graphing calculator table for input-output pairs for the function $f(x) = 4^x$ is displayed in Fig. 36. Some of the outputs are approximations. Write the entries of the table in standard decimal form.

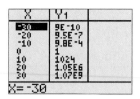

Figure 36 Input-output pairs for $f(x) = 4^x$

84. Recall that for an exponential function $f(x) = b^x$, the base b is a positive number not equal to 1. In this exercise, you will explore what happens if we try to define a function where the base is negative. Consider $f(x) = (-4)^x$.

    a. Explain why $f\left(\dfrac{1}{2}\right)$ is undefined.

    b. Explain why $f\left(\dfrac{1}{4}\right)$ is undefined.

    c. List three more values for x that result in undefined outputs.

85. The graphs of the exponential functions $f(x) = -ab^x$ and $g(x) = ab^x$ are reflections of each other across the x-axis. Explain why this makes sense.

86. Explain how to sketch the graph of a function of the form $f(x) = ab^x$, where $b > 0$. Include in your discussion the effect of a value for a or b on the graph.

4.4 FINDING EQUATIONS FOR EXPONENTIAL FUNCTIONS

Never stop learning; knowledge doubles every fourteen months. —Anthony J. D'Angelo, *The College Blue Book*

Objectives

▸ Use the base multiplier property to find an exponential equation.

▸ Solve an equation of the form $ab^n = k$ for the base b.

▸ Use two points to find an exponential equation.

In this section we will discuss two ways to find an exponential equation in two variables. For the first way, we will use the base multiplier property.

Note

Recall that we discussed the base multiplier property in Section 4.3.

Base Multiplier Property

For an exponential function of the form $y = ab^x$, if the value of the independent variable increases by 1, then the value of the dependent variable is multiplied by the base b.

Table 15 Solutions of an Exponential Equation

x	$f(x)$
0	3
1	6
2	12
3	24
4	48

Example 1 Finding an Equation for an Exponential Curve

An exponential curve contains the points listed in Table 15. Find an equation for the curve.

Solution

For $f(x) = ab^x$, recall that the y-intercept is $(0, a)$. From Table 15 we see that the y-intercept is $(0, 3)$, so $a = 3$. As the value of x increases by 1, the value of y is multiplied by 2. By the base multiplier property, we know that $b = 2$. Therefore, an

equation for the curve is

$$f(x) = 3(2)^x.$$

We check our result using a graphing calculator table (see Fig. 37).

Figure 37 Verify the exponential equation $f(x) = 3(2)^x$

Graphing Calculator

See Section B.13.

For Example 2, it will be helpful to review the slope addition property from Section 1.4.

Slope Addition Property

For a linear function $y = mx + b$, if the value of the independent variable increases by 1, then the value of the dependent variable changes by the slope m.

Example 2 Linear versus Exponential Functions

1. Find a possible equation for a function whose input-output pairs are listed in Table 16.
2. Find a possible equation for a function whose input-output pairs are listed in Table 17.

Solution

1. As the value of x increases by 1 throughout Table 16, the value of y is multiplied by $\frac{1}{3}$. This suggests that there is an exponential function $f(x) = a\left(\frac{1}{3}\right)^x$ that contains the points in Table 16. Since the y-intercept is $(0, 162)$, we have $f(x) = 162\left(\frac{1}{3}\right)^x$.

2. As the value of x increases by 1 throughout Table 17, the value of y changes by adding -4. This suggests that there is a linear function $g(x) = -4x + b$ that contains the points in Table 17. Since the y-intercept is $(0, 50)$, we have $g(x) = -4x + 50$.

Table 16 Input-Output Pairs for f

x	$f(x)$
0	162
1	54
2	18
3	6
4	2

Table 17 Input-Output Pairs for g

x	$g(x)$
0	50
1	46
2	42
3	38
4	34

So far, we have discussed how to find an equation of an exponential curve that contains points whose x-coordinates are *consecutive* integers. Later in this section we will discuss how to find an equation of an exponential curve that contains two given points such as $(2, 5)$ and $(5, 63)$, where the x-coordinates are *not* consecutive integers. To use this method, we first need to discuss how to solve equations of the form $ab^n = k$ for the base b.

Example 3 Solving Equations

Solve each equation for b.

1. $b^2 = 25$
2. $b^3 = 8$
3. $2b^4 = 32$
4. $10b^5 = 90$
5. $b^6 = -28$

Solution

1. $b^2 = 25$

   $b = \pm 5$ since $(-5)^2 = 25$ and $5^2 = 25$

Note

Recall that $b = \pm 5$ means $b = -5$ or $b = 5$.

2. $b^3 = 8$

   $b = 2$ since $2^3 = 8$

3. $2b^4 = 32$

   $b^4 = 16$ Divide both sides by 2.

   $b = \pm 2$ since $(-2)^4 = 16$ and $2^4 = 16$

   We can check that both -2 and 2 satisfy the equation $2b^4 = 32$.

4. $10b^5 = 90$

   $b^5 = 9$ Divide both sides by 10.

   $b = 9^{1/5}$

   $b \approx 1.55$ since $1.55^5 \approx 9$

   We can check that 1.55 approximately satisfies the equation $10b^5 = 90$ by using a graphing calculator to check that $10(1.55)^5 \approx 90$ (see Fig. 38).

Figure 38 Checking that 1.55 approximately satisfies $10b^5 = 90$

5. The equation $b^6 = -28$ has no real number solutions since an even number exponent gives a positive number. ■

The problems in Example 3 suggest how to solve equations of the form $b^n = k$ for b:

1. If n is odd, the solution is $k^{1/n}$.

2. If n is even and $k \geq 0$, the solution is $\pm k^{1/n}$.

3. If n is even and $k < 0$, there is no real number solution.

Example 4   Solving Equations

Solve each equation.

1. $5.42b^6 - 3.19 = 43.74$

2. $\dfrac{b^9}{b^4} = \dfrac{70}{3}$

Solution

1. $5.42b^6 - 3.19 = 43.74$

   $5.42b^6 = 43.74 + 3.19$ Add 3.19 to both sides.

   $5.42b^6 = 46.93$

   $b^6 = \dfrac{46.93}{5.42}$ Divide both sides by 5.42.

   $b = \pm\left(\dfrac{46.93}{5.42}\right)^{1/6}$

   $b \approx \pm 1.43$

2. $\dfrac{b^9}{b^4} = \dfrac{70}{3}$

   $b^5 = \dfrac{70}{3}$ $\dfrac{b^m}{b^n} = b^{m-n}$

   $b = \left(\dfrac{70}{3}\right)^{1/5}$

   $b \approx 1.88$ ■

Knowing how to solve equations in *one* variable will help us find equations in *two* variables for exponential functions.

Example 5 Finding an Equation for an Exponential Curve

Find the equation $y = ab^x$ for the exponential curve that contains the points $(0, 3)$ and $(4, 70)$.

Solution

Since the y-intercept is $(0, 3)$, we know the equation has the form $y = 3b^x$. Next, we substitute $(4, 70)$ in the equation $y = 3b^x$ and solve for b:

$$70 = 3b^4$$

$$3b^4 = 70$$

$$b^4 = \frac{70}{3} \qquad \text{Divide both sides by 3.}$$

$$b = \left(\frac{70}{3}\right)^{1/4} \qquad \text{The base of an exponential function is positive.}$$

$$b \approx 2.20$$

So, our equation is $y = 3(2.20)^x$, which contains the given point $(0, 3)$. Since we rounded the value b, the graph of the equation comes close to, but does not pass through, the given point $(4, 70)$.

We use a graphing calculator to verify our work (see Figs. 39 and 40). ■

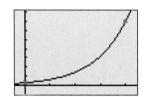

Figure 39 Verify work with axes on

Graphing Calculator

To turn axes off, see Section B.24.

In Example 6, we will find an equation of a curve that approximates the exponential curve containing two given points. Neither point will be the y-intercept.

Example 6 Finding an Equation for an Exponential Curve

Find an equation that approximates the exponential curve that passes through $(2, 5)$ and $(5, 63)$.

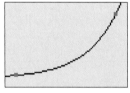

Figure 40 Verify work with axes off to see $(0, 3)$ more clearly

Solution

Since both of the given points must satisfy the equation $y = ab^x$, we have the following system of equations:

$$5 = ab^2$$
$$63 = ab^5$$

It will be slightly easier to solve this system if we switch the equations so the equation with the greater exponent of b is listed first.

$$63 = ab^5$$
$$5 = ab^2$$

Note that if $A = B$ and $C = D$, then it follows that $\dfrac{A}{C} = \dfrac{B}{D}$ if $C \neq 0$ and $D \neq 0$.

If this fact is applied to the system above, we get the following result for nonzero a and b:

$$\frac{63}{5} = \frac{ab^5}{ab^2}$$

By then applying the exponential property $\dfrac{b^m}{b^n} = b^{m-n}$ to the right side of the equation, we have an equation in terms of b (and not a).

$$\frac{63}{5} = \frac{b^3}{1}$$

We can now solve for b by finding the cube root of $\dfrac{63}{5}$.

$$b^3 = \frac{63}{5}$$

$$b = \left(\frac{63}{5}\right)^{1/3}$$

$$\approx 2.33$$

So, 2.33 can be substituted for the constant b in the equation $y = ab^x$:

$$y \approx a(2.33)^x$$

Since $(2, 5)$ comes close to satisfying the equation $y = a(2.33)^x$, we can now find an approximate value for a.

$$5 = a(2.33)^2$$

$$a = \frac{5}{(2.33)^2}$$

$$\approx 0.92$$

So, an equation that approximates the exponential curve that passes through $(2, 5)$ and $(5, 63)$ is $y = 0.92(2.33)^x$.

We use a graphing calculator to verify our work (see Fig. 41). ∎

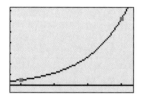

Figure 41 Check that the curve contains $(2, 5)$ and $(5, 63)$

EXPLORATION

Comparing three ways to find exponential equations

An exponential curve contains the points listed in Table 18.

1. Use the point $(0, 5)$ and one other point in Table 18 to find an equation for the curve (see Example 5).
2. Use two points in Table 18 other than $(0, 5)$ to find an equation for the curve (see Example 6).
3. Use the base multiplier property to find an equation for the curve. [**Hint**: First find $f(1)$ by recognizing a pattern.]
4. Compare your equations from problems 1, 2, and 3.
5. An exponential curve contains the points listed in Table 19. Which method would you use to find an equation that approximates the exponential curve? Explain. Also, find the equation.

Table 18 Solutions for an Exponential Equation

x	f(x)
0	5
2	20
4	80
6	320
8	1280

Table 19 Solutions for an Exponential Equation

x	f(x)
0	400
3	200
6	100
9	50
12	25

Tips on Succeeding in This Course

It can be helpful to meet with a friend from class and discuss what happened in class that day. Not only can you ask questions *of* each other, but you will learn just as much by explaining concepts *to* each other. Explaining a concept to someone else forces you to clarify your own understanding of the concept.

Key Points OF THIS SECTION

Solving $b^n = k$ for b

When solving an equation of the form $b^n = k$ for b:

- If n is odd, the solution is $k^{1/n}$.
- If n is even and $k \geq 0$, the solution is $\pm k^{1/n}$.
- If n is even and $k < 0$, there is no real number solution.

You have two ways to find an equation of an exponential model.

- Use the base multiplier property.
- Use two points.

Finding an equation

To find an equation of an exponential curve that contains two given points:

- If one of the points is the y-intercept, say $(0, 7)$, then the equation is of the form $y = 7b^x$. The value of b can be found by substituting the values of x and y of the other point into $y = 7b^x$ and then solving for b.
- If neither point is the y-intercept, see Example 6.

Using two points to find an equation

After finding an exponential equation, use a graphing calculator table or graph to verify your work.

Verifying work

HOMEWORK 4.4

SSM
Tutor Center

MathPro 4
MathPro 4

MathPro 5
MathPro 5

CDs/Videos

www

www

---

1. Some values of functions f, g, h, and k are provided in Table 20. Find a possible equation for each function. Verify your results using a graphing calculator table.

Table 20 Values of Four Functions (Exercise 1)

x	f(x)	g(x)	h(x)	k(x)
0	4	36	5	250
1	8	12	50	50
2	16	4	500	10
3	32	$\frac{4}{3}$	5,000	2
4	64	$\frac{4}{9}$	50,000	$\frac{2}{5}$

2. Some values of functions f, g, h, and k are provided in Table 21. Find a possible equation for each function. Verify your results using a graphing calculator table. [**Hint:** Use linear or exponential equations.]

Table 21 Values of Four Functions (Exercise 2)

x	f(x)	g(x)	h(x)	k(x)
0	3	19	2	2500
1	12	13	9	500
2	48	7	16	100
3	192	1	23	20
4	768	−5	30	4

3. Some values of functions f, g, h, and k are provided in Table 22. Find a possible equation for each function. Verify your results using a graphing calculator table.

Table 22 Values of Four Functions (Exercise 3)

x	f(x)	g(x)	h(x)	k(x)
0	100	100	2	2
1	50	50	6	6
2	25	0	10	18
3	12.5	−50	14	54
4	6.25	−100	18	162

4. The graph of a function passes through the points $(0, 3)$ and $(2, 12)$. Could the function be linear, exponential, either of these types, or neither of these types?

Solve each equation. Verify that your results satisfy the equation.

5. $b^2 = 16$

6. $b^3 = 27$

7. $b^4 = 81$

8. $b^5 = 100{,}000$

9. $3b^5 = 96$

10. $5b^2 = 8$

11. $5b^4 = 100$

12. $44b^3 = 12$

13. $3.6b^3 = 42.5$

14. $1.7b^4 = 86.4$

15. $32.7b^6 + 4 = 20.2$

16. $2.1b^5 - 8.2 = 0$

17. $\frac{1}{4}b^3 - \frac{1}{2} = \frac{9}{4}$

18. $\frac{1}{6}b^4 + \frac{5}{3} = \frac{11}{2}$

19. $\dfrac{b^6}{b^2} = 81$

20. $\dfrac{b^{10}}{b^3} = 2187$

21. $\dfrac{b^8}{b^3} = \dfrac{79}{5}$

22. $\dfrac{b^9}{b^6} = \dfrac{2}{9}$

Solve or simplify, whichever is appropriate.

23. $\dfrac{b^7}{b^2}$

24. $\dfrac{b^8}{b^4} = \dfrac{65}{3}$

25. $\dfrac{b^7}{b^2} = 76$

26. $\dfrac{b^8}{b^4}$

27. $\dfrac{8b^3}{6b^{-1}}$

28. $\dfrac{10b^{-7}}{15b^{-2}}$

29. $\dfrac{8b^3}{6b^{-1}} = \dfrac{3}{7}$

30. $\dfrac{10b^{-7}}{15b^{-2}} = \dfrac{4}{7}$

Find an equation of the exponential curve that passes through the given pair of points. Then verify your equation using your graphing calculator.

31. $(0, 4)$ and $(1, 8)$

32. $(0, 5)$ and $(1, 15)$

33. $(0, 3)$ and $(5, 100)$

34. $(0, 8)$ and $(4, 0.7)$

35. $(0, 87)$ and $(6, 14)$

36. $(0, 13)$ and $(7, 256)$

37. $(0, 7.4)$ and $(3, 1.3)$

38. $(0, 2.1)$ and $(5, 9.7)$

39. $(0, 5.5)$ and $(2, 73.9)$

40. $(0, 97.2)$ and $(4, 17.1)$

41. (0, 39.18) and (15, 3.66)

42. (0, 12.94) and (20, 357.03)

43. Find an equation of the exponential curve sketched in Fig. 42. [**Hint**: Choose two points whose coordinates appear to be integers.]

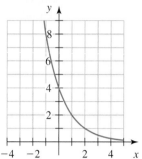

Figure 42 Graph of
f—exercise 43

44. Find an equation of the exponential curve sketched in Fig. 43.

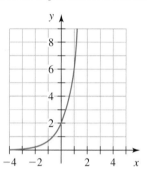

Figure 43 Graph of
f—exercise 44

Find an equation of the exponential curve that passes through the given pair of points. Then verify your equation using your graphing calculator.

45. (1, 4) and (2, 12)   **46.** (2, 5) and (3, 10)

47. (3, 4) and (5, 9)   **48.** (2, 7) and (5, 1)

49. (10, 329) and (30, 26)   **50.** (11, 8) and (17, 492)

51. (5, 8.1) and (9, 2.4)   **52.** (1, 3.5) and (5, 5)

53. (2, 73.8) and (7, 13.2)

54. (4, 6.3) and (10, 250.8)

55. (13, 24.71) and (21, 897.35)

56. (8, 39.43) and (12, 6.52)

57. Find an equation of the exponential curve sketched in Fig. 44.

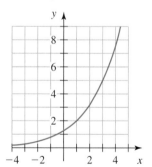

Figure 44 Graph of
f—exercise 57

58. Find an equation of the exponential curve sketched in Fig. 45.

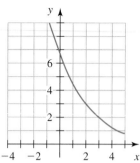

Figure 45 Graph of
f—exercise 58

59. Let f be a linear function and g be an exponential function. Assume that the graphs of f and g both contain the points (0, 2) and (1, 6).

   a. Find an equation for f.

   b. Find an equation for g.

   c. Use your graphing calculator to draw the graphs of f and g in the same viewing window.

60. In this exercise you will compare the linear function $f(x) = 2x + 100$ to the exponential function $g(x) = 3(2)^x$.

   a. Find the y-intercept of each function.

   b. For both functions f and g, describe what happens to the value of y as the value of x increases by 1.

   c. Based on your responses to parts a and b, which function's outputs will eventually dominate the other's outputs? Explain.

   d. Use a graphing calculator table to verify your responses to parts a–c. To do this, enter the functions and set up a table as indicated in Figs. 46 and 47, respectively.

Figure 46 Enter the
functions

Figure 47 Set up a
table

61. Solve the system. [**Hint**: Think graphically.]

$$y = 6(4)^x$$
$$y = 6\left(\frac{1}{3}\right)^x$$

62. Solve the system.

$$y = 7(3)^x$$
$$y = 4(3)^x$$

63. Is it possible for a linear function and an exponential function to have the indicated number of intersection points? If so, give equations for the two functions. [**Hint**: First sketch some graphs.]

   a. 3 intersection points

   b. 2 intersection points

   c. 1 intersection point

   d. 0 intersection points

64. a. Is there an exponential curve that passes through the points (0, 5) and (1, 5)? If so, find the equation of the exponential curve. If not, explain.

b. Is there an exponential curve that passes through the points (0, 3) and (7, 3)? If so, find the equation of the exponential curve. If not, explain.

c. Is there an exponential curve that passes through two given points that have the same *y*-coordinate? Explain.

65. Describe the base multiplier property and explain why it makes sense. Include an example in your explanation.

66. Describe how to find the equation of an exponential curve that contains two given points. Include in your description both the case where one of the points is the *y*-intercept and the case where neither of the points is the *y*-intercept.

4.5 USING EXPONENTIAL FUNCTIONS TO MODEL DATA

Objectives

▸ Find an equation of an exponential model.

▸ Make estimates and predictions using an exponential model.

▸ For a model $f(t) = ab^t$, know the meaning of the coefficient a and the base b in terms of the situation being modeled.

In Section 4.4 we found exponential equations. In this section we use this skill to model data.

Example 1   Modeling with an Exponential Function

Suppose that a peach has 3 million bacteria on it at noon on Monday and that a bacterium divides into two bacteria every hour, on average (see Fig. 48).

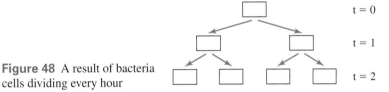

Figure 48 A result of bacteria cells dividing every hour

Let $B = f(t)$ represent the number of bacteria (in millions) on the peach at t hours after noon, Monday.

1. Find an equation for f.

2. Predict the number of bacteria at noon on Tuesday.

Solution

1. We complete a table of values of f based on the assumption that a bacterium divides into two bacteria every hour (see Table 23).

As the value of t increases by 1, the value of B changes by greater and greater amounts, so it would *not* be appropriate to model the data using a linear function. Note, though, that as the value of t increases by 1, the value of B is multiplied by 2, so we *can* model the situation using an *exponential* model of the form $f(t) = a(2)^t$. The B-intercept is (0, 3), so $f(t) = 3(2)^t$.

We use a graphing calculator table and graph to verify our work (see Figs. 49 and 50).

Table 23 Values of a Bacteria Model

t (hours)	$B = f(t)$ (millions)
0	3
1	6
2	12
3	24
4	48

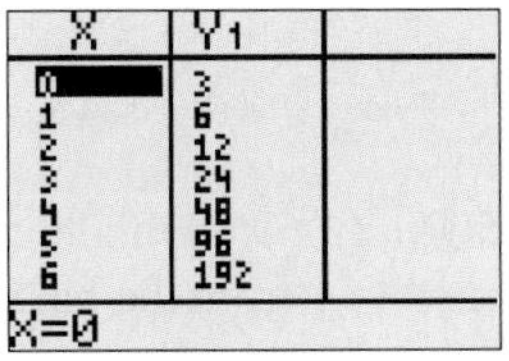

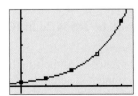

Figure 49 Table for bacteria model

Figure 50 Graph of bacteria model

2. We use $t = 24$ to represent noon on Tuesday. We substitute 24 for t in our equation for f.

$$f(24) = 3(2)^{24} = 50,331,648$$

According to the model, there would be 50,331,648 million bacteria. To omit writing "million," we must add six zeros to 50,331,648. This would be 50,331,648,000,000. There would be about 50 trillion bacteria at noon on Tuesday. ∎

Example 2 Modeling with an Exponential Function

A person invests $5000 in an account that earns 6% interest compounded annually. The term **compounded annually** means that the interest earned each year equals 6% of the sum of the $5000 and any interest earned in previous years (all of which becomes part of the investment).

1. Let $V = f(t)$ represent the value (in dollars) of the account at t years after the money is invested. Find an equation for f.

2. What will be the value after 10 years?

Solution

1. Each year, the investment value is equal to the last year's value (100% of it) plus 6% of the last year's value. So the value is equal to 106% of the last year's value. For example, after one year, the value will be 106% of $5000, or $1.06(5000) = 5300$ dollars. At two years, the value will be $1.06(5300) = 5618$ dollars. See Table 24.

   As the value of t increases by 1, the value of V is multiplied by 1.06. So, f is the exponential function $f(t) = a(1.06)^t$. Since the V-intercept is $(0, 5000)$, we have $f(t) = 5000(1.06)^t$.

Table 24 Values of a Compound Interest (6%) Account

t	$V = f(t)$
0	5000.00
1	$5000.00(1.06) = 5300.00$
2	$5300.00(1.06) = 5618.00$
3	$5618.00(1.06) = 5955.08$
4	$5955.08(1.06) \approx 6312.38$

Our new coalition has decided you can have a gallon of milk now, if you give us 2 lbs of cheese, 1 pint of cream, and a chocolate cake later.

2. To find the value in 10 years, we substitute 10 for t.

$$f(10) = 5000(1.06)^{10} \approx 8954.24$$

The value will be $8954.24 in 10 years. ∎

In Example 2, we used the function $f(t) = 5000(1.06)^t$ to model the value of the 6% compounded interest account. Note that subtracting 1 from the base 1.06 gives the interest rate in decimal form:

$$b - 1 = 1.06 - 1 = 0.06 = \text{interest rate (decimal form)}$$

Example 3   Modeling with an Exponential Function

The worst nuclear accident in the world occurred in Chernobyl, Ukraine on April 26, 1986. Immediately after the disaster, 31 people died from acute radiation sickness. It is predicted that about 17,000 people will die from radiogenic cancer, mostly due to the release of the radioactive element cesium-137 (Anspaugh, et al.).*

Cesium-137 has a **half-life** of 30 years, which means that every 30 years the number of cesium-137 atoms is reduced to half. Let $P = f(t)$ represent the percent of the cesium-137 that still remains at t years since 1986.

1. Find an equation for f.
2. Describe the meaning of the base of f.
3. What percent of the cesium-137 will remain in 2009?

Solution

1. We discuss two methods for finding an equation for f.

   Method 1   At time $t = 0$, 100% of the cesium-137 is present. At time $t = 30$, there will be $\frac{1}{2}(100) = 50$ percent. At time $t = 60$, there will be $\frac{1}{2} \cdot \frac{1}{2}(100) = 25$ percent. We organize these results and one more in Table 25.

Table 25 Percentages of Cesium-137 That Remain

Year t	Percent P
0	$100 = 100\left(\frac{1}{2}\right)^0$
30	$100 \cdot \frac{1}{2} = 100 \cdot \left(\frac{1}{2}\right)^1$
60	$100 \cdot \frac{1}{2} \cdot \frac{1}{2} = 100\left(\frac{1}{2}\right)^2$
90	$100 \cdot \frac{1}{2} \cdot \frac{1}{2} \cdot \frac{1}{2} = 100\left(\frac{1}{2}\right)^3$
t	$100\left(\frac{1}{2}\right)^{t/30}$

From Table 25 we see that the situation can be modeled well using an exponential function. Furthermore, we note that each exponent in the second column of the table is equal to the value of t in the first column divided by 30. The equation for f is

$$f(t) = 100\left(\frac{1}{2}\right)^{t/30}$$

We can use a graphing calculator table to verify our equation (see Fig. 51). We can write this equation in $f(t) = ab^t$ form:

$$f(t) = 100\left(\frac{1}{2}\right)^{\frac{t}{30}} = 100\left(\frac{1}{2}\right)^{\frac{1}{30} \cdot t} = 100\left(\left(\frac{1}{2}\right)^{\frac{1}{30}}\right)^t$$

X	Y1
0	100
30	50
60	25
90	12.5
120	6.25
150	3.125
180	1.5625

X=0

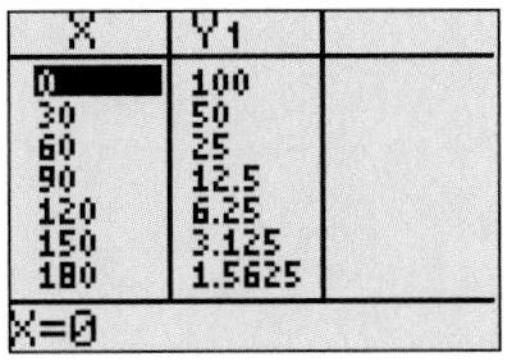

Figure 51 Table for
$f(t) = 100\left(\frac{1}{2}\right)^{t/30}$

Since $\left(\frac{1}{2}\right)^{1/30} \approx 0.977$, we can write

$$f(t) = 100(0.977)^t$$

Method 2   Instead of recognizing a pattern from a table, we can find an equation for f using the points $(0, 100)$ and $(30, 50)$. Since the P-intercept is $(0, 100)$, we have

$$P = f(t) = 100b^t$$

---

*Chernobyl, www.hyperphysics.phy-astr.gsu.edu/htbase/nucene/cherno2.html; Chernobyl, www.dhushara.com/book/explod/cher/cher.htm. Both accessed 6/9/02.

We use the point (30, 50) to find a value for b.

$$50 = 100b^{30}$$

$$b^{30} = \frac{50}{100} \qquad \text{Divide both sides by 100.}$$

$$b^{30} = \frac{1}{2} \qquad \text{Reduce } \frac{50}{100}.$$

$$b = \left(\frac{1}{2}\right)^{1/30} \qquad \text{The base of an exponential function is positive.}$$

$$\approx 0.977$$

So, an equation for f is $f(t) = 100(0.977)^t$, the same equation we found earlier.

2. The base of f is 0.977. Each year 97.7% of the previous year's cesium-137 is present. In other words, the cesium-137 decays by 2.3% each year.

3. Since $2009 - 1986 = 23$, we substitute 23 for t in the equation $f(t) = 100(0.977)^t$.

$$f(23) = 100(0.977)^{23} \approx 58.56$$

In 2009 about 58.6% of the cesium-137 will still remain. ∎

In Example 3 we worked with the half-life of cesium-137. Every radioactive element decays exponentially and has a half-life.

DEFINITION half-life

If a quantity decays exponentially, the **half-life** is the amount of time for the quantity to be reduced to half.

From Example 3 we know that the half-life of cesium-137 is 30 years. The half-lives of some radioactive elements are much less than that of cesium-137. For example, polonium-214 has a half-life of 0.164 second. On the other hand, uranium-238 has a half-life of 4.5 billion years!

In Example 3 we used the P-intercept and another point to find the cesium-137 model. In Example 4, we use two nonintercept points to find a model.

Example 4 Modeling with an Exponential Function

The Dow Jones Industrial Average is a measure of the strength of companies on the New York Stock Exchange (see Table 26).

Year	Dow Jones Average	Year	Dow Jones Average
1980	839	1988	1939
1982	875	1990	2753
1984	1259	1992	3169
1986	1547	1994	3754
		1996*	5117

Table 26 Dow Jones Industrial Averages

Source: *BigCharts—Historical Quotes, www.bigcharts.marketwatch.com/historical. Accessed 6/10/02.*

Let $A = f(t)$ represent the Dow Average at t years since 1980.

1. Find an equation for f.

2. Estimate the percent rate of growth of the Dow Average.

---

*The Dow Averages are for the start of each given year.

Solution

1. We use a graphing calculator to view a scattergram of the data (see Fig. 52).

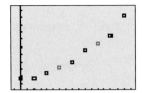

Figure 52 Dow Jones scattergram

Since the points "bend upward," the scattergram suggests that we can better model the data using an exponential function than a linear function. To find an equation for an exponential model, we substitute the coordinates of the points (6, 1547) and (12, 3169) into the equation $f(t) = ab^t$.

$$3169 = ab^{12}$$
$$1547 = ab^{6}$$

Note

If we imagine an exponential curve that contains the points (6, 1547) and (12, 3169), it appears that the curve might come close to the other data points. If this didn't turn out to be the case, we could try another pair of points (not necessarily data points).

Next, we divide both sides of the first equation by the same sides in the second equation and solve for b.

$$\frac{3169}{1547} = \frac{ab^{12}}{ab^{6}}, \quad \text{where } a \neq 0 \text{ and } b \neq 0$$

$$\frac{3169}{1547} = b^{6} \qquad\qquad \frac{b^{m}}{b^{n}} = b^{m-n}$$

$$b = \left(\frac{3169}{1547}\right)^{1/6}$$

$$\approx 1.127$$

So, an equation is $f(t) = a(1.127)^t$. To find a, we substitute the coordinates of (6, 1547) into the equation.

$$1547 = a(1.127)^{6}$$

$$a = \frac{1547}{1.127^{6}}$$

$$\approx 755.00$$

The equation is $f(t) = 755.00(1.127)^t$. The graph in Fig. 53 shows that our exponential model fits the data quite well.

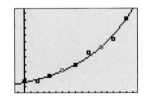

Figure 53 Check how well the model fits the data

2. The base of $f(t) = 755.00(1.127)^t$ is 1.127. So, the percent rate of growth of the Dow Average from 1980 to 1996 was about 12.7%. ∎

Graphing Calculator

See Section B.16.

In Example 4 we used two points to find the exponential model $f(t) = 755.00(1.127)^t$. We can also find an exponential model using a graphing calculator's *exponential regression*. Exponential regression gives the model $r(t) = 776.79(1.124)^t$. In Fig. 54, we see that the graphs of the models are so similar that when using ZoomStat there appears to be just one curve.

We call the equation $r(t) = 776.79(1.124)^t$ the **exponential regression equation** for the given data and refer to its graph as an **exponential regression curve**.

Note that the coefficients of both models are approximately equal, as are the bases. In particular, the base 1.127 of f estimates that the Dow Average grows exponentially by 12.7% each year. The base 1.124 of r estimates that the Dow Average grows exponentially by 12.4% each year.

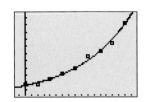

Figure 54 Comparing the fit of models f and r

Meaning of the Base of an Exponential Model

Suppose $f(t) = ab^t$, $a > 0$, models a quantity at time t. Then the percent rate of growth is constant, and in particular:

- If $b > 1$, then $b - 1$ represents the percent rate of growth (decimal form) of the quantity per unit time.
- If $0 < b < 1$, then $1 - b$ represents the percent rate of decay (decimal form) of the quantity per unit time.

In Example 2 we found the compound interest model $f(t) = 5000(1.06)^t$. Since $1.06 - 1 = 0.06$, the base 1.06 indicates that the value grows exponentially by 6% per year.

In Example 3 we found the cesium-137 model $f(t) = 100(0.977)^t$. Since $1 - 0.977 = 0.023$, the base 0.977 indicates that the mass decays exponentially by 2.3% per year.

Table 27 Dow Jones Industrial Averages

Year	Dow Jones Average
1997	6448
1998	7908
1999	9181
2000	11,497
2001	10,787

Source: *BigCharts—Historical Quotes, www.bigcharts.marketwatch.com/historical. Accessed 6/10/02.*

Note

The stock market was closed for one week following the terrorist attacks on September 11, 2001. The Average fell by 685 the first day the stock market reopened. By October 11, the Average had rebounded by 489 points.

Example 5   Using a Model to Analyze a Situation

The exponential regression equation for the Dow Jones Industrial Average $r(t)$ is

$$r(t) = 776.79(1.124)^t$$

where t is the number of years since 1980.

1. During the late 1990s the percent rate of growth of the Dow Average was greater than that from 1980 to 1996 (see Table 27).

    Compute by how much the model r underestimates the Dow Average for the years shown in Table 27.

2. Due to a recession, the Dow Average generally declined during 2001. The Average was 10,022 on January 1, 2002. Use the model r to estimate the Average on that day and describe what your result might imply.

Solution

1. First, we use our model to estimate the Average in 1997 by evaluating r at 17:

$$r(17) = 776.79(1.124)^{17} \approx 5666.71$$

Our estimate is 5667. Next, we compute the difference of the true value 6448 and our estimate 5667:

$$6448 - 5667 = 781$$

So, our model underestimates the Average in 1997 by 781. We perform similar calculations for 1998, 1999, 2000, and 2001 and list these calculations in Table 28.

Table 28 Underestimates of Dow Averages

Year	Dow Jones Average	Model Estimate	Difference in the Average and the Estimate
1997	6448	5667	781
1998	7908	6369	1539
1999	9181	7159	2022
2000	11,497	8047	3450
2001	10,787	9045	1742

So, r underestimates the Average for each of the years. In Fig. 55 we see that the graph of r is indeed below the data points for the years from 1997 to 2001.

Each estimate by the model r being so much smaller than the true Average suggests that either the economy was really doing well during this period or many companies' stock prices overestimated the true values of the companies.

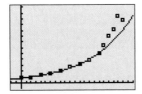

Figure 55 The scattergram
for 1980–2001 and the
graph of r

2. We find the estimate for 2002 by evaluating r at 22:

$$r(22) = 776.79(1.124)^{22} \approx 10,166.29$$

The overestimate 10,166 is quite close to the true Average 10,022. The model r and a scattergram including the data point for 2002 are shown in Fig. 56. Figure 56 suggests that although the Averages can wildly swing up and down in the short run, the percent rate of growth in the long run will tend to be about 12.4%. This is why many investors get a good return on investments in stocks, provided that they make informed choices and invest the money for a long time. ■

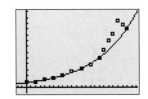

Figure 56 The scattergram for
1980–2002 and the graph of r

At this point in the course, we have a choice between modeling some data with a linear function or an exponential function. For some true-to-life situations, we will know which type of model to use based on our knowledge of the situation. For other situations, we can decide by viewing a scattergram of the data. Here, we review the four-step modeling process:

Four-Step Modeling Process

1. Create a scattergram of the data. Decide whether a line or an exponential curve comes close to the points.

2. Find an equation for your function.

3. Verify that your equation has a graph that comes close to the points in the scattergram. If it doesn't, check for calculation errors or try using different points to find the equation. An alternative is to reconsider your choice of model in step one.

4. Use your equation of the model to draw conclusions, make estimates, and/or make predictions.

EXPLORATION
Comparing a linear model to an exponential model

The world population in 1950 was 2.5 billion. In 1987, it was 5.0 billion.

1. First, we assume that the world population is growing exponentially. Let $E(t)$ represent the world population (in billions) at t years since 1950. Find a formula for E.

2. Now we assume that the world population is growing linearly. Let $L(t)$ represent the world population (in billions) at t years since 1950. Find a formula for L.

3. Use your formulas for E and L to make two predictions of the world population for each of the following years.
   a. 2010   **b.** 2050   **c.** 2150

4. Compare the graphs of E and L using the window settings shown in Fig. 57.

5. Would there be much difference in the world population if it grows exponentially or linearly in the short run? in the long run? Explain.

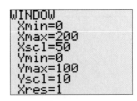

Figure 57 Compare the graphs of
the exponential and linear models

EXPLORATION
Looking ahead: Inverse function

A company's profit for 2000 was $5 million. Each year, the profit increases by $2 million. Let $p = f(t)$ represent the profit (in millions of dollars) for the year that is t years after 2000.

1. Find an equation for f. Write the equation using $f(t)$ notation. Also, write the equation in terms of p and t.
2. Use an equation for f to predict when the company will have a profit of $21 million.
3. In problem 1 you found the equation $p = 2t + 5$. Solve this equation for t.
4. Substitute 21 for p in your equation from problem 3 and solve for t.
5. Compare your results for problems 2 and 4.
6. Enter the equation you found in problem 3 in a graphing calculator to help you complete Table 29.

Table 29 From Profit to Year	
Profit (millions of dollars) p	Years since 2000 t
21	
22	
23	
24	
25	

7. In problems 2 and 4 you used two different methods for finding when the company will have a profit of $21 million. If the company wants to know when it might attain 15 different profit levels, which method would be best to use? Explain.

Tips on Succeeding in This Course

Even if you don't know how to do one step of a problem, you can still show the instructor that you understand the other parts of the problem. Depending on how your instructor grades tests, you may earn partial credit if you pick some number (which would probably be incorrect) to be the result for the one step and then show what you would do with that number in the remaining steps of the solution.

For example, suppose you want to find an equation of the line that passes through the points $(4, 7)$ and $(5, 9)$, and you forget how to find slope. You could write:

I've drawn a blank on finding slope. However, assuming that the slope is 3, then

$$y = 3x + b$$
$$7 = 3(4) + b$$
$$7 = 12 + b$$
$$7 - 12 = b$$
$$-5 = b$$
$$b = -5$$

Therefore, $y = 3x - 5$.

You could also point out that you know your result is incorrect as the graph of $y = 3x - 5$ does not pass through the point $(5, 9)$. Also, seeing your result (with graph) may jog your memory about finding slope so you can go back and do the problem correctly.

Key Points
OF THIS SECTION

We can find an equation of an exponential model by

* using the base multiplier property;
* using two points; and
* exponential regression.

If a quantity decays exponentially, the half-life is the amount of time for the quantity to be reduced to half.

Assume that $f(t) = ab^t$, $a > 0$, is an exponential model, where $f(t)$ represents a quantity at time t.

* The coefficient a represents the amount of the quantity present at time $t = 0$.
* If $b > 1$, then the quantity grows exponentially at a rate of $b - 1$ percent (decimal form) per unit of time.
* If $0 < b < 1$, then the quantity decays exponentially at a rate of $1 - b$ percent (decimal form) per unit of time.

Three ways to find an equation of a model

Half-life

Properties of an exponential model

HOMEWORK 4.5

SSM
Tutor Center

MathPro 4
MathPro 4

MathPro 5
MathPro 5

CDs/Videos

www

1. Suppose a rumor is spreading in the United States that chocolate causes cancer. Assume that 40 people have heard the rumor as of today and that each day the total number of people who have heard the rumor triples. Let $f(t)$ represent the total number of people who have heard the rumor at t days since today.

   a. Find an equation for f.

   b. How many people will have heard the rumor 10 days from now?

   c. How many people will have heard the rumor 15 days from now? Has model breakdown occurred? [**Hint**: The U.S. population was about 287 million in 2002.]

2. Suppose a flu epidemic has broken out at your school. Assume that on February 10, a total of 20 people come down with the flu and that each day the total number of people who have gotten the flu doubles. Let $f(t)$ represent the total number of people who have gotten the flu by the day that is t days since February 10.

   a. Find an equation for f.

   b. Find $f(15)$. What does your result mean in terms of the flu epidemic?

3. Laserdisc player sales were $108 million in 1995. Since then, sales have been approximately halving each year. Let $f(t)$ represent the sales (in millions of dollars) for the year that is t years since 1995.

   a. Find an equation for f.

   b. Predict laserdisc player sales in 2008.

4. In 1996, doctors prescribed a total of $40 million worth of oxycontin, a powerful painkiller. Doctors are clearly impressed with its effectiveness as sales have approximately doubled every year since then. One problem with the drug being so effective is that some patients are abusing it. An official of the Drug Enforcement Administration said that no other prescription drug in the last 20 years has been illegally abused by so many people so soon after it appeared (*The New York Times*).

   a. Let $f(t)$ represent oxycontin sales (in millions of dollars) for the year that is t years since 1996. Find an equation for f.

   b. Predict oxycontin sales in 2004.

5. Google™ is a search engine, which provides a way to search the Internet for information. Google's name is a play on the word "googol," which refers to the number 1 followed by one hundred zeroes. Many companies including Yahoo!®, Netscape®, and AOL® rely on Google to power searches on their web sites.

   a. In 2002 Google's index comprised 2 billion web pages. According to founders Larry Page and Sergey Brin, the size of the index doubles every year (*Charlie Rose*, PBS, 7/27/01). Let $f(t)$ represent the number of web pages (in billions) in Google's index at t years since 2002. Find a formula for f.

   b. Predict the number of web pages that Google's index will comprise in 2010.

   c. If the web pages in Google's index in 2010 were printed and stacked in one pile, what would be the height (in miles) of that pile? [**Hint**: A stack of 500 pages of multipurpose computer paper has height 2 inches.]

6. In 1992 there were 2 deaths caused by airbags in cars. From 1992 to 1996, the number of deaths approximately doubled each year. Let $n = f(t)$ represent the number of deaths caused by airbags during the year that is t years since 1992.

   a. Find an equation for $f(t)$.

   b. One major reason that the number of deaths caused by airbags has doubled each year is that each year they have been installed in more makes and models of cars. By 1998 most manufacturers had included airbags in most of their cars. Estimate the number of deaths caused by airbags in 1998.

c. Explain why the exponential model you have derived will probably break down after 1998.

d. Sketch a qualitative graph that describes the relationship between t and n for all of time, past and future.

7. Someone invests $3000 in an account at 8% interest compounded annually. Let $f(t)$ represent the value (in dollars) of the account at t years after investing the $3000.

a. Find an equation for f.

b. What is the base b of your model $f(t) = ab^t$? What does it mean in terms of the account?

c. What is the coefficient a of your model $f(t) = ab^t$? What does it mean in terms of the account?

8. A person invests $7000 at 10% interest compounded annually. Let $f(t)$ represent the value (in dollars) of the account at t years after depositing the $7000.

a. Find an equation for f.

b. What will be the value in 10 years? Explain why the value has more than doubled, even though the investment earned 10% for 10 years.

9. Suppose that someone invests $4000 in stocks today and the value of the stocks doubles every 6 years. Let $f(t)$ represent the value (in dollars) of the investment at t years from now.

a. Find an equation for f.

b. Find the value of the investment 20 years from now.

10. A person invests $5000 at 6% interest compounded annually for 3 years and then invests the balance (the $5000 plus the interest earned) in an account at 8% interest for 5 years. Find the value of the investment after the 8 years.

11. In this exercise you will explore two types of interest-bearing accounts.

a. Suppose that $800 is deposited into a savings account that earns 3% interest compounded annually. Let $C(t)$ represent the value (in dollars) of the 3% annual compounded interest account at t years after depositing $800. Find a formula for $C(t)$.

b. Now suppose that $800 is deposited into a savings account that earns 3% simple interest. The term **simple interest** means that the interest earned each year is 3% of the $800 only. Let $S(t)$ represent the value (in dollars) of the 3% simple interest account at t years after depositing $800. Find a formula for $S(t)$. [**Hint**: Each year the balance increases by $800 \cdot 0.03 = 24$ dollars.]

c. Find $C(1)$, $C(2)$, $S(1)$, and $S(2)$. Explain in terms of the situation why it makes sense that $C(1)$ is equal to $S(1)$, but $C(2)$ is not equal to $S(2)$.

d. Compare $C(20)$ and $S(20)$. What does your comparison mean in terms of the accounts?

12. On Monday 20 people receive a prank e-mail "warning" them that there is going to be a gasoline shortage soon. On Tuesday 40 more people receive the e-mail.

a. First, assume that the number of people receiving the e-mail is growing exponentially. Let $E(t)$ represent the number of people who receive the e-mail on the day that is t days since Monday. Find an equation for E.

b. Now assume that the number of people receiving the e-mail is growing linearly. Let $L(t)$ represent the number of people who receive the e-mail on the day that is t days since Monday. Find an equation for L.

c. Compare $E(7)$ and $L(7)$. What does your comparison mean in terms of the situation?

d. Compare $E(28)$ and $L(28)$. What does your comparison mean in terms of the situation?

e. Use your graphing calculator to draw the graphs of E and L on the same coordinate system. Compare the graphs.

f. Is there much difference in the number of people who will receive the e-mail if it grows exponentially or linearly? Explain.

13. If the average number of children per couple is a little over 2, a country's population will remain constant. Although most countries have increasing populations, Italy's population is actually decreasing. In fact, the average number of children per couple in Italy is only 1.2. If this continues, Italy's population will decay exponentially, with a half-life of 40 years. In 2002 the population was about 57.7 million.

a. Let $f(t)$ represent Italy's population (in millions) at t years since 2002. Find an equation for f.

b. Find $f(16)$. What does your result mean in terms of Italy's population?

14. A storage tank contains radium, a radioactive element with a half-life of 1600 years. Let $f(t)$ represent the percentage of radium that remains in the tank at t years since the element was placed in the tank.

a. Find an equation for f.

b. What percentage of the radium will remain in the tank after 100 years?

c. What percentage of the radium will remain in the tank after 3200 years? Explain how you can find this result without using the equation for f.

15. A storage tank contains cobalt-60, a radioactive element with a half-life of 5.3 years. Let $g(t)$ represent the percentage of cobalt-60 that remains in the tank at t years since the element was placed in the tank.

a. Find an equation for g.

b. What percentage of the cobalt-60 will remain in the tank after 28 years?

c. Cobalt-60 decays by what percentage per year?

16. Do you wait very long for art to download from the Web? The amount of time you wait depends on (among other things) the speed of your modem. The speed of modems manufactured in 1997 is 56.6 K. (The term 56.6 K stands for 56,600 bits per second.) Industry estimates that a new version of a technological product will be twice as fast as the previous version that was available 1.5 years previously. Let $f(t)$ represent the speed of a modem (in thousands of bits per second) manufactured at t years since 1997.

a. Find an equation for f.

b. Predict the speed of modems that will be manufactured in 2007.

c. Suppose that a modem made 10 years from now will take 1 minute to download some art. How long would it take to download the art using a modem manufactured today? [**Hint**: Find the ratio of the modem speeds.]

Consider the scattergram of data and the graph of the model $f(t) = ab^t$ in the indicated figure. Sketch the graph of an exponential model that better describes the data and then explain how you would adjust the values of a and b of the original model in order to better describe the data.

17. See Fig. 58.

18. See Fig. 59.

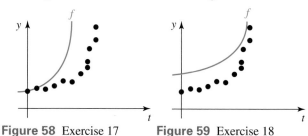

Figure 58 Exercise 17   **Figure 59** Exercise 18

19. The speed of a computer depends on (among other things) its chip speed. The speeds of various chips in megahertz (MHz) and when they were first available on the market are listed in Table 30.

Table 30 Chip Speeds

Type of Chip	Year	Clock Speed (MHz)
4004	1971	0.1
8080	1974	2
80186	1982	12
386	1985	16
486	1992	50
Pentium I	1995	120
Pentium II	1997	300
Pentium III	2000	1000

Source: *History of the Intel Microprocessor, www.i-probe.com/i-probe/ip_intel.html. Accessed 3/30/02.*

Let $f(t)$ represent the speed (in MHz) of a chip introduced into the market at t years since 1971.

a. Use a graphing calculator to draw a scattergram of the data. Is it better to model the data using a linear or an exponential function? Explain.

b. Find an equation for f.

c. If f is linear, find the slope. If f is exponential, find the base b of $f(t) = ab^t$. What does it mean in terms of chip speeds?

d. Suppose that you purchase a computer today. How many times faster will your computer be than a computer manufactured 10 years ago? [**Hint**: Find the ratio of the two chip speeds.]

This is the fastest one in the store.

e. Although it is true that the speed of a computer depends on its chip speed, explain why the ratio of the two chip speeds in part d may not equal the ratio of the computer speeds. [**Hint**: What does "depend" mean in this context?]

20. The total values of purchases using credit cards in the United States from Thanksgiving to Christmas are listed for various years in Table 31.

Table 31 Christmas Season Credit Card Purchases

Year	Credit Card Purchases (billions of dollars)
1987	37.0
1989	48.9
1991	59.8
1993	79.1
1995	116.3
1997	109.5
1999	126.5
2001	165.8

Source: *CardWeb.com, Inc., www.cardweb.com. Accessed 3/30/02.*

Let $f(t)$ represent the total value of credit card purchases (in billions of dollars) from Thanksgiving to Christmas during the year that is t years since 1987.

a. Use the data for 1987 and 2001 to find an equation for f.

b. The data for 1993 and 1995 give the exponential model:

$$f(t) = 24.89(1.2126)^t$$

The data for 1995 and 1997 give the exponential model:

$$f(t) = 147.99(0.9703)^t$$

In part a you found yet another possible model. Use your graphing calculator to determine which of these three equations seems to best model the credit card data. For the three pairs of data points discussed in this exercise, is it possible to see which two points will yield the best exponential model by inspecting the scattergram without deriving the three equations? If so, explain how.

21. The world populations for various years are provided in Table 32.

Table 32 World Populations

Year	Population (billions)
1930	2.070
1940	2.295
1950	2.500
1960	3.050
1970	3.700
1980	4.454
1990	5.279
2000	6.080
2002	6.215

Source: *The Cambridge Factfinder; Census Bureau—POP clocks, www.census.gov/main/www/popclock.html. Accessed 3/30/02.*

Let $f(t)$ represent the world population (in billions) at t years since 1900.

a. Use a graphing calculator to draw a scattergram of the data. Is it better to use a linear or an exponential function to model the data? Explain.

b. Find an equation for f.

c. What is the base b of your model $f(t) = ab^t$? What does it mean in terms of world population?

d. What is the coefficient a of your model $f(t) = ab^t$? What does it mean in terms of world population?

e. Find $f(110)$. What does your result mean in terms of world population?

22. The world enrollments in the Mormon Church for various years are listed in Table 33. Let $f(t)$ represent the enrollment (in millions) in the Mormon Church at t years since 1900.

Table 33 World Enrollments in the Mormon Church	
Year	Enrollment (millions)
1947	1.0
1971	2.7
1980	4.64
1990	7.57
1996	9.43
2002	11.07

Source: *Church of Jesus Christ of Latter-Day Saints*

a. Perform the first three steps of the four-step modeling process to find an equation for f. [**Modeling Process:** 1. Scattergram 2. Equation 3. Verify 4. Estimate/predict]

b. Find $f(108)$. What does your result mean in terms of Mormon enrollment?

c. A reasonable model for the world population is $f(t) = 1.197(1.0163)^t$, where $f(t)$ represents the population (in billions) at t years since 1900. Compare the base of the Mormon model with the base of the world population model. What does your comparison tell you? Explain.

23. The number of Starbucks® stores has substantially increased since the mid 1980s (see Table 34).

Table 34 Numbers of Starbucks Stores	
Year	Number of Stores
1987	17
1989	55
1991	116
1993	272
1995	676
1997	1412
1999	2135
2001	4709

Source: *Data found by Hudi B., Spring Semester, 2002: Timeline and History, www.starbucks.com/aboutus/timeline.asp. Accessed 5/24/02.*

Let $f(t)$ represent the number of stores at t years since 1980.

a. Perform the first three steps of the four-step modeling process to find an equation for f. [**Modeling Process**: 1. Scattergram 2. Equation 3. Verify 4. Estimate/predict]

b. Predict the number of stores in 2008.

c. What is the percentage rate of growth of stores?

24. The numbers of homes that have sold for more than $1 million has greatly increased in the last decade (see Table 35).

Table 35 Numbers of Homes Sold for Over $1 Million	
Year	Number (in thousands)
1995	2.5
1996	3.4
1997	4.9
1998	7.2
1999	10.3
2000	15.6

Source: *DataQuick Information Systems*

Let $n = g(t)$ represent the number of homes (in thousands) that sell for over $1 million in the year that is t years since 1990.

a. Perform the first three steps of the four-step modeling process to find an equation for g.

b. Predict how many homes will be sold for over $1 million in 2007.

c. Find the n-intercept of g. What does it mean in terms of the situation?

25. From 1790 to 1860, the U.S. population grew rapidly (see Table 36).

a. Complete the third column of Table 36. The first entry is 1.36, since $\dfrac{1800 \text{ population}}{1790 \text{ population}} = \dfrac{5.3}{3.9} \approx 1.36$.

Table 36 U.S. Populations

Year	Population (millions)	Population Ratio (current to previous)
1790	3.9	[leave blank]
1800	5.3	1.36
1810	7.2	
1820	9.6	
1830	12.9	
1840	17.1	
1850	23.2	
1860	31.4	

Source: *U.S. Census Bureau*

b. What do you observe about the ratios in the third column?

c. Based on your observation in part b, is it better to use an exponential function or a linear function to model the data? Explain.

d. Let $f(t)$ represent the U.S. population (in millions) at t years since 1790. Find an equation for an exponential function that models the data from 1790 to 1860.

e. Complete the third column of Table 37.

Table 37 U.S. Populations

Year	Population (millions)	Population Ratio (current to previous)
1860	31.4	[leave blank]
1870	39.8	
1880	50.2	
1890	62.9	
1900	76.0	

Source: *U.S. Census Bureau*

f. Is it likely that f gives reasonable population estimates after 1860? Explain.

g. Use f to predict the population in 2002. The actual population was 286.7 million. What is the error in your estimate?

26. Laptop computer sales nearly doubled from 1989 to 1994 (see Table 38).

Table 38 Laptop Computer Sales

Year	Sales (million laptops)	Sales Ratio (current to previous)
1989	0.97	[leave blank]
1990	1.11	1.14
1991	1.26	
1992	1.45	
1993	1.66	
1994	1.90	

Source: Where Sales Will Soar, *Thayer C. Taylor*, as reported by *Dave T., Spring Semester, 1999*

a. Complete the third column of Table 38. The first entry is 1.14, since $\dfrac{1990 \text{ sales}}{1989 \text{ sales}} = \dfrac{1.11}{0.97} \approx 1.14.$

b. What do you observe about the ratios in the third column? What is the percent growth each year in laptop sales from 1989 to 1994?

c. Based on your observation in part b, is it better to use an exponential or linear model? Explain.

d. Let $f(t)$ represent the laptop sales (in millions) at t years since 1989. Use the base multiplier property to find an equation for f.

e. Now find the regression exponential equation for f. Compare the coefficient and the base of the regression equation with your result in part d. [**Graphing Calculator:** See Section B.16.]

27. The amounts of the federal debt for various years are listed in Table 39.

Table 39 Federal Debt Amounts

Year	Federal Debt (billions of dollars)
1965	322
1970	381
1975	542
1980	909
1985	1817
1990	3206
1993	4351
1994	4644

Source: *U.S. Office of Management and Budget*

a. Let $D = f(t)$ represent the federal debt (in billions of dollars) at t years since 1950. Find an equation for a function f that models the data well for the years 1965–1994.

b. From 1994 to 1997, the debt increased by smaller and smaller amounts to the point where in 1998 the United States ran a surplus. Explain why this means that model breakdown will occur if we use f to make predictions for any of the years from 1995 to 1998.

c. To get an idea of what *would* have happened if the federal debt had continued to grow exponentially, respond to the following questions by assuming that f will model the debt well for future years.

 i. Use f to predict the federal debt at 2009.

 ii. If the federal debt were paid off in 2009 by each person in the United States contributing an equal amount of money, how much would each person have to pay? Assume that the population will be about 297 million in 2009.*

 iii. Use f to predict the federal debt at 2050.

 iv. If the federal debt were paid off in 2050 by each person in the United States contributing an equal amount of money, how much would each person have to pay? Assume that the population will be about 394 million in 2050.† Explain why the United States wants to reduce or eliminate the debt now, rather than later.

28. In 1989 an institute at Stanford University performed research that included comparing the economic strength of a country with the percentage of the population that is involved in producing agriculture (see Table 40). **Gross national product** (GNP) measures the amount of product produced by a country.

---

*Source: U.S. Bureau of the Census, *Statistical Abstract of the United States*, 118th ed. (Washington, DC, 1998), 9.

†Ibid.

Table 40 Percent of Population in Agriculture versus GNP

Country	Percent of Population in Agriculture	GNP per Person (dollars)*
United States	1	14,100
Great Britain	1	9,300
West Germany	3	11,400
Canada	5	12,300
Australia	6	11,400
France	8	10,500
Japan	9	10,100
Italy	10	6,400
Argentina	12	2,050
Soviet Union	16	6,700
Mexico	33	2,700
South Korea	36	1,950
Brazil	37	1,900
Nigeria	49	700
Egypt	49	600
Pakistan	51	400
Indonesia	55	500
China	56	350
India	61	300
Bangladesh	83	200

Source: Population and Resources in a Changing World *by Kingsley Davis et al.*

*The GNPs are for 1983.

a. Let $f(p)$ represent the GNP per person (in dollars) for a country where p percent of the population is involved with agriculture. Use your graphing calculator to find the regression equation for f. [**Graphing Calculator:** See Section B.16.]

b. Use your graphing calculator to sketch a scattergram of the data as well as your model on the same coordinate system. Does the model fit the data well?

c. What is the base b of your function $f(p) = ab^p$? What does the base mean in terms of the situation? Explain why this makes sense in terms of productivity.

d. Which country's data point is farthest from the regression curve? What does the position of the point in relation to the other data points and the regression curve suggest about the workers in this country?

29. Explain how to find an exponential model for a given situation. Also, explain how to use the model to make an estimate or prediction for the situation.

Taking it to the lab

For each lab assignment, consult with your instructor on whether to organize your responses as a numbered list or to write them in a paragraph.

STRINGED INSTRUMENT LAB

Stringed instruments (like guitars, banjos, and basses) often have thin metal strips called **frets** across the neck and underneath the strings of the instrument. These frets are precisely placed so the instruments produce the 12 chromatic notes of our Western musical scale. In this lab you will discover where the frets of a stringed instrument must be placed in order to produce the 12 chromatic notes. What you learn can also be applied to determine where violinists and cellists must put their fingers to produce the 12 chromatic notes.

Materials

You will need the following materials:

1. a stringed instrument with frets
2. a meter stick or tape measure

Note

There are many bridges that allow for each of the instruments's strings to have slightly different open string lengths. Be consistent with the string you use for this lab.

Recording of Data

Measure the length (in centimeters) of one of the strings of your instrument from the *nut* to the *bridge* (see Fig. 60). This is the length of an *open string*. Then measure the length (in centimeters) of the same string from the 12th fret to the bridge.

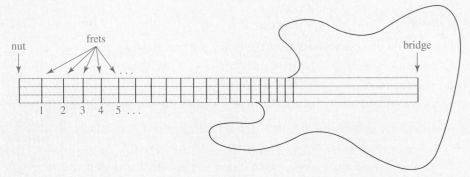

Figure 60 A bass guitar

Analyzing the Data

1. Compare the sound produced by plucking the open string and plucking the string when it is pressed just before the 12th fret (toward the nut). The higher-pitched note is called the **octave** of the lower note. How do distances between the nut and the bridge, and between the 12th fret and the bridge, compare?

2. You probably found that when you halve the length of an open string, you can produce the octave of the open string. Since there are 12 notes in the Western chromatic musical scale from a note to its octave, this means that the 12th fret should be placed in the middle of the nut and the bridge. Where should the 24th fret be placed to achieve the next octave?

3. Complete Table 41.

4. Let $f(n)$ represent the distance (in centimeters) from the nth fret to the bridge. Based on the entries in Table 41, is f linear, exponential, or neither? Explain.

5. Find an equation for f.

6. Use your equation for f to find the distance between the fifth fret and the bridge.

7. Use a calculator table to find the appropriate distances for all the frets of your instrument. Compare these values with the actual distances.

8. The author owns a 1974 Fender® Jazz bass with one of its strings having open string length 86.5 cm. Explain how the Fender music technician knew where to put the frets.

9. In this lab you observed that an octave is achieved by halving a string. Explain why it follows that the frets of an instrument are closer together when they are closer to the bridge.

Note

Many stringed instruments have fewer than 24 frets.

Table 41 Fret Positions of a Bass Guitar	
Number of Frets	Length of String from Fret to Bridge (centimeters)
0	
12	
24	
36	
48	
n	

COOLING WATER LAB

In this lab you will relate the temperature of some heated water to the amount of time that the water has been cooling.

In case the necessary measuring devices are not available for this experiment, some sample data are listed in Table 42.

Table 42 Temperatures of Water*		
Time (minutes)	Temperature (°C)	Difference between Temperature and Room Temperature* (°C)
0	83.49	
5	71.22	
10	63.09	
15	57.23	
20	52.65	
25	48.83	
30	45.63	

Source: *Data collected by the author*

*Room temperature $= 21.7°C$

Materials

In order to do this lab you will need the following materials:

1. some hot water
2. a coffee cup
3. a temperature probe that measures temperatures in degrees Celsius. Or, you can measure temperatures in degrees Fahrenheit and use the equation $C = \dfrac{F - 32}{1.8}$ to convert from degrees Fahrenheit to degrees Celsius. If you are using a thermometer for your temperature probe, make sure it can handle the temperature of your hot water, otherwise it may break or even explode.
4. a timing device

Preparation

Prepare some hot water in a coffee cup. Set the cup on a counter or table. Choose an environment where the temperature of the air will not change much during the 30-minute experiment.

Recording of Data

Record the temperature of the air at the start and then again at the end of the experiment. If the room temperatures at the beginning and end of the experiment are significantly different, redo the experiment in an environment where the room temperature will not change much. Also, record the temperature of the cooling water every minute for 30 minutes.

Analyzing the Data

1. Find the average of the room temperature in degrees Celsius at the start and end of the experiment. This average will be referred to as *the* room temperature.
   If you are using the sample data in Table 42, the room temperature when it was collected was 21.7 degrees Celsius.
2. Display the water temperature data in the first two columns of a table like Table 42.
3. For each temperature reading of the water, compute the difference between the temperature of the water and the room temperature. Enter these differences in the third column of your table.
4. Let $D = f(t)$ represent the difference between the temperatures of the water and the room at t minutes after the water is allowed to cool. Use pencil and paper to create a scattergram comparing D to t for your cooling water data.
5. Use a graphing calculator to draw a scattergram of the data. Which will best model the data, a linear function or an exponential function? Explain.
6. Find an equation for f.
7. If your model is linear, what is the meaning of the slope? If your model is exponential, what is the meaning of the base?
8. Estimate the temperature of the water at 22 minutes.
9. Estimate the temperature of the water at one hour.
10. What will be the temperature of the water when it stops cooling? Does $f(t)$ predict when this temperature will be reached? According to $f(t)$, how much time will it take for this to happen? Is this a reasonable prediction? Explain.

EXPONENTIAL LAB: TOPIC OF YOUR CHOICE

Your objective in this lab is to find some true-to-life data that appears to be exponential and then model it with an exponential function. Your function should model a situation that has not been discussed in this text. Your first task will be to find some data. Almanacs, newspapers, magazines, scientific journals, and the Internet are good resources. Or, you can conduct an experiment to obtain your data.

Analyzing the Data

1. What two variables did you explore? Did you make any false starts? If so, what variables were you exploring then? Explain why you chose new variables.

2. Which variable is the dependent variable? Which variable is the independent variable?

3. Does it make sense to you why an exponential function would best model your situation? Explain.

4. State the source of your data. If you conducted an experiment, provide a careful description with specific details of how you ran your experiment.

5. Include a table of your data.

6. Use your graphing calculator to draw a scattergram of your data.

7. Find an equation for your exponential model.

8. Use your graphing calculator to graph your exponential model and your scattergram in the same viewing window. Also sketch the graph of your model and scattergram using pencil and paper.

9. What is the base b of your exponential model? What does it tell you about the situation you are modeling?

10. What is the coefficient a of your exponential model? What does it tell you about the situation you are modeling?

11. Find any intercepts of your exponential model. What do they represent in terms of the situation being modeled?

12. Make a prediction or estimate about your dependent variable based on a specific value for your independent variable. Interpret your prediction in terms of the situation being modeled.

13. Comment on your lab experience.

    a. For example, you might address whether this lab was enjoyable, insightful, and so on.

    b. Were you surprised by any of your findings? If so which ones?

    c. How would you improve your lab process for this lab if you did it again?

    d. How would you improve your process if you had more time and money?

Chapter Summary

Key Points
OF THIS CHAPTER

For the following definitions, n is a counting number.

Definitions of exponents

- $b^0 = 1, b \neq 0$

- $b^{-n} = \dfrac{1}{b^n}, b \neq 0$

- If $b \geq 0$, then $b^{1/n}$ is the nonnegative number whose nth power is b.

- If $b < 0$ and n is odd, then $b^{1/n}$ is the (negative) number whose nth power is b.

For the following definition, m and n are counting numbers and b is any real number for which $b^{1/n}$ is a real number.

- $b^{m/n} = \left(b^{1/n}\right)^m = \left(b^m\right)^{1/n}$

- $b^{-m/n} = \dfrac{1}{b^{m/n}}, b \neq 0$

For the following properties, m and n are rational numbers and b and c are any real numbers for which b^m, b^n, and c^n are real numbers.

Properties of exponents

- $b^m b^n = b^{m+n}$

Properties of exponents (Continued)

- $\dfrac{b^m}{b^n} = b^{m-n}, b \neq 0$

- $(bc)^n = b^n c^n$

- $\left(\dfrac{b}{c}\right)^n = \dfrac{b^n}{c^n}, c \neq 0$

- $(b^m)^n = b^{mn}$

Simplifying an exponential expression

An exponential expression is simplified if:

1. there are no parentheses;
2. each constant appears as a base as few times as possible. (For example, for $b \neq 0$,
$$b^2 + \frac{b^4}{b^3} = b^2 + b);$$
3. each numerical expression (such as 3^2 or $4 \cdot 6$) has been calculated and each numerical fraction has been reduced; and
4. each exponent is positive.

Scientific and standard decimal notations

To write the number $N \times 10^k$ in standard decimal notation, move the decimal point of the number N as follows:

- If k is positive, move the decimal point k places to the right.
- If k is negative, move the decimal point k places to the left.

To write a number in scientific notation, count the number of places k that the decimal point needs to be moved so that the new number N is between 1 and 10 (or -1 and -10).

- If the decimal point is moved to the left, then the scientific notation of the original number is $N \times 10^k$.
- If the decimal point is moved to the right, then the scientific notation of the original number is $N \times 10^{-k}$.

Properties of an exponential function

For the exponential function $f(x) = ab^x$:

- If the value of x increases by 1, the value of y is multiplied by the base b.
- The y-intercept is $(0, a)$.
- The domain is the set of all real numbers.
- If $a > 0$, then the graph of f is above the x-axis and the range is the set of all positive numbers.
- If $a < 0$, then the graph of f is below the x-axis and the range is the set of all negative numbers.
- If $a > 0$ and $b > 1$, then f is an increasing function.
- If $a > 0$ and $0 < b < 1$, then f is a decreasing function.
- The x-axis is a horizontal asymptote.

Reflection property

The graphs of $f(x) = -ab^x$ and $g(x) = ab^x$ are mirror reflections of each other across the x-axis.

Solving an equation in one variable

When solving an equation of the form $b^n = k$ for b:

- if n is odd, the solution is $k^{1/n}$;
- if n is even and $k \geq 0$, the solution is $\pm k^{1/n}$; and
- if n is even and $k < 0$, there is no real number solution.

Finding an equation in two variables

We can find an equation of an exponential model by:

- using the base multiplier property;
- using two points; and
- exponential regression.

Assume that $f(t) = ab^t$, $a > 0$, is an exponential model, where $f(t)$ represents a quantity at time t.

Properties of an exponential model

- The coefficient a represents the amount of the quantity present at time $t = 0$.
- If $b > 1$, then the quantity grows exponentially at a rate of $b - 1$ percent (decimal form) per unit of time.
- If $0 < b < 1$, then the quantity decays exponentially at a rate of $1 - b$ percent (decimal form) per unit of time.

CHAPTER 4 REVIEW EXERCISES

Simplify the expression. Assume that b and c are positive.

1. $\dfrac{2^{-400}}{2^{-405}}$

2. $(8b^{-3}c^5)(6b^{-9}c^{-2})$

3. $\dfrac{4b^{-3}c^{12}}{16b^{-4}c^3}$

4. $(2b^{-5}c^{-2})^3(3b^4c^{-6})^{-2}$

5. $(6b + c)^0 - (4b)^0$

6. $\dfrac{(20b^{-2}c^{-9})(27b^5c^3)}{(18b^3c^{-1})(30b^{-1}c^{-4})}$

7. $(37b^{-3}c^4)^{-97}(37b^{-3}c^4)^{97}$

8. $32^{4/5}$

9. $16^{-3/4}$

10. $\dfrac{b^{-1/3}}{b^{4/3}}$

11. $\dfrac{(16b^8c^{-4})^{1/4}}{(25b^{-6}c^4)^{3/2}}$

12. $\left(\dfrac{32b^2c^5}{2b^{-6}c^1}\right)^{1/4}$

13. $(8^{2/3}b^{-1/3}c^{3/4})(64^{-1/3}b^{1/2}c^{-5/2})$

Simplify the expression. Assume that n is a counting number.

14. $b^{2n-1}b^{4n+3}$

15. $\dfrac{b^{n/2}}{b^{n/3}}$

16. Use properties of exponents to show why $3^{2x} = 9^x$.

Let $f(x) = (8)^x$.

17. $f(2)$

18. $f(0)$

19. $f(-2)$

20. $f\left(\dfrac{2}{3}\right)$

21. Find x when $f(x) = 2$.

22. Find x when $f(x) = \dfrac{1}{8}$.

Write the number in standard decimal form.

23. 4.4487×10^7

24. 3.85×10^{-5}

Write each number in scientific notation.

25. 54,698,201

26. -0.00897

Graph each function by hand.

27. $f(x) = 2(3)^x$

28. $g(x) = 12\left(\dfrac{1}{2}\right)^x$

29. $h(x) = -3(2)^x$

30. $k(x) = -18\left(\dfrac{1}{3}\right)^x$

Solve each equation.

31. $b^3 = 8$

32. $2b^5 = 60$

33. $3.9b^7 = 283.5$

34. $5b^4 - 13 = 67$

35. $\dfrac{1}{3}b^2 - \dfrac{1}{5} = \dfrac{2}{3}$

36. Some values of the functions f, g, h, and k are provided in Table 43. For each function, determine whether the given values suggest that the function is linear, exponential, or neither. If the function could be linear or exponential, find a possible equation for it.

Table 43 Values of Four Functions (Exercises 36–40)				
x	$f(x)$	$g(x)$	$h(x)$	$k(x)$
1	30	5	2	96
2	26	15	3	48
3	22	45	6	24
4	18	135	11	12
5	14	405	18	6

For exercises 37–40, refer to Table 43.

37. Find $f(4)$.

38. Find $h(3)$.

39. Find x when $g(x) = 5$.

40. Find x when $k(x) = 6$.

Find an equation of the exponential curve that passes through the given pair of points.

41. $(0, 2)$ and $(5, 3)$

42. $(0, 3.8)$ and $(4, 113.2)$

43. $(3, 30)$ and $(9, 7)$

44. $(5, 6.9)$ and $(20, 78.6)$

45. Consider the scattergram of data and the graph of the model $f(x) = ab^t$ in Fig. 61. Sketch the graph of an exponential model that better describes the data and then explain how you would adjust the values of a and b of the original model in order to better describe the data.

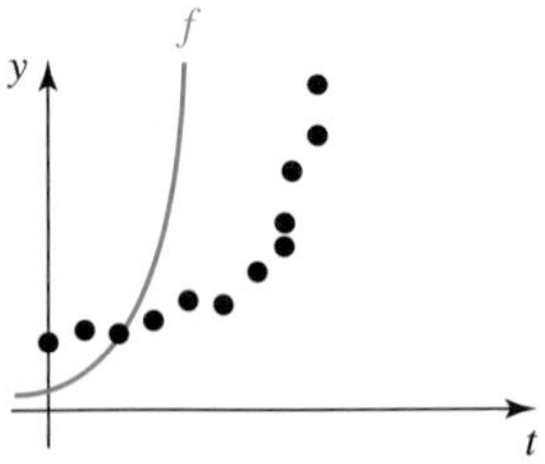

Figure 61 Exercise 45

46. Suppose that $2000 is deposited into an account that earns 7% interest compounded annually. Let $f(t)$ represent the value (in dollars) of the account at t years after depositing the $2000.

    a. Find an equation for f.

    b. Find the value of the account at 5 years.

47. A corporation's annual total sales have doubled every 7 years. The total sales in 2000 was $500,000.

    a. Let $g(t)$ represent the total sales (in thousands of dollars) for the year that is t years since 1990. Find an equation for g.

    b. Predict the corporation's total sales in 2009.

48. A storage tank contains carbon-14, a radioactive element with a half-life of 5730 years. Let $f(t)$ represent the percent of carbon-14 that remains in the tank at t years since the element was placed in the tank.

    a. Find an equation for f.

    b. Use your equation for f to predict the percentage of carbon-14 remaining in the tank after 100 years.

49. Table 44 shows the number of Kohl's® stores for various years.

Table 44 Numbers of Kohl's Stores

Year	Number of Stores
1996	150
1997	182
1998	213
1999	259
2000	320
2001	382

Source: *SEC filings*

a. Let $f(t)$ represent the number of stores at t years since 1990. Use a graphing calculator to draw a scattergram of the data. Is it better to use a linear or an exponential function to model the data? Explain.

b. Find an equation for f.

c. What is the coefficient a of your model $f(t) = ab^t$? What does it mean in terms of the number of stores?

d. What is the base b of your model $f(t) = ab^t$? What does it mean in terms of the number of stores?

e. Find $f(18)$. What does your result mean in terms of the number of stores?

50. As the political momentum against the tobacco industry has increased over the past few years, so have the number of lawsuits filed against tobacco companies (see Table 45).

Table 45 Numbers of Lawsuits Filed against Tobacco Companies

Year	Number of Lawsuits
1993	49
1994	73
1995	200
1996	352
1997	733

Source: *R.J. Reynolds Tobacco Co.*

a. Let $f(t)$ represent the number of lawsuits filed during the year that is t years since 1993. Find an equation for f.

b. Predict the number of lawsuits that will be filed during 2008.

c. Some experts believe that many of the filed cases will disappear or fail. Suppose that only 10% of the filed cases go to court and that the tobacco industry loses half of these cases and pays about $10 million for each lost case.* Predict how much money the tobacco industry will pay for lost cases during the year 2008.

d. On June 17, 1998, a bill introduced by Senator John McCain was defeated. This bill would have required the tobacco industry to pay $516 billion. For months before this bill's defeat, tobacco industry executives said that they would much rather fight cases in court than be required to pay money due to such a bill. If your model turns out to be accurate through 2008 (see part c), are the tobacco industry executives suggesting a strategy that would be good for their industry?

CHAPTER 4 TEST

Compute each expression. Show work to demonstrate that you can find the result without using a graphing calculator.

1. $32^{2/5}$

2. $-8^{-4/3}$

Simplify the expression.

3. $(2b^3c^8)^3$

4. $\left(\dfrac{4b^{-3}c}{25b^5c^{-9}}\right)^0$

5. $\dfrac{b^{1/2}}{b^{1/3}}$

6. $\dfrac{25b^{-9}c^{-8}}{35b^{-10}c^{-3}}$

7. $\left(\dfrac{6b(b^3c^{-2})}{3b^2c^5}\right)^2$

8. $\dfrac{(25b^8c^{-6})^{3/2}}{(7b^{-2})(2c^3)^{-1}}$

9. Use properties of exponents to show that $8^{x/3}2^{x+3} = 8(4)^x$.

*Some verdicts are for substantially more than $10 million. In fact, the tobacco industry paid out $35 billion during 1997 to settle lawsuits brought by four states: Minnesota, Florida, Mississippi, and New York.

Sketch the graph of the function.

10. $f(x) = -5(2)^x$

11. $f(x) = 18\left(\dfrac{1}{3}\right)^x$

12. For each graph in Fig. 62, find an equation of an exponential function that could fit the graph.

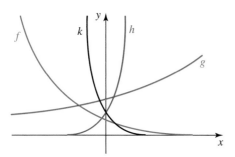

Figure 62 Exercise 12

13. Some values of a function f are provided in Table 46. Find a possible equation for f in terms of t.

Table 46 Values of a Function f (Exercise 13)	
t	$f(t)$
0	160
1	80
2	40
3	20
4	10
5	5

14. Solve for b: $3b^6 = 79$.

Find an equation of an exponential curve that passes through the given pair of points.

15. $(0, 70)$ and $(6, 20)$

16. $(4, 9)$ and $(7, 50)$

For exercises 17–19, refer to Fig. 63.

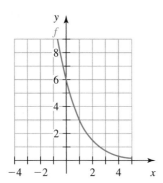

Figure 63 Exercises 17–19

17. Find $f(0)$.

18. Find x where $f(x) = 3$.

19. Find a possible equation for f.

20. On March 1 there are 40 leaves on a tree. For a while, the total number of leaves doubles each week. Let $f(t)$ represent the total number of leaves on the tree at t weeks since March 1.

   a. Find an equation for f.

   b. Find $f(6)$. What does your result mean in terms of the leaves?

   c. Find $f(52)$. What does your result mean in terms of the leaves?

21. In recent years the chances for treating infertility have greatly increased. Such treatments also increase the chances of a woman having a multiple birth (triplets or more). Due to these treatments and an increasing population, there has been a substantial increase in the number of multiple births, including septuplets (7 babies) and octuplets (8 babies). See Table 47.

Table 47 Numbers of Multiple Births	
Year	Number of Multiple Births*
1971	1000
1975	1100
1980	1300
1985	1900
1990	3300
1995	5000
2000	7325

Source: *U.S. Department of Health and Human Services*

*Triplets or more

   a. Let $f(t)$ represent the number of multiple births (triplets of more) in the United States at t years since 1970. Find an equation for f.

   b. Estimate the percent growth in multiple births (triplets or more) each year.

   c. What is the coefficient a of your model $f(t) = ab^t$? What does it mean in terms of the situation?

   d. Predict the number of multiple births in 2008.

Logarithmic Functions

Never regard study as a duty, but as the enviable opportunity to learn to know the liberating influence of beauty in the realm of the spirit for your own personal joy and to the profit of the community to which your later work belongs.

—*Albert Einstein*

5.1 INVERSE FUNCTIONS

Objectives

▸ Know the meaning of *inverse of a function, invertible function,* and *one-to-one function.*

▸ Know the *reflection property.*

▸ Graph the inverse of a function.

▸ Use the inverse of a model to make predictions.

▸ Find an equation for the inverse of a function.

In this chapter we will study a type of function that has a special relationship to a given function. It is called the *inverse* of a function. We will begin this section with an example that will allow us to develop a definition of inverse function. Then we will describe such functions by first using tables, then graphs, and finally, equations.

Although the United States and several other countries use the Fahrenheit scale (°F) to measure temperature, most countries use the Celsius scale (°C). A comparison of the two scales is shown in Table 1.

Table 1 Celsius and Fahrenheit Equivalent Readings	
Celsius	Fahrenheit
0	32
20	68
40	104
60	140
80	176
100	212

It's 30 degrees here.

Same here!

If an American visiting France hears that the local temperature will reach a high of 20°C, it would be helpful to be able to convert 20°C to 68°F. There is a function g that converts Celsius inputs to Fahrenheit outputs (see Fig. 1).

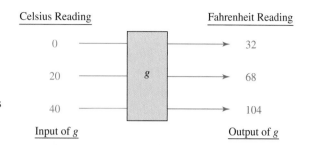

Celsius Reading · Fahrenheit Reading

0 → 32

20 → g → 68

40 → 104

Input of g · Output of g

Figure 1 The function g converts Celsius readings to Fahrenheit readings

If a European visiting the United States hears that the local temperature will reach a high of 68°F, it would be helpful to be able to convert 68°F to 20°C. If we reverse the arrows of Fig. 1, we have an input-output diagram of a new relation (see Fig. 2). For each Fahrenheit reading, there is exactly one Celsius reading, so this new relation is also a function. We call this function the *inverse* of g "g^{-1}" (read "g-inverse").

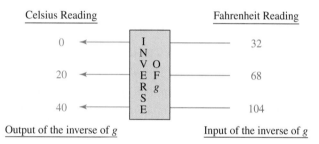

Celsius Reading · Fahrenheit Reading

0 ← 32

20 ← INVERSE OF g ← 68

40 ← 104

Output of the inverse of g · Input of the inverse of g

Figure 2 The inverse of g converts Fahrenheit readings back to Celsius readings

There are two key observations we can make about g^{-1}:

1. g^{-1} sends outputs of g to inputs of g. For example, g sends the input 0 to the output 32 and g^{-1} sends 32 to 0 (see Figs. 1 and 2). Using symbols, we write

$$g(0) = 32 \quad \text{and} \quad g^{-1}(32) = 0$$

We say that these two statements are **equivalent**, which means that one statement implies the other and vice versa.

2. g^{-1} undoes g. For example, g sends 0 to 32 and g^{-1} undoes this action by sending 32 *back* to 0 (see Fig. 3).

The inverse of a function is not necessarily a function. If the inverse of a function f is also a function we say that f is **invertible** and use "f^{-1}" as a name for the inverse of f.

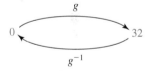

Figure 3 g^{-1} undoes g

Note

Recall that $a^{-1} = \dfrac{1}{a}$. Now it becomes important to keep in mind that $f^{-1}(x)$ is *not* $\dfrac{1}{f(x)}$. It is possible to express $\dfrac{1}{f(x)}$ as $[\,f(x)]^{-1}$, if so desired.

Property of an Inverse Function

For an invertible function f, the following statements are equivalent:

$$f(a) = b \quad \text{and} \quad f^{-1}(b) = a$$

In words, if f sends a to b, then f^{-1} sends b to a. If f^{-1} sends b to a, then f sends a to b.

Example 1 Evaluating an Inverse Function

Let f be an invertible function where $f(2) = 5$. Find $f^{-1}(5)$.

Solution

Since f sends 2 to 5, we know that f^{-1} sends 5 back to 2. So, $f^{-1}(5) = 2$. ∎

Table 2 Input-Output Values of f	
x	$f(x)$
0	1
1	3
2	9
3	27
4	81

Example 2 Evaluating f and f^{-1}

Some values for an invertible function f are shown in Table 2.

1. Find $f(3)$.
2. Find $f^{-1}(9)$.

Solution

1. $f(3) = 27$.
2. Since f sends 2 to 9, we conclude that f^{-1} sends 9 back to 2. Therefore, $f^{-1}(9) = 2$. ∎

All linear functions with nonzero slope and all exponential functions are invertible. (We will see why at end of this section.) Therefore, we can use the notation f^{-1} whenever describing the inverse of either of these two types of functions.

Example 3 Finding Input-Output Values for an Inverse Function

Let $f(x) = 16\left(\dfrac{1}{2}\right)^x$.

1. Find five input-output values for f^{-1}.
2. Find $f^{-1}(8)$.

Solution

1. We begin by finding input-output values for f (see Table 3). Since f^{-1} sends outputs of f to inputs of f, we conclude that f^{-1} sends 16 to 0, 8 to 1, 4 to 2, 2 to 3, and 1 to 4. We list these results from smallest input to largest input in Table 4.
2. From Table 4 we see that f^{-1} sends the input 8 to the output 1, so $f^{-1}(8) = 1$.

Table 3 Input-Output Values of f	
x	$f(x)$
0	16
1	8
2	4
3	2
4	1

Table 4 Input-Output Values of f^{-1}	
x	$f^{-1}(x)$
1	4
2	3
4	2
8	1
16	0

∎

Let's return to the function g that converts Celsius readings to Fahrenheit readings. Consider the input-output diagrams for g and g^{-1} in Figs. 4 and 5. We've discussed that to find the inverse of g we reverse the arrow for g. But also notice that if we reverse the arrow for g^{-1}, we get the arrow for g. This suggests that the inverse of g^{-1} is g, which is true. So, g and g^{-1} are inverses of each other.

Figure 4 g sends values of C to values of F

Figure 5 g^{-1} sends values of F to values of C

---

f and f^{-1} are Inverses of Each Other

If f is an invertible function, then

- f^{-1} is invertible, and
- f and f^{-1} are inverses of each other.

So far, we have described inverse functions by using tables. We now describe them by using graphs.

Example 4   Comparing the Graphs of a Function and Its Inverse

Sketch the graphs of $f(x) = 2^x$, f^{-1}, and $y = x$ on the same set of axes.

Solution

We list some input-output values of f and f^{-1} in Tables 5 and 6.

Table 5 Input-Output Pairs for f	
x	$f(x)$
-3	$\frac{1}{8}$
-2	$\frac{1}{4}$
-1	$\frac{1}{2}$
0	1
1	2
2	4

Table 6 Input-Output Pairs for f^{-1}	
x	$f^{-1}(x)$
$\frac{1}{8}$	-3
$\frac{1}{4}$	-2
$\frac{1}{2}$	-1
1	0
2	1
4	2

From Tables 5 and 6 it is clear that if a point (a, b) is on the graph of f, then the point (b, a) is on the graph of f^{-1}. This observation will lead us to a key step in graphing inverses in future problems.

Next, we plot the input-output pairs of f using blue dots, plot the input-output pairs of f^{-1} using red dots, and sketch the line $y = x$ (see Fig. 6).

Note that if we were to draw the blue dots in wet ink and fold the paper along the line $y = x$, the ink would make dots where the red dots are. We say that the red dots are the reflection of the blue dots across the line $y = x$.

In Fig. 7 we sketch a graph of $f(x) = 2^x$ using a blue exponential curve. By reflecting all the blue points on the graph of f across the line $y = x$, we obtain the red graph of f^{-1}.

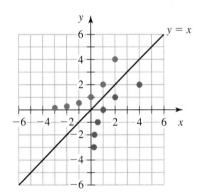

Figure 6 Some points of the graph of f (in blue), some points of the graph of f^{-1} (in red), and the line $y = x$

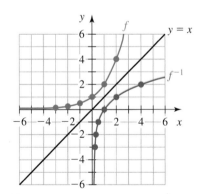

Figure 7 The graphs of f, f^{-1} and $y = x$

Note that the graph of f^{-1} is the reflection of the graph of $f(x) = 2^x$ across the line $y = x$. ∎

Reflection Property

For an invertible function f, the graph of f^{-1} is the reflection of the graph of f across the line $y = x$.

In Example 4 we saw that if a point (a, b) is on the graph of f, then the point (b, a) is on the graph of the inverse of f.

Graphing an Inverse Function

For an invertible function f, we sketch the graph of f^{-1} by the following steps:

1. Sketch the graph of f.
2. Choose several points that lie on the graph of f.
3. For each point (a, b) chosen in step 2, plot the point (b, a).
4. Sketch the curve that contains the points plotted in step 3.

Example 5   Graphing an Inverse Function

Let $f(x) = \dfrac{1}{3}x - 1$. Sketch the graph of f, f^{-1}, and $y = x$ on the same set of axes.

Solution

We apply the four steps to graph the inverse function.

Step 1. Sketch the graph of f: See Fig. 8.

Step 2. Choose several points that lie on the graph of f: $(-6, -3)$, $(-3, -2)$, $(0, -1)$, $(3, 0)$, and $(6, 1)$.

Step 3. For each point (a, b) chosen in step 2, plot the point (b, a): We plot $(-3, -6)$, $(-2, -3)$, $(-1, 0)$, $(0, 3)$, and $(1, 6)$ in Fig. 9.

Step 4. The points from step 3 lie on a line. We sketch the line that contains these points. See Fig. 9.

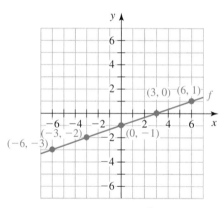

Figure 8 Graph of $f(x) = \dfrac{1}{3}x - 1$

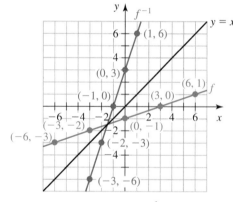

Figure 9 Graph of $f(x) = \dfrac{1}{3}x - 1$, $y = x$, and f^{-1}

So far, we have described inverse functions by using tables and graphs. We now describe them by using equations.

Example 6   Finding the Equation of the Inverse of a Model

Note

You may have found this equation in exercise 85 in Homework 2.3.

Let's return one last time to the temperature model g. An equation for g is

$$g(C) = 1.8C + 32$$

where $F = g(C)$ is the Fahrenheit reading corresponding to a Celsius reading of C degrees.

1. Find an equation for g^{-1}.
2. Find $g(25)$. What does this result mean in terms of the situation?
3. Find $g^{-1}(59)$. What does this result mean in terms of the situation?

Solution

1. Since g sends values of C to values of F, g^{-1} sends values of F to values of C (see Figs. 10 and 11).

$$C \xrightarrow{\hspace{0.4cm}} \boxed{g} \xrightarrow{\hspace{0.4cm}} F$$

Figure 10 g sends values of C to values of F

$$F \xrightarrow{\hspace{0.4cm}} \boxed{g^{-1}} \xrightarrow{\hspace{0.4cm}} C$$

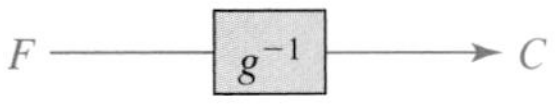

Figure 11 g^{-1} sends values of F to values of C

To find an equation for g^{-1}, we want to write C in terms of F. Here are three steps to find an equation for g^{-1}:

Step 1. We replace $g(C)$ with F: $F = 1.8C + 32$.

Step 2. We solve the equation for C:

$$F = 1.8C + 32$$
$$F - 32 = 1.8C$$
$$1.8C = F - 32$$
$$C = \frac{F - 32}{1.8}$$

Step 3. Now that we have found an equation for g^{-1}, we replace C with $g^{-1}(F)$:

$$g^{-1}(F) = \frac{F - 32}{1.8}$$

Note

$g^{-1}(F) = C$ because g^{-1} sends values of F to values of C (see Fig. 11).

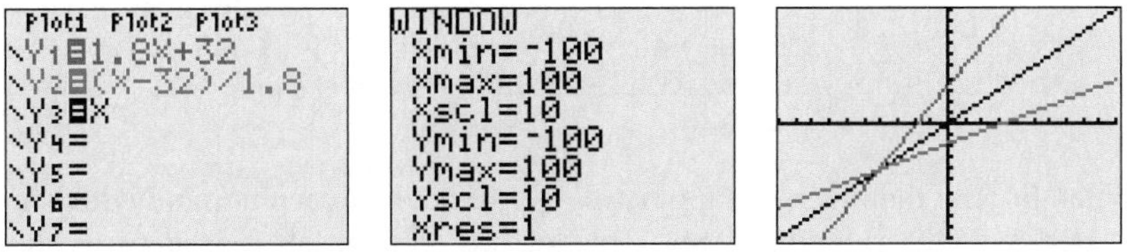

We can verify our work by checking that the graph of g^{-1} is the reflection of the graph of g across the line $y = x$ (see Fig. 12).

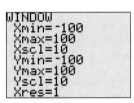
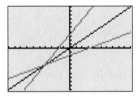

Figure 12 Check that the graph of g^{-1} is the reflection of g across the line $y = x$

2. $g(25) = 1.8(25) + 32 = 77$. Since g sends values of C to values of F, this means that $F = 77$ when $C = 25$. So, 77°F is equivalent to 25°C.

3. $g^{-1}(59) = \dfrac{59 - 32}{1.8} = 15$. Since g^{-1} sends values of F to values of C, this means that $C = 15$ when $F = 59$. So, 15°C is equivalent to 59°F. ∎

In Example 6 we performed three steps to find an equation of the inverse of a model.

Three-Step Process to Find the Inverse of a Model

To find the inverse of an invertible model f where $p = f(t)$:

1. Replace $f(t)$ with p.
2. Solve for t.
3. Replace t with $f^{-1}(p)$.

We can take three similar steps followed by one more step to find an equation of the inverse of a function that is not a model.

Example 7 Finding the Inverse of a Function That Is Not a Model

Find the inverse of $f(x) = 2x - 3$.

Solution

Step 1. Substitute y for $f(x)$: $y = 2x - 3$

Step 2. Solve for x:

$$y = 2x - 3$$
$$y + 3 = 2x$$
$$2x = y + 3$$
$$x = \frac{y + 3}{2}$$

Step 3. Replace x with $f^{-1}(y)$: $f^{-1}(y) = \frac{y + 3}{2}$

Step 4. When a function is not a model, we usually want the input variable to be x. So, we rewrite the equation $f^{-1}(y) = \frac{y + 3}{2}$ in terms of x:

$$f^{-1}(x) = \frac{x + 3}{2}$$

We can use ZStandard followed by ZSquare to verify our work by checking that the graph of f^{-1} is the reflection of the graph of f across the line $y = x$ (see Fig. 13).

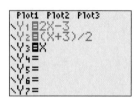
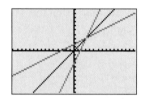

Figure 13 Check that the graph of f^{-1} is the reflection of f across the line $y = x$

When finding the inverse of a function f, we need to keep in mind whether or not f is a model. If the function is not a model, we perform all four steps as shown in Example 7. However, if the function is a model, we perform only the first three steps so that the variables retain their original meaning.

Four-Step Process to Find the Inverse of a Function That Is Not a Model

Let f be an invertible function that is not a model. To find the inverse of f where $y = f(x)$:

1. Replace $f(x)$ with y.
2. Solve for x.
3. Replace x with $f^{-1}(y)$.
4. Write the equation for f^{-1} in terms of x.

So far, we have used tables and graphs to describe the inverse of an exponential function such as $f(x) = 2^x$. Can we find an equation of the inverse of $f(x) = 2^x$? To find such an equation we would have to solve the equation $y = 2^x$ for x. We cannot do this using familiar operations such as adding, subtracting, and so on. Nonetheless, the inverse of an exponential function is a powerful tool that we will continue to explore throughout the rest of this chapter.

The inverse of a function is not necessarily a function. For example, consider $f(x) = x^2$. Note that $f(-2) = 4$ and $f(2) = 4$. So, f sends both of the inputs -2 and 2 to the output 4 (see Fig. 14). Therefore, the inverse of f sends the input 4 to the *two* outputs −2 and 2 (see Fig. 15). Thus, the inverse of f is not a function. Note that the inverse of f is not a function because f sends different inputs to the *same* output.

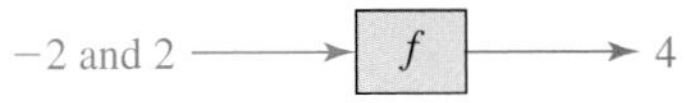

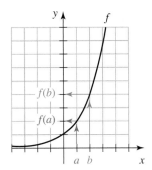

−2 and 2 ⟶ f ⟶ 4

Figure 14 f sends −2 and 2 to 4

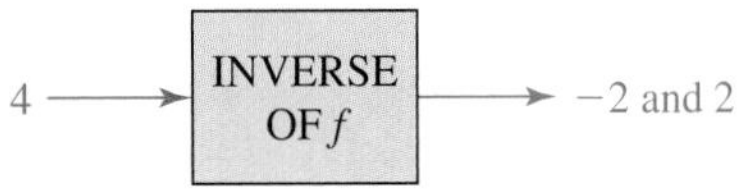

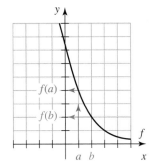

4 ⟶ INVERSE OF f ⟶ −2 and 2

Figure 15   The inverse of f sends 4 to −2 and 2

If a function sends different inputs to different outputs, we say that the function is **one-to-one**. A one-to-one function is invertible. All exponential functions are one-to-one and, hence, invertible. To see why, note that if f is an increasing exponential function, then $a < b$ implies that $f(a) < f(b)$ (see Fig. 16). If f is a decreasing exponential function, then $a < b$ implies that $f(a) > f(b)$ (see Fig. 17). Both of these implications show that if two inputs a and b are different, then their outputs $f(a)$ and $f(b)$ are different, too.

Figure 16   Graph of an increasing exponential function

Figure 17   Graph of a decreasing exponential function

Using similar approaches, we could show that all linear functions with nonzero slope are one-to-one and, hence, invertible. Or, we could show that such a function is invertible by finding the equation of the inverse of the function; you are asked to do this in exercise 77.

EXPLORATION
Looking ahead: A logarithm is an exponent

1. **a.** Solve $2^x = 16$.
   b. Solve $2^x = 32$.
   c. Approximate the solution to $2^x = 24$ by using a trial-and-error approach. Your result should be correct up to four decimal places. [**Hint:** Parts a and b should suggest a reasonable first guess. Then you can use calculator tables to speed up the trial-and-error process.]
2. Approximate the solution to $3^x = 15$ using trial and error. (We call the solution $\log_3(15)$, where *log* is shorthand for *logarithm*.)
3. Approximate the solution to $10^x = 500$. (We call the solution $\log_{10}(500)$.)

Tips on Succeeding in This Course

At various times throughout the course, you can improve your understanding of algebra by reviewing material that you have learned so far. Your review should include solving problems, revisiting explorations, and reconsidering key points from previous sections.

Key Points
OF THIS SECTION

Invertible function	If the inverse of a function f is also a function we say that f is *invertible* and use "f^{-1}" as a name for the inverse of f.
Meaning of an inverse function	For an invertible function f, the following statements are equivalent: $f(a) = b$ and $f^{-1}(b) = a$. In words, if f sends a to b, then f^{-1} sends b to a. If f^{-1} sends b to a, then f sends a to b.
Reflection property	For an invertible function f, the graph of f^{-1} is the reflection of the graph of f across the line $y = x$.
Sketching the graph of f^{-1}	For an invertible function f, we sketch the graph of f^{-1} by the following steps: **1.** Sketch the graph of f. **2.** Choose several points that lie on the graph of f. **3.** For each point (a, b) chosen from step 2, plot the point (b, a). **4.** Sketch the curve that contains the points plotted in step 3.
Finding the inverse of a model	To find the inverse of an invertible *model* f where $p = f(t)$: **1.** Replace $f(t)$ with p. **2.** Solve for t. **3.** Replace t with $f^{-1}(p)$.
Finding the inverse of a function that is not a model	Let f be an invertible function that is *not* a model. To find the inverse of f where $y = f(x)$: **1.** Replace $f(x)$ with y. **2.** Solve for x. **3.** Replace x with $f^{-1}(y)$. **4.** Write the equation for f^{-1} in terms of x.
One-to-one function	A *one-to-one* function sends different inputs to different outputs. A one-to-one function is *invertible*.

HOMEWORK 5.1

SSM
Tutor Center

MathPro 4
MathPro 4

MathPro 5
MathPro 5

CDs/Videos

www
www

1. Let f be an invertible function where $f(4) = 7$. Find $f^{-1}(7)$.

2. Let g be an invertible function where $g^{-1}(3) = 2$. Find $g(2)$.

Some values of an invertible function f are given in Table 7. For exercises 3–8, refer to this table.

Table 7 Values of f
(Exercises 3–8)

x	$f(x)$
2	10
3	8
4	6
5	4
6	2

3. Find $f(4)$.

4. Find $f(2)$.

5. Find $f^{-1}(4)$.

6. Find $f^{-1}(2)$.

7. Use a table to describe five input-output values of f^{-1}.

8. Find x when $f(x) = 2$.

Some values of an invertible function g are given in Table 8. For exercises 9–14, refer to this table.

Table 8 Values of g
(Exercises 9–14)

x	$g(x)$
1	2
2	6
3	18
4	54
5	162
6	486

9. Find $g(2)$.

10. Find $g(6)$.

11. Find $g^{-1}(2)$.

12. Find $g^{-1}(6)$.

13. Use a table to describe six input-output values of g^{-1}.

14. Find x when $g(x) = 6$.

15. Complete Table 9 by using the table of values for f to complete the table of values for f^{-1}.

Table 9 Finding Values for f^{-1} (Exercise 15)

x	f(x)	x	$f^{-1}(x)$
1	34	4	
2	28	10	
3	22	16	
4	16	22	
5	10	28	
6	4	34	

16. Complete Table 10 by using the table of values for f to complete the table of values for f^{-1}.

Table 10 Finding Values for f^{-1} (Exercise 16)

x	f(x)	x	$f^{-1}(x)$
2	5	2	
3	2	3	
4	3	4	
5	6	5	
6	4	6	

Let $f(x) = 3(2)^x$.

17. Use a table to describe five input-output values for f^{-1}.

18. Find $f(2)$.

19. Find $f(3)$.

20. Find $f^{-1}(24)$.

21. Find $f^{-1}(3)$.

22. Find x when $f(x) = 6$.

Sketch the graph of the given function, its inverse, and $y = x$ on the same set of axes. Label each graph with f, f^{-1}, or $y = x$.

23. $f(x) = 3^x$

24. $f(x) = 2^x$

25. $f(x) = 3(2)^x$

26. $f(x) = 2(3)^x$

27. $f(x) = 2x$

28. $f(x) = 3x$

29. $f(x) = \frac{1}{3}x$

30. $f(x) = \frac{1}{2}x$

31. $f(x) = 3x - 2$

32. $f(x) = 2x - 3$

33. $f(x) = \frac{1}{2}x + 1$

34. $f(x) = \frac{2}{3}x - 1$

35. $f(x) = 4\left(\frac{1}{2}\right)^x$

36. $f(x) = 6\left(\frac{1}{3}\right)^x$

37. $f(x) = \left(\frac{1}{3}\right)^x$

38. $f(x) = 2\left(\frac{1}{2}\right)^x$

39. In exercise 7 of Homework 2.2, you found an equation close to $p = f(t) = 0.81t - 45.86$ that models the percentage p of births out of wedlock at t years since 1900 (see Table 11).

Table 11 Births out of Wedlock

Year	Percent
1970	10.7
1975	14.3
1980	18.4
1985	22.0
1990	28.0
1995	32.2
2000	33.2

a. Find an equation for the inverse function of f.

b. Find $f(100)$. What does it mean in terms of the percentage of births out of wedlock?

c. Find $f^{-1}(100)$. What does it mean in terms of the percentage of births out of wedlock?

40. In exercise 87 of Homework 2.3, you may have found the equation $n = f(t) = 0.29t + 3.36$, where n represents the number of bald eagle male-female pairs (in thousands) at t years since 1990 (see Table 12).

Table 12 Numbers of Bald Eagle Pairs

Year	Eagle Pairs (thousands)
1993	4.2
1994	4.5
1995	4.9
1996	5.1
1997	5.3
1998	5.7

a. Find an equation for f^{-1}.

b. Find $f(9)$. What does your result mean in terms of eagle pairs?

c. Find $f^{-1}(9)$. What does your result mean in terms of eagle pairs?

d. What is the slope of f^{-1}? What does it mean in terms of eagle pairs? [**Hint:** f^{-1} sends outputs of f to inputs of f.]

41. The number of cremations in the United States has steadily increased over the past two decades (see Table 13).

Table 13 Percentages of Bodies That Are Cremated

Year	Percent
1980	9.7
1985	14.0
1990	17.6
1995	21.4
2000	25.5

Source: *Cremation Association of North America Victimization Survey*

Let p represent the percentage of bodies that are cremated at t years since 1900.

a. Use your graphing calculator to draw a scattergram. Is it better to use a linear or exponential function to model the situation?

b. Find an equation for f.

c. Find an equation for f^{-1}.

d. Use f to predict when half of people who die will be cremated.

e. Now use f^{-1} to predict when half of people who die will be cremated.

f. Compare your results in parts d and e.

42. In exercise 20 in Section 2.4, you may have found the equation $f(t) = 6.03t + 79.79$, where $p = f(t)$ represents the median price (in thousands of dollars) for homes sold during the year that is t years after 1990 (see Table 14).

Table 14 Median Sales Prices of Homes

Year	Median Price (thousand dollars)
1995	110.5
1996	115.8
1997	121.8
1998	128.4
1999	133.3
2000	139.0
2001	147.5

a. Find an equation for f^{-1}.

b. Use f^{-1} to predict when the median sales price will be 200 thousand dollars.

c. Find $f^{-1}(170)$, $f^{-1}(180)$, and $f^{-1}(190)$. Is f^{-1} an increasing function or a decreasing function? Explain why this makes sense in terms of median sales prices of homes.

43. The typical list prices for records and CDs for various years are listed in Table 15.

Table 15 List Prices for Records or CDs

Year	Price (dollars)
1965	4.98
1975	7.98
1985	12.98
1995	15.98
2002	18.98

Source: *USA Today research,*
www.newsengine.com

Let $p = f(t)$ represent the typical list price (in dollars) for records and CDs at t years since 1900.

a. Use your graphing calculator to draw a scattergram of the data. Is it better to use a linear or exponential function to model the data? Explain.

b. Find an equation for f.

c. Find an equation for f^{-1}.

d. Find $f(109)$. What does your result mean in terms of the situation?

e. Find $f^{-1}(25)$. What does your result mean in terms of the situation?

44. The percentage of adults over age 25 who say they are heavier than the recommended weight range for their height and frame has increased over the past 12 years (see Table 16).

Table 16 Percentages of Adults Who Say They Are Overweight

Year	Percent
1990	64
1992	66
1994	69
1996	74
1998	76
2000	79
2002	80

Source: *Harris Interactive,*
www.harrisinteractive.com. Accessed 4/2/02.

Let $p = f(t)$ represent the percentage of adults who say they are overweight at t years since 1900.

a. Use your graphing calculator to draw a scattergram. Does it appear that the data can best be modeled using a linear or an exponential function? Explain.

b. Find an equation for f.

c. Find an equation for f^{-1}.

d. Find $f(100)$. What does your result mean in terms of adults over age 25?

e. Find $f^{-1}(100)$. What does your result mean in terms of adults over age 25?

Find the inverse of each given function. Use a graphing calculator to verify your work by graphing f, f^{-1}, and $y = x$ on the same set of axes.

45. $f(x) = x - 3$

46. $f(x) = x + 6$

47. $f(x) = x + 8$

48. $f(x) = x - 7$

49. $f(x) = 4x$

50. $f(x) = -5x$

51. $f(x) = \dfrac{x}{7}$

52. $f(x) = -\dfrac{x}{2}$

53. $f(x) = 4x + 5$

54. $f(x) = -2x + 7$

55. $f(x) = -6x - 2$

56. $f(x) = 3x - 8$

57. $f(x) = 0.4x - 7.9$

58. $f(x) = -2.9x + 4.5$

59. $f(x) = -72.83x + 99.23$

60. $f(x) = -3.28x - 70.52$

61. $f(x) = \dfrac{7}{3}x + 1$

62. $f(x) = -\dfrac{2}{5}x - 8$

63. $f(x) = -\dfrac{5}{6}x - 3$

64. $f(x) = \dfrac{8}{5}x + 4$

65. $f(x) = \dfrac{6x - 2}{5}$

66. $f(x) = \dfrac{x - 7}{4}$

67. $f(x) = -3(2x - 5)$

68. $f(x) = 5(3x + 1)$

69. $f(x) = 7 - 8(x + 1)$

70. $f(x) = 2(x - 1) + 5$

71. $f(x) = x$

72. $f(x) = -x$

73. Explain why it makes sense that the function $g(x) = x - 5$ is the inverse of the function $f(x) = x + 5$.

74. Explain why it makes sense that the function $g(x) = \dfrac{x}{3}$ is the inverse of the function $f(x) = 3x$.

75. Is the function $f(x) = 3$ one-to-one? Is it invertible? Explain.

76. Is the function $f(x) = x^4$ one-to-one? Is it invertible? Explain.

77. In this exercise you will show that all linear functions with nonzero slope are invertible.

a. Let $f(x) = mx + b$, where $m \neq 0$. Find an equation for the inverse of f.

b. Explain why your work in part a shows that the inverse of f is a function.

78. Explain how to find the inverse of an invertible linear function. Also, explain the meaning of an inverse function.

5.2 LOGARITHMIC FUNCTIONS

Objectives

▸ Know the meaning of a *logarithm* and a *logarithmic function*.

▸ Find logarithms and evaluate logarithmic functions.

▸ Graph logarithmic functions.

In this section we continue to explore the inverse of an exponential function. Consider the function $f(x) = 2^x$. Input-output values for f and f^{-1} are shown in Tables 17 and 18.

Table 17 Input-Output Pairs for $f(x) = 2^x$	
x	$f(x)$
0	2^0
1	2^1
2	2^2
3	2^3
4	2^4

Table 18 Input-Output Pairs for f^{-1}	
x	$f^{-1}(x)$
2^0	0
2^1	1
2^2	2
2^3	3
2^4	4

Note

Function names like f and g tell us nothing about the functions. The name f^{-1} tells us that the function is the inverse of a function named f. The name $\log_2$ tells us more, that the function is the inverse of another function, *and* that the other function is the function $f(x) = 2^x$.

From Table 18 we see that $f^{-1}(2^3) = 3$. Note that the output 3 is the *exponent* of the input 2^3. In fact, all of the outputs in Table 18 are *exponents* of the corresponding inputs. We know that $f^{-1}(2^6) = 6$ since 6 is the exponent of 2^6. Likewise, $f^{-1}(2^7) = 7$.

For $f(x) = 2^x$, the function f^{-1} is so useful that we give it a specific name and call it $\log_2$ (read "logarithm base 2"). So, we write $\log_2(2^3) = 3$, $\log_2(2^6) = 6$, and so on.

To find $\log_2(32)$, we first write 32 as a power of 2:

$$\log_2(32) = \log_2(2^5) = 5$$

So, $\log_2(32) = 5$ because $2^5 = 32$. Likewise, $\log_2(16) = 4$ because $2^4 = 16$. In general,

$$\log_2(a) = k \quad \text{if} \quad 2^k = a$$

We define $\log_3$ in a similar way:

$$\log_3(a) = k \quad \text{if} \quad 3^k = a$$

For example, $\log_3(9) = 2$ since $3^2 = 9$. Note that the logarithm 2 is the *exponent* on the base 3 that gives 9.

DEFINITION *Logarithm*

For $b > 0$, $b \neq 1$, and $a > 0$,

$$\log_b(a) \text{ is the number } k \text{ so that } b^k = a.$$

In words: $\log_b(a)$ is the exponent on the base b that gives a. Also, we call b the **base** of the logarithm.

In short, a logarithm is an exponent.

Example 1   Evaluating a Logarithmic Function

Evaluate each logarithmic function.

1. $\log_7(49)$ **2.** $\log_2(16)$ **3.** $\log_3(81)$

4. $\log_5(125)$ **5.** $\log_4(64)$ **6.** $\log_{10}(100{,}000)$

Solution

1. $\log_7(49) = 2$ since $7^2 = 49$.
2. $\log_2(16) = 4$ since $2^4 = 16$.
3. $\log_3(81) = 4$ since $3^4 = 81$.
4. $\log_5(125) = 3$ since $5^3 = 125$.
5. $\log_4(64) = 3$ since $4^3 = 64$.
6. $\log_{10}(100{,}000) = 5$ since $10^5 = 100{,}000$. ■

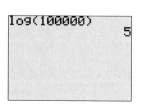

Figure 18 Use a graphing calculator to find that $\log(100{,}000) = 5$

We often write "$\log(a)$" as shorthand for $\log_{10}(a)$. For example, instead of writing $\log_{10}(100{,}000)$, we write $\log(100{,}000)$. We can use the "log" key on a calculator to find base 10 logarithms (see Fig. 18).

It is important to keep in mind that a logarithm is an exponent. In Example 2, we use properties of exponents to find more logarithms.

Example 2 Finding Logarithms

Find each logarithm.

1. $\log_7(7)$
2. $\log_4(1)$
3. $\log_5\left(\dfrac{1}{5}\right)$
4. $\log_3\left(\dfrac{1}{9}\right)$
5. $\log_6(\sqrt{6})$
6. $\log(0.001)$

Solution

1. $\log_7(7) = 1$, since $7^1 = 7$.
2. $\log_4(1) = 0$, since $4^0 = 1$.
3. $\log_5\left(\dfrac{1}{5}\right) = -1$, since $5^{-1} = \dfrac{1}{5}$.
4. $\log_3\left(\dfrac{1}{9}\right) = -2$, since $3^{-2} = \dfrac{1}{3^2} = \dfrac{1}{9}$.
5. $\log_6(\sqrt{6}) = \dfrac{1}{2}$, since $6^{\frac{1}{2}} = \sqrt{6}$.
6. Remember that "log" is shorthand for $\log_{10}$, so $\log(0.001) = -3$, since
$$10^{-3} = \frac{1}{10^3} = \frac{1}{1000} = 0.001.$$ ■

Note

We assume that $b > 0$ and $b \neq 1$.

In Example 2, we found that $\log_7(7) = 1$. In general, $\log_b(b) = 1$, since $b^1 = b$. We also found that $\log_4(1) = 0$. In general, $\log_b(1) = 0$, since $b^0 = 1$.

Properties of Logarithmic Functions

For $b > 0$ and $b \neq 1$,

- $\log_b(b) = 1$
- $\log_b(1) = 0$

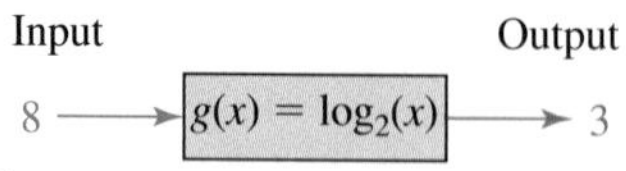

Input   Output

$8 \longrightarrow \boxed{g(x) = \log_2(x)} \longrightarrow 3$

Figure 19 Illustration of $g(8) = \log_2(8) = 3$

Recall that $\log_2$ is the name of a function. When we write $\log_2(8) = 3$, we mean that the function $\log_2$ sends the input 8 to the output 3 (see Fig. 19).

The nonpositive numbers are not in the domain of $\log_2(x)$. For example, consider $\log_2(0)$. Note that no exponent on the base 2 gives 0. For another example, consider $\log_2(-8)$. Note that no exponent on the base 2 gives a negative number. In general, the domain of $g(x) = \log_b(x)$ is the set of all positive real numbers.

The function $\log_2(x)$ is the inverse of $f(x) = 2^x$. From our work in Section 5.1, we can conclude that $f(x) = 2^x$ is also the inverse of $\log_2(x)$.

Logarithmic and Exponential Functions are Inverses of Each Other

- For the exponential function $f(x) = b^x$, $f^{-1}(x) = \log_b(x)$.
- For the logarithmic function $g(x) = \log_b(x)$, $g^{-1}(x) = b^x$.

In words, $\log_b(x)$ and $f(x) = b^x$ are inverse functions of each other.

Example 3   Finding an Inverse Function

Find the inverse function of each function.

1. $f(x) = 4^x$
2. $h(x) = \log_9(x)$

Solution

1. $f^{-1}(x) = \log_4(x)$
2. $h^{-1}(x) = 9^x$ ■

Example 4   Evaluating f and f^{-1}

Let $f(x) = 3^x$.

1. Find $f(4)$.
2. Find $f^{-1}(9)$.

Solution

1. $f(4) = 3^4 = 81$
2. $f^{-1}(9) = \log_3(9) = 2$ ■

Because a logarithmic function is the inverse of an exponential function, we can use the four-step method discussed in Section 5.1 to help us graph a logarithmic function.

Example 5   Graphing a Logarithmic Function

Sketch the graph of $y = \log_3(x)$.

Solution

Note that the inverse of $f(x) = 3^x$ is $f^{-1}(x) = \log_3(x)$. So, we can apply the four-step method to graph the inverse of f.

Step 1. Sketch the graph of f: See Fig. 20.

Step 2. Choose several points that lie on the graph of f: $\left(-1, \dfrac{1}{3}\right)$, $(0, 1)$, $(1, 3)$, and $(2, 9)$.

Step 3. For each point (a, b) chosen in step 2, plot the point (b, a): We plot $\left(\dfrac{1}{3}, -1\right)$, $(1, 0)$, $(3, 1)$, and $(9, 2)$ in Fig. 21.

Step 4. We sketch a curve that contains the points plotted in step 3 (see Fig. 21). The curve is the graph of $y = \log_3(x)$.

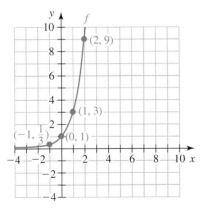

Figure 20 Graph of $f(x) = 3^x$

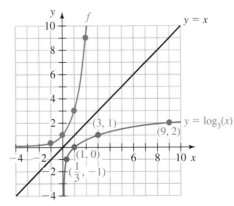

Figure 21 Graphs of $f(x) = 3^x$, $y = \log_3(x)$, and $y = x$

■

To graph a logarithmic function $y = \log_b(x)$, we use the four-step method to sketch the inverse of $f(x) = b^x$.

Scientists often use logarithms to rescale measurements of objects or phenomena when the measurements tend to be very small (e.g., 3.2×10^{-8}) and/or very large (e.g., 7.9×10^{13}). For example, scientists use logarithms to rescale measurements of amplitudes of earthquakes, pH of solutions, and noise levels of sounds.

The energy released by an earthquake is sometimes measured using a *Richter scale*. The *Richter number*, R, of an earthquake is given by

$$R = \log\left(\frac{A}{A_0}\right)$$

where A is the amplitude of the shockwave of the earthquake and A_0, called the *reference amplitude*, is the amplitude of the shockwave of the smallest earth movement that can be recorded on a seismograph.

Note

You will study the pH of a solution and the noise level of a sound in exercises 76 and 77, respectively.

Note

The *amplitude* of a wave is half the difference of the height of the highest point on the wave minus the height of the lowest point on the wave.

Example 6 Richter Numbers

In 1906 an earthquake in San Francisco had an amplitude 2×10^8 times the reference amplitude A_0. In 1989 an earthquake in San Francisco had an amplitude 8×10^6 times A_0.

1. Find the Richter number for each earthquake.
2. Compare the ratio of the Richter numbers to the ratio of the amplitudes for the 1906 and 1989 earthquakes.

Solution

1. The Richter number for the 1906 earthquake is

$$R = \log\left(\frac{2 \times 10^8 A_0}{A_0}\right)$$
$$= \log(2 \times 10^8)$$
$$\approx 8.3$$

A Mean Trick to Play
on a
Seismographer

Oh NO!!

Hee hee!

The Richter number for the 1989 earthquake is

$$R = \log\left(\frac{8 \times 10^6 A_0}{A_0}\right)$$
$$= \log(8 \times 10^6)$$
$$\approx 6.9$$

2. The ratio of the Richter numbers for the 1906 earthquake and the 1989 earthquake is

$$\frac{8.3}{6.9} \approx 1.2$$

The ratio of the amplitudes for the 1906 earthquake and the 1989 earthquake is

$$\frac{2 \times 10^8 A_0}{8 \times 10^6 A_0} = 25$$

So, the ratio of 1.2 for the Richter numbers means that the 1906 earthquake released 25 times the energy of the 1989 earthquake. ■

EXPLORATION
Looking ahead: Power property for logarithms

1. Use your calculator to compare $\log(3^2)$ and $2 \log(3)$. (Remember "log" stands for $\log_{10}$.)
2. Use your calculator to compare $\log(7^4)$ and $4 \log(7)$.
3. Use a calculator table to compare values for $f(x) = \log(x^3)$ and $g(x) = 3 \log(x)$. Also compare the graphs of f and g in the same viewing window.
4. Use a calculator table to compare values for $f(x) = \log(x^5)$ and $g(x) = 5 \log(x)$. Also compare the graphs of f and g in the same viewing window.
5. What do problems 1–4 suggest about $\log(x^p)$? Test your conjecture.

Tips on Succeeding in This Course

If you have trouble solving a problem, try exploring aspects of the problem that may not lead directly to a solution. For example, suppose you draw a blank on how to solve the equation $2x + 1 = 11$. Here are some strategies you could try:

1. You could substitute a pattern of numbers for x and try to recognize a pattern in the $2x + 1$ values.
2. You could solve the simpler equation $x + 1 = 11$ and reflect on how your work relates to solving the equation $2x + 1 = 11$.
3. You could use your graphing calculator to generate a table of values for the function $y = 2x + 1$ and then examine the numbers in the table.

People who are successful at mathematics are willing to tinker with aspects of a problem, even if they do not immediately see how to solve it. Some problems, in fact, have taken years, even centuries, for mathematicians to solve ... not that the problems in this text will take anywhere near that long!

Key Points
OF THIS SECTION

$\log_b(a)$ is an exponent, namely the exponent on the base b that gives a.

A logarithm is an exponent

A *logarithmic function, base b*, is a function of the form $g(x) = \log_b(x)$, where $b > 0$ and $b \neq 1$. The domain is the set of positive real numbers.

Logarithmic function

For the exponential function $f(x) = b^x$, $f^{-1}(x) = \log_b(x)$.

Inverse function

For the logarithmic function $g(x) = \log_b(x)$, $g^{-1}(x) = b^x$.

$\log(x)$ is shorthand for $\log_{10}(x)$.

Meaning of $\log(x)$

Graphing $y = \log_b(x)$

To graph a logarithmic function $y = \log_b(x)$, use the four-step method to graph the inverse of $f(x) = b^x$.

Richter number

The *Richter number*, R, of an earthquake is given by

$$R = \log\left(\frac{A}{A_0}\right)$$

where A is the amplitude of the shockwave of the earthquake and A_0, called the reference amplitude, is the amplitude of the shockwave of the smallest earth movement detectable on a seismograph.

HOMEWORK 5.2

 SSM Tutor Center   MathPro 4 MathPro 4   MathPro 5 MathPro 5   CDs/Videos   www www

Find the logarithm.

1. $\log_9(81)$ **2.** $\log_7(49)$

3. $\log_3(27)$ **4.** $\log_3(9)$

5. $\log_2(32)$ **6.** $\log_3(243)$

7. $\log_6(216)$ **8.** $\log_5(125)$

9. $\log_6(36)$ **10.** $\log_2(128)$

11. $\log(100)$ **12.** $\log(1000)$

13. $\log_7(343)$ **14.** $\log_2(64)$

15. $\log_5(15{,}625)$ **16.** $\log_4(16{,}384)$

17. $\log_4\left(\dfrac{1}{4}\right)$ **18.** $\log_3\left(\dfrac{1}{3}\right)$

19. $\log_2\left(\dfrac{1}{8}\right)$ **20.** $\log_3\left(\dfrac{1}{81}\right)$

21. $\log\left(\dfrac{1}{10}\right)$ **22.** $\log(0.01)$

23. $\log_5(1)$ **24.** $\log_8(8)$

25. $\log_9(3)$ **26.** $\log_{36}(6)$

27. $\log_8(2)$ **28.** $\log_{32}(2)$

29. $\log_7(\sqrt{7})$ **30.** $\log_2(\sqrt{2})$

31. $\log_5(\sqrt[4]{5})$ **32.** $\log_7(\sqrt[3]{7})$

33. $\log_2(\log_2(16))$ **34.** $\log_2(\log_3(81))$

35. $\log(\log(10))$ **36.** $\log_3(\log_3(\sqrt[3]{3}))$

37. $\log_b(b)$ **38.** $\log_b(1)$

39. $\log_b(b^2)$ **40.** $\log_b\left(\dfrac{1}{b^3}\right)$

41. $\log_b(\sqrt{b})$ **42.** $\log_b(\sqrt[5]{b})$

43. $\log_b(\log_b(b))$ **44.** $\log_b(\log_b(b^b))$

For each function, find a formula for the inverse function.

45. $f(x) = 3^x$

46. $g(x) = 8^x$

47. $h(x) = 10^x$ **48.** $g(x) = \left(\dfrac{1}{3}\right)^x$

49. $f(x) = \log_5(x)$ **50.** $g(x) = \log_4(x)$

51. $h(x) = \log(x)$ **52.** $f(x) = \log_{\frac{1}{2}}(x)$

Let $f(x) = 2^x$.

53. Find $f(2)$.   **54.** Find $f(4)$.

55. Find $f^{-1}(2)$.   **56.** Find $f^{-1}(4)$.

Let $g(x) = \log_3(x)$.

57. Find $g(3)$.   **58.** Find $g(81)$.

59. Find $g^{-1}(3)$.   **60.** Find $g^{-1}(2)$.

For exercises 61–64, refer to the values of the function $f(x) = 3^x$ listed in Table 19.

Table 19 Values of $f(x) = 3^x$
(Exercises 61–64)

x	$f(x)$
0	1
1	3
2	9
3	27
4	81

61. Find $f(1)$.   **62.** Find $f(3)$.

63. Find $f^{-1}(1)$. Also, write your result as a logarithm.

64. Find $f^{-1}(3)$. Also, write your result as a logarithm.

Sketch the graph of the function.

65. $y = \log_2(x)$ **66.** $y = \log_4(x)$

67. $y = \log(x)$ **68.** $y = \log_5(x)$

69. $y = \log_{\frac{1}{2}}(x)$ **70.** $y = \log_{\frac{1}{3}}(x)$

71. Describe the inverse of $f(x) = 5^x$ using a table, a graph, and an equation.

72. Describe the inverse of $f(x) = \left(\dfrac{1}{4}\right)^x$ using a table, a graph, and an equation.

73. a. Solve the equation $5^x = 25$.

   b. Find $\log_5(25)$.

   c. Explain why the results to parts a and b are the same.

74. a. Find $\log(100)$.

   b. Complete Table 20.

Table 20 Values of $\log(x)$
(Exercise 74)

x	$\log(x)$
0.001	
0.01	
0.1	
1	
10	
100	
1000	

   c. Examine the entries in Table 20 and describe all patterns that you observe.

75. In 1920 an earthquake in Gansu, China, killed 180,000 people. In 1980 an earthquake in Naples, Italy, killed 4500 people. The Gansu earthquake's amplitude was 4.0×10^8 times the reference amplitude A_0, and the Naples' earthquake was 1.6×10^7 times A_0.

   a. Find the Richter number for the Gansu earthquake.

   b. Find the Richter number for the Naples earthquake.

   c. Find the ratio of the Gansu earthquake's Richter number to the Naples earthquake's Richter number.

   d. Find the ratio of the Gansu earthquake's amplitude to the Naples earthquake's amplitude.

   e. In your own words, compare the energy released by the two earthquakes.

76. The acidity or basicity of a solution is measured using a *pH scale*. The pH of a solution is given by

$$pH = -\log(H^+)$$

where H^+ is the hydrogen ion concentration of the solution in moles per liter. Distilled water has a pH of 7. Acidic solutions have a pH less than 7, and basic solutions have a pH greater than 7. Most solutions have a pH between 1 and 14. Find the pH of each solution listed in Table 21 and determine whether the solution is acidic or basic.

Table 21 Hydrogen Ion Concentrations of Some Solutions

Solution	Hydrogen Ion Concentration (moles per liter)
Vinegar	1.6×10^{-3}
Human blood	6.3×10^{-8}
Shampoo	7.4×10^{-10}
Orange juice	6.3×10^{-4}
Hydrochloric acid	2.5×10^{-2}

77. The loudness of sound can be measured using a *decibel scale*. The sound level L (in decibels) of a sound is given by

$$L = 10 \log\left(\frac{I}{I_0}\right)$$

where I is the intensity of the sound (watts per square meter, W/m^2) and $I_0 = 10^{-12}$ W/m^2. The constant I_0 is the approximate intensity of the softest sound that can be heard by humans. Find the decibel readings for the sounds listed in Table 22.

Table 22 Examples of Sound Intensities

Sound	Intensity of Sound (W/m^2)
Faintest sound heard by humans	10^{-12}
Whisper	10^{-10}
Inside a running car	10^{-8}
Conversation	10^{-6}
Noisy street corner	10^{-4}
Soft rock concert	10^{-2}
Threshold of pain	1

78. Give an example of an exponential function f of the form $f(x) = b^x$ and list five input-output pairs for f. Then list five input-output pairs for f^{-1}. What is another name for f^{-1}?

79. Explain how to find a logarithm. Also, explain why a logarithmic function is the inverse of a function.

5.3 PROPERTIES OF LOGARITHMS

Objectives

▸ Convert expressions in logarithmic form to exponential form and vice-versa.

▸ Know the *power property* for logarithms.

▸ Use properties of logarithms to solve exponential and logarithmic equations.

In Section 5.2 we discussed how to find logarithms. For example,

$$\log_2(8) = 3 \quad \text{since} \quad 2^3 = 8$$

We say the equation $\log_2(8) = 3$ is in *logarithmic* form and the equation $2^3 = 8$ is in *exponential* form. The forms $\log_2(8) = 3$ and $2^3 = 8$ are equivalent.

Here are more examples of equations in equivalent logarithmic and exponential forms.

Logarithmic form	*Exponential form*
$\log_2(16) = 4$	$2^4 = 16$
$\log_3(9) = 2$	$3^2 = 9$
$\log_5(125) = 3$	$5^3 = 125$
$\log(100,000) = 5$	$10^5 = 100,000$

Exponential/Logarithmic Forms Property

For $a > 0, b > 0$, and $b \neq 1$,

$$\log_b(a) = c \text{ and } b^c = a \text{ are equivalent.}$$

The equation $\log_b(a) = c$ is in **logarithmic form** and the equation $b^c = a$ is in **exponential form**. Either form can replace the other when solving a problem.

Example 1   Solving an Equation in Logarithmic Form

Solve for x.

1. $\log_4(x) = 3$
2. $\log(x) = 2.16$
3. $\log_2(3x - 1) = 5$
4. $\log_3(x^4) = 2$

Solution

1. We write $\log_4(x) = 3$ in exponential form and solve for x.

$$4^3 = x$$
$$x = 64$$

2. We write $\log(x) = 2.16$ in exponential form and solve for x.

$$10^{2.16} = x$$
$$x \approx 144.5440$$

3. We write $\log_2(3x - 1) = 5$ in exponential form and solve for x.

$$2^5 = 3x - 1$$
$$3x - 1 = 32$$
$$3x = 33$$
$$x = 11$$

4. We write $\log_3(x^4) = 2$ in exponential form and solve for x.

$$3^2 = x^4$$
$$x^4 = 9$$
$$x = \pm 9^{1/4}$$
$$x \approx \pm 1.7321$$

In most cases throughout this chapter, we will round approximate solutions to the ten-thousandth place.

The first step in solving each of the logarithmic equations in Example 1 involves writing the equation in exponential form. For each equation in Example 2, we perform the same first step, but now to solve for the *base* of a logarithm.

Example 2   Solving for the Base of a Logarithm

Solve for b.

1. $\log_b(81) = 4$
2. $\log_b(67) = 5$

Solution

1. We write $\log_b(81) = 4$ in exponential form and solve for b.

$$b^4 = 81$$
$$b = 81^{1/4} \qquad \text{The base of a logarithm is positive.}$$
$$b = 3$$

2. We write $\log_b(67) = 5$ in exponential form and solve for b.

$$b^5 = 67$$
$$b = 67^{1/5}$$
$$b \approx 2.3185 \qquad \blacksquare$$

There is an important property called the *power property* for logarithms that will help us solve exponential equations.

Power Property

For $x > 0$, $b > 0$, and $b \neq 1$,

$$\log_b(x^p) = p\log_b(x)$$

For example, $\log_3(x^5) = 5\log_3(x)$. Also, $\log_2(x^7) = 7\log_2(x)$.
A proof of the power property for logarithms follows.
Let $k = \log_b(x^p)$. The exponential form of this equation is $b^k = x^p$. Using the exponent $\dfrac{1}{p}$ on both sides and simplifying gives:

$$(b^k)^{1/p} = (x^p)^{1/p}$$
$$b^{k\left(\frac{1}{p}\right)} = x^{p\left(\frac{1}{p}\right)} \qquad (b^m)^n = b^{mn} \quad \text{if } b > 0$$
$$b^{k/p} = x$$

The logarithmic form of this equation is:

$$\log_b(x) = \frac{k}{p}$$

Multiplying both sides by p and simplifying gives:

$$p\log_b(x) = p\left(\frac{k}{p}\right)$$
$$p\log_b(x) = k$$
$$k = p\log_b(x)$$

But $k = \log_b(x^p)$, so by substitution we have:

$$\log_b(x^p) = p\log_b(x)$$

This statement is what we set out to prove.

Example 3   Solving Exponential Equations

Solve each equation.

1. $2^x = 12$ **2.** $3(4)^x = 71$ **3.** $5^{3x-1} = 17$

Solution

1. If $2^x = 12$, it follows that

$$\log(2^x) = \log(12)$$

Note

We say that we "Take the log of both sides."

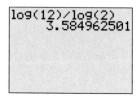
Next, we use the power property on the left side and then solve for x.

$$x \log(2) = \log(12) \qquad \text{Power property}$$

$$x = \frac{\log(12)}{\log(2)} \qquad \text{Divide by } \log(2) \text{ on each side.}$$

$$x \approx 3.5850$$

We check that 3.5850 approximately satisfies the equation $2^x = 12$.

$$2^{3.5850} \approx 12.0003 \approx 12$$

2.
$$3(4)^x = 71$$

$$4^x = \frac{71}{3} \qquad \text{Divide each side by 3.}$$

$$\log(4^x) = \log\left(\frac{71}{3}\right) \qquad \text{Take the log of both sides.}$$

$$x \log(4) = \log\left(\frac{71}{3}\right) \qquad \text{Power property}$$

$$x = \frac{\log\left(\dfrac{71}{3}\right)}{\log(4)} \qquad \text{Divide each side by } \log(4).$$

$$x \approx 2.2824$$

We check that 2.2824 approximately satisfies the equation $3(4)^x = 71$.

$$3(4)^{2.2824} \approx 71.0008 \approx 71$$

3.
$$5^{3x-1} = 17$$

$$\log(5^{3x-1}) = \log(17) \qquad \text{Take the log of both sides.}$$

$$(3x - 1)\log(5) = \log(17) \qquad \text{Power property}$$

$$3x - 1 = \frac{\log(17)}{\log(5)} \qquad \text{Divide each side by } \log(5).$$

$$3x = \frac{\log(17)}{\log(5)} + 1 \qquad \text{Add 1 to both sides.}$$

$$x = \frac{\dfrac{\log(17)}{\log(5)} + 1}{3} \qquad \text{Divide both sides by 3.}$$

$$x \approx 0.9201 \qquad\qquad\qquad\qquad\qquad\qquad \blacksquare$$

In problem 2 in Example 3, we solved $3(4)^x = 71$ by first dividing both sides of the equation by 3. Next, we took the log of both sides of the equation, and then used the power property to help solve for x.

Since $3(4)^x \neq (3 \cdot 4)^x$, we *cannot* begin the process of solving $3(4)^x = 71$ by saying

$$3(4)^x = 71$$

$$\log[3(4)^x] = \log(71)$$

$$x \log(3 \cdot 4) = \log(71) \qquad \text{Cannot do this since } 3(4)^x \neq (3 \cdot 4)^x.$$

That is why we begin by dividing both sides of $3(4)^x = 71$ by 3.

In general,

$$\log_b(ax^p) \neq p \log(ax)$$

EXPLORATION
Comparing the power property with other statements

Consider the following equations, where $x > 0$, $a > 0$, $b > 0$, and $b \neq 1$:

$$\log_b(x^p) = p \log_b(x)$$
$$\log_b[a(x^p)] = p \log_b(ax)$$
$$\log_b[(ax)^p] = p \log_b(ax)$$

1. Which equation(s) is true in general? Explain why in terms of the power property.
2. Show that the other equation(s) is false by using the substitutions $b = 10$, $a = 10$, $x = 10$, and $p = 2$.
3. Three students tried to solve $5(4)^x = 30$. Which students, if any, solved the equation correctly? If any errors were made, describe them and where they occurred.

Student 1's work	*Student 2's work*	*Student 3's work*
$5(4)^x = 30$	$5(4)^x = 30$	$5(4)^x = 30$
$4^x = 6$	$\log[5(4)^x] = \log(30)$	$20^x = 30$
$\log(4^x) = \log(6)$	$x \log[5(4)] = \log(30)$	$\log(20^x) = \log(30)$
$x \log(4) = \log(6)$	$x \log(20) = \log(30)$	$x \log(20) = \log(30)$
$x = \dfrac{\log(6)}{\log(4)}$	$x = \dfrac{\log(30)}{\log(20)}$	$x = \dfrac{\log(30)}{\log(20)}$
$x \approx 1.2925$	$x \approx 1.1353$	$x \approx 1.1353$

Tips on Succeeding in This Course

Have you ever had trouble solving a problem and then returned to the problem hours later and found it easy to solve? By taking a break, you can return to the exercise with a different perspective and renewed energy. You've also given your unconscious mind a chance to reflect on the problem while taking your break. You can strategically put this phenomenon to your advantage by allocating time at two different points in your day to complete your homework assignment.

Key Points
OF THIS SECTION

The equations $\log_b(a) = c$ and $b^c = a$ are equivalent. Either can be written in the place of the other.	**Equivalent equations**
For an equation of the form $\log_b(x) = k$, we can solve for b or for x by writing the equation in exponential form.	**Solving $\log_b(x) = k$**
For $x > 0$, $b > 0$, and $b \neq 1$, $$\log_b(x^p) = p \log_b(x) \qquad \text{Power property}$$	**Power property**
We solve some equations of the form $ab^x = c$ for x by first dividing both sides by a and then taking the log of both sides of the equation. Next, we use the power property for logarithms.	**Solving $ab^x = c$ for x**

HOMEWORK 5.3

SSM Tutor Center   MathPro 4   MathPro 5   CDs/Videos   www

Write the equation in exponential form. Assume that all constants are positive and not equal to 1.

1. $\log_3(243) = 5$

2. $\log_2(32) = 5$

3. $\log(100) = 2$

4. $\log_5\left(\dfrac{1}{25}\right) = -2$

5. $\log_b(a) = c$

6. $\log_r(s) = t$

7. $\log(m) = n$

8. $\log_w(y) = z$

Write the equation in logarithmic form. Assume that all constants are positive and not equal to 1.

9. $5^3 = 125$

10. $2^5 = 32$

11. $10^3 = 1000$

12. $7^2 = 49$

13. $y^w = x$

14. $r^s = t$

15. $10^p = q$

16. $b^x = y$

Solve for x.

17. $\log_4(x) = 2$

18. $\log_2(x) = 3$

19. $\log_2(x) = -4$

20. $\log(x) = 5$

21. $\log(x) = -2$

22. $\log_8(x) = \dfrac{1}{3}$

23. $\log_4(x) = 0$

24. $\log_{2.8}(x) = 3.4$

25. $\log_8(4x) = 1$

26. $\log_5(4x + 1) = 3$

27. $\log_6(x^3) = 2$

28. $\log(x^6) = 4$

29. $\log_2(\log_3(x)) = 3$

30. $\log_3(\log_2(x)) = -1$

Solve for b.

31. $\log_b(49) = 2$

32. $\log_b(16) = 2$

33. $\log_b(8) = 3$

34. $\log_b(125) = 3$

35. $\log_b(16) = 4$

36. $\log_b(80) = 7$

37. $\log_b(3.6) = 6$

38. $\log_b(9.34) = 3$

Solve for x. Verify your result by checking that it approximately satisfies the original equation.

39. $4^x = 9$

40. $10^x = 50$

41. $5(4^x) = 80$

42. $3(2^x) = 17$

43. $-11(7^x) = -5400$

44. $-19 = -2(3^x)$

45. $8 + 5(2^x) = 79$

46. $20 = -3 + 4(2^x)$

47. $3.83(2.18^x) = 170.91$

48. $8 = 3^{7x-1}$

49. $2^{4x+5} = 17$

50. $5^x + 7 = 50 - 3(5^x)$

51. $4^{3x}4^{2x-1} = 100$

52. $\dfrac{2^{5x}}{2^{2x+3}} = 39$

53. $3^x = -8$

54. $1^x = 13$

Solve.

55. $\log_b(81) = 4$

56. $3^x = 11$

57. $\log_3(x) = 4$

58. $50 = 2^x$

59. $\log_4(x) = 3$

60. $\log_b(12) = 5$

61. $\log_7(x) = 1$

62. $2(6)^x - 17 = 3$

63. $7^x = 0$

64. $\log_{1.9}(x) = 2.2$

65. $\log_{27}(x - 1) = \dfrac{1}{3}$

66. $\log_b(80) = 7$

67. $\log_2\left(\dfrac{x}{2}\right) = -1$

68. $7 - 3(5^x) = -1000$

69. $\log_b(2) = 5$

70. $5^{4x-1} = 1$

71. A student has made an error in trying to solve the equation $3(8^x) = 7$. Find the error in the student's work shown below.

$$3(8^x) = 7$$
$$\log[3(8^x)] = \log(7)$$
$$x \log[3(8)] = \log(7)$$
$$x \log(24) = \log(7)$$
$$x = \frac{\log(7)}{\log(24)}$$
$$x \approx 0.6123$$

72. A student has made an error in trying to solve the equation $2(3^x) = 10$. Find the error in the student's work shown below.

$$2(3^x) = 10$$
$$6^x = 10$$
$$\log(6^x) = \log(10)$$
$$x \log(6) = 1$$
$$x = \frac{1}{\log(6)}$$
$$x \approx 1.2851$$

In exercises 73–76, assume that $b > 0$, $b \neq 1$, and that the constants have values for which the equation has exactly one real number solution.

73. Solve $ab^x = c$ for x.

74. Solve $ab^x + d = c$ for x.

75. Solve $ab^{kx} + d = c$ for x.

76. Solve $ab^{kx-p} + d = c$ for x.

Let $f(x) = 4^x$.

77. Find $f(4)$.

78. Find $f^{-1}(16)$.

79. Find x when $f(x) = 3$.

80. Find x when $f^{-1}(x) = 3$.

Let $g(x) = \log_2(x)$.

81. Find $g(8)$.

82. Find $g^{-1}(4)$.

83. Find a when $g(a) = 5$.

84. Find a when $g^{-1}(a) = 5$.

85. The following incorrect work shows that the logarithm of any positive number is 0. What error(s) is made? Assume that $b > 0$, $b \neq 1$, and $k > 0$.

$$\log_b(k) = \log_b(k \cdot 1) = \log_b(k \cdot 5^0) = 0 \cdot \log_b(5k) = 0$$

86. Describe the power property. Include an example in your description. Does the power property imply that $x^p = px$? Explain.

87. Describe how to use the power property to solve an exponential equation.

5.4 USING THE POWER PROPERTY WITH EXPONENTIAL MODELS TO MAKE PREDICTIONS

Math is a way of life. Without it everything would be a mess. —*Christina A., student*

Objectives

▶ Use the power property with exponential models to make predictions.
▶ Use the half-life of carbon-14 to date artifacts and fossils.

In Section 4.5 we used exponential functions to model data. In this section we use the power property with an exponential function to make predictions for the independent variable of the function.

Example 1   Using the Power Property to Make a Prediction

A person invests $7000 in a compound interest account with a yearly interest rate of 6%. When will the balance be $10,000?

Solution

Let $B = f(t)$ represent the balance (in thousands of dollars) after t years or any fraction thereof. From our work with compound interest accounts in Section 4.5, we know that we can model the situation well by using an exponential model of the form $f(t) = ab^t$. The B-intercept is $(0, 7)$ (for $7000 when $t = 0$), so $a = 7$ and $f(t) = 7b^t$. By the end of each year, the account has increased by 6% of the previous year's balance, so $b = 1.06$. Thus,

$$f(t) = 7(1.06)^t$$

To find when the balance is $10,000 ($B = 10$), we substitute 10 for $f(t)$ and solve for t.

$$10 = 7(1.06)^t$$

$$\frac{10}{7} = 1.06^t \qquad \text{Divide each side by 7.}$$

$$\log\left(\frac{10}{7}\right) = \log(1.06^t) \qquad \text{Take the log of both sides.}$$

$$\log\left(\frac{10}{7}\right) = t\log(1.06) \qquad \text{Power property}$$

$$t = \frac{\log\left(\frac{10}{7}\right)}{\log(1.06)} \qquad \text{Divide each side by } \log(1.06).$$

$$t \approx 6.1212$$

So, it will take about 6 years and 45 days for the balance to reach $10,000.   ■

   In Example 1 we used the function $f(t) = 7(1.06)^t$ to make a prediction for the independent variable t by first substituting 10 for the dependent variable $f(t)$ and then dividing both sides of the equation by the coefficient 7. Next, we took the log of both sides of the equation, and then used the power property to help solve for t.

Example 2   Using the Power Property to Make a Prediction

The infant mortality rate is the number of deaths of infants under one year old per 1000 births.* In 1915 the rate was almost 100 deaths per 1000 infants, or 1 death per 10 infants. The infant mortality rate has substantially decreased since then (see Table 23).

Table 23 Infant Mortality Rates

Year	Rate (number of deaths) per 1000 infants)
1915	99.9
1920	85.8
1930	64.6
1940	47.0
1950	29.2
1960	26.0
1970	20.0
1980	12.6
1990	9.2
2000	6.9

Source: *National Center for Health Statistics*

*The rate does not include fetal deaths.

1. Let $I = f(t)$ represent the infant mortality rate (number of deaths per 1000 infants) at t years since 1900. Find an equation for f.

2. What is the percent rate of decay for infant mortality rates?

3. Find $f^{-1}(5)$. What does this mean in terms of infant mortality rates?

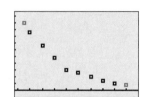

Figure 23 Scattergram of data

Solution

1. By viewing the scattergram in Fig. 23, we see that the points "bend" and that an exponential function will model the data better than a linear function.

   We use the points $(15, 99.9)$ and $(100, 6.9)$ to find an equation of the form $I = ab^t$. We substitute the coordinates of the two chosen points into $f(t) = ab^t$:

   $$6.9 = ab^{100}$$
   $$99.9 = ab^{15}$$

   We divide the left sides and the right sides of the equations and solve for b:

   $$\frac{6.9}{99.9} = \frac{ab^{100}}{ab^{15}}, \quad \text{where } a \text{ and } b \text{ are nonzero}$$

   $$\frac{6.9}{99.9} = b^{85}$$

   $$b = \left(\frac{6.9}{99.9}\right)^{1/85}$$

   $$b \approx 0.969$$

   So, $f(t) = a(0.969)^t$. We substitute the coordinates of $(100, 6.9)$ into the equation $f(t) = a(0.969)^t$ and solve for a:

   $$6.9 = a(0.969)^{100}$$

   $$a = \frac{6.9}{0.969^{100}}$$

   $$a \approx 160.87$$

   Thus, $f(t) = 160.87(0.969)^t$. In Fig. 24, we see that the model appears to fit the data very well.

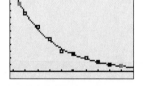

Figure 24 Verifying the model

2. The base b is 0.969. Since $1 - 0.969 = 0.031$, we conclude that the model estimates that the infant mortality rates decayed by 3.1% per year.

3. Since f sends values of t to values of I, f^{-1} sends values of I to values of t (see Figs. 25 and 26).

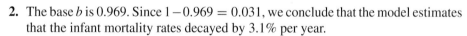

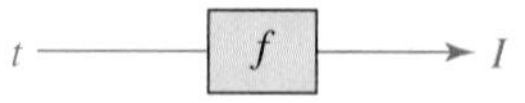

Figure 25 f sends values of t to values of I 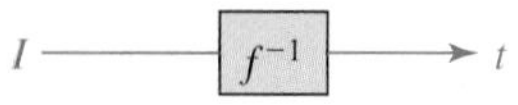   **Figure 26** f^{-1} sends values of I to values of t

Therefore, $f^{-1}(5)$ represents the year (since 1900) when the infant mortality rate will be 5 deaths per 1000 infants. To find the year, we substitute 5 for $f(t)$ in the equation $f(t) = 160.87(0.969)^t$ and solve for t.

$$5 = 160.87(0.969)^t$$

$$\frac{5}{160.87} = 0.969^t$$

$$\log\left(\frac{5}{160.87}\right) = \log(0.969^t)$$

$$\log\left(\frac{5}{160.87}\right) = t\log(0.969)$$

$$t = \frac{\log\left(\dfrac{5}{160.87}\right)}{\log(0.969)}$$

$$t \approx 110.23$$

The model predicts that the infant mortality rate will be 5 deaths per 1000 infants by 2010. ∎

Before performing some more modeling in Example 3, it will be helpful to know a fact related to the base multiplier property. For an exponential function $y = f(t) = ab^t$, we know by the base multiplier property that as the value of t increases by 1, the value of y is multiplied by the base b. Also, if the value of y is multiplied by M in going from $t = 0$ to $t = k$, then the value of y will continue to be multiplied by M each time t is increased by k. To see this, recall that $f(0) = a$ and note that:

$$f(k) = ab^k = aM, \text{ so } b^k = M$$

We can use the fact $b^k = M$ to show that if t is increased by k a total of n times, then y is multiplied by M a total of n times.

$$f(nk) = ab^{nk} = a(b^k)^n = a(M)^n$$

We use this idea in Example 3.

Example 3 Using the Power Property to Estimate Doubling Time

In exercise 19 in Section 4.5 on the speeds of computer chips, you may have found $f(t) = 0.10(1.37)^t$, where $f(t)$ represents the chip speed (in MHz) at t years since 1971 (see Table 24). A rule of thumb for estimating how quickly technological products improve is that they double in speed every 2 years.* Use f to estimate the doubling time.

Table 24 Chip Speeds

Year	Clock Speed (MHz)
1971	0.1
1974	2
1982	12
1985	16
1992	50
1995	120
1997	300
2000	1000

Solution

In 1971, the chip speed was 0.10 MHz. We find when the speed was $2(0.10) = 0.20$ MHz by substituting 0.20 for $f(t)$ in the equation $f(t) = 0.10(1.37)^t$ and solving for t.

$$0.20 = 0.10(1.37)^t$$
$$2 = 1.37^t$$
$$\log(2) = \log(1.37^t)$$
$$\log(2) = t\log(1.37)$$
$$t = \frac{\log(2)}{\log(1.37)}$$
$$t \approx 2.20$$

According to the exponential function f, it took 2.20 years (about 2 years and 2 months) for the 1971 chip speed to be doubled. By the discussion preceding this example, we can conclude that the chip speed will double *every* 2.20 years. So, according to the model, the doubling time is about two months more than the rule of thumb estimate of 2 years. ∎

Recall that the half-life of an element is the amount of time it takes for the number of atoms to be reduced to half as much. All living organisms are, in part, composed of the elements carbon-12 and carbon-14. After an animal or plant dies, its carbon-14

*In 1965, Gordon Moore used modeling to show that the number of transistors per integrated circuit had been doubling about every two years. He predicted that this rule of thumb (now referred to as "Moore's Law") would model the situation well until at least 1975. In fact, the model still accurately describes the situation to date (Silicon Showcase, www.intel.com/research/silicon/mooreslaw.htm. Accessed 6/22/02.).

decays exponentially with a half-life of 5730 years. However, the amount of carbon-12 remains constant. Since scientists know the ratio of carbon-14 to carbon-12 of *living* organisms, they can *carbon date* an artifact by measuring the decreased ratio of carbon-14 to carbon-12.

Example 4   Using the Power Property to Make an Estimate

In 1997, four finely crafted but ancient spears were found in a coal mine excavation 60 miles from Hanover, Germany. The spears were used to hunt horses who frequented a nearby ancient lake. What's extraordinary about these 6.5 foot-long spears is that they have turned scientific theory upside down about when humans began hunting. Until this discovery, scientists had believed that humans began hunting some 40,000 years ago.

If only 9.676×10^{-20} percent of the carbon-14 still remains in the spears, when were our early ancestors using them to hunt?

Solution

Let $P = f(t)$ represent the percentage of carbon-14 that still remains t years after the spears were made. Since the percentage is halved every 5730 years, we will find an exponential equation of the form

$$f(t) = ab^t$$

At time $t = 0$, 100% (all) of the carbon-14 remained, so the P-intercept is $(0, 100)$. Therefore, $a = 100$ and $f(t) = 100b^t$. At time $t = 5730$, $\frac{1}{2}(100) = 50$ percent of the carbon-14 remained. So, we substitute the coordinates of the point $(5730, 50)$ into the equation $f(t) = 100b^t$ and solve for t.

$$50 = 100b^{5730}$$
$$0.5 = b^{5730} \qquad \text{Divide by 100.}$$
$$b = 0.5^{1/5730}$$
$$b \approx 0.999879 \qquad \text{The base of an exponential function is positive.}$$

Note

We use more digits than usual for the base, as even a small change in the base greatly affects estimates made for events well into the past (or future).

Therefore, the equation is $f(t) = 100(0.999879)^t$.

We now estimate the age of the spears by substituting 9.676×10^{-20} for $f(t)$ and solving for t.

$$9.676 \times 10^{-20} = 100(0.999879)^t$$
$$\frac{9.676 \times 10^{-20}}{10^2} = 0.999879^t \qquad \text{Divide by } 100 = 10^2.$$
$$9.676 \times 10^{-22} = 0.999879^t \qquad \frac{b^m}{b^n} = b^{m-n}$$
$$\log(9.676 \times 10^{-22}) = \log(0.999879^t)$$
$$\log(9.676 \times 10^{-22}) = t \log(0.999879)$$
$$t = \frac{\log(9.676 \times 10^{-22})}{\log(0.999879)}$$
$$t \approx 399{,}870.21$$

So, the age of the spears is approximately 400,000 years. Due to the discovery and the carbon dating of the spears, scientists have adjusted their theory about when humans first hunted from 40,000 years to 400,000 years ago. ■

EXPLORATION
Finding an equation of the inverse for an exponential model

In Example 1 we found the model $B = f(t) = 7(1.06)^t$, where B is the balance (in thousands of dollars) of an account at 6% interest compounded annually, where the starting balance is $7000. In this exploration, you will find the equation for the inverse function f^{-1}.

1. Substitute B for $f(t)$ in the equation $f(t) = 7(1.06)^t$.

2. Solve your equation for t.

3. Replace t with $f^{-1}(B)$ in your equation. You now have an equation for f^{-1}.

4. Use your equation for f^{-1} to find $f^{-1}(12)$. What does your result mean in terms of the balance of the account?

5. Use a graphing calculator to help you complete Table 25.

6. How can the information in your completed table help an investor?

Table 25 Values of f^{-1}

B	$f^{-1}(B)$
7	
8	
9	
10	
11	
12	
13	
14	

EXPLORATION
Looking ahead: Product and quotient properties for logarithms

1. **a.** Use your calculator to compare $\log(2 \cdot 3)$ and $\log(2) + \log(3)$.

   b. Use your calculator to compare $\log(7 \cdot 2)$ and $\log(7) + \log(2)$.

   c. Use your calculator to compare $\log(4 \cdot 6)$ and $\log(4) + \log(6)$.

   d. What do parts a–c of this exploration suggest about $\log(xy)$? Test your conjecture. (Your observation is referred to as the *product property* for logarithms.)

2. Form a conjecture about how $\log\left(\dfrac{x}{y}\right)$ can be expressed in terms of two logarithms. [**Hint**: Choose specific values for x and y. Then compute $\log\left(\dfrac{x}{y}\right)$, $\log(x)$, and $\log(y)$, and compare the values.] Test your conjecture. (Your observation is referred to as the *quotient property* for logarithms.)

Tips on Succeeding in This Course

Many students believe that success in math courses depends mostly on their mathematical ability. In reality, students' success depends mostly on their *commitment* to learning math.

Key Points
OF THIS SECTION

To make a prediction for the independent variable t of an exponential model of the form $f(t) = ab^t$, substitute a value for $f(t)$ and then divide each side of the equation by the coefficient a. Next, take the log of both sides and then use the power property to solve for t.

Making a prediction

Scientists use the half-life of carbon-14 (5730 years) and the percentage of carbon-14 remaining in an artifact to estimate the age of the artifact.

Estimating the age of an artifact

HOMEWORK 5.4

SSM
Tutor Center

MathPro 4
MathPro 4

MathPro 5
MathPro 5

CDs/Videos

www
www

1. A person invests $2000 in an account at 5% interest compounded annually. Let $V = f(t)$ represent the value (in dollars) of the account after t years or any fraction thereof.

   a. Find an equation for f.

   b. What is the V-intercept? What does it mean in terms of the account balance?

   c. When will the balance be $3000?

2. A person invests $4000 in an account at 7% interest compounded annually. When will the value of the investment be doubled?

3. In exercise 21 of Homework 4.5, you may have found an equation close to $f(t) = 1.197(1.0163)^t$, where $f(t)$ models the world population (in billions) at t years since 1900 (see Table 26).

Table 26 World Populations

Year	World Population (billions)
1930	2.070
1940	2.295
1950	2.500
1960	3.050
1970	3.700
1980	4.454
1990	5.279
2000	6.080
2002	6.215

   a. Many experts believe that the world can support about 10 billion people.* Predict when the population will reach 10 billion.

   b. Use pencil and paper to sketch a graph of the model f for the years from 1900 to 2100. Assuming that the Earth can support at most 10 billion people, for what future years is there model breakdown? Sketch the graph of a function that would better model the world population.

4. In exercise 6 of Homework 4.5, you may have found the equation $f(t) = 2(2)^t$, where $f(t)$ represents the number of deaths due to air bags during the year that is t years since 1992.

   a. Recall that there were only 2 deaths during 1992. To get an idea of how quickly an exponential function grows, use f to predict during which single year the number of deaths would equal the U.S. population. Assume that the United States population is its 2002 population, 287 million.

   b. What assumptions were made in part a? Explain how these assumptions affected your prediction.

5. In exercise 23 of Homework 4.5, you found an equation close to $f(t) = 1.46(1.48)^t$, where $f(t)$ represents the number of Starbucks stores at t years since 1980 (see Table 27).

   a. What is the coefficient a of the model $f(t) = ab^t$? What does it mean in terms of the situation?

Table 27 Numbers of Starbucks stores

Year	Number of Stores
1987	17
1989	55
1991	116
1993	272
1995	676
1997	1412
1999	2135
2001	4709

   b. The first Starbucks store opened in 1971 in Seattle, Washington. Do you think the number of stores grew exponentially from 1971 to 2001? Explain.

   c. What is the base b of the model $f(t) = ab^t$? What does it mean in terms of the situation?

   d. Use f to predict when there will be an average of 1000 Starbucks stores per state.

   e. Use f to predict when there will be one Starbucks store for every household. Assume that there are 105 million households in the United States.

6. The illicit use of methamphetamine (or "meth") has skyrocketed in rural areas. The chemical by-products of creating the drug are a big environmental hazard, and toxic spill cleanup crews are often called in to clean up abandoned meth labs. Table 28 shows the number of meth labs cleaned by toxic spill crews in Washington State alone.

Table 28 Numbers of Methamphetamine Labs Cleaned by Toxic Spill Crews in Washington

Year	Number of Labs
1995	42
1996	146
1997	208
1998	350
1999	792
2000	1458

Source: *Washington State Dept. of Ecology*

Let $f(t)$ represent the number of methamphetamine labs cleaned by toxic spill crews in the year that is t years since 1995.

   a. Use a graphing calculator to draw a scattergram of the data. Is it better to use a linear or an exponential function to model the data? Explain.

   b. Find an equation for f.

   c. What is the base b of the function $f(t) = ab^t$? What does it tell you about the number of labs? Now consider the *actual* number of labs in Table 28. During which year did the number of labs increase by the greatest percent?

---

*Estimates fall most frequently in the range of 4 billion to 16 billion. Source: Joel E. Cohen, *How Many People Can the Earth Support?* (New York: W.W. Norton & Company, 1995), 215.

d. Use f to predict in which year the number of meth labs cleaned by toxic spill crews will equal the number of households in Washington. Assume there are 2.2 million households in Washington.*

7. The percentages of U.S. food expenditures devoted to food away from home has greatly increased in the past century (see Table 29).

Table 29 Percentages of Food Expenditures Away from Home	
Year	Percent
1869	5
1899	9
1929	17
1959	26
1989	46
2000	47

Source: *Economic Research Service, USDA*

Let $f(t)$ represent the percentage of U.S. food expenditures devoted to food away from home in the year that is t years since 1800.

a. Use a graphing calculator to draw a scattergram of the data. Is it better to use a linear or an exponential function to model the data? Explain.

b. Find an equation for f.

c. Find $f(100)$. What does your result mean in terms of the situation?

d. Find $f^{-1}(100)$. What does your result mean in terms of the situation?

e. Find t when $f(t) = 100$. What does your result mean in terms of the situation?

8. The number of female competitors in the Olympic games has greatly increased during the past four decades (see Table 30).

Table 30 Numbers of Female Competitors in the Olympic Games	
Year	Number of Female Competitors
1960	610
1964	683
1968	781
1972	1070
1976	1251
1980	1088
1984	1620
1988	2186
1992	2710
1996	3684
2000	3906

Source: Guinness Book of Answers *as reported by Jennifer S., Spring Semester, 1999; for 2000 data: CNN/SI-Olympics, www.sportsillustrated.cnn.com/ olympics/news/1999/03/31/sydney_women. Accessed 4/2/02.*

a. Let $f(t)$ represent the number of female competitors at t years since 1960. Use a graphing calculator to draw a scattergram of the data. Is it better to use a linear or an exponential function to model the data? Explain.

b. Find an equation for f.

c. Find $f(48)$. What does your result mean in terms of female competitors?

d. Find $f^{-1}(7000)$. What does your result mean in terms of female competitors?

9. In a seven-year study of ten of the nation's most selective colleges and universities, researchers found that a student applicant has a better chance of being accepted to a college through early decision than by regular decision. Table 31 shows a comparison of SAT scores (out of 1600) and acceptance rates by both systems.

Table 31 Percentages of Applicants Accepted by Early Decision and Regular Decision		Percent	
SAT score Group	SAT score Used to Represent SAT Group	Early Decision	Regular Decision
1100–1190	1145	21	10
1200–1290	1245	35	17
1300–1390	1345	52	31
1400–1490	1445	70	48
1500–1600	1550	93	72

Source: *Professor Christopher Avery, Kennedy School of Government, Harvard University*

For students who score s points, let $E(s)$ and $R(s)$ be the percentages of early decision and regular decision applicants, respectively, that are accepted.

a. Find regression equations for E and R.

b. For early decision applicants who score 1425 points, what percentage get accepted? What about for regular decision applicants who score 1425 points?

c. For what score do half of early decision applicants get accepted? What about for regular decision applicants?

d. The study concluded that students who apply for early decision have the equivalent of 100 points added to their SAT score compared with applying for regular decision. What do your results from part c suggest the equivalent number of added points to be?

e. Use your graphing calculator to find the intersection point(s) of the graphs of E and R. What does this point(s) represent in terms of the situation?

It's true your newborn applying to our college is planning pretty far ahead. But by applying to our ultra-early decision system, your child need only score 1 point on the SAT to be accepted!

---

*There were approximately 2.2 million households in Washington in 2002.

10. Allstate Life Insurance offers a $250,000 life insurance policy for men who are smokers. The monthly rates for men of various ages are listed in Table 32. Although the rates for various ages are adjusted every five years, a person's rate will always be determined by age at the time of purchase.

Table 32 Allstate Monthly Insurance Rates for Male Smokers

Issue Age	Monthly Rates (dollars)
30	50.85
35	61.74
40	92.27
45	134.97
50	193.25
55	290.18
60	444.54
65	674.06

Source: *Allstate Life Insurance Company*

a. Let $f(t)$ represent the monthly rate (in dollars) for a man t years of age who smokes. Perform the first three steps of the four-step modeling process to find an equation for f.

b. Find $f(25)$ and $f(70)$. What do these outputs of f mean in terms of Allstate's rates, according to the model? Explain in terms of the life insurance described why it makes sense that $f(70)$ for such a life insurance policy is so much larger than $f(25)$.

c. Find $f^{-1}(500)$. What does your result mean in terms of Allstate's rates?

d. What is the base b of your model $f(t) = ab^t$? What does it tell you in terms of this policy's rates?

11. Suppose that a rumor is spreading in the United States that the airlines will soon be giving away free promotional tickets for flights within the United States. Assume that 30 people have heard the rumor as of today and that each day the total number of people who have heard the rumor doubles. Let $f(t)$ represent the total number of people who have heard the rumor at t days since today.

a. Find an equation for f.

b. Predict when all Americans will have heard the rumor. Assume that the United States population is its 2002 population, 287 million.

12. There are 4 million bacteria on a peach at noon on Tuesday. Assume that a bacterium divides into two bacteria every hour, on average. Let $f(t)$ represent the number of bacteria (in millions) on the peach at t hours after Tuesday noon.

a. Find an equation for f.

b. Find $f(24)$. What does your result mean in terms of the number of bacteria?

c. Find $f^{-1}(8000)$. What does your result mean in terms of the number of bacteria?

13. According to the U.S. Occupational Safety and Health Administration (OSHA) standard, an average person can listen to 8 hours of sound per day at a sound level of 90 decibels without experiencing hearing loss. Recall from exercise 77 in Section 5.2 that a decibel is a unit for the measurement of the intensity of sound. For each increase of 5 decibels, the exposure time must be cut in half. For example, an average person can listen to 4 hours of sound per day at a sound level of 95 decibels without experiencing hearing loss.

Some examples of sound being made at various decibel readings are listed in Table 33.

Table 33 Examples of Decibel Readings

Sound Level (decibels)	Example
0	Faintest sound heard by humans
20	Whisper
40	Inside a running car
60	Conversation
80	Noisy street corner
100	Soft rock concert
120	Threshold of pain

Source: Math and Music

a. Let $T = f(d)$ represent the number of hours of safe exposure time in one day to a sound at a level of d decibels *above 90 decibels*. Find an equation for f.

b. Many rock bands (e.g., Linkin Park, blink−182) play at about 114 decibels. Use your equation for f to predict how long they can play at concerts so that fans who attend a lot of concerts do not experience hearing loss.* Based on your result, do you think that these types of fans experience hearing loss? You may assume that most people are not wearing earplugs.

c. If you include the warm-up band's act, many rock concerts last about 3 hours. At what sound level should the bands play so that fans who attend a lot of concerts will not experience hearing loss?

14. A mammal's fossil bones are found in a cave excavation. Recall that the half-life of carbon-14 is 5730 years. If 5% of the carbon-14 remains in the bones, estimate when the mammal died.

---

*A one-time overexposure may result in temporary hearing loss, but probably not permanent hearing loss.

15. A bowl made of pinewood is found at an archeological site. The half-life of carbon-14 is 5730 years. If 90% of the carbon-14 remains in the wood, estimate the age of the wood.

16. An archeologist discovers a tool made of wood.

    a. If 50% of the wood's carbon-14 remains, how old is the wood? Explain how you can find this result without using an equation. Recall that the half-life of carbon-14 is 5730 years.

    b. If 25% of the wood's carbon-14 remains, how old is the wood? Explain how you can find this result without using an equation.

    c. If 10% of the wood's carbon-14 remains, how old is the wood? First, guess an approximate age without solving an equation. Explain how you decided on your estimate. Then, use an equation to find the result.

17. A person drinks alcohol at a party. After her last drink, the alcohol level of her blood soon reaches a maximum of 0.28 milligram alcohol per milliliter of blood. If the half-life of alcohol in her blood is 2 hours, how long must she wait before driving at the legal limit of 0.08 milligram alcohol per milliliter of blood?*

18. A person drinks a cup of coffee. We assume that the caffeine enters his bloodstream immediately, and that there was no caffeine in his bloodstream prior to drinking the coffee. The half-life of caffeine in a person's bloodstream is about 6 hours. A cup of coffee contains about 100 milligrams of caffeine.

    a. Let $f(t)$ represent the number of milligrams of caffeine in the person's bloodstream at t hours after drinking the cup of coffee. Find an equation for f.

    b. The person drinks the cup of coffee at 8 A.M. and goes to bed at 11 P.M. Use f to predict the amount of caffeine in his bloodstream when he goes to bed.

    c. Suppose the person drinks another cup of coffee 24 hours after the first cup. How much caffeine will be in his bloodstream from these 2 cups of coffee just after drinking the second cup? Explain how you can find this result without using an equation.

    d. Now assume that the person drinks the cup of coffee at 8 A.M. and then drinks a cup of coffee every morning at 8 A.M. from then on. Sketch a qualitative graph that describes the relationship between caffeine in the person's bloodstream and time. Describe any assumptions that you make.

19. A storage tank contains radium, which has a half-life of 1600 years.

    a. Let $f(t)$ represent the percentage of the radium that remains in the tank at t years after the element was placed in the tank. Find a formula for f.

    b. Predict when 10% of the radium will remain.

20. A storage tank contains a liquid radioactive element with a half-life of 100 years. It will be relatively safe for the contents of the tank to leak from the tank when 0.01% of the radioactive element remains. How long must the tank remain intact for this storage procedure to be safe?

21. Describe the four-step modeling process in your own words. If an exponential function is chosen to model a situation, in which of the four steps do we use the power property? Explain.

5.5 MORE PROPERTIES OF LOGARITHMS

Objectives

▸ Know *product*, *quotient*, and *change of base properties* for logarithms.

▸ Use properties of logarithms to simplify expressions and solve equations.

▸ Use a calculator to evaluate a logarithm with base other than 10.

In Section 5.3 we studied some properties of logarithms. In this section we will discuss three more. We begin with the *product* property.

Product Property for Logarithms

For $x > 0$, $y > 0$, $b > 0$, and $b \neq 1$,

$$\log_b(x) + \log_b(y) = \log_b(xy)$$

We can use the product property to add two logarithms that have the same base. For example, for $x > 0$, $\log_3(5) + \log_3(x) = \log_3(5x)$. A proof of the product property for logarithms follows.

Let $m = \log_b(x)$ and $n = \log_b(y)$. Writing both equations in exponential form, we have:

$$x = b^m$$
$$y = b^n$$

---

*There is no safe way to drive after drinking. Even one drink can make you an unsafe driver. Also, the half-life of alcohol in a person's blood can vary by body type, sex, health status, and other factors.

Multiplying yields:

$$xy = (b^m)(b^n)$$
$$= b^{m+n}$$

Writing $xy = b^{m+n}$ in logarithmic form we have:

$$m + n = \log_b(xy)$$

Substituting $\log_b(x)$ for m and $\log_b(y)$ for n yields:

$$\log_b(x) + \log_b(y) = \log_b(xy)$$

This statement is what we set out to prove.

Example 1   Product Property

Simplify. Write each sum of logarithms as a single logarithm.

1. $\log_b(x) + \log_b(x - 1)$ **2.** $3 \log_b(x) + \log_b(6x)$

Solution

1. $\log_b(x) + \log_b(x - 1) = \log_b[x(x - 1)]$ Product property

                          $= \log_b(x^2 - x)$ Distributive law

2. $3 \log_b(x) + \log_b(6x) = \log_b(x^3) + \log_b(6x)$ Power property

                          $= \log_b[x^3(6x)]$ Product property

                          $= \log_b(6x^4)$ ■

In order to apply the product property, the coefficient of each logarithm must be 1. So, in problem 2 of Example 1, we first applied the power property to get coefficients of 1:

$$3 \log_b(x) + \log_b(6x) = \log_b(x^3) + \log_b(6x)$$

Then, we applied the product property.

We use the product property to simplify the sum of two logarithms with the same base. We use the *quotient* property to simplify the difference of two logarithms with the same base.

Quotient Property for Logarithms

For $x > 0$, $y > 0$, $b > 0$, and $b \neq 1$,

$$\log_b(x) - \log_b(y) = \log_b\left(\frac{x}{y}\right)$$

For example, for $x > 0$, $\log_4(x) - \log_4(7) = \log_4\left(\frac{x}{7}\right)$. You are asked to prove the quotient property in exercise 57.

Example 2   Product and Quotient Properties

Simplify. Write each result as a single logarithm with a coefficient of 1.

1. $\log_b(6x^7) - \log_b(3x^2)$ **2.** $\log_b(x) + 3 \log_b(5x) - 5 \log_b(2x)$

Solution

1. $\log_b(6x^7) - \log_b(3x^2) = \log_b\left(\dfrac{6x^7}{3x^2}\right)$ Quotient property

                          $= \log_b(2x^5)$ $\dfrac{b^m}{b^n} = b^{m-n}$

2. $\log_b(x) + 3\log_b(5x) - 5\log_b(2x)$

$$= \log_b(x) + \log_b(5x)^3 - \log_b(2x)^5 \qquad \text{Power property}$$

$$= \log_b[x(5x)^3] - \log_b(2x)^5 \qquad \text{Product property}$$

$$= \log_b \frac{x(5x)^3}{(2x)^5} \qquad \text{Quotient property}$$

$$= \log_b \frac{125x^4}{32x^5} \qquad (bc)^n = b^n c^n$$

$$= \log_b \frac{125}{32x} \qquad \text{Simplify.} \qquad \blacksquare$$

We can use the power, product, and quotient properties to solve logarithmic equations.

Example 3   Solving Logarithmic Equations

Solve each equation.

1. $2\log_5(3x) + 4\log_5(2x) = 3$ **2.** $5\log_7(x) - 2\log_7(3x) = 2$

Solution

1. $2\log_5(3x) + 4\log_5(2x) = 3$

$$\log_5(3x)^2 + \log_5(2x)^4 = 3 \qquad \text{Power property}$$

$$\log_5[(3x)^2(2x)^4] = 3 \qquad \text{Product property}$$

$$\log_5[9x^2(16x^4)] = 3 \qquad (bc)^n = b^n c^n$$

$$\log_5(144x^6) = 3 \qquad b^m b^n = b^{m+n}$$

$$5^3 = 144x^6 \qquad \text{Write in exponential form.}$$

$$x^6 = \frac{125}{144} \qquad \text{Divide both sides by 144.}$$

Although there is a negative 6th root of $\frac{125}{144}$, the original equation contains $4\log_5(2x)$ and the domain of a logarithmic function is the set of *positive* numbers. So $2x$ must be positive and, hence, x must be positive.

$$x = \left(\frac{125}{144}\right)^{1/6}$$

$$x \approx 0.9767$$

2. $5\log_7(x) - 2\log_7(3x) = 2$

$$\log_7(x^5) - \log_7(3x)^2 = 2 \qquad \text{Power property}$$

$$\log_7 \frac{x^5}{(3x)^2} = 2 \qquad \text{Quotient property}$$

$$\log_7 \frac{x^5}{9x^2} = 2 \qquad (bc)^n = b^n c^n$$

$$\log_7 \frac{x^3}{9} = 2 \qquad \frac{b^m}{b^n} = b^{m-n}$$

$$7^2 = \frac{x^3}{9} \qquad \text{Write in exponential form.}$$

$$x^3 = 441 \qquad \text{Multiply both sides by 9.}$$

$$x = 441^{1/3} \qquad \text{Find the cube root.}$$

$$x \approx 7.6117 \qquad \blacksquare$$

Note that we solve each equation in Example 3 by first applying the power property so that the coefficient of each logarithm is 1. Next we combine the logarithms on one side of the equation by using the product property or the quotient property. Then, we solve the equation using techniques discussed in Section 5.3.

Recall that the "log" key on a calculator finds logarithms base 10. We use the *change of base* property to find logarithms for bases other than 10.

Change of Base Property

For a and b positive and not equal to 1, $x > 0$,

$$\log_b(x) = \frac{\log_a(x)}{\log_a(b)}$$

For example, $\log_3(5) = \dfrac{\log_2(5)}{\log_2(3)}$. Also, $\log_3(5) = \dfrac{\log_4(5)}{\log_4(3)}$ and $\log_3(5) = \dfrac{\log(5)}{\log(3)}$.

Note that we are free to write a logarithm in terms of any new base, including base 10.

Now we prove the change of base property. Let $k = \log_b(x)$. In exponential form, we have

$$b^k = x$$

Next, we take $\log_a$ of both sides and solve for k.

$$\log_a(b^k) = \log_a(x)$$

$$k \log_a(b) = \log_a(x) \qquad \text{Power property}$$

$$k = \frac{\log_a(x)}{\log_a(b)} \qquad \text{Divide each side by } \log_a(b).$$

But $k = \log_b(x)$, so by substitution we have:

$$\log_b(x) = \frac{\log_a(x)}{\log_a(b)}$$

which is what we set out to prove.

In Example 4 we find a logarithm by converting to $\log_{10}$ and then using the log key on a calculator.

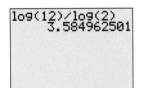

Figure 27 Compute $\dfrac{\log(12)}{\log(2)}$

Example 4 Converting to $\log_{10}$

Find $\log_2(12)$.

Solution

By the change of base property, we have $\log_2(12) = \dfrac{\log(12)}{\log(2)}$. Using the log key on a calculator, we compute that $\dfrac{\log(12)}{\log(2)} \approx 3.5850$ (see Fig. 27). So, $\log_2(12) \approx 3.5850$. $\blacksquare$

In Section 5.2 we sketched the graph of a logarithmic function. From now on we can use a graphing calculator to verify such a graph by converting the logarithmic function to $\log_{10}$.

Example 5   Using a Graphing Calculator to Graph a Logarithmic Function

Use a graphing calculator to draw a graph of $y = \log_3(x)$.

Solution

By the change of base property, we have $y = \log_3(x) = \dfrac{\log(x)}{\log(3)}$. Using the log key on a graphing calculator, we enter the function and draw the graph (see Fig. 28). This graph verifies the graph that we sketched by hand in Example 5 in Section 5.2.   ■

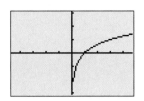

Figure 28 Graph of $y = \dfrac{\log(x)}{\log(3)}$

Example 6   Change of Base Property

Write $\dfrac{\log_7(x)}{\log_7(4)}$ as a single logarithm.

Solution

By the change of base property, we have $\dfrac{\log_7(x)}{\log_7(4)} = \log_4(x)$.   ■

The quotient property tells us that a difference of logarithms is equal to a logarithm of a quotient:

$$\log_b(x) - \log_b(y) = \log_b\left(\frac{x}{y}\right)$$

Note

Throughout this discussion, we assume that x, y, a, and b are positive and a and b are not equal to 1.

The change of base property tells us that a logarithm is equal to a quotient of logarithms (with "new" base):

$$\log_b(x) = \frac{\log_a(x)}{\log_a(b)}$$

It is easy to confuse the change of base property and the quotient property. Beware! In general,

$$\log_b(x) - \log_b(y) \neq \frac{\log_b(x)}{\log_b(y)}$$

and

$$\log_b\left(\frac{x}{y}\right) \neq \frac{\log_b(x)}{\log_b(y)}$$

EXPLORATION

Function of a sum

1. For each statement, substitute 2 for x and 3 for y and use a calculator to help you decide whether the result is a true or false statement.
   a. $\log(x + y) = \log(x) + \log(y)$
   b. $2^{x+y} = 2^x + 2^y$
   c. $(x + y)^2 = x^2 + y^2$
   d. $\sqrt{x + y} = \sqrt{x} + \sqrt{y}$

2. All of the statements in problem 1 are of the form $f(x + y) = f(x) + f(y)$. Is the statement $f(x + y) = f(x) + f(y)$ true for every function f? Explain.

3. According to the distributive law, $a(x + y) = ax + ay$. Explain why this statement is true for all values of a, but the statement $f(x + y) = f(x) + f(y)$ is not true for all functions f.

4. Give an example of a function f so that the statement $f(x + y) = f(x) + f(y)$ *is* true.

Tips on Succeeding in This Course

If you get flustered during a test, consider closing your eyes, taking a couple of deep breaths, and thinking about something pleasant, or nothing at all. This break from working on a test may give you some perspective and help you relax.

Key Points OF THIS SECTION

Properties of logarithms	For positive x, y, a, and b; $a \neq 1$; and $b \neq 1$;
	• $\log_b(x) + \log_b(y) = \log_b(xy)$ Product property
	• $\log_b(x) - \log_b(y) = \log_b\left(\dfrac{x}{y}\right)$ Quotient property
	• $\log_b(x) = \dfrac{\log_a(x)}{\log_a(b)}$ Change of base property
Finding $\log_b(a)$ where $b \neq 10$	A calculator can be used to find a logarithm of base other than 10 by using the change of base property for logarithms.

HOMEWORK 5.5

SSM Tutor Center   MathPro 4   MathPro 5   CDs/Videos   www

Simplify. Write your result as a single logarithm with a coefficient of 1.

1. $\log_b(2x) + \log_b(x)$

2. $\log_b(x + 1) + \log_b(x)$

3. $\log_b(10x) - \log_b(2)$

4. $\log_b(x^5) - \log_b(x^4)$

5. $3\log_b(x) + \log_b(3x)$

6. $5\log_b(x) + 2\log_b(3x)$

7. $4\log_b(2x) + 7\log_b(x)$

8. $2\log_b(x) - \log_b(3x)$

9. $\log_b(8x) - 4\log_b(2x)$

10. $\log_b(4x) - 2\log_b(2x)$

11. $\log_b(x^3) + 3\log_b(x^2)$

12. $3\log_b(5x^4) - 2\log_b(5x^2)$

13. $2\log_b(3x^2) - 4\log_b(x^3)$

14. $2\log_b(2x^2) + 3\log_b(2x^6)$

Solve.

15. $\log_6(x) + \log_6(4x) = 2$

16. $\log_2\left(\dfrac{x}{2}\right) + \log_2(2x) = 4$

17. $\log_4(x^2) + \log_4(x) = 3$

18. $\log_2(10x^5) - \log_2(5x^3) = 5$

19. $3\log_7(x^2) + \log_7(x^3) = 3$

20. $\log(5x^5) - 2\log(x^2) = 2$

21. $3\log(5x) + 4\log(2x) = 3$

22. $2\log_6(3x) + 3\log_6(2x) = 5$

23. $5\log_4(2x^2) - 2\log_4(2x^3) = 3$

24. $3\log_2(4x^3) - 2\log_2(9x^3) = 4$

Simplify by writing your result as a single logarithm with a coefficient of 1, or solve, whichever is appropriate.

25. $\log_2(x^4) + \log_2(x^3)$

26. $\log_2(x^9) - \log_2(x^5) = 5$

27. $\log_2(x^4) + \log_2(x^3) = 4$

28. $\log_2(x^9) - \log_2(x^5)$

29. $4\log_9(2x^5) + 2\log_9(x^4) = 2$

30. $2\log_3(4x^7) - 3\log_3(2x^2)$

31. $4\log_9(2x^5) + 2\log_9(x^4)$

32. $2\log_3(4x^7) - 3\log_3(2x^2) = 1$

Evaluate.

33. $\log_3(7)$

34. $\log_2(11)$

35. $\log_6(1000)$

36. $\log_{17}(5)$

37. $\log_5(41.2)$

38. $\log_{12}(2.88)$

39. $\log_8\left(\dfrac{1}{70}\right)$

40. $\log_{\frac{1}{2}}(7)$

Write the expression as a single logarithm.

41. $\dfrac{\log_2(x)}{\log_2(7)}$

42. $\dfrac{\log_4(x)}{\log_4(5)}$

43. $\dfrac{\log_b(r)}{\log_b(s)}$

44. $\dfrac{\log_b(8)}{\log_b(2)}$

45. Three students tried to solve the equation $3(2^x) = 7$. The work of each student is shown. Decide which student(s) solved the equation correctly.

Student 1's work

$$3(2^x) = 7$$

$$2^x = \frac{7}{3}$$

$$\log(2^x) = \log\left(\frac{7}{3}\right)$$

$$x\log(2) = \log\left(\frac{7}{3}\right)$$

$$x = \frac{\log\left(\dfrac{7}{3}\right)}{\log(2)}$$

Student 2's work

$$3(2^x) = 7$$

$$\log[3(2^x)] = \log(7)$$

$$\log(3) + \log(2^x) = \log(7)$$

$$\log(2^x) = \log(7) - \log(3)$$

$$x\log(2) = \log(7) - \log(3)$$

$$x = \frac{\log(7) - \log(3)}{\log(2)}$$

Student 3's work

$$3(2^x) = 7$$

$$2^x = \frac{7}{3}$$

$$x = \log_2\left(\frac{7}{3}\right)$$

46. The statement

$$\frac{a(5)}{a(3)} = \frac{5}{3}, \text{ where } a \text{ is a nonzero real number}$$

is a true statement. Is the statement

$$\frac{\log_b(5)}{\log_b(3)} = \frac{5}{3}, \text{ where } b > 0 \text{ and } b \neq 1$$

a true statement? Explain.

Let $f(x) = \log_5(x)$. *Find each output.*

47. $f(13)$

48. $f(59)$

49. $f\left(\dfrac{1}{2}\right)$

50. $f(25)$

Let $g(x) = \log_{12}(x)$. *Find each output.*

51. $g(17)$

52. $g(50)$

53. $g(37)$

54. $g(1)$

55. Which of the following expressions are equal?

$$\log_b(b^2) \qquad \log_b\left(\frac{b^6}{b^4}\right) \qquad \log_b(b^6) \qquad 2$$

$$\log_b(b^6) - \log_b(b^4) \qquad \frac{\log_b(b^6)}{\log_b(b^4)}$$

56. Clearly $\log_b(x) - \log_b(x) = 0$. Use a logarithmic property to write $\log_b(x) - \log_b(x)$ in another form to show the fact that $\log_b(1) = 0$. Assume that $b > 0$ and $b \neq 1$.

57. Prove the quotient property for logarithms. [**Hint**: Try to find a creative way to use the product property followed by the power property, with the expression $\log_b\left(\dfrac{x}{y}\right)$.]

58. List the properties for logarithms discussed in this section and Section 5.3. Also, explain how each property can be used. Give examples to illustrate your points.

59. a. Simplify $\log_2(x^3) + \log_2(x^5)$.

b. Solve $\log_2(x^3) + \log_2(x^5) = 7$.

c. Compare the process of simplifying an expression with solving an equation.

d. Explain how simplifying an expression can help when solving an equation.

5.6 NATURAL LOGARITHMS

Objectives

▸ Know the meaning of a *natural logarithm*.

▸ Use a calculator to evaluate a natural logarithm.

▸ Use properties of natural logarithms to simplify expressions and solve equations.

In this section we discuss a logarithm with a special base called "*e*," where *e* represents an irrational number:

$$e \approx 2.718281828459045\ldots$$

To the nearest ten thousandth, we have $e = 2.7183$.

Many equations for useful models contain *e*. For example, the equation for one type of "bell curve" is

$$f(x) = \frac{e^{-0.5x^2}}{\sqrt{2\pi}}$$

Note that the graph of f has the shape of a bell (see Fig. 29). Bell curves can be used to model heights of women (or men), IQs of people, widths of trunks of redwood trees, and many other quantities.

Recall that $\log_b(a)$ is the exponent on the base b that gives a. So $\log_e(a)$ is the exponent on the base e that gives a. We call $\log_e(a)$ a *natural logarithm* and we write $\ln(a)$ as shorthand for $\log_e(a)$.

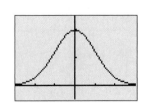

Figure 29 Graph of
$$f(x) = \frac{e^{-0.5x^2}}{\sqrt{2\pi}}$$

DEFINITION Natural Logarithm

A **natural logarithm** is a logarithm with base e. We represent $\log_e(a)$ as $\ln(a)$.

Also recall that $\log_b(a) = c$ and $b^c = a$ are equivalent forms. In terms of base e, this means that $\ln(a) = c$ and $e^c = a$ are equivalent forms.

Exponential/Logarithmic Forms Property

For $a > 0$

$$\ln(a) = c \text{ and } e^c = a \text{ are equivalent.}$$

Graphing Calculator

To compute ln(50):

Press [LN] [50] [)] [ENTER].

To compute $e^{3.9120}$:

Press [2nd] [LN] [3.9120] [)]

[ENTER].

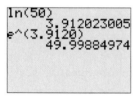

Figure 30 Computing ln(50)

The equation $\ln(a) = c$ is in **logarithmic form** and the equation $e^c = a$ is in **exponential form**. Either form can replace the other when solving a problem.

There likely is a key on your graphing calculator labeled "ln" or "LN" that will give you the natural logarithm of a number.

Example 1   Finding a Natural Logarithm

Use a calculator to find $\ln(50)$.

Solution

We find that $\ln(50) \approx 3.9120$ (see Fig. 30). This means that $e^{3.9120} \approx 50$. We also check that $e^{3.9120} \approx 50$. ■

We can find the natural logarithm of powers of e without using a calculator.

Example 2   Finding a Natural Logarithm

Find $\ln(e^5)$.

Solution

$\ln(e^5) = \log_e(e^5) = 5$ ■

From our work in Example 2, we see that for the function $\ln(x)$, the input e^5 leads to the output 5. Recall that for *any* logarithmic function, the domain is the set of positive numbers.

In Example 3 we solve a logarithmic equation and an exponential equation.

Example 3   Solving Equations

Solve the equation.

1. $\ln(x) = 4$ **2.** $5e^{x-1} = 100$

Solution

1. We write $\ln(x) = 4$ in the exponential form $x = e^4$ (≈ 54.5982; see Fig. 31).

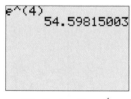

Figure 31 Computing e^4

2.

$$5e^{x-1} = 100$$

$$e^{x-1} = 20$$

$$\ln(20) = x - 1 \qquad \text{Write in logarithmic form.}$$

$$\ln(20) + 1 = x$$

$$x = \ln(20) + 1$$

$$x \approx 3.9957$$

We check that 3.9957 approximately satisfies the equation $5e^{x-1} = 100$:

$$5e^{3.9957-1} \approx 99.9968 \approx 100$$ ■

Next, we list properties for $\log_b(x)$ and the corresponding properties for $\ln(x)$. Assume that $x > 0$, $y > 0$, $b > 0$, and $b \neq 1$.

$\log_b(1) = 0$	$\ln(1) = 0$	The natural logarithm of 1 is 0.
$\log_b(b) = 1$	$\ln(e) = 1$	The natural logarithm of e is 1.
$\log_b(x^p) = p\log_b(x)$	$\ln(x^p) = p\ln(x)$	Power property
$\log_b(x) + \log_b(y) = \log_b(xy)$	$\ln(x) + \ln(y) = \ln(xy)$	Product property
$\log_b(x) - \log_b(y) = \log_b\left(\dfrac{x}{y}\right)$	$\ln(x) - \ln(y) = \ln\left(\dfrac{x}{y}\right)$	Quotient property

We can use the power property for natural logarithms to solve exponential equations.

Example 4   Solving an Equation

Solve $2(5)^x + 3 = 63$.

Solution

$$2(5)^x + 3 = 63$$
$$2(5)^x = 60$$
$$5^x = 30$$

$$\ln(5^x) = \ln(30) \qquad \text{Take the natural logarithm of both sides.}$$
$$x\ln(5) = \ln(30) \qquad \text{Power property}$$
$$x = \frac{\ln(30)}{\ln(5)}$$
$$x \approx 2.1133$$

We check that 2.1133 approximately satisfies the equation $2(5)^x + 3 = 63$:

$$2(5)^{2.1133} + 3 \approx 63.0017 \approx 63 \qquad \blacksquare$$

In Example 4 we used ln to solve an exponential equation. In Section 5.3 we used log to solve exponential equations. It does not matter whether we use ln or log to solve an exponential equation such as $2(5)^x + 3 = 63$. Both methods require about the same amount of work and give the same result.

In Example 5 we use the power property and the quotient property to simplify a logarithmic expression.

Example 5   Power and Quotient Properties

Write $5\ln(x) - 3\ln(2x)$ as a single logarithm with a coefficient of 1. Simplify the result.

Solution

$$5\ln(x) - 3\ln(2x) = \ln(x^5) - \ln(2x)^3 \qquad \text{Power property}$$
$$= \ln\frac{x^5}{(2x)^3} \qquad \text{Quotient property}$$
$$= \ln\frac{x^5}{8x^3}$$
$$= \ln\frac{x^2}{8}$$

We verify our work by comparing tables for $y = 5\ln(x) - 3\ln(2x)$ and $y = \ln\dfrac{x^2}{8}$ (see Fig. 32).

Figure 32 Verifying that
$$5\ln(x) - 3\ln(2x) = \ln\frac{x^2}{8}$$

Example 6 Solving an Equation

Solve $3\ln(4x) + \ln(5x) = 7$.

Solution

$$3\ln(4x) + \ln(5x) = 7$$

$$\ln(4x)^3 + \ln(5x) = 7 \qquad \text{Power property}$$

$$\ln[(4x)^3(5x)] = 7 \qquad \text{Product property}$$

$$\ln[64x^3(5x)] = 7$$

$$\ln(320x^4) = 7$$

$$e^7 = 320x^4 \qquad \text{Write in exponential form.}$$

$$x^4 = \frac{e^7}{320}$$

Although there is a negative 4th root of $\dfrac{e^7}{320}$, the original equation contains $3\ln(4x)$ and the domain of a (natural) logarithm function is the set of *positive* numbers. So $4x$ must be positive and, hence, x must be positive.

$$x = \left(\frac{e^7}{320}\right)^{1/4}$$

$$x \approx 1.3606$$

EXPLORATION
Newton's Law of Cooling

A hot potato is taken out of an oven and allowed to cool to room temperature. Let p represent the temperature (in Fahrenheit degrees) of the potato t minutes after being removed from the oven.

1. Sketch a qualitative graph that describes the relationship between t and p.

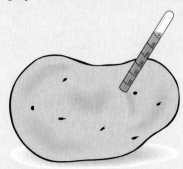

2. Newton's Law of Cooling states that

$$p - r = ae^{-kt}$$

where r represents room temperature (in Fahrenheit degrees) and a and k are constants. The room temperature is $70°F$. Substitute $r = 70$ into the equation.

3. The temperature of the potato was $350°F$ when it was removed from the oven. Find the value of the constant a and substitute it into your equation.

4. The temperature of the potato was $200°F$ after 5 minutes. Find the value of k and substitute it into your equation.

5. Isolate p on one side of your equation. Then use your graphing calculator to draw the graph of your equation. Compare your graph with your sketch in problem 1.

6. What will be the temperature of the potato after a long time? Explain.

7. At what temperature can a potato be comfortably eaten? How long will it take for the potato to reach this temperature?

EXPLORATION

Looking ahead: Significance of a, h, and k for $y = a(x - h)^2 + k$

1. Use ZStandard followed by ZSquare to draw a graph of $y = x^2$.

2. In this problem you will explore the graphical significance of the constant k for functions of the form $y = x^2 + k$. Graph

$$y = x^2 + 1$$
$$y = x^2 + 2$$
$$y = x^2 + 3$$
$$y = x^2 + 4$$

in order, and state in terms of k how you could "move" the graph of $y = x^2$ to get each of these graphs.

Do the same with:

$$y = x^2 - 1$$
$$y = x^2 - 2$$
$$y = x^2 - 3$$
$$y = x^2 - 4$$

3. In this problem, you will explore the graphical significance of the constant h for functions of the form $y = (x - h)^2$. Graph

$$y = (x - 1)^2$$
$$y = (x - 2)^2$$
$$y = (x - 3)^2$$
$$y = (x - 4)^2$$

in order, and state in terms of h how you could "move" the graph of $y = x^2$ to get each of these graphs.

Do the same with:

$$y = (x + 1)^2$$
$$y = (x + 2)^2$$
$$y = (x + 3)^2$$
$$y = (x + 4)^2$$

4. In this problem, you will explore the graphical significance of the constant a for functions of the form $y = ax^2$. You should have a pretty good idea of how to go about this based on your recent explorations of the constants h and k. Do this now and describe what you find. Don't forget to try negative values of a as well as values of a between 0 and 1.

5. **a.** Graph

$$y = x^2$$
$$y = 0.5x^2$$
$$y = -0.5x^2$$
$$y = -0.5(x+1)^2$$
$$y = -0.5(x+1)^2 - 6$$

in order, and explain how these graphs relate to your observations in problems 2, 3, and 4.

b. Refer to your graph of $y = -0.5(x+1)^2 - 6$ and find the coordinates of the vertex. Compare these coordinates to the equation $y = -0.5(x+1)^2 - 6$. What do you notice?

6. Summarize your findings about a, h, and k in terms of how you could move or adjust the graph of $y = x^2$ to get the graph of $y = a(x-h)^2 + k$. Also discuss how the coordinates of the vertex are related to a, h, and k. If you are unsure, continue exploring.

Key Points
OF THIS SECTION

Approximation for e	To the nearest ten-thousandth, $e = 2.7183$.
Meaning of natural logarithm	A **natural logarithm** is a logarithm with base e. We write $\ln(x)$ to represent $\log_e(x)$. For $a > 0$, $\ln(a)$ is the exponent on the base e that gives a.
Equivalent forms	For $a > 0$, $\ln(a) = c$ and $e^c = a$ are equivalent forms.
Properties of natural logarithm	$\ln(1) = 0$
	$\ln(e) = 1$
	For $x > 0$ and $y > 0$,

- $\ln(x^p) = p\ln(x)$ Power property
- $\ln(x) + \ln(y) = \ln(xy)$ Product property
- $\ln(x) - \ln(y) = \ln\left(\dfrac{x}{y}\right)$ Quotient property

HOMEWORK 5.6

 SSM Tutor Center   MathPro 4   MathPro 5   CDs/Videos   www

Use a calculator to find the natural logarithm.

1. $\ln(7)$ **2.** $\ln(15)$ **3.** $\ln(54.8)$

4. $\ln(37.28)$ **5.** $\ln(0.8)$ **6.** $\ln(4.2)$

7. $\ln\left(\dfrac{1}{2}\right)$ **8.** $\ln\left(\dfrac{5}{8}\right)$

Find the natural logarithm. Verify your result using a graphing calculator.

9. $\ln(e^4)$ **10.** $\ln(e^6)$

11. $\ln(e)$ **12.** $\ln(1)$

13. $\ln\left(\dfrac{1}{e}\right)$ **14.** $\ln\left(\dfrac{1}{e^2}\right)$

15. $\dfrac{1}{2}\ln(e^6)$ **16.** $\ln(\sqrt{e})$

Solve the equation. Check your result.

17. $\ln(x) = 2$ **18.** $\ln(x) = 5$

19. $e^x = 20$

20. $e^x = 17$

21. $\ln(x + 5) = 3$

22. $\ln(4x) = 5$

23. $e^{2x} = 71$

24. $e^{x-4} = 10$

25. $7e^x = 44$

26. $3e^x = 85$

27. $5\ln(3x) = 5$

28. $2\ln(x - 1) = 0$

29. $4e^{3x-1} = 68$

30. $7e^{2x+10} = 100$

31. $4^x = 90$

32. $2^x = 27$

33. $3.1^x = 49.8$

34. $2.4^x = 63.5$

35. $3(6^x) - 1 = 97$

36. $2(4^{2x+1}) = 84$

Simplify. Write each expression as a single logarithm with a co-efficient of 1. Use graphing calculator tables or graphs to verify your result.

37. $\ln(4x) + \ln(3x)$

38. $\ln(8x^2) - \ln(4x)$

39. $\ln(25x^4) - \ln(5x^3)$

40. $\ln(6x^3) + \ln(2x^4)$

41. $2\ln(3x^4) + 3\ln(2x)$

42. $4\ln(3x^2) - 5\ln(2x^3)$

43. $3\ln(3x^3) - 2\ln(3x^2)$

44. $2\ln(2x^3) + 3\ln(2x^5)$

45. $\ln(2x + 1) + \ln(3)$

46. $\ln(6) + \ln(4x - 2)$

Solve the equation. Check your result.

47. $\ln(3x) + \ln(x) = 4$

48. $\ln(2x) + \ln(5x) = 6$

49. $\ln(4x^5) - 2\ln(x^2) = 5$

50. $2\ln(4x^3) - \ln(8x^5) = 1$

51. $5\ln(x^4) + 2\ln(4x^3) = 8$

52. $4\ln(2x^5) - 3\ln(2x^3) = 9$

Simplify by writing your result as a single logarithm with a coef-ficient of 1, or solve, whichever is appropriate.

53. $\ln(x^8) - \ln(x^3)$

54. $2\ln(4x^6) + 3\ln(2x^2) = 7$

55. $\ln(x^8) - \ln(x^3) = 4$

56. $2\ln(4x^6) + 3\ln(2x^2)$

57. Explain in your own words why $\ln(e) = 1$.

58. Explain in your own words why $\ln(1) = 0$.

59. Assume that the equation $ae^{bx} = c$ has a solution for x, where $a \neq 0$ and $b \neq 0$. Solve for x.

60. Assume that the equation $ae^{bx+d} + k = c$ has a solution for x, where $a \neq 0$ and $b \neq 0$. Solve for x.

61. Which expressions are equal? Assume that $x > 0$ and $x \neq 1$.

$$3\ln(x) \qquad \ln(x^7) - \ln(x^4) \qquad \frac{\ln(x^7)}{\ln(x^4)} \qquad 2\ln(x)\ln(x)$$

$$\ln(x^3) \qquad \ln(3x)$$

62. a. Solve $e^x = 30$ by writing the equation in logarithmic form.

  b. Solve $e^x = 30$ by taking the natural logarithm of both sides of the equation.

  c. Compare your results in parts a and b.

63. A person buys a cup of coffee and drinks it in the store. The temperature of the coffee y (in Fahrenheit degrees) is given by

$$y = 70 + 137e^{-0.06t}$$

where t is the number of minutes since the person bought the coffee.

a. What is the temperature of the coffee when the person bought it?

b. If the person will begin drinking the coffee when it reaches 180°F, how much time must he wait after buying the coffee?

c. Use a graphing calculator table or graph to estimate the room temperature of the *store*.

64. A person makes a cup of tea. The temperature of the tea y (in Fahrenheit degrees) is given by

$$y = 68 + 132e^{-0.05t}$$

where t is the number of minutes since the person made the tea.

a. What is the temperature of the tea when the person made it?

b. If the person waits 5 minutes to begin drinking the tea, what is the temperature of the tea then?

c. The tea is *lukewarm* at a temperature of about 98.6°F. If the person gets distracted and lets the tea sit until it is lukewarm, how much time has gone by since she made the tea?

65. A cable is hanging between two poles that are 20 feet apart (see Fig. 33). The height of the cable (in feet) is given by

$$h(x) = 10(e^{0.03x} + e^{-0.03x}), \qquad -10 \le x \le 10$$

where x is the horizontal position (in feet) as indicated in Fig. 33.

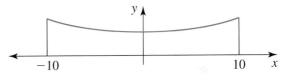

Figure 33 Exercise 65

a. Find the height of the cable at either pole.

b. Find $h(6)$. What does your result mean in terms of the situation?

c. Find the least height of the cable.

66. A cable is hanging between two poles that are 30 feet apart (see Fig. 34). The height of the cable (in feet) is given by

$$h(x) = 20(e^{0.05x} + e^{-0.05x}), \qquad -15 \le x \le 15$$

where x is the horizontal position (in feet) as indicated in Fig. 34.

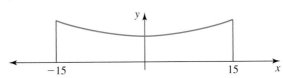

Figure 34 Exercise 66

a. Find $h(-8)$. What does your result mean in terms of the situation?

b. Find the least height of the cable.

c. Explain why there is model breakdown for $x < -15$ and for $x > 15$.

67. Explain how to use the power property for natural logarithms to solve an exponential equation in one variable.

Taking it to the lab

For each lab assignment, consult with your instructor on whether to organize your responses as a numbered list or to write them in a paragraph.

CHINA AND INDIA POPULATIONS WEB LAB

In this lab you will compare China's and India's populations.*

Collecting the Data

For each country, you will find populations for every five years since 1950 plus a projection for the current year.

1. Go to http://www.census.gov/ipc/www/idbagg.html. Select "Total Mid-Year Population, Area, Density." Select "Display mode" and submit query. If you can't find the site, try http://esa.un.org/unpp/.
2. For variable group, select "Population." Accept the defaults on aggregation options. Select China and also type "China" into the "User region name" box. Select the current year and also enter 1950 into the "From" box, enter the current year into the "To" box, and enter 5 into the "By" box. Then submit your query.
3. Record the data or print the screen.
4. Go back and collect data for India's populations.

Analyzing the Data

1. Include tables of data for China's populations and India's populations.
2. Define the variables for your models. Which variables are the independent variables? Which variables are the dependent variables? If you round the data, describe how.
3. Use your graphing calculator to draw scattergrams of the data. Copy the scattergrams on paper. For each population, discuss whether a linear or exponential function is the best choice of model. Explain.
4. For each population, find an equation of a model.
5. For each population, use your graphing calculator to graph your model and your scattergram in the same viewing window. Also sketch the graph of your model and scattergram by hand.
6. Which country has the largest current population? Use your models to estimate the difference in the current populations.
7. Use "intersect" on your graphing calculator to predict when the populations will be equal.
8. The world population in 2002 was 6.215 billion. Use a model to predict when India's population will reach that size. Use a model to predict when China's population will reach that size. Explain in terms of the types of functions used to model the populations why the two results are so different.
9. Use your models to predict when the sum of the populations of China and India will reach 6.215 billion. Explain how you found your result.

FOLDING PAPER LAB

In this lab, you will investigate the thickness of a sheet of paper by folding it many times.

---

*Adapted from a lab written by Cheryl Gregory, College of San Mateo, CA.

Materials

In order to do this lab, you will need the following materials:

1. an $8\frac{1}{2}$-inch by 11-inch piece of paper
2. a ruler

Preparation

Very carefully fold the piece of paper in half six times, each time without unfolding.

Recording of Data

Measure the thickness of the folded paper.

Analyzing the Data

1. What is the thickness of the folded paper (after 6 folds)? Include units.
2. Use your answer to part 1 to estimate the thickness of the paper when it is unfolded. Include units.
3. Let $f(n)$ represent the thickness of the paper if it has been folded n times. Find an equation for f. What are the units for $f(n)$?
4. Can you fold the paper a seventh time? If not, use f to predict the thickness if it could be folded seven times. If yes, keep folding the paper until you cannot fold it anymore and predict the thickness if it could be folded one more time.
5. How thick would the folded paper be if you could fold it 15 times? Would the folded paper be taller or shorter than you?
6. After how many folds would the folded paper be at least as tall as a football field is long. (A football field is 120 yards long if you include the end zones.)
7. After how many folds would the thickness of the folded paper match the distance to the moon?
8. The above situation seems limited because of one's inability to fold the paper many times. Instead of folding a piece of paper, consider cutting a piece of paper in half and then stacking the two halves. Next, cut the stack of papers and restack the two piles of papers. Cutting the stack in two each time and then restacking achieves the same thickness as folding the paper in two. For cutting and stacking, can the result described in part 7 be achieved? Explain.

EXPONENTIAL/LOGARITHMIC LAB: TOPIC OF YOUR CHOICE

Repeat the Exponential Lab: Topic of Your Choice in Chapter 4, but with a different situation. Also, make a prediction or estimate about your independent variable based on a specific value for your dependent variable. Interpret your prediction in terms of the situation being modeled.

Chapter Summary

Key Points
OF THIS CHAPTER

For an invertible function f, we sketch the graph of f^{-1} by the following steps: **Sketching the graph of f^{-1}**

1. Sketch the graph of f.
2. Choose several points that lie on the graph of f.
3. For each point (a, b) chosen from step 2, plot the point (b, a).
4. Sketch the curve that contains the points plotted in step 3.

Finding the inverse of a model

To find the inverse of an invertible *model* f where $p = f(t)$:

1. Replace $f(t)$ with p.
2. Solve for t.
3. Replace t with $f^{-1}(p)$.

Finding the inverse of a function that is not a model

Let f be an invertible function that is *not* a model. To find the inverse of f where $y = f(x)$:

1. Replace $f(x)$ with y.
2. Solve for x.
3. Replace x with $f^{-1}(y)$.
4. Write the equation for f^{-1} in terms of x.

Meaning of $\log_b(a)$

$\log_b(a)$ is an exponent, namely the exponent on the base b that gives a.

Inverse functions

For the exponential function $f(x) = b^x$, $f^{-1}(x) = \log_b(x)$.

For the logarithmic function $g(x) = \log_b(x)$, $g^{-1}(x) = b^x$.

Meaning of $\log(x)$

The notation $\log(x)$ is shorthand for $\log_{10}(x)$.

Properties of $\log_b$

For positive x, y, a, and b, $a \neq 1$, and $b \neq 1$:

- $\log_b(b) = 1$

- $\log_b(1) = 0$

- $\log_b(x^p) = p \log_b(x)$ Power property

- $\log_b(x) + \log_b(y) = \log_b(xy)$ Product property

- $\log_b(x) - \log_b(y) = \log_b\left(\dfrac{x}{y}\right)$ Quotient property

- $\log_b(x) = \dfrac{\log_a(x)}{\log_a(b)}$ Change of base property

Equivalent equation

The equations $\log_b(a) = c$ and $b^c = a$ are equivalent. Either can be written in the place of the other.

Solving equations

We solve an equation of the form $\log_b(x) = k$ for b or for x by writing the equation in exponential form.

We solve some equations of the form $ab^x = c$ for x by first dividing both sides by a and then taking the log of both sides of the equation. Next, we use the power property for logarithms.

Carbon dating

Scientists measure the ratio of carbon-14 to carbon-12 in artifacts and fossils to estimate their ages. The half-life of carbon-14 is 5730 years.

Approximation for e

To the nearest ten-thousandth, $e = 2.7183$.

Meaning of a natural logarithm

A **natural logarithm** is a logarithm with base e. We write $\ln(x)$ to represent $\log_e(x)$.

For $a > 0$,

- $\ln(a)$ is the exponent on the base e that gives a.

- $\ln(a) = c$ and $e^c = a$ are equivalent forms.

$\ln(1) = 0$

$\ln(e) = 1$

For $x > 0$ and $y > 0$,

- $\ln(x^p) = p\ln(x)$ Power property
- $\ln(x) + \ln(y) = \ln(xy)$ Product property
- $\ln(x) - \ln(y) = \ln\left(\dfrac{x}{y}\right)$ Quotient property

Properties of ln

CHAPTER 5 REVIEW EXERCISES

Find the inverse of the function.

1. $f(x) = 3x$

2. $g(x) = \dfrac{4x - 7}{8}$

3. $h(x) = 3^x$

4. $f(x) = \log_4(x)$

5. $g(x) = 10^x$

6. $h(x) = \log(x)$

For exercises 7–10, refer to Table 34.

Table 34 Values of f
(Exercises 7–10)

x	$f(x)$
0	1
1	2
2	4
3	3
4	0

7. Find $f(0)$.

8. Find $f(2)$.

9. Find $f^{-1}(0)$.

10. Find $f^{-1}(2)$.

Sketch the graph of f, f^{-1}, and $y = x$ on the same set of axes.

11. $f(x) = 2x - 3$

12. $g(x) = 3^x$

Evaluate the logarithmic function at the given value. Explain.

13. $\log_5(25)$

14. $\log(100{,}000)$

15. $\log_3\left(\dfrac{1}{9}\right)$

16. $\ln\left(\dfrac{1}{e^3}\right)$

17. $\log_6(\sqrt{6})$

18. $\log_4(\sqrt[3]{4})$

19. $\log(0.001)$

20. $\log_3(7)$

21. $\ln(5)$

22. $\log_b(b)$

23. $\log_b(b^7)$

24. $\log_b(1)$

25. Write the equation $d^t = k$ in logarithmic form.

26. Write the equation $\log_y(w) = r$ in exponential form.

27. Sketch the graph of $y = \log_4(x)$.

Solve.

28. $6(2)^x = 30$

29. $\log_3(x) = 4$

30. $4e^x = 75$

31. $4.3(9.8)^x - 3.3 = 8.2$

32. $\log_b(83) = 6$

33. $-3\ln(x) + 7 = 1$

34. $\log_2(x^2) + \log_2(2x) = 4$

35. $\log_5(x) = 3.17$

36. $5(4)^{3x-7} = 40$

37. $e^{3x-8} = 12$

38. $\log_b(32) = 5$

39. $5\log_6(2x) - 3\log_6(4x) = 2$

40. $\ln(3x + 1) = 2$

41. $2\log_3(5x) + 4\log_3(x) = 5$

42. $\ln(4x^8) - \ln(2x^7) = 5$

43. $\log_2(\log_2(x)) = 2$

44. $\dfrac{5^{6x}}{5^{4x}} = 48$

Simplify. Write your result as a single logarithm with a coefficient of 1.

45. $\log_b(x) + \log_b(6x) - \log_b(2x)$

46. $3\log_b(2x) + 2\log_b(3x)$

47. $3\ln(4x^3) + 3\ln(2x^5)$

48. $4\log_b(3x^2) - 2\log_b(9x^5)$

49. $\ln(2x^7) - 4\ln(2x^3)$

50. $\dfrac{\log_b(w)}{\log_b(y)}$

51. Which of the following expressions are equal?

$$\log_b(b^5) - \log_b(b^2) \qquad \dfrac{\log_b(b^5)}{\log_b(b^2)}$$

$$\log_b(b^3) \qquad 3 \qquad \log_b(b^5) \qquad \log_b\left(\dfrac{b^5}{b^2}\right)$$

52. Suppose that \$8000 is deposited into an account that earns 5% interest compounded annually. Let $f(t)$ represent the balance (in thousands of dollars) in the account after t years or any fraction thereof.

  a. Find a formula for f.

  b. Find the balance in the account after 9 years.

  c. After how many years will the balance in the account be doubled?

53. On April 1, a tree has 30 leaves. Each week, the number of leaves quadruples (four times as great). Let $f(t)$ represent the number of leaves on the tree at t weeks since April 1.

   a. Find a formula for f.

   b. Find $f(5)$. What does your result mean in terms of the number of leaves on the tree?

   c. Find $f^{-1}(100,000)$. What does your result mean in terms of the number of leaves on the tree?

54. Before they came up with the idea of selling juice drinks, Tom First and Tom Scott lived in Nantucket and worked odd jobs like tending bar, shucking scallops, and even shampooing dogs to pay their rent. Nantucket Nectars® juice-drink sales have increased rapidly since the company's beginning in 1990 (see Table 35).

Table 35 Nantucket Nectar Juice-drink Sales	
Year	Sales (million dollars)
1990	0.1
1991	0.2
1992	0.5
1993	1.0
1994	6.0
1995	15.0
1996	30.0
1997	60.0

Source: New York Times

   a. Let $f(t)$ represent the sales (in millions of dollars) at t years since 1990. Use a graphing calculator to draw a scattergram of the data. Is it better to use a linear or an exponential function to model the data? Explain.

   b. Find an equation for f.

   c. What is the base b of the function $f(t) = ab^t$? What does it tell you about the sales? Now, consider the *actual* sales in Table 35. During which year did the company experience the greatest percent growth in sales?

   d. Use your model to predict when sales will be $1 billion.

55. A student ran an experiment to investigate the relationship between the length of a rubber band and the weight applied to one end of it. He set up the experiment by hooking one end of a rubber band on a horizontal pole supported by two chairs. He attached a plastic bag to the other end of the rubber band. Then, he recorded the lengths of the rubber band with various numbers of tape cassettes in the plastic bag (see Table 36).

Table 36 Lengths of a Rubber Band Stretched by Tape Cassettes	
Number of Cassettes	Length of Rubber Band (inches)
0	10.00
1	12.00
2	15.38
3	19.50
4	28.31
5	33.50
6	45.45
7	64.15

Source: *Experiment run by Michael S., Fall Semester, 1998*

   a. Let $f(n)$ represent the length (in inches) of the rubber band stretched by n cassettes. Find a formula for f.

   b. What is the base b of your function $f(n) = ab^n$? What does it mean in terms of rubber band lengths?

   c. What is the coefficient a of your function $f(n) = ab^n$. What does it mean in terms of rubber band lengths?

   d. Use f to estimate the length of the rubber band stretched by 8 cassettes. Describe two scenarios in which model breakdown might occur for your estimate.

   e. Use f to estimate the number of cassettes needed to stretch the rubber band to 139 inches. If model breakdown occurs for your estimate made in part d, does that imply that model breakdown occurs for your estimate made in this part? Explain.

CHAPTER 5 TEST

Find the inverse of the function.

1. $g(x) = 2x - 9$ **2.** $h(x) = 4^x$

3. $f(x) = \log_5(x)$

4. The graph of a function f is sketched in Fig. 35. Sketch the graph of f^{-1}.

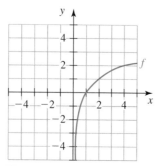

Figure 35 Exercise 4

Evaluate.

5. $\log_2(16)$ **6.** $\log_4\left(\dfrac{1}{64}\right)$

7. $\log_7(10)$ **8.** $\log(0.1)$

9. $\log_2(\log(10,000))$ **10.** $\log_b(1)$

11. $\log_b(b^3)$ **12.** $\log_b(\sqrt{b})$

13. $\ln\left(\dfrac{1}{e^2}\right)$ **14.** $\log_b(\log_b(b))$

Solve.

15. $\log_b(50) = 4$

16. $6(2)^x - 9 = 23$

17. $\log_3(x) + \log_3(2x) = 5$

18. $2e^{3x-1} = 54$

19. $2\log_4(3x^5) - 3\log_4(x^2) = 3$

20. $7\ln(x - 2) - 1 = 4$

21. Write $s^t = w$ in logarithmic form.

22. Write $\ln(k) = p$ in exponential form.

Simplify. Write your result as a single logarithm with a coefficient of 1.

23. $\log_b(x^3) + \log_b(5x)$

24. $4\ln(x^3) - 5\ln(2x^2)$

25. An archeologist discovers an artifact made of wood. Let $f(t)$ represent the percent of carbon-14 that remains at t years since the artifact was made. Recall that the half-life of carbon-14 is 5730 years.

   a. Find a formula for f.

   b. If 40% of the carbon-14 remains, how old is the wood?

26. In New York City, the number of arrests for possession of marijuana has increased dramatically (see Table 37).

Table 37 Numbers of Arrests for Possession of Marijuana

Year	Number of Arrests (thousands)
1992	1.0
1993	2.4
1994	3.9
1995	7.0
1996	11.0
1997	19.8

Source: New York Times

   a. Let $f(t)$ represent the number of arrests (in thousands) for possession of marijuana at t years since 1990. Find a formula for f.

   b. What is the coefficient a of the function $f(t) = ab^t$? What does it mean in terms of the number of arrests?

   c. What is the base b of the function $f(t) = ab^t$? What does it mean in terms of the number of arrests?

   d. According to the model f, when will there be 1 million arrests? In your opinion, is this possible? Explain.

Cumulative Review of Chapters 1–5

Solve.

1. $2(4)^{5x-1} = 17$

2. $\log_3(x - 5) = 4$

3. $3b^7 - 18 = 7$

4. $\dfrac{7}{12}x - \dfrac{2}{3} = \dfrac{5}{4}x$

5. $8 + 2e^x = 15$

6. $\log_b(93) = 4$

7. $4\log_5(3x^2) + 3\log_5(6x^4) = 3$

8. $7 - 3(4x - 2) = 5x + 1$

9. $2\ln(12x^9) - \ln(3x^3) = 5$

Solve each system.

10.
   $x = 2y - 5$
   $4x - 5y = -14$

11.
   $3(2 - 4x) = -10 - 2y$
   $2x - 3y = -8$

12.
   $\dfrac{1}{2}x - \dfrac{3}{4}y = -12$
   $\dfrac{5}{6}x + \dfrac{1}{3}y = -1$

13. Solve the inequality $8x - 3 \geq -3(4x - 5)$. Describe the solution set with inequality notation, interval notation, and a graph.

Simplify.

14. $(4b^{-3}c^2)^3(5b^{-7}c^{-1})^2$

15. $\dfrac{8b^{1/3}c^{-1/2}}{6b^{-1/2}c^{3/4}}$

16. $\left(\dfrac{2b^{-5}c^8}{3b^{-2}c^7}\right)^3$

Simplify. Write your result as a single logarithm with a coefficient of 1.

17. $4\log_b(2x^7) - 2\log_b(7x^2)$

18. $3\ln(4x^6) + 4\ln(3x^2)$

For exercises 19–25, refer to Table 38.

Table 38 Values for Four Functions (Exercises 19–25)

x	$f(x)$	x	$g(x)$	x	$h(x)$	x	$k(x)$
0	5	0	25	4	83	3	160
1	15	1	28	5	76	4	80
2	45	2	31	6	69	5	40
3	135	3	34	7	62	6	20
4	405	4	37	8	55	7	10
5	1215	5	40	9	48	8	5

19. Find an equation for f.

20. Find an equation for g.

21. Find the slope of h.

22. Find $g(5)$.

23. Find x when $k(x) = 5$.

24. Find $f^{-1}(5)$.

25. Find the y-intercept of k.

Sketch the graph of the function.

26. $y = 8\left(\dfrac{1}{2}\right)^x$

27. $y = -\dfrac{2}{5}x + 3$

28. $y = -3(2)^x$

29. $4x - 3y = 12$

30. $y = \log_2(x)$

31. Find an equation of a line that contains the points $(-4, 7)$ and $(5, -3)$.

32. Find an equation of an exponential curve that contains the points $(0, 8.1)$ and $(20, 719.3)$.

33. Find an equation of an exponential curve that contains the points $(3, 85)$ and $(7, 13)$.

Let $f(x) = 2(3)^x$.

34. Find $f(-4)$.

35. Find x when $f(x) = 16$.

36. Find the y-intercept of f.

37. Sketch a graph of f.

38. Sketch a graph of f^{-1}.

39. Find $f^{-1}(35)$.

Find the logarithm.

40. $\log_3\left(\dfrac{1}{81}\right)$

41. $\log_3(\log_2(8))$

42. $\log_b(\sqrt[7]{b})$

43. $\log_8(73)$

For exercises 44–47, refer to Fig. 36.

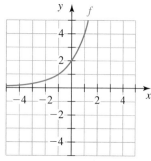

Figure 36 Exercises 44–47

44. Estimate $f(-1)$.

45. Find an equation for f.

46. Sketch the graph of f^{-1}.

47. Estimate $f^{-1}(2)$.

Find the inverse of the function.

48. $f(x) = 4x - 6$

49. $g(x) = 8^x$

50. Sketch a graph of a relation that is not a function. Next, create a table of five ordered pairs of the relation. Explain why your relation is not a function.

51. Compare the function $f(x) = 3x + 2$ to the function $g(x) = 2(3)^x$.
    a. Find the y-intercept of each function.
    b. For both of the functions f and g, describe what happens to the value of y as the value of x increases by 1.
    c. For a large input value for x, which function will have a greater output value for y? Explain.
    d. Use pencil and paper to sketch the graphs of f and g on the same coordinate system.

52. Let f be the linear function and g be the exponential function whose graphs contain the points $(4, 3)$ and $(7, 8)$.
    a. Find an equation for f.
    b. Find an equation for g.
    c. Use your graphing calculator to draw the graphs of f and g in the same viewing window.

53. Let $f(x) = 3x$ and $g(x) = 3^x$.
    a. Find $f(2)$ and $g(2)$.
    b. Find an equation for f^{-1} and an equation for g^{-1}.
    c. Find $f^{-1}(81)$ and $g^{-1}(81)$.

54. One U-Haul office rents pickup trucks for a one-day fee of $19.95 plus $0.69 per mile. One Budget location charges a one-day fee of $29.95 plus $0.45 per mile.
    a. Let $U(x)$ represent U-Haul's charge (in dollars) and $B(x)$ represent Budget's charge (in dollars), both for driving x miles in one day. Find equations for U and B.
    b. Find the slopes of U and B. What do your results represent in terms of the situation?
    c. For how many miles driven is the one-day charge at U-Haul equal to the one-day charge at Budget?
    d. Solve the inequality $U(x) < B(x)$. What does your result mean in terms of the situation?

55. During the 1990s, the syphilis rate in the United States decreased rapidly (see Table 39).

Table 39 Syphilis Rates in the United States

Year	Syphilis Rate (number of cases per 100,000 people)	Year	Syphilis Rate (number of cases per 100,000 people)
1990	20.1	1995	6.2
1991	16.7	1996	4.3
1992	13.1	1997	3.2
1993	10.2	1998	2.6
1994	8.0	1999	2.4
		2000	2.2

Source: New York Times

Let $s = f(t)$ represent the syphilis rate (number of cases per 100,000 people) at t years since 1990.

a. Perform the first three steps of the modeling process to find an equation for f.

b. What is the s-intercept of f? What does it mean in terms of the situation?

c. What is the percent rate of decay? What does this mean in terms of the situation?

d. Use f to predict when there will be 1 syphilis case per 100,000 people.

e. Find $f^{-1}(0.1)$. What does your result mean in terms of the situation?

f. Federal health officials have developed a plan designed to eliminate syphilis from the United States by 2004. Use f to predict the syphilis rate in 2004. How many *cases* is this? Use 289 million as an estimate of the U.S. population in 2004.

56. The number of people undergoing laser eye surgery has increased over the past several years (see Table 40).

Table 40 Numbers of People Undergoing Laser Eye Surgery

Year	Number of People (millions)
1997	0.1
1998	0.3
1999	0.6
2000	0.8
2001	1.1

Source: *Market Scope*

Let $n = f(t)$ represent the number of people (in millions) undergoing laser eye surgery in the year that is t years since 1990.

a. Use a graphing calculator to draw a scattergram of the data. Is it better to use a linear or an exponential function to model the data? Explain.

b. Find an equation for f.

c. What is the slope of f? What does it mean in terms of the situation?

d. What is the t-intercept of f? What does it mean in terms of the situation?

e. What is the n-intercept of f? What does it mean in terms of the situation?

f. Find $f(18)$. What does it mean in terms of the situation?

g. Find an equation for f^{-1}.

h. Find $f^{-1}(4)$. What does it mean in terms of the situation?

Polynomial Functions

"No other field can offer, to such an extent as mathematics, the joy of discovery, which is perhaps the greatest human joy."

—*Rósza Péter, "Mathematics is Beautiful,"* Mathematics Intelligencer *12 (Winter 1990): 62*

6.1 SKETCHING GRAPHS OF QUADRATIC FUNCTIONS IN VERTEX FORM

Objectives

▸ Know the definition of a *quadratic function*.

▸ Sketch the graph of a quadratic function in vertex form.

So far, we have studied linear, exponential, and logarithmic functions. In this section we turn our attention to *quadratic* functions.

Here, we list several quadratic functions:

$$f(x) = x^2, \quad h(x) = -2(x+4)^2, \quad g(x) = 3(x-2)^2 + 7, \quad k(x) = -\frac{5}{2}(x-3)^2 - 2$$

DEFINITION *Quadratic Function*

A **quadratic function** is a function whose equation can be put into the form
$$f(x) = a(x - h)^2 + k,$$
where $a \neq 0$.

In this section we sketch the graph of a quadratic function.

Example 1 Sketching the Graph of a Quadratic Function

Sketch the graph of $f(x) = x^2$.

Solution

We list input-output pairs of f in Table 1. For example, $f(-3) = (-3)^2 = 9$. So, $(-3, 9)$ is an input-output pair. We plot the found points in Fig. 1.

Then, in Fig. 1, we sketch a curve that contains the plotted points.

Table 1 Input-Output Pairs of $f(x) = x^2$	
x	y
-3	9
-2	4
-1	1
0	0
1	1
2	4
3	9

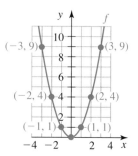

Figure 1 Graph of $y = x^2$

Recall that all points that lie on the graph of an equation satisfy the equation. All points that do not lie on the graph do not satisfy the equation.

The curve sketched in Fig. 1 is called a *parabola*. A **parabola** is the graph of a quadratic function. Some examples of parabolas are sketched in Fig. 2.

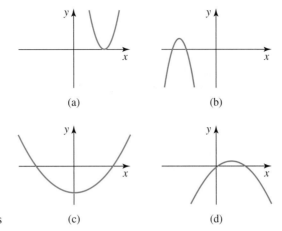

(a)   (b)

(c)   (d)

Figure 2 Sketches of parabolas

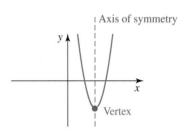

There is a difference particularly worth noting between parabolas and other curves that we have studied. A parabola that *opens upwards* (see graphs (a) and (c) in Fig. 2) has a lowest point called a **minimum point**. A parabola that *opens downward* (see graphs (b) and (d) in Fig. 2) has a highest point called a **maximum point**. The minimum point or maximum point of a parabola is called the **vertex** of the parabola, or the vertex of the function. In contrast, the curves that we studied in earlier chapters do not have a lowest or highest point.

The vertical line that passes through the vertex is called the **axis of symmetry** (see Fig. 3). The axis of symmetry divides the parabola into two parts: the points that lie to the left of the axis of symmetry and the points that lie to the right of it. The two parts are mirror reflections of each other across the axis of symmetry.

Figure 3 The axis of symmetry and vertex of a parabola

Example 2   Vertical Stretching

Compare the graph of $g(x) = 2x^2$ and the graph of $f(x) = x^2$.

Solution

We list input-output pairs of f and g in Table 2 and sketch the graphs of f and g in Fig. 4. For example, $g(-2) = 2(-2)^2 = 2(4) = 8$. Therefore, $(-2, 8)$ is a point on the graph of g.

Table 2 Input-Output Pairs of $f(x) = x^2$ and $g(x) = 2x^2$

x	$f(x) = x^2$	$g(x) = 2x^2$
−3	9	18
−2	4	8
−1	1	2
0	0	0
1	1	2
2	4	8
3	9	18

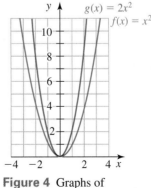

Figure 4 Graphs of $f(x) = x^2$ and $g(x) = 2x^2$

For each value of x, the value of y is twice as large for $g(x) = 2x^2$ as it is for $f(x) = x^2$. Therefore, the graph of g appears narrower (steeper) than the graph of f. Also, notice that the vertex for both functions is the point $(0, 0)$. ■

Example 3 Reflecting a Graph across the *x*-axis

Compare the graph of $f(x) = \frac{1}{2}x^2$ and the graph of $g(x) = -\frac{1}{2}x^2$.

Solution

We list input-output pairs of f and g in Table 3 and sketch the graphs of f and g in Fig. 5.

Table 3 Input-Output Pairs of $f(x) = \frac{1}{2}x^2$ and $g(x) = -\frac{1}{2}x^2$

x	$f(x) = \frac{1}{2}x^2$	$g(x) = -\frac{1}{2}x^2$
−3	4.5	−4.5
−2	2	−2
−1	0.5	−0.5
0	0	0
1	0.5	−0.5
2	2	−2
3	4.5	−4.5

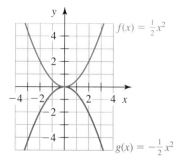

Figure 5 Graphs of $f(x) = \frac{1}{2}x^2$ and $g(x) = -\frac{1}{2}x^2$

Note that the graph of $g(x) = -\frac{1}{2}x^2$ is the reflection across the *x*-axis of the graph of $f(x) = \frac{1}{2}x^2$. To see why, note that for each value of *x*, the value of *y* for $g(x) = -\frac{1}{2}x^2$ is the opposite of the value of *y* for $f(x) = \frac{1}{2}x^2$ (see Table 3). ∎

Our observations made in Examples 2 and 3 suggest the following properties.

Graphs of Parabolas of the Form $f(x) = ax^2$

- The graph of $f(x) = ax^2$ is a parabola with vertex $(0, 0)$.
- If $|a|$ is a large number, then the parabola is narrow (steep).
- If a is near zero, the parabola is wide (not steep).
- If $a > 0$, then the parabola opens upwards.
- If $a < 0$, then the parabola opens downwards.
- The graph of $y = -ax^2$ is the reflection across the *x*-axis of the graph of $f(x) = ax^2$.

Next, we investigate graphs of functions of the form $f(x) = x^2 + k$.

Example 4 Vertical Translation (Up-Down Shifts)

Compare the graphs of $f(x) = x^2 - 3$, $g(x) = x^2$, and $h(x) = x^2 + 3$.

Solution

We list input-output pairs of f, g, and h in Table 4 and sketch the graphs of the functions in Fig. 6.

Table 4 Input-Output Pairs of $f(x) = x^2 - 3$, $g(x) = x^2$, and $h(x) = x^2 + 3$

x	$f(x) = x^2 - 3$	$g(x) = x^2$	$h(x) = x^2 + 3$
−3	6	9	12
−2	1	4	7
−1	−2	1	4
0	−3	0	3
1	−2	1	4
2	1	4	7
3	6	9	12

Figure 6 Graphs of $f(x) = x^2 - 3$, $g(x) = x^2$, and $h(x) = x^2 + 3$

For each value of x, the values of y for $h(x) = x^2 + 3$ are 3 more than the values of y for $g(x) = x^2$, which are 3 more than the values of y for $f(x) = x^2 - 3$.

To sketch the graph of $h(x) = x^2 + 3$, we *translate* (move) the graph of $g(x) = x^2$ up 3 units. To sketch the graph of $f(x) = x^2 - 3$, we translate the graph of $g(x) = x^2$ down 3 units. ■

Example 5 Horizontal Translation (Left-Right Shifts)

Compare the graph of $g(x) = (x - 5)^2$ to the graph of $f(x) = x^2$.

Solution

We list input-output pairs of g in Table 5 and sketch the graphs of f and g in Fig. 7.

Table 5 Input-Output Pairs of $g(x) = (x - 5)^2$

x	y
−1	36
0	25
1	16
2	9
3	4
4	1
5	0
6	1
7	4
8	9

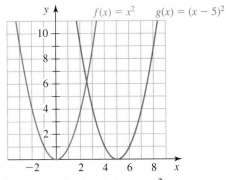

Figure 7 Graphs of $f(x) = x^2$ and $g(x) = (x - 5)^2$

To graph $g(x) = (x - 5)^2$, we translate the graph of $f(x) = x^2$ five units to the right. To see why this makes sense, we solve each equation for x:

$$f(x) = x^2 \qquad\qquad g(x) = (x - 5)^2$$
$$x^2 = y \qquad\qquad (x - 5)^2 = y$$
$$x = \pm y^{1/2} \qquad\qquad x - 5 = \pm y^{1/2}$$
$$x = \pm y^{1/2} + 5$$

For each value of y, the value of x for g is 5 more than the value of x for f. Therefore, the graph of $g(x) = (x - 5)^2$ lies 5 units to the right of the graph of $f(x) = x^2$. ■

To graph $h(x) = (x + 5)^2$, we translate the graph of $f(x) = x^2$ five units to the left.

Three-Step Method for Graphing a Quadratic Function

To sketch the graph of $f(x) = a(x - h)^2 + k, a \neq 0$:

1. Sketch the graph of $y = ax^2$.
2. Translate the graph (from step 1) to the right h units if $h > 0$ and to the left $|h|$ units if $h < 0$.
3. Translate the graph (from step 2) up k units if $k > 0$ and down $|k|$ units if $k < 0$.

Recall that the domain of a function is the set of all inputs and the range of a function is the set of all outputs.

Example 6 Sketching the Graph of a Quadratic Function

Sketch the graph of $f(x) = -2(x + 6)^2 - 2$. Also, find the domain and range of f.

Solution

First, we write the equation for f in $f(x) = a(x - h)^2 + k$ form:

$$f(x) = -2(x - (-6))^2 + (-2)$$

Then, we follow the three-step graphing method.

Step 1. We sketch the graph of $y = -2x^2$ in Fig. 8.

Step 2. Since $h = -6$, we imagine translating the graph (from step 1) left 6 units.

Step 3. Since $k = -2$, we translate the graph (from step 2) down 2 units.

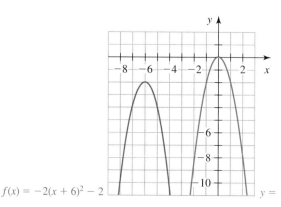

$f(x) = -2(x + 6)^2 - 2$ $y = -2x^2$

Figure 8 Graphs of $y = -2x^2$ and $f(x) = -2(x + 6)^2 - 2$

Table 6 Input-Output Pairs of $f(x) = -2(x + 6)^2 - 2$

x	$f(x) = -2(x+6)^2 - 2$
-8	-10
-7	-4
-6	-2
-5	-4
-4	-10

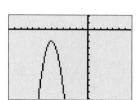

Figure 9 Verify the graph of $f(x) = -2(x + 6)^2 - 2$

We check that the input-output pairs of f listed in Table 6 are points on our sketched parabola.

Also, we use a graphing calculator to verify our sketch (see Fig. 9).

Since we can compute a value for $-2(x + 6)^2 - 2$ for any real number x, the domain of $f(x) = -2(x + 6)^2 - 2$ is the set of all real numbers. From the graph of f, we see that the vertex $(-6, -2)$ is the maximum point and that values of y are less than or equal to -2. So, the range of f is the set of numbers y where $y \leq -2$. ■

In Fig. 8 we see that the vertex of $f(x) = -2(x + 6)^2 - 2$ is the point $(-6, -2)$. This makes sense, since the vertex $(0, 0)$ of $y = x^2$ has been translated 6 units to the left and down 2 units.

Vertex of a Quadratic Function

The vertex of the quadratic function $f(x) = a(x - h)^2 + k$ is the point (h, k). We say that the equation is in **vertex form**.

For example, the vertex of the function $f(x) = -3(x - 1)^2 + 5$ is $(1, 5)$. To find the vertex of the function $g(x) = 6(x + 2)^2 - 7$, we write the equation as $g(x) = 6(x - (-2))^2 + (-7)$. So, the vertex is $(-2, -7)$.

Example 7   Sketching the Graph of a Quadratic Function

Sketch the graph of $h(x) = \dfrac{1}{3}(x - 5)^2 + 3$. Also, find the domain and range of h.

Solution

We follow the three-step graphing method:

Step 1. We sketch the graph of $y = \dfrac{1}{3}x^2$ in Fig. 10.

Step 2. Since $h = 5$, we imagine translating the graph (from step 1) right 5 units.

Step 3. Since $k = 3$, we translate the graph (from step 2) up 3 units.

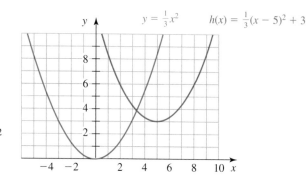

$$y = \frac{1}{3}x^2 \qquad h(x) = \frac{1}{3}(x - 5)^2 + 3$$

Figure 10 Graphs of $y = \frac{1}{3}x^2$

and $h(x) = \frac{1}{3}(x - 5)^2 + 3$

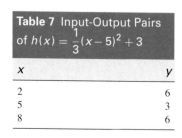

Table 7 Input-Output Pairs of $h(x) = \frac{1}{3}(x - 5)^2 + 3$

x	y
2	6
5	3
8	6

Note

In Table 7, we choose to use inputs 2, 5, and 8 to avoid fractional outputs.

We check that the input-output pairs of h listed in Table 7 are points on our sketched parabola. Also, from the equation $h(x) = \frac{1}{3}(x-5)^2 + 3$, we see that the vertex is $(5, 3)$, which matches with our sketched parabola's vertex. Finally, we can use a graphing calculator to verify our work (see Fig. 11).

Since we can compute a value for $\frac{1}{3}(x - 5)^2 + 3$ for any real number x, the domain of $h(x) = \frac{1}{3}(x - 5)^2 + 3$ is the set of all real numbers. From the graph of h, we see that the vertex $(5, 3)$ is the minimum point and that values of y are greater than or equal to 3. So, the range of f is the set of numbers y where $y \geq 3$. ■

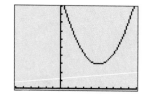

Figure 11 Graph of $h(x) = \frac{1}{3}(x - 5)^2 + 3$

 EXPLORATION

Drawing families of parabolas

1. On your calculator, graph eight parabolas to make a design like the one in Fig. 12. List the equations of your parabolas.

2. Now make a design like the one in Fig. 13. List the equations of your parabolas.

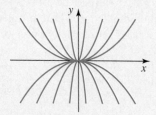

Figure 12 A family of parabolas

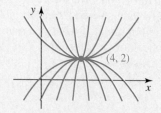

(4, 2)

Figure 13 A family of parabolas with vertex (4, 2)

3. For a quadratic function of the form $f(x) = a(x - h)^2 + k$, summarize what you have learned about a, h, and k from this exploration and from the rest of this section.

 EXPLORATION

Looking ahead: Finding an equation using trial and error

1. In this problem you will use your graphing calculator to graph a parabola like the one in Fig. 14. Use a function of the form $f(x) = a(x - h)^2 + k$, where a, h, and k are real number constants. What is the equation for f that works? You should be able to determine the values for h and k by inspecting the graph. You can find a through trial and error.

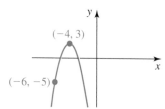

(−4, 3)

(−6, −5)

Figure 14 Problem 1

2. Table 8 shows the U.S. populations for various years. Let $f(t)$ represent the U.S. population (in millions) at t years since 0 A.D.

Table 8 U.S. Population, 1800–2002

Year	Population (millions)	Year	Population (millions)
1800	5.3	1920	106.5
1820	9.6	1940	132.1
1840	17.1	1960	179.3
1860	31.5	1980	226.5
1880	50.3	2000	281.4
1900	76.1	2002	286.7

Source: *U.S. Census Bureau*

In this problem you will use your graphing calculator to draw a scattergram and then to find an equation for a parabola that comes close to the points in the scattergram. From your work so far, you know that the equation can be of the form $f(t) = a(t - h)^2 + k$. So your task will be to find appropriate values for a, h, and k.

a. Use your graphing calculator to sketch a scattergram of the U.S. population data.

b. Guess values for a, h, and k. Substitute your guesses for a, h, and k into the vertex equation $f(t) = a(t - h)^2 + k$.

c. Graph your vertex equation and your scattergram in the same viewing window to get an idea of how well your quadratic model fits the data.

d. Now get better values for a, h, and k through trial and error. When you are satisfied with your values for a, h, and k, write the equation of the parabola you have found.

Tips on Succeeding in This Course

After you think you have solved a problem, it is wise to reread the problem to make sure you have answered its question(s). You should also reread the problem with the solution in mind. If what you read seems to make sense, then you've provided another check of your result(s).

Key Points OF THIS SECTION

Quadratic function

A **quadratic function** is a function whose equation can be put into the vertex form $f(x) = a(x - h)^2 + k$, where $a \neq 0$.

Properties of a quadratic function

For a quadratic function of the form $f(x) = a(x - h)^2 + k$:

- The graph of f is a parabola with vertex (h, k).
- If $a > 0$, the parabola opens upward and the vertex is the minimum point of the function f.
- If $a < 0$, the parabola opens downward and the vertex is the maximum point of the function f.
- If $|a|$ is large, the parabola is narrow.
- If a is near zero, the parabola is wide.
- The graph of f can be found by translating the graph of $y = ax^2$ horizontally by $|h|$ units (right if $h > 0$, left if $h < 0$) and vertically by $|k|$ units (up if $k > 0$, down if $k < 0$.)

HOMEWORK 6.1

 SSM
Tutor Center

MathPro 4
MathPro 4

MathPro 5
MathPro 5

 CDs/Videos

 www

Sketch the graph of the function. Give the coordinates for the vertex. Verify your result using your graphing calculator.

1. $y = 3x^2$

2. $y = -3x^2$

3. $y = -\dfrac{1}{4}x^2$

4. $y = \dfrac{1}{3}x^2$

5. $y = x^2 - 1$

6. $y = x^2 + 4$

7. $y = -x^2 + 5$

8. $y = -x^2 - 2$

9. $y = (x + 5)^2$

10. $y = (x - 4)^2$

11. $y = -(x - 1)^2$

12. $y = -(x + 3)^2$

13. $y = (x - 4)^2 - 6$

14. $y = -(x + 2)^2 + 5$

15. $y = 2(x - 6)^2$

16. $y = -3(x - 4)^2 + 1$

17. $y = 3(x + 5)^2 - 7$

18. $y = -2(x - 1)^2 + 9$

19. $y = 2(x + 7)^2 - 3$

20. $y = 3(x - 2)^2 - 6$

21. $y = -\dfrac{1}{3}(x + 4)^2 - 1$

22. $y = \dfrac{1}{3}(x + 6)^2 - 2$

23. $y = -\dfrac{1}{2}(x + 6)^2 + 3$

24. $y = \dfrac{1}{2}(x - 2)^2 - 8$

Sketch the graph of the function. Verify your result using your graphing calculator. Also, find the domain and range of the function.

25. $f(x) = x^2 - 4$

26. $f(x) = x^2 + 1$

27. $f(x) = (x + 6)^2$

28. $f(x) = (x - 2)^2$

29. $f(x) = -3(x - 5)^2 + 2$

30. $f(x) = 2(x + 1)^2 + 4$

31. Four functions of the form $y = a(x - h)^2 + k$ are sketched in Fig. 15. Describe the signs of the constants a, h, and k for each of these functions.

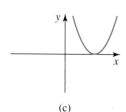

(a)

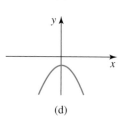

(b)

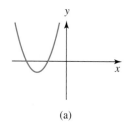

(c)

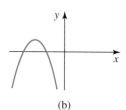

(d)

Figure 15 Exercise 31

32. For a quadratic function $f(x) = a(x - h)^2 + k$, for what values of a, h, and k does f have a maximum point? For what values of a, h, and k does f have a minimum point? Describe the maximum or minimum point in terms of a, h, and k.

33. Use your graphing calculator to graph a family of parabolas similar to the family shown in Fig. 16. List the equations of your parabolas.

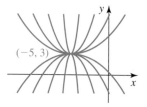

Figure 16 A family of parabolas with vertex $(-5, 3)$

34. Use your graphing calculator to graph a family of parabolas similar to the family shown in Fig. 17. List the equations of your parabolas.

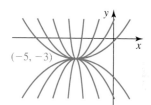

Figure 17 A family of parabolas with vertex $(-5, -3)$

35. Find an equation of the function f sketched in Fig. 18. [**Hint:** For $f(x) = a(x - h)^2 + k$, $a \neq 0$, you can determine h and k from the vertex. Find a value for a using trial and error. Or, you can find a by using the fact that a point on a graph satisfies a graph's equation.]

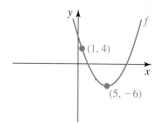

Figure 18 Exercise 35

36. A graph of the function $f(x) = 2.1(x - 2.73)^2 - 3.71$ is shown in Fig. 19. Find an equation for the function g also sketched in Fig. 19.

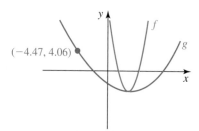

Figure 19 Exercise 36

37. A graph of the function $f(x) = 2.1(x - 2.73)^2 - 3.71$ is sketched in Fig. 20. Find an equation for the function g also sketched in Fig. 20. Assume that the graphs of f and g have the same "shape."

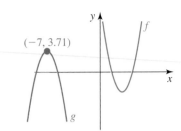

Figure 20 Exercise 37

38. Find equations of the four functions that produce the design shown in Fig. 21.

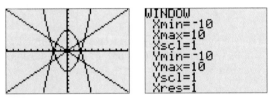

Figure 21 A design created by graphs of functions

For each exercise, decide if it is possible for a parabola to have the indicated number of x-intercepts. If it is possible, find an equation of such a parabola. If it is not possible, explain why.

39. no x-intercepts

40. one x-intercept

41. two x-intercepts

42. three x-intercepts

43. Table 9 shows the percentages of Americans in support of abortion rights (pro-choice) for various years.

Table 9 Percentages of Americans Who Are Pro-Choice

Year	Percent
1995	56
1996	52
1997	49
1998	48
1999	48
2000	48
2001	50

Source: *Gallup Poll Topics,*
www.frtl.org/abortion/
gallup%20poll%20topics.htm

Let $f(t)$ represent the percentage of Americans who are pro-choice at t years since 1990.

a. Use a graphing calculator to draw a scattergram of the data. Will it be best to model the data with a linear, exponential, or quadratic function? Explain.

b. Use your graphing calculator to draw the graph of the function $f(t) = 0.56(t - 8.86)^2 + 47.49$. Does the parabola come close to the points in the scattergram?

c. What is the vertex of the parabola? Verify your result using TRACE. What does the vertex represent in terms of the situation?

d. Use TRACE to estimate/predict when 57% of Americans were/will be pro-choice.

e. Since 1973, when abortion became legal in the United States, the percentage of Americans who have been pro-choice has varied from about 48% to about 58%. Use TRACE to predict the percentage of Americans in 2008

that will be pro-choice. In your opinion, is it likely that this prediction will turn out to be correct? Explain.

44. The numbers of wildfires in the United States for various years are shown in Table 10.

Table 10 U.S. Wildfires

Year	Number of Wildfires (thousands)
1996	115
1997	90
1998	81
1999	94
2000	123

Source: *Wildland Fire Statistics,*
www.nifc.gov/stats/wildlandfirestats.html

Let $f(t)$ represent the number of wildfires (in thousands) in the year that is t years since 1995.

a. Use a graphing calculator to draw a scattergram of the data. Will it be best to model the data with a linear, exponential, or quadratic function? Explain.

b. Use your graphing calculator to draw the graph of the function $f(t) = 9.29(t - 2.89)^2 + 81.97$. Does the parabola come close to the points in the scattergram?

c. What is the vertex of the parabola? Verify your result using TRACE. What does the vertex represent in terms of the situation? Does the model underestimate or overestimate the number of wildfires in that year? Explain.

d. Use TRACE to predict the number of wildfires in 2007.

e. Use TRACE to predict when there will be 1 million wildfires.

45. Solve the system by finding ordered pairs that satisfy both equations. [**Hint:** Graph both equations on the same coordinate system.]

$$y = 2(x - 2)^2 + 5$$
$$y = -3(x - 2)^2 + 5$$

46. Use ZDecimal to draw a graph of $f(x) = 0.7x^2 + 2x - 1$. Then use TRACE to complete Table 11. Verify your table entries using a graphing calculator table.

Table 11 Values of $f(x) = 0.7x^2 + 2x - 1$

x	$f(x)$
−2	
−1	
0	
1	
2	

47. Use your graphing calculator to sketch a graph of the function $f(x) = x^2 + 1$.

    a. Use ZDecimal and TRACE to find three points that lie on the graph of f. Show that each point is a solution of the equation $f(x) = x^2 + 1$.

    b. Choose three points that do not lie on the graph of f. Show that each point is not a solution of the equation $f(x) = x^2 + 1$.

48. a. Find equations of a parabola and a line that don't intersect.

    b. Find equations of a parabola and a line that intersect at exactly one point. (Challenge: Do this for a line that is neither horizontal nor vertical.)

    c. Find equations of a parabola and a line that intersect at exactly two points.

    d. Is it possible for a parabola and a line to intersect at more than two points? Explain.

49. a. Find two equations of the form $y = a(x - h)^2 + k, a \neq 0$, whose graphs don't intersect.

    b. Find two equations of the form $y = a(x - h)^2 + k, a \neq 0$, whose graphs intersect at one point.

    c. Find two equations of the form $y = a(x - h)^2 + k, a \neq 0$, whose graphs intersect at two points.

    d. Is it possible for two equations of the form $y = a(x - h)^2 + k, a \neq 0$, to have graphs that intersect at more than two points? Explain.

50. Find equations of a parabola and two perpendicular lines, neither of which intersects the parabola.

51. Use your graphing calculator to graph $y = x^2$.

    a. Use the WINDOW settings displayed in Fig. 22. (You can get these settings using ZDecimal.)

    b. Use the WINDOW settings displayed in Fig. 23. Compare what you see with what you see in part a.

    c. Use the WINDOW settings displayed in Fig. 24. Compare what you see with what you see in part a.

    d. Explain your results in parts b and c.

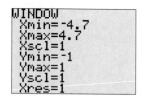

Figure 22 Window for Exercise 51a   **Figure 23** Window for Exercise 51b

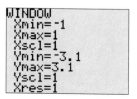

Figure 24 Window for Exercise 51c

52. Use your graphing calculator to graph each function. Record WINDOW settings that allow you to see the graph, including the vertex, on your calculator screen.

    a. $f(x) = 1000x^2$

    b. $g(x) = 1000(x - 1000)^2$

    c. $h(x) = (x + 10{,}000)^2 + 10{,}000$

53. A student uses ZStandard on his graphing calculator to graph $f(x) = 0.0001x^2 + 5$. He is confused because he thought that the graph should be a parabola, but his calculator screen displays a horizontal line. What would you tell him?

54. The vertex of a parabola is $(5, 3)$ and the parabola passes through the point $(8, 10)$. Find a third point that lies on the parabola.

55. For each part, sketch on the same coordinate system the graph of $f(x) = a(x - h)^2 + k$ as shown in Fig. 25 and the given function. Be sure to label which graph is which.

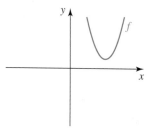

Figure 25 Exercise 55

    a. $g(x) = a(x - 2h)^2 + k$ **b.** $p(x) = a(x - h)^2 + 2k$

    c. $q(x) = 2a(x - h)^2 + k$ **d.** $l(x) = 2a(x - 2h)^2 + 2k$

56. Find two other quadratic functions that have the same domain and range as the function $f(x) = -2(x + 4)^2 + 7$.

57. Describe how to sketch the graph of a function of the form $f(x) = a(x - h)^2 + k, a \neq 0$. Include in your description the effects that the values for a, h, and k have on the graph.

58. To sketch the graph of the quadratic function $y = a(x - h)^2 + k$ we translate the graph of $y = a(x - h)^2$ up k units if $k > 0$ and down $|k|$ units if $k < 0$. Explain.

6.2 EXPANDING AND FACTORING POLYNOMIALS

Objectives

▸ Know the meaning of *polynomial*.

▸ Know that factoring a polynomial is the reverse process of expanding a polynomial.

▸ Expand and factor polynomials.

▸ Convert quadratic functions in vertex form to standard form.

▸ Use the formula for the square of a sum, square of a difference, or difference of two squares to expand a polynomial.

In this chapter and Chapter 7 we will work with *polynomial expressions*, or *polynomials* for short.

Here are some examples of polynomials:

$$x^2 - 9, \qquad 5x^3 - 4x^2 + 9x - 7, \qquad 4, \qquad 2x^{99} + 3x + 1, \qquad -7x^5$$

A polynomial can be written in the form

$$a_n x^n + a_{n-1} x^{n-1} + a_{n-2} x^{n-2} + \cdots + a_2 x^2 + a_1 x + a_0$$

Note

The polynomial a_0 has degree 0 if $a_0 \neq 0$; the degree is undefined for the polynomial 0.

where $a_n, a_{n-1}, a_{n-2}, \ldots, a_2, a_1, a_0$ are real number constants and n is a whole number. If $a_n \neq 0$, then n is the *degree* of the polynomial.

Here are the names for and examples of polynomials with degrees $n = 1$, $n = 2$, and $n = 3$:

Degree	Name	Example
$n = 1$	linear polynomial	$2x + 9, \quad -5x$
$n = 2$	quadratic polynomial	$-3x^2 + 7x + 1, \quad 4x^2 - 9$
$n = 3$	cubic polynomial	$8x^3 - 5x^2 - 4x + 2, \quad 7x^3 + x$

We begin our work with polynomials by reviewing the distributive law:

$$a(b + c) = ab + ac$$

For example, $5(x + 3) = 5x + 15$. Here, we have *expanded* $5(x + 3)$ into $5x + 15$. If we do the reverse, that is, write $5x + 15$ as $5(x + 3)$, we say we have *factored* $5x + 15$ into $5(x + 3)$. See Fig. 26.

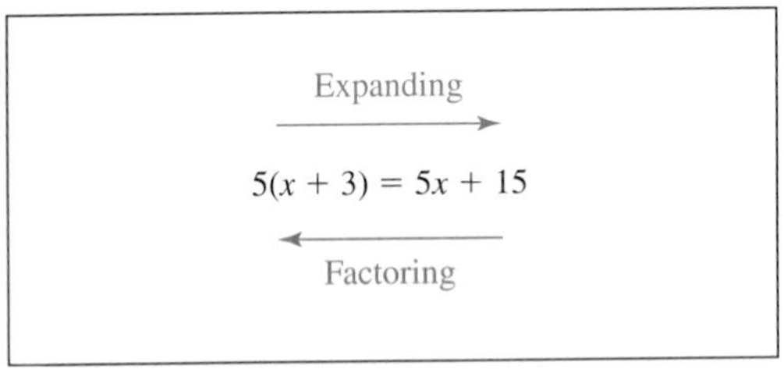

Figure 26 Expanding versus factoring

We **factor** a polynomial by writing it as a product of two or more factors. We **expand** a factored polynomial by performing the indicated multiplication.

Now, we factor $3x + 6$. Note that 3 is a common factor of $3x$ and 6:

$$3x + 6 = 3x + 3(2)$$

We use the distributive law to "factor out" the common factor 3:

$$3x + 6 = 3x + 3(2) = 3(x + 2)$$

To factor $8x + 12$, we could write $2(4x + 6)$, since 2 is a common factor of $8x$ and 12. However, to *completely factor* $8x + 12$, we factor out 4, the *greatest common factor (GCF)*:

$$8x + 12 = 4(2x + 3)$$

When we factor polynomials in this course, our intent is to factor completely.

Example 1 Expanding and Factoring Polynomials

1. Expand $2x(x - 3)$.
2. Expand $-3x(x - 6)$.
3. Factor $2x^2 - 10x$.
4. Factor $-8x^2 - 12x$.

Solution

1. $2x(x - 3) = 2x(x) - 2x(3) = 2x^2 - 6x$

We can verify our result by comparing graphing calculator tables for $y = 2x(x - 3)$ and $y = 2x^2 - 6x$ (see Fig. 27).

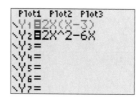

Figure 27 Comparing tables for $y = 2x(x - 3)$ and $y = 2x^2 - 6x$

2. $-3x(x - 6) = -3x(x) - (-3x)6 = -3x^2 + 18x$

3. To factor $2x^2 - 10x$, note that $2x$ is the GCF of $2x^2$ and $10x$:

$$2x^2 - 10x = 2x(x) - 2x(5)$$
$$= 2x(x - 5) \qquad \text{Factor out the GCF.}$$

We can check our work by expanding the result:

$$2x(x - 5) = 2x(x) - 2x(5) = 2x^2 - 10x$$

Since $2x^2 - 10x$ is the original polynomial, we know that our work is correct.

4. For $-8x^2 - 12x$, the GCF is $4x$. If the first term has a negative coefficient, we usually factor out the opposite of the GCF. Here, we factor out $-4x$:

$$-8x^2 - 12x = -4x(2x) + (-4x)(3)$$
$$= -4x(2x + 3) \qquad \text{Factor out } -4x. \qquad ■$$

Throughout the rest of this section we will focus on expanding polynomials. In Sections 6.3 and 6.4 we will factor more types of polynomials.

We can expand $(a + b)(c + d)$ using the distributive law three times:

$$(a + b)(c + d) = (a + b)c + (a + b)d \qquad \text{Apply the distributive law once.}$$
$$= ac + bc + ad + bd \qquad \text{Apply the distributive law twice more.}$$
$$= ac + ad + bc + bd \qquad \text{Commutative law}$$

By examining the result $ac + ad + bc + bd$, we see that we can expand $(a + b)(c + d)$ in one step by adding the four products formed by multiplying each term in the first sum by each term in the second sum:

$$(a + b)(c + d) = ac + ad + bc + bd$$

Example 2   **Expanding Polynomials**

Expand each polynomial.

1. $(x + 2)(x + 3)$

2. $(2x + 5)(3x - 8)$

3. $4(x^3 - 2)(x^2 - 3)$

Solution

1. $(x + 2)(x + 3) = x(x) + x(3) + 2(x) + 2(3)$
$$= x^2 + 3x + 2x + 6$$
$$= x^2 + 5x + 6$$

We verify our work by comparing graphing calculator tables for $y = (x+2)(x+3)$ and $y = x^2 + 5x + 6$ (see Fig. 28).

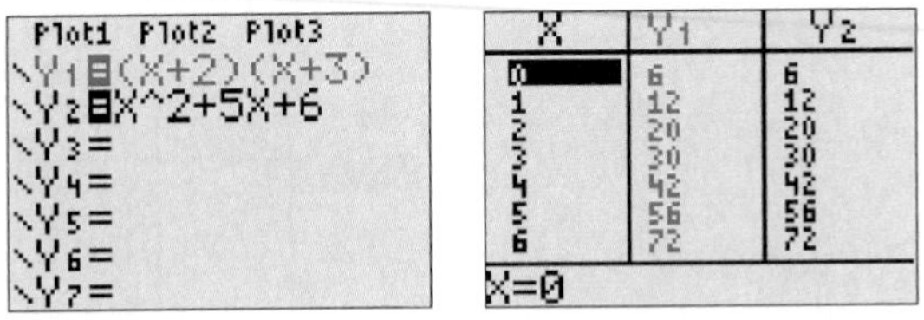

Figure 28 Comparing tables for $y = (x + 2)(x + 3)$ and $y = x^2 + 5x + 6$

2. $(2x + 5)(3x - 8) = (2x + 5)(3x + (-8))$

$\qquad = 2x(3x) + 2x(-8) + 5(3x) + 5(-8)$

$\qquad = 6x^2 - 16x + 15x - 40$

$\qquad = 6x^2 - x - 40$

3. $4(x^3 - 2)(x^2 - 3) = 4(x^3 \cdot x^2 - 3x^3 - 2x^2 + 6)$

$\qquad = 4(x^5 - 3x^3 - 2x^2 + 6)$

$\qquad = 4x^5 - 12x^3 - 8x^2 + 24$ ■

We refer to a polynomial as a *monomial*, *binomial*, or *trinomial*, depending on whether it has one, two, or three nonzero terms, respectively:

Name	Examples	Meaning
monomial	$3x^2$, $-5x$, x, 2	one term
binomial	$4x^2 - 8$, $3x^2 + 5x$, $7x + 1$	two terms
trinomial	$7x^2 - 3x + 4$, $x^2 + 6x - 9$	three terms

In problems 1 and 2 in Example 2, we found the product of two binomials. We'll now discuss how to find the product of a binomial and a trinomial.

We can expand the product $(a + b)(c + d + e)$ by applying the distributive law four times:

$(a + b)(c + d + e) = (a + b)c + (a + b)d + (a + b)e$ ⟶ Apply the distributive law.

$\qquad = ac + bc + ad + bd + ae + be$ ⟶ Apply the distributive law three more times.

$\qquad = ac + ad + ae + bc + bd + be$ ⟶ Commutative law

By examining the result $ac + ad + ae + bc + bd + be$, we see that we can effectively expand $(a + b)(c + d + e)$ in one step by adding the six products formed by multiplying each term in the first sum by each term in the second sum:

$$(a + b)(c + d + e) = ac + ad + ae + bc + bd + be$$

We can expand the product of two polynomials of any degrees by adding the products formed by multiplying each term in one polynomial by each term in the other polynomial.

Example 3 Expanding Polynomials

Expand.

1. $(2x + 3)(x^2 - 4x + 5)$

2. $(x^2 + 3x - 2)(x^2 - x + 4)$

Solution

1. $(2x + 3)(x^2 - 4x + 5) = 2x \cdot x^2 - 2x \cdot 4x + 2x \cdot 5 + 3 \cdot x^2 - 3 \cdot 4x + 3 \cdot 5$

$$= 2x^3 - 8x^2 + 10x + 3x^2 - 12x + 15$$
$$= 2x^3 - 5x^2 - 2x + 15$$

We use graphing calculator tables to verify our work (see Fig. 29).

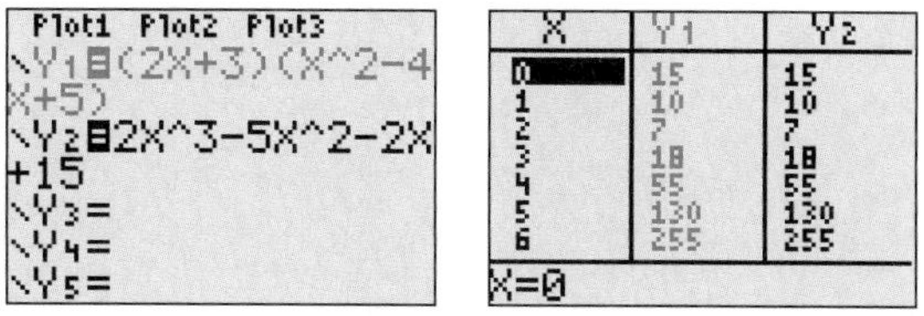

Figure 29 Verifying that $(2x + 3)(x^2 - 4x + 5) = 2x^3 - 5x^2 - 2x + 15$

2. We add the nine products formed by multiplying each term in the first trinomial by each term in the second trinomial.

$$(x^2 + 3x - 2)(x^2 - x + 4) = x^2 \cdot x^2 - x^2 \cdot x + x^2 \cdot 4 + 3x \cdot x^2 - 3x \cdot x$$
$$+ 3x \cdot 4 - 2 \cdot x^2 + 2 \cdot x - 2 \cdot 4$$
$$= x^4 - x^3 + 4x^2 + 3x^3 - 3x^2 + 12x - 2x^2 + 2x - 8$$
$$= x^4 + 2x^3 - x^2 + 14x - 8$$

■

Next, we discuss how to find the square of a sum such as $(x + 3)^2$. Here we expand $(x + 3)^2$:

$$(x + 3)^2 = (x + 3)(x + 3) \qquad b^2 = bb$$
$$= x^2 + 3x + 3x + 9$$
$$= x^2 + 6x + 9$$

> **Note**
>
> "Expand" also means to perform exponentiation, since exponentiation is repeated multiplication.

Now we expand $(x + k)^2$:

$$(x + k)^2 = (x + k)(x + k) \qquad b^2 = bb$$
$$= x^2 + kx + kx + k^2$$
$$= x^2 + 2kx + k^2$$

So, $(x + k)^2 = x^2 + 2kx + k^2$. We can use similar steps to find the formula $(x - k)^2 = x^2 - 2kx + k^2$.

Square of a Sum Formula and Square of a Difference Formula

$$(x + k)^2 = x^2 + 2kx + k^2 \qquad \text{Square of a sum}$$
$$(x - k)^2 = x^2 - 2kx + k^2 \qquad \text{Square of a difference}$$

Beware! In general, $(x + k)^2 \neq x^2 + k^2$ and $(x - k)^2 \neq x^2 - k^2$.
Now, we expand $(x + 3)^2$ using the square of a sum formula:

$$(x + 3)^2 = x^2 + 2(3)x + 3^2$$
$$= x^2 + 6x + 9$$

> **Note**
>
> It *is* true that $(xk)^2 = x^2 k^2$ and that $\left(\dfrac{x}{k}\right)^2 = \dfrac{x^2}{k^2}$, $k \neq 0$.

Note that the result is the same as our result from writing $(x + 3)^2$ as $(x + 3)(x + 3)$ and then expanding.

Example 4   Expanding Polynomials

Expand each polynomial.

1. $(x - 8)^2$ **2.** $(3x + 5)^2$ **3.** $3x(x - 1)^2$

Solution

1. $(x - 8)^2 = x^2 - 2(8)x + 8^2$ $(x - k)^2 = x^2 - 2kx + k^2$

$= x^2 - 16x + 64$

2. $(3x + 5)^2 = (3x)^2 + 2(5)(3x) + 5^2$ $(x + k)^2 = x^2 + 2kx + k^2$

$= 9x^2 + 30x + 25$

3. $3x(x - 1)^2 = 3x(x^2 - 2(1)x + 1^2)$ $(x - k)^2 = x^2 - 2kx + k^2$

$= 3x(x^2 - 2x + 1)$

$= 3x^3 - 6x^2 + 3x$ Distributive law ∎

In Section 6.1 we sketched the graph of a quadratic function in vertex form, such as $f(x) = 2(x + 5)^2 + 3$. Here, we expand the right side of this equation:

$$f(x) = 2(x + 5)^2 + 3$$

$$= 2(x^2 + 10x + 25) + 3 \qquad (x + k)^2 = x^2 + 2kx + k^2$$

$$= 2x^2 + 20x + 50 + 3 \qquad \text{Distributive law}$$

$$= 2x^2 + 20x + 53$$

Note

By assuming that the graph of a quadratic equation in standard form is a parabola, you will discover how to convert an equation from standard form to vertex form in exercise 68 of Section 6.6.

The function $f(x) = 2x^2 + 20x + 53$ is now in *standard form*. Any quadratic function in vertex form can be written in standard form. Also, it turns out that *any* function in the form $f(x) = ax^2 + bx + c, a \neq 0$, can be written in vertex form, so such a function is quadratic and its graph is a parabola.

Standard Form of a Quadratic Function

Let $f(x) = ax^2 + bx + c$, where $a \neq 0$. Then f is a quadratic function and its graph is a parabola. We say that the equation is in **standard form**.

Example 5   Writing a Quadratic Function in Standard Form

Write $f(x) = -3(x - 4)^2 + 8$ in standard form.

Solution

$$f(x) = -3(x - 4)^2 + 8$$

$$= -3(x^2 - 8x + 16) + 8 \qquad (x - k)^2 = x^2 - 2kx + k^2$$

$$= -3x^2 + 24x - 48 + 8 \qquad \text{Distributive law}$$

$$= -3x^2 + 24x - 40$$

We verify our result by comparing graphs of $y = -3(x - 4)^2 + 8$ and $y = -3x^2 + 24x - 40$ in ZStandard windows (see Fig. 30).

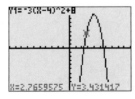
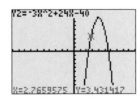

Figure 30 Graphs of $y = -3(x - 4)^2 + 8$ and $y = -3x^2 + 24x - 40$ ∎

Now, we expand the expression $(x + k)(x - k)$:

$$(x + k)(x - k) = x^2 + kx - kx - k^2$$
$$= x^2 - k^2$$

We see that the product of $x + k$ and $x - k$ is a difference of two squares $x^2 - k^2$.

Difference of Two Squares Formula

$$(x + k)(x - k) = x^2 - k^2$$

We can use the difference of two squares formula to help us expand (and factor) quadratic polynomials.

Example 6   Expanding Polynomials

Expand each polynomial.

1. $(3x + 5)(3x - 5)$ **2.** $-2(x - 1)(x + 1)$

3. $\left(\dfrac{1}{2}x + \dfrac{1}{3}\right)\left(\dfrac{1}{2}x - \dfrac{1}{3}\right)$

Solution

1. $(3x + 5)(3x - 5) = (3x)^2 - 5^2 = 9x^2 - 25$

2. $-2(x - 1)(x + 1) = -2(x^2 - 1) = -2x^2 + 2$

3. $\left(\dfrac{1}{2}x + \dfrac{1}{3}\right)\left(\dfrac{1}{2}x - \dfrac{1}{3}\right) = \left(\dfrac{1}{2}x\right)^2 - \left(\dfrac{1}{3}\right)^2 = \dfrac{1}{4}x^2 - \dfrac{1}{9}$ ■

So far, we have discussed how to factor binomials of the form $ax^2 + bx$ and $bx + c$. In Section 6.3 we discuss how to factor other types of binomials as well as quadratic trinomials.

EXPLORATION
Looking ahead: Factoring the difference of two squares

1. Expand $2(x + 4)$. Then factor $2x + 8$. Compare expanding $2(x + 4)$ with factoring $2x + 8$.

2. Expand each expression.
   - **a.** $(x - 3)(x + 3)$
   - **b.** $(x - 4)(x + 4)$
   - **c.** $(x - 5)(x + 5)$
   - **d.** $(2x - 7)(2x + 7)$

3. Factor each expression.
   - **a.** $x^2 - 9$
   - **b.** $x^2 - 36$
   - **c.** $16x^2 - 25$
   - **d.** $4x^2 - 9$

4. Describe in general how to factor the difference of two squares.

EXPLORATION
Looking ahead: Factoring trinomials

1. Complete Table 12.

Table 12 Expanding Some Quadratic Polynomials

Factored Expression	Last Terms	Expanded Expression	Coefficient of x	Constant Term
$(x+3)(x+4)$	3 and 4	$x^2 + 7x + 12$	7	12
$(x+5)(x-3)$	5 and −3	$x^2 + 2x - 15$	2	−15
$(x-2)(x+6)$				
$(x-4)(x-3)$				

2. For each row of Table 12, what connection do you notice between the last terms of the factored expression and the coefficient of x of the expanded expression? Explain why this happens.

3. For each row of Table 12, what connection do you notice between the last terms of the factored expression and the constant term of the expanded expression? Explain why this happens.

4. **a.** Do the observations that you have made in this exploration apply to the expression $(2x + 3)(x + 4)$? If so, show that they do. If not, explain why not in terms of how you expand $(2x + 3)(x + 4)$.

   b. Do the observations that you have made in this exploration apply to the expression $(x + 5)(3x + 4)$? If so, show that they do. If not, explain why not in terms of how you expand $(x + 5)(3x + 4)$.

   c. Discuss, in general, when your observations apply and when they do not.

Tips on Succeeding in This Course

How often do you get confused by class notes you had written earlier the same day, even though the class activities made sense to you? If this happens a lot, you can counteract this by reviewing your notes as soon after class as possible. Even reviewing your notes for just a few minutes between classes will help. This will increase the likelihood of you remembering what you learned in class, and will give you the opportunity to add additional comments to your notes while the class experience is still fresh in your mind. Teaming with a classmate to review notes can also be helpful.

Key Points
OF THIS SECTION

Factoring and expanding a polynomial

We *factor* a polynomial by writing it as a product. We *expand* a product by performing the multiplication.

Expanding the product of two polynomials

We can expand the product of two polynomials of any degrees by adding the products formed by multiplying each term in one polynomial by each term in the other polynomial.

Vertex and standard forms

A function in vertex form $f(x) = a(x - h)^2 + k$ can be written in standard form $f(x) = ax^2 + bx + c$ and vice versa. A function that can be written in either form (with $a \neq 0$) is a quadratic function and its graph is a parabola.

We can expand some quadratic polynomials using the following formulas:

$$(x + k)^2 = x^2 + 2kx + k^2 \quad \text{Square of a sum}$$

$$(x - k)^2 = x^2 - 2kx + k^2 \quad \text{Square of a difference}$$

$$(x + k)(x - k) = x^2 - k^2 \quad \text{Difference of two squares}$$

Formulas for expanding polynomials

HOMEWORK 6.2

 SSM
Tutor Center

 MathPro 4
MathPro 4   **MathPro 5**
MathPro 5

 CDs/Videos   www

Expand.

1. $x(x - 1)$

2. $4x(x - 5)$

3. $-7x(x - 2)$

4. $-2x(3 - x)$

5. $-3x(8x + 5)$

6. $6x(2x - 7)$

Factor. Verify that you have factored correctly by expanding your factored polynomial.

7. $x^2 + 4x$

8. $3x^2 - 6x$

9. $-27x^2 + 36x$

10. $-x^2 - x$

11. $-25x^2 + 35x$

12. $39x^2 - 26x$

13. $-4x^2 - 8x$

14. $-40x^2 + 15x$

15. $3.8x^2 + 4.7x$

16. $17.29x^2 - 5.64x$

Factor the right side of each equation. Verify your result by comparing graphing calculator tables of your result and the original function.

17. $f(x) = x^2 - 7x$

18. $g(x) = -5x^2 + 10x$

19. $h(x) = 24x^2 - 30x$

20. $k(x) = -18x^2 - 12x$

21. $f(x) = -2x^2 - 8x$

22. $g(x) = 12x - 24x^2$

23. $h(x) = 2.5x^2 - 6.2x$

24. $k(x) = 3.71x^2 + 8.44x$

25. A student attempts to factor the polynomial $12x^2 + 18x$:

$$12x^2 + 18x = 2x(6x + 9)$$

The student finds that graphing calculator tables for $y = 12x^2 + 18x$ and $y = 2x(6x + 9)$ are the same. Is the factoring correct? Explain.

26. A student attempts to factor the polynomial $14x^2 - 7x$. The student's work is shown below. What would you tell the student?

$$14x^2 - 7x = 7x(2x - 1)$$
$$= 14x^2 - 7x$$

27. Three students tried to factor the polynomial $-16x^2 + 24x$. The work of each student is shown. Decide which student(s) factored the polynomial correctly.

Student 1's work

$$-16x^2 + 24x = 8x(-2x + 3)$$
$$= 8x(3 - 2x)$$

Student 2's work

$$-16x^2 + 24x = -8x(2x - 3)$$

Student 3's work

$$-16x^2 + 24x = -4x(4x - 6)$$
$$= -4x(2)(2x - 3)$$
$$= -8x(2x - 3)$$

28. Two students both attempt to expand $(x + 5)^2$. Did one, both, or neither of these students expand the polynomial correctly? If any errors were made, describe them and where they occurred. Also, verify your decision by comparing graphing calculator tables or graphs for $y = (x + 5)^2$, $y = x^2 + 25$, and $y = x^2 + 10x + 25$.

Student 1's work

$$(x + 5)^2 = x^2 + 5^2$$
$$= x^2 + 25$$

Student 2's work

$$(x + 5)^2 = (x + 5)(x + 5)$$
$$= x^2 + 5x + 5x + 5^2$$
$$= x^2 + 10x + 25$$

Expand.

29. $2x(7x - 4)$

30. $3x(6x + 1)$

31. $(x + 2)(x + 4)$

32. $(x + 6)(x + 3)$

33. $(x - 3)(x + 6)$

34. $(x - 5)(x - 4)$

35. $(x + 8)(x + 8)$

36. $(x - 10)(x - 10)$

37. $(x - 7)^2$

38. $(x + 4)^2$

39. $(x + 5)(x - 5)$

40. $(x - 9)(x + 9)$

41. $(3x + 5)(4x + 1)$

42. $(2x + 7)(5x - 4)$

43. $(4x - 1)(6x - 5)$

44. $(7x - 2)(3x + 6)$

45. $(x^2 - 4)(x + 2)$

46. $(x^2 + 3)(x^2 - 5)$

47. $(x^3 + 6)(x^2 + 3)$

48. $(x^2 - 1)(x^4 - 7)$

49. $(2x + 5)^2$

50. $(7x - 9)^2$

51. $(3x + 2)(3x - 2)$

52. $(4x - 6)(4x + 6)$

53. $-x(1 - x)$

54. $-5x(x - 4)$

55. $-(x - 1)^2$

56. $3(x - 9)(x - 2)$

57. $3x(5x - 1)^2$

58. $x(x + 2)(x - 4)$

59. $-2x(4x + 5)^2$

60. $-3x(2x + 1)(2x - 1)$

61. $(2.1x - 3.8)(1.8x + 5.6)$

62. $(1.9x + 4.7)^2$

63. $(x + 5)(x^2 + x + 2)$

64. $(x + 3)(x^2 + 4x + 1)$

65. $(2x - 3)(3x^2 + x - 4)$

66. $(3x + 5)(2x^2 - 4x - 1)$

67. $(x + 4)(x^2 - 4x + 16)$

68. $(x - 3)(x^2 + 3x + 9)$

69. $(x^2 + 2x + 3)(x^2 + x + 2)$

70. $(x^2 + 4x + 1)(x^2 + 3x + 1)$

71. $(2x^2 + x - 3)(x^2 - 2x + 1)$

72. $(3x^2 - 5x + 2)(x^2 - 2x + 1)$

73. $(x + 1)(x + 2)(x + 3)$

74. $(x - 5)(x + 2)(x + 4)$

75. $\left(\frac{1}{5}x - 2\right)\left(\frac{1}{5}x + 2\right)$ **76.** $\left(x - \frac{1}{3}\right)^2$

77. $(x - 5)^2 - (x + 5)^2$ **78.** $(x + 5)^2 - (x - 5)^2$

79. $(x - 4)^2 + (x - 4)^2$ **80.** $(x + 4)^2 - (x + 4)^2$

Write the quadratic function in standard form. Use graphing calculator tables or graphs to verify your result.

81. $f(x) = (x + 6)^2$ **82.** $g(x) = (x - 5)^2$

83. $h(x) = 2(x + 3)^2 + 1$ **84.** $k(x) = -3(x + 2)^2 - 5$

85. $p(x) = 4(x - 5)^2 + 2$ **86.** $f(x) = -2(x - 4)^2 - 8$

87. $g(x) = -4(x - 1)^2 - 1$ **88.** $h(x) = 3(x + 1)^2 - 1$

89. $k(x) = 1.5(x + 2.8)^2 - 3.7$

90. $p(x) = -2.7(x - 1.1)^2 + 4.4$

Expand the right side of each equation to help you decide whether each function is linear or quadratic. Use your graphing calculator to verify your decision.

91. $f(x) = 5x(x - 2)$ **92.** $g(x) = -3(x - 5)$

93. $h(x) = (x + 6)(x - 6)$ **94.** $k(x) = x^2 - (x + 1)^2$

95. $p(x) = (x + 6)^2 - (x - 6)^2$

96. $q(x) = (x + 2)^2 - (x + 2)$

Let $f(x) = x^2 - 3x$. Find each output.

97. $f(4)$ **98.** $f(3)$ **99.** $f(2.8)$

100. $f\left(\frac{1}{3}\right)$ **101.** $f(a)$ **102.** $f(2a)$

103. $f(a + 1)$ **104.** $f(a - 1)$

Let $f(x) = -2x^2 + 5x - 1$. Find each output.

105. $f(0)$ **106.** $f(3)$ **107.** $f\left(\frac{1}{2}\right)$

108. $f(4.5)$ **109.** $f(a)$ **110.** $f(3a)$

111. $f(a + 3)$ **112.** $f(a - 3)$

113. Describe the various types of polynomial expansions that have been discussed in this section. Also, explain how to recognize which type of expansion to use for a given polynomial.

6.3 FACTORING QUADRATIC POLYNOMIALS

Objective

▸ Factor quadratic polynomials.

In Example 1 of Section 6.2 (problems 3 and 4), we factored quadratic polynomials of the form $ax^2 + bx + c$, where $c = 0$. To see how to factor quadratic polynomials where $c \neq 0$, consider the following expansion:

$$\overset{\text{last terms}}{}$$

$$(x + 2)(x + 3) = x^2 + 2x + 3x + 6$$
$$= x^2 + 5x + 6$$

For $x^2 + 5x + 6$, notice that the coefficient of x is 5, which is the sum of the *last terms*, 2 and 3, of $(x + 2)(x + 3)$. Also, the constant term is 6, which is the product of 2 and 3.

Now, we expand the general polynomial $(x + p)(x + q)$:

$$(x + p)(x + q) = x^2 + px + qx + pq$$
$$= x^2 + (p + q)x + pq$$

In the result, we see that the coefficient of x is the sum of the last terms p and q and the constant term is the product of the last terms p and q.

Working backwards, we can factor some quadratic polynomials of the form $ax^2 + bx + c$ with $a = 1$, such as $x^2 + 10x + 21$:

$$x^2 + 10x + 21 = (x + p)(x + q)$$

We know that p and q are two integers whose sum is 10 and whose product is 21. For each pair of integers whose product is 21, we find the sum:

Note

The integers are the numbers $0, 1, -1, 2, -2, 3, -3, \ldots$

Product $= 21$	**Sum $= 10$?**	
$1(21) = 21$	$1 + 21 = 22$	
$3(7) = 21$	$3 + 7 = 10$	Success!
$-1(-21) = 21$	$-1 + (-21) = -22$	
$-3(-7) = 21$	$-3 + (-7) = -10$	

Since $3(7) = 21$ and $3 + 7 = 10$, we conclude that p and q are 3 and 7:

$$x^2 + 10x + 21 = (x + 3)(x + 7)$$

Note

Since $cd = dc$, we could also write $x^2 + 10x + 21 = (x + 7)(x + 3)$.

Factoring $ax^2 + bx + c$, with $a = 1$

If $x^2 + bx + c = (x + p)(x + q)$, then $p + q = b$ and $pq = c$.

In words: If $x^2 + bx + c$ can be factored as a product of two binomials, then there are two integers whose sum is b and whose product is c. The two integers are the last terms of the factors $(x + p)$ and $(x + q)$.

Example 1 Factoring $ax^2 + bx + c$, with $a = 1$

Factor.

1. $x^2 + 11x + 24$ **2.** $x^2 - x - 12$

Solution

1. To factor $x^2 + 11x + 24$, we find two integers whose sum is 11 and whose product is 24. To begin, we list pairs of integers whose product is 24 and compute each sum:

Product $= 24$	**Sum $= 11$?**	
$1(24) = 24$	$1 + 24 = 25$	
$2(12) = 24$	$2 + 12 = 14$	
$3(8) = 24$	$3 + 8 = 11$	Success!
$4(6) = 24$	$4 + 6 = 10$	

Note

We can rule out negative last terms in the factors since the middle term $11x$ of $x^2 + 11x + 24$ has positive coefficient 11.

Since $3(8) = 24$ and $3 + 8 = 11$, we conclude that the last terms of the factors are 3 and 8:

$$x^2 + 11x + 24 = (x + 3)(x + 8)$$

We check the result by expanding $(x + 3)(x + 8)$:

$$(x + 3)(x + 8) = x^2 + 3x + 8x + 24 = x^2 + 11x + 24$$

2. To factor $x^2 - x - 12$, we find two integers whose sum is -1 and whose product is -12. For each pair of integers whose product is -12, we find their sum:

Product $= -12$	**Sum $= -1$?**	
$1(-12) = -12$	$1 + (-12) = -11$	
$2(-6) = -12$	$2 + (-6) = -4$	
$3(-4) = -12$	$3 + (-4) = -1$	Success!
$4(-3) = -12$	$4 + (-3) = 1$	
$6(-2) = -12$	$6 + (-2) = 4$	
$12(-1) = -12$	$12 + (-1) = 11$	

Since $3(-4) = -12$ and $3 + (-4) = -1$, we know that the last terms of the factors are 3 and -4:

$$x^2 - x - 12 = (x + 3)(x + (-4)) = (x + 3)(x - 4)$$

We check our result by comparing graphing calculator tables for $y = x^2 - x - 12$ and $y = (x + 3)(x - 4)$. See Fig. 31.

Figure 31 Comparing tables for $y = x^2 - x - 12$ and $y = (x + 3)(x - 4)$

Not all polynomials can be factored. If a polynomial cannot be factored, it is **prime**. In problem 1 of Example 2, we show that the polynomial $x^2 + x + 6$ is prime.

Example 2   Factoring Trinomials

Factor.

1. $x^2 + x + 6$ **2.** $-9x^2 + 45x - 36$

Solution

1. To factor $x^2 + x + 6$, we want to find two integers with product 6 and sum 1.

Product = 6	*Sum = 1?*
$1(6) = 6$	$1 + 6 = 7$
$2(3) = 6$	$2 + 3 = 5$
$-1(-6) = 6$	$-1 + (-6) = -7$
$-2(-3) = 6$	$-2 + (-3) = -5$

Since there are no two integers with product 6 and sum 1, we conclude that the polynomial $x^2 + x + 6$ is prime.

2. For $-9x^2 + 45x - 36$, the GCF is 9. Since the first term $-9x^2$ has a negative coefficient, we factor out -9:

$$-9x^2 + 45x - 36 = -9(x^2 - 5x + 4)$$

To factor completely, we find two integers whose product is 4 and whose sum is -5. The integers are -1 and -4 since $-1(-4) = 4$ and $-1 + (-4) = -5$. Therefore, we have

$$-9x^2 + 45x - 36 = -9(x^2 - 5x + 4) = -9(x - 1)(x - 4)$$

We use graphing calculator tables to verify our work (see Fig. 32).

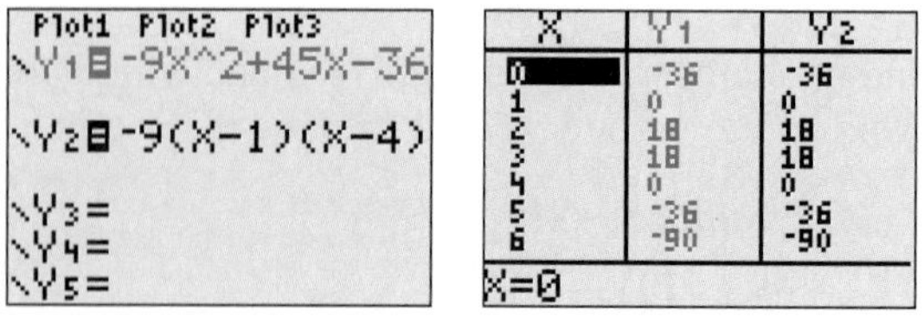

Figure 32 Verifying that $-9x^2 + 45x - 36 = -9(x - 1)(x - 4)$

Since the graphing calculator tables in Fig. 32 are the same, this suggests that our result $-9(x - 1)(x - 4)$ is equivalent to the original polynomial $-9x^2 + 45x - 36$. It does *not* suggest that we have *completely* factored the polynomial. For example, the polynomials $-9(x^2 - 5x + 4)$ and $-9x^2 + 45x - 36$ have the same tables, but $-9(x^2 - 5x + 4)$ is not completely factored.

If the terms of a polynomial have a GCF not equal to one, we factor out the GCF (or the opposite of the GCF) first and then factor the result further, if possible.

Now, consider the expansion of $(2x + 1)(x + 3)$:

$$(2x + 1)(x + 3) = 2x^2 + 6x + x + 3$$
$$= 2x^2 + 7x + 3$$

Note that for $(2x + 1)(x + 3)$, the sum of the last terms, 1 and 3, does *not* equal the coefficient of x, 7, in $2x^2 + 7x + 3$:

$$1 + 3 \neq 7$$

Beware! If $ax^2 + bx + c$ can be factored as a product of two binomials and $a \neq 1$, it is *not* true that b is equal to the sum of the last terms of the factored polynomials. We discuss how to factor such polynomials in Example 3.

Example 3   Factoring $ax^2 + bx + c$, with $a \neq 1$

Factor each polynomial.

1. $3x^2 + 14x + 8$ **2.** $6x^2 - 39x + 45$

Solution

1. If we can factor $3x^2 + 14x + 8$, the result is of the form

$$(3x + ?)(x + ?)$$

The product of the last terms is 8, so the last terms are 1 and 8, or 2 and 4. We decide by expanding:

$$(3x + 1)(x + 8) = 3x^2 + 25x + 8$$
$$(3x + 8)(x + 1) = 3x^2 + 11x + 8$$
$$(3x + 2)(x + 4) = 3x^2 + 14x + 8 \qquad \text{Success!}$$
$$(3x + 4)(x + 2) = 3x^2 + 10x + 8$$

Note

We can rule out negative last terms in the factors since the middle term $14x$ of $3x^2 + 14x + 8$ has positive coefficient 14.

So, $3x^2 + 14x + 8 = (3x + 2)(x + 4)$.

2. To factor $6x^2 - 39x + 45$, we first factor out the GCF 3:

$$6x^2 - 39x + 45 = 3(2x^2 - 13x + 15)$$

If we can factor further, the desired result is of the form

$$3(2x - ?)(x - ?)$$

The product of the last terms is 15, so the last terms are -1 and -15, or -3 and -5. We decide by expanding:

$$(2x - 1)(x - 15) = 2x^2 - 31x + 15$$
$$(2x - 15)(x - 1) = 2x^2 - 17x + 15$$
$$(2x - 3)(x - 5) = 2x^2 - 13x + 15 \qquad \text{Success!}$$
$$(2x - 5)(x - 3) = 2x^2 - 11x + 15$$

Note

We can rule out positive last terms in the factors since the middle term $-13x$ of $2x^2 - 13x + 15$ has negative coefficient -13.

Therefore, $6x^2 - 39x + 45 = 3(2x^2 - 13x + 15) = 3(2x - 3)(x - 5)$.

We use graphing calculator tables to verify our work (see Fig. 33).

Figure 33 Verifying that $6x^2 - 39x + 45 = 3(2x - 3)(x - 5)$

Factoring $ax^2 + bx + c$, with $a \neq 1$, by Trial and Error

If $ax^2 + bx + c = (mx + p)(nx + q)$, then $mn = a$ and $pq = c$.

In words: If $ax^2 + bx + c$ can be factored as a product of two binomials, then the product of the coefficients of the first terms is equal to a and the product of the last terms is equal to c. We expand the possible products to find the factored polynomial.

Example 4 Factoring $ax^2 + bx + c$, with $a \neq 1$

Factor $6x^2 + 83x - 14$.

Solution
If we can factor $6x^2 + 83x - 14$, the result is of one of the four forms:

$$(6x + ?)(x - ?) \qquad (6x - ?)(x + ?) \qquad (3x + ?)(2x - ?) \qquad (3x - ?)(2x + ?)$$

The product of the last terms is -14, so the last terms are 1 and -14, -1 and 14, 2 and -7, or -2 and 7, where each pair can be written in either order.

Rather than expand all 16 possible products, we note that the coefficient 83 of the middle term of $6x^2 + 83x - 14$ is large and so it is reasonable to first expand products where the first terms are $6x$ and x and the last terms are 1 and -14, or -1 and 14:

Note

We can rule out $(6x - 14)(x + 1)$ and $(6x + 14)(x - 1)$ because both $6x - 14$ and $6x + 14$ are divisible by 2 and $6x^2 + 83x - 14$ is not.

$$(6x + 1)(x - 14) = 6x^2 - 83x - 14$$

$$(6x - 1)(x + 14) = 6x^2 + 83x - 14 \qquad \text{Success!}$$

So, $6x^2 + 83x - 14 = (6x - 1)(x + 14)$. ∎

Sometimes we can factor a quadratic polynomial of the form $ax^2 + bx + c$, where $b = 0$. In some of these cases, it will be helpful to use the *difference of two squares* formula:

$$x^2 - k^2 = (x + k)(x - k)$$

Beware! A polynomial of the form $x^2 + k^2$, $k \neq 0$, is prime. Some students think that this polynomial can be factored as $(x + k)^2$, but recall that $(x + k)^2$ equals $x^2 + 2kx + k^2$, not $x^2 + k^2$.

Example 5 Using the Difference of Two Squares Formula

1. Factor $x^2 - 49$.   **2.** Factor $36x^2 - 25$.

3. Factor $12 - 3x^2$.   **4.** Factor $4x^2 + 36$.

Solution

1. $x^2 - 49 = x^2 - 7^2 = (x + 7)(x - 7)$

2. $36x^2 - 25 = (6x)^2 - 5^2 = (6x + 5)(6x - 5)$

3. $12 - 3x^2 = -3(x^2 - 4) = -3(x + 2)(x - 2)$

4. $4x^2 + 36 = 4(x^2 + 9)$. We have completely factored $4x^2 + 36$ as $4(x^2 + 9)$ since a polynomial of the form $x^2 + k^2$, $k \neq 0$, is prime. (Note that $4x^2 + 36$ *is* a sum of squares and it *is not* prime, but $x^2 + 9$ is prime.) ∎

EXPLORATION
Factors of an expression and x-intercepts of a function

In this exploration you will explore a connection between factors of a polynomial of the form $x^2 + bx + c$ and x-intercepts of the polynomial function $f(x) = x^2 + bx + c$.

1. Factor $x^2 + x - 2$.

2. Use ZDecimal to draw a graph of the function $f(x) = x^2 + x - 2$. What are the x-intercepts?

3. What connection do you notice between your result in problem 1 and the x-intercepts of f? Explain why this connection makes sense.

4. The graph of a function $g(x) = x^2 + bx + c$ is sketched in the indicated figure. Use the graph to help you factor $x^2 + bx + c$. Then, find the values of b and c.

   a. Fig. 34   b. Fig. 35

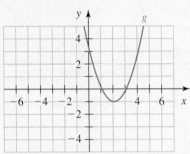

Figure 34 Graph of $g(x) = x^2 + bx + c$

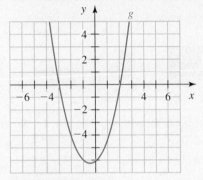

Figure 35 Graph of $g(x) = x^2 + bx + c$

Tips on Succeeding in This Course

While there are many things you can do to enhance your learning, there is no substitute for solving problems. Your mathematical ability will respond to solving problems in much the same way as your muscles respond to weight lifting. Muscles greatly increase in strength when you work out intensely, frequently, and consistently.

Just as with building muscles, to learn math, you must be an *active* participant. No amount of watching weight lifters lift, reading about weight lifting techniques, or conditioning yourself psychologically can replace working out by lifting weights. The same is true with learning math. No amount of watching your instructor do problems, reading your text, or listening to a tutor can replace "working out" by solving problems.

Key Points
OF THIS SECTION

Factoring $x^2 + bx + c$	If $x^2 + bx + c$ can be factored as a product of two binomials, we factor it by first finding two integers, p and q, whose product is c and whose sum is b. Then we write: $$x^2 + bx + c = (x + p)(x + q)$$
Factoring out the GCF	If the GCF of the terms of a polynomial is not 1, the first step in factoring completely is to factor out the GCF (or the opposite of the GCF).
Factoring $ax^2 + bx + c$, with $a \neq 1$	If the GCF of the terms of $ax^2 + bx + c$ is 1 and $a \neq 1$, then samples of factoring such a polynomial are shown in problem 1 of Example 3 and Example 4.
Difference of two squares	We factor the difference of two squares using: $x^2 - k^2 = (x + k)(x - k)$.
The polynomial $x^2 + k^2$	The polynomial $x^2 + k^2$, $k \neq 0$, is prime.

HOMEWORK 6.3

 SSM Tutor Center   **MathPro 4** MathPro 4   **MathPro 5** MathPro 5   CDs/Videos   www

Factor. Verify that you have factored correctly by expanding your factored polynomial.

1. $x^2 + 7x + 12$

2. $x^2 + 17x + 30$

3. $x^2 - 21x + 20$

4. $x^2 + 2x - 15$

5. $x^2 + 2x - 10$

6. $x^2 + 4x + 8$

7. $3x^2 - 3x - 18$

8. $2x^2 + 12x + 18$

9. $-5x^2 + 20x + 60$

10. $2x^2 - 20x + 50$

11. $6x^2 + 36x + 54$

12. $2x^2 - 4x + 2$

Factor. Verify your result using graphing calculator tables or graphs.

13. $x^2 - 16$

14. $x^2 - 81$

15. $x^2 - 1$

16. $x^2 + 49$

17. $25 - x^2$

18. $4 - x^2$

19. $100x^2 - 1$

20. $5x^2 - 45$

21. $-x^2 + 49$

22. $-4x^2 + 25$

23. $x^2 + 4$

24. $36x^2 - 9$

25. $64x^2 - 49$

26. $2x^2 + 18$

27. $x^2 - \dfrac{1}{9}$

28. $x^2 - \dfrac{1}{16}$

Factor. Verify that you have factored correctly by expanding your factored polynomial.

29. $3x^2 + 10x + 8$

30. $6x^2 + 5x + 1$

31. $2x^2 - 13x + 15$

32. $2x^2 - 2x - 3$

33. $9x^2 + 9x + 2$

34. $3x^2 - 6x + 10$

35. $6x^2 + 7x - 12$

36. $-10x^2 - 32x - 24$

37. $9x^2 - 15x + 4$

38. $-12x^2 + 3x + 9$

39. $20x^2 + 30x - 140$

40. $6x^2 + 34x - 12$

Factor. Verify your result using graphing calculator tables or graphs.

41. $x^2 + 11x + 30$

42. $x^2 - 25$

43. $x^2 - 3x + 2$

44. $4x^2 - 64$

45. $15x^2 - 27x$

46. $x^2 + 4x + 8$

47. $10x^2 + 23x + 12$

48. $-4x^2 + 24x - 32$

49. $x^2 - 7x - 24$

50. $25x^2 - 1$

51. $x^2 - 15x - 100$

52. $x^2 + 5x - 24$

53. $-3x^2 - 75$

54. $20x - 15x^2$

55. $x^2 + 2x - 48$

56. $10x^2 - 90$

57. $12x^2 - 7x - 5$

58. $x^2 - 3x - 18$

59. $7x^2 + 14x + 21$

60. $x^2 + 25$

61. $81x^2 - 16$

62. $-9 + 4x^2$

63. $6x^2 - 3x - 4$

64. $3x^2 - 7x$

65. $8x^2 - 2x - 15$

66. $50 - 72x^2$

67. $x^2 - 100$

68. $4x^2 + 4x - 15$

69. $x^2 - 14x + 49$

70. $x^2 - \dfrac{1}{25}$

71. $9x^2 - 18x + 8$

72. $8x^2 + 12x + 5$

73. $4x^2 + 79x - 20$

74. $36 - 25x^2$

75. $2x^2 - 22x + 48$

76. $6x^2 - 24$

77. $10x^2 + x - 24$

78. $16x^2 - 12x$

79. $36x^2 + 48x + 16$

80. $12x^2 - 36x + 32$

81. A student tries to factor the polynomial $4x^2 + 8x + 3$:

$$4x^2 + 8x + 3 = 4x(x + 2) + 3$$

Did the student factor the polynomial correctly? Explain.

82. A student tries to factor the polynomial $3x^2 - 9x - 30$:

$$3x^2 - 9x - 30 = (3x - 15)(x + 2) = 3(x - 5)(x + 2)$$

Did the student factor the polynomial correctly? Explain.

83. A student tries to factor the polynomial $4x^2 + 100$:

$$4x^2 + 100 = 4(x^2 + 25) = 4(x + 5)(x + 5) = 4(x + 5)^2$$

Did the student factor the polynomial correctly? Explain.

84. Which of the following equations describe the same function?

Use graphing calculator tables or graphs to verify your work.

$$y = 2(x - 2)(x - 6) \qquad y = 2(x^2 - 8x + 12)$$
$$y = 2x^2 - 16x + 24 \qquad y = (x - 2)(2x - 12)$$
$$y = (x - 2)(2x - 6) \qquad y = (2x - 2)(x - 6)$$
$$y = 2(x - 4)^2 - 8 \qquad y = 2x^2 - 40$$
$$y = (2x - 4)(x - 6)$$

85. Describe the various factoring techniques addressed in this section and Section 6.2. Also, give an example to illustrate each technique. Finally, explain how to recognize polynomials to which each technique applies.

6.4 FACTORING POLYNOMIALS

The great thing about mathematics is that, unlike liberal arts, it is objective rather than subjective. It is great to be able to work on a problem and know that there is a correct and an incorrect answer. —*Colleen S., student*

Objectives

▸ Factor polynomials by factoring out the GCF.

▸ Factor polynomials by grouping.

▸ Know a factoring strategy.

So far, we have factored linear and quadratic polynomials. In this section we will discuss how to factor polynomials with degree at least two. Recall that if the terms of an expression have a GCF not equal to one, we factor out the GCF first and then factor the result further, if possible.

Example 1 Factoring out the GCF

Factor.

1. $4x^3 + 12x^2 - 40x$

2. $18x^5 - 32x^3$

Solution

1. To factor $4x^3 + 12x^2 - 40x$, we first factor out the GCF $4x$:

$$4x^3 + 12x^2 - 40x = 4x(x^2 + 3x - 10) \qquad \text{Factor out } 4x.$$
$$= 4x(x + 5)(x - 2) \qquad \text{Find two integers whose product is } -10 \text{ and whose sum is 3.}$$

2. To factor $18x^5 - 32x^3$, we first factor out the GCF $2x^3$:

$$18x^5 - 32x^3 = 2x^3(9x^2 - 16) \qquad \text{Factor out } 2x^3.$$
$$= 2x^3(3x + 4)(3x - 4) \qquad a^2 - b^2 = (a + b)(a - b)$$

We use graphing calculator tables to verify our work (see Fig. 36).

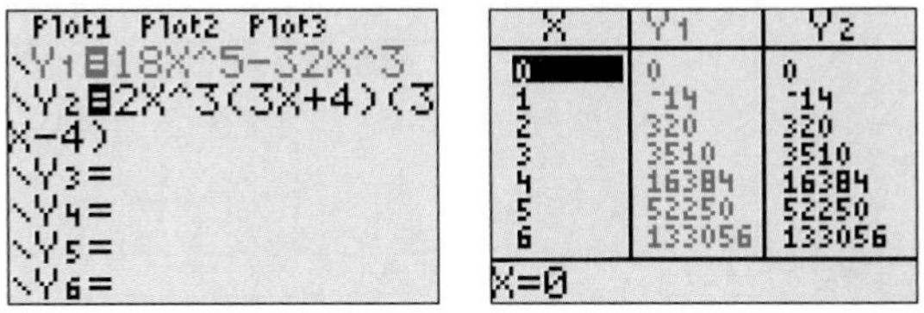

Figure 36 Verifying that
$18x^5 - 32x^3 = 2x^3(3x + 4)(3x - 4)$

So far, we have factored out a GCF when the GCF is a monomial. For example, here we factor out the monomial x:

$$x^2(x) + 5(x) = (x^2 + 5)(x)$$

We can also factor out a GCF when it is a binomial. For example, here we factor out the binomial $x - 2$:

$$x^2(x - 2) + 5(x - 2) = (x^2 + 5)(x - 2)$$

We can factor some polynomials with four terms by first factoring the first two terms and the last two terms. For example,

$$x^3 - 2x^2 + 5x - 10 = x^2(x - 2) + 5(x - 2)$$
$$= (x^2 + 5)(x - 2) \qquad \text{Factor out } (x - 2).$$

We call this method *factoring by grouping*. Note that $x^2(x - 2) + 5(x - 2)$ is *not* factored; it is a sum, not a product. It is only when we factor out the binomial $x - 2$ to get $(x^2 + 5)(x - 2)$ that we have a factored expression.

Example 2 Factoring by Grouping

Factor $3x^3 - 12x^2 - 2x + 8$.

Solution

We begin by factoring the first two terms and the last two terms:

$$3x^3 - 12x^2 - 2x + 8 = 3x^2(x - 4) - 2(x - 4)$$
$$= (3x^2 - 2)(x - 4) \qquad \text{Factor out } (x - 4). \qquad \blacksquare$$

Factoring by Grouping

If a polynomial with four terms can be factored by grouping, we factor it by:

1. Factoring the first two terms and the last two terms.
2. Factoring out the binomial GCF.

Example 3 Factoring by Grouping

Factor $x^3 + 3x^2 - 25x - 75$.

Solution

$$x^3 + 3x^2 - 25x - 75 = x^2(x + 3) - 25(x + 3) \Bigg\}$$
$$= (x^2 - 25)(x + 3) \qquad \Bigg\} \quad \text{Factor by grouping.}$$
$$= (x + 5)(x - 5)(x + 3) \qquad a^2 - b^2 = (a + b)(a - b)$$

We use graphing calculator tables to verify our work (see Fig. 37).

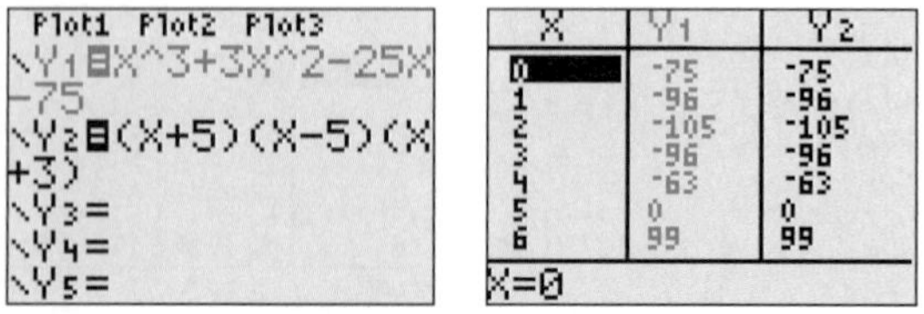

Plot1 Plot2 Plot3
\Y1■X^3+3X^2-25X
-75
\Y2■(X+5)(X-5)(X
+3)
\Y3=
\Y4=
\Y5=

X	Y1	Y2
0	-75	-75
1	-96	-96
2	-105	-105
3	-96	-96
4	-63	-63
5	0	0
6	99	99

X=0

Figure 37 Verifying that $x^3 + 3x^2 - 25x - 75 = (x + 5)(x - 5)(x + 3)$

$\blacksquare$

In this section and Section 6.3, we have discussed several factoring methods. It is important to develop a strategy to help us be thorough, yet efficient, in factoring polynomials.

Five-Step Factoring Strategy

Here is a five-step method to factor many polynomials. Steps 2–4 can be applied to the entire polynomial or to a factor of the polynomial.

1. If the GCF is not 1, factor it out.
2. For a binomial, try using the difference of two squares formula.
3. For a trinomial $ax^2 + bx + c$:
    a. If $a = 1$, then try to find two integers whose product is c and whose sum is b.
    b. If $a \neq 1$, then try to factor using the trial and error method.
4. For an expression with four terms, try factoring by grouping.
5. Continue applying steps 2–4 until the polynomial is completely factored.

Example 4   Factoring Polynomials

Factor.

1. $-12x^3 + 75x$
2. $-x^2 + 9x - 20$
3. $2x^3 - 30x^2 + 12x^4$
4. $12x^3 + 16x^2 - 27x - 36$

Solution

1. To factor $-12x^3 + 75x$, we first factor out $-3x$:

$$-12x^3 + 75x = -3x(4x^2 - 25) \qquad \text{Factor out } -3x.$$
$$= -3x(2x + 5)(2x - 5) \quad a^2 - b^2 = (a + b)(a - b)$$

2. For $-x^2 + 9x - 20$, the GCF is 1. Since the first term $-x^2$ has a negative coefficient, we first factor out -1:

$$-x^2 + 9x - 20 = -(x^2 - 9x + 20) \quad \text{Factor out } -1.$$
$$= -(x - 5)(x - 4) \quad \begin{array}{l}\text{Find two integers whose product}\\ \text{is 20 and whose sum is } -9.\end{array}$$

3. To factor $2x^3 - 30x^2 + 12x^4$, we first write the polynomial in *descending powers of x*:

$$2x^3 - 30x^2 + 12x^4 = 12x^4 + 2x^3 - 30x^2 \quad \text{Rearrange the terms.}$$
$$= 2x^2(6x^2 + x - 15) \quad \text{Factor out } 2x^2.$$
$$= 2x^2(2x - 3)(3x + 5) \quad \text{Trial and error method}$$

4. Since $12x^3 + 16x^2 - 27x - 36$ has four terms, we try to factor by grouping:

$$12x^3 + 16x^2 - 27x - 36 = 4x^2(3x + 4) - 9(3x + 4) \quad \Big\} \quad \text{Factor by grouping.}$$
$$= (4x^2 - 9)(3x + 4)$$
$$= (2x + 3)(2x - 3)(3x + 4) \quad a^2 - b^2 = (a + b)(a - b)$$

■

EXPLORATION
Factoring polynomials

1.   A student tries to factor $x^2 + 9$:

$$x^2 + 9 = (x + 3)(x + 3)$$

Expand $(x + 3)(x + 3)$ to show that the work is incorrect. Then correctly factor $x^2 + 9$.

2.   A student tries to factor $4x^2 - 16$:

$$4x^2 - 16 = (2x + 4)(2x - 4)$$

Explain why the work is not correct. Then correctly factor the polynomial.

3.   A student tries to factor $x^3 - 3x^2 + 2x - 6$:

$$x^3 - 3x^2 + 2x - 6 = x^2(x - 3) + 2(x - 3)$$

Explain why the student has not succeeded in factoring the given expression. Then correctly factor it.

4.   A student tries to factor $2x^2 - x - 6$. Since the product of -3 and 2 is -6 and the sum of -3 and 2 is -1, the student does the following work:

$$2x^2 - x - 6 = (2x - 3)(x + 2)$$

Expand $(2x - 3)(x + 2)$ to show that the work is incorrect. Explain what is wrong with the student's reasoning. Then correctly factor the polynomial.

EXPLORATION
Looking ahead: Zero factor property

1.   What can you say about A or B if $AB = 0$?
2.   What can you say about A or B if $A(B - 1) = 0$?
3.   What can you say about x if $x(x - 1) = 0$?
4.   Solve the equation $x^2 - x = 0$. [**Hint**: Does this have something to do with problem 3?]
5.   Solve $2x^2 - 6x = 0$.
6.   Solve $x^2 - 8x + 15 = 0$.

Tips on Succeeding in This Course

To improve your effectiveness at studying for your courses, consider taking stock of when and where you can study at your best. Tracy, a student who lives in a sometimes distracting household, completes her assignments at the campus library just after attending her classes. Gerome, a morning person, gets up early so that he can study before classes. Being consistent in the time and location for studying can help, too. Studies have shown that after repeating a daily activity for about 21 days, the activity becomes habit. So, even if it takes some willpower to shuffle your schedule so that you can study at your prime time and location, things will start to feel comfortable and familiar in three weeks or maybe less.

Key Points

OF THIS SECTION

If a polynomial with four terms can be factored by grouping, we factor it by:

Factor by grouping

1. Factoring the first two terms and the last two terms.

2. Factoring out the binomial GCF.

Here is a five-step method to factor many polynomials. Steps 2–4 can be applied to the entire polynomial or to a factor of the polynomial.

Five-step method for factoring many polynomials

1. If the GCF is not 1, factor it out.

2. For a binomial, try using the difference of two squares formula.

3. For a trinomial $ax^2 + bx + c$:
   a. If $a = 1$, then try to find two integers whose product is c and whose sum is b.
   b. If $a \neq 1$, then try to factor using the trial and error method.

4. For an expression with four terms, try factoring by grouping.

5. Continue applying steps 2–4 until the polynomial is completely factored.

HOMEWORK 6.4

 SSM Tutor Center   MathPro 4 MathPro 4   MathPro 5 MathPro 5   CDs/Videos   www www

For each polynomial, first factor out the GCF. Then continue factoring until your result is completely factored.

1. $3x^3 - 75x$

2. $8x^3 - 32x$

3. $63x^3 - 7x$

4. $32x^3 - 98x$

5. $8x^5 - 72x^3$

6. $9x^4 - 36x^2$

7. $-75x^4 + 27x^2$

8. $-28x^5 + 7x^3$

9. $2x^3 + 8x^2 + 6x$

10. $5x^3 + 30x^2 + 40x$

11. $4x^4 + 12x^3 - 40x^2$

12. $2x^4 - 10x^3 + 12x^2$

13. $-2x^3 + 20x^2 - 50x$

14. $-6x^3 - 12x^2 - 6x$

15. $6x^5 + 33x^4 + 45x^3$

16. $12x^4 + 20x^3 + 9x^2$

17. $80x^3 + 140x^2 - 150x$

18. $60x^3 - 55x^2 + 10x$

19. $-30x^6 - 5x^5 + 5x^4$

20. $-24x^4 + 54x^3 - 27x^2$

Factor each polynomial by grouping. Then continue factoring until your result is completely factored.

21. $x^3 + 3x^2 + 4x + 12$

22. $x^3 + 2x^2 + 5x + 10$

23. $x^3 - 4x^2 + 3x - 12$

24. $x^3 - x^2 - 2x + 2$

25. $3x^3 - 15x^2 - 4x + 20$

26. $6x^3 + 12x^2 - 3x - 6$

27. $2x^3 + x^2 - 32x - 16$

28. $3x^3 - 2x^2 - 12x + 8$

29. $x^3 - 5x^2 - 4x + 20$

30. $x^3 + 2x^2 - 9x - 18$

31. $4x^3 - 12x^2 - 9x + 27$

32. $9x^3 + 18x^2 - 16x - 32$

33. $12x^3 + 4x^2 - 3x - 1$

34. $18x^3 + 27x^2 - 2x - 3$

Factor.

35. $2x^3 - 50x$

36. $6x^2 - x - 15$

37. $x^3 + 4x^2 - 7x - 28$

38. $x^2 - 10x + 25$

39. $x^2 - 81$

40. $x^3 + 2x^2 + 3x + 6$

41. $8x^2 - 10x + 3$

42. $49 - x^2$

43. $9x^2 - 15x$

44. $x^2 - 14x + 24$

45. $3x^4 + 18x^3 + 24x^2$

46. $8 - 2x^2 + x^3 - 4x$

47. $14x + 49 + x^2$

48. $3x^5 - 75x^3$

49. $-4x^4 - 20x^3 - 24x^2$

50. $9x^3 + 6x^2 + x$

51. $x^2 + x$

52. $x^2 + 15x + 56$

53. $16x - x^3$

54. $-x^2 - 10x - 24$

55. $4x^3 + 20x^2 - x - 5$

56. $45x^3 + 12x^2 - 12x$

57. $x^3 - 100x$

58. $x^3 - x^2 - x + 1$

59. $15 - 6x^2 + x^3 - 5x$

60. $5x - 25x^3$

61. $5x^4 - 35x^3 + 60x^2$

62. $7x^3 - 28x^2$

63. $26x^2 + 12x + 12x^3$

64. $4x^5 - 12x^4 - 40x^3$

65. $4x^2 - 16x + 16$

66. $36 + x^2 - 13x$

67. $-x^2 - 2x - 1$

68. $x^3 - 5x^2 - 9x + 45$

69. $8x^4 - 6x^5$

70. $3x^3 - 3x$

71. $x^3 - 4x^2 - 9x + 9$

72. $12x^2 - 22x^3 + 8x^4$

73. $20x^5 + 20x^4 - 15x^3$

74. $4x^3 - 8x^2 - 96x$

75. A student tries to factor $x^3 + 5x^2 - 3x - 15$:
$$x^3 + 5x^2 - 3x - 15 = x^2(x + 5) - 3(x + 5)$$
What would you tell the student?

76. Two students try to factor $6x^3 - 2x^2 - 15x + 5$. Did both, one, or neither student factor the expression correctly?

Student 1's work
$$6x^3 - 2x^2 - 15x + 5 = 2x^2(3x - 1) - 5(3x - 1)$$
$$= (2x^2 - 5)(3x - 1)$$

Student 2's work
$$6x^3 - 2x^2 - 15x + 5 = 6x^3 - 15x - 2x^2 + 5$$
$$= 3x(2x^2 - 5) - (2x^2 - 5)$$
$$= (3x - 1)(2x^2 - 5)$$

77. A student tries to factor $16x^2 - 36$:

$$16x^2 - 36 = (4x + 6)(4x - 6)$$

The student then checks that the input-output values for $y = 16x^2 - 36$ and $y = (4x - 6)(4x + 6)$ are the same. Explain why the student's work is incorrect even though the input-output values for the two functions are the same.

78. Describe in your own words how to factor a polynomial by grouping.

79. Describe in your own words a strategy for factoring polynomials.

6.5 USING FACTORING TO SOLVE POLYNOMIAL EQUATIONS

Objectives

▶ Know the *zero factor property*.

▶ Use factoring to solve polynomial equations.

▶ Use graphing calculator tables and graphs to verify work in solving polynomial equations.

▶ Find *x*-intercepts of polynomial functions.

We begin this section by using factoring to solve *quadratic equations*. A **quadratic equation in one variable** is an equation that can be put into the form

$$ax^2 + bx + c = 0$$

where a, b, and c are constants and $a \neq 0$. The connection between solving a quadratic equation and factoring an expression lies in the zero factor property.

Zero Factor Property

Let A and B be real numbers.

$$\text{If } AB = 0, \text{ then } A = 0 \text{ or } B = 0$$

In words: If the product of two numbers is zero, then at least one of the numbers is zero.

Example 1 Solving Quadratic Equations

Solve each equation.

1. $x(x + 1) = 0$

2. $x^2 - 2x - 8 = 0$

Solution

1. $x(x + 1) = 0$

$x = 0$	or	$x + 1 = 0$	Zero factor property
$x = 0$	or	$x = -1$	Add -1 on both sides of $x + 1 = 0$.

We check that both 0 and -1 satisfy the original equation:

Check $x = 0$	***Check*** $x = -1$
$x(x + 1) = 0$	$x(x + 1) = 0$
$0(0 + 1) \overset{?}{=} 0$	$-1(-1 + 1) \overset{?}{=} 0$
$0(1) \overset{?}{=} 0$	$-1(0) \overset{?}{=} 0$
$0 \overset{?}{=} 0$	$0 \overset{?}{=} 0$
true	true

So, for $x(x + 1) = 0$, the solutions are 0 and -1.

2. $x^2 - 2x - 8 = 0$

$(x + 2)(x - 4) = 0$ Factor the left side.

$x + 2 = 0$ or $x - 4 = 0$ Zero factor property

$x = -2$ or $x = 4$

We check that both -2 and 4 satisfy the original equation:

Check $x = -2$	**Check $x = 4$**
$x^2 - 2x - 8 = 0$	$x^2 - 2x - 8 = 0$
$(-2)^2 - 2(-2) - 8 \overset{?}{=} 0$	$4^2 - 2(4) - 8 \overset{?}{=} 0$
$4 + 4 - 8 \overset{?}{=} 0$	$16 - 8 - 8 \overset{?}{=} 0$
$0 \overset{?}{=} 0$	$0 \overset{?}{=} 0$
true	true

So, the solutions are -2 and 4.   ■

The key step in solving a quadratic equation of the form $ax^2 + bx + c = 0$ is to factor the left side of the equation so that we can apply the zero factor property.

Example 2   Finding *x*-intercepts of a Quadratic Function

Find the x-intercepts of the function $f(x) = x^2 - 7x + 10$.

Solution

To find the x-intercepts, we substitute 0 for $f(x)$ and solve for x:

$0 = x^2 - 7x + 10$ Substitute 0 for $f(x)$.

$0 = (x - 5)(x - 2)$ Factor the right side.

$x - 5 = 0$ or $x - 2 = 0$ Zero factor property

$x = 5$ or $x = 2$

So, the x-intercepts are $(2, 0)$ and $(5, 0)$. We verify our work using a graphing calculator (see Fig. 38).

Figure 38 The x-intercepts are $(2, 0)$ and $(5, 0)$

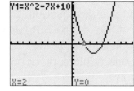
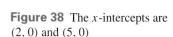
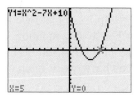

       ■

In Example 2 we found the x-intercepts of $f(x) = x^2 - 7x + 10$ by solving the equation $x^2 - 7x + 10 = 0$. In general, we find the x-intercepts of $f(x) = ax^2 + bx + c$ by solving the equation $ax^2 + bx + c = 0$.

Connection between *x*-intercepts and Solutions

Let f be a function. If k is a real number solution of the equation $f(x) = 0$, then $(k, 0)$ is an x-intercept of the function f. Also, if $(k, 0)$ is an x-intercept of f, then k is a solution of $f(x) = 0$.

Note

This property holds true for *any* function f, including a quadratic function.

This property suggests that we can verify our solutions to a quadratic equation $ax^2 + bx + c = 0$ by using a graphing calculator to find the x-intercepts of the function $f(x) = ax^2 + bx + c$.

Example 3 Solving Quadratic Equations

Solve each equation.

1. $2x^2 + 13x + 15 = 0$ **2.** $3x^2 - 30x + 75 = 0$

Solution

1.
$$2x^2 + 13x + 15 = 0$$

$(2x + 3)(x + 5) = 0$ Factor the left side.

$2x + 3 = 0$ or $x + 5 = 0$ Zero factor property

$2x = -3$ or $x = -5$

$x = -\dfrac{3}{2}$ or $x = -5$

We verify our work by checking that the x-intercepts of $f(x) = 2x^2 + 13x + 15$ are $\left(-\dfrac{3}{2}, 0\right)$ and $(-5, 0)$. See Fig. 39.

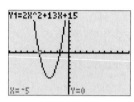
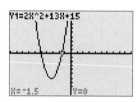

Figure 39 The x-intercepts are $(-5, 0)$ and $\left(-\dfrac{3}{2}, 0\right)$

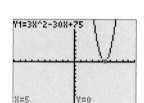

Figure 40 Check that the x-intercept of $f(x) = 3x^2 - 30x + 75$ is $(5, 0)$

2.
$$3x^2 - 30x + 75 = 0$$

$x^2 - 10x + 25 = 0$ Divide both sides by 3.

$(x - 5)(x - 5) = 0$ Factor the left side.

$x - 5 = 0$ Zero factor property

$x = 5$

Since using the zero factor property yields the *one* equation $x - 5 = 0$, there is one solution, 5. We can verify our result by checking that the x-intercept of $f(x) = 3x^2 - 30x + 75$ is $(5, 0)$. See Fig. 40. ∎

In Example 3 we found that the equation $2x^2 + 13x + 15 = 0$ has two solutions, whereas the equation $3x^2 - 30x + 75 = 0$ has one solution. What are the possible numbers of solutions for a quadratic equation? To decide, note that a quadratic function can have two x-intercepts, one x-intercept, or no x-intercepts (see Fig. 41).

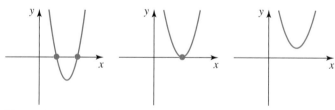

Figure 41 A quadratic function can have two, one, or no x-intercepts

Since the number of real number solutions of an equation $ax^2 + bx + c = 0$ is equal to the number of x-intercepts of the function $f(x) = ax^2 + bx + c$, we conclude that the solution set of such an equation may contain two real numbers, one real number, or no real numbers. We will solve quadratic equations that have no real number solutions in Chapter 7.

If a quadratic equation is not in the form $ax^2 + bx + c = 0$, we put it in this form before trying to factor the nonzero side of the equation.

Example 4   Solving Quadratic Equations

Solve each equation.

1. $2x^2 - 8x = 5x - 20$ **2.** $(x + 2)(x - 4) = 7$

Solution

1. $2x^2 - 8x = 5x - 20$

$2x^2 - 13x + 20 = 0$ Write in $ax^2 + bx + c = 0$ form.

$(2x - 5)(x - 4) = 0$ Factor the left side.

$2x - 5 = 0$ or $x - 4 = 0$

$x = \dfrac{5}{2}$ or $x = 4$

To verify that $\dfrac{5}{2}$ is a solution of $2x^2 - 8x = 5x - 20$, we can use a graphing calculator table to check that for input $\dfrac{5}{2}$ the output for $y = 2x^2 - 8x$ is equal to the output for $y = 5x - 20$. We do similarly for input $x = 4$ (see Fig. 42).

Figure 42 Comparing outputs for $y = 2x^2 - 8x$ and $y = 5x - 20$

2. Although the left side of $(x + 2)(x - 4) = 7$ is factored, the right side is not zero. First, we expand the left side of the equation.

$(x + 2)(x - 4) = 7$

$x^2 - 2x - 8 = 7$ Expand the left side.

$x^2 - 2x - 15 = 0$ Write in $ax^2 + bx + c = 0$ form.

$(x - 5)(x + 3) = 0$ Factor the left side.

$x - 5 = 0$ or $x + 3 = 0$ Zero factor property

$x = 5$ or $x = -3$

Therefore, the solutions are -3 and 5.   ■

Example 5   Finding an Input and an Output
of a Quadratic Function

Let $f(x) = x^2 - 3x - 23$.

1. Find $f(5)$.
2. Find x when $f(x) = 5$.

Solution

1. $f(5) = 5^2 - 3(5) - 23 = -13$
2. We substitute 5 for $f(x)$ in the equation $f(x) = x^2 - 3x - 23$.

$$5 = x^2 - 3x - 23$$
$$0 = x^2 - 3x - 28$$
$$0 = (x - 7)(x + 4)$$
$$x - 7 = 0 \quad \text{or} \quad x + 4 = 0$$
$$x = 7 \quad \text{or} \quad x = -4$$

Next, we verify that $f(7) = 5$ and $f(-4) = 5$.

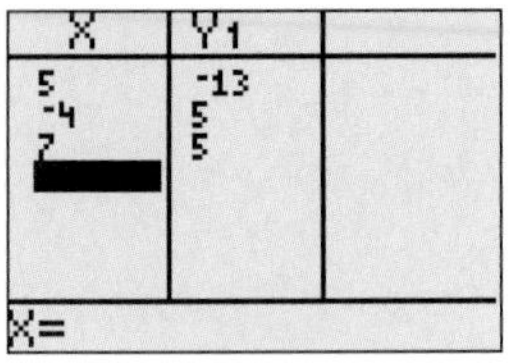

Figure 43 Verify our work

Check that $f(7) = 5$	**Check that $f(-4) = 5$**
$f(x) = x^2 - 3x - 23$	$f(x) = x^2 - 3x - 23$
$f(7) = 7^2 - 3(7) - 23$	$f(-4) = (-4)^2 - 3(-4) - 23$
$= 5$	$= 5$

or, we can verify our work in problems 1 and 2 by using a graphing calculator table (see Fig. 43). ∎

Recall that the inverse of a function g sends each output of g to its corresponding input. From our work in Example 5, we see that the function $f(x) = x^2 - 3x - 23$ is not invertible, since the output 5 corresponds to not one input, but two, -4 and 7. It turns out that no quadratic function is invertible.

So far, we have solved quadratic equations. We will now solve some *cubic equations*. A **cubic equation in one variable** is an equation that can be put into the form

$$ax^3 + bx^2 + cx + d = 0$$

where $a, b, c,$ and d are constants and $a \neq 0$. We can solve some cubic equations by using the zero factor property. Here we restate this property for a product of *three* numbers that is equal to zero.

Zero Factor Property

Let A, B, and C be real numbers.

$$\text{If } ABC = 0, \text{ then } A = 0, B = 0, \text{ or } C = 0.$$

In words, if the product of three numbers is zero, then at least one of the numbers is zero.

Example 6 Solving a Cubic Equation

Solve $2x^3 = 42x + 8x^2$.

Solution

$$2x^3 = 42x + 8x^2$$

	Write the equation in the form $ax^3 + bx^2 + cx + d = 0$.
$2x^3 - 8x^2 - 42x = 0$	
$2x(x^2 - 4x - 21) = 0$	Factor out the GCF on the left side.
$2x(x - 7)(x + 3) = 0$	Completely factor the left side.
$2x = 0 \quad \text{or} \quad x - 7 = 0 \quad \text{or} \quad x + 3 = 0$	Zero factor property
$x = 0 \quad \text{or} \qquad x = 7 \quad \text{or} \qquad x = -3$	

So, the solutions are $-3, 0,$ and 7. ∎

Note that solving a cubic equation is similar to solving a quadratic equation. We now summarize the steps used to solve either type of equation.

Solving Quadratic and Cubic Equations

If an equation can be solved by factoring, we solve it by the following steps:

1. Write the equation so that one side of the equation is equal to zero.
2. Factor the nonzero side of the equation.
3. Apply the zero factor property.
4. Solve the equations that result from applying the zero factor property.

Example 7   Finding *x*-intercepts of a Cubic Function

Find the *x*-intercepts of the function $f(x) = x^3 - 5x^2 - 4x + 20$.

Solution

We substitute 0 for $f(x)$ and solve for x.

$$x^3 - 5x^2 - 4x + 20 = 0$$

$$\left.\begin{array}{r} x^2(x-5) - 4(x-5) = 0 \\ (x^2 - 4)(x-5) = 0 \end{array}\right\} \quad \text{Factor by grouping.}$$

$$(x-2)(x+2)(x-5) = 0 \qquad \text{Difference of two squares formula}$$

$$\begin{array}{ccccc} x-2=0 & \text{or} & x+2=0 & \text{or} & x-5=0 \quad \text{Zero factor property} \\ x=2 & \text{or} & x=-2 & \text{or} & x=5 \end{array}$$

So, the *x*-intercepts are $(-2, 0)$, $(2, 0)$, and $(5, 0)$. We use a graphing calculator to verify our work (see Fig. 44).

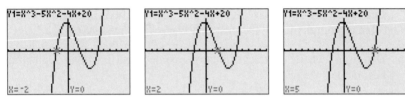

Figure 44 The *x*-intercepts are $(-2, 0)$, $(2, 0)$, and $(5, 0)$

In Example 7 we worked with the *cubic function* $f(x) = x^3 - 5x^2 - 4x + 20$. If the equation of a function f can be put into the form $f(x) = ax^3 + bx^2 + cx + d$, where $a \neq 0$, then f is called a **cubic function**. Some typical graphs of cubic functions are shown in Fig. 45.

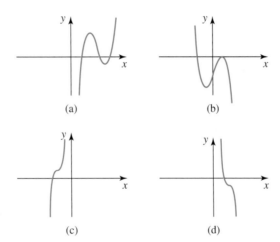

Figure 45 Typical graphs of cubic functions

Note that the cubic function sketched in graph (a) in Fig. 45 has exactly three *x*-intercepts, the cubic function sketched in graph (b) has exactly two *x*-intercepts and both cubic functions sketched in graphs (c) and (d) have exactly one *x*-intercept. It turns out that any cubic function has exactly one, two, or three *x*-intercepts.

Since the number of real number solutions of a cubic equation $ax^3 + bx^2 + cx + d = 0$ is equal to the number of *x*-intercepts of the function $f(x) = ax^3 + bx^2 + cx + d$, we conclude that the solution set of such an equation may contain one, two, or three real numbers.

In Example 8 we use a quadratic function to model a situation.

Example 8 Modeling with a Quadratic Function

Between 1994 and 1998, the number of people receiving food stamps declined (see Table 13).

Table 13 Average Monthly Participation in the Food Stamp Program

Year	Average Monthly Participation (millions of people)
1990	20.0
1991	22.5
1992	25.2
1993	26.6
1994	27.3
1995	26.6
1996	25.4
1997	22.7
1998	19.8

Source: New York Times

Let $f(t)$ represent the average monthly participation (in millions of people) in the food stamp program during the year that is t years since 1990.

1. Determine whether a linear function, exponential function, or quadratic function best models the data.
2. Draw a graph of the model $f(t) = -\frac{1}{2}t^2 + 4t + 19$ and the scattergram of the data in the same viewing window. Is f a reasonable model?
3. Predict in what year(s) the average monthly participation was/will be 9 million people.
4. Predict the average monthly participation in 2008.

Solution

1. The scattergram in Fig. 46 suggests that the data can be best modeled using a quadratic function.
2. We sketch the graph of f and the scattergram of the data in the same viewing window (see Fig. 47). It appears that f models the data quite well.

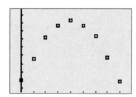

Figure 46 Food stamp scattergram

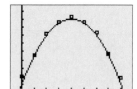

Figure 47 Verify the model

3. To predict when the average monthly participation will be 9 million people, we substitute 9 for $f(t)$ in $f(t) = -\frac{1}{2}t^2 + 4t + 19$ and solve for t.

$$9 = -\frac{1}{2}t^2 + 4t + 19$$

$$18 = -t^2 + 8t + 38 \qquad \text{Multiply both sides by 2.}$$

$$t^2 - 8t - 20 = 0$$

$$(t - 10)(t + 2) = 0$$

$$t - 10 = 0 \quad \text{or} \quad t + 2 = 0$$

$$t = 10 \quad \text{or} \quad t = -2$$

To verify our result, we enter $y = -\frac{1}{2}t^2 + 4t + 19$ in a graphing calculator and check that the inputs -2 and 10 lead to the output 9 (see Fig. 48).

So, according to the model, the average monthly participation would be 9 million people in 1988 and 2000.

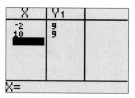

Figure 48 Verify the result

4. To predict the average monthly participation in 2008, we find $f(18)$:

$$f(18) = -\frac{1}{2}(18)^2 + 4(18) + 19 = -71$$

Since it does not make sense to have a negative number of people, we conclude that model breakdown has occurred. ∎

EXPLORATION
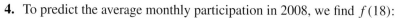
Finding equations of quadratic functions

1. Use ZStandard to sketch graphs of $f(x) = (x-2)(x+3)$, $g(x) = 2(x-2)(x+3)$, $h(x) = \frac{1}{2}(x-2)(x+3)$, and $k(x) = -(x-2)(x+3)$.

   a. What do you notice about the x-intercepts of f, g, h, and k?

   b. For a function of the form $y = a(x-2)(x+3)$, describe the effect that the value of a has on the graph of the function. Sketch more graphs with varying values of a if you are unsure.

2. Find a possible equation for the function sketched in Fig. 49. Use a graphing calculator to verify your work.

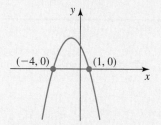

Figure 49 Find a possible equation (problem 2)

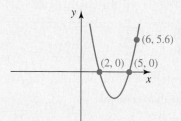

Figure 50 Find an equation (problem 3)

3. Find an equation for the function sketched in Fig. 50. Use a graphing calculator to verify your work. [**Hint**: A point on the graph of an equation satisfies the equation.]

Tips on Succeeding in This Course

When you have a question during class time, do you ask it? Many students are reluctant to ask questions for fear of slowing the class down or for feeling embarrassed that they are unclear about some concept. If you tend to shy away from asking questions, keep in mind that the main idea of school is for you to be able to learn through open communication with your instructor and other students in your class.

If you are confused about some concept, it's highly likely that lots of other students in your class are confused too. If you ask your question, everyone else who is confused will be grateful that you did so. Most instructors want students to ask questions. It helps an instructor to know when students are understanding the material and when they are having trouble.

Key Points
OF THIS SECTION

Quadratic equation	A **quadratic equation in one variable** is an equation that can be put into the form $ax^2 + bx + c = 0$, where a, b, and c are constants and $a \neq 0$.
Cubic equation	A **cubic equation in one variable** is an equation that can be put into the form $ax^3 + bx^2 + cx + d = 0$, where a, b, c, and d are constants and $a \neq 0$.
Zero factor property	If $ABC = 0$, then $A = 0$, $B = 0$, or $C = 0$.
Solving an equation by factoring	If an equation can be solved by factoring, we solve it by the following steps: 1. Write the equation so that one side of the equation is equal to zero. 2. Factor the nonzero side of the equation. 3. Apply the zero factor property. 4. Solve the equations that result from applying the zero factor property.
Solutions and x-intercepts	Let f be a function. If k is a solution of the equation $f(x) = 0$, then $(k, 0)$ is an x-intercept of f. If $(k, 0)$ is an x-intercept of f, then k is a solution of the equation $f(x) = 0$.
Finding x-intercepts	To find the x-coordinate of each x-intercept of a function f, substitute 0 for $f(x)$ and solve for x.
Solution set of a quadratic equation	The solution set of a quadratic equation may contain either two real numbers, one real number, or no real numbers.
Solution set of a cubic equation	The solution set of a cubic equation may contain one, two, or three real numbers.

HOMEWORK 6.5

SSM
Tutor Center

MathPro 4
MathPro 4

MathPro 5
MathPro 5

CDs/Videos

www

Solve. Verify your results by checking that they satisfy the equation.

1. $x^2 - 3x = 0$

2. $x^2 - 8x = 0$

3. $4x^2 - 6x = 0$

4. $x^2 - 6x + 5 = 0$

5. $x^2 + 8x + 15 = 0$

6. $x^2 - 81 = 0$

7. $x^2 - 4 = 0$

8. $x^2 - x - 12 = 0$

9. $7x^2 + 6x + 1 = 1$

10. $5 - 2x = x^2 + 5$

11. $3x^2 = 6x$

12. $3x^2 + 3x - 90 = 0$

13. $9x = -2x^2 + 5$

14. $7x^2 = 5x$

15. $9x^2 - 12x + 4 = 0$

16. $5x^2 - 80 = 0$

17. $3x^2 = 12$

18. $2x(x - 2) = 5x$

19. $16x^2 = 25$

20. $3x^2 = 48$

Solve. Verify your results by using graphing calculator tables or graphs.

21. $x(5x + 3) - 4 = -4$

22. $2x^2 = 3x$

23. $3x(x - 2) = 4x$

24. $x^2 = 25$

25. $9x^2 - 49 = 0$

26. $4x^2 = 100$

27. $x^2 = 1$

28. $25x^2 + 20x + 4 = 0$

29. $64 = 48x - 9x^2$

30. $10x^2 = 30x - 20$

31. $4x^2 - 8x = 32$

32. $x(x - 3) = 4$

33. $x^2 - 12x + 36 = 0$

34. $49 - 14x = -x^2$

35. $x^2 - \dfrac{1}{25} = 0$

36. $x^2 - \dfrac{1}{49} = 0$

37. $12x^2 - 2x - 2 = 0$

38. $2x^2 = 3x + 5$

39. $\dfrac{1}{4}x^2 - \dfrac{1}{2}x = 6$

40. $\dfrac{1}{4}x^2 = x + 3$

41. $6x^3 - 24x = 0$

42. $x^3 + 3x^2 - 4x - 12 = 0$

43. $4x^3 - 2x^2 - 36x + 18 = 0$

44. $2x^3 - 6x^2 - 36x = 0$

45. $4x^3 = 7x^2$

46. $36x = 24x^2 - 4x^3$

47. $4x^3 - 9x = 20x^2 - 45$

48. $20x^2 = -2x^3 - 50x$

49. $9x^3 - 12 = 4x - 27x^2$

50. $2x^2 + 25x = x^3 + 50$

51. $18x^3 + 3x^2 = 6x$

52. $3x^2 - 4x = 12 - x^3$

53. $-17x = 28 - 3x^2$

54. $4x(x - 3) = 3x(x - 4)$

55. $(x + 2)(x + 5) = 40$

56. $x(x - 3) + 7 = 10 - 3(x - 2)$

57. $4x(x - 1) - 24 = 3x(x - 2)$

58. $2x^2 + 2(x - 1) = x(1 - x)$

Factor or solve, whichever is appropriate.

59. $x^2 + 5x + 6$

60. $25x^2 - 64 = 0$

61. $x^2 + 5x + 6 = 0$

62. $25x^2 - 64$

63. $x^2 - 7x + 10 = 0$

64. $3x^3 + 8x^2 + 4x = 0$

65. $x^2 - 7x + 10$

66. $3x^3 + 8x^2 + 4x$

67. Is the following statement true or false? Explain.

$$\text{If } ab = 20, \text{ then } a = 2 \text{ or } b = 10$$

68. A student tries to solve the equation $(x - 3)(x - 7) = 12$. First, explain why the student's work is incorrect. Then, solve the equation.

$$(x - 3)(x - 7) = 12$$
$$x - 3 = 2 \quad \text{or} \quad x - 7 = 6$$
$$x = 5 \quad \text{or} \quad x = 13$$

69. A student tries to solve $x^3 + 4x^2 - 9x - 36 = 0$. What would you tell the student?

$$x^3 + 4x^2 - 9x - 36 = 0$$
$$x^2(x + 4) - 9(x + 4) = 0$$
$$x + 4 = 0$$
$$x = -4$$

70. A student tries to solve $2x^2 - 18x + 36 = 0$. Is the student's work correct? Explain.

$$2x^2 - 18x + 36 = 0$$
$$2(x^2 - 9x + 18) = 0$$
$$2(x - 3)(x - 6) = 0$$
$$x = 2, x = 3, \text{ or } x = 6$$

Find the x-intercept(s), if any, for the function. Verify your intercept(s) graphically using your graphing calculator.

71. $f(x) = x^2 - 9x + 20$

72. $f(x) = x^2 + 4x - 21$

73. $f(x) = x^2 - 16$

74. $f(x) = x^2 - 36$

75. $f(x) = 2x^2 - 20x + 32$

76. $f(x) = -4x^2 - 8x$

77. $f(x) = 36x^2 - 25$

78. $f(x) = 27x^2 - 12$

79. $f(x) = 12x^2 - 7x - 10$

80. $f(x) = 2x^2 + 5x + 2$

81. $f(x) = 3x^2 - 30x$

82. $f(x) = 6x^2 + 28x - 10$

83. $f(x) = 2x^3 - 2x^2 - 84x$

84. $f(x) = 5x^3 - 25x^2 + 30x$

85. $f(x) = x^3 + 2x^2 - x - 2$

86. $f(x) = 4x^3 - 12x^2 - 9x + 27$

87. Sales data for Corona® beer are shown in Table 14.

Table 14 Sales of Corona Beer

Year	Sales (millions of cases)
1987	24
1989	16
1991	13
1993	15
1995	23
1997	38
1998	55
1999	70

Source: New York Times

Let $C(t)$ represent sales (in millions of cases) of Corona beer during the year that is t years since 1980.

a. Use a graphing calculator to draw a scattergram of the data. Can the data best be modeled using a linear function, an exponential function, or a quadratic function? Explain.

b. Use your graphing calculator to draw the graph of the function $f(t) = t^2 - 22t + 130$ and the scattergram of the data in the same viewing window. Is f a reasonable model?

c. Estimate when sales were 109 million cases. Verify your result using a graphing calculator graph or table.

d. Predict the sales for 2008. Verify your result using a graphing calculator graph or table.

88. In exercise 44 of Homework 6.1, you may have modeled the number of wildfires in the United States with respect to time (see Table 15).

Table 15 U.S. Wildfires

Year	Number of Wildfires (thousands)
1996	115
1997	90
1998	81
1999	94
2000	123

Let $f(t)$ represent the number of wildfires (in thousands) in the year that is t years since 1995. A possible equation for f is:

$$f(t) = 9.29t^2 - 53.71t + 159.60$$

a. Use a graphing calculator to draw the graph of f and the scattergram of the data in the same viewing window. Is f a reasonable model?

b. Find t so that $f(t) = 159.6$. What does your result mean in terms of the situation?

c. The average size of a wildfire in the 1990s was 34.3 acres. Use this estimate and f to predict the area of land that will be burned from wildfires in 2009. To get an idea of the size of your result, assume that all of the wildfires will occur in Colorado and compute what percent of the state's 104,091 square miles of land will burn. [**Hint**: There are 640 acres in a square mile.] Colorado is the eighth largest state in area of land.

89. In exercise 43 of Homework 6.1, you may have modeled the percentages of Americans who are in favor of abortion rights (pro-choice). See Table 16.

Table 16 Percentages of Americans Who Are Pro-Choice

Year	Percent
1995	56
1996	52
1997	49
1998	48
1999	48
2000	48
2001	50

Let $f(t)$ represent the percentage of Americans who are pro-choice at t years since 1990. A possible equation for f is:

$$f(t) = \frac{4}{7}t^2 - 10t + 91$$

a. Use a graphing calculator to draw the graph of f and the scattergram of the data in the same viewing window. Is f a reasonable model?

b. Use f to predict when 63% of Americans will be pro-choice.

c. Since 1973, when abortion became legal in the United States, the percentage of Americans who are pro-choice has varied from about 48% to about 58%. Use f to predict when 91% of Americans will be pro-choice. In your opinion, is it likely that this prediction will turn out to be correct? Explain.

90. A company's profit can be modeled by

$$P(t) = \frac{1}{4}t^2 - 2t + 15$$

where $P(t)$ represents the profit (in millions of dollars) for the year that is t years since 2000.

a. Use P to predict when the profit will be $12 million. Verify your result using a graphing calculator graph or table.

b. Use P to predict when the profit will be $15 million. Verify your result using a graphing calculator graph or table.

Let $f(x) = x^2 - x - 6$.

91. Find $f(9)$.   **92.** Find $f(0)$.

93. Find $f(-7)$.   **94.** Find $f(-12)$.

95. Find x when $f(x) = 14$.

96. Find x when $f(x) = 0$.

97. Find x when $f(x) = -6$.

98. Find x when $f(x) = 6$.

For exercises 99–106, refer to Fig. 51.

99. Estimate $f(-2)$.   **100.** Estimate $f(0)$.

101. Estimate $f(1)$.   **102.** Estimate $f(4)$.

103. Estimate a when $f(a) = 4$.

104. Estimate a when $f(a) = 0$.

105. Estimate a when $f(a) = -1$.

106. Estimate a when $f(a) = -2$.

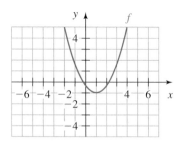

Figure 51 A sketch of the graph of f (Exercises 99–106)

Values for a quadratic function f are listed in Table 17. For exercises 107–114, refer to this table.

107. Find $f(0)$.   **108.** Find $f(3)$.

109. Find $f(4)$.   **110.** Find $f(6)$.

111. Find x when $f(x) = 19$.

112. Find x when $f(x) = 3$.

113. Find x when $f(x) = 1$.

114. Find x when $f(x) = 0$.

Table 17 Some Values of a Quadratic Function f (Exercises 107–114)

x	$f(x)$
0	19
1	9
2	3
3	1
4	3
5	9
6	19

115. For this exercise, consider the quadratic function f described in Table 17.

a. Find x when $f(x) = 9$.

b. Explain why f does not have an inverse function.

116. The values for a quadratic function f are listed in Table 18. Estimate the x-intercept(s) and the y-intercept(s).

Table 18 Values of a Function f (Exercise 116)

x	$f(x)$
-3	5.35
-2	0.17
-1	-3.61
0	-5.99
1	-6.97
2	-6.55
3	-4.73
4	-1.51
5	3.11
6	9.13

117. Give an example of a function whose x-intercepts are $(-5, 0)$ and $(1, 0)$. Verify your work using a graphing calculator.

118. Give three examples of quadratic functions where each function's only x-intercept is $(2, 0)$.

119. Give an example of a quadratic equation whose solutions are -3 and 1.

120. Give an example of a quadratic equation that has 4 as its only solution.

121. The graph of a function h is sketched in Fig. 52. Find a possible equation for h. Verify your equation using a graphing calculator.

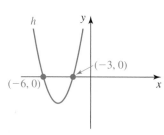

Figure 52 Exercise 121

122. The graph of a function g is sketched in Fig. 53. Find a possible equation for g. Verify your equation using a graphing calculator.

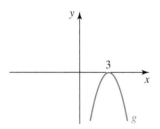

Figure 53 Exercise 122

123. Two students attempt to solve the equation $x^2 = x$. Which student's work is correct? Explain the error made by the other student. [**Hint**: Under what conditions can we divide both sides of an equation by x?]

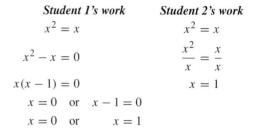

124. Explain why a quadratic equation in one variable cannot have three solutions.

125. Explain how to solve a quadratic equation in one variable.

6.6 SKETCHING GRAPHS OF QUADRATIC FUNCTIONS IN STANDARD FORM

Objectives

▸ Know the meaning of *symmetric points*.

▸ Find the x-coordinate of the vertex of a parabola by averaging the x-coordinates of a pair of symmetric points.

▸ Sketch the graph of a quadratic function in standard form.

In Section 6.1 we sketched graphs of quadratic functions in vertex form $f(x) = a(x - h)^2 + k$. Now, we sketch graphs of quadratic functions in standard form $f(x) = ax^2 + bx + c$.

To begin, consider the parabola sketched in Fig. 54. Recall that the axis of symmetry divides the parabola into two parts: the points that lie to the left of the axis of symmetry and the points that lie to the right of it. The two parts are mirror reflections of each other across the axis of symmetry. For any point on one part, there is a *symmetric point* on the other part with equal y-coordinate that is the "mirror reflection" of the first point.

For example, the point $(1, 6)$ and the point $(5, 6)$ are symmetric points. Since the vertex lies on the axis of symmetry, the x-coordinate of the vertex, 3, is equal to the average of the x-coordinates of the points $(1, 6)$ and $(5, 6)$:

$$\frac{1 + 5}{2} = 3$$

The average of the x-coordinates of *any* two symmetric points is equal to the x-coordinate of the vertex. For another example, we find the average of the x-coordinates of the y-intercept $(0, 11)$ and its symmetric point $(6, 11)$:

$$\frac{0 + 6}{2} = 3$$

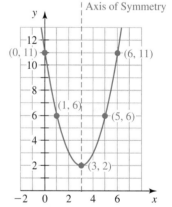

Figure 54 Locating symmetric points on a parabola

Note

The only point that does not have a symmetric point is the vertex, which lies on the axis of symmetry.

Note

The average of a and b is $\dfrac{a + b}{2}$. On the number line, the average lies between a and b, an equal distance from a and b.

Finding the *x*-coordinate of the Vertex from Symmetric Points

The x-coordinate of the vertex of a parabola is equal to the average of the x-coordinates of a pair of symmetric points.

Example 1   Find the *x*-coordinate of a Vertex

Find the x-coordinate of the vertex of the parabola sketched in the indicated figure.

1. Fig. 55   **2.** Fig. 56

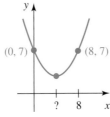

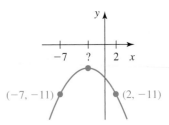

Figure 55 Problem 1   **Figure 56** Problem 2

Solution

1. The x-coordinates of the symmetric points $(0, 7)$ and $(8, 7)$ are 0 and 8. Since $\dfrac{0+8}{2} = 4$, the x-coordinate of the vertex must be 4.

2. The x-coordinates of the symmetric points $(-7, -11)$ and $(2, -11)$ are -7 and 2. Since $\dfrac{-7+2}{2} = -2.5$, the x-coordinate of the vertex must be -2.5. ■

Averaging the x-coordinates of the y-intercept and its symmetric point to find the x-coordinate of the vertex is a key step in sketching the graph of a quadratic function in standard form.

Note

Although we can find the x-coordinate of the vertex using *any* pair of symmetric points, the computations are easier using the y-intercept and its symmetric point.

Example 2   Sketching the Graph of a Quadratic Function

Sketch the graph of $g(x) = x^2 - 4x + 7$.

Solution

First, we find the y-intercept of $g(x) = x^2 - 4x + 7$.

$$g(0) = 0^2 - 4(0) + 7 = 7$$

The y-intercept is $(0, 7)$. See Fig. 57.

Next, we find the symmetric point to the y-intercept. Since symmetric points have the same height, we know the y-coordinate of the symmetric point is 7. We find the x-coordinate by substituting 7 for y in the equation $g(x) = x^2 - 4x + 7$ and solving for x:

$$7 = x^2 - 4x + 7$$
$$0 = x^2 - 4x$$
$$0 = x(x - 4)$$
$$x = 0 \quad \text{or} \quad x - 4 = 0$$
$$x = 0 \quad \text{or} \quad x = 4$$

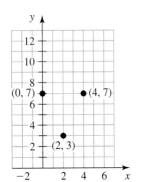

Figure 57 The vertex and two symmetric points

Therefore, the points that have height 7 are $(0, 7)$ and $(4, 7)$; these are symmetric points. To find the x-coordinate of the vertex, we average the x-coordinates of the symmetric points:

$$x\text{-coordinate of the vertex} = \frac{0+4}{2} = 2$$

To find the y-coordinate of the vertex, we find $g(2)$:

$$g(2) = 2^2 - 4(2) + 7 = 4 - 8 + 7 = 3$$

So the vertex is $(2, 3)$.

We can find another pair of symmetric points on the graph by evaluating g for the values of x that are 3 units from $x = 2$ on either side, namely at $x = -1$ and $x = 5$. At $x = 5$,

$$g(5) = 5^2 - 4(5) + 7 = 25 - 20 + 7 = 12$$

Thus, the graph passes through $(5, 12)$ and its symmetric point $(-1, 12)$. We plot these points to assist us in sketching a graph of g (see Fig. 58).

We use a graphing calculator to verify our sketch (see Fig. 59). In particular, we check that the vertex (minimum point) is at $(2, 3)$.

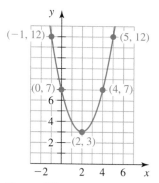

Figure 58 Graph of g using 5 points

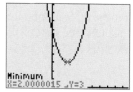

Figure 59 Verify the sketch of the graph of g

Note

The "minimum" choice on a graphing calculator will closely approximate the vertex. See Section B.19.

Graphing a Quadratic Function in Standard Form

To sketch a graph of a quadratic function $y = ax^2 + bx + c = 0$ with $b \neq 0$:

1. Find the y-intercept.
2. Find the y-intercept's symmetric point.
3. Average the x-coordinates of the two symmetric points to find the x-coordinate of the vertex.
4. Find the y-coordinate of the vertex.
5. Depending on how accurate your sketch is to be, find additional points on the parabola, as needed.
6. Sketch a parabola to contain the found points.

We use the above steps to graph a quadratic function $f(x) = ax^2 + bx + c$, provided $b \neq 0$. If $b = 0$, the y-intercept is the vertex and, therefore, does not have a symmetric point. In this case, however, $f(x) = ax^2 + 0x + c = ax^2 + c$ is in vertex form and we can readily use the methods of Section 6.1 to sketch the graph.

Example 3   Sketching the Graph of a Quadratic Function

Sketch the graph of $f(x) = -0.9x^2 - 5.8x - 5.7$.

Solution

We find the y-intercept by finding $f(0)$:

$$f(0) = -0.9(0)^2 - 5.8(0) - 5.7 = -5.7$$

So, the y-intercept is $(0, -5.7)$.

Next, we find the symmetric point for the y-intercept. We find it by substituting -5.7 for y and solving for x:

$$-5.7 = -0.9x^2 - 5.8x - 5.7$$
$$0 = -0.9x^2 - 5.8x$$
$$0 = -x(0.9x + 5.8)$$
$$-x = 0 \quad \text{or} \quad 0.9x + 5.8 = 0$$
$$x = 0 \quad \text{or} \quad x = \frac{-5.8}{0.9} \approx -6.44$$

With the same y-coordinate as the y-intercept, the symmetric point is approximately $(-6.44, -5.7)$.

We find the approximate x-coordinate of the vertex by averaging the x-coordinates of the points $(0, -5.7)$ and $(-6.44, -5.7)$:

$$\frac{0 + (-6.44)}{2} = -3.22$$

We find the approximate y-coordinate of the vertex by computing $f(-3.22)$:

$$f(-3.22) = -0.9(-3.22)^2 - 5.8(-3.22) - 5.7 \approx 3.64$$

So, the vertex is approximately $(-3.22, 3.64)$. See Fig. 60.

Although we could find and plot additional points, we can sketch a fairly accurate graph of f from the three points already found (see Fig. 61). We use a graphing calculator to verify our sketch (see Fig. 62). In particular, we check that the approximate vertex (maximum point) is $(-3.22, 3.64)$.

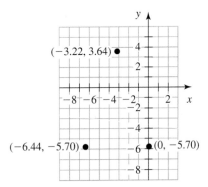

Figure 60 The vertex and two symmetric points

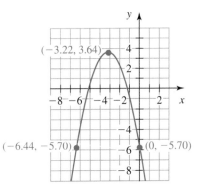

Figure 61 Graph of f using 3 points

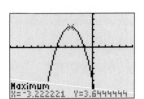

Figure 62 Graph of $f(x) = -0.9x^2 - 5.8x - 5.7$

In this section we located the vertex to help ourselves graph quadratic functions in standard form. Recall that the vertex is also the minimum or maximum point of the parabola. If the vertex is the maximum point, we call the y-coordinate of the vertex the **maximum value** of the function (see Fig. 63). If the vertex is the minimum point, we call the y-coordinate of the vertex the **minimum value** of the function (see Fig. 64).

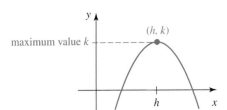

Figure 63 Quadratic function with a maximum value k but no minimum value

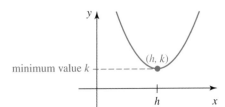

Figure 64 Quadratic function with a minimum value k but no maximum value

It turns out that if we convert a quadratic function in standard form $f(x) = ax^2 + bx + c$ into vertex form $f(x) = a(x - h)^2 + k$, the constant a has the same value for both forms. So, what we have determined about the graphical significance of a for quadratic functions in vertex form also applies to quadratic functions in standard form.

Maximum or Minimum Value of a Function

For a quadratic function $f(x) = ax^2 + bx + c$ with vertex (h, k):

- If $a < 0$, then the parabola opens downward and the maximum value of the function is k (see Fig. 63).
- If $a > 0$, then the parabola opens upward and the minimum value of the function is k (see Fig. 64).

In Example 4 we find the vertex of a quadratic model to estimate the maximum value of a quantity.

Example 4   Modeling with a Quadratic Function

A firework is launched into the air. The firework's height $h = f(t)$ (in feet) at time t seconds is modeled well by the function $f(t) = -16t^2 + 200t + 4$. When should the firework explode so that it goes off at the greatest height? What is that height?

Solution

The function f is of the form $f(t) = at^2 + bt + c$ with $a = -16 < 0$, so the graph of f is a parabola that opens downward. Therefore, the vertex is the maximum point.

To find the vertex, we begin by finding the h-coordinate of the h-intercept:

$$f(0) = -16(0)^2 + 200(0) + 4 = 4$$

So, the h-intercept is $(0, 4)$. To find the symmetric point for the h-intercept we substitute 4 for $f(t)$ and solve for t:

$$4 = -16t^2 + 200t + 4$$
$$16t^2 - 200t = 0$$
$$4t(4t - 50) = 0$$
$$4t = 0 \quad \text{or} \quad 4t - 50 = 0$$
$$t = 0 \quad \text{or} \quad t = \frac{50}{4}$$
$$t = 0 \quad \text{or} \quad t = 12.5$$

With the same h-coordinate as the h-intercept, the symmetric point is $(12.5, 4)$.

We find the t-coordinate of the vertex by averaging the t-coordinates of the symmetric points:

$$\frac{0 + 12.5}{2} = 6.25$$

We find the h-coordinate of the vertex by computing $f(6.25)$:

$$f(6.25) = -16(6.25)^2 + 200(6.25) + 4 = 629$$

So, the vertex is $(6.25, 629)$. We verify our work using "maximum" on a graphing calculator (see Fig. 65).

Graphing Calculator

See Section B.19.

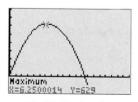

Figure 65 Verify that the vertex is $(6.25, 629)$

A vertex of $(6.25, 629)$ means that the firework would reach a maximum height of 629 feet at 6.25 seconds. The firework should explode at 6.25 seconds to go off at the maximum height. ∎

Note that the function $f(t) = -16t^2 + 200t + 4$ describes the height of the firework in terms of time, *not* in terms of horizontal distance. So, the graph of f does *not* describe the path of the firework. The path of the firework can be modeled by a quadratic function that describes the height in terms of horizontal distance.

EXPLORATION
Finding the x-coordinate of the vertex

1. Find the x-coordinate of the vertex for the function $f(x) = 2x^2 + 3x + 5$. Verify your result by using "minimum" on a graphing calculator.

2. Find the x-coordinate of the vertex for a quadratic function of the form $g(x) = ax^2 + bx + c$, where $b \neq 0$. [**Hint**: Perform the same steps as you did in problem 1.]

3. Use your result from problem 2 to find the x-coordinate of the vertex of $h(x) = 3x^2 - 7x + 1$. Verify your result by using "minimum" on a graphing calculator.

Tips on Succeeding in This Course

When taking a test, it is best to quickly scan the test items, pick the items with which you feel most comfortable, and complete those problems first. By doing so, you will have warmed up, feel more confident, and probably do better on the rest of the test. Also, you will probably have a better idea of how to allot your time on the remaining problems.

Key Points
OF THIS SECTION

x-coordinate of the vertex

The x-coordinate of the vertex of a parabola is equal to the average of the x-coordinates of a pair of symmetric points.

Sketching the graph of a quadratic function

To sketch a graph of a quadratic function $y = ax^2 + bx + c$ with $b \neq 0$:
1. Find the y-intercept.
2. Find the y-intercept's symmetric point.
3. Average the x-coordinates of the two symmetric points to find the x-coordinate of the vertex.
4. Find the y-coordinate of the vertex.
5. Depending on how accurate your sketch is to be, find additional points on the parabola, as needed.
6. Sketch a parabola to contain the found points.

Maximum or minimum value

For a quadratic function $f(x) = ax^2 + bx + c$ with vertex (h, k):

- If $a < 0$, then the parabola opens downward and the maximum value of the function is k.
- If $a > 0$, then the parabola opens upward and the minimum value of the function is k.

HOMEWORK 6.6

SSM
Tutor Center

 MathPro 4
MathPro 4

 MathPro 5
MathPro 5

CDs/Videos

www
www

Find the x-coordinate of the vertex of the parabola sketched in the figure.

1. Fig. 66

2. Fig. 67

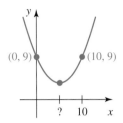

Figure 66 Exercise 1

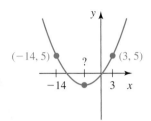

Figure 67 Exercise 2

Find the x-coordinate of the vertex of a parabola passing through the given points.

3. $(0, 8)$ and $(6, 8)$

4. $(0, 5)$ and $(8, 5)$

5. $(0, -3)$ and $(-4, -3)$

6. $(0, -6)$ and $(-9, -6)$

7. $(0, 2)$ and $(7, 2)$

8. $(0, -7)$ and $(15, -7)$

9. $(2, 9)$ and $(8, 9)$

10. $(3, -5)$ and $(6, -5)$

11. $(-4, 2)$ and $(6, 2)$

12. $(-9, -5)$ and $(-2, -5)$

A parabola has the given vertex and y-intercept. Find another point on the parabola.

13. vertex $(2, 5)$ and y-intercept $(0, 9)$

14. vertex $(-3, -8)$ and y-intercept $(0, 4)$

A parabola has the given vertex and contains the given point. Find another point on the parabola.

15. vertex $(3, 7)$ and point $(1, 1)$

16. vertex $(-5, -6)$ and point $(2, 3)$

Sketch by hand the graph of the function. Give the coordinates for the vertex. Verify your sketch using your graphing calculator.

17. $y = x^2 - 6x + 7$

18. $y = x^2 - 4x + 5$

19. $y = x^2 + 8x + 9$

20. $y = x^2 + 2x - 7$

21. $y = -x^2 + 8x - 10$

22. $y = -2x^2 + 12x - 9$

23. $y = 3x^2 + 6x - 4$

24. $y = 2x^2 - 10x + 1$

25. $y = -3x^2 + 12x - 5$

26. $y = -2x^2 + 10x - 3$

27. $y = -4x^2 - 9x - 5$

28. $y = -2x^2 + 5x + 3$

29. $y = 2x^2 - 7x + 7$

30. $y = -3x^2 - 2x - 4$

31. $4x^2 - y + 6 = 8x$

32. $6x^2 = 3y - 24x - 15$

33. $y = 2(x - 3)^2 + 15$

34. $y = -3(x + 6)^2 + 2$

35. $y = x^2 - 6$

36. $y = 3 - x^2$

37. $y = 2.8x^2 - 8.7x + 4$

38. $y = -1.6x^2 - 4.8x + 3$

39. $y = 3.9x^2 + 6.9x - 3.4$

40. $y = -2.4x^2 + 6.1x - 7.8$

41. $3.6y - 26.3x = 8.3x^2 - 7.1$

42. $5.3 - 2.1y = 9.8x^2 - 3.4x + 8.3$

A quadratic function has the given x-intercepts. What is the x-coordinate of the vertex?

43. $(2, 0)$ and $(6, 0)$

44. $(-4, 0)$ and $(3, 0)$

45. $(-9, 0)$ and $(4, 0)$

46. $(-7, 0)$ and $(-3, 0)$

Find the x-intercepts and y-intercept of the function. Next, find the vertex. Then sketch by hand the graph of the function. Verify your result by using your graphing calculator.

47. $y = 5x^2 - 10x$

48. $y = -2x^2 - 8x$

49. $y = x^2 - 4x$

50. $y = -3x^2 + 9x$

51. $y = x^2 - 10x + 24$

52. $y = x^2 + 5x - 14$

53. $y = x^2 - 8x + 7$

54. $y = x^2 - 5x - 24$

55. $y = x^2 - 9$

56. $y = x^2 - 1$

57. $y = 2x^2 - 11x - 21$

58. $y = 12x^2 - 26x + 10$

59. A baseball is hit by a batter. The height $h(t)$ (in feet) of the ball after t seconds is given by:

$$h(t) = -16t^2 + 140t + 3$$

a. What is the height of the ball when the batter makes contact?

b. What is the maximum height of the ball? When does it reach that height?

c. Sketch a graph of h.

60. A person on the edge of a cliff throws a stone so that it hits the ground near the base of the cliff. The height of the stone $h(t)$ (in feet above the base of the cliff) after t seconds is given by:

$$h(t) = -16t^2 + 30t + 200$$

a. Find the vertex of h. What does the vertex mean in terms of the stone?

b. Estimate the height of the cliff. State any assumptions that you make.

c. Did the person throw the stone upwards or downwards? Explain.

61. The profit $P(t)$ (in millions of dollars) of a company for the year that is t years since 1990 is given by

$$P(t) = 0.09t^2 - 1.65t + 9.72$$

  a. In what year was the profit the least? What was the profit?

  b. Sketch a graph of P.

62. A company's revenue can be modeled by the function

$$r(t) = -0.07t^2 + 1.70t + 4.45$$

where $r(t)$ represents the revenue (in millions of dollars) for the year that is t years since 1990.

  a. What was the company's revenue in 1990?

  b. When was the company's revenue at a maximum? What was the maximum revenue?

63. The median ages at which men were first married are shown in Table 19.

Table 19 Median Ages at First Marriages for Men

Year	Median Age
1940	24.3
1950	22.8
1960	22.8
1970	23.2
1980	24.5
1990	26.3
2000	26.8

Source: Who We Are, A Portrait of America, *Sam Roberts*

Let $f(t)$ represent the median age at which men were first married at t years since 1900. A possible equation for f is

$$f(t) = 0.00248t^2 - 0.29t + 31.48$$

What was the minimum median age at which men were first married? What was the year?

64. Are you afraid of traveling by airplane? Most people worry about midair engine failure or an unsuccessful takeoff or landing. But few worry about two planes colliding while on the ground.

  In recent years, the Federal Aviation Administration has focused on minimizing the number of *midair* collisions, but pilots, accident investigators, and air controllers say that the most overlooked safety hazard is the collision of planes while on the runways. Pilots say that the anxiety of takeoff and landing does not compare to the stress of taxiing planes along complex ground routes in a congested airport, especially at night or in poor weather.

  Let $f(t)$ represent the number of runway accidents during the year that is t years since 1990. A reasonable equation for f is

$$f(t) = -4.70t^2 + 85.11t - 62.76$$

  a. Estimate the year when runways accidents were at a maximum. Estimate the number of accidents in that year.

  b. According to the model, has safety on the runways improved or declined in recent years? Explain.

65. Let $f(t)$ represent the sales of premium sports cars (in thousands) in the United States for the year that is t years since 1990. A reasonable equation for f is

$$f(t) = -5.07t^2 + 102.93t - 460.40$$

When were the sales of premium sports cars at a maximum? What was that maximum?

66. Suppose that $f(x) = ax^2 + bx + c$, where $a > 0$ and $g(x)$ is a linear function. If $f(2) = g(2)$ and $f(5) = g(5)$, which is larger, $f(4)$ or $g(4)$? Explain. [**Hint**: Think graphically.]

67. a. Find the x-coordinate of the vertex of $f(x) = x^2 + 4x - 12$ by averaging the x-coordinates of the y-intercept and its symmetric point.

  b. Find the x-coordinate of the vertex of $f(x) = x^2 + 4x - 12$ by averaging the x-coordinates of the x-intercepts.

  c. Compare the methods you used in parts a and b. Are your results the same?

  d. Which method from parts a and b is easier to use to find the x-coordinate of the vertex of $g(x) = 54x^2 - 195x - 216$? Explain.

  e. Which method(s) can be used to find the x-coordinate of the vertex of $h(x) = x^2 + 4x + 6$? Explain.

  f. Summarize your findings in this exercise.

68. In this exercise you will discover how to convert the standard form of a quadratic function to its vertex form.

  a. Find the vertex of $f(x) = 3x^2 - 6x + 7$.

  b. Recall that if (h, k) is the vertex of a quadratic function, then $f(x) = a(x - h)^2 + k$. Use your result from part a to determine h and k for $f(x) = 3x^2 - 6x + 7$.

  c. Substitute the values you found for h and k from part b into $f(x) = a(x - h)^2 + k$.

  d. Compare $f(x) = 3x^2 - 6x + 7$ with your result in part c. Determine the value of a. [**Hint**: You may see this immediately. If not, expand your result in part c and compare again with $f(x) = 3x^2 - 6x + 7$.]

  e. Substitute your value for a into your result from part c.

  f. Verify your result graphically by comparing the graph of $f(x) = 3x^2 - 6x + 7$ with the graph of $f(x) = a(x - h)^2 + k$ for the values you found for a, h, and k.

69. Input-output pairs for four quadratic functions f, g, h, and k are listed in Table 20. For each function, decide whether $(3, 2)$ is the vertex. If $(3, 2)$ is not the vertex, estimate the coordinates of the vertex.

Table 20 Values of Four Quadratic Functions

x	$f(x)$	$g(x)$	$h(x)$	$k(x)$
1	10	8.2	19.3	11.6
2	4	2.9	7.3	4.4
3	2	2	2	2
4	4	5.5	3.4	4.4
5	10	13.4	11.5	11.6

70. Explain how to sketch the graph of a quadratic function $f(x) = ax^2 + bx + c$, where $b \neq 0$.

71. The x-coordinate of the vertex of a quadratic function is equal to the average of the x-coordinates of two symmetric points. Explain. Include a sketch of a graph of a quadratic function in your explanation.

CHAPTER SUMMARY

Key Points
OF THIS CHAPTER

For a function of the form $f(x) = a(x - h)^2 + k, a \neq 0$:

Properties of a quadratic function

- The graph of f is a parabola with vertex (h, k).

- If $a > 0$, the parabola opens upward and the vertex is the minimum point of the function f.

- If $a < 0$, the parabola opens downward and the vertex is the maximum point of the function f.

- If $|a|$ is large, the parabola will be narrow.

- If a is near zero, the parabola will be wide.

- The graph of f can be found by translating the graph of $y = ax^2$ horizontally by $|h|$ units (right if $h > 0$, left if $h < 0$) and vertically by $|k|$ units (up if $k > 0$, down if $k < 0$.)

We *factor* a polynomial by writing it as a product. We *expand* a product by performing the multiplication.

Factoring and expanding a polynomial

A function in vertex form $f(x) = a(x - h)^2 + k$ can be written in standard form $f(x) = ax^2 + bx + c$ and vice versa. A function that can be written in either form with $a \neq 0$ is a quadratic function and its graph is a parabola.

Vertex and standard forms

We can expand/factor some quadratic polynomials using the formulas:

Formulas for expanding and factoring polynomials

$$(x + k)^2 = x^2 + 2kx + k^2 \qquad \text{Square of a sum}$$
$$(x - k)^2 = x^2 - 2kx + k^2 \qquad \text{Square of a difference}$$
$$(x + k)(x - k) = x^2 - k^2 \qquad \text{Difference of two squares}$$

If $x^2 + bx + c$ can be factored as a product of two binomials, we factor it by first finding two integers, p and q, whose product is c and whose sum is b. Then we write:

Factoring $x^2 + bx + c$

$$x^2 + bx + c = (x + p)(x + q)$$

If $ax^2 + bx + c$ can be factored as $(mx + p)(nx + q)$, then $mn = a$ and $pq = c$.

Factoring $ax^2 + bx + c$

The polynomial $x^2 + k^2, k \neq 0$, is prime.

The polynomial $x^2 + k$

If a polynomial with four terms can be factored by grouping, we factor it by:

Factor by grouping

1. Factoring the first two terms and the last two terms.

2. Factoring out the binomial GCF.

Here is a five-step method for factoring many polynomials. Steps 2–4 can be applied to the entire polynomial or to a factor of the polynomial.

Five-step method for factoring many polynomials

1. If the GCF is not 1, factor it out.

2. For a binomial, try using the difference of two squares formula.

3. For a trinomial $ax^2 + bx + c$:
   a. If $a = 1$, then try to find two integers whose product is c and whose sum is b.
   b. If $a \neq 1$, then try to factor using the trial and error method.

4. For an expression with four terms, try factoring by grouping.

5. Continue applying steps 2–4 until the polynomial is completely factored.

Solving an equation	If an equation can be solved by factoring, we solve it by the following steps:
	1. Write the equation so that one side of the equation is equal to zero.
	2. Factor the nonzero side of the equation.
	3. Apply the zero factor property.
	4. Solve the equations that result from applying the zero factor property.
Solutions and x-intercepts	Let f be a function. If k is a solution of the equation $f(x) = 0$, then $(k, 0)$ is an x-intercept of f. If $(k, 0)$ is an x-intercept of f, then k is a solution of the equation $f(x) = 0$.
Finding x-intercepts	To find the x-coordinate of each x-intercept of a function f, substitute 0 for $f(x)$ in the equation for f and solve for x.
Sketching the graph of a quadratic function	To sketch a graph of a quadratic function $y = ax^2 + bx + c$ with $b \neq 0$:
	1. Find the y-intercept.
	2. Find the y-intercept's symmetric point.
	3. Average the x-coordinates of the two symmetric points to find the x-coordinate of the vertex.
	4. Find the y-coordinate of the vertex.
	5. Depending on how accurate your sketch is to be, find additional points on the parabola, as needed.
	6. Sketch a parabola to contain the found points.

CHAPTER 6 REVIEW EXERCISES

Sketch by hand the graph of each function.

1. $y = -3x^2$

2. $y = 3(x + 1)^2 - 4$

3. $y = -2(x - 3)^2 + 5$

4. $y = \frac{1}{2}(x + 4)^2 - 2$

5. A function of the form $y = a(x - h)^2 + k$ has been sketched in Fig. 68. Describe the signs of the constants a, h, and k for this function.

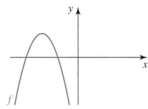

Figure 68 Exercise 5

6. Find an equation for the function sketched in Fig. 69.

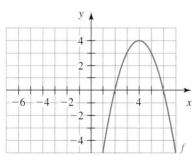

Figure 69 Exercise 6

Expand.

7. $-3x(x + 2)$

8. $(x + 4)^2$

9. $(x + 2)(x - 5)$

10. $3x(2x - 5)^2$

11. $-(2x - 3)(5x + 1)$

12. $-2x(4x - 7)(4x + 7)$

13. $\left(\frac{1}{5}x - \frac{1}{8}\right)\left(\frac{1}{5}x + \frac{1}{8}\right)$

14. $(x^2 - 3)(x^2 - 7)$

15. $(x + 2)(x^2 - 5x + 4)$

16. $(3x^2 - 2x + 1)(x^2 + 3x - 2)$

17. Which of the following equations describe the same function?
$$y = 3x^2 + 9x - 12 \qquad y = 3(x - 2)^2 + 9x$$
$$y = 3x(x + 3) - 12 \qquad y = 3(x + 4)(x - 1)$$
$$y = (3x + 4)(x - 1) \qquad y = (x + 4)(3x - 3)$$

18. Convert the function $f(x) = -2(x + 1)^2 - 3$ to standard form.

Factor.

19. $-4x^2 + 12x$

20. $x^2 - 2x - 24$

21. $7x^2 - 63$

22. $x^2 - \frac{1}{4}$

23. $3x^2 - 6x - 45$

24. $2.4x^2 - 7.9x$

25. $2x^2 - 16x + 32$

26. $8x^2 + 14x - 15$

27. $6x^2 - 33x + 36$

28. $5x^3 - 45x$

29. $3x^3 - 27x^2 + 42x$

30. $4x^3 + 8x^2 - 9x - 18$

Solve.

31. $x^2 - 2x - 24 = 0$

32. $x^2 - 25 = 0$

33. $x^2 - \dfrac{1}{81} = 0$

34. $3x^2 - 7x + 2 = 0$

35. $10x^2 = 100x$

36. $-3x^2 + 10x = -5x^2 + x + 5$

37. $3x^2 + 30x = 6x^3$

38. $x^3 - 4x = 12 - 3x^2$

39. $3x(x - 7) + 13 = 2x^2 - 7$

40. $20x^2 - 24x = 4x^2 - 9$

Find the x-intercepts of the function.

41. $f(x) = 6x^2 - 7x - 5$

42. $f(x) = 64x^2 - 49$

Let $f(x) = 5x^2 - x - 4$.

43. Find $f(0)$.

44. Find $f\left(\dfrac{1}{5}\right)$.

45. Find $f(a + 2)$.

46. Find x when $f(x) = 0$.

47. Find x when $f(x) = 2$.

48. Find x when $f(x) = -4$.

49. Find the x-coordinate of the vertex of a parabola that contains the points $(-2, 7)$ and $(9, 7)$.

Sketch by hand the graph of each function. Determine the y-intercept and vertex for each function.

50. $y = -2x^2 + 8x + 5$

51. $y = 9 - x^2$

52. $y = 3x^2 - 6x$

53. $1.7x + 2.6x^2 + y = 6.7x^2 - 10x + 2.1$

54. Give an example of a quadratic function whose only x-intercept is $(5, 0)$.

55. Give an example of a quadratic function whose x-intercepts are $(2, 0)$ and $(6, 0)$.

56. A batter hits a baseball into the air. The height $h(t)$ (in feet) of the baseball after t seconds is given by
$$h(t) = -16t^2 + 100t + 3.$$

   a. Find the maximum height of the baseball. When does it reach that height?

   b. A fielder catches the ball at a height of 3 feet. How many seconds after the batter hit the ball did the fielder have to get into position?

   c. Sketch a graph of h.

57. The median ages at which women were first married are shown in Table 21.

Table 21 Median Ages at First Marriages for Women	
Year	Median Age
1940	21.5
1950	20.3
1960	20.3
1970	20.8
1980	22.0
1990	24.0
2000	25.1

Source: Who We Are, A Portrait of America, *Sam Roberts*

Let $f(t)$ represent the median age at which women were first married at t years since 1900. A possible equation for f is $f(t) = 0.0027t^2 - 0.31t + 29.29$. What was the minimum median age at which women were first married? What was the year?

CHAPTER 6 TEST

1. Use pencil and paper to sketch the graph of the function $f(x) = -2(x + 5)^2 - 1$.

2. What can you say about the values of a, h, and k if the equation of the parabola pictured in Fig. 70 is $y = a(x - h)^2 + k$?

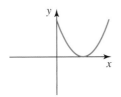

Figure 70 Exercise 2

3. Give an example of a quadratic function that has vertex $(2, -7)$ and has no x-intercepts.

Expand.

4. $-2(x - 2)(x + 7)$

5. $(3x - 7)(3x + 7)$

6. $2x(5x + 8)^2$

7. $(2x^2 - x + 3)(x^2 + 2x - 1)$

8. Write $f(x) = -2(x + 6)^2 + 11$ in standard form.

Factor.

9. $8x^2 + 13x - 6$

10. $2x^3 - 12x^2 + 18x$

Solve.

11. $x^2 - 3x - 10 = 0$

12. $(2x - 7)(x - 3) = 10$

13. $25x^2 - 16 = 0$

14. $2x^3 + 3x^2 = 18x + 27$

15. If the point $(1, 3)$ lies on a parabola with vertex $(4, 7)$, find another point that lies on the parabola.

16. a. Find the x-intercepts of the function $f(x) = x^2 - 5x - 24$.

b. Find the vertex of f.

c. Sketch by hand the graph of f. Indicate the x-intercepts and vertex on the graph.

Let $f(x) = 3x^2 + 10x - 8$.

17. Find $f(0)$.

18. Find $f(-3)$.

19. Find x when $f(x) = 0$.

20. Find x when $f(x) = 5$.

For exercises 21–25, refer to Fig. 71.

21. Estimate $f(0)$.

22. Estimate $f(-3)$.

23. Estimate x when $f(x) = 3$.

24. Estimate x when $f(x) = 1$.

25. Estimate x when $f(x) = -3$.

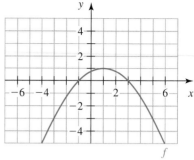

Figure 71 Exercises 21–25

26. The profit $P(t)$ (in millions of dollars) of a company for the year that is t years since 1995 is given by

$$P(t) = 0.2t^2 - 2.6t + 14.1$$

a. In what year(s) was the profit $14.1 million?

b. In what year was the profit the least? What was the profit?

c. Sketch by hand the graph of P.

Using Quadratic Functions to Model Data

Like any other art form, math takes practice and a lot of attention. I guess more like a girlfriend.

—*Jose C., student*

7.1 SOLVING QUADRATIC EQUATIONS BY EXTRACTING SQUARE ROOTS

Objectives

▸ Know the *product property* and *quotient property* for square roots.

▸ Rationalize the denominator of a fraction.

▸ Solve quadratic equations of the form $a(x - h)^2 + k = 0$.

In Chapter 6 we used factoring to solve quadratic equations of the form $ax^2 + bx + c = 0$. However, not all expressions $ax^2 + bx + c$ can be factored. In this section we discuss how to solve any quadratic equation of the form $a(x - h)^2 + k = 0$. In Section 7.2 we discuss how to write any quadratic equation in this form, so we will have a method by which to solve *any* quadratic equation.

Recall from Section 4.2 that we can represent a principal nth root such as $k^{1/n}$ as $\sqrt[n]{k}$, where $k \geq 0$ if n is even. We call $\sqrt[n]{k}$ a *radical* and refer to k as the *radicand* of $\sqrt[n]{k}$. In this section we restrict our study to principal square roots $\sqrt{k}$.

The radicand k must be nonnegative for the radical $\sqrt{k}$ to be a real number. For example, $\sqrt{-4}$ is not a real number.

Recall that for $k \geq 0$

$$\sqrt{k} = k^{1/2}$$

We can use exponential properties to prove the *product property* for square roots. Assume that $a \geq 0$ and $b \geq 0$. Then

$$\sqrt{ab} = (ab)^{1/2} \qquad \sqrt{k} = k^{1/2}$$
$$= a^{1/2}b^{1/2} \qquad (ab)^n = a^n b^n$$
$$= \sqrt{a}\sqrt{b}$$

So, $\sqrt{ab} = \sqrt{a}\sqrt{b}$. We can use a similar approach to prove the *quotient property* for square roots:

$$\sqrt{\frac{a}{b}} = \frac{\sqrt{a}}{\sqrt{b}}$$

where $a \geq 0$ and $b > 0$.

Square Root Properties

For $a \geq 0$ and $b \geq 0$,

$$\sqrt{ab} = \sqrt{a}\sqrt{b} \qquad \text{Product property}$$

For $a \geq 0$ and $b > 0$,

$$\sqrt{\frac{a}{b}} = \frac{\sqrt{a}}{\sqrt{b}} \qquad \text{Quotient property}$$

Note

You will prove the quotient property in exercise 73.

We can use the product property and quotient property to simplify square roots. We do so by writing the radicand as a product (or quotient) of a number and one of the *perfect squares* 4, 9, 16, 25, 36, ...

Example 1 Simplifying Radical Expressions

Simplify each radical.

1. $\sqrt{20}$
2. $\sqrt{\dfrac{x}{9}}$

Graphing Calculator

To compute $\sqrt{20}$:

Press [2nd] [x^2] [20] [)] [ENTER].

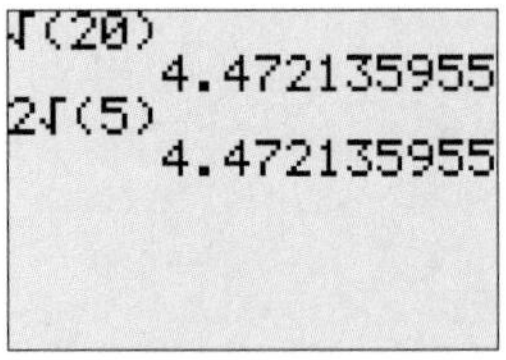

```
√(20)
       4.472135955
2√(5)
       4.472135955
```

Figure 1 Verify that $\sqrt{20} = 2\sqrt{5}$

Solution

1. $\sqrt{20} = \sqrt{4 \cdot 5}$ 4 is a perfect square.

   $\quad\ = \sqrt{4}\sqrt{5}$ $\sqrt{ab} = \sqrt{a}\sqrt{b}$

   $\quad\ = 2\sqrt{5}$ $\sqrt{4} = 2$

   We verify our work using a graphing calculator (see Fig. 1).

2. $\sqrt{\dfrac{x}{9}} = \dfrac{\sqrt{x}}{\sqrt{9}}$ $\sqrt{\dfrac{a}{b}} = \dfrac{\sqrt{a}}{\sqrt{b}}$

   $\quad\ = \dfrac{\sqrt{x}}{3}$ $\sqrt{9} = 3$

A radical $\sqrt{k}$ with a counting number k, $k \geq 2$, is *simplified* if k does not have any perfect square factors other than 1. For an expression such as $\sqrt{\dfrac{3}{5}}$, we simplify by leaving no radicand as a fraction.

For an expression such as $\dfrac{7}{\sqrt{2}}$, we simplify by leaving no denominator as a radical. We call this process *rationalizing the denominator*.

Example 2 Simplifying Radical Expressions

Simplify each expression.

1. $\dfrac{2}{\sqrt{3}}$
2. $\sqrt{\dfrac{5}{18}}$

Solution

1. Since $\sqrt{3}\sqrt{3} = \sqrt{3 \cdot 3} = \sqrt{9} = 3$, we rationalize the denominator of $\dfrac{2}{\sqrt{3}}$ by multiplying by $\dfrac{\sqrt{3}}{\sqrt{3}}$.

$$\frac{2}{\sqrt{3}} = \frac{2}{\sqrt{3}} \cdot 1 \qquad a \cdot 1 = a$$

$$= \frac{2}{\sqrt{3}} \cdot \frac{\sqrt{3}}{\sqrt{3}} \qquad \frac{\sqrt{3}}{\sqrt{3}} = 1$$

$$= \frac{2\sqrt{3}}{\sqrt{9}} \qquad \sqrt{a}\sqrt{b} = \sqrt{ab}$$

$$= \frac{2\sqrt{3}}{3} \qquad \sqrt{9} = 3$$

2. $\sqrt{\dfrac{5}{18}} = \dfrac{\sqrt{5}}{\sqrt{18}}$ $\qquad$ $\sqrt{\dfrac{a}{b}} = \dfrac{\sqrt{a}}{\sqrt{b}}$

$\qquad = \dfrac{\sqrt{5}}{3\sqrt{2}}$ $\qquad$ $\sqrt{18} = \sqrt{9 \cdot 2} = \sqrt{9}\sqrt{2} = 3\sqrt{2}$

$\qquad = \dfrac{\sqrt{5}}{3\sqrt{2}} \cdot \dfrac{\sqrt{2}}{\sqrt{2}}$ $\qquad$ Rationalize the denominator.

$\qquad = \dfrac{\sqrt{10}}{3\sqrt{4}}$

$\qquad = \dfrac{\sqrt{10}}{6}$

We verify our work using a graphing calculator (see Fig. 2). ■

```
√(5/18)
           .5270462767
√(10)/6
           .5270462767
```

Figure 2 Verify that $\sqrt{\dfrac{5}{18}} = \dfrac{\sqrt{10}}{6}$

As shown in problem 1 in Example 2, to rationalize the denominator of a fraction of the form $\dfrac{p}{\sqrt{q}}$, $q > 0$, we multiply the fraction by 1 in the form $\dfrac{\sqrt{q}}{\sqrt{q}}$.

We now turn our attention to solving quadratic equations of the form

$$x^2 = k$$

where $k \geq 0$. From our work in Section 4.4, we know the solution(s) to the equation $x^2 = k$ is

$$\pm k^{1/2} = \pm\sqrt{k}$$

Solving Equations of the Form $x^2 = k$

Let k be a nonnegative constant. Then $x^2 = k$ is equivalent to
$$x = \pm\sqrt{k}$$

Example 3 Solving Quadratic Equations

Solve each equation.

1. $x^2 = 25$ $\qquad\qquad$ **2.** $x^2 = 7$ $\qquad\qquad$ **3.** $x^2 - 45 = 0$

4. $x^2 = -4$ $\qquad\qquad$ **5.** $3x^2 - 11 = 0$

Solution

1. $x^2 = 25$

$\qquad x = \pm\sqrt{25}$ $\qquad$ The solutions of $x^2 = k$ are $\pm\sqrt{k}$.

$\qquad\;\; = \pm 5$

2. $x^2 = 7$

$\qquad\;\; = \pm\sqrt{7}$

3. $x^2 - 45 = 0$

$\qquad x^2 = 45$ $\qquad$ Add 45 to both sides.

$\qquad x = \pm\sqrt{45}$

$\qquad\;\; = \pm\sqrt{9 \cdot 5}$ $\qquad$ 9 is a perfect square.

$\qquad\;\; = \pm 3\sqrt{5}$

The approximate solutions are -6.71 and 6.71. We can check that each approximate solution approximately satisfies the original equation $x^2 - 45 = 0$.

$$(-6.71)^2 - 45 = 0.0241 \approx 0 \quad \text{and} \quad (6.71)^2 - 45 = 0.0241 \approx 0$$

4. Since the square of a number is nonnegative, we conclude that there are no real number solutions for $x^2 = -4$.

5. $3x^2 - 11 = 0$

$$3x^2 = 11 \qquad \text{Add 11 to both sides.}$$

$$x^2 = \frac{11}{3} \qquad \text{Divide both sides by 3.}$$

$$x = \pm\sqrt{\frac{11}{3}}$$

$$= \pm\frac{\sqrt{11}}{\sqrt{3}} \qquad\qquad \sqrt{\frac{a}{b}} = \frac{\sqrt{a}}{\sqrt{b}}$$

$$= \pm\frac{\sqrt{11}}{\sqrt{3}} \cdot \frac{\sqrt{3}}{\sqrt{3}} \qquad \text{Rationalize the denominator.}$$

$$= \pm\frac{\sqrt{33}}{3}$$

We enter $y = 3x^2 - 11$ in a graphing calculator and check that for both inputs $-\dfrac{\sqrt{33}}{3}$ and $\dfrac{\sqrt{33}}{3}$, the output is 0 (see Fig. 3).

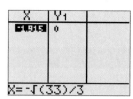 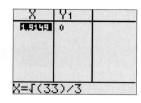

Figure 3 Verify that both $-\dfrac{\sqrt{33}}{3}$ and $\dfrac{\sqrt{33}}{3}$ satisfy the equation $3x^2 - 11 = 0$

Since we used square *roots* to solve the quadratic equations in Example 3, we say we solved them by *extracting square roots*. We can solve an equation such as $(x - 3)^2 = 5$ by extracting square roots. To solve this equation, first let $B = x - 3$.

$$(x - 3)^2 = 5$$

$$B^2 = 5 \qquad \text{Substitute } B \text{ for } x - 3.$$

$$B = \pm\sqrt{5} \qquad \text{Solve for } B.$$

Next, substitute $x - 3$ for B:

$$x - 3 = \pm\sqrt{5}$$

$$x = 3 \pm \sqrt{5}$$

In the future, we streamline the work by writing

$$(x - 3)^2 = 5$$

$$x - 3 = \pm\sqrt{5}$$

$$x = 3 \pm \sqrt{5}$$

Example 4   Solving Quadratic Equations

Solve each equation.

1. $(2x + 5)^2 = 36$

2. $2(4x - 3)^2 + 1 = 11$

3. $(5x + 2)^2 = 0$

4. $4(6x - 11)^2 + 8 = 0$

Solution

1. $(2x + 5)^2 = 36$

$$2x + 5 = \pm\sqrt{36}$$
$$2x + 5 = \pm 6$$

$2x + 5 = -6$ or $2x + 5 = 6$

$2x = -11$ or $2x = 1$

$x = -\dfrac{11}{2}$ or $x = \dfrac{1}{2}$

2. $2(4x - 3)^2 + 1 = 11$

$$2(4x - 3)^2 = 10$$
$$(4x - 3)^2 = 5$$
$$4x - 3 = \pm\sqrt{5}$$
$$4x = 3 \pm \sqrt{5}$$
$$x = \frac{3 \pm \sqrt{5}}{4}$$

We verify each result by storing it as x and checking that $2(4x - 3)^2 + 1$ is equal to 11 (see Fig. 4).

Graphing Calculator

See Section B.20.

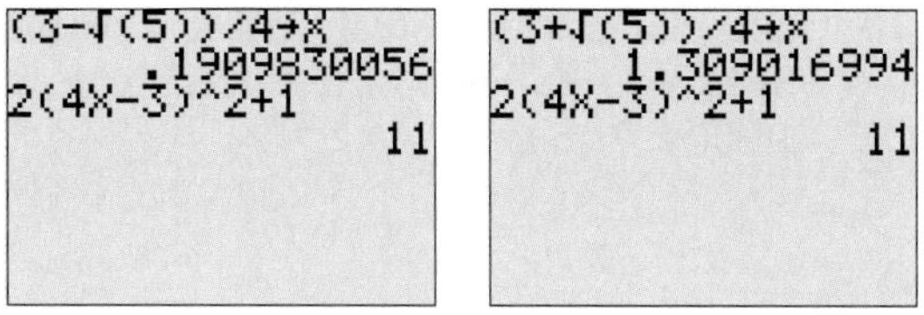

Figure 4 Check that both results satisfy $2(4x - 3)^2 + 1 = 11$

3. $(5x + 2)^2 = 0$

$$5x + 2 = \pm\sqrt{0}$$
$$5x + 2 = 0$$
$$5x = -2$$
$$x = -\frac{2}{5}$$

4. $4(6x - 11)^2 + 8 = 0$

$$4(6x - 11)^2 = -8$$
$$(6x - 11)^2 = -2$$

Since the square of a number is nonnegative, we conclude that there are no real number solutions. ■

Recall that a quadratic equation has two, one, or no real number solutions. Problems 2, 3, and 4 in Example 4 serve as respective examples of these three types of equations.

Example 5   Finding *x*-intercepts

Find the approximate x-intercepts of the function $f(x) = -2(x - 7)^2 + 20$.

Solution

To find the x-intercepts, we substitute 0 for $f(x)$:

$$-2(x - 7)^2 + 20 = 0$$
$$-2(x - 7)^2 = -20$$
$$(x - 7)^2 = 10$$
$$x - 7 = \pm\sqrt{10}$$
$$x = 7 \pm \sqrt{10}$$

$x = 7 - \sqrt{10}$ or $x = 7 + \sqrt{10}$

$x \approx 3.84$ or $x \approx 10.16$

The approximate x-intercepts are $(3.84, 0)$ and $(10.16, 0)$. We verify our work by using "intersect" to locate approximate intersection points of the function f and the x-axis ($y = 0$). See Fig. 5.

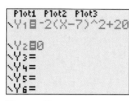 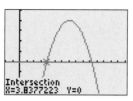 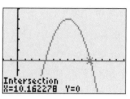

Figure 5 Use "intersect" to verify the approximate x-intercepts

EXPLORATION

Deriving a formula to solve quadratic equations in $a(x-h)^2 + k = p$ form

1. Solve $2(x-5)^2 + 7 = 10$.

2. Solve the equation $a(x-h)^2 + k = p$ for x. Assume that $a \neq 0$ and $\dfrac{p-k}{a} \geq 0$.
   [**Hint:** Follow the same steps as in problem 1.]

3. Use your result from problem 2 to solve the equation $3(x-4)^2 + 2 = 7$. [**Hint:** Substitute the appropriate values for a, h, k, and p into your formula from problem 2.]

EXPLORATION

Looking ahead: Perfect square trinomials

1. Consider the true statement

$$(x + k)^2 = x^2 + (2k)x + k^2$$

   Note that for the trinomial on the right side, the coefficient of x is $2k$ and the constant term is k^2. Describe how to use operations such as adding, subtracting, multiplying, dividing, and/or squaring to change $2k$ into k^2.

2. Show that your description in problem 1 works for the true statement

$$(x + 3)^2 = x^2 + 6x + 9$$

3. Expand $(x - k)^2$. Make a description similar to the one you made in problem 1. Show that your description works for the true statement

$$(x - 4)^2 = x^2 - 8x + 16$$

4. Find a value for c so that the trinomial can be factored as the square of a binomial $(x + k)^2$. Then factor the trinomial.

   a. $x^2 + 8x + c$ **b.** $x^2 + 10x + c$

   c. $x^2 - 14x + c$ **d.** $x^2 - 18x + c$

Tips on Succeeding in This Course

One way to improve your chances of success in this course is to do what you are doing now—read the text. Class time is a great opportunity to be introduced to new concepts and to see how they fit together with previously learned ones. However, there is usually not enough time to address details as well as a textbook can. In this way, a textbook can serve as a supplement to what you learn during class time.

Key Points
OF THIS SECTION

For $a \geq 0$ and $b \geq 0$,

$$\sqrt{ab} = \sqrt{a}\sqrt{b}$$

Product property

For $a \geq 0$ and $b > 0$,

$$\sqrt{\frac{a}{b}} = \frac{\sqrt{a}}{\sqrt{b}}$$

Quotient property

Let k be a nonnegative constant. Then $x^2 = k$ is equivalent to

$$x = \pm\sqrt{k}$$

Solving $x^2 = k$

A radical $\sqrt{k}$ with a counting number k, $k \geq 2$, is *simplified* if k does not have any perfect square factors other than 1.

Simplifying $\sqrt{k}$

For an expression such as $\sqrt{\frac{2}{7}}$, we simplify by leaving no radicand as a fraction.

Simplifying $\sqrt{\frac{p}{q}}$

For an expression such as $\frac{5}{\sqrt{6}}$, we simplify by leaving no denominator as a radical. We call this process rationalizing the denominator.

Simplifying $\frac{p}{\sqrt{q}}$

To rationalize the denominator of a fraction of the form $\frac{p}{\sqrt{q}}$, $q > 0$, we multiply the fraction by 1 in the form $\frac{\sqrt{q}}{\sqrt{q}}$.

Rationalizing the denominator

HOMEWORK 7.1

SSM Tutor Center   MathPro 4 — MathPro 4   MathPro 5 — MathPro 5   CDs/Videos   www

Simplify.

1. $\sqrt{16}$

2. $\sqrt{49}$

3. $\sqrt{8}$

4. $\sqrt{18}$

5. $\sqrt{12}$

6. $\sqrt{75}$

7. $\sqrt{\dfrac{4}{9}}$

8. $\sqrt{\dfrac{25}{81}}$

9. $\sqrt{\dfrac{6}{49}}$

10. $\sqrt{\dfrac{13}{100}}$

11. $\dfrac{5}{\sqrt{2}}$

12. $\dfrac{4}{\sqrt{7}}$

13. $\dfrac{9}{\sqrt{6}}$

14. $\dfrac{15}{\sqrt{10}}$

15. $\dfrac{3}{\sqrt{32}}$

16. $\dfrac{8}{\sqrt{60}}$

17. $\sqrt{\dfrac{3}{2}}$

18. $\sqrt{\dfrac{7}{5}}$

19. $\sqrt{\dfrac{11}{20}}$

20. $\sqrt{\dfrac{7}{90}}$

Solve.

21. $x^2 = 36$

22. $x^2 = 9$

23. $x^2 = 5$

24. $x^2 = 11$

25. $x^2 - 3 = 0$

26. $19 - x^2 = 0$

27. $x^2 = 32$

28. $x^2 = 24$

29. $x^2 = 300$

30. $x^2 = 98$

31. $x^2 - 48 = 0$

32. $0 = x^2 - 45$

33. $x^2 = -9$

34. $x^2 = -5$

35. $x^2 + 25 = 0$

36. $0 = x^2 + 57$

37. $4x^2 = 36$

38. $7x^2 = 21$

39. $5x^2 = 8$

40. $2x^2 = 13$

41. $3x^2 - 14 = 0$

42. $33 - 5x^2 = 0$

43. $(x - 4)^2 = 25$

44. $(x + 1)^2 = 81$

45. $(8x + 3)^2 = 36$

46. $(5x - 1)^2 = 4$

47. $(9x - 7)^2 = 0$

48. $(2x + 3)^2 = 0$

49. $(8x - 3)^2 + 6 = 90$

50. $(4x + 9)^2 - 3 = 36$

51. $3(4x + 7)^2 = -9$

52. $-6(3x - 8)^2 = 12$

53. $5(x - 6)^2 + 3 = 23$

54. $-2(x + 4)^2 - 8 = -24$

55. $-3(4x - 9)^2 + 5 = -31$

56. $7(2x + 1)^2 + 4 = 39$

Find the x-intercepts of each function.

57. $f(x) = x^2 - 17$

58. $g(x) = 4x^2 - 24$

59. $h(x) = (x - 2)^2 - 4$

60. $k(x) = (x + 6)^2 - 25$

61. $f(x) = -3(x + 1)^2 + 12$

62. $g(x) = 6(x - 3)^2 - 54$

63. $h(x) = 2(x + 5)^2 + 1$

64. $k(x) = -4(x - 2)^2 - 8$

Find the approximate x-intercepts of the function. Also, find the vertex. Then, sketch a graph of the function by hand. Use a graphing calculator to verify your work.

65. $f(x) = (x-5)^2 - 9$

66. $g(x) = (x+4)^2 - 36$

67. $h(x) = -2(x+7)^2 + 9$

68. $f(x) = 3(x-2)^2 - 49$

69. $g(x) = x^2 - 31$

70. $h(x) = -2x^2 + 10$

71. A student tries to solve the equation $x^2 - 10x + 25 = 0$:

$$x^2 - 10x + 25 = 0$$
$$x^2 = 10x - 25$$
$$x = \pm\sqrt{10x - 25}$$

Did the student correctly solve it? Explain. If the student did not correctly solve the equation, show how to solve it.

72. A student attempts to solve the equation $x^2 + 6x + 9 = 49$:

$$x^2 + 6x + 9 = 49$$
$$(x+3)^2 = 49$$
$$x + 3 = \pm 7$$
$$x = -3 \pm 7$$
$$x = -10 \quad \text{or} \quad x = 4$$

Did the student correctly solve it? Explain. If the student did not correctly solve the equation, show how to solve it.

73. Assume that $a \geq 0$ and $b > 0$. Show that $\sqrt{\dfrac{a}{b}} = \dfrac{\sqrt{a}}{\sqrt{b}}$.

74. Quadratic equations have two, one, or no real number solutions. Give an example (different from the ones in the text) of each type of equation. Then solve each of your equations.

75. What forms of quadratic equations can you solve using extraction of square roots? Using factoring? For example, quadratic equations of the form $x^2 + 2bx + b^2 = 0$ can be solved by factoring. Include examples in your discussion.

7.2 SOLVING QUADRATIC EQUATIONS BY COMPLETING THE SQUARE

Objective

▸ Solve quadratic equations by completing the square.

In Section 7.1 we solved equations of the form $(x-h)^2 = p$ for x by extracting square roots. In this section we discuss how to write an equation of the form $ax^2 + bx + c = 0$, $a \neq 0$, in the form $(x-h)^2 = p$. We call this method *completing the square*. We can solve *any* quadratic equation by putting it into the form $ax^2 + bx + c = 0$, completing the square, and then extracting square roots.

We begin our study of completing the square by expanding the square of a sum:

$$(x+3)^2 = x^2 + 6x + 9$$

We call $x^2 + 6x + 9$ a *perfect square trinomial*. A **perfect square trinomial** is the trinomial that results from expanding an expression like $(ax + b)^2$.

For $x^2 + 6x + 9$, there is a special connection between the 6 and the 9. If we divide the 6 by 2 and then square the result, we get 9:

$$x^2 + 6x + 9$$

$$\left(\frac{6}{2}\right)^2 = 3^2 = 9$$

This is no coincidence. For example, consider the expansion $(x - 4)^2 = x^2 - 8x + 16$. If we divide -8 by 2 and then square the result, we get 16:

$$x^2 - 8x + 16$$

$$\left(\frac{-8}{2}\right)^2 = (-4)^2 = 16$$

In general, we expand $(x + k)^2$:

$$(x+k)^2 = x^2 + (2k)x + k^2$$

$$\left(\frac{2k}{2}\right)^2 = k^2$$

For $x^2 + (2k)x + k^2$, we see that if we divide the coefficient of x by 2 and then square the result, we get the constant term k^2.

Perfect Square Trinomial Property

For a perfect square trinomial $x^2 + bx + c$, dividing b by 2 and then squaring the result gives c:

$$x^2 + bx + c$$

$$\left(\frac{b}{2}\right)^2 = c$$

Note

The coefficient b may be negative. For example, the property applies for $x^2 - 12x + 36$. Here $b = -12$.

Note that this property is for a perfect square trinomial $ax^2 + bx + c$, where $a = 1$.

Example 1   Perfect Square Trinomials

Find the value of c so that the expression is a perfect square trinomial. Then factor the perfect square trinomial.

1. $x^2 + 10x + c$ **2.** $x^2 - 9x + c$ **3.** $x^2 + \dfrac{5}{3}x + c$

Solution

1. We divide 10 by 2 and then square the result:

$$\left(\frac{10}{2}\right)^2 = 5^2 = 25 = c$$

The expression is $x^2 + 10x + 25$ and its factored form is $(x + 5)^2$.

2. We divide -9 by 2 and then square the result:

$$\left(\frac{-9}{2}\right)^2 = \frac{81}{4} = c$$

The expression is $x^2 - 9x + \dfrac{81}{4}$ and its factored form is $\left(x - \dfrac{9}{2}\right)^2$.

3. Dividing by 2 is the same as multiplying by $\dfrac{1}{2}$. So we multiply $\dfrac{5}{3}$ by $\dfrac{1}{2}$ and square the result:

$$\left(\frac{5}{3} \cdot \frac{1}{2}\right)^2 = \left(\frac{5}{6}\right)^2 = \frac{25}{36} = c$$

The expression is $x^2 + \dfrac{5}{3}x + \dfrac{25}{36}$ and its factored form is $\left(x + \dfrac{5}{6}\right)^2$.   ∎

We can solve equations that can be put into the form $x^2 + bx + c = 0$ by writing the equation in the form $(x - h)^2 = p$, and then extracting square roots.

Example 2   Solving Quadratic Equations

Solve each equation.

1. $x^2 - 8x + 16 = 7$ **2.** $x^2 + 6x = -4$

Solution

1. $x^2 - 8x + 16 = 7$

$\qquad (x - 4)^2 = 7$ Factor the left side.

$\qquad x - 4 = \pm\sqrt{7}$ Extract square roots.

$\qquad x = 4 \pm \sqrt{7}$

4-√(7)
      1.354248689
4+√(7)
      6.645751311

Figure 6 Computing $4 - \sqrt{7}$ and $4 + \sqrt{7}$

The approximate solutions are 1.35 and 6.65 (see Fig. 6).

We can check that each approximate solution approximately satisfies the original equation.

2. Since $\left(\dfrac{6}{2}\right)^2 = 3^2 = 9$, we add 9 to both sides of $x^2 + 6x = -4$ so that the left side will be a perfect square trinomial.

$$x^2 + 6x = -4$$

$$x^2 + 6x + 9 = -4 + 9 \qquad \text{Add 9 to both sides.}$$

$$(x + 3)^2 = 5 \qquad \text{Factor the left side.}$$

$$x + 3 = \pm\sqrt{5} \qquad \text{Extract square roots.}$$

$$x = -3 \pm \sqrt{5}$$

To check, we enter $y = x^2 + 6x$ in a graphing calculator and find that for both inputs $-3 - \sqrt{5}$ and $-3 + \sqrt{5}$, the output is -4 (see Fig. 7).

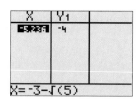
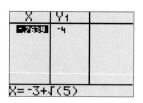

Figure 7 Verify that both $-3 - \sqrt{5}$ and $-3 + \sqrt{5}$ satisfy the equation $x^2 + 6x = -4$

In problem 2 of Example 2, we added 9 to both sides of $x^2 + 6x = -4$ and then factored the left side to get $(x + 3)^2 = 5$. By adding 9 to both sides of the equation, we completed the square for $x^2 + 6x$.

Example 3 Solving by Completing the Square

Solve each equation.

1. $x(x + 10) + 30 = -3$ **2.** $x^2 - 5x + 1 = 0$

Solution

1. To solve $x(x + 10) + 30 = -3$, we first put the equation in the form $x^2 + bx = p$ before completing the square.

$$x(x + 10) + 30 = -3$$

$$x^2 + 10x + 30 = -3 \qquad \text{Distributive law}$$

$$x^2 + 10x = -33 \qquad \text{Subtract 30 from both sides.}$$

$$x^2 + 10x + 25 = -33 + 25 \qquad \text{Add } \left(\dfrac{10}{2}\right)^2 = 5^2 = 25 \text{ to both sides.}$$

$$(x + 5)^2 = -8 \qquad \text{Factor the left side.}$$

Since the square of a number is positive, there are no real number solutions.

2. $\quad x^2 - 5x + 1 = 0$

$$x^2 - 5x = -1 \qquad \text{Subtract 1 from both sides.}$$

$$x^2 - 5x + \dfrac{25}{4} = -1 + \dfrac{25}{4} \qquad \text{Add } \left(\dfrac{-5}{2}\right)^2 = \dfrac{25}{4} \text{ to both sides.}$$

$$\left(x - \dfrac{5}{2}\right)^2 = -\dfrac{4}{4} + \dfrac{25}{4} \qquad \text{Find a least common denominator.}$$

$$\left(x - \dfrac{5}{2}\right)^2 = \dfrac{21}{4} \qquad \dfrac{a}{b} + \dfrac{c}{b} = \dfrac{a + c}{b}$$

$$x - \dfrac{5}{2} = \pm\sqrt{\dfrac{21}{4}} \qquad \text{Extract square roots.}$$

$$x - \frac{5}{2} = \pm\frac{\sqrt{21}}{2} \qquad \sqrt{\frac{a}{b}} = \frac{\sqrt{a}}{\sqrt{b}}$$

$$x = \frac{5}{2} \pm \frac{\sqrt{21}}{2}$$

$$x = \frac{5 \pm \sqrt{21}}{2}$$

We use a graphing calculator to verify our work by storing an approximation of each solution as x and checking that $x^2 - 5x + 1$ is approximately 0 (see Fig. 8).

Figure 8 Verify that both $\frac{5 - \sqrt{21}}{2}$ and $\frac{5 + \sqrt{21}}{2}$ satisfy the equation $x^2 - 5x + 1 = 0$

Graphing Calculator

For calculator entry, the numerator of each fraction must be in parentheses. For storing a valve, see Section B.20.

So far, we have worked only with expressions of the form $ax^2 + bx + c$ where $a = 1$. Consider the expansion of $(2x + 5)^2$:

$$(2x + 5)^2 = 4x^2 + 20x + 25$$

Dividing 20 by 2 and then squaring the result does *not* give 25:

$$\left(\frac{20}{2}\right)^2 = 10^2 = 100 \neq 25$$

So, the perfect square trinomial property given for $ax^2 + bx + c$ where $a = 1$, does not extend to trinomials with $a \neq 1$. However, when solving an equation of the form $ax^2 + bx + c = 0$ with $a \neq 1$, we can first divide both sides by a to obtain an equation involving "$1x^2$," one to which we can apply the property.

Example 4 Solving by Completing the Square

Solve the equation $3x^2 + 7x = 5$ by completing the square.

Solution

$$3x^2 + 7x = 5$$

$$x^2 + \frac{7}{3}x = \frac{5}{3} \qquad \text{Divide both sides by 3.}$$

$$x^2 + \frac{7}{3}x + \frac{49}{36} = \frac{5}{3} + \frac{49}{36} \qquad \text{Add } \left(\frac{7}{3} \cdot \frac{1}{2}\right)^2 = \left(\frac{7}{6}\right)^2 = \frac{49}{36}.$$

$$\left(x + \frac{7}{6}\right)^2 = \frac{60}{36} + \frac{49}{36} \qquad \text{Get a least common denominator.}$$

$$\left(x + \frac{7}{6}\right)^2 = \frac{109}{36} \qquad \text{Add the fractions.}$$

$$x + \frac{7}{6} = \pm\sqrt{\frac{109}{36}} \qquad \text{Extract square roots.}$$

$$x + \frac{7}{6} = \pm\frac{\sqrt{109}}{6} \qquad \sqrt{\frac{a}{b}} = \frac{\sqrt{a}}{\sqrt{b}}, b > 0$$

$$x = -\frac{7}{6} \pm \frac{\sqrt{109}}{6}$$

$$x = \frac{-7 \pm \sqrt{109}}{6}$$

We use a graphing calculator to verify our work by storing an approximation of each solution as x and checking that $3x^2 + 7x$ is approximately 5 (see Fig. 9).

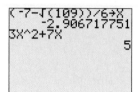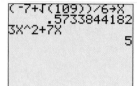

Figure 9 Verify that $\dfrac{-7 - \sqrt{109}}{6}$ and $\dfrac{-7 + \sqrt{109}}{6}$ satisfy the equation $3x^2 + 7x = 5$

■

The first step in solving a quadratic equation involves deciding whether to solve the equation using factoring or by completing the square. For some quadratic equations, it is easier to solve by factoring. However, not all equations can be solved by factoring, whereas *any* equation can be solved by completing the square.

Solve by Completing the Square

To solve a quadratic equation by completing the square:

1. Write the equation in the form $ax^2 + bx = p$.
2. If $a \neq 1$, divide both sides of the equation by a: $x^2 + \dfrac{b}{a}x = \dfrac{p}{a}$.
3. Complete the square for the expression $x^2 + \dfrac{b}{a}x$.
4. Solve the equation by extracting roots.

 EXPLORATION

Looking ahead: Deriving a formula to solve quadratic equations of the form $ax^2 + bx + c = 0$

1. Solve the equation $2x^2 + 9x + 3 = 0$.
2. Solve the quadratic equation $ax^2 + bx + c = 0$ for x. [**Hint**: Follow the same steps as in problem 1.]
3. Use your result from problem 2 to solve the equation $3x^2 + 11x + 5 = 0$. [**Hint**: Substitute the appropriate values for a, b, and c into your formula from problem 2.]

Tips on Succeeding in This Course

It is always a good idea to take notes during classroom activities. Not only will you have something to refer to when later doing the homework, you will also have something to help you prepare for tests. Also, the process of taking notes makes you even more involved with the material, which means that you probably will increase both your understanding and retention of it.

Key Points
OF THIS SECTION

Perfect square trinomial

For a perfect square trinomial $x^2 + bx + c$, dividing b by 2 and then squaring the result gives c.

Completing the square

To solve a quadratic equation by completing the square:

1. Write the equation in the form $ax^2 + bx = p$.

2. If $a \neq 1$, divide both sides of the equation by a: $x^2 + \dfrac{b}{a}x = \dfrac{p}{a}$.

3. Complete the square for the expression $x^2 + \dfrac{b}{a}x$.

4. Solve the equation by extracting roots.

We can solve *any* quadratic equation by completing the square.

When completing the square can be used

HOMEWORK 7.2

SSM
Tutor Center

MathPro 4
MathPro 4

MathPro 5
MathPro 5

CDs/Videos

www
www

Find the value of c for which each expression is a perfect square trinomial. Then factor the perfect square trinomial.

1. $x^2 + 12x + c$

2. $x^2 - 20x + c$

3. $x^2 - 14x + c$

4. $x^2 + 18x + c$

5. $x^2 - 7x + c$

6. $x^2 - 11x + c$

7. $x^2 + 3x + c$

8. $x^2 + 5x + c$

9. $x^2 + \dfrac{1}{2}x + c$

10. $x^2 - \dfrac{1}{7}x + c$

11. $x^2 - \dfrac{4}{5}x + c$

12. $x^2 + \dfrac{3}{4}x + c$

Solve by completing the square.

13. $x^2 + 6x = 1$

14. $x^2 + 8x = 3$

15. $x^2 + 12x = 2$

16. $x^2 - 4x = 6$

17. $x^2 - 2x = 10$

18. $x^2 + 14x = 9$

19. $x^2 - 18x - 3 = 0$

20. $x^2 - 10x - 2 = 0$

21. $x^2 + 18x - 7 = 2$

22. $x^2 - 6x + 1 = 5$

23. $x^2 + 11 = 3 - 4x$

24. $x^2 + 5 = 9 + 20x$

25. $x^2 - 7x = 3$

26. $x^2 - 3x = 12$

27. $x^2 + 5x + 8 = 12$

28. $x^2 + 9x - 4 = 5$

29. $x + 8 = 1 - x^2$

30. $-x + 3 = 10 - x^2$

31. $x^2 - \dfrac{5}{2}x = \dfrac{1}{2}$

32. $x^2 - \dfrac{4}{3}x = \dfrac{5}{3}$

33. $2x^2 - 8x = 3$

34. $3x^2 - 12x = 1$

35. $5x^2 + 15x + 2 = -7$

36. $4x^2 + 8x + 6 = 1$

37. $3x^2 + 4x = 5$

38. $6x^2 - 2x = 7$

39. $2x^2 - x + 4 = 11$

40. $3x^2 + x - 3 = 2$

41. $5x^2 + 7x + 3 = 0$

42. $2x^2 - 3x - 7 = 0$

43. $8x^2 - 5x = 4$

44. $6x^2 - 7x + 8 = 0$

45. $2x(x - 3) = 5 - x$

46. $3x(x + 4) = 7$

47. $x(x - 1) + 3 = 5(x + 1)$

48. $x(x + 6) = -2(x - 1)$

49. A student tries to solve the equation $4x^2 + 6x = 1$ by completing the square:

$$4x^2 + 6x = 1$$
$$4x^2 + 6x + 9 = 1 + 9$$
$$(2x + 3)^2 = 10$$
$$2x + 3 = \pm\sqrt{10}$$
$$2x = -3 \pm \sqrt{10}$$
$$x = \dfrac{-3 \pm \sqrt{10}}{2}$$

Did the student correctly solve the equation? If so, explain how the student can verify the work. If not, explain the error(s) and show how to solve the equation.

50. Find nonzero values of a, b, and c so that the equation $ax^2 + bx + c = 0$ does not have any real number solutions. Your equation should be different from those in the text. [**Hint**: Begin with an appropriate equation of the form $(x - h)^2 = k$ and expand the left side of the equation.]

51. Let $f(x) = x^2 + 6x + 13$.
   a. Find x when $f(x) = 3$.
   b. Find x when $f(x) = 4$.
   c. Find x when $f(x) = 5$.

52. Give an example of an equation that can be solved by factoring. Solve the equation by factoring and then solve it by completing the square. Which process was easier? Explain.

Find the x-intercept(s), if any, of the function.

53. $f(x) = x^2 - 8x + 3$

54. $g(x) = x^2 - 2x + 1$

55. $h(x) = x^2 + 4x + 5$

56. $f(x) = x^2 + 6x + 10$

57. $g(x) = x^2 + 10x + 25$

58. $h(x) = x^2 - 12x + 1$

59. Compare the methods of solving a quadratic equation by factoring, by extracting square roots, and by completing the square. Describe the methods as well as their advantages and disadvantages.

60. Explain how to solve a quadratic equation by completing the square.

7.3 SOLVING QUADRATIC EQUATIONS BY USING THE QUADRATIC FORMULA

Objectives

▶ Solve quadratic equations by using the quadratic formula.

▶ Use the discriminant to determine the number of real number solutions of a quadratic equation.

Completing the square can be used to solve *any* quadratic equation. However, this process can be somewhat involved for some equations. In this section we use completing the square to find the *quadratic formula*, which we can use to solve quadratic equations fairly easily.

Example 1 Deriving the Quadratic Formula

Solve $ax^2 + bx + c = 0$ for x, where $a \neq 0$.

Solution

We solve the equation by completing the square.

$$ax^2 + bx + c = 0$$

$$x^2 + \frac{b}{a}x + \frac{c}{a} = 0 \qquad \text{Divide both sides by } a.$$

$$x^2 + \frac{b}{a}x = -\frac{c}{a} \qquad \text{Subtract } \frac{c}{a} \text{ from both sides.}$$

$$x^2 + \frac{b}{a}x + \frac{b^2}{4a^2} = -\frac{c}{a} + \frac{b^2}{4a^2} \qquad \text{Add } \left(\frac{b}{a} \cdot \frac{1}{2}\right)^2 = \left(\frac{b}{2a}\right)^2 = \frac{b^2}{4a^2}.$$

$$\left(x + \frac{b}{2a}\right)^2 = -\frac{c}{a} \cdot \frac{4a}{4a} + \frac{b^2}{4a^2} \qquad \begin{array}{l}\text{Factor the left side. Find a common} \\ \text{denominator for the right side.}\end{array}$$

$$\left(x + \frac{b}{2a}\right)^2 = \frac{b^2 - 4ac}{4a^2}$$

$$x + \frac{b}{2a} = \pm\sqrt{\frac{b^2 - 4ac}{4a^2}} \qquad \text{Extract square roots.}$$

$$x = -\frac{b}{2a} \pm \frac{\sqrt{b^2 - 4ac}}{2a} \qquad \sqrt{\frac{A}{B}} = \frac{\sqrt{A}}{\sqrt{B}}$$

$$x = \frac{-b \pm \sqrt{b^2 - 4ac}}{2a} \qquad\qquad \blacksquare$$

In Example 1 we found a formula (the last line) for the real number solutions of a quadratic equation. We can use this formula to help us solve a quadratic equation.

Quadratic Formula

If a quadratic equation $ax^2 + bx + c = 0$ has real number solutions, they are given by

$$\frac{-b \pm \sqrt{b^2 - 4ac}}{2a}$$

Example 2 Solving by Using the Quadratic Formula

Solve $x^2 - 6x + 8 = 0$.

Solution

Comparing $x^2 - 6x + 8 = 0$ with $ax^2 + bx + c = 0$, we see that $a = 1$, $b = -6$, and $c = 8$. We substitute these values for a, b, and c in the quadratic formula:

$$x = \frac{-(-6) \pm \sqrt{(-6)^2 - 4(1)(8)}}{2(1)}$$

$$x = \frac{6 \pm \sqrt{4}}{2}$$

$$x = \frac{6 \pm 2}{2}$$

$$x = \frac{6-2}{2} \quad \text{or} \quad x = \frac{6+2}{2}$$
$$x = 2 \quad \text{or} \quad x = 4$$

The solutions are 2 and 4. ∎

Note that instead of using the quadratic formula, we can solve $x^2 - 6x + 8 = 0$ by factoring:

$$x^2 - 6x + 8 = 0$$
$$(x-2)(x-4) = 0$$
$$x - 2 = 0 \quad \text{or} \quad x - 4 = 0$$
$$x = 2 \quad \text{or} \quad x = 4$$

A benefit of the quadratic formula is that we can use it to solve equations that are difficult to solve by factoring. We solve such an equation in Example 3.

Example 3 Solving by Using the Quadratic Formula

Solve $2x^2 = 10x - 3$.

Solution

First, we write $2x^2 = 10x - 3$ in the form $ax^2 + bx + c = 0$:

$$2x^2 - 10x + 3 = 0$$

So, $a = 2$, $b = -10$, and $c = 3$. By the quadratic formula,

$$x = \frac{-(-10) \pm \sqrt{(-10)^2 - 4(2)(3)}}{2(2)}$$

$$x = \frac{10 \pm \sqrt{76}}{4}$$

$$x = \frac{10 \pm 2\sqrt{19}}{4} \qquad \sqrt{76} = \sqrt{4 \cdot 19} = \sqrt{4}\sqrt{19} = 2\sqrt{19}$$

$$x = \frac{10 - 2\sqrt{19}}{4} \quad \text{or} \quad x = \frac{10 + 2\sqrt{19}}{4}$$

$$x = \frac{2(5 - \sqrt{19})}{4} \quad \text{or} \quad x = \frac{2(5 + \sqrt{19})}{4} \qquad \text{Factor out a 2.}$$

$$x = \frac{5 - \sqrt{19}}{2} \quad \text{or} \quad x = \frac{5 + \sqrt{19}}{2} \qquad \text{Reduce.}$$

The solutions are $\dfrac{5 \pm \sqrt{19}}{2}$. ∎

If we are interested in finding approximate solutions to $2x^2 = 10x - 3$, we could stop at the step

$$x = \frac{-(-10) \pm \sqrt{(-10)^2 - 4(2)(3)}}{2(2)}$$

and use a graphing calculator to find the approximate solutions 0.32 and 4.68 (see Fig. 10).

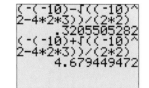

Figure 10 Approximate solutions of $2x^2 = 10x - 3$

Example 4 Solving by Using the Quadratic Formula

Solve $x(x + 1) = -1$.

Solution

First, we write $x(x + 1) = -1$ in the form $ax^2 + bx + c = 0$:

$$x(x + 1) = -1$$
$$x^2 + x = -1$$
$$x^2 + x + 1 = 0$$

So, $a = 1$, $b = 1$, and $c = 1$. Using the quadratic formula gives:

$$x = \frac{-1 \pm \sqrt{1^2 - 4(1)(1)}}{2(1)}$$

$$x = \frac{-1 \pm \sqrt{-3}}{2}$$

Since the square root of a negative number is not a real number, we conclude that there are no real number solutions. ■

Recall that a quadratic equation can have either two, one, or no real number solutions. How does this fact relate to the quadratic formula?

$$x = \frac{-b \pm \sqrt{b^2 - 4ac}}{2a}$$

The answer lies with the number $b^2 - 4ac$, known as the *discriminant*. If the discriminant is positive, then there are two real number solutions (see Example 3). If the discriminant is negative then there are no real number solutions (see Example 4). Finally, if the discriminant is 0, then the quadratic formula gives

$$x = \frac{-b \pm \sqrt{0}}{2a} = \frac{-b}{2a}$$

and, therefore, there is one real number solution.

Determining the Number of Real Number Solutions

For the quadratic equation $ax^2 + bx + c = 0$, the **discriminant** is $b^2 - 4ac$. Also:

- If $b^2 - 4ac > 0$, there are two real number solutions.
- If $b^2 - 4ac = 0$, there is one real number solution.
- If $b^2 - 4ac < 0$, there are no real number solutions.

Example 5   Determining the Number of Real Number Solutions

Determine the number of real number solutions for the equation $2x^2 - 3x + 5 = 0$.

Solution
Since $b^2 - 4ac = (-3)^2 - 4(2)(5) = -31 < 0$, we conclude that the quadratic equation $2x^2 - 3x + 5 = 0$ has no real number solutions. ■

We can also use the discriminant to determine the number of points on a parabola at a given height.

Example 6   Finding the Number of Points at a Given Height

For $f(x) = x^2 - 6x + 12$, find the number of points that lie on the graph of f at the indicated height.

1. $y = 5$ **2.** $y = 3$ **3.** $y = 1$

Solution
1. We substitute 5 for $f(x)$ in the equation $f(x) = x^2 - 6x + 12$.

$$x^2 - 6x + 12 = 5$$

$$x^2 - 6x + 7 = 0$$

Since $b^2 - 4ac = (-6)^2 - 4(1)(7) = 8 > 0$, we conclude that there are two solutions to the equation $x^2 - 6x + 12 = 5$, which means that there are two (symmetric) points that have height $y = 5$.

2. We substitute 3 for $f(x)$:

$$x^2 - 6x + 12 = 3$$
$$x^2 - 6x + 9 = 0$$

Since $b^2 - 4ac = (-6)^2 - 4(1)(9) = 0$, we conclude that there is one solution to the equation $x^2 - 6x + 12 = 3$, which means that there is one point with height $y = 3$. Since the point does not have a symmetric point, it must be the vertex of the parabola.

3. We substitute 1 for $f(x)$:

$$x^2 - 6x + 12 = 1$$
$$x^2 - 6x + 11 = 0$$

Since $b^2 - 4ac = (-6)^2 - 4(1)(11) = -8 < 0$, we conclude that there are no real number solutions to the equation $x^2 - 6x + 12 = 1$ which means that there are no points on the parabola with height $y = 1$. ∎

In Example 6 we found that the parabola $f(x) = x^2 - 6x + 12$ has exactly two points at height $y = 5$, exactly one point at height $y = 3$, and no points at height $y = 1$. We indicate the three points on the graph of f in Fig. 11.

In Example 7 we make a prediction for the dependent variable of a quadratic model by using the quadratic formula.

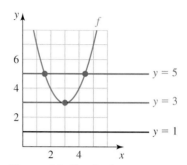

Figure 11 Graph of $f(x) = x^2 - 6x + 12$

Example 7   Modeling with a Quadratic Function

Sales of bottled water increased during the 1990s (see Table 1).

It just tastes better!!

Table 1 Bottled Water Consumptions

Year	Bottled Water Consumption (billions of gallons)
1990	2.24
1992	2.42
1994	2.90
1996	3.45
1998	4.15
2000	5.03
2001	5.50

Source: *Beverage Marketing Corporation*

Let $f(t)$ represent the bottled water consumption (in billions of gallons) during the year that is t years since 1990. A possible equation for f is

$$f(t) = 0.0191t^2 + 0.089t + 2.22$$

1. Verify that f models the data well.

2. Predict when sales will reach 10 billion gallons.

Solution

1. We draw the graph of f and the scattergram of the data in the same viewing window (see Fig. 12). It appears that f is a reasonable model.

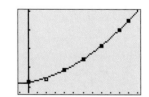

Figure 12 Check how well f models the data

2. To predict when sales will be 10 billion gallons, we substitute 10 for $f(t)$ and solve for t:

$$10 = 0.0191t^2 + 0.089t + 2.22$$
$$0 = 0.0191t^2 + 0.089t - 7.78$$

$$t = \frac{-0.089 \pm \sqrt{0.089^2 - 4(0.0191)(-7.78)}}{2(0.0191)}$$

$$t \approx -22.65 \quad \text{or} \quad t \approx 17.99$$

Figure 13 Verify the results

We verify the results by entering $y = 0.0191t^2 + 0.089t + 2.22$ in a graphing calculator and checking that the inputs -22.65 and 17.99 lead to outputs of about 10 (see Fig. 13).

The values $t = -22.65 \approx -23$ and $t = 17.99 \approx 18$ represent the years $1990 - 23 = 1967$ and $1990 + 18 = 2008$. The estimate of 1967 is a result of model breakdown, as a little research would show that bottled water consumption in 1967 was much less than 10 billion gallons. Therefore, we predict that it will be 2008 when bottled water consumption reaches 10 billion gallons. ∎

In Example 7 we used the model

$$f(t) = 0.0191t^2 + 0.089t + 2.22$$

to make a prediction. To make a prediction for the dependent variable $f(t)$, we substitute a value for t and solve for $f(t)$. To make a prediction for the independent variable t, we substitute a value for $f(t)$ and solve for t using the quadratic formula.

EXPLORATION
Comparing methods of solving quadratic equations

1. Solve the equation $x^2 + 5x + 6 = 0$ by factoring.
2. Solve the equation $x^2 + 5x + 6 = 0$ by completing the square.
3. Solve the equation $x^2 + 5x + 6 = 0$ by using the quadratic formula.
4. Compare your results from problems 1, 2, and 3. Which method was easiest?
5. Try solving the equation $x^2 + 4x - 7 = 0$ by each of the three methods. What do you find?
6. Compare the methods of solving quadratic equations by factoring, by completing the square, and by using the quadratic formula. What are the advantages and disadvantages of each method?

EXPLORATION
Looking ahead: Finding an equation of a parabola

In this exploration, you will find an equation for the parabola that passes through the points $(1, 1)$, $(2, 3)$, and $(3, 9)$.

1. Since $(1, 1)$ lies on the parabola, this ordered pair should satisfy the equation $y = ax^2 + bx + c$. Find the equation that results from substituting 1 for x and 1 for y. This equation will be in terms of a, b, and c. Find another equation using $(2, 3)$. Finally, find a third equation using the ordered pair $(3, 9)$.
2. You should now have three equations, each in terms of a, b, and c. Choose any two of these equations and combine them to eliminate c. Then choose another pair of equations and, again, combine them to eliminate c.
3. You should now have two equations, both in terms of a and b, forming a system of two equations in two unknowns. Solve this system using substitution or elimination.
4. You should now know the values of a and b. Find c by substituting the values of a and b into one of the three equations found in problem 1 of this exploration.
5. You should now know the values of a, b, and c. Substitute these values into the equation $y = ax^2 + bx + c$ to obtain an equation of the parabola.
6. Verify that the graph of your equation passes through the points $(1, 1)$, $(2, 3)$, and $(3, 9)$ by using a graphing calculator table or graph.

Key Points
OF THIS SECTION

If the quadratic equation $ax^2 + bx + c = 0$ has real number solutions, they are given by

$$x = \frac{-b \pm \sqrt{b^2 - 4ac}}{2a}$$

Quadratic formula and number of real number solutions

Also:
- If $b^2 - 4ac > 0$, there are two real number solutions.
- If $b^2 - 4ac = 0$, there is one real number solution.
- If $b^2 - 4ac < 0$, there are no real number solutions.

Any quadratic equation can be solved by using the quadratic formula.

When the quadratic formula can be used

We can simplify a fraction such as $\dfrac{3 + 3\sqrt{5}}{6}$ by factoring the numerator and then reducing the fraction:

Simplifying by factoring and reducing

$$\frac{3 + 3\sqrt{5}}{6} = \frac{3(1 + \sqrt{5})}{6} = \frac{1 + \sqrt{5}}{2}$$

To make a prediction for the dependent variable of a quadratic model, substitute the chosen value for the independent variable in the equation, and then solve for the dependent variable.

Predicting for the dependent variable

To make a prediction for the independent variable of a quadratic model, substitute the chosen value for the dependent variable in the equation, and then solve for the independent variable, usually by using the quadratic formula.

Predicting for the independent variable

HOMEWORK 7.3

 SSM Tutor Center

 MathPro 4 MathPro 5 CDs/Videos

 www www

Solve. Use the quadratic formula.

1. $2x^2 + 5x - 2 = 0$
2. $5x^2 + 5x - 1 = 0$
3. $3x^2 + 7x - 2 = 0$
4. $2x^2 + 7x - 3 = 0$
5. $2x^2 - 5x + 1 = 0$
6. $3x^2 - x - 1 = 0$
7. $3x^2 - 2x + 8 = 0$
8. $x^2 - 3x + 15 = 0$
9. $2x^2 + 5x = 3$
10. $4x^2 - 7x = 2$
11. $3x^2 - 17 = 0$
12. $4x^2 - 81 = 0$
13. $2x^2 = -5x$
14. $5x^2 = 3$
15. $4x^2 = 2x + 3$
16. $x^2 = 3 - 4x$
17. $-3x(x - 1) = 2$
18. $2x(3x - 1) = -3$
19. $3(x^2 - 2) = 5x$
20. $2x(x + 3) - 1 = 2(x + 1)$
21. $3(x + 4) = -2x(x - 1)$
22. $5x(4x - 3) = -2$
23. $3 - 6x = x^2$
24. $7x + 6 = 3x^2$

Find approximate solutions for each equation.

25. $2x^2 - 5x - 4 = 0$
26. $-3x^2 = 9x + 2$
27. $2.8x^2 - 7.1x = 4.4$
28. $3.98x^2 = 2.17x + 3.68$
29. $-5.4x(x + 9.8) + 4.1 = 3.2 - 6.9x$
30. $7.1x(x - 4.9) = 3.2$

Solve. Use the method of your choice.

31. $x^2 = 25$
32. $2x^2 + 3x + 9 = 0$
33. $x(x - 8) = -16$
34. $(x + 1)^2 = 81$
35. $4x^2 = 80$
36. $x^2 + 2x - 24 = 0$
37. $(4x + 3)^2 + 2 = 22$
38. $3x = x(x - 1) + 1$
39. $2x(3x - 4) = x - 5$
40. $5x^2 - 8x + 2 = 0$
41. $9x^2 - 5x = 0$
42. $x^2 - 18x = -81$
43. $5x^2 = 3x + 8$
44. $100x^2 = 49$
45. $3x^2 = 27x$
46. $3x^2 - 7x + 2 = 0$
47. $2(2x^2 + 1) = 3x$
48. $3x^2 = 6x + 2$
49. $x^2 + 12x + 36 = 0$
50. $16x^2 - 20x = -4$
51. $24x^2 - 18x - 60 = 0$
52. $2x^2 = 9$
53. $(x - 4)^2 - 3 = 7$
54. $6x^2 - 7x + 3 = 0$
55. $x(x + 1) = 7 - x$
56. $2x^2 - 72 = 0$
57. $2x(3x - 1) = 3(x - 2)$
58. $x^2 - 17x + 16 = 0$
59. $2x(x - 3) = 80$
60. $x^2 - 18x + 45 = 0$
61. $5x^2 = 35x$
62. $12x^2 - 75 = 0$
63. $50x = 5x^2 + 105$
64. $5x^2 = 13$
65. $(3x - 1)^2 = 7$
66. $3(x^2 + 10x) = -75$

67. $(x - 2)(x - 5) = 1$

68. $(x + 3)(x - 2) = 25$

69. $(x - 1)^2 + (x + 2)^2 = 0$

70. $(x - 1)^2 + (x - 3)^2 = 3$

Expand or solve, whichever is appropriate.

71. $(x + 2)(x - 5)$

72. $2x(3x - 5) = 4$

73. $(x + 2)(x - 5) = 3$

74. $2x(3x - 5)$

75. $-4(x - 2)^2 + 3 = -1$

76. $(x + 3)^2$

77. $-4(x - 2)^2 + 3$

78. $(x + 3)^2 = 7$

79. Let $f(x) = x^2 - 4x + 8$. Find the number of points that lie on the graph of f at the indicated height.

    a. $y = 3$ **b.** $y = 4$ **c.** $y = 5$

    d. Use your graphing calculator to draw a graph of f and sketch the graph on paper. Then explain why you found the number of points to be 0, 1, and 2 for parts a, b, and c, respectively.

80. Let $g(x) = -x^2 + 6x - 2$. Find the number of points that lie on the graph of g at the indicated height.

    a. $y = 6$ **b.** $y = 7$ **c.** $y = 8$

    d. Use your graphing calculator to draw a graph of g and sketch the graph on paper. Then explain why you found the number of points to be 2, 1, and 0 for parts a, b, and c, respectively.

81. Let $f(x) = x^2 - 6x + 7$. Find the approximate coordinates of any points on the graph of f at height $y = 2$. Then find the vertex of f. Finally, sketch a graph of f.

82. Let $g(x) = x^2 + 8x + 6$. Find the approximate coordinates of any points on the graph of g at height $y = -5$. Then find the vertex of g. Finally, sketch a graph of g.

83. The number of charter schools—new, special-purpose schools that are not required to conform with traditional curricula—increased greatly in the United States in the 1990s (see Table 2).

Table 2 Charter Schools in the United States	
Year	Number of Charter Schools
1992	2
1994	98
1996	425
1998	1092
2000	2000
2001	2372

Source: *U.S. Department of Education, www.ed.gov/pdfdocs/4yrrpt.pdf. CER Education Reform Update, www.edreform.com/press/ 2001/010917.html. Both accessed 4/4/02.*

Let $f(t)$ represent the number of charter schools in the United States at t years since 1990. A possible equation for f is

$$f(t) = 30.32t^2 - 122.15t + 109.75$$

    a. Draw the graph of f and the scattergram of the data in the same viewing window. Does f model the data reasonably well?

    b. Predict the number of charter schools in 2007. Verify your result using a graphing calculator graph or table.

    c. When will there be an average of 200 charter schools per state? Verify your result using a graphing calculator graph or table.

84. The numbers of new U.S. hotel openings for various years are shown in Table 3.

Table 3 New Hotel Openings	
Year	Number of Openings
1997	1476
1998	1519
1999	1402
2000	1246
2001	1047

Source: *Lodging Econometrics*

Let $f(t)$ represent the number of new U.S. hotel openings in the year that is t years since 1990. A possible equation for f is

$$f(t) = -37.36t^2 + 559.33t - 595.31$$

    a. Draw the graph of f and the scattergram of the data in the same viewing window. Does f model the data reasonably well?

    b. Find t when $f(t) = 700$. What does your result mean in terms of the situation?

    c. Find the t-intercepts of f. What do the t-intercepts mean in terms of the situation?

85. The numbers of convictions of police officers for the past five years are listed in Table 4.

Table 4 Convictions of Police Officers	
Year	Convictions
1994	143
1995	135
1996	83
1997	149
1998	246

Source: *Federal Bureau of Investigation*

Let $f(t)$ represent the number of police officers convicted during the year that is t years since 1990. A possible equation for f is

$$f(t) = 23.43t^2 - 259.14t + 815.77$$

Give me all your money!!

    a. Draw the graph of f and the scattergram of the data in the same viewing window. Does f model the data reasonably well?

    b. Find $f(13)$. Verify your result using a graphing calculator graph or table. What does your result mean in terms of convictions of police officers?

c. Find t when $f(t) = 2000$. Verify your result using a graphing calculator graph or table. What does your result mean in terms of convictions of police officers?

86. A person throws a stone into the air. The height $h(t)$ (in feet) after t seconds is given by

$$h(t) = -16t^2 + 52t + 4$$

a. What is the height of the stone after 3 seconds?

b. When is the stone at a height of 30 feet?

c. When does the stone reach the ground?

87. A student tries to solve $2x^2 + 5x = 1$:

$$x = \frac{-5 \pm \sqrt{5^2 - 4(2)(1)}}{2(2)}$$

$$x = \frac{-5 \pm \sqrt{17}}{4}$$

Did the student correctly solve the equation? If so, show how to verify the work. If not, describe the error(s) and show how to solve the equation.

88. A student tries to solve $3x^2 + 2x - 4 = 0$:

$$x = \frac{-2 \pm \sqrt{2^2 - 4(3)(-4)}}{2(3)}$$

$$x = \frac{-2 \pm \sqrt{52}}{6}$$

$$x = \frac{-2 \pm 2\sqrt{13}}{6}$$

$$x = \frac{-1 \pm 2\sqrt{13}}{3}$$

Did the student correctly solve the equation? If so, show how to verify the work. If not, describe the error(s) and show how to solve the equation.

Find the approximate x-intercept(s), if any, of the function.

89. $f(x) = x^2 - 4x - 6$

90. $g(x) = x^2 - 81$

91. $h(x) = 3x^2 - 2x + 5$

92. $g(x) = 4x^2 - 28x$

93. The quadratic formula gives the solutions of any equation of the form $ax^2 + bx + c = 0$ with $a \neq 0$.

a. Find a "linear formula" that gives the solution of *any* equation of the form $mx + b = 0$ with $m \neq 0$.

b. Use your linear formula to solve $7x + 21 = 0$. Verify your result by solving $7x + 21 = 0$ in the usual way.

94. a. Use factoring to solve $3x^2 + 2x = 0$.

b. Use factoring to solve $ax^2 + bx = 0$ for x.

c. Use the formula you found in part b to solve $3x^2 + 2x = 0$. Compare your results with your results from part a.

d. Use the quadratic formula to solve $3x^2 + 2x = 0$. Compare your results with your results from part a.

e. Use the quadratic formula to solve $ax^2 + bx = 0$ for x. Compare your results with your results from part b.

f. Use the formula you found in part e to solve the equation $7x^2 - 3x = 0$.

*Find the nonzero value(s) of the constant k so that the equation has exactly one solution for the variable x. [**Hint:** What do you know about the discriminant?]*

95. $kx^2 - 12x + 4 = 0$

96. $kx^2 + 20x + 25 = 0$

97. $2x^2 + kx + 8 = 0$

98. $5x^2 + kx + 5 = 0$

99. $9x^2 - 6x + k = 0$

100. $4x^2 - 28x + k = 0$

101. Solve the equation $x^2 - x - 20 = 0$ by factoring, by completing the square, and by using the quadratic formula. Compare your results.

102. Explain how to determine whether to solve a quadratic equation by factoring, by extracting square roots, by completing the square, or by using the quadratic formula. Give examples and, for each example, describe the advantages with the method you chose to use and the disadvantages with the other methods.

103. Describe how to solve a quadratic equation using the quadratic formula.

7.4 SOLVING SYSTEMS OF THREE LINEAR EQUATIONS TO FIND QUADRATIC FUNCTIONS

If I had to select one quality, one personal characteristic that I regard as being most highly correlated with success, whatever the field, I would pick the trait of persistence. Determination. The will to endure to the end, to get knocked down seventy times and get up off the floor saying, "Here comes number seventy-one!" —*Richard M. DeVos*

Objective

▶ Find an equation of a parabola that contains three given points.

In this section we discuss how to find an equation for a parabola that contains three given points. In Section 7.5 we use this skill to find an equation of a quadratic model.

Example 1 Finding an Equation of a Parabola

Find an equation of a parabola that contains the points $(1, 1)$, $(2, 3)$, and $(3, 9)$.

Solution

Our goal is to find values of the constants a, b, and c for the equation $y = ax^2 + bx + c$. Since the three given points lie on the parabola, each of these points satisfies the equation $y = ax^2 + bx + c$.

$$1 = a(1)^2 + b(1) + c \qquad \text{Substitute (1, 1) into } y = ax^2 + bx + c.$$
$$3 = a(2)^2 + b(2) + c \qquad \text{Substitute (2, 3) into } y = ax^2 + bx + c.$$
$$9 = a(3)^2 + b(3) + c \qquad \text{Substitute (3, 9) into } y = ax^2 + bx + c.$$

We can simplify these equations.

$$a + b + c = 1 \qquad\qquad (1)$$
$$4a + 2b + c = 3 \qquad\qquad (2)$$
$$9a + 3b + c = 9 \qquad\qquad (3)$$

We can find values for a, b, and c using elimination. It is easiest to eliminate c. We begin by multiplying both sides of equation (1) by -1.

$$-a - b - c = -1 \qquad\qquad (4)$$

Adding the left sides and adding the right sides of equations (2) and (4) gives:

$$3a + b = 2 \qquad\qquad (5)$$

Adding the left sides and adding the right sides of equations (3) and (4) gives:

$$8a + 2b = 8 \qquad\qquad (6)$$
$$4a + b = 4 \qquad \text{Divide both sides by 2.} \qquad (7)$$

Equations (5) and (7) form a system in two variables:

$$3a + b = 2 \qquad\qquad (8)$$
$$4a + b = 4 \qquad\qquad (9)$$

To eliminate b, we multiply both sides of equation (8) by -1 and add each side to the corresponding side of equation (9).

$$a = 2$$

Next, we substitute 2 for a in the equation $3a + b = 2$.

$$3(2) + b = 2$$
$$b = -4$$

Then, we substitute 2 for a and -4 for b in the equation $a + b + c = 1$.

Note

We would also get $c = 3$ if we made the substitutions in equation (2) or equation (3).

$$2 + (-4) + c = 1$$
$$c = 3$$

Therefore, $a = 2$, $b = -4$, and $c = 3$. So, the equation of the parabola is

$$y = 2x^2 - 4x + 3$$

We use a graphing calculator table as well as a scattergram and graph to check that the parabola $y = 2x^2 - 4x + 3$ contains the points (1, 1), (2, 3), and (3, 9). See Fig. 14.

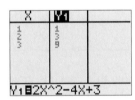
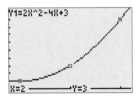

Figure 14 Verify that the parabola $y = 2x^2 - 4x + 3$ contains the given points

In Example 1 we substituted three given points into the equation $y = ax^2 + bx + c$ to obtain the *system of three equations in three unknowns*:

$$a + b + c = 1$$
$$4a + 2b + c = 3$$
$$9a + 3b + c = 9$$

Next, we used elimination three times to find the values $a = 2$, $b = -4$, and $c = 3$. We say that the *solution* of the system is $(2, -4, 3)$. A **solution** of a system of three equations satisfies *all* three equations.

By finding the values $a = 2$, $b = -4$, and $c = 3$, we concluded that an equation of a parabola containing the points $(1, 1)$, $(2, 3)$, and $(3, 9)$ is

$$y = 2x^2 - 4x + 3.$$

To find a linear equation $y = mx + b$, we need *two* points to find the *two* constants m and b. To find an exponential equation $y = ab^x$, we need *two* points to find the *two* constants a and b. To find a quadratic equation $y = ax^2 + bx + c$ we need *three* points to find the *three* constants a, b, and c.

The process of finding an equation of a parabola is considerably easier if one of the three given points is the y-intercept, as we shall see in Example 2.

Note

In exercise 34 you will show that two points do *not* determine a parabola $y = ax^2 + bx + c$.

Example 2 Finding an Equation When One of the Given Points is the y-intercept

Find the equation of a parabola that contains the points $(0, 1)$, $(3, 7)$, and $(4, 5)$.

Solution

We begin by substituting the given points into $y = ax^2 + bx + c$.

$$1 = a(0)^2 + b(0) + c \qquad \text{Substitute } (0, 1).$$
$$7 = a(3)^2 + b(3) + c \qquad \text{Substitute } (3, 7).$$
$$5 = a(4)^2 + b(4) + c \qquad \text{Substitute } (4, 5).$$

Next, we simplify the equations

$$c = 1 \tag{10}$$
$$9a + 3b + c = 7 \tag{11}$$
$$16a + 4b + c = 5 \tag{12}$$

Since $c = 1$, we substitute 1 for c in equations (11) and (12).

$$9a + 3b + 1 = 7$$
$$16a + 4b + 1 = 5$$

Simplifying these equations gives:

$$9a + 3b = 6 \tag{13}$$
$$16a + 4b = 4 \tag{14}$$

To eliminate b, we multiply both sides of equation (13) by -4, and both sides of equation (14) by 3.

$$-36a - 12b = -24 \tag{15}$$
$$48a + 12b = 12 \tag{16}$$

Adding the left sides and adding the right sides of equations (15) and (16) gives

$$12a = -12$$
$$a = -1$$

Next, we substitute -1 for a in the equation $9a + 3b = 6$ and solve for b:

$$9(-1) + 3b = 6$$
$$3b = 15$$
$$b = 5$$

Therefore, $a = -1$, $b = 5$, and $c = 1$, and the equation of the parabola is

$$y = -x^2 + 5x + 1$$

We use a graphing calculator table as well as a scattergram and graph to verify that the graph of $y = -x^2 + 5x + 1$ contains the points $(0, 1)$, $(3, 7)$, and $(4, 5)$. (See Fig. 15.)

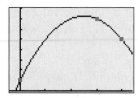

Figure 15 Verify that the parabola contains the given points

■

In Example 2 we were able to find an equation of the desired parabola using elimination once, rather than three times. We need only use elimination once when one of the three given points is the *y*-intercept.

EXPLORATION

For any three points, is there a quadratic function that contains them?

1. Find the values of a, b, and c for the function $f(x) = ax^2 + bx + c$, where the graph of f contains the points $(0, 1)$, $(1, 4)$, and $(2, 7)$. What type of function is f? Why did this happen?

2. Do the same for the points $(0, 1)$, $(0, 8)$, and $(1, 4)$. What happens? Is there a function $f(x) = ax^2 + bx + c$ whose graph contains these points? Explain.

3. What must be true of three points so that there is a quadratic function whose graph contains the points? Give an example of three such points, plot them, and sketch the graph of the quadratic function that contains them. Then find the equation and use a graphing calculator to view the graph. Compare the two graphs.

Tips on Succeeding in This Course

When learning a definition or property, try to create an example. While studying the material in this section, you could select three points and determine whether there is a parabola that contains the chosen points and, if so, you could find an equation of the parabola. Creating examples will shed light on many details of a concept and also personalize the information.

Key Points
OF THIS SECTION

Finding an equation of a parabola

To find an equation of a parabola that contains three given points:

• Obtain a system of three equations in three unknowns by substituting the three given points into the equation $y = ax^2 + bx + c$.

• If none of the three points is the *y*-intercept, solve the system by using elimination three times. If one of the points is the *y*-intercept, solve the system using elimination once.

• Substitute the found values for a, b, and c into the equation $y = ax^2 + bx + c$.

HOMEWORK 7.4

SSM
Tutor Center

MathPro 4
MathPro 4

MathPro 5
MathPro 5

CDs/Videos

www

Find an equation of a parabola that contains the given points. Use your graphing calculator to verify that the graph of your equation contains the points.

1. $(1, 6)$, $(2, 11)$, $(3, 18)$

2. $(1, 1)$, $(2, 5)$, $(3, 15)$

3. $(1, 5)$, $(2, 11)$, $(3, 19)$

4. $(1, 5)$, $(2, 8)$, $(3, 15)$

5. $(1, 9)$, $(2, 7)$, $(4, -15)$

6. $(1, 4)$, $(2, 3)$, $(3, 0)$

7. $(2, 2)$, $(3, 11)$, $(4, 24)$

8. $(2, 3)$, $(3, -2)$, $(4, -11)$

9. $(1, -3), (3, 9), (5, 29)$

10. $(2, -1), (4, 19), (5, 38)$

11. $(3, 7), (4, 0), (5, -11)$

12. $(2, 4), (4, 30), (5, 49)$

13. $(3, 2), (4, 16), (5, 36)$

14. $(4, -3), (5, 2), (6, 9)$

15. $(2, -5), (4, 3), (5, 13)$

16. $(1, 3), (2, 3), (4, -9)$

17. $(1, -1), (2, -1), (3, 3)$

18. $(1, 3), (2, 7), (3, 13)$

19. $(0, 4), (2, 8), (3, 1)$

20. $(0, 0), (1, 4), (2, 14)$

21. $(0, -1), (1, 3), (2, 13)$

22. $(0, 5), (2, 13), (3, 26)$

23. $(0, 17), (2, 11), (3, 2)$

24. $(0, -3), (1, 1), (2, 1)$

25. $(1, 1), (2, 4), (3, 9)$

26. $(1, -1), (2, -4), (3, -9)$

27. Find the values of a, b, and c for the function $f(x) = ax^2 + bx + c$, where the graph of f contains the points $(1, 1)$, $(2, 2)$, and $(3, 3)$. What type of function is f?

28. Find the values of a, b, and c for the function $g(x) = ax^2 + bx + c$, where the graph of g contains the points $(1, 2)$, $(2, 5)$, and $(3, 8)$. What type of function is g?

29. Find an equation of a parabola with vertex $(5, -7)$ that contains the point $(8, 11)$. [**Hint**: Try using the vertex form $y = a(x - h)^2 + k$.]

30. Find an equation of a parabola with vertex $(-5, 8)$ that contains the point $(-7, -4)$.

31. The graph of a quadratic function has y-intercept $(0, 4)$ and x-intercepts $(1, 0)$ and $(2, 0)$. Find an equation for the function.

32. The graph of a quadratic function has y-intercept $(0, 8)$ and x-intercepts $(-4, 0)$ and $(2, 0)$. Find an equation for the function.

33. Find equations of a linear function, quadratic function, and exponential function so that the graph of each equation contains the points $(0, 2)$ and $(1, 4)$. Use your graphing calculator to verify your work.

34. In this exercise, you will show that two points do not determine a quadratic function.

   a. Plot the points $(1, 4)$ and $(3, 4)$ on a coordinate system.

   b. Sketch three parabolas that all contain both of the points $(1, 4)$ and $(3, 4)$. Find the vertex of each of your parabolas.

   c. Write an equation for each of your three sketched parabolas. [**Hint**: You may want to use the vertex form $y = a(x - h)^2 + k$.]

   d. Explain why two points do not determine a quadratic function.

35. Find an equation of the parabola sketched in Fig. 16. [**Hint**: Choose three points whose coordinates appear to be integers.]

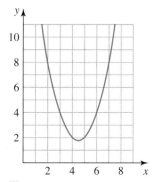

Figure 16 Exercise 35

36. Find an equation of the parabola sketched in Fig. 17.

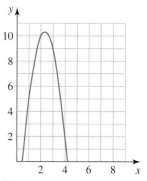

Figure 17 Exercise 36

37. Find an equation of the parabola sketched in Fig. 18.

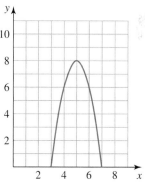

Figure 18 Exercise 37

38. Find an equation of the parabola sketched in Fig. 19.

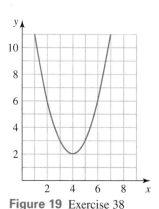

Figure 19 Exercise 38

39. Choose three points that do not all lie on a line and of which no two are in line vertically. Find an equation of a parabola that contains the points. [**Hint**: To make the calculations easier, choose one of the points so that it is the y-intercept.]

40. Describe how to find an equation for a parabola that contains three points that do not all lie on a line and of which no two are in line vertically.

Table 5 Property Crimes	
Year	Number of Property Crimes (millions)
1984	10.61
1986	11.72
1988	12.36
1990	12.66
1992	12.51
1994	12.13
1996	11.79
1998	10.95
2000	9.90

Source: *Federal Bureau of Investigation*

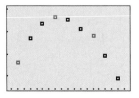

Figure 20 Property crime scattergram

7.5 SOLVING SYSTEMS OF THREE LINEAR EQUATIONS TO FIND QUADRATIC MODELS

Objectives

▸ Find an equation for a quadratic model.

▸ Determine whether a linear function, exponential function, or a quadratic function can be used to model a situation.

Now that we can find an equation of a parabola that contains three given points, we are ready to find the equation of a quadratic model.

Example 1   Finding an Equation of a Quadratic Model

Although the number of property crimes increased significantly from 1984 to 1991, it actually decreased from 1991 to 2000 (see Table 5). Property crimes consist of burglaries, larceny theft, and motor vehicle theft. Find an equation of a function that models the data well.

Solution

Let $f(t)$ represent the number of property crimes (in millions) during the year that is t years since 1980. A scattergram of the data is shown in Fig. 20. Note that the data appears to be quadratically related.

Through practice, we envision that a parabola containing the data points $(4, 10.61)$, $(10, 12.66)$, and $(16, 11.79)$ may be close to the other data points. To find an equation for this parabola, we substitute the three ordered pairs into the standard form $f(t) = at^2 + bt + c$.

$$10.61 = a(4)^2 + b(4) + c$$
$$12.66 = a(10)^2 + b(10) + c$$
$$11.79 = a(16)^2 + b(16) + c$$

We can simplify these equations:

$$16a + 4b + c = 10.61 \tag{17}$$
$$100a + 10b + c = 12.66 \tag{18}$$
$$256a + 16b + c = 11.79 \tag{19}$$

Now, we eliminate c. The equation produced by multiplying both sides of equation (17) by -1 and adding the left sides and adding the right sides of the result and equation (18) is:

$$84a + 6b = 2.05 \tag{20}$$

The equation that results from multiplying both sides of equation (18) by -1 and adding the left sides and adding the right sides of the result and equation (19) is:

$$156a + 6b = -0.87 \tag{21}$$

Equations (20) and (21) form a system in two variables:

$$84a + 6b = 2.05 \tag{22}$$
$$156a + 6b = -0.87 \tag{23}$$

To eliminate b, we first multiply both sides of equation (23) by -1 and then add the left sides and add the right sides of the result and equation (22) to obtain:

$$-72a = 2.92$$
$$a \approx -0.0406$$

We now substitute -0.0406 for a in $84a + 6b = 2.05$ and solve for b.

$$84(-0.0406) + 6b = 2.05$$
$$-3.4104 + 6b = 2.05$$
$$6b = 5.4604$$
$$b \approx 0.91$$

Then we substitute -0.0406 for a and 0.91 for b in $16a + 4b + c = 10.61$ and solve for c.

$$16(-0.0406) + 4(0.91) + c = 10.61$$
$$2.9904 + c = 10.61$$
$$c \approx 7.62$$

Finally, we substitute our approximate values for a, b, and c in the general equation $f(t) = at^2 + bt + c$ to obtain our quadratic model:

$$f(t) = -0.0406t^2 + 0.91t + 7.62$$

We verify the equation by observing that the graph of f appears to contain the points $(4, 10.61)$, $(10, 12.66)$, and $(16, 11.79)$. See Fig. 21.

Since the graph appears to come close to the other data points, we conclude that f is a reasonable model of the situation. ∎

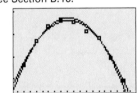

Figure 21 Verifying the property crime model

In Example 1 we found the quadratic model

$$f(t) = -0.0406t^2 + 0.91t + 7.62$$

by using three data points. We can decide on three "good" points to use by visualizing a parabola that comes close to the data points and choosing three points (not necessarily data points) that lie on or close to the parabola.

There is another way. Instead of choosing three points and solving a system of equations, we can use a graphing calculator to find the *quadratic regression equation*, which for Example 1 is

$$r(t) = -0.0357t^2 + 0.80t + 8.12$$

In Fig. 22 we see that both of the models f and r fit the data well.

Graphing Calculator

See Section B.16.

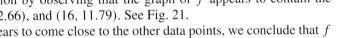

Figure 22 Comparing the graphs of f and r

Example 2 Determining Which Model to Use

Sales of DVD players in the United States have increased over the years (see Table 6). Let $f(t)$ represent the sales (in millions) of DVD players during the year that is t years since 1990. Find an equation for f.

Table 6 Sales of DVD Players

Year	DVD Player Sales (millions)
1997	0.3
1998	0.9
1999	3.6
2000	9.9
2001	16.0

Source: *DVD Entertainment Group*

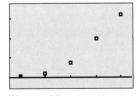

Figure 23 DVD player scattergram

Solution

First, we draw a scattergram of the data (see Fig. 23). Since the points suggest a curve that "bends," we will not use a linear function to model the data. To decide between using an exponential model or a quadratic model, we use a graphing calculator to find the exponential regression equation and the quadratic regression equation:

$$f(t) = 0.000246(2.82)^t \qquad \text{Exponential regression equation}$$
$$g(t) = 1.04t^2 - 14.73t + 52.17 \qquad \text{Quadratic regression equation}$$

Next, we see how well each regression model fits the data (see Figs. 24 and 25).

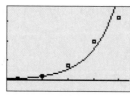

Figure 24 Exponential regression model

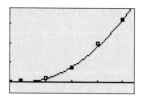

Figure 25 Quadratic regression model

It appears that the quadratic model fits the data much better than the exponential model. We conclude that the quadratic function

$$g(t) = 1.04t^2 - 14.73t + 52.17$$

is the best model for DVD player sales that we can find at this time.

Note

In Section 7.6 we take a closer look at the DVD quadratic model.

Note that the first step of the modeling process now involves deciding whether a situation can be modeled by a linear function, an exponential function, or a quadratic function. We outline the process again.

Four-Step Modeling Process

1. Create a scattergram of the data. Decide whether any of a line, an exponential curve, or a parabola comes close to the points.
2. Find an equation for your function.
3. Verify that your equation has a graph that comes close to the points in the scattergram. If it doesn't, check for calculation errors or try using different points to find the equation. An alternative is to reconsider your choice of model in step one.
4. Use your equation of the model to draw conclusions, make estimates, and/or make predictions.

We will perform activities from step 4 in Section 7.6.

 EXPLORATION
Choosing three "good" points to find a quadratic model

Table 7 Paid Vacation Days and Holidays

Years of Service	Days Off
1	9.4
3	11.2
5	13.6
10	16.6
15	18.8
20	20.4
25	21.6
30	21.9

Source: USA Today

Table 7 lists the average numbers of paid vacation days and holidays for full-time workers at medium-to-large companies for various years of experience. Let D represent the average number of paid vacation days and holidays in one year for someone who has worked at a company for t years.

1. Use your calculator to create a scattergram for the vacation data. Which of a linear function, an exponential function, or a quadratic function would seem to best model the data? Explain.
2. Use the three data points $(1, 9.4)$, $(15, 18.8)$, and $(30, 21.9)$ to find an equation $D = at^2 + bt + c$ of the parabola that comes close to the data points in your scattergram.

3. Draw the graph of your quadratic model and your scattergram in the same viewing window to verify that the parabola passes through the points $(1, 9.4)$, $(15, 18.8)$, and $(30, 21.9)$. Does your quadratic function seem to be a reasonable model for the vacation data?

4. The first three rows in Table 7 give the data points $(1, 9.4)$, $(3, 11.2)$, and $(5, 13.6)$. Had you used these three points you would have found the equation $D = 0.075t^2 + 0.60t + 8.73$. Compare its graph to the graph you drew in problem 3. Explain why the graphs look so different.

5. In the future, you will encounter other data sets that can be modeled well using a quadratic function. Describe a general "game plan" for deciding which three points to use to find a quadratic model.

Key Points
OF THIS SECTION

When performing step 1 of the modeling process, you must now decide whether any of a linear function, an exponential function, or a quadratic function is suitable for modeling the situation.

Determining which model to use

To find an equation of a quadratic function to model some data:

Finding an equation for a quadratic model

- Inspect a scattergram of the data, imagine a parabola that comes close to the data points, and choose three points that lie on or close to the parabola.

- Substitute the three chosen points into the equation $y = at^2 + bt + c$ to obtain a system of three equations in three unknowns a, b, and c.

- If none of the three points is the y-intercept, solve the system by using elimination three times. If one of the points is the y-intercept, solve the system using elimination once.

- Substitute the found values for a, b, and c into the equation $y = at^2 + bt + c$.

HOMEWORK 7.5

SSM
Tutor Center

MathPro 4
MathPro 4

MathPro 5
MathPro 5

CDs/Videos

www
www

1. Four scattergrams of data are sketched in Fig. 26. Decide whether a linear function, exponential function, quadratic function, or none of these types of functions would be reasonable for modeling the data.

2. Make a sketch of each scattergram in Fig. 26. Then sketch the graph of the function you would use to model the data.

3. A student believes that the data listed in Table 8 suggests a quadratic relationship since the values of y increase and then decrease. What would you tell the student?

(a)

(b)

(c)

(d)

Figure 26 Scattergrams of data—exercise 1

Table 8 Is There a Quadratic Relationship? (Exercise 3)

x	y
0	3
1	4
2	7
3	12
4	20
5	35
6	20
7	12
8	7
9	4
10	3

4. A student uses the points $(2, 2.5)$, $(3, 4.1)$, and $(4, 6.4)$ to find an equation for a quadratic function to model the data in Table 9. Did the student make a good selection of points? If so, explain and then find the equation for those points. If not, explain and then find an equation using a better choice of points.

Table 9 A Student Models Some Data (Exercise 4)

x	y
2	2.5
3	4.1
4	6.4
5	7.5
6	8.0
7	7.8
8	7.1
9	5.8
10	3.9
11	1.4
12	−1.7

5. The percentages of major U.S. firms that perform drug tests on employees and/or job applicants for various years are shown in Table 10.

Table 10 Percentages of Firms That Perform Drug Tests

Year	Percent
1987	22
1990	51
1992	72
1995	78
1998	74
2001	67

Source: *American Management Association*

Let $f(t)$ represent the percentage of firms that perform drug tests on employees and/or job applicants at t years since 1980. Find and verify an equation for f.

6. Gasoline prices have generally increased since 1980. However, when inflation is taken into account, the adjusted prices actually declined from 1980 to 1995 (see Table 11).

Table 11 Inflation-Adjusted Prices of Gasoline

Year	Adjusted Price (dollars per gallon)*
1980	2.25
1985	1.75
1990	1.40
1995	1.36
2000	1.60
2002	1.75

Source: *Energy Information Administration*
Prices are described in 2002 dollars

Let $f(t)$ represent the adjusted price (in 2002 dollars) of a gallon of gasoline at t years since 1980. Find and verify an equation for f.

7. The number of people waiting for organ transplants has greatly increased during the past decade (see Table 12).

Table 12 Number of People Waiting for Organ Transplants

Year	Number of People Waiting (thousands)
1988	16
1990	22
1992	30
1994	37
1996	50
1998	64
1999	72

Source: *United Network for Organ Sharing*

Let $f(t)$ represent the number of people (in thousands) waiting for organ transplants at t years since 1988.

a. Find a linear equation, an exponential equation, and a quadratic equation for f. Compare how well the models fit the data.

b. For the years before 1987, which of the three models likely gives better estimates of the number of people waiting for organ transplants? [**Hint**: Use a graphing calculator to sketch graphs of the three equations. If you used ZoomStat to form your window, now use ZOOM OUT.]

8. The number of Home Depot® stores has increased over the years (see Table 13).

Table 13 Numbers of Home Depot Stores

Year	Number of Stores
1985	50
1988	96
1991	174
1994	340
1997	624
2000	1123
2001	1319

Source: New York Times

Let $f(t)$ represent the number of stores at t years since 1985.

a. Find a quadratic model for f and an exponential model for f. Compare how well the models fit the data.

b. For years before 1987, which of the two models likely gives better estimates of the number of stores?

9. Table 14 gives the involvement rate (number of fatal crashes per 100 million miles driven) for various drivers' ages. For example, 16-year-olds are involved in 17.7 fatal crashes per 100 million miles driven. One way to think of this statistic is that when this nation's 16-year-olds have driven a total of 100 million miles, then this group will have been involved in a total of about 18 accidents where someone died.

Table 14 Drivers' Ages and Involvement Rates*

Age of Driver	Involvement Rate (number of fatal vehicle crashes per 100 million miles)
16	17.7
18	9.5
20	6.2
25	4.1
35	2.8
45	2.4
55	3.0
65	3.8
75	8.0
79	16.3

Source: *Insurance Institute for Highway Safety*

Although young drivers tend to get into more accidents than elderly drivers, both groups tend to be involved in more fatal crashes than middle-aged people.

Let $n = f(t)$ represent the involvement rate for drivers who are t years old. Perform the first three steps of the four-step modeling process. [**Modeling Process:** 1. Scattergram 2. Equation 3. Verify 4. Estimate/predict.]

10. July is the most popular month for Americans to take vacations (see Table 15).

Table 15 Percentages of Americans Who Vacation by Months

Month	Number of Months Since July t	Percent
July	0	43
August	1	34
September	2	21
October	3	15
November	4	11
December	5	11
January	6	3
February	7	3
March	8	4
April	9	8
May	10	20
June	11	30

Source: *2001 American Express Leisure Travel Index*

Let $f(t)$ represent the percentage of Americans who take a vacation in the month that is t months since July.

a. Use the data for August, October, and May to find a quadratic equation for f.

b. The data for July, August, and September lead to the equation $f(t) = -2t^2 - 7t + 43$. The data for November, December, and January lead to the equation $f(t) = -4t^2 + 36t - 69$. The data for December, March, and May lead to the equation $f(t) = 2.07t^2 - 29.20t + 105.33$. In part a, you found yet another equation for f. Use your graphing calculator to determine which of these four triples of months gives the best model for the data. Could you have guessed this before deriving the equations? If so, explain how.

11. Table 16 shows the U.S. populations for various years.

Table 16 U.S. Populations

Year	Population (millions)	Year	Population (millions)
1790	3.9	1910	92.4
1800	5.3	1920	106.5
1810	7.2	1930	123.2
1820	9.6	1940	132.1
1830	12.9	1950	151.3
1840	17.1	1960	179.3
1850	23.3	1970	203.3
1860	31.5	1980	226.5
1870	39.9	1990	248.7
1880	50.3	2000	281.4
1890	63.1	2002	286.7
1900	76.1		

Source: *U.S. Census Bureau*

Let $f(t)$ represent the U.S. population (in millions) at t years since 1790. Find and verify an equation for f.

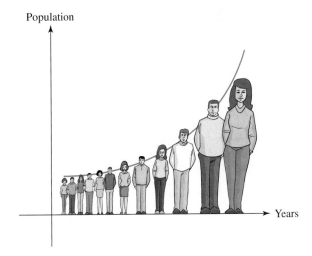

Population

Years

12. The percentages of Americans of various age groups who say they do volunteer service are listed in Table 17.

Table 17 Percentages of Americans Who Say They Volunteer

Age	Percent
18–24	38
25–34	51
35–54	55
55–64	48
65–74	45
over 74	34

Source: *Gallup Survey*

Let $f(A)$ represent the percentage of Americans at age A who say they volunteer.

a. Create a scattergram of the volunteer data. [**Hint**: Use $A = 21$ to represent Americans from 18 to 24 years of age. Use $A = 29.5$ to represent Americans from 25 to 34. In other words, use the midpoint of each group of ages for A. When choosing an age to represent Americans over age 74, keep in mind that there are many more people close to 74 than close to an older age, such as 100.]

b. Find and verify an equation for f.

13. During a baseball game, a batter hits a pitched baseball. He hits the ball when it is 4 feet above home plate. The baseball travels directly over the pitcher's mound and then directly over second base. Heights of the ball and lengths of parts of the field are shown in Fig. 27. Let $f(d)$ represent the height (in feet) of the baseball at a horizontal distance of d feet from home plate.

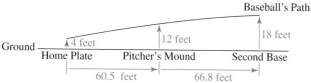

Figure 27 Heights of a baseball and lengths of parts of a baseball field

a. Thinking in terms of the baseball's trajectory (path), decide which of our three types of functions will best model the relationship between height and horizontal distance of the baseball. Explain.

b. Find and verify an equation for f.

14. The percentages of California's population who are foreign-born and the percentages who were born in other U.S. states for various years are listed in Table 18.

Table 18 Californians Not Originally from California

| | Percent | |
Year	Foreign-Born	Born in Other U.S. States
1930	18.9	47.0
1940	13.4	50.0
1950	10.0	53.0
1960	8.5	51.0
1970	8.8	47.9
1980	15.1	39.5
1990	21.7	31.8
2000	25.9	23.5

Source: *William Frey analysis of U.S. Census sources*

a. Let $f(t)$ and $g(t)$ represent the percentages of California's population that are foreign-born and born in other U.S. states, respectively, at t years since 1900. Find regression equations for f and g and verify these equations.

b. Use "intersect" on a graphing calculator to find the intersection points of the graphs of f and g. What do these points mean in terms of the situation?

15. Corona® beer and Heineken® beer have been the top two in imported beer sales for many years (see Tables 19 and 20).

Table 19 Sales of Corona Beer

Year	Corona Sales (millions of cases)
1987	24
1989	16
1991	13
1993	15
1995	23
1997	38
1999	70

Source: New York Times

Table 20 Sales of Heineken Beer

Year	Heineken Sales (millions of cases)
1992	29
1993	31
1994	34
1995	35
1996	38
1997	40
1998	43
1999	47.2

Source: New York Times

a. Let $C(t)$ and $H(t)$ represent the sales (in millions of cases) of Corona beer and Heineken beer, respectively, during the year that is t years since 1980. Find an equation for C and an equation for H. [**Hint**: Be sure to do step one of the four-step modeling process.]

b. Use "intersect" on a graphing calculator to find the intersection points of the graphs of C and H. What do these points mean in terms of the situation?

16. Describe how to find an equation for a quadratic function that can be used to model data whose scattergrams suggest a quadratic relationship.

7.6 MODELING WITH QUADRATIC FUNCTIONS

Objectives

▶ Make estimates and predictions using a quadratic model.
▶ Estimate the maximum or minimum value of a quantity.
▶ Make estimates and predictions using a system of two quadratic equations.

In Section 7.5 we discussed how to find an equation of a quadratic model. In this section we use such an equation to make estimates and predictions.

Example 1 Making Estimates and Predictions

In Example 2 of Section 7.5, we found the equation $g(t) = 1.04t^2 - 14.73t + 52.17$ where $S = g(t)$ represents the sales (in millions) of DVD players during the year that is t years since 1990 (see Table 21).

1. Find $g(19)$. What does the result mean in terms of DVD players?
2. Find t when $g(t) = 100$. What does the result mean in terms of DVD players?
3. In what years does there seem to be model breakdown?

Table 21 DVD Player Sales	
Year	Sales (millions)
1997	0.3
1998	0.9
1999	3.6
2000	9.9
2001	16.0

Solution

1. $g(19) = 1.04(19)^2 - 14.73(19) + 52.17 = 147.74$

   So, about 148 million DVD players will be sold in 2009 according to the model.

2. To find t when $g(t) = 100$, we substitute 100 for $g(t)$ and solve for t:

   $$1.04t^2 - 14.73t + 52.17 = 100 \qquad \text{Substitute 100 for } g(t).$$
   $$1.04t^2 - 14.73 - 47.83 = 0 \qquad \text{Subtract 100 from both sides.}$$

   Next, we apply the quadratic formula.

   $$t = \frac{-(-14.73) \pm \sqrt{(-14.73)^2 - 4(1.04)(-47.83)}}{2(1.04)}$$

   $$t \approx -2.72 \quad \text{or} \quad t \approx 16.89$$

   We can use a graphing calculator to verify our work (see Fig. 28).

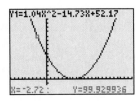 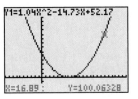

Figure 28 Verify that $t \approx -2.72$ and $t \approx 16.89$

   The found values of t represent the years 1987 and 2007. Model breakdown occurs for 1987 as DVD players were not available for sale then. So, we predict that there will be sales of 100 million DVD players in 2007.

3. To determine for what years there is model breakdown, we consider the graph of f. Note that in problem 2, we found two symmetric points:

   $$(-2.72, 100) \quad \text{and} \quad (16.89, 100)$$

   See Fig. 29.

> **Note**
> Coordinates of points in Example 1 have been rounded to the nearest hundredth.

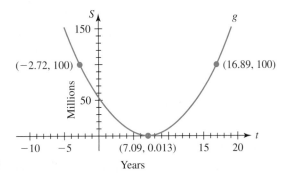

Figure 29 Graph of $g(t) = 1.04t^2 - 14.73t + 52.17$

> **Note**
> Due to the scaling of the vertical axis, the vertex *appears* to be on the t-axis, but it *is* above it.

We find the t-coordinate of the vertex by averaging the t-coordinates of the symmetric points:

$$t\text{-coordinate of vertex} = \frac{-2.72 + 16.89}{2} \approx 7.09$$

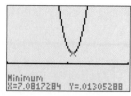

Figure 30 Verify that the vertex is approximately (7.09, 0.01)

Next, we find $g(7.09)$.

$$g(7.09) = 1.04(7.09)^2 - 14.73(7.09) + 52.17 \approx 0.013$$

So, the vertex is approximately $(7.09, 0.01)$. We verify our work using "minimum" on a graphing calculator (see Fig. 30).

The portion of the parabola that lies to the left of the vertex suggests that DVD player sales decreased for the years before 1997, which is false, as a little research would show. We conclude that model breakdown occurs for years up to 1997. ∎

In Example 1 we used a quadratic model to make predictions. Recall that to make predictions for the dependent variable we substitute a value for the independent variable and solve for the dependent variable. To make predictions for the independent variable, we substitute a value for the dependent variable and solve the equation, usually by using the quadratic formula.

Also in Example 1 we found the vertex of

$$g(t) = 1.04t^2 - 14.73t + 52.17$$

to help us decide when model breakdown occurs. We found the t-coordinate of the vertex by averaging the t-coordinates of two symmetric points.

In Example 2 we find the vertex of a quadratic function to help us determine the maximum value of the function.

Example 2   Making Estimates and Predictions

Table 22 lists the percentage of workers who use computers on the job by age groups.

Table 22 Workers Who Use Computers on the Job		
Age Group	Age Used to Represent Age Group	Percent Who Use Computers at Work
18–25	21.5	34.4
25–29	27.0	48.3
30–39	34.5	50.7
40–49	44.5	51.3
50–59	54.5	43.9
60+	62.5	27.2

Source: *National Center for Education Statistics*

Let $p = f(a)$ represent the percentage of workers who use computers at age a years.

1. Find a formula for a function that provides a reasonable model for the computer data.

2. Estimate the age(s) at which half the workers use computers on the job.

3. Estimate the age of workers who are *most likely* to use computers on the job (maximum percentage). What percentage of workers at this age use computers on the job?

4. Use f to estimate the percentage of 22-year-old workers who use computers on the job.

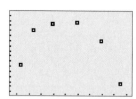

Figure 31 Computer worker scattergram

Solution

1. We begin by drawing a scattergram of the data (see Fig. 31). It looks like a parabola would fit the data well. We can use a graphing calculator to find the quadratic regression model.

$$f(a) = -0.051a^2 + 4.09a - 28.14$$

To verify this result, we draw the scattergram and the graph of f in the same viewing window (see Fig. 32). It appears that f is a reasonable model for the data.

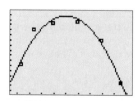

Figure 32 Verify computer worker model

2. Half of the workers is 50%, so to find the age(s) we can solve the equation:

$$50 = -0.051a^2 + 4.09a - 28.14$$

To solve, we rewrite the equation as

$$-0.051a^2 + 4.09a - 78.14 = 0$$

and then apply the quadratic formula.

$$a = \frac{-4.09 \pm \sqrt{4.09^2 - 4(-0.051)(-78.14)}}{2(-0.051)}$$

$$a \approx 31.40 \quad \text{or} \quad a \approx 48.80$$

So, according to our model, half of the 31-year-old workers and half of the 49-year-old workers use computers on the job. We can verify these results using a graphing calculator table or graph.

3. For the model $f(a) = -0.051a^2 + 4.09a - 28.14$, the coefficient of the square term $-0.051a^2$ is -0.051, a negative number. Therefore, the parabola opens downward and the parabola has a maximum point at the vertex. We can find the age when workers are most likely to use computers by finding the vertex of the graph.

We see from our work in problem 2 that $(31.40, 50)$ and $(48.80, 50)$ are symmetric points. The average of the a-coordinates is

$$\frac{31.40 + 48.80}{2} = 40.10,$$

the a-coordinate of the vertex. Since $40.10 \approx 40$, and

$$f(40) = -0.051(40)^2 + 4.09(40) - 28.14 = 53.86,$$

we see that the vertex is about $(40, 53.9)$. So, according to our model, about 53.9% of 40-year-old workers use computers on the job, the highest percentage for any age group. We can verify our computations using "maximum" on a graphing calculator.

4. Since

$$f(22) = -0.051(22)^2 + 4.09(22) - 28.14 \approx 37.16$$

we estimate that about 37.2% of 22-year-old workers use computers at work. We can verify this computation using a graphing calculator table or graph. ∎

Consider the quadratic model $f(x) = ax^2 + bx + c$ with vertex (h, k). Recall that if $a < 0$, then the parabola opens downward and the model has maximum value k (see Fig. 33). If $a > 0$, then the parabola opens upward and the model has minimum value k (see Fig. 34).

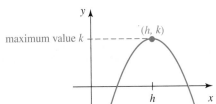

Figure 33 Quadratic function with a maximum value k but no minimum value

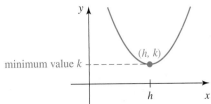

Figure 34 Quadratic function with a minimum value k but no maximum value

In Example 3 we make an estimate by working with a system of two quadratic equations.

Example 3   Modeling with a System of Quadratic Equations

In September, 1997, Circuit City® stores announced that in 1998 they would begin selling *Divx*, a way to watch a movie on a pay-per-view basis without having to return the disc to a rental store. Divx critics believed that consumers would prefer to purchase DVDs rather than rent Divxs, as there has been an increasing trend of consumers buying rather than renting videos (see Table 23).

Note

In 1999, Circuit City announced that it was abandoning its plans to sell Divx.

Year	Video Rentals (billions of dollars)	Video Sales (billions of dollars)
1991	8.4	3.6
1992	9.1	4.0
1994	9.5	5.5
1996	9.3	7.3
2000	8.25	10.8

Table 23 Video Rentals and Video Sales

Source: *Adams Media Research*

Let $r(t)$ and $s(t)$ represent the revenues (in billions of dollars) in video rentals and video sales, respectively, during the year that is t years since 1990. Scattergrams of the data suggest using quadratic functions to model revenues in video rentals and video sales. Regression equations for r and s are:

$$M = r(t) = -0.054t^2 + 0.56t + 8.03$$
$$M = s(t) = 0.019t^2 + 0.60t + 2.86$$

Estimate when video sales revenue overtook video rentals revenue.

Solution

To find when video sales revenue equaled video rentals revenue, we substitute $-0.054t^2 + 0.56t + 8.03$ for M in the equation $M = 0.019t^2 + 0.60t + 2.86$.

$$-0.054t^2 + 0.56t + 8.03 = 0.019t^2 + 0.60t + 2.86$$
$$-0.073t^2 - 0.04t + 5.17 = 0$$

The quadratic formula gives:

$$t = \frac{-(-0.04) \pm \sqrt{(-0.04)^2 - 4(-0.073)(5.17)}}{2(-0.073)}$$

$$t \approx -8.69 \quad \text{or} \quad t \approx 8.15$$

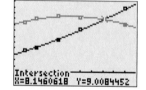

Figure 35 Verify using "intersect"

Our found values of t represent 1981 and 1998. The year 1981 is not relevant to our current study as we are interested in what happened between 1996 and 2000. So, video sales revenue was approximately equal to video rentals revenue in 1998 according to our models. We verify our work using "intersect" on a graphing calculator (see Fig. 35).

Note that the graphs of r and s suggest that video sales revenue has been greater than video rentals revenue since 1998. ■

In Example 3 we used substitution with a system of two quadratic equations to estimate when video sales revenue was equal to video rentals revenue. We can use substitution to solve any system of two quadratic equations, where the equations are in standard form $y = ax^2 + bx + c$.

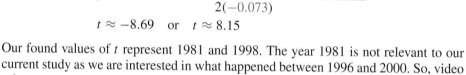

EXPLORATION
Modeling differences of quantities

Refer to Example 3 for background information on videos.

1. Complete Table 24.

Table 24 Differences between Video Rentals Revenues and Video Sales Revenues

Year	Video Rentals Revenue (billions of dollars)	Video Sales Revenue (billions of dollars)	Difference between Video Rentals Revenues and Video Sales Revenues (billions of dollars)
1991	8.4	3.6	$8.4 - 3.6 = 4.8$
1992	9.1	4.0	
1994	9.5	5.5	
1996	9.3	7.3	
2000	8.25	10.8	

2. Let $D(t)$ represent the difference between video rentals revenue and video sales revenue (in billions of dollars) during the year that is t years since 1990. Perform the first three steps of the four-step modeling process to find an equation for D. Use the regression choice on your graphing calculator to find the equation.

3. Which of the following statements is correct? Explain.

$$D(t) = r(t) + s(t) \qquad D(t) = r(t) - s(t) \qquad D(t) = s(t) - r(t)$$

4. In Example 3 we used the following equations for r and s:

$$M = r(t) = -0.054t^2 + 0.56t + 8.03$$
$$M = s(t) = 0.019t^2 + 0.60t + 2.86$$

Substitute $-0.054t^2 + 0.56t + 8.03$ for $r(t)$ and $0.019t^2 + 0.60t + 2.86$ for $s(t)$ in the equation $D(t) = r(t) - s(t)$ to find an equation for $D(t)$. Simplify your equation and compare it with the equation you found in problem 2.

5. Find t when $D(t) = 0$. What does your result mean in terms of video rentals and sales? Compare your result with the result in Example 3.

Tips on Succeeding in This Course

If you get an answer to a test item and realize your answer is incorrect but do not have time to work a correct solution, it is a good idea to write on your test paper that you know your answer is incorrect and what you would do to correct your error if you had more time. If you are not sure how to correct your error, you can at least acknowledge that you know your answer is incorrect and state why you know. In either case, you are demonstrating ability to analyze the correctness of a solution. Depending on how your instructor grades tests, you may earn more points.

Key Points

OF THIS SECTION

For a quadratic model $f(t) = at^2 + bt + c$ with vertex (h, k):

- If $a < 0$, the parabola opens downward and k is a maximum value of the model.
- If $a > 0$, the parabola opens upward and k is a minimum value of the model.

Maximum and minimum values

We can use substitution to solve any system of two quadratic equations that are both in standard form $y = ax^2 + bx + c$.

Solving a system of quadratic equations

HOMEWORK 7.6

 SSM Tutor Center

MathPro 4 MathPro 4

MathPro 5 MathPro 5

 CDs/Videos

 www

1. In exercise 5 of Homework 7.5, you found an equation close to $f(t) = -0.66t^2 + 21.49t - 95.85$, where $f(t)$ represents the percentage of firms that perform drug tests on employees and/or job applicants at t years since 1980 (see Table 25).

   a. Find the t-intercepts of the model. What do the t-intercepts mean in terms of the situation?

   b. For what values of t is there model breakdown for certain? Which years are represented by these values?

   c. Estimate the maximum percentage of firms that performed drug tests. Estimate the year that this percentage occurred.

Table 25 Percentages of Firms That Perform Drug Tests

Year	Percent
1987	22
1990	51
1992	72
1995	78
1998	74
2001	67

2. In exercise 6 of Homework 7.5, you found an equation close to $f(t) = 0.0051t^2 - 0.136t + 2.27$, where $f(t)$ represents the adjusted price (in 2002 dollars) of a gallon of gasoline at t years since 1980 (see Table 26).

Table 26 Inflation-Adjusted Prices of Gasoline

Year	Adjusted Price (dollars per gallon)*
1980	2.25
1985	1.75
1990	1.40
1995	1.36
2000	1.60
2002	1.75

*Prices are in 2002 dollars

a. Use f to predict the price of gasoline in 2008.

b. Use f to predict when the price of gasoline will be $3 per gallon.

c. Estimate the minimum adjusted price of gasoline. Estimate the year that this adjusted price occurred.

3. In exercise 9 of Homework 7.5, you found an equation close to $f(t) = 0.013t^2 - 1.19t + 28.24$, where $f(t)$ represents the involvement rate (number of fatal crashes per 100 million miles driven) for drivers who are t years old (see Table 27).

Table 27 Drivers' Ages and Involvement Rates

Age of Driver	Involvement Rate
16	17.7
18	9.5
20	6.2
25	4.1
35	2.8
45	2.4
55	3.0
65	3.8
75	8.0
79	16.3

a. Estimate the involvement rate for drivers at your age (or an age of your choice).

b. What age driver has the minimum involvement rate? What is that minimum involvement rate according to the model?

4. A *householder* is the person in whose name a house, condominium, or apartment is owned or rented. The percentages of householders who own a home for various age groups are listed in Table 28.

Table 28 Percentages of Householders Who Own a Home

Age Group	Age Used to Represent Age Group	Percent Who Own a Home
15–24	19.5	18
25–34	29.5	46
35–44	39.5	66
45–54	49.5	75
55–64	59.5	80
65–74	69.5	81
75–84	79.5	77

Source: *U.S. Census Bureau*

Let $f(t)$ represent the percentage of householders who own a home at age a years. A reasonable equation for f is $f(t) = -0.031t^2 + 4.03t - 47.05$.

a. Use f to estimate the percentage of 22-year-old householders who own a home.

b. Use f to estimate the age(s) at which half the householders own homes.

c. Use f to estimate the age of householders who are most likely to own a home. What percentage of householders at this age own homes?

5. In exercise 13 of Homework 7.5, you found the equation $f(d) = -0.000333d^2 + 0.152d + 4$, where $f(d)$ represents the height (in feet) of a baseball at a horizontal distance of d feet from home plate. Recall that the batter hit the baseball so that it went directly over the pitcher's mound and second base.

a. Find the maximum height of the baseball.

b. If the baseball goes over the *home run wall* without bouncing on the ground before going over, we say the batter hit a *home run*.* If the home run wall is 10 feet high and 400 feet from home plate, was the hit described in the exercise a home run? Assume that no player interferes with the path of the ball.

6. A tennis ball is tossed upwards and its height and the elapsed time are recorded. The data from the experiment are given in Table 29.

Table 29 Heights of Tennis Ball at Elapsed Times

Time (in seconds)	Height (in feet)
0.00	1.6097
0.02	1.8690
0.04	2.0886
0.06	2.2363
0.10	2.2435
0.12	2.1138
0.14	1.9734

Source: *Data collected by the author*

Let $h = f(t)$ represent the height (in feet) of the tennis ball at t seconds.

a. Perform the first three steps of the modeling process. Use the regression choice on your graphing calculator to find an equation.

---

*The ball must go over the wall in *fair territory* to be a home run. If a baseball goes directly over the pitcher's mound and second base, it will be in fair territory.

b. When does the tennis ball reach the ground?

c. Find the maximum height of the tennis ball. When did it reach this height?

d. Find the domain and the range of your model.

e. Assume that after the toss the tennis ball bounces on the ground several times. Sketch a qualitative graph that describes the relationship between the height of the tennis ball and the time until the tennis ball comes to rest.

7. The percentage of households that have cable television has increased over the past two decades (see Table 30).

Table 30 Percentages of Households with Cable Television

Year	Percent
1980	21.7
1985	40.8
1990	54.1
1995	61.5
2000	64.7

Source: Paul Kagan Associates, Inc. *Travis T., student, contributed the data*

Let $f(t)$ represent the percentage of households with cable television at t years since 1980. A reasonable equation for f is $f(t) = -0.108t^2 + 4.29t + 21.83$.

a. Find $f(30)$. What does your result mean in terms of households with cable television?

b. Find t when $f(t) = 30$. What does your result mean in terms of households with cable television?

c. According to the model, what will be the maximum percentage? In what year will this happen?

d. For what years is there likely model breakdown? [**Hint:** Consider the past and the future.]

8. In exercise 12 of Homework 7.5, you found an equation close to $f(A) = -0.02A^2 + 1.86A + 9.90$, where $f(A)$ represents the percentage of Americans at age A who say they do volunteer work (see Table 31).

Table 31 Percentages Who Say They Volunteer

Age	Percent
18–24	38
25–34	51
35–54	55
55–64	48
65–74	45
over 74	34

a. Find $f(50)$. What does your result mean in terms of the situation?

b. Find A where $f(A) = 50$. What does your result mean in terms of the situation?

c. For what age is the percentage who say they volunteer the greatest? What is that maximum percentage?

9. In exercise 15 of Homework 7.5, beer sales are modeled by the system:

$$s = C(t) = 0.90t^2 - 19.94t + 121.38$$
$$s = H(t) = 2.48t - 1.30$$

where $C(t)$ and $H(t)$ represent beer sales (in millions of cases) of Corona and Heineken, respectively, at t years since 1980. In what year(s) were the sales for the two brands of beer about the same?

10. Throughout the past five decades, women usually have been younger than men in their first marriage. Table 32 provides the median ages at which men and women were first married in various years.

Table 32 Median Ages at First Marriages

Year	Women	Men
1940	21.5	24.3
1950	20.3	22.8
1960	20.3	22.8
1970	20.8	23.2
1980	22.0	24.5
1990	24.0	26.3
2000	25.1	26.8

Source: Who We Are, A Portrait of America, *Sam Roberts*

Let $W(t)$ represent the median age at which women were first married and $M(t)$ represent the median age at which men were first married at t years since 1900.

a. Perform the first three steps of the modeling process to find a model for men's median marrying age and another model for women's median marrying age. Find each equation by using the regression choice on a graphing calculator.

b. Predict when the median age(s) at which women and men first marry will be equal, if ever.

11. In exercise 11 of Homework 7.5, you found an equation close to $P = f(t) = 0.0067t^2 - 0.12t + 6.39$, where $f(t)$ represents the U.S. population (in millions) at t years since 1790 (see Table 33).

Table 33 U.S. Populations

Year	Population (millions)	Year	Population (millions)
1790	3.9	1910	92.4
1800	5.3	1920	106.5
1810	7.2	1930	123.2
1820	9.6	1940	132.1
1830	12.9	1950	151.3
1840	17.1	1960	179.3
1850	23.3	1970	203.3
1860	31.5	1980	226.5
1870	39.9	1990	248.7
1880	50.3	2000	281.4
1890	63.1	2002	286.7
1900	76.1		

a. Find $f(217)$. What does your result mean in terms of the population?

b. Find t where $f(t) = 300$. What does your result mean in terms of the population?

c. Use pencil and paper to sketch a graph of f.

d. For what values of t is there model breakdown? Which years are represented by these values?

e. Sketch a qualitative graph that describes the relationship between t and P for all of time, past and future.

12. Let $U(t)$ represent the U.S. population (in millions) and $W(t)$ represent the world population (in millions) at t years since 1930. See Tables 33 and 34. Possible equations for U and W are:

$$U(t) = 0.0067t^2 + 1.76t + 121.29$$
$$W(t) = 1945.09(1.0163)^t$$

Table 34 World Populations

Year	World Population (millions)
1930	2070
1940	2295
1950	2500
1960	3050
1970	3700
1980	4454
1990	5279
2000	6080
2002	6215

Compare the graphs of U and W and determine which curve appears to be steeper at the value of t that represents the current year. Explain why your observation makes sense in terms of what you know about industrialized countries.

13. The function

$$Q(t) = 0.0067t^2 - 0.12t + 6.39$$

is a reasonable model of the U.S. population (in millions), where t is the number of years since 1790. However, in exercise 25 of Homework 4.5, you found that the U.S. population (in millions) can be modeled extremely well from 1790 to 1860 by an exponential model such as

$$E(t) = 3.9(1.030)^t$$

where t is the number of years since 1790.

a. Calculate what the U.S. population would have been in 2002, if it had continued to increase after 1860 according to the model E. Compare this value to the actual value of 286.7 million.

b. Use a graphing calculator table to compare the predictions of the models Q and E in the years 2100, 2200, 2300, 2400, and 2500. Is there much difference in these predictions? Explain. If the U.S. population had continued to increase exponentially from 1860 to 2500, guess what problems there would be for the population in 2500. Explain.

14. The numbers of airplane collisions on runways for various years are listed in Table 35.

Table 35 Accidents on Runways

Year	Number of Runway Accidents
1994	202
1995	245
1996	285
1997	292
1998	325
1999	321

Source: *Federal Aviation Administration*

Let $f(t)$ represent the number of runway accidents during the year that is t years since 1990.

a. In exercise 64 in Section 6.6, you may have modeled the runway data using a quadratic function. Explain why it is better to use a quadratic model rather than an exponential model or a linear model. Then, find the quadratic regression equation.

b. Use a graphing calculator table to help you complete the third column of Table 36. Then complete the fourth column.

Table 36 Accidents on Runways

Year	Actual Number of Runway Accidents	Model's Estimate of Number of Runway Accidents	Difference Between Model's Estimate and Actual Value
1994	202		
1995	245		
1996	285		
1997	292		
1998	325		
1999	321		

c. For which of the years from 1994 to 1999 does the model give the best estimate of the number of runway accidents? For which year does the model give the worst estimate?

15. The percentages of public schools with Internet access for various years are shown in Table 37.

Table 37 Public Schools with Internet Access

Year	Percent
1994	35
1995	50
1996	65
1997	78
1998	89
1999	95
2000	98

Source: *Internet Access in U.S. Public Schools and Classrooms: 1994–2000, http://nces.ed.gov/pubs2001/2001071.pdf. Accessed 4/2/02*

Let $p = f(t)$ represent the percentage of public schools with Internet access at t years since 1990.

a. Perform the first three steps of the modeling process to find an equation for f.

b. Find the t-intercepts of f. What do the t-intercepts mean in terms of the situation?

c. If f is linear, find the slope. If f is exponential, find the base b of $f(t) = ab^t$. If f is quadratic, find the vertex of f. What does your result mean in terms of the situation?

d. For what values of *t* does model breakdown either likely occur or occur for certain? Which years are represented by these values?

e. Use pencil and paper to sketch a graph of *f*. On the same set of axes, also sketch a nonquadratic model that you think better describes the situation.

16. The numbers of prisoners under the sentence of death ("on death row") in the United States are listed by year in Table 38.

Table 38 Numbers of Prisoners on Death Row

Year	Number of Prisoners on Death Row
1995	3064
1996	3242
1997	3328
1998	3465
1999	3540
2000	3593

Source: *Bureau of Justice Statistics Prisoners Under Sentence of Death Trends Chart, www.ojp.usdoj.gov/bjs/glance/dr.htm. Accessed 4/4/02*

Let $n = f(t)$ represent the number of prisoners on death row at *t* years since 1990.

a. Find an equation for *f*.

b. Find the *n*-intercept of *f*. What does it represent in terms of the situation?

c. Find the value(s) of *t* when $f(t) = 2800$. What does your result(s) represent in terms of the situation?

d. If *f* is linear, find the slope. If *f* is exponential, find the base *b* of $f(t) = ab^t$. If *f* is quadratic, find the vertex of *f*. What does it represent in terms of the situation?

17. Harnessing wind energy is much more affordable and cleaner than producing many other traditional energy sources such as oil and coal. Most people do not realize the high potential of wind energy. For example, the combined potential wind energy in North Dakota, South Dakota, and Texas is enough to provide electricity for the entire nation. Table 39 lists the worldwide wind-generating capacities in thousands of megawatts (thousand MW) for various years.

Table 39 Worldwide Wind-Generating Capacities

Year	Wind-Generating Capacity (thousand MW)
1990	1.9
1992	2.5
1994	3.7
1996	6.1
1998	9.6
2000	17.8
2001	23.3

Source: *Worldwatch Press Brief. www.worldwatch.org/alerts/981229a.html. Accessed 4/27/02*

Let $f(t)$ represent the wind-generating capacity (in thousand MW) at *t* years since 1990.

a. Find a linear regression equation, an exponential regression equation, and a quadratic regression equation for *f*. Compare how well the models fit the data.

b. For the years before 1990, which of the models likely gives better estimates of the wind-generating capacity?

c. One MW of wind-generating capacity will meet the electricity needs of about 1000 people in an industrial country. Use your exponential model to predict when wind energy could meet the electricity needs of the entire world. Assume that the world population is 6.2 billion. [**Hint**: Note that the units of $f(t)$ are *thousands* of MW.]

d. Now use your quadratic model to predict when wind energy could meet the electricity needs of the entire world.

e. Explain why the year you predicted in part d is so much later than the year you predicted in part c.

OK, Timmy, trim the sails—we're approaching the freeway !

18. In exercise 14 of Homework 7.5, the percentages of California's population $f(t)$ and $g(t)$ who are foreign-born and born in other U.S. states, respectively, are modeled by the system

$$p = f(t) = 0.0111t^2 - 1.32t + 48.24$$
$$p = g(t) = -0.0117t^2 + 1.17t + 22.79$$

where *t* represents the number of years since 1900 (see Table 40).

Table 40 Californians Not Originally from California

	Percent	
Year	Foreign-Born	Born in Other U.S. States
1930	18.9	47.0
1940	13.4	50.0
1950	10.0	53.0
1960	8.5	51.0
1970	8.8	47.9
1980	15.1	39.5
1990	21.7	31.8
2000	25.9	23.5

a. In what year are the percentages of foreign-born and those born in other U.S. states about the same? What is that percentage? What percentage of California's population are originally from California for that year?

b. Let $h(t)$ represent the percentage of California's population who are originally from California. Which of the following statements is correct? Explain.

$$h(t) = f(t) - g(t) \qquad h(t) = g(t) - f(t)$$
$$h(t) = 100 - f(t) - g(t) \qquad h(t) = 100 + f(t) + g(t)$$

c. Substitute $0.0111t^2 - 1.32t + 48.24$ for $f(t)$ and $-0.0117t^2 + 1.17t + 22.79$ for $g(t)$ in the equation you

chose in part b and simplify the right side of the equation. You should now have an equation for h in terms of t.

d. Find $h(110)$. What does your result mean in terms of the situation?

e. Use your graphing calculator to sketch a graph of h. Is the curve increasing or decreasing for $t \geq 30$? What does this mean in terms of the situation?

19. For a quadratic model, discuss how to find intercepts, the vertex, the maximum or minimum value, how to make predictions for the dependent or independent variable, and how to determine values of the independent variable where model breakdown occurs.

Taking it to the lab

For each lab assignment, consult with your instructor on whether to organize your responses as a numbered list or to write them in a paragraph.

WATER FLOW LAB

In this experiment you will fill a cylinder with water and allow the water to flow out of a small hole at the bottom. The point of this experiment will be to explore the relationship between the volume of water that flows out of the small hole and the amount of time that has elapsed.

In case of a lack of the necessary measuring devices or a lack of class time to devote to running this experiment, some data are listed in Table 41.

Table 41 Heights of Water in a Cylinder with Radius 1.5 Inches

Time (seconds)	Height (inches)
0	30
9	25
18	20
29	15
42	10
58	5
79	1

Source: *Data collected by the author*

Materials

If you are going to perform your own experiment, you will need at least three people and the following items:

1. A timing device, a tape measure or ruler, and a marker.

2. You will need a transparent cylinder or some other water container that has uniform cross sections. One end of the cylinder should have a hole that is large enough so that water flows out rather than drips out. The cylinder should be large enough and the hole small enough so that it takes at least a minute for a full container to drain. For example, a cylinder that is four feet long with a diameter of 3 inches works well with a hole that has a diameter of about $\frac{1}{8}$ inch.

3. You may also need a bucket to catch the water that flows out of the container, depending on where you are performing this experiment.

Preparation

Make about 8 equally spaced marks along the cylinder. Fill the cylinder with water, keeping the hole at the bottom of the cylinder sealed until you are ready to begin timing. While two people are preparing the cylinder and water, a third person should prepare to record the times it takes for the water level to reach the various marks on the cylinder.

Recording of Data

The height of each mark and the amount of time (from the start of the experiment) that it takes for the water level to reach each mark should be recorded. The radius of the cylinder should also be recorded.

Analyzing the Data

1. If you collect your own water flow data, display these data in a table. If not, then use the data in Table 41.
2. Let $H = f(t)$ represent the height (in inches) of the water at t seconds after the water begins to flow out of the cylinder. Create a scattergram of the water flow data.
3. Find an equation for f. The water flow is likely to be a bit erratic at the end, so it's probably best to avoid using a data point that corresponds to a water level of zero, if you recorded such a data point.
4. How well does your function model the data?
5. Find the H-intercept of f. What does this point represent in terms of water in the cylinder? Does model breakdown occur before the H-intercept, after the H-intercept, or neither? Explain.
6. Find the t-intercept(s) of f. What does such a point represent in terms of the water in the cylinder? Does model breakdown occur before the t-intercept, after the t-intercept, both, or neither? Explain. If there is no t-intercept(s), what does your model imply?
7. Find the vertex of your model. Does model breakdown occur before the vertex, after the vertex, or neither? Explain.
8. Use your model to estimate the height of the water at 20 seconds.
9. Use your model to estimate how many seconds it took for the water level to reach a height of 7 inches.

Taking It One Step Further

10. For a cylinder of radius R and height H, the volume of the cylinder can be described by the equation $V = \pi R^2 H$. If R and H are in inches, then V is in cubic inches. Substitute your value for R in the equation $V = \pi R^2 H$.
11. You have $H = at^2 + bt + c$ and $V = \pi R^2 H$, where a, b, c, and R are known constants. Substitute $at^2 + bt + c$ for H in the equation $V = \pi R^2 H$ to find an equation that describes volume of water in terms of the time elapsed since the start of the experiment.
12. Use your result from part 11 to estimate the volume of water in the cylinder at $t = 10$ seconds.
13. How much water flowed out of the cylinder during the time span from $t = 10$ to $t = 20$? From $t = 20$ to $t = 30$? Explain why the two amounts are not equal.

PROJECTILE LAB

In this lab you will toss a softball vertically into the air and record the softball's height at various times.

Materials

If you are going to perform your own experiment, you will need the following materials:

1. a softball (or some other object)
2. a CBR unit, or a CBL unit with a Vernier motion detector probe

In case the necessary measuring devices are not available or there is not enough class time to devote to this experiment, some data are listed in Table 42.

Table 42 Heights of a Softball			
Time (in seconds)	Height (feet)	Time (in seconds)	Height (feet)
0.00	3.3994	0.32	4.4545
0.02	3.5650	0.34	4.4077
0.04	3.7271	0.36	4.3393
0.06	1.5000	0.38	4.2925
0.08	4.0620	0.40	4.2024
0.10	4.1340	0.42	4.0980
0.12	4.1916	0.44	3.9900
0.14	4.2709	0.46	3.8711
0.16	4.3429	0.48	3.7379
0.18	4.4005	0.50	3.5903
0.20	4.4473	0.52	3.4354
0.22	4.4797	0.54	3.2662
0.24	4.4977	0.56	3.0861
0.26	4.5049	0.58	2.9061
0.28	4.5013	0.60	2.7044
0.30	4.4797		

Source: *Data collected by the author*

Analyzing the Data

1. If you collected your own softball data, display these data in a table. If not, use the data in Table 42.
2. Let $h = f(t)$ represent the height (in feet) of the softball at t seconds. Use your graphing calculator to draw a scattergram of the softball data.
3. If you are using the data in Table 42, should the graph of your model come close to (0.06, 1.5)? If you performed your own experiment, should the graph of your model come close to all of your data points? Explain.
4. Find an equation for f.
5. Use your graphing calculator to draw a graph of f and your scattergram in the same viewing window. Use what you see on your screen to help you sketch a graph of your model and scattergram on paper.
6. Use your equation for f to estimate when the softball reached the ground.
7. What is the h-intercept of f? What does your result mean in terms of the height of the softball?

Note

Remember, if you think model breakdown occurs, say so, say where, and explain why.

8. Use your equation for f to estimate the height of the softball at 0.7 second.
9. Use your equation for f to estimate the height of the softball at 10 seconds.
10. For what values of t is there model breakdown? Explain.
11. Use your equation for f to estimate when the softball reached its maximum height. What was the maximum height?

QUADRATIC LAB: TOPIC OF YOUR CHOICE

Your objective in this lab is to find a quadratic function to model a true-to-life situation (a situation that has not already been discussed in this text). Your first task will be to find some data. Almanacs, newspapers, magazines, scientific journals, and the Internet are good resources for data. Or, you can conduct an experiment to obtain your data.

Analyzing the Data

1. What two variables did you explore? Did you make any false starts? If so, what variables were you exploring then? Explain why you chose to pick new variables.
2. Does it make sense that a quadratic function best models your situation? Explain.
3. Which variable is the dependent variable? Which variable is the independent variable?

4. State the source of your data. If you conducted an experiment, provide a careful description with specific details of how you ran your experiment.

5. Include a table of your data and create a scattergram. Draw some conclusions.

6. Find an equation for your quadratic model.

7. Use pencil and paper to sketch a graph of your quadratic model and your scattergram on the same coordinate system.

8. Find the intercepts of your quadratic model. What do they represent in terms of the situation being modeled? Has model breakdown occurred at the intercepts?

9. Find the vertex of your model. What does it represent in terms of the situation you chose to model? Has model breakdown occurred at the vertex?

10. Make a prediction or estimate about your dependent variable based on a specific value for your independent variable. What does your prediction mean in terms of the situation being modeled?

11. Make a prediction or estimate about your independent variable based on a specific value for your dependent variable. What does your prediction mean in terms of the situation being modeled?

12. Comment on your lab experience.

   a. For example, you might address whether this lab was enjoyable, insightful, and so on.

   b. Were you surprised by any of your findings? If so, which ones?

   c. How would you improve your process for this lab if you did it again?

   d. How would you improve your process if you had more time and money?

CHAPTER SUMMARY

Key Points OF THIS CHAPTER

If $a \geq 0$ and $b \geq 0$, then $\sqrt{ab} = \sqrt{a}\sqrt{b}$.	**Product property**
If $a \geq 0$ and $b > 0$, then $\sqrt{\dfrac{a}{b}} = \dfrac{\sqrt{a}}{\sqrt{b}}$.	**Quotient property**
For $k \geq 0$, if $x^2 = k$, then $x = \pm\sqrt{k}$.	**Solving $x^2 = k$**
A radical $\sqrt{k}$ with a counting number k, $k \geq 2$, is *simplified* if k does not have any perfect square factors other than 1.	**Simplifying $\sqrt{k}$**
For an expression such as $\sqrt{\dfrac{2}{7}}$, we simplify by leaving no radicand as a fraction.	**Simplifying $\sqrt{\dfrac{p}{q}}$**
For an expression such as $\dfrac{5}{\sqrt{6}}$, we simplify by leaving no denominator as a radical. We call this process *rationalizing the denominator*.	**Simplifying $\dfrac{p}{\sqrt{q}}$**
To rationalize the denominator of a fraction of the form $\dfrac{p}{\sqrt{q}}$, $q > 0$, we multiply the fraction by 1 in the form $\dfrac{\sqrt{q}}{\sqrt{q}}$.	**Rationalizing the denominator**
For a perfect square trinomial $x^2 + bx + c$, dividing b by 2 and then squaring the result gives c.	**Perfect square trinomial**
To solve a quadratic equation of the form $ax^2 + bx = p$ with $a \neq 1$ by completing the square, we first divide both sides of the equation by a before completing the square.	**Completing the square**
We can solve *any* quadratic equation by using the quadratic formula. We can also do so by completing the square.	**Solving a quadratic equation**

Quadratic formula and number of solutions	If the equation $ax^2 + bx + c = 0$, $a \neq 0$, has real number solutions, they are given by

$$x = \frac{-b \pm \sqrt{b^2 - 4ac}}{2a}$$

Also:
- If $b^2 - 4ac > 0$, there are two real number solutions.
- If $b^2 - 4ac = 0$, there is one real number solution.
- If $b^2 - 4ac < 0$, there are no real number solutions.

Factoring and reducing	We simplify a fraction such as $\dfrac{6 + 2\sqrt{3}}{4}$ by factoring the numerator and then reducing the fraction:

$$\frac{6 + 2\sqrt{3}}{4} = \frac{2(3 + \sqrt{3})}{4} = \frac{3 + \sqrt{3}}{2}$$

Determining which model to use	When performing step 1 of the modeling process, you must now decide whether any of a linear function, an exponential function, or a quadratic function is suitable for modeling the situation.
Finding an equation of a quadratic model	To find an equation of a quadratic function to model some data:

1. Inspect a scattergram of the data, imagine a parabola that comes close to the data points, and choose three points that lie on or close to the parabola.
2. Substitute the three chosen points into the equation $y = at^2 + bt + c$ to obtain a system of three equations in three unknowns a, b, and c.
3. If none of the three points is the y-intercept, solve the system by using elimination three times. If one of the points is the y-intercept, solve the system using elimination once.
4. Substitute the found values for a, b, and c into the equation $y = at^2 + bt + c$.

Predicting for the dependent variable	To make a prediction for the dependent variable of a quadratic model, substitute the chosen value for the independent variable in the equation, and then solve for the dependent variable.
Predicting for the independent variable	To make a prediction for the independent variable of a quadratic model, substitute the chosen value for the dependent variable in the equation, and then solve for the independent variable, usually by using the quadratic formula.
Maximum or minimum value	The vertex of a parabola $y = ax^2 + bx + c$ is either the maximum point or the minimum point.

- If the vertex is the maximum point, then the maximum value of the function is the y-coordinate of the vertex.
- If the vertex is the minimum point, then the minimum value of the function is the y-coordinate of the vertex.

CHAPTER 7 REVIEW EXERCISES

Simplify.

1. $\sqrt{72}$

2. $\sqrt{\dfrac{3}{5}}$

3. $\sqrt{\dfrac{50}{49}}$

4. $\sqrt{\dfrac{49}{100}}$

Solve.

5. $6x^2 - 3x - 2 = 0$

6. $5x^2 = 7$

7. $5(x - 3)^2 + 4 = 7$

8. $x^2 = 98$

9. $(x + 1)(x - 7) = 4$

10. $-2(x + 4)^2 = 9$

11. $3x^2 = 6x$

12. $x^2 - 2x - 24 = 0$

13. $4x^2 - 25 = 0$

14. $3x^2 - 7x + 2 = 0$

15. $25x^2 - 9 = 0$

16. $3x^2 - 1 = x^2 - 5x + 1$

17. $2x^2 = 4 - 5x$

18. $4x - x^2 = 1$

19. $-3x^2 + x = -5x^2 + x + 5$

20. $2x(2x - 5) + 13 = 3x^2 + 5$

21. $3x(x - 7) + 13 = 2x^2 - 7$

22. $2(x^2 - x) = 3(x^2 - 2x) + 5$

23. $(x + 2)^2 + (x - 3)^2 = 2$

24. $5(5x^2 - 8) = 9$

Find approximate solutions of the equation.

25. $2.7x^2 - 5.1x = 9.8$

26. $1.7(x^2 - 2.3) = 3.4 - 2.8x$

Solve by completing the square.

27. $x^2 + 6x - 4 = 0$

28. $8x^2 + 12x + 8 = 0$

29. $2x^2 = -3x + 6$

30. $x(x - 3) + 2 = 2x(x + 1) + 1$

Find the x-intercepts.

31. $f(x) = 3x^2 - 10$

32. $g(x) = -2(x + 3)^2 + 5$

33. $h(x) = 3x^2 + 2x - 2$

34. $k(x) = -5x^2 + 3x - 1$

35. $f(x) = 2x^2 - 3x - 4$

36. $g(x) = -3x^2 + 5x + 7$

37. Solve $x^2 - 2x - 8 = 0$ by factoring, completing the square, and using the quadratic formula.

38. Find the value(s) of k so that the equation $3x^2 + kx + 12 = 0$ has exactly one solution.

39. Let $f(x) = 3x^2 - 6x + 7$.
   a. Find x when $f(x) = 3$.
   b. Find x when $f(x) = 4$.
   c. Find x when $f(x) = 5$.
   d. Discuss in terms of the graph of f why you found 0, 1, and 2 values of x for parts a, b, and c, respectively.

40. The vertex of a parabola is $(-4, 3)$ and the parabola passes through the point $(-3, 6)$. Find an equation of the parabola.

Find an equation of the parabola that passes through the three given points.

41. $(1, 3), (2, 6), (3, 13)$

42. $(1, 4), (3, -2), (4, -11)$

43. $(2, 9), (3, 18), (5, 48)$

44. $(0, 5), (2, 3), (4, -15)$

45. $(0, 7), (1, 8), (3, -8)$

46. $(0, 3), (2, 13), (3, 24)$

47. Find equations of a linear function, an exponential function, and a quadratic function so that the graph of each of these functions contains both of the points $(0, 4)$ and $(1, 2)$.

48. Find an equation of the parabola sketched in Fig. 36.

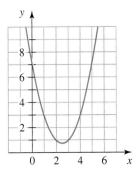

Figure 36 Exercise 48

49. The percentages of Americans who reported voting in the 2000 election, for various age groups, are listed in Table 43.

Table 43 Percentages of Americans Who Voted		
Age Group	Age Used to Represent Age Group	Percent Who Voted
18–24	19.5	32.3
25–34	29.5	43.7
35–44	39.5	55.0
45–54	49.5	62.3
55–64	59.5	66.8
65–74	69.5	69.9

Source: *U.S. Census Bureau*

Let $f(t)$ represent the percentage of Americans who voted at age a years.

   a. Find an equation for f.
   b. Use f to estimate the percentage of 18-year-old Americans who voted.
   c. Estimate the age(s) at which half of Americans voted.
   d. Estimate the age of Americans who are most likely to vote.
   e. When candidates run for presidential office, most tend to focus on issues that concern older people, such as Social Security benefits and Medicare. In your opinion, why do they tend to do this?

50. As difficult as it is to get into Princeton University, the good news is that once you are in, you likely will get a degree. In fact, students who put in a so-so effort increasingly get good grades according to a report released on February 12, 1998, by the Dean of the College at Princeton (*The New York Times*). Although there is no evidence that the caliber of students has improved over past decades, the percentages of both A's and B's have increased. The percentages of A's are listed in Table 44.

Table 44 Percentages of A's at Princeton	
Year	Percent
1973–77	30.1
1978–82	31.6
1983–87	32.8
1988–92	36.9
1993–97	43.4

Source: *Princeton University*

Let $f(t)$ represent the percentage of A's during the year that is t years since 1970.

a. Create a scattergram of the Princeton data. [**Hint**: Use the midpoint of each group of years for t. For example, use $t = 5$ to represent the years between 1973 and 1977.]

b. Find an equation for a function that provides a reasonable model for the Princeton data.

c. Find $f(37)$. What does your result mean in terms of grades at Princeton?

d. Find t when $f(t) = 100$. What does your result mean in terms of grades at Princeton?

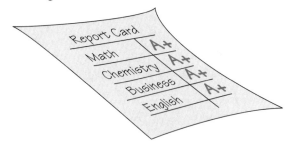

CHAPTER 7 TEST

Simplify.

1. $\sqrt{32}$

2. $\dfrac{7}{\sqrt{2}}$

3. $\sqrt{\dfrac{20}{75}}$

Solve.

4. $x^2 - 3x - 10 = 0$

5. $6x^2 = 100$

6. $-3x^2 + x - 4 = 0$

7. $4(x - 3)^2 + 1 = 7$

8. $3x^2 - 21x = 0$

9. $x^2 - 81 = 0$

10. $(x - 3)(x + 5) = 6$

11. $-5(x + 4)^2 = 10$

12. $2x(x + 5) = 4x - 3$

13. $9x^2 = 16 + 24x$

14. Find approximate solutions for $3.7x^2 = 2.4 - 5.9x$.

Solve by completing the square.

15. $x^2 - 8x - 2 = 0$

16. $2(x^2 - 4) = -3x$

17. Find the x-intercepts of $f(x) = 3x^2 - 8x + 1$.

18. Find the approximate x-intercepts of $f(x) = -2(x - 3)^2 + 5$. Also, find the vertex. Then, sketch by hand a graph of f.

19. Find the nonzero value(s) of a so that the equation $ax^2 - 4x + 4a = 0$ has exactly one solution.

20. Find an equation of the parabola that passes through the points $(1, 4)$, $(2, 9)$, and $(3, 16)$.

21. Find an equation of the parabola with vertex $(5, 3)$ that contains the point $(3, 11)$.

22. Let $f(x) = x^2 - 6x + 11$. Find the number of points that lie on the graph of f at the indicated height.

a. $y = 1$ b. $y = 2$ c. $y = 3$

23. The percentages of television shows that are at least an hour long for various years are shown in Table 45.

Table 45 Television Shows That Are at Least an Hour Long	
Year*	Percent
1998	48
1999	53
2000	57
2001	59
2002	57

Source: *American Demographics and Inside.com*

Each year indicates when the season ended.

Let f represent the percentage of television shows that are at least one hour long at t years since 1990.

a. Find an equation for f.

b. Find $f(15)$. What does your result mean in terms of the situation?

c. Predict when half of all televisions shows will be at least one hour long.

d. Find the t-intercepts of f. What do the t-intercepts mean in terms of the situation?

e. For what years is there model breakdown for certain?

24. A batter hits a baseball. Let $f(t)$ represent the height (in feet) of the baseball at t seconds after the batter hit the ball. An equation for f is $f(t) = -16x^2 + 80x + 3$. At what time is the ball at its maximum height? What is that height?

Cumulative Review of Chapters 1–7

Solve.

1. $5x^2 = 24$

2. $\dfrac{5}{6}x + \dfrac{1}{3} = \dfrac{7}{2}$

3. $\log_4(x + 3) = 2$

4. $3e^x - 5 = 49$

5. $5b^6 + 4 = 82$

6. $3x^2 - 5x - 4 = 0$

7. $3(2)^{4x-5} = 95$

8. $2(3x^2 - 10) = -7x$

9. $\log_b(65) = 4$

10. $3x^2 + 5x^2 = 12x + 20$

11. $2(5x + 2) - 1 = 9(x - 3)$

12. $3\log_2(4x^4) - 2\log_2(4x^3) = 5$

13. $2x(3x - 4) + 5 = 3 - x^2$

14. $3\ln(2x^4) + 2\ln(3x^7) = 8$

15. $5(x - 3)^2 + 4 = 13$

16. Solve $2x^2 + 3x - 6 = 0$ by completing the square.

Solve each system.

17. $4x - 7y = -10$
$3x + 4y = 11$

18. $y = 3x - 1$
$2x - 3y = -11$

19. $\dfrac{1}{2}x - y = \dfrac{5}{2}$
$\dfrac{2}{5}x - \dfrac{3}{5}y = \dfrac{6}{5}$

20. Solve the inequality $2(3x - 4) < 5 - 3(6x + 5)$. Describe the solution set with inequality notation, interval notation, and a graph.

Simplify.

21. $(2b^4 c^{-5})^3 (3b^{-1}c^{-2})^4$

22. $\dfrac{10b^{-3/4}c^{2/5}}{15b^{-1/2}c^{3/5}}$

23. $\left(\dfrac{6b^8 c^{-3}}{8b^{-4}c^{-1}}\right)^3$

Simplify. Write your result as a single logarithm with a coefficient of 1.

24. $3\ln(2x^7) + 2\ln(4x^5)$

25. $3\log_b(4x^8) - 4\log_b(2x^3)$

Expand.

26. $(3x - 4)^2$

27. $(5x - 7)(5x + 7)$

28. $-3x(x^2 - 5)(x^3 + 8)$

29. $(2x - 3)(x^2 + 4x - 5)$

30. Write $f(x) = -2(x - 5)^2 + 3$ in standard form.

Factor.

31. $81x^2 - 25$

32. $x^3 - 13x^2 + 40x$

33. $8x^2 + 22x - 21$

34. $x^3 + 4x^2 - 9x - 36$

For exercises 35–41, refer to Table 46.

Table 46 Values for Four Functions (Exercises 35–41)

x	f(x)	x	g(x)	x	h(x)	x	k(x)
0	20	1	4	4	7	0	-8
1	17	2	12	5	14	1	-4
2	14	3	36	6	28	2	0
3	11	4	108	7	56	3	4
4	8	5	324	8	112	4	8
5	5	6	972	9	224	5	12

35. Find a possible equation for f.

36. Find a possible equation for g.

37. Find the slope of k.

38. Find $g(4)$.

39. Find x when $g(x) = 4$.

40. Find $h^{-1}(7)$.

41. Find the x-intercept of k.

Sketch the graph of the function.

42. $y = -2(x + 4)^2 + 3$

43. $y = \log_3(x)$

44. $2x - 5y = 20$

45. $y = \dfrac{4}{3}x - 2$

46. $y = 2(3)^x$

47. $y = x^2 + 4x - 5$

48. $y = 8\left(\dfrac{1}{4}\right)^x$

49. Find the slope of the line that contains the points $(-5, -2)$ and $(-3, 4)$.

50. Find the equation of the line that contains the point $(-2, 6)$ and is perpendicular to the line $3x - 4y = 5$.

51. Find an equation of an exponential curve that contains the points $(0, 78)$ and $(9, 13)$.

52. Find an equation of an exponential curve that contains the points $(2, 27)$ and $(5, 83)$.

53. Find an equation of a parabola that contains the points $(1, -1)$, $(2, 4)$, and $(4, 20)$.

54. Let f be the linear function, g be the exponential function, and h be a quadratic function whose graphs contain the points $(0, 3)$ and $(1, 6)$.

   a. Find possible equations for f, g, and h.

   b. Use your graphing calculator to draw the graphs of f, g, and h in the same viewing window.

Let $f(x) = -x^2 + 6x - 5$.

55. Find x when $f(x) = 3$.

56. Find x when $f(x) = 4$.

57. Find x when $f(x) = 5$.

58. Find the x-intercepts.

59. Find the y-intercept.

60. Sketch a graph of f.

Find the logarithm.

61. $\log_2(16)$

62. $\log_b(b^3)$

63. $\log_5\left(\dfrac{1}{25}\right)$

64. $\log_9(2)$

65. The graph of a function f is sketched in Fig. 37. Sketch the graph of f^{-1}.

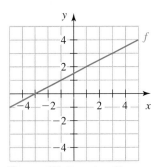

Figure 37 Exercise 65

Find the inverse of the function.

66. $g(x) = 3^x$

67. $f(x) = \dfrac{2}{5}x + 1$

68. An increasing number of companies are offering their employees stock options (see Table 47). This means that workers have the right to buy company stock at a set price for a set period of time.

Table 47 Companies Offering Stock Options to at Least Half of Their Employees

Year	Percent of Companies Offering Stock Options
1994	19
1995	23
1996	27
1997	30
1998	35
1999	39

Source: William M. Mercer

Let $f(t)$ represent the percentage of companies offering stock options to at least half of their employees at t years since 1990.

a. Perform the first three steps of the modeling process to find an equation for f.

b. Find $f(18)$. What does your result mean in terms of the situation?

c. Find t when $f(t) = 100$. What does your result mean in terms of the situation?

d. Find the t-intercept. What does your result mean in terms of the situation?

e. For what values of t is model breakdown certain? Explain.

69. Skateboarding hardware and apparel sales revenues for various years are shown in Table 48.

Table 48 Skateboarding Hardware and Apparel Sales Revenues

Year	Revenue (billions of dollars)
1996	0.52
1997	0.60
1998	0.76
1999	0.90
2000	1.15
2001	1.40

Source: *TransWorld SKATEboarding Magazine*

Let $f(t)$ represent hardware and apparel sales revenues (in billions of dollars) at t years since 1990.

a. Find a linear equation, an exponential equation, and a quadratic equation for f.

b. Discuss how well each model fits the data.

c. Given that sales revenues declined during the early 1990s, which is the best model for the years 1990–2001?

d. Estimate the minimum sales revenues for the years 1990–2001.

e. Use the exponential model to estimate the percent rate of growth in sales revenues for the years 1996–2001.

f. Use the quadratic model to predict when sales revenues will reach $5 billion. Then use the exponential model to predict when sales revenues will reach $5 billion. Explain in terms of the graphs of the models why the year predicted by the exponential model is before the year predicted by the quadratic model.

70. Although grade school students in Hartford used to have the lowest test scores in the state of Connecticut, their scores have increased due to reform measures. In particular, their mathematics test scores have improved (see Table 49).

Table 49 Hartford and Connecticut Mathematics Test Scores

	Percent of Students Scoring Above Goals	
Year	Hartford Students	Connecticut Average
1994	20	57
1996	26	59
1998	29	61
2000	35	63

Source: *Connecticut Department of Education*

Let $h(t)$ and $c(t)$ represent the percentages of Hartford and Connecticut grade school students, respectively, scoring on mathematics tests above goals at t years since 1990.

a. Find equations for h and for c.

b. Compare the rate of change of percentage of students scoring above goals for students in Hartford versus students in Connecticut.

c. Find $h(18)$ and $c(18)$. What do these results mean in terms of the situation?

d. Using your models, predict when the percentage of Hartford grade school students who score above goals will be equal to the percentage for Connecticut grade school students who score above goals.

Rational Functions

Individually [mathematicians] are very different in their mathematical personalities, the kind of mathematics they like, and the way that they do mathematics, but they are alike in one respect. Almost without exception, they love their subject, are happy in their choice of a career, and consider that they are exceptionally lucky in being able to do for a living what they would do for fun.
—*Constance Reid, "Becoming a Mathematician," 1990.*

8.1 FINDING THE DOMAINS OF RATIONAL FUNCTIONS AND SIMPLIFYING RATIONAL EXPRESSIONS

Objectives

▶ Know the meaning of *rational function* and *vertical asymptote*.

▶ Find the domain of a rational function.

▶ Simplify a rational expression.

▶ Know the connection between vertical asymptotes of a rational function and the domain of such a function.

In chapters 6 and 7 we worked with polynomials. In this section we work with *rational expressions*, which are ratios of polynomials. The name "rational" refers to *ratio*. Here are some examples:

$$\frac{x^3 - 3x + 6}{x^7 - 8}, \quad \frac{-2x^2 + 17}{5x - 1}, \quad -\frac{2}{7x^3}$$

DEFINITION *Rational Function*

A **rational function** is a function whose equation can be put into the form

$$f(x) = \frac{P(x)}{Q(x)}$$

where $P(x)$ and $Q(x)$ are polynomials, and $Q(x)$ is nonzero.

Note

"$Q(x)$ is nonzero" means that $Q(x)$ is not the zero polynomial, 0. $Q(x)$ could be, say, $x - 3$, even though $x - 3$ does have the value 0 when $x = 3$.

Consider the rational function $f(x) = \frac{5}{x}$. Note that $f(0)$ is undefined since $\frac{5}{0}$ is undefined. So, 0 is *not* in the domain of f. Since division by nonzero numbers is defined, the domain of f is the set of real numbers except 0.

In Fig. 1 we display a graphing calculator graph of $f(x) = \frac{5}{x}$. Note that the x-axis appears to be a horizontal asymptote of f, which is, in fact, true. Also note that the

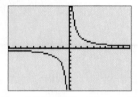

Figure 1 Use ZStandard to draw a graph of $f(x) = \dfrac{5}{x}$

graph of f appears to get arbitrarily close to, but never intersects, the y-axis, which is also true. We say that the y-axis is a *vertical asymptote* of f. Just as with a horizontal asymptote, a vertical asymptote is *not* part of the graph of a function.

Example 1   Domain of a Rational Function

Find the domain of each function.

1. $f(x) = \dfrac{2}{x-3}$ **2.** $g(x) = \dfrac{x-5}{x^2-4}$ **3.** $h(x) = \dfrac{x+1}{2x^2+7x+3}$

Solution

1. For $f(x) = \dfrac{2}{x-3}$, the number 3 is not in the domain, since $\dfrac{2}{3-3}$ is not defined. Since 3 is the only value of x that leads to division by zero, the domain is the set of real numbers except 3. In Fig. 2 we draw a graph of f and build a table of input-output pairs of f.

 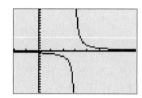 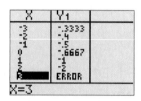

Figure 2 f has a vertical asymptote at $x = 3$, and 3 is not in the domain of f

From the graph of f, it appears that f has a vertical asymptote at $x = 3$, which is fact. Also, the "ERROR" message across from $x = 3$ in the table supports the idea that the value 3 for x leads to division by zero.

2. Since division by zero is undefined, we find numbers *not* in the domain of g by setting the denominator of the right side of the equation $g(x) = \dfrac{x-5}{x^2-4}$ equal to zero and solving the equation.

$$x^2 - 4 = 0$$
$$(x+2)(x-2) = 0$$
$$x + 2 = 0 \quad \text{or} \quad x - 2 = 0$$
$$x = -2 \quad \text{or} \quad x = 2$$

The numbers -2 and 2 are *not* in the domain since these values for x lead to divisions by zero. The domain of g is the set of real numbers except -2 and 2. In Fig. 3 we draw a graph of g and build a table of input-output pairs of g.

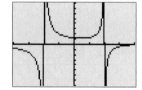 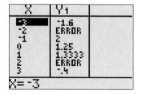

Figure 3 g has vertical asymptotes at $x = -2$ and $x = 2$; also, -2 and 2 are not in the domain of g

From the graph, it appears that g has vertical asymptotes at $x = -2$ and $x = 2$. The "ERROR" messages across from $x = -2$ and $x = 2$ in the table support the idea that the values -2 and 2 for x lead to divisions by zero.

3. We find the numbers *not* in the domain by setting the denominator of the right side

of $h(x) = \dfrac{x + 1}{2x^2 + 7x + 3}$ equal to zero and solving for x.

$$2x^2 + 7x + 3 = 0$$
$$(x + 3)(2x + 1) = 0$$
$$x + 3 = 0 \quad \text{or} \quad 2x + 1 = 0$$
$$x = -3 \quad \text{or} \quad 2x = -1$$
$$x = -3 \quad \text{or} \quad x = -\frac{1}{2}$$

The numbers -3 and $-\dfrac{1}{2}$ are *not* in the domain since these values of x lead to

divisions by zero. The domain of h is the set of real numbers except -3 and $-\dfrac{1}{2}$.

In Fig. 4 we draw a graph of h and build a table of input-output pairs of h.

Figure 4 h has vertical asymptotes at $x = -3$ and $x = -\dfrac{1}{2}$; also, -3 and $-\dfrac{1}{2}$ are
not in the domain of h

From the graph, it appears that h has vertical asymptotes at $x = -3$ and $x = -\dfrac{1}{2}$.

The "ERROR" messages in the table are due to divisions by zero when $x = -3$

and $x = -\dfrac{1}{2}$. ■

Our work in problems 1–3 in Example 1 suggests the following property of rational
functions.

Domain of a Rational Function

The domain of a rational function $f(x) = \dfrac{P(x)}{Q(x)}$ is the set of real numbers except
for those numbers that when substituted for x give $Q(x) = 0$.

Example 2   Domain of a Rational Function

Find the domain of the function.

1. $f(x) = \dfrac{x - 4}{x^2 + 5x + 2}$ **2.** $g(x) = \dfrac{x^2 - 5x + 1}{x^2 + 10}$

Solution

1. We set the denominator of the right side of $f(x) = \dfrac{x - 4}{x^2 + 5x + 2}$ equal to zero.

$$x^2 + 5x + 2 = 0$$

Applying the quadratic formula gives:

$$x = \frac{-5 \pm \sqrt{5^2 - 4(1)(2)}}{2(1)}$$
$$= \frac{-5 \pm \sqrt{17}}{2}$$

The domain of f is the set of real numbers except $\dfrac{-5 - \sqrt{17}}{2}$ and $\dfrac{-5 + \sqrt{17}}{2}$.

2. We set the denominator of the right side of $g(x) = \dfrac{x^2 - 5x + 1}{x^2 + 10}$ equal to zero and solve.

$$x^2 + 10 = 0$$
$$x^2 = -10$$

Since a real number squared is nonnegative, we conclude that the solution set is the empty set. For the function g, no values of x lead to division by zero and, therefore, the domain is the set of (all) real numbers. ∎

Now we will discuss how to reduce rational expressions. Recall that we reduce fractions by the two-step procedure of using factoring and then using the property $\dfrac{a}{a} = 1$, when $a \neq 0$. For example:

$$\frac{8}{12} = \frac{4(2)}{4(3)} = \frac{4}{4} \cdot \frac{2}{3} = 1 \cdot \frac{2}{3} = \frac{2}{3}$$

Note that 4 is the GCF of the numerator and the denominator of $\dfrac{8}{12}$.

We reduce the rational expression $\dfrac{2x + 6}{x^2 - 9}$ in a similar manner:

$$\frac{2x + 6}{x^2 - 9} = \frac{2(x + 3)}{(x - 3)(x + 3)} \qquad \text{Factor the numerator and the denominator.}$$

$$= \frac{2}{x - 3} \cdot \frac{x + 3}{x + 3} \qquad x + 3 \text{ is the GCF of the numerator and the denominator.}$$

$$= \frac{2}{x - 3} \cdot 1 \qquad \frac{x + 3}{x + 3} = 1 \text{ when } x \neq -3$$

$$= \frac{2}{x - 3}$$

When we write

$$\frac{2x + 6}{x^2 - 9} = \frac{2}{x - 3}$$

Note

The expressions are not equivalent at $x = -3$ since $\dfrac{2(-3) + 6}{(-3)^2 - 9}$ gives a zero in the denominator while $\dfrac{2}{-3 - 3} = \dfrac{2}{-6} = -\dfrac{1}{3}$ is a real number.

we mean that the two expressions give the same result for each real number substituted for x except -3 or 3. We say that the two expressions are *equivalent* for all values of x except -3 or 3.

Throughout the rest of this text you may assume that all values of a variable that would leave an expression undefined are excluded.

Reducing a Rational Expression

To reduce a rational expression:

1. Factor the numerator and denominator.

2. Use the property $\dfrac{AB}{AC} = \dfrac{B}{C}$, where A and C are nonzero and A is the GCF of the numerator and the denominator.

Throughout the rest of this chapter you may assume that the form $\dfrac{A}{B}$ represents a rational expression.

For the expression $\dfrac{AB}{AC}$, note that the polynomial A is a factor both of the numerator and the denominator. The expression

$$\frac{5(x+2)}{3(x+2)}$$

can be reduced to $\dfrac{5}{3}$ when $x \neq -2$ since $x + 2$ is a factor of both the numerator and the denominator. However, the expression

$$\frac{5(x+2)+1}{3(x+2)}$$

cannot be reduced. Note in particular that $x + 2$ is not a factor of both the numerator and the denominator.

If an expression cannot be reduced, we say it is in *lowest terms*. We *simplify* a rational expression by reducing it so that it is in lowest terms. We also write the numerator and the denominator in factored form.

> **Note**
>
> To simplify $\dfrac{5(x+2)+1}{3(x+2)}$, we simplify the numerator to get $\dfrac{5x+11}{3(x+2)}$. In this form, it is clear that we cannot reduce the rational expression.

> **Note**
>
> A "simplified" rational expression can mean that the numerator and the denominator are expanded or factored. Consult your instructor for his/her preference.

Example 3   Simplifying the Right Side of an Equation

Simplify the right side of the equation $f(x) = \dfrac{2x^2 - 6x - 20}{2x^2 - 50}$.

Solution

$$
\begin{aligned}
f(x) &= \frac{2x^2 - 6x - 20}{2x^2 - 50} \\[2mm]
&= \left.\frac{2(x^2 - 3x - 10)}{2(x^2 - 25)}\right\} \quad \text{Factor the numerator and the} \\[2mm]
&= \frac{2(x-5)(x+2)}{2(x-5)(x+5)} \quad \text{denominator of the right side.} \\[2mm]
&= \frac{x+2}{x+5} \qquad\qquad\quad \text{Reduce.}
\end{aligned}
$$

So, we can describe f by the equation $f(x) = \dfrac{x+2}{x+5}$, where the domain is all real numbers except -5 and 5.

To perform a check, we substitute a value, say 3, for x in both of the expressions $\dfrac{2x^2 - 6x - 20}{2x^2 - 50}$ and $\dfrac{x+2}{x+5}$ and check that the results are equal:

$$\frac{2(3)^2 - 6(3) - 20}{2(3)^2 - 50} = \frac{5}{8}, \qquad \frac{3+2}{3+5} = \frac{5}{8}$$

We perform a more convincing check by comparing graphing calculator graphs for the functions $y = \dfrac{2x^2 - 6x - 20}{2x^2 - 50}$ and $y = \dfrac{x+2}{x+5}$. See Fig. 5.

> **Note**
>
> To hand-sketch a graph of $f(x) = \dfrac{2x^2 - 6x - 20}{2x^2 - 50}$, we use an open circle at the point $(5, 0.7)$ to indicate that the point is not part of the graph.

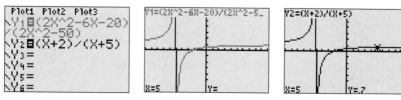

Figure 5 Compare graphs for $y = \dfrac{2x^2 - 6x - 20}{2x^2 - 50}$ and $y = \dfrac{x+2}{x+5}$

Note that the graphs of the functions are the same for $x \neq -5$ and $x \neq 5$.   ■

Problems 1–3 in Example 1 suggest that if a rational function has a vertical asymptote at $x = k$, then k is not in the domain of the function, which is, in fact, true.

In Example 3 we wrote f in the form

$$f(x) = \frac{2(x-5)(x+2)}{2(x-5)(x+5)}$$

by factoring the numerator and denominator of the right side of the original equation for f. From this form, we see that the domain of f is the set of real numbers except -5 and 5. Yet, $x = -5$ is the only vertical asymptote of f (see Fig. 5). The reason $x = 5$ is not a vertical asymptote has to do with the fact that both the numerator and the denominator of f contain the factor $x - 5$.

Domain and Vertical Asymptote(s) of a Rational Function

If a rational function g has a vertical asymptote $x = k$, then k is not in the domain of g. If k is not in the domain of a rational function h, then $x = k$ may or may not be a vertical asymptote of h.

Example 4   Simplifying the Right Side of an Equation

Simplify the right side of the equation $f(x) = \dfrac{x^3 - 5x^2 + 6x}{x^3 - 2x^2 - 9x + 18}$.

Solution

$$\begin{aligned}
f(x) &= \frac{x^3 - 5x^2 + 6x}{x^3 - 2x^2 - 9x + 18} \\[2mm]
&= \frac{x(x^2 - 5x + 6)}{x^2(x-2) - 9(x-2)} \\[2mm]
&= \frac{x(x-3)(x-2)}{(x^2 - 9)(x-2)} \\[2mm]
&= \frac{x(x-3)(x-2)}{(x+3)(x-3)(x-2)} \\[2mm]
&= \frac{x}{x+3}
\end{aligned}$$

Factor the numerator and the denominator.

Reduce.

So, we can describe f by the equation $f(x) = \dfrac{x}{x+3}$, where the domain is all real numbers except $-3, 2$, and 3. We verify our work in Fig. 6.

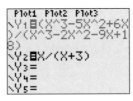 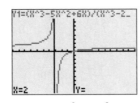 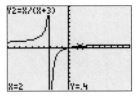

Figure 6 Compare graphs for $y = \dfrac{x^3 - 5x^2 + 6x}{x^3 - 2x^2 - 9x + 18}$ and $y = \dfrac{x}{x+3}$

■

The distributive law applies to the expression $-(b - a)$ as follows

$$-(b-a) = -1(b-a) = (-1)b - (-1)a = -b + a = a - b$$

or

$$-(b - a) = a - b$$

The statement

$$a - b = -(b - a)$$

is a useful tool in reducing rational expressions, as you shall see in Example 5.

Example 5

Simplify each expression.

1. $\dfrac{7-x}{x-7}$　　　　　　　**2.** $\dfrac{x^2-x-12}{4-x}$

Solution

1. $\dfrac{7-x}{x-7} = \dfrac{-(x-7)}{x-7}$　　　　　$a-b=-(b-a)$

　　　$= -1$　　　　　　　　　Reduce.

2. $\dfrac{x^2-x-12}{4-x} = \dfrac{(x-4)(x+3)}{4-x}$　　　Factor the numerator.

　　　$= \dfrac{(x-4)(x+3)}{-(x-4)}$　　　$a-b=-(b-a)$

　　　$= -\dfrac{(x-4)(x+3)}{x-4}$　　　$\dfrac{a}{-b}=-\dfrac{a}{b}$

　　　$= -(x+3)$　　　　　Reduce.

　　　$= -x-3$　　■

EXPLORATION

Connection between the domain and vertical asymptotes

1. Find the domain of $f(x) = \dfrac{6}{x}$. Use ZStandard to draw a graph of f. Explain why it makes sense that f does not have a y-intercept. Find any vertical asymptotes of f.

2. Use ZStandard to graph $f(x) = \dfrac{6}{x}$ and $g(x) = \dfrac{6}{x-4}$ on the same viewing screen. Describe how you can translate the graph of f to get the graph of g.

3. What is the vertical asymptote of $g(x) = \dfrac{6}{x-4}$? What do you observe about the connection between the vertical asymptote of g and the domain of g?

4. Use ZStandard to graph $h(x) = \dfrac{7x-7}{x^2+2x-8}$. What are the vertical asymptote(s)? What is the connection between the vertical asymptote(s) of h and the domain of h?

5. Find the domain of the function $f(x) = \dfrac{x-2}{2-x}$. Use ZDecimal to graph f. Explain why the graph is so different from the graphs you drew in problems 1, 2, and 4. [**Hint**: Simplify the right side of the equation for f.] Also, describe what happens when you use TRACE to try to find the value of y when $x = 2$. Explain.

6. Describe the connection between the vertical asymptote(s) of the function $g(x) = \dfrac{5x-10}{x^2-7x+10}$ and the domain of g. [**Hint**: Simplify the right side of the equation for g.]

7. True or False? Explain.

a. If a rational function f has a vertical asymptote $x = k$, then k is not in the domain of f.

b. If k is not in the domain of a rational function g, then $x = k$ is a vertical asymptote of g.

Tips on Succeeding in This Course

If you are having difficulty doing an exercise, don't panic! It is a good idea to reread the exercise and reflect on what you have already sorted out about the problem—what you know, and where you want to go. Your solution to the problem may be just around the corner.

Key Points OF THIS SECTION

Rational function	A **rational function** is a function whose equation can be put into the form $$f(x) = \frac{P(x)}{Q(x)}$$ where $P(x)$ and $Q(x)$ are polynomials, and $Q(x)$ is nonzero.
Domain of a rational function	The domain of a rational function $f(x) = \dfrac{P(x)}{Q(x)}$ is the set of real numbers except for those numbers that when substituted for x give $Q(x) = 0$.
Vertical asymptote and domain	If a rational function g has a vertical asymptote $x = k$, then k is not in the domain of g. If k is not in the domain of a rational function h, then $x = k$ may or may not be a vertical asymptote of h.
Reducing a rational expression	To reduce a rational expression: 1. Factor the numerator and denominator. 2. Use the property $\dfrac{AB}{AC} = \dfrac{B}{C}$, where A and C are nonzero and A is the GCF of the numerator and the denominator.

HOMEWORK 8.1

SSM
Tutor Center

MathPro 4
MathPro 4

MathPro 5
MathPro 5

CDs/Videos

www

Find the domain for each function. Verify your result using a graphing calculator table or graph.

1. $f(x) = \dfrac{8}{x}$

2. $f(x) = \dfrac{x}{8}$

3. $f(x) = \dfrac{x}{2}$

4. $f(x) = \dfrac{9}{x}$

5. $f(x) = \dfrac{x-5}{x+3}$

6. $f(x) = \dfrac{x+4}{x-9}$

7. $f(x) = \dfrac{2-x}{x-2}$

8. $f(x) = \dfrac{x-4}{4-x}$

9. $f(x) = \dfrac{x-3}{2x+1}$

10. $f(x) = \dfrac{x+4}{5x-7}$

11. $f(x) = \dfrac{x+5}{(x+8)(x-4)}$

12. $f(x) = \dfrac{(x+2)(x-7)}{(x-1)(x-5)}$

13. $f(x) = \dfrac{x+9}{x^2-3x-10}$

14. $f(x) = \dfrac{x^2-3x-10}{x^2+5x-24}$

15. $f(x) = \dfrac{x^2-25}{x^2-9}$

16. $f(x) = \dfrac{x-3}{x^2-1}$

17. $f(x) = \dfrac{x+2}{4x^2-25}$

18. $f(x) = -\dfrac{7}{81x^2-49}$

19. $f(x) = \dfrac{2}{x^2+1}$

20. $f(x) = \dfrac{x+8}{x^2+4}$

21. $f(x) = \dfrac{x-10}{2x^2-7x-15}$

22. $f(x) = \dfrac{x}{6x^2+7x-20}$

23. $f(x) = \dfrac{x+3}{x^2-3x+6}$

24. $f(x) = -\dfrac{5}{3x^2-x+4}$

25. $f(x) = \dfrac{x^2+5x-1}{3x^2-2x-7}$

26. $f(x) = \dfrac{x+14}{6x^2+4x-1}$

27. $f(x) = \dfrac{x-7}{4x^3-8x^2-9x+18}$

28. $f(x) = \dfrac{5x+10}{x^3+5x^2-4x-20}$

29. $f(x) = \dfrac{3}{2(x-1)+3(x-2)}$

30. $f(x) = \dfrac{2}{3(x+4)+5(x+1)}$

31. $f(x) = \dfrac{4x-7}{(x+2)(x-3)+(x+1)(x+5)}$

32. $f(x) = \dfrac{x+1}{(x-2)^2-(x-3)^2}$

Simplify the right side of the equation. Verify your result by comparing graphing calculator tables or graphs for the original function and its simplified form.

33. $f(x) = \dfrac{x^2}{x^6}$

34. $f(x) = \dfrac{x^8}{x^3}$

35. $f(x) = \dfrac{20x^7}{15x^4}$

36. $f(x) = \dfrac{12x^3}{16x^{10}}$

37. $f(x) = \dfrac{4x - 28}{5x - 35}$

38. $f(x) = \dfrac{-4x - 20}{6x + 30}$

39. $f(x) = \dfrac{x^2 + 7x + 10}{x^2 - 7x - 18}$

40. $f(x) = \dfrac{x^2 + 8x + 12}{x^2 + 9x + 18}$

41. $f(x) = \dfrac{5x + 10}{x^2 - 4}$

42. $f(x) = \dfrac{6x - 4}{9x^2 - 4}$

43. $f(x) = \dfrac{12x + 15}{16x^2 - 25}$

44. $f(x) = \dfrac{x^2 - x - 12}{x^2 - 9}$

45. $f(x) = \dfrac{x^2 - 49}{x^2 - 14x + 49}$

46. $f(x) = \dfrac{4x^2 + 4x - 8}{6x^2 - 12x + 6}$

47. $f(x) = \dfrac{2x^2 - 50}{6x^2 - 24x - 30}$

48. $f(x) = \dfrac{12x^2 - 75}{6x^2 - 39x + 60}$

49. $f(x) = \dfrac{2x^2 - 7x - 4}{2x^2 + 9x + 4}$

50. $f(x) = \dfrac{x - 2}{-7x + 14}$

51. $f(x) = \dfrac{x - 5}{5 - x}$

52. $f(x) = \dfrac{2 - x}{x - 2}$

53. $f(x) = \dfrac{4x - 12}{18 - 6x}$

54. $f(x) = \dfrac{-2x + 10}{4x - 20}$

55. $f(x) = \dfrac{6x - 18}{9 - x^2}$

56. $f(x) = \dfrac{9 - x}{x^2 - 81}$

57. $f(x) = \dfrac{x^2 + 2x - 35}{-x^2 + 3x + 10}$

58. $f(x) = \dfrac{36 - x^2}{x^2 - 8x + 12}$

59. $f(x) = \dfrac{6x^2 - 5x - 6}{15 - 10x}$

60. $f(x) = \dfrac{25 - 4x^2}{2x^2 - 21x + 40}$

61. $f(x) = \dfrac{3x^3 + 21x^2 + 36x}{x^2 - 9}$

62. $f(x) = \dfrac{x^2 - 7x - 30}{2x^3 - 8x^2 - 42x}$

63. $f(x) = \dfrac{x^3 - x^2 - 4x + 4}{3x^2 + 3x - 6}$

64. $f(x) = \dfrac{3x^2 + 11x + 6}{3x^3 + 2x^2 - 27x - 18}$

65. A person plans to drive at a constant speed from San Francisco to Los Angeles. The driving time (in hours) $T(s)$ is given by the equation

$$T(s) = \dfrac{420}{s}$$

where s is the constant speed (in miles per hour).

a. Find $T(50)$. What does your result mean in terms of the length of the trip?

b. Find $T(55)$, $T(60)$, $T(65)$, and $T(70)$.

c. Is T an increasing function or a decreasing function for $s > 0$? Explain why this makes sense in terms of the trip.

Use ZStandard and then ZOOM OUT once or twice to verify your answer.

66. Some students agree to share equally in the expense of renting a large beach house for $1200 during their spring break.

a. What is the per-student expense if 10 students rent the house?

b. What is the per-student expense if 12 students rent the house?

c. Let $p(n)$ represent the per-student expense (in dollars) when n students rent the house. Find an equation for p. [**Hint**: Reflect on your work in parts a and b.]

d. Use ZStandard and then ZOOM OUT twice to draw a graph of p. Is p a decreasing function or an increasing function for $n > 0$? Explain why this makes sense in terms of the per-student expense.

e. Find $p(15)$. What does your result mean in terms of the situation?

67. Give examples of three rational functions, each of whose domains are the set of real numbers except -3 and 3.

68. Give examples of three rational functions, each of whose domains are the set of real numbers except $\dfrac{1}{2}$ and $\dfrac{1}{3}$.

69. A graphing calculator table for a rational function is shown in Fig. 7. List seven members of the domain. List seven members of the range.

X	Y1
-3	50.4
-2	25.2
-1	16.8
0	12.6
1	10.08
2	8.4
3	7.2

X= -3

Figure 7 Exercise 69

70. A student decides that the domain of the function

$$f(x) = \dfrac{6}{x(x - 3) + 2(x - 3)}$$

is the set of real numbers except 3. After viewing the table shown in Fig. 8, the student believes that his work is correct. What would you tell this student?

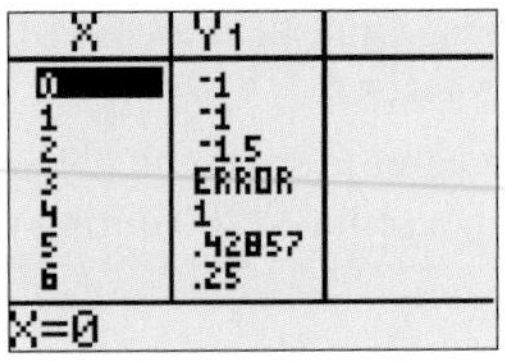

Figure 8 Exercise 70

71. A student tries to reduce the expression $\dfrac{2(x+4)+3}{(x+4)(x-1)}$ as follows:

$$\frac{2(x+4)+3}{(x+4)(x-1)} = \frac{2+3}{x-1} = \frac{5}{x-1}$$

Now, substitute 3 for x in the original expression and in the student's result. What can you conclude about the student's work based on the results of the substitutions you made? Explain.

72. A student tries to reduce the expression $\dfrac{(x-2)(x-1)}{4(x-2)+5}$ as follows:

$$\frac{(x-2)(x-1)}{4(x-2)+5} = \frac{x-1}{4+5} = \frac{x-1}{9}$$

Next, the student substitutes 3 for x in the original expression,

with the result:

$$\frac{(3-2)(3-1)}{4(3-2)+5} = \frac{2}{9}, \qquad \frac{3-1}{9} = \frac{2}{9}$$

The student concludes that the result $\dfrac{x-1}{9}$ is correct. What would you tell the student?

Let $f(x) = \dfrac{(x-1)(x-2)}{(x-3)(x-4)}$. *Find each indicated output.*

73. $f(1)$

74. $f(2)$

75. $f(3)$

76. $f(4)$

77. A student states that the domain of the function

$$f(x) = \frac{(x-2)(x-4)}{(x-5)(x-1)}$$

is the set of real numbers except 1, 2, 4, and 5. What would you tell the student?

78. Describe how to find the domain of a rational function. If you believe that a number is not in the domain, describe how you can verify that this is the case.

79. Describe how to reduce a rational expression. Explain how you can check that the result and the original expression are equivalent.

8.2 MULTIPLYING AND DIVIDING RATIONAL EXPRESSIONS

Objective

▸ Multiply and divide rational expressions.

In this section we multiply and divide rational expressions. To find the product $\dfrac{2}{3} \cdot \dfrac{5}{7}$, we use the property $\dfrac{a}{b} \cdot \dfrac{c}{d} = \dfrac{ac}{bd}$, where b and d are nonzero:

$$\frac{2}{3} \cdot \frac{5}{7} = \frac{2 \cdot 5}{3 \cdot 7} = \frac{10}{21}$$

We multiply more complicated rational expressions in a similar way.

Multiplying Rational Expressions

If $\dfrac{A}{B}$ and $\dfrac{C}{D}$ are rational expressions with B and D nonzero, then

$$\frac{A}{B} \cdot \frac{C}{D} = \frac{AC}{BD}$$

In words: To multiply two rational expressions, multiply their numerators and multiply their denominators.

Example 1   Multiplying Rational Expressions

Find the indicated product $\dfrac{4x^3}{7x-1} \cdot \dfrac{3x+5}{2x}$. Simplify the result.

Solution

$$\frac{4x^3}{7x-1} \cdot \frac{3x+5}{2x} = \frac{4x^3(3x+5)}{2x(7x-1)} \qquad \frac{A}{B} \cdot \frac{C}{D} = \frac{AC}{BD}$$

$$= \frac{2x^2(3x+5)}{7x-1} \qquad \text{Reduce.}$$

We can verify our result by comparing tables for $y = \dfrac{4x^3}{7x-1} \cdot \dfrac{3x+5}{2x}$ and
$y = \dfrac{2x^2(3x+5)}{7x-1}$. See Fig. 9.

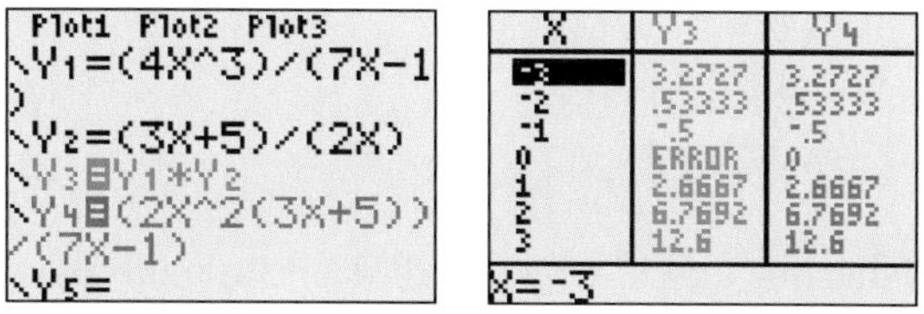

Graphing Calculator

To enter a function using Y_n references, see Section B.25.

Figure 9 Compare the tables
for $y = \dfrac{4x^3}{7x-1} \cdot \dfrac{3x+5}{2x}$ and
$y = \dfrac{2x^2(3x+5)}{7x-1}$

Note

0 is in the domain of one function, but not the other. Why?

To find products of rational expressions, it is usually easiest to start by factoring, if possible, the numerators and denominators.

Example 2   Multiplying Rational Expressions

Find the product $\dfrac{x^2-9}{2x^2-x-10} \cdot \dfrac{4x^2-25}{x^2+4x-21}$. Simplify the result.

Solution

We begin by factoring the numerators and denominators.

$$\frac{x^2-9}{2x^2-x-10} \cdot \frac{4x^2-25}{x^2+4x-21} = \frac{(x-3)(x+3)}{(2x-5)(x+2)} \cdot \frac{(2x-5)(2x+5)}{(x-3)(x+7)}$$

$$= \frac{(x-3)(2x-5)(x+3)(2x+5)}{(x-3)(2x-5)(x+2)(x+7)} \qquad \frac{A}{B} \cdot \frac{C}{D} = \frac{AC}{BD}$$

$$= \frac{(x+3)(2x+5)}{(x+2)(x+7)} \qquad\qquad \text{Reduce.}$$

To multiply two rational expressions:

1. Factor the numerators and the denominators.
2. Multiply using the property $\dfrac{A}{B} \cdot \dfrac{C}{D} = \dfrac{AC}{BD}$, where B and D are nonzero.
3. Simplify the result.

If $f(x)$ and $g(x)$ are functions, then we can form the **product function, $P(x)$,** where $P(x) = f(x)g(x)$.

Example 3   Finding a Product Function

Let $P(x) = f(x)g(x)$, where $f(x) = \dfrac{35x^2-25x}{x^2-36}$ and $g(x) = \dfrac{6-x}{15x^4}$. Find an equation for P and simplify the right side of the equation.

Solution

$$P(x) = f(x)g(x)$$

$$= \frac{35x^2-25x}{x^2-36} \cdot \frac{6-x}{15x^4}$$

$$= \frac{5x(7x-5)}{(x-6)(x+6)} \cdot \frac{-(x-6)}{15x^4} \qquad 6-x = -(x-6)$$

$$= \frac{-5x(x-6)(7x-5)}{15x^4(x-6)(x+6)}$$

$$= -\frac{7x-5}{3x^3(x+6)} \qquad\qquad \text{Reduce.}$$

Now we turn our attention to dividing rational expressions. To find the quotient $\frac{2}{3} \div \frac{5}{7}$, we use the property $\frac{a}{b} \div \frac{c}{d} = \frac{a}{b} \cdot \frac{d}{c}$, where b, c, and d are nonzero.

$$\frac{2}{3} \div \frac{5}{7} = \frac{2}{3} \cdot \frac{7}{5}$$
$$= \frac{14}{15}$$

Note that dividing by $\frac{5}{7}$ is the same as multiplying by $\frac{7}{5}$. We call $\frac{7}{5}$ the *reciprocal* of $\frac{5}{7}$.

Dividing Rational Expressions

If $\frac{A}{B}$ and $\frac{C}{D}$ are rational expressions and B, C, and D are nonzero, then

$$\frac{A}{B} \div \frac{C}{D} = \frac{A}{B} \cdot \frac{D}{C}$$

In words: The quotient of two rational expressions is the product of the first rational expression and the reciprocal of the second rational expression.

Example 4 Dividing Two Rational Expressions

Find the quotient $\frac{6x^2}{x-3} \div \frac{4x^7}{x+1}$.

Solution

$$\frac{6x^2}{x-3} \div \frac{4x^7}{x+1} = \frac{6x^2}{x-3} \cdot \frac{x+1}{4x^7} \qquad \frac{A}{B} \div \frac{C}{D} = \frac{A}{B} \cdot \frac{D}{C}$$
$$= \frac{6x^2(x+1)}{4x^7(x-3)}$$
$$= \frac{3(x+1)}{2x^5(x-3)} \qquad \text{Reduce.}$$

We verify our work by comparing graphing calculator graphs for $y = \frac{6x^2}{x-3} \div \frac{4x^7}{x+1}$ and $y = \frac{3(x+1)}{2x^5(x-3)}$. See Fig. 10.

Graphing Calculator

To view the portion of the graph near the vertical asymptote $x = 3$, use WINDOW settings of $X_{min} = 2.9$ and $X_{max} = 3.1$.

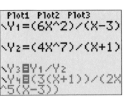 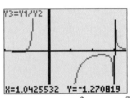 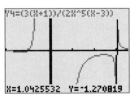

Figure 10 Compare the graphs of $y = \frac{6x^2}{x-3} \div \frac{4x^7}{x+1}$ and $y = \frac{3(x+1)}{2x^5(x-3)}$

To divide two rational expressions:

1. Write the quotient as a product using the property $\frac{A}{B} \div \frac{C}{D} = \frac{A}{B} \cdot \frac{D}{C}$, where B, C, and D are nonzero.

2. Find the product.

3. Simplify.

In Example 3 we combined two rational functions to form a product function. In Example 5 we combine two rational functions to form a *quotient function*.

Example 5   Finding a Quotient Function

Let $Q(x) = f(x) \div g(x)$, where $f(x) = \dfrac{81x^2 - 49}{3x^2 + 16x + 5}$ and $g(x) = \dfrac{7 - 9x}{18x + 6}$. Find an equation for Q and simplify the right side of the equation.

Solution

$Q(x) = f(x) \div g(x)$

$= \dfrac{81x^2 - 49}{3x^2 + 16x + 5} \div \dfrac{7 - 9x}{18x + 6}$

$= \dfrac{81x^2 - 49}{3x^2 + 16x + 5} \cdot \dfrac{18x + 6}{7 - 9x}$ $\dfrac{A}{B} \div \dfrac{C}{D} = \dfrac{A}{B} \cdot \dfrac{D}{C}$

$= \dfrac{(9x - 7)(9x + 7)}{(3x + 1)(x + 5)} \cdot \dfrac{6(3x + 1)}{-(9x - 7)}$ Factor numerators and denominators.

$= -\dfrac{6(9x - 7)(3x + 1)(9x + 7)}{(9x - 7)(3x + 1)(x + 5)}$ $\dfrac{a}{-b} = -\dfrac{a}{b}$

$= -\dfrac{6(9x + 7)}{x + 5}$ ■

If $f(x)$ and $g(x)$ are functions, then we can form the **quotient function**, $Q(x)$, where $Q(x) = f(x) \div g(x)$.

In Example 6 we combine three rational expressions using two operations.

Example 6   Combining Three Rational Expressions

Perform the indicated operations:

$$\left(\dfrac{x^2 - 2x - 48}{x^2 + 8x + 16} \div \dfrac{3x^2 - 9x}{x^2 - 16} \right) \cdot \dfrac{6x + 24}{5x + 30}$$

Solution

$$\left(\dfrac{x^2 - 2x - 48}{x^2 + 8x + 16} \div \dfrac{3x^2 - 9x}{x^2 - 16} \right) \cdot \dfrac{6x + 24}{5x + 30}$$

$$= \left(\dfrac{x^2 - 2x - 48}{x^2 + 8x + 16} \cdot \dfrac{x^2 - 16}{3x^2 - 9x} \right) \cdot \dfrac{6x + 24}{5x + 30}$$

$$= \left(\dfrac{(x - 8)(x + 6)}{(x + 4)^2} \cdot \dfrac{(x - 4)(x + 4)}{3x(x - 3)} \right) \cdot \dfrac{6(x + 4)}{5(x + 6)}$$

$$= \dfrac{(x + 6)(x + 4)(x - 8)(x - 4)}{3x(x + 4)^2(x - 3)} \cdot \dfrac{6(x + 4)}{5(x + 6)}$$

$$= \dfrac{6(x + 6)(x + 4)^2(x - 8)(x - 4)}{15x(x + 6)(x + 4)^2(x - 3)}$$

$$= \dfrac{2(x - 8)(x - 4)}{5x(x - 3)}$$ ■

EXPLORATION
Looking ahead: Adding and subtracting rational expressions

1. Compare graphing calculator tables for $y = \dfrac{2}{x} + \dfrac{3}{x}$ and $y = \dfrac{5}{x}$. What does this comparison suggest about performing the addition $\dfrac{A}{B} + \dfrac{C}{B}$?

2. Compare graphing calculator tables for $y = \dfrac{5}{x} - \dfrac{2}{x}$ and $y = \dfrac{3}{x}$. What does this comparison suggest about performing the subtraction $\dfrac{A}{B} - \dfrac{C}{B}$?

3. Use a graphing calculator table to find outputs of the function $y = \dfrac{x}{5} - \dfrac{x+10}{5}$. What do you notice about the outputs? Explain why it makes sense that the outputs are like they are.

4. Compare graphing calculator tables for $y = \dfrac{2}{x} + \dfrac{x}{3}$ and $y = \dfrac{2+x}{x+3}$. Is the following statement, in general, true or false? Explain. [**Hint**: In what way is this sum different from the sum and differences in problems 1–3?]

$$\frac{A}{B} + \frac{C}{D} = \frac{A+C}{B+D}, \quad \text{where } B, D, \text{ and } B+D \text{ are nonzero}$$

5. State any concepts suggested by this exploration.

Key Points
OF THIS SECTION

Multiplying two rational expressions

To multiply rational expressions $\dfrac{A}{B}$ and $\dfrac{C}{D}$, where B and D are nonzero:

1. Factor the numerators and denominators.

2. Multiply using the property $\dfrac{A}{B} \cdot \dfrac{C}{D} = \dfrac{AC}{BD}$.

3. Simplify.

Dividing two rational expressions

To divide rational expressions $\dfrac{A}{B}$ and $\dfrac{C}{D}$, where B, C, and D are nonzero:

1. Rewrite the division as a multiplication using $\dfrac{A}{B} \div \dfrac{C}{D} = \dfrac{A}{B} \cdot \dfrac{D}{C}$.

2. Factor the numerators and denominators.

3. Multiply using the property $\dfrac{A}{B} \cdot \dfrac{D}{C} = \dfrac{AD}{BC}$.

4. Simplify.

HOMEWORK 8.2

SSM
Tutor Center

MathPro 4
MathPro 4

MathPro 5
MathPro 5

CDs/Videos

www

Perform the indicated operation. Simplify the result.

1. $\dfrac{5}{x} \cdot \dfrac{2}{x}$

2. $\dfrac{x}{9} \cdot \dfrac{3}{x}$

3. $\dfrac{7x^5}{2} \div \dfrac{5x^3}{6}$

4. $\dfrac{14}{8x^9} \div \dfrac{21}{6x^5}$

5. $\dfrac{5}{4x-8} \cdot \dfrac{3x-6}{10}$

6. $\dfrac{5x-20}{21x^6} \cdot \dfrac{7x^2}{3x^2-12x}$

7. $\dfrac{6x-18}{5x} \div \dfrac{5x-15}{x^3}$

8. $\dfrac{20x^3}{7x-14} \div \dfrac{50x^8}{2x-4}$

9. $\dfrac{(x-3)(x+8)}{(x+2)(x+3)} \cdot \dfrac{4(x+2)}{3(x-3)}$

10. $\dfrac{15(x-6)(x+1)}{8(x-3)} \cdot \dfrac{6(x-3)^2}{5(x-6)}$

11. $\dfrac{(x+8)(x+9)}{3(x-3)(x-5)} \div \dfrac{2(x+8)}{12(x-5)}$

12. $\dfrac{12(x+2)^2}{3(x-5)} \div \dfrac{16(x+2)(x-2)}{9(x-5)}$

13. $\dfrac{3x+15}{x^7} \div \dfrac{x+5}{x^2}$

14. $\dfrac{4x^6}{10x-25} \div \dfrac{6x^{10}}{6x-15}$

15. $\dfrac{x^2+10x+21}{x-9} \cdot \dfrac{2x-18}{x^2-9}$

16. $\dfrac{6x^2-3x}{9x^7} \cdot \dfrac{12x^4}{4x^2-4x+1}$

17. $\dfrac{x^2+3x+2}{3x-3} \div \dfrac{x^2-x-6}{6x-6}$

18. $\dfrac{x^2-36}{2x^2-50} \div \dfrac{2x+12}{3x-15}$

19. $\dfrac{2x-12}{x+1} \div \dfrac{18-3x}{4x+4}$

20. $\dfrac{2x+4}{x-4} \div \dfrac{x+2}{16-x^2}$

21. $\dfrac{2x^2-32}{x^2-2x-24} \cdot \dfrac{x^2-7x+6}{x+6}$

22. $\dfrac{3x+6}{5x} \cdot \dfrac{x^2+4}{x^2+10x+16}$

23. $\dfrac{4-x}{x^2+10x+25} \cdot \dfrac{25-x^2}{3x^2-9x-12}$

24. $\dfrac{x^2-9}{5x+5} \cdot \dfrac{2x+2}{-x+3}$

25. $\dfrac{-x+3}{x^2-16} \cdot \dfrac{x^2+8x+16}{x^2-2x-3}$

26. $\dfrac{-x^2+9}{-x^2+49} \div \dfrac{2x-6}{5x-35}$

27. $\dfrac{-x^2+7x-10}{2x^2+5x-12} \div \dfrac{-x^2+4}{8x^2-18}$

28. $\dfrac{x^2+7x}{6x-30} \cdot \dfrac{x^2-2x-15}{x^2-49}$

29. $\dfrac{-4x-6}{36-x^2} \cdot \dfrac{4x+24}{6x^2+x-12}$

30. $\dfrac{7-21x}{x^2-8x+15} \cdot \dfrac{x^2-x-20}{6x-2}$

31. $\dfrac{x^2-16}{2x+6} \cdot \dfrac{x+3}{x-4}$

32. $\dfrac{9x^2}{x^2+12x+36} \cdot \dfrac{x^2+6x}{12x}$

33. $\dfrac{x^2-25}{4x+20} \div \dfrac{15-3x}{12x^2}$

34. $\dfrac{x^2-16}{x^2-10x+25} \div \dfrac{3x-12}{x^2-3x-10}$

35. $\dfrac{x^2-8x+16}{-8x^2+16x} \cdot \dfrac{x^2-4x+4}{x^2-16}$

36. $\dfrac{x^2-9}{2x+16} \cdot \dfrac{x+8}{5x-15}$

37. $\dfrac{8x^2+14x+5}{6x^2+24} \cdot \dfrac{4x-8}{-6x-3}$

38. $\dfrac{x^2-16}{x^2} \cdot \dfrac{x^2-4x}{x^2-x-12}$

39. $\dfrac{x^2-9}{x^2} \div \dfrac{x^5+3x^4}{x+2}$

40. $\dfrac{3x^2-5x}{x^2-10x+25} \div \dfrac{4x^2-100}{6x^2-7x-5}$

41. $\left(3x^2+5x-12\right) \div \dfrac{9x^2-12x}{x+2}$

42. $\left(x^2-9x+14\right) \div \dfrac{x^2-49}{3x+6}$

43. $\dfrac{12x^3-20x^2-8x}{2x-6} \cdot \dfrac{5x-15}{x^2-4}$

44. $\dfrac{x^2-6x}{x^2+8x+16} \div \dfrac{x^2-36}{2x^3-2x^2-40x}$

45. $\dfrac{x^2+4x-5}{x^3+6x^2-4x-24} \cdot \dfrac{x^2+8x+12}{x^2+10x+25}$

46. $\dfrac{x^2-3x-10}{7x+21} \div \dfrac{x^3+2x^2-25x-50}{4x+12}$

Perform the indicated operations. Simplify the result.

47. $\left(\dfrac{20x^7}{x^2-9} \div \dfrac{x^2-14x+24}{5x-15}\right) \cdot \dfrac{x^2+x-6}{8x^{13}}$

48. $\left(\dfrac{8x^2+10x-3}{-2x^2-8x} \cdot \dfrac{x^2+x}{16x^2-1}\right) \div \dfrac{-10x-15}{8x^5}$

49. $\dfrac{12x^3}{x^2-4} \div \left(\dfrac{22x^6}{-6x+12} \cdot \dfrac{x}{11x+22}\right)$

50. $\dfrac{3-x}{20x^5} \div \left(\dfrac{x^2-9}{30x^2} \div \dfrac{8x^2+6x}{3x}\right)$

51. $\left(\left(\dfrac{x-4}{x+5}\right)^2 \cdot \left(\dfrac{x+5}{x-1}\right)^2\right) \div \left(\dfrac{x-4}{x-1}\right)^2$

52. $\dfrac{3x+6}{4x+20} \div \left(\dfrac{x^2-4}{x^2-25} \div \dfrac{x-2}{x-5}\right)^2$

Let $f(x) = \dfrac{x^2-6x-16}{x^2+3x-40}$ *and* $g(x) = \dfrac{x^2-64}{x^2-3x-10}$.

53. Find $f(x)g(x)$.   **54.** Find $f(x) \div g(x)$.

55. Find $g(x) \div f(x)$.

Let $f(x) = \dfrac{1-x^2}{x^2-3x-28}$ *and* $g(x) = \dfrac{x^2-8x+7}{x^2+5x+4}$.

56. Find $f(x)g(x)$.   **57.** Find $f(x) \div g(x)$.

58. Find $g(x) \div f(x)$.

59. A student tries to perform the multiplication $\dfrac{x+2}{x+5} \cdot \dfrac{x+2}{x-5}$:

$$\dfrac{x+2}{x+5} \cdot \dfrac{x+2}{x-5} = \dfrac{x^2+4}{x^2-25}$$

Find a value to substitute for x to show that the student's work is incorrect. Then perform the multiplication correctly. Use a graphing calculator table to verify your work.

60. Perform the indicated operations.

a. $\dfrac{1}{x} \div \dfrac{1}{x}$

b. $\dfrac{1}{x} \div \left(\dfrac{1}{x} \div \dfrac{1}{x}\right)$

c. $\dfrac{1}{x} \div \left(\dfrac{1}{x} \div \left(\dfrac{1}{x} \div \dfrac{1}{x}\right)\right)$

d. $\dfrac{1}{x} \div \left(\dfrac{1}{x} \div \left(\dfrac{1}{x} \div \left(\dfrac{1}{x} \div \dfrac{1}{x}\right)\right)\right)$

e. $\underbrace{\dfrac{1}{x} \div \left(\dfrac{1}{x} \div \left(\dfrac{1}{x} \div \cdots \div \left(\dfrac{1}{x} \div \left(\dfrac{1}{x} \div \dfrac{1}{x}\right)\right)\cdots\right)\right)}_{n \text{ division symbols}}$

61. In exercise 7 of Homework 7.6 you modeled the percentages of households that have cable television. In Table 1 we list the *numbers* of households that have cable television.

Table 1 Numbers of Households with Cable Television

Year	Number of Households with Cable Television (millions)	Total Number of Households (millions)
1980	17.5	80.8
1985	35.4	86.8
1990	50.5	93.3
1995	60.9	99.0
2000	67.7	104.7

Source: *Paul Kagan Associates, Inc.*

The number of households with cable television $c(t)$ and the total number of households $h(t)$, both in millions, can be modeled by the system

$$c(t) = -0.077t^2 + 4.06t + 17.38$$
$$h(t) = 1.20t + 80.92$$

Where t represents the number of years since 1980.

a. Use c to estimate the number of households that had cable television in 1999.

b. Use h to estimate the total number of households in 1999.

c. Estimate the percentage of households that had cable television in 1999.

d. Let $p(t)$ represent the percentage of households that have cable television at t years since 1980. Use your equations for c and h to build an equation for p.

e. Find $p(19)$. Compare your result to your result in part c.

f. Find $p(18)$. What does your result mean in terms of the situation?

g. In exercise 7 of Homework 7.6 you found the following equation for p:

$$p(t) = -0.108t^2 + 4.29t + 21.83$$

Use a graphing calculator to compare the graphs of this equation and the equation you found in part d for $0 \le t \le 20$. Then, ZOOM OUT and compare the graphs.

62. In Example 8 in Section 6.5, we modeled the number of people who receive food stamps. A reasonable model is

$$F(t) = -0.46t^2 + 3.68t + 19.71$$

where $F(t)$ represents the average monthly participation (in millions) in the food stamp program during the year that is t years since 1990 (see Table 2). The U.S. population (in millions) $U(t)$ can be modeled by

$$U(t) = 0.0067t^2 + 2.56t + 250.39$$

where t is the number of years since 1990.

Table 2 Average Monthly Participation in the Food Stamp Program

Year	Average Monthly Participation (millions of people)
1990	20.0
1991	22.5
1992	25.2
1993	26.6
1994	27.3
1995	26.6
1996	25.4
1997	22.7
1998	19.8

a. Use F to estimate the average monthly participation in 1999.

b. Use U to estimate the U.S. population in 1999.

c. Estimate the percentage of Americans who received food stamps in 1999. Describe any assumptions you have made.

d. Let $P(t)$ represent the percentage of Americans receiving food stamps during the year that is t years since 1990. Find an equation for P. [**Hint**: Use the equations for F and U to build an equation for P.] Describe any assumptions you have made.

e. Find $P(9)$. Compare your result to part c.

f. Find $P(10)$. What does your result mean in terms of the situation?

63. Find two rational expressions whose product is $\dfrac{x^2 + 10x + 25}{x^2 - 10x + 25}$ and whose quotient is 1.

64. Find two rational expressions whose product is 1 and whose quotient is $\dfrac{x^2 - 4x + 4}{x^2 - 8x + 16}$.

65. Describe how to multiply two rational expressions. Also describe how to divide two rational expressions.

8.3 ADDING AND SUBTRACTING RATIONAL EXPRESSIONS

Objective

▸ Add and subtract rational expressions.

In Section 8.2 we multiplied and divided rational expressions. In this section we add and subtract rational expressions.

We perform the addition $\dfrac{2}{7} + \dfrac{3}{7}$ by using the property $\dfrac{a}{b} + \dfrac{c}{b} = \dfrac{a+c}{b}$, where $b \ne 0$:

$$\frac{2}{7} + \frac{3}{7} = \frac{2+3}{7}$$
$$= \frac{5}{7}$$

Note that the expressions $\dfrac{2}{7}$ and $\dfrac{3}{7}$ have a *common denominator* of 7. We add more complicated rational expressions with a common denominator in a similar way.

Adding Rational Expressions with a Common Denominator

If $\dfrac{A}{B}$ and $\dfrac{C}{B}$ are rational expressions where B is nonzero, then

$$\frac{A}{B} + \frac{C}{B} = \frac{A + C}{B}$$

In words: To add two rational expressions with a common denominator, add the numerators to get the numerator of the sum. The denominator of the sum is the common denominator.

After adding two rational expressions, it may be possible to reduce the result.

Example 1   Adding Two Rational Expressions

Perform the indicated addition: $\dfrac{x^2 + 5x}{x^2 - 9} + \dfrac{6}{x^2 - 9}$

Solution

$$\frac{x^2 + 5x}{x^2 - 9} + \frac{6}{x^2 - 9} = \frac{x^2 + 5x + 6}{x^2 - 9} \qquad \frac{A}{B} + \frac{C}{B} = \frac{A + C}{B}$$

$$= \frac{(x + 2)(x + 3)}{(x - 3)(x + 3)} \qquad \text{Factor the numerator and the denominator.}$$

$$= \frac{x + 2}{x - 3} \qquad \text{Reduce.} \qquad \blacksquare$$

To find a sum such as $\dfrac{1}{4} + \dfrac{5}{6}$, where the denominators of the two fractions are different, we use the fact $\dfrac{a}{a} = 1$ when $a \neq 0$ to rewrite the fractions with the LCD.

To find the LCD, we first prime factor the denominators:

$$4 = 2 \cdot 2$$
$$6 = 2 \cdot 3$$

Next, we count the most times that each prime appears in any one factorization. The most that 2 appears is twice. The most that 3 appears is once. Then we find the product of that many 2s and 3s. So, the LCD is $2 \cdot 2 \cdot 3 = 12$.

Now we write the fractions so that each one has denominator 12.

$$\frac{1}{4} + \frac{5}{6} = \frac{1}{4} \cdot 1 + \frac{5}{6} \cdot 1$$

$$= \frac{1}{4} \cdot \frac{3}{3} + \frac{5}{6} \cdot \frac{2}{2}$$

$$= \frac{3}{12} + \frac{10}{12}$$

$$= \frac{13}{12}$$

Although we could have written the fractions with a common denominator of 24, 36, 48, or any other multiple of 12, using the LCD 12 yields the result more efficiently.

To perform the addition

$$\frac{3}{(x - 1)(x - 2)} + \frac{5}{(x - 1)(x - 2)^2}$$

we note that the first fraction's denominator has only one $x - 2$ factor whereas the second fraction's denominator has two $x - 2$ factors. If we multiply $\dfrac{3}{(x - 1)(x - 2)}$ by $\dfrac{x - 2}{x - 2} = 1$ (when $x \neq 2$), then we will have two fractions with a common denominator:

$$\frac{3}{(x - 1)(x - 2)} + \frac{5}{(x - 1)(x - 2)^2} = \frac{3}{(x - 1)(x - 2)} \cdot \frac{x - 2}{x - 2} + \frac{5}{(x - 1)(x - 2)^2}$$

$$= \frac{3x - 6}{(x - 1)(x - 2)^2} + \frac{5}{(x - 1)(x - 2)^2}$$

$$= \frac{3x - 1}{(x - 1)(x - 2)^2}$$

The expression $(x - 1)(x - 2)^2$ is the LCD of the two original fractions being added.

Example 2 Adding Two Rational Expressions

Perform the indicated addition.

1. $\dfrac{5}{6x^3} + \dfrac{7}{8x}$

2. $\dfrac{2}{x + 3} + \dfrac{4}{x - 5}$

Solution

1. $\dfrac{5}{6x^3} + \dfrac{7}{8x} = \dfrac{5}{6x^3} \cdot \dfrac{4}{4} + \dfrac{7}{8x} \cdot \dfrac{3x^2}{3x^2}$ Multiply by 1 to get the LCD, $24x^3$.

$$= \frac{20}{24x^3} + \frac{21x^2}{24x^3}$$ Find the products.

$$= \frac{21x^2 + 20}{24x^3}$$ $\dfrac{A}{B} + \dfrac{C}{B} = \dfrac{A + C}{B}$

Note

The comparison does not tell us whether the result is in lowest terms.

Note that the result is in lowest terms. We verify our result by comparing graphing calculator graphs for $y = \dfrac{5}{6x^3} + \dfrac{7}{8x}$ and $y = \dfrac{21x^2 + 20}{24x^3}$ (see Fig. 11).

 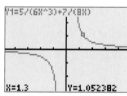 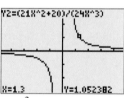

Figure 11 Compare graphs for $y = \dfrac{5}{6x^3} + \dfrac{7}{8x}$ and $y = \dfrac{21x^2 + 20}{24x^3}$

2. $\dfrac{2}{x + 3} + \dfrac{4}{x - 5}$

$$= \frac{2}{x + 3} \cdot \frac{x - 5}{x - 5} + \frac{4}{x - 5} \cdot \frac{x + 3}{x + 3}$$ Multiply by 1 to get the LCD, $(x - 5)(x + 3)$.

$$= \frac{2(x - 5)}{(x - 5)(x + 3)} + \frac{4(x + 3)}{(x - 5)(x + 3)}$$ $\dfrac{A}{B} \cdot \dfrac{C}{D} = \dfrac{AC}{BD}$

$$= \frac{2(x - 5) + 4(x + 3)}{(x - 5)(x + 3)}$$ Add the numerators.

$$= \frac{2x - 10 + 4x + 12}{(x - 5)(x + 3)}$$

$$= \frac{6x + 2}{(x - 5)(x + 3)}$$

$$= \frac{2(3x + 1)}{(x - 5)(x + 3)}$$

Note that the result is in lowest terms. We can verify the result by comparing graphing calculator tables of $y = \dfrac{2}{x+3} + \dfrac{4}{x-5}$ and $y = \dfrac{2(3x+1)}{(x-5)(x+3)}$. ■

For the sum of two expressions with different denominators, we factor the denominators, if possible, to help us find the LCD.

Example 3 Adding Two Rational Expressions

Perform the addition: $\dfrac{-9}{x^2-9} + \dfrac{x}{2x+6}$

Solution

$$\dfrac{-9}{x^2-9} + \dfrac{x}{2x+6}$$

$$= \dfrac{-9}{(x-3)(x+3)} + \dfrac{x}{2(x+3)} \qquad \text{Factor the denominators.}$$

$$= \dfrac{-9}{(x-3)(x+3)} \cdot \dfrac{2}{2} + \dfrac{x}{2(x+3)} \cdot \dfrac{x-3}{x-3} \qquad \begin{array}{l}\text{Multiply by 1 to get the LCD,} \\ 2(x-3)(x+3).\end{array}$$

$$= \dfrac{-18}{2(x-3)(x+3)} + \dfrac{x(x-3)}{2(x+3)(x-3)}$$

$$= \dfrac{-18 + x(x-3)}{2(x-3)(x+3)} \qquad \text{Add the numerators.}$$

$$= \dfrac{x^2 - 3x - 18}{2(x-3)(x+3)} \qquad \text{Simplify the numerator.}$$

$$= \dfrac{(x-6)(x+3)}{2(x-3)(x+3)} \qquad \text{Factor the numerator.}$$

$$= \dfrac{x-6}{2(x-3)} \qquad \text{Reduce.}$$

We can verify our result by comparing graphs for $y = \dfrac{-9}{x^2-9} + \dfrac{x}{2x+6}$ and $y = \dfrac{x-6}{2(x-3)}$ (see Fig. 12).

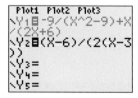 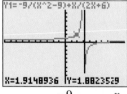 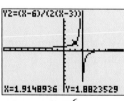

Figure 12 Compare graphs for $y = \dfrac{-9}{x^2-9} + \dfrac{x}{2x+6}$ and $y = \dfrac{x-6}{2(x-3)}$

■

To add two rational expressions with different denominators:

1. Factor, if possible, the denominators of the expressions. Identify the LCD.

2. Use the property $\dfrac{A}{A} = 1$ when $A \neq 0$ to write equivalent expressions having the LCD.

3. Add the expressions using the property $\dfrac{A}{B} + \dfrac{C}{B} = \dfrac{A+C}{B}$, where B is nonzero.

4. Simplify.

If $f(x)$ and $g(x)$ are two functions, then we can form the **sum function**, $S(x)$, where $S(x) = f(x) + g(x)$.

Example 4 Finding a Sum Function

Let $S(x) = f(x) + g(x)$, where $f(x) = \dfrac{3}{12x^3 - 22x^2 + 6x}$ and $g(x) = \dfrac{x+1}{30x^2 - 10x}$.
Find an equation for S and simplify the right side of the equation.

Solution

$$S(x) = f(x) + g(x)$$

$$= \frac{3}{12x^3 - 22x^2 + 6x} + \frac{x+1}{30x^2 - 10x}$$

$$= \frac{3}{2x(6x^2 - 11x + 3)} + \frac{x+1}{10x(3x - 1)}$$

$$= \frac{3}{2x(3x - 1)(2x - 3)} + \frac{x+1}{10x(3x - 1)}$$

$$= \frac{3}{2x(3x - 1)(2x - 3)} \cdot \frac{5}{5} + \frac{x+1}{10x(3x - 1)} \cdot \frac{2x - 3}{2x - 3}$$

$$= \frac{15}{10x(3x - 1)(2x - 3)} + \frac{(x+1)(2x - 3)}{10x(3x - 1)(2x - 3)}$$

$$= \frac{15 + (x+1)(2x - 3)}{10x(3x - 1)(2x - 3)}$$

$$= \frac{15 + 2x^2 - x - 3}{10x(3x - 1)(2x - 3)}$$

$$= \frac{2x^2 - x + 12}{10x(3x - 1)(2x - 3)}$$

Since $2x^2 - x + 12$ is prime, the right side of the equation $S(x) = \dfrac{2x^2 - x + 12}{10x(3x - 1)(2x - 3)}$
is in lowest terms, so we are done. ∎

We subtract two fractions having a common denominator, such as $\dfrac{5}{7} - \dfrac{3}{7}$, by using
the property $\dfrac{a}{b} - \dfrac{c}{b} = \dfrac{a - c}{b}$, where $b \neq 0$:

$$\frac{5}{7} - \frac{3}{7} = \frac{5 - 3}{7}$$

$$= \frac{2}{7}$$

We subtract more complicated rational expressions with a common denominator in a similar way.

Subtracting Rational Expressions with a Common Denominator

If $\dfrac{A}{B}$ and $\dfrac{C}{B}$ are rational expressions with B nonzero, then

$$\frac{A}{B} - \frac{C}{B} = \frac{A - C}{B}$$

In words: To subtract two rational expressions with a common denominator, subtract the numerators for the numerator of the difference. The denominator of the difference is the common denominator.

Example 5 Subtracting Two Rational Expressions

Perform the subtraction: $\dfrac{x^2}{x+1} - \dfrac{x+2}{x+1}$

Solution

$$\frac{x^2}{x+1} - \frac{x+2}{x+1} = \frac{x^2 - (x+2)}{x+1} \qquad \frac{A}{B} - \frac{C}{B} = \frac{A-C}{B}$$

$$= \frac{x^2 - x - 2}{x+1}$$

$$= \frac{(x-2)(x+1)}{x+1} \qquad \text{Factor the numerator.}$$

$$= x - 2 \qquad \text{Reduce.} \qquad ■$$

It is a common student error to write:

$$\frac{x^2}{x+1} - \frac{x+2}{x+1} = \frac{x^2 - x + 2}{x+1} \qquad \text{Incorrect}$$

This first step is incorrect. We must subtract the *entire* numerator:

$$\frac{x^2}{x+1} - \frac{x+2}{x+1} = \frac{x^2 - (x+2)}{x+1} = \frac{x^2 - x - 2}{x+1}$$

Note

See Example 5 for the rest of the work.

To find the difference of rational expressions with different denominators, we find equivalent expressions having the LCD.

Example 6 Subtracting Two Rational Expressions

Perform the subtraction: $\dfrac{3x-1}{2x^2 - 7x - 4} - \dfrac{5}{x^2 - 8x + 16}$

Solution

$$\frac{3x-1}{2x^2 - 7x - 4} - \frac{5}{x^2 - 8x + 16}$$

$$= \frac{3x-1}{(2x+1)(x-4)} - \frac{5}{(x-4)^2} \qquad \text{Factor denominators.}$$

$$= \frac{3x-1}{(2x+1)(x-4)} \cdot \frac{x-4}{x-4} - \frac{5}{(x-4)^2} \cdot \frac{2x+1}{2x+1} \qquad \begin{array}{l}\text{Multiply by 1 to get the} \\ \text{LCD, } (x-4)^2(2x+1).\end{array}$$

$$= \frac{(3x-1)(x-4) - 5(2x+1)}{(x-4)^2(2x+1)} \qquad \text{Subtract the numerators.}$$

$$= \frac{3x^2 - 13x + 4 - 10x - 5}{(x-4)^2(2x+1)}$$

$$= \frac{3x^2 - 23x - 1}{(x-4)^2(2x+1)}$$

Since $3x^2 - 23x - 1$ is prime, the result is in lowest terms, so we are done. ■

In Example 7 we use the property $a - b = -(b - a)$ to help us find the LCD of two expressions.

Example 7 Adding Two Rational Expressions

Perform the addition: $\dfrac{x}{x-2} + \dfrac{3}{2-x}$

Solution

$$\frac{x}{x-2} + \frac{3}{2-x} = \frac{x}{x-2} + \frac{3}{-(x-2)} \qquad a - b = -(b-a)$$

$$= \frac{x}{x-2} - \frac{3}{x-2} \qquad \frac{a}{-b} = -\frac{a}{b}$$

$$= \frac{x-3}{x-2} \qquad ■$$

In Example 8 we combine three rational expressions by performing two operations.

Example 8 Combining Three Rational Expressions

Perform the indicated operations: $\left(\dfrac{x+2}{x^2-x} - \dfrac{6}{x^2-1} \right) + \dfrac{3}{x^2+x}$

Solution

$$\left(\frac{x+2}{x^2-x} - \frac{6}{x^2-1} \right) + \frac{3}{x^2+x}$$

$$= \left(\frac{x+2}{x(x-1)} - \frac{6}{(x-1)(x+1)} \right) + \frac{3}{x(x+1)}$$

$$= \left(\frac{x+2}{x(x-1)} \cdot \frac{x+1}{x+1} - \frac{6}{(x-1)(x+1)} \cdot \frac{x}{x} \right) + \frac{3}{x(x+1)} \cdot \frac{x-1}{x-1}$$

$$= \frac{(x+2)(x+1) - 6x}{x(x-1)(x+1)} + \frac{3(x-1)}{x(x-1)(x+1)}$$

$$= \frac{(x+2)(x+1) - 6x + 3(x-1)}{x(x-1)(x+1)}$$

$$= \frac{x^2 + 3x + 2 - 6x + 3x - 3}{x(x-1)(x+1)}$$

$$= \frac{x^2 - 1}{x(x-1)(x+1)}$$

$$= \frac{(x-1)(x+1)}{x(x-1)(x+1)}$$

$$= \frac{1}{x} \qquad \blacksquare$$

If $f(x)$ and $g(x)$ are two functions, then we can form a **difference function**, $D(x)$, where $D(x) = f(x) - g(x)$.

Example 9 Finding a Difference Function

Let $D(x) = f(x) - g(x)$, where $f(x) = \dfrac{x-1}{x+1}$ and $g(x) = \dfrac{x+1}{x-1}$. Find an equation for D and simplify the right side of the equation.

Solution

$$D(x) = f(x) - g(x)$$

$$= \frac{x-1}{x+1} - \frac{x+1}{x-1}$$

$$= \frac{x-1}{x+1} \cdot \frac{x-1}{x-1} - \frac{x+1}{x-1} \cdot \frac{x+1}{x+1}$$

$$= \frac{(x-1)^2 - (x+1)^2}{(x+1)(x-1)}$$

$$= \frac{x^2 - 2x + 1 - (x^2 + 2x + 1)}{(x-1)(x+1)}$$

$$= \frac{x^2 - 2x + 1 - x^2 - 2x - 1}{(x-1)(x+1)}$$

$$= \frac{-4x}{(x-1)(x+1)}$$

We can check our result by comparing graphs for $y = \dfrac{x-1}{x+1} - \dfrac{x+1}{x-1}$ and $y = \dfrac{-4x}{(x-1)(x+1)}$ (see Fig. 13).

 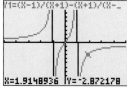 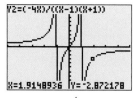

Figure 13 Compare graphs for $y = \dfrac{x-1}{x+1} - \dfrac{x+1}{x-1}$ and $y = \dfrac{-4x}{(x-1)(x+1)}$

EXPLORATION
Adding and subtracting rational expressions

For problems 1–3, a student tries to simplify an expression. If the result is correct, decide whether there is a more efficient way to simplify. If the result is incorrect, describe the error and then simplify the result correctly.

1. Simplify $\dfrac{5}{x+1} + \dfrac{2}{x+3}$.

$$\dfrac{5}{x+1} + \dfrac{2}{x+3} = \left(\dfrac{5}{x+1} + \dfrac{2}{x+3}\right) \cdot (x+1)(x+3)$$

$$= \dfrac{5}{x+1} \cdot (x+1)(x+3) + \dfrac{2}{x+3} \cdot (x+1)(x+3)$$

$$= 5(x+3) + 2(x+1)$$

$$= 7x + 17$$

2. Simplify $\dfrac{4}{(x-2)(x+3)} + \dfrac{1}{x-2}$.

$$\dfrac{4}{(x-2)(x+3)} + \dfrac{1}{x-2} = \dfrac{4}{(x-2)(x+3)} \cdot \dfrac{x-2}{x-2} + \dfrac{1}{x-2} \cdot \dfrac{(x-2)(x+3)}{(x-2)(x+3)}$$

$$= \dfrac{4(x-2) + (x-2)(x+3)}{(x-2)^2(x+3)}$$

$$= \dfrac{4x - 8 + x^2 + x - 6}{(x-2)^2(x+3)}$$

$$= \dfrac{x^2 + 5x - 14}{(x-2)^2(x+3)}$$

$$= \dfrac{(x-2)(x+7)}{(x-2)^2(x+3)}$$

$$= \dfrac{x+7}{(x-2)(x+3)}$$

3. Simplify $\dfrac{5x}{x-7} - \dfrac{3x+4}{x-7}$.

$$\dfrac{5x}{x-7} - \dfrac{3x+4}{x-7} = \dfrac{5x - 3x + 4}{x-7}$$

$$= \dfrac{2x+4}{x-7}$$

Tips on Succeeding in This Course

Consider writing, after each class meeting, a summary of what you have learned. Your summaries will increase your understanding as well as your memory of concepts and procedures. They will also serve as good reference for quizzes and exams.

Key Points
OF THIS SECTION

Adding or subtracting two rational expressions

To add or subtract two rational expressions with different denominators:

1. Factor, if possible, the denominators of the expressions. Identify the LCD.

2. Use the property $\dfrac{a}{a} = 1$ when $a \neq 0$ to write equivalent expressions having the LCD.

3. Add the expressions using the property $\dfrac{A}{B} + \dfrac{C}{B} = \dfrac{A + C}{B}$, where $B \neq 0$. Subtract the expressions using the property $\dfrac{A}{B} - \dfrac{C}{B} = \dfrac{A - C}{B}$, where $B \neq 0$.

4. Simplify.

Subtracting an entire numerator

When subtracting, be sure to subtract the *entire* numerator. For example:

$$\frac{3x^2}{x+7} - \frac{x+2}{x+7} = \frac{3x^2 - (x+2)}{x+7} = \frac{3x^2 - x - 2}{x+7}$$

HOMEWORK 8.3

SSM
Tutor Center

MathPro 4
MathPro 4

MathPro 5
MathPro 5

CDs/Videos

www

Perform the indicated operation. Simplify the result.

1. $\dfrac{3x}{x-1} + \dfrac{2x+1}{x-1}$

2. $\dfrac{x}{x^2-4} + \dfrac{2}{x^2-4}$

3. $\dfrac{3x^2+5x}{x^2+10x+21} - \dfrac{2x^2+7x+15}{x^2+10x+21}$

4. $\dfrac{2x^2-4x}{3x^2-6x} - \dfrac{x^2+x-6}{3x^2-6x}$

5. $\dfrac{2}{x} - \dfrac{4}{x^2}$

6. $\dfrac{4}{x^5} - \dfrac{7}{x^3}$

7. $\dfrac{5}{10x^6} + \dfrac{3}{12x^4}$

8. $\dfrac{3}{14x^2} + \dfrac{4}{21x^9}$

9. $\dfrac{3}{x+1} + \dfrac{4}{x-2}$

10. $\dfrac{3}{x-2} + \dfrac{2}{x+3}$

11. $\dfrac{6}{(x+4)(x-6)} - \dfrac{4}{(x-1)(x+4)}$

12. $\dfrac{2}{(x+1)(x-2)} - \dfrac{3}{(x-2)(x+3)}$

13. $\dfrac{5}{3x-6} - \dfrac{2}{5x+15}$

14. $\dfrac{x}{8x-4} - \dfrac{2}{3x-6}$

15. $\dfrac{3}{x^2-25} + \dfrac{5}{x^2-5x}$

16. $\dfrac{-2}{x^2-2x-3} + \dfrac{3}{x^2-9}$

17. $\dfrac{2}{x^2-9} + \dfrac{3}{x^2-7x+12}$

18. $\dfrac{4}{x^2-2x} + \dfrac{1}{x^2-5x+6}$

19. $2 + \dfrac{x-3}{x+1}$

20. $\dfrac{x+2}{x-5} + 3$

21. $2 - \dfrac{2x+4}{x^2+3x+2}$

22. $\dfrac{6x^2+2x-4}{x^2-1} - 5$

23. $\dfrac{8}{x-6} - \dfrac{4}{6-x}$

24. $\dfrac{x}{x-3} + \dfrac{2}{3-x}$

25. $\dfrac{2x+1}{x^2-4x-21} + \dfrac{3}{14-2x}$

26. $\dfrac{x+1}{x^2-36} + \dfrac{x-2}{6-x}$

27. $\dfrac{-2x}{7-2x} - \dfrac{x+1}{4x^2-49}$

28. $\dfrac{4}{25-4x^2} - \dfrac{6}{2x^2+11x+15}$

29. $\dfrac{x-1}{x+2} + \dfrac{x+2}{x-1}$

30. $\dfrac{x+2}{x-1} - \dfrac{x+1}{x-2}$

31. $\dfrac{x+3}{x-5} + \dfrac{x-1}{x+4}$

32. $\dfrac{x-5}{x-3} - \dfrac{x+3}{x+5}$

33. $\dfrac{x+4}{2x+10} - \dfrac{5}{x^2-25}$

34. $\dfrac{3x-2}{x^2-x-2} - \dfrac{x-1}{x^2-1}$

35. $\dfrac{x+2}{(x-4)(x+3)^2} + \dfrac{x-1}{(x-4)(x+1)(x+3)}$

36. $\dfrac{x}{(x+1)(x+2)(x+3)} - \dfrac{7}{(x+2)(x+3)(x+4)}$

37. $\dfrac{x+2}{x^2-4}+\dfrac{3x}{x^2-2x}$

38. $\dfrac{x+3}{x^2-x}-\dfrac{12}{x^2+x-2}$

39. $\dfrac{x-1}{4x^2+20x+25}-\dfrac{x+4}{6x^2+17x+5}$

40. $\dfrac{x-3}{2x^2+11x-6}-\dfrac{x}{2x^2+7x-4}$

41. $\dfrac{3x-1}{x^2+4x+4}-\dfrac{2x+1}{3x^2+5x-2}$

42. $\dfrac{x-3}{3x^2-x-4}+\dfrac{2x+5}{6x^2+x-12}$

43. $\dfrac{x-1}{6x^2-24x}+\dfrac{5}{3x^3-6x^2-24x}$

44. $\dfrac{3}{2x^3+14x^2+20x}-\dfrac{x+2}{14x^2+28x}$

Perform the indicated operations. Simplify your result.

45. $\left(\dfrac{2}{x^2-4}+\dfrac{3}{x+2}\right)-\dfrac{1}{2x-4}$

46. $\left(\dfrac{5}{3x-9}-\dfrac{2x-1}{x^2-9}\right)+\dfrac{4}{x+3}$

47. $\dfrac{3}{x+1}-\left(\dfrac{2x-3}{x^2+6x+5}+\dfrac{2}{x+5}\right)$

48. $\dfrac{2x+11}{x^2+x-6}-\left(\dfrac{2}{x+3}+\dfrac{3}{2-x}\right)$

49. $\dfrac{4x+5}{x+2}+\left(\dfrac{3x+15}{x^2-4}\cdot\dfrac{x^2-2x}{x^2+7x+10}\right)$

50. $\dfrac{2x-7}{x-5}-\left(\dfrac{2x+10}{x^2+9x+20}\div\dfrac{x^2-25}{3x+12}\right)$

51. $\dfrac{5x+5}{3x+6}\cdot\left(\dfrac{x^2+4x}{x^2+2x+1}+\dfrac{4}{x^2+2x+1}\right)$

52. $\dfrac{x^2-16}{2x-12}\div\left(\dfrac{x^2}{x^2-36}-\dfrac{2x+8}{x^2-36}\right)$

Let $f(x)=\dfrac{x+3}{x-4}$ and $g(x)=\dfrac{x+4}{x-3}$.

53. Find $f(x)+g(x)$.　　**54.** Find $f(x)-g(x)$.

55. Find $g(x)-f(x)$.

Let $f(x)=\dfrac{x-2}{x^2-2x-8}$ and $g(x)=\dfrac{x+1}{3x+6}$.

56. Find $f(x)+g(x)$.　　**57.** Find $f(x)-g(x)$.

58. Find $g(x)-f(x)$.

59. A student tries to find the sum $\dfrac{2}{x+1}+\dfrac{3}{x+2}$. Describe the error and then perform the addition correctly.

$$\dfrac{2}{x+1}+\dfrac{3}{x+2}=\dfrac{2}{x+1}\cdot\dfrac{1}{x+2}+\dfrac{3}{x+2}\cdot\dfrac{1}{x+1}$$

$$=\dfrac{2}{(x+1)(x+2)}+\dfrac{3}{(x+2)(x+1)}$$

$$=\dfrac{5}{(x+1)(x+2)}$$

60. A student tries to find the difference $\dfrac{6x}{x+4}-\dfrac{3x+2}{x+4}$. Describe the error and then perform the subtraction correctly.

$$\dfrac{6x}{x+4}-\dfrac{3x+2}{x+4}=\dfrac{6x-(3x+2)}{x+4}$$

$$=\dfrac{6x-3x+2}{x+4}$$

$$=\dfrac{3x+2}{x+4}$$

61. A student tries to find the difference $\dfrac{9x}{x-3}-\dfrac{5x+1}{x-3}$. Describe the error and then perform the subtraction correctly.

$$\dfrac{9x}{x-3}-\dfrac{5x+1}{x-3}=\dfrac{9x-5x+1}{x-3}$$

$$=\dfrac{4x+1}{x-3}$$

62. Two students try to find the sum $\dfrac{3}{x-4}+\dfrac{2}{4-x}$. Compare the two methods. Are both methods correct? Explain. Discuss why student A's method is shorter.

Student A's Work

$$\dfrac{3}{x-4}+\dfrac{2}{4-x}=\dfrac{3}{x-4}+\dfrac{2}{-(x-4)}$$

$$=\dfrac{3}{x-4}-\dfrac{2}{x-4}$$

$$=\dfrac{1}{x-4}$$

Student B's Work

$$\dfrac{3}{x-4}+\dfrac{2}{4-x}=\dfrac{3}{x-4}\cdot\dfrac{4-x}{4-x}+\dfrac{2}{4-x}\cdot\dfrac{x-4}{x-4}$$

$$=\dfrac{3(4-x)+2(x-4)}{(x-4)(4-x)}$$

$$=\dfrac{12-3x+2x-8}{(x-4)(4-x)}$$

$$=\dfrac{4-x}{(x-4)(4-x)}$$

$$=\dfrac{1}{x-4}$$

63. Find two rational expressions whose sum is $\dfrac{4x}{x-1}$ and whose difference is $\dfrac{2x}{x-1}$.

64. Find two rational expressions whose sum is $\dfrac{x+5}{x-5}$ and whose product is $\dfrac{2x+6}{x^2-10x+25}$.

65. Describe how to add two rational expressions that have different denominators. Then describe how to subtract two such expressions.

8.4 SIMPLIFYING COMPLEX RATIONAL EXPRESSIONS

Objective

▶ Simplify complex rational expressions.

In this section we discuss how to simplify *complex rational expressions*. Here are some examples of such expressions:

$$\frac{\dfrac{x^2}{2} - \dfrac{3}{x^3}}{\dfrac{x}{6} + \dfrac{7}{x^2}}, \qquad \frac{\dfrac{3x}{x-1}}{\dfrac{x^3}{x-2}}, \qquad \frac{\dfrac{5x}{x^2-2x+1}}{x+4}$$

A **complex rational expression** is a rational expression whose numerator or denominator (or both) is a rational expression.

Here we find the values of two numerical complex rational expressions:

$$\frac{2}{\dfrac{2}{2}} = \frac{2}{1} = 2 \qquad \frac{\dfrac{2}{2}}{2} = \frac{1}{2}$$

From these two examples, we see that it is important to keep track of the main division bar (the one in bold) of the complex fraction.

We will discuss two methods for simplifying complex rational expressions. Ask your instructor whether you are required to know method 1, method 2, or both methods. In case the choice is yours to make, compare the use of method 1 in Examples 1 and 2 with the use of method 2 in Examples 3 and 4. The complex rational expressions in these examples are simplified by both methods, so you can get a sense of the advantages/disadvantages of each method.

Method 1: Writing a complex rational expression as a quotient of two rational expressions.

Note that an expression in the form $\dfrac{R}{S}$ can also be written in the form $R \div S$. We use this idea to help simplify a complex rational expression. If $\dfrac{A}{B}$ and $\dfrac{C}{D}$ are rational expressions, and B, C, and D are nonzero, then:

$$\frac{\dfrac{A}{B}}{\dfrac{C}{D}} = \frac{A}{B} \div \frac{C}{D}$$

$$= \frac{A}{B} \cdot \frac{D}{C}$$

$$= \frac{AD}{BC}$$

Note

Note that *AD* and *BC* are polynomials, since the product of two polynomials is a polynomial.

We *simplify* a complex rational expression by writing it as a rational expression $\dfrac{P}{Q}$, with $\dfrac{P}{Q}$ in lowest terms.

Example 1 Simplifying a Complex Rational Expression Using Method 1

Simplify the complex rational expression.

1. $\dfrac{\dfrac{12}{x}}{\dfrac{8}{x^3}}$

2. $\dfrac{\dfrac{x^2-9}{x^2+2x+1}}{\dfrac{2x-6}{4x+4}}$

Solution

1. $\dfrac{\dfrac{12}{x}}{\dfrac{8}{x^3}} = \dfrac{12}{x} \div \dfrac{8}{x^3}$ $\dfrac{R}{S} = R \div S$

$\quad = \dfrac{12}{x} \cdot \dfrac{x^3}{8}$ $\dfrac{A}{B} \div \dfrac{C}{D} = \dfrac{A}{B} \cdot \dfrac{D}{C}$

$\quad = \dfrac{12x^3}{8x}$ Multiply numerators; multiply denominators.

$\quad = \dfrac{3x^2}{2}$ Reduce.

We verify our work by comparing graphing calculator tables for $y = \dfrac{\dfrac{12}{x}}{\dfrac{8}{x^3}}$ and $y = \dfrac{3x^2}{2}$ (see Fig. 14).

Figure 14 Compare tables

The tables support our conclusion that the original expression and our result are equivalent expressions for values of x not equal to zero.

2. $\dfrac{\dfrac{x^2 - 9}{x^2 + 2x + 1}}{\dfrac{2x - 6}{4x + 4}} = \dfrac{x^2 - 9}{x^2 + 2x + 1} \div \dfrac{2x - 6}{4x + 4}$ $\dfrac{R}{S} = R \div S$

$\quad = \dfrac{x^2 - 9}{x^2 + 2x + 1} \cdot \dfrac{4x + 4}{2x - 6}$ $\dfrac{A}{B} \div \dfrac{C}{D} = \dfrac{A}{B} \cdot \dfrac{D}{C}$

$\quad = \dfrac{(x - 3)(x + 3)}{(x + 1)^2} \cdot \dfrac{4(x + 1)}{2(x - 3)}$ Factor numerators and denominators.

$\quad = \dfrac{4(x - 3)(x + 1)(x + 3)}{2(x - 3)(x + 1)^2}$ Multiply numerators; multiply denominators.

$\quad = \dfrac{2(x + 3)}{x + 1}$ Reduce. ∎

Next we simplify a complex rational expression that has two rational expressions in the numerator and two rational expressions in the denominator.

Example 2   Simplifying a Complex Rational Expression Using Method 1

Simplify $\dfrac{\dfrac{3}{2x} + \dfrac{1}{x^2}}{\dfrac{2}{5} - \dfrac{1}{x}}$.

Solution

We write both the numerator and the denominator as a fraction and then simplify as before.

$$\frac{\dfrac{3}{2x}+\dfrac{1}{x^2}}{\dfrac{2}{5}-\dfrac{1}{x}} = \frac{\left.\begin{array}{c}\dfrac{3}{2x}\cdot\dfrac{x}{x}+\dfrac{1}{x^2}\cdot\dfrac{2}{2}\end{array}\right\}}{\left.\begin{array}{c}\dfrac{2}{5}\cdot\dfrac{x}{x}-\dfrac{1}{x}\cdot\dfrac{5}{5}\end{array}\right\}}$$

Multiply by 1 to get a common denominator.

Multiply by 1 to get a common denominator.

$$= \frac{\dfrac{3x}{2x^2}+\dfrac{2}{2x^2}}{\dfrac{2x}{5x}-\dfrac{5}{5x}}$$

$$= \frac{\dfrac{3x+2}{2x^2}}{\dfrac{2x-5}{5x}} \qquad \frac{A}{B}+\frac{C}{B}=\frac{A+C}{B}$$

$$= \frac{3x+2}{2x^2}\div\frac{2x-5}{5x} \qquad \frac{R}{S}=R\div S$$

$$= \frac{3x+2}{2x^2}\cdot\frac{5x}{2x-5} \qquad \frac{A}{B}\div\frac{C}{D}=\frac{A}{B}\cdot\frac{D}{C}$$

$$= \frac{5x(3x+2)}{2x^2(2x-5)} \qquad \text{Multiply numerators; multiply denominators.}$$

$$= \frac{5(3x+2)}{2x(2x-5)} \qquad \text{Reduce.}$$

Since our result is in lowest terms, we are done. ∎

Using Method 1 to Simplify a Complex Rational Expression

1. Write both the numerator and the denominator as a fraction.

2. Use the following property to write the complex rational expression as the quotient of two rational expressions:

$$\frac{\dfrac{A}{B}}{\dfrac{C}{D}} = \frac{A}{B}\div\frac{C}{D}, \quad \text{where } B, C, \text{ and } D \text{ are nonzero}$$

3. Divide the rational expressions.

Method 2: Multiplying by $\dfrac{\text{LCD}}{\text{LCD}}$

By this method, we simplify a complex rational expression by first finding the LCD of all of the fractions appearing in the numerator and denominator. Then, we multiply by 1 in the form $\dfrac{\text{LCD}}{\text{LCD}}$.

In Example 3 we simplify the same complex rational expressions that we simplified in Example 1, but now we use method 2.

Example 3   Simplifying a Complex Rational Expression Using Method 2

Simplify the complex rational expression.

1. $\dfrac{\dfrac{12}{x}}{\dfrac{8}{x^3}}$

2. $\dfrac{\dfrac{x^2 - 9}{x^2 + 2x + 1}}{\dfrac{2x - 6}{4x + 4}}$

Solution

1. The LCD of $\dfrac{12}{x}$ and $\dfrac{8}{x^3}$ is x^3. So, we multiply the complex rational expression by 1 in the form $\dfrac{x^3}{x^3}$.

$$\dfrac{\dfrac{12}{x}}{\dfrac{8}{x^3}} = \dfrac{\dfrac{12}{x} \cdot \dfrac{x^3}{1}}{\dfrac{8}{x^3} \cdot \dfrac{x^3}{1}} \qquad \text{Multiply by } \dfrac{\text{LCD}}{\text{LCD}}.$$

$$= \dfrac{\dfrac{12x^3}{x}}{\dfrac{8x^3}{x^3}} \qquad \text{Simplify.}$$

$$= \dfrac{12x^2}{8} \qquad \begin{array}{l}\text{Reduce the fractions in the}\\ \text{numerator and the denominator.}\end{array}$$

$$= \dfrac{3x^2}{2} \qquad \text{Reduce.}$$

The result is the same as our result in problem 1 of Example 1.

2. $\dfrac{\dfrac{x^2 - 9}{x^2 + 2x + 1}}{\dfrac{2x - 6}{4x + 4}} = \dfrac{\dfrac{(x - 3)(x + 3)}{(x + 1)^2}}{\dfrac{2(x - 3)}{4(x + 1)}} \qquad \begin{array}{l}\text{Factor the numerators and}\\ \text{denominators of the fractions.}\end{array}$

$$= \dfrac{\dfrac{(x - 3)(x + 3)}{(x + 1)^2} \cdot \dfrac{4(x + 1)^2}{1}}{\dfrac{2(x - 3)}{4(x + 1)} \cdot \dfrac{4(x + 1)^2}{1}} \qquad \text{Multiply by } \dfrac{\text{LCD}}{\text{LCD}}.$$

$$= \dfrac{4(x - 3)(x + 3)}{2(x - 3)(x + 1)}$$

$$= \dfrac{2(x + 3)}{x + 1} \qquad \text{Reduce.}$$

The result is the same as our result in problem 2 of Example 1. ■

In comparing our work in Examples 1 and 3, we see that methods 1 and 2 required about the same number of steps. One advantage that method 1 has over method 2 for *these* complex rational expressions is that method 1 does not require finding an LCD.

In Example 4 we simplify the same expression as in Example 2, but now by using method 2.

Example 4 Simplifying a Complex Rational Expression Using Method 2

Simplify $\dfrac{\dfrac{3}{2x} + \dfrac{1}{x^2}}{\dfrac{2}{5} - \dfrac{1}{x}}$.

Solution

Note that the LCD of the rational expressions in the numerator and the denominator is $10x^2$. To simplify, we multiply by $\dfrac{10x^2}{10x^2}$.

$$\frac{\dfrac{3}{2x} + \dfrac{1}{x^2}}{\dfrac{2}{5} - \dfrac{1}{x}} = \frac{\dfrac{3}{2x} + \dfrac{1}{x^2}}{\dfrac{2}{5} - \dfrac{1}{x}} \cdot \frac{10x^2}{10x^2} \qquad \text{Multiply by } \frac{\text{LCD}}{\text{LCD}}.$$

$$= \frac{\dfrac{3}{2x} \cdot \dfrac{10x^2}{1} + \dfrac{1}{x^2} \cdot \dfrac{10x^2}{1}}{\dfrac{2}{5} \cdot \dfrac{10x^2}{1} - \dfrac{1}{x} \cdot \dfrac{10x^2}{1}} \qquad \text{Distributive law}$$

$$= \frac{15x + 10}{4x^2 - 10x} \qquad \text{Simplify.}$$

$$= \frac{5(3x + 2)}{2x(2x - 5)} \qquad \text{Factor the numerator and the denominator.}$$

Note that the result is the same as our result in Example 2. ∎

In comparing our work in Examples 2 and 4, we see that method 2 required fewer steps than method 1. In general, when the numerator or denominator, or both, contains two rational expressions, it is usually easier to use method 2.

Using Method 2 to Simplify a Complex Rational Expression

1. Find the LCD of all the fractions appearing in the numerator and denominator.
2. Multiply by 1 in the form $\dfrac{\text{LCD}}{\text{LCD}}$.
3. Reduce the numerator and the denominator to polynomials.
4. Reduce the rational expression.

In Example 5 we form the quotient function of two rational functions.

Example 5 Finding a Quotient Function Using Method 2

Let $Q(x) = \dfrac{f(x)}{g(x)}$, where $f(x) = 2 - \dfrac{5}{x+2}$ and $g(x) = \dfrac{x}{x+2} + \dfrac{x+1}{x^2 - 4x - 12}$.
Find an equation for Q and simplify the right side of the equation.

Solution

Since the denominator (and the numerator) contains two rational expressions, we choose to use method 2.

$$Q(x) = f(x) \div g(x)$$

$$= \frac{2 - \dfrac{5}{x+2}}{\dfrac{x}{x+2} + \dfrac{x+1}{x^2 - 4x - 12}}$$

$$= \frac{2 - \dfrac{5}{x+2}}{\dfrac{x}{x+2} + \dfrac{x+1}{(x-6)(x+2)}} \qquad \text{Factor } x^2 - 4x - 12.$$

$$= \frac{2 - \dfrac{5}{x+2}}{\dfrac{x}{x+2} + \dfrac{x+1}{(x-6)(x+2)}} \cdot \frac{(x-6)(x+2)}{(x-6)(x+2)} \qquad \text{Multiply by } \dfrac{\text{LCD}}{\text{LCD}}.$$

$$= \frac{2(x-6)(x+2) - \dfrac{5}{x+2} \cdot \dfrac{(x-6)(x+2)}{1}}{\dfrac{x}{x+2} \cdot \dfrac{(x-6)(x+2)}{1} + \dfrac{x+1}{(x-6)(x+2)} \cdot \dfrac{(x-6)(x+2)}{1}} \qquad \text{Distributive law}$$

$$= \frac{2(x-6)(x+2) - 5(x-6)}{x(x-6) + (x+1)} \qquad \text{Simplify.}$$

$$= \frac{2x^2 - 8x - 24 - 5x + 30}{x^2 - 6x + x + 1} \qquad \text{Expand.}$$

$$= \frac{2x^2 - 13x + 6}{x^2 - 5x + 1} \qquad \text{Simplify.}$$

$$= \frac{(2x-1)(x-6)}{x^2 - 5x + 1} \qquad \text{Factor the numerator.}$$

We verify our work in Fig. 15.

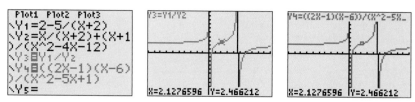

Figure 15 Verify our work

Our work in Example 5 suggests that if two functions f and g are rational, then the quotient function $\dfrac{f(x)}{g(x)}$ is rational. This is, indeed, true if the numerator of $g(x)$ is nonzero.

EXPLORATION
Looking ahead: Solving rational equations

1. Solve $\dfrac{1}{2} + \dfrac{x}{3} = \dfrac{5}{6}$.

2. Solve $\dfrac{1}{x-2} + \dfrac{2}{3} = \dfrac{5}{x-2}$. [**Hint**: Multiply both sides by the LCD of all three fractions.]

3. Solve $\dfrac{x}{x-2} - 5 = \dfrac{2}{x-2}$. Then, check whether your result satisfies the equation. What is the solution set? Explain.

4. The equations in problems 1–3 are called *rational equations*. What does your work in problem 3 suggest that you should always do when solving a rational equation?

Key Points
OF THIS SECTION

Complex rational expression	A complex rational expression is a rational expression whose numerator or denominator (or both) is a rational expression.
Method 1	Using Method 1 to Simplify a Complex Rational Expression: 1. Write both the numerator and the denominator as a fraction. 2. Use the following property to write the complex rational expression as the quotient of two rational expressions: $$\dfrac{\dfrac{A}{B}}{\dfrac{C}{D}} = \dfrac{A}{B} \div \dfrac{C}{D}, \text{ where } B, C, \text{ and } D \text{ are nonzero}$$ 3. Divide the rational expressions.
Method 2	Using Method 2 to Simplify a Complex Rational Expression: 1. Find the LCD of all the fractions appearing in the numerator and denominator. 2. Multiply by 1 in the form $\dfrac{\text{LCD}}{\text{LCD}}$. 3. Reduce the numerator and the denominator to polynomials. 4. Reduce the rational expression.

HOMEWORK 8.4

 SSM Tutor Center   **MathPro 4** MathPro 4   **MathPro 5** MathPro 5   CDs/Videos   www

Simplify each complex rational expression.

1. $\dfrac{\dfrac{2}{x}}{\dfrac{3}{x}}$

2. $\dfrac{\dfrac{x}{8}}{\dfrac{x}{6}}$

3. $\dfrac{\dfrac{7}{x^2}}{\dfrac{21}{x^5}}$

4. $\dfrac{\dfrac{8}{x^8}}{\dfrac{24}{x^2}}$

5. $\dfrac{\dfrac{9x^3}{16}}{\dfrac{12x^7}{20}}$

6. $\dfrac{\dfrac{18}{22x^9}}{\dfrac{4}{33x^4}}$

7. $\dfrac{\dfrac{x^2 + 2x + 1}{x - 5}}{\dfrac{5x + 5}{x^2 - 11x + 30}}$

8. $\dfrac{\dfrac{6x - 24}{5x - 10}}{\dfrac{3x^2 - 12x}{2x - 4}}$

9. $\dfrac{\dfrac{x^2 - 49}{3x^2 - 9x}}{\dfrac{x^2 - 5x - 14}{7x - 21}}$

10. $\dfrac{\dfrac{25x^2 - 4}{9x^2 - 16}}{\dfrac{25x^2 - 20x + 4}{9x^2 - 24x + 16}}$

11. $\dfrac{\dfrac{6}{x^2} - \dfrac{1}{x^2}}{\dfrac{2}{x^3} + \dfrac{8}{x^3}}$

12. $\dfrac{\dfrac{3x^2}{x-1} + \dfrac{5x^2}{x-1}}{\dfrac{4x}{x^2-1} - \dfrac{2x}{x^2-1}}$

13. $\dfrac{\dfrac{2}{x^3} - \dfrac{3}{x}}{\dfrac{5}{x^3} + \dfrac{4}{x^2}}$

14. $\dfrac{\dfrac{7}{2x^2} - \dfrac{3}{4x^3}}{\dfrac{2}{3x} + \dfrac{1}{5}}$

15. $\dfrac{\dfrac{4}{3x^2} - \dfrac{6}{3x^2}}{\dfrac{9}{5x} - \dfrac{7}{5x}}$

16. $\dfrac{\dfrac{3}{x} - \dfrac{1}{x^2}}{\dfrac{3}{x^2} - \dfrac{1}{x^3}}$

17. $\dfrac{4 + \dfrac{3}{x}}{\dfrac{2}{x} - 3}$

18. $\dfrac{\dfrac{3}{x} + 5}{\dfrac{2}{x} + 4}$

19. $\dfrac{2 - \dfrac{1}{x}}{4 - \dfrac{1}{x^2}}$

20. $\dfrac{\dfrac{3}{4x^2} - 2}{4 - \dfrac{5}{8x^2}}$

21. $\dfrac{\dfrac{1}{x} - \dfrac{8}{x^2} + \dfrac{15}{x^3}}{\dfrac{1}{x} - \dfrac{5}{x^2}}$

22. $\dfrac{\dfrac{4}{x} - \dfrac{1}{x^2} + \dfrac{3}{x^3}}{\dfrac{2}{x} - \dfrac{5}{x^2}}$

23. $\dfrac{\dfrac{x}{x-4} - \dfrac{2x}{x+1}}{\dfrac{x}{x+1} - \dfrac{2x}{x-4}}$

24. $\dfrac{\dfrac{2}{x-3} + \dfrac{5}{x-2}}{\dfrac{3}{x-2} - \dfrac{4}{x-3}}$

25. $\dfrac{x + \dfrac{2}{x-4}}{x - \dfrac{3}{x-4}}$

26. $\dfrac{\dfrac{3}{x+1} + 2}{4 - \dfrac{5}{x-1}}$

27. $\dfrac{\dfrac{1}{x+3} - \dfrac{1}{x}}{3}$

28. $\dfrac{\dfrac{2}{x+4} - \dfrac{2}{x}}{4}$

29. $\dfrac{\dfrac{1}{(x+2)^2} - \dfrac{1}{x^2}}{2}$

30. $\dfrac{\dfrac{5}{(x+3)^2} - \dfrac{5}{x^2}}{3}$

31. $\dfrac{\dfrac{6}{2x-8} + \dfrac{10}{x^2-4x}}{\dfrac{1}{x^2-x-12} - \dfrac{2}{x^2-16}}$

32. $\dfrac{\dfrac{x+7}{x^2+7x+10} - \dfrac{6}{x^2+2x}}{\dfrac{x+1}{x^2+7x+10} + \dfrac{6}{x^2+5x}}$

Let $h(x) = \dfrac{f(x)}{g(x)}$. Find an equation for h and simplify the right side of the equation.

33. $f(x) = \dfrac{2}{x} + \dfrac{x}{2}$ and $g(x) = \dfrac{3}{x} + \dfrac{2}{x}$

34. $f(x) = \dfrac{4}{x^2}$ and $g(x) = \dfrac{2}{x}$

35. $f(x) = \dfrac{5x+10}{x^2-6x+9}$ and $g(x) = \dfrac{4x+8}{x^2-4x+3}$

36. $f(x) = \dfrac{5x-10}{2x^2+x-15}$ and $g(x) = \dfrac{7x-14}{x^2+5x+6}$

37. $f(x) = 1 - \dfrac{1}{x-1}$ and $g(x) = 1 + \dfrac{1}{x+1}$

38. $f(x) = \dfrac{6x}{x^2-9} + \dfrac{x-3}{x+3}$ and $g(x) = \dfrac{x+3}{x-3} - \dfrac{6x}{x^2-9}$

*Simplify the expression. [**Hint**: Recall that $x^{-n} = \dfrac{1}{x^n}$ when $x \neq 0$.]*

39. $\dfrac{x^{-1} + x^{-2}}{x^{-2} - x^{-1}}$

40. $\dfrac{2x^{-2} - 4x^{-3}}{8x^{-1} - 6x^{-2}}$

41. $\dfrac{x - x^{-1}}{x^{-1} + x^{-2}}$

42. $\dfrac{3x + 2x^{-1}}{6x - 4x^{-1}}$

43. A student tries to simplify a complex rational expression. The work is shown below. Describe any error and then simplify the expression.

$$\dfrac{x}{\dfrac{1}{x} + \dfrac{1}{2}} = x \div \left(\dfrac{1}{x} + \dfrac{1}{2} \right)$$

$$= x \left(\dfrac{x}{1} + \dfrac{2}{1} \right)$$

$$= x \cdot (x + 2)$$

$$= x^2 + 2x$$

44. Describe how to simplify a complex rational expression.

45. Simplify the expression using method 1 and then simplify it again using method 2. Decide which method you prefer for this expression. Explain.

$$\dfrac{\dfrac{6}{x^2} - \dfrac{5}{x}}{\dfrac{2}{x^2} + \dfrac{3}{2x}}$$

46. Simplify the expression using method 1 and then simplify it again using method 2. Decide which method you prefer for this expression. Explain.

$$\dfrac{\dfrac{x^2-3x-28}{3x-6}}{\dfrac{x^2-14x+49}{6x-12}}$$

8.5 SOLVING RATIONAL EQUATIONS

By taking intermediate algebra, I realized that I have to be ready mentally to take on this course. I do better when I play the piano before I go to class. —*Ena C., student*

Objectives

▸ Solve rational equations.

▸ Distinguish between simplifying a rational expression and solving a rational equation.

In Sections 8.1–8.4 we have worked with rational expressions. In this section we solve *rational equations*. Here are some examples of rational equations:

$$\frac{x+2}{x-3} + \frac{2}{x-5} = \frac{x+1}{x-3}, \qquad \frac{5}{x} = 9, \qquad x^2 + \frac{3}{x^2-6x+9} = \frac{x-5}{x-3}$$

A **rational equation** is an equation where each side can be written as a rational expression.

A key step in solving a rational equation is to put the equation in a form where there are no denominators (other than 1) on either side of the equation. We do so by multiplying both sides of the equation by the LCD of all the rational expressions in the equation. This is sometimes called "clearing the equation of fractions."

Example 1   Solving a Rational Equation

Solve the equation $\dfrac{2}{x} + 5 = \dfrac{8}{x} - 1$.

Solution

$$\frac{2}{x} + 5 = \frac{8}{x} - 1$$

$$x\left(\frac{2}{x} + 5\right) = x\left(\frac{8}{x} - 1\right) \qquad \text{Multiply both sides by the LCD, } x.$$

$$x \cdot \frac{2}{x} + x \cdot 5 = x \cdot \frac{8}{x} - x \cdot 1 \qquad \text{Distributive law}$$

$$2 + 5x = 8 - x \qquad \text{Simplify.}$$

$$6x = 6$$

$$x = 1$$

We check that 1 satisfies the original equation:

$$\frac{2}{x} + 5 = \frac{8}{x} - 1$$

$$\frac{2}{1} + 5 \stackrel{?}{=} \frac{8}{1} - 1$$

$$7 \stackrel{?}{=} 7$$

true ∎

With rational equations, it is possible to take the usual steps for solving equations, yet arrive at values for *x* that are *not* solutions. So, with rational equations, it is doubly important that we confirm that each of our results satisfies the equation.

Example 2   Solving a Rational Equation

Solve the equation $2 - \dfrac{1}{x-2} = \dfrac{x-3}{x-2}$.

Solution

$$2 - \frac{1}{x-2} = \frac{x-3}{x-2}$$

$$(x-2)\left(2 - \frac{1}{x-2}\right) = (x-2) \cdot \frac{x-3}{x-2} \qquad \text{Multiply both sides by the LCD, } x-2.$$

$$(x-2) \cdot 2 - (x-2) \cdot \frac{1}{x-2} = (x-2) \cdot \frac{x-3}{x-2} \qquad \text{Distributive law}$$

$$(x-2) \cdot 2 - 1 = x - 3 \qquad \text{Simplify.}$$

$$2x - 4 - 1 = x - 3 \qquad \text{Distributive law}$$

$$x = 2$$

We check whether 2 satisfies the original equation:

$$2 - \frac{1}{x-2} = \frac{x-3}{x-2}$$

$$2 - \frac{1}{2-2} \stackrel{?}{=} \frac{2-3}{2-2}$$

$$2 - \frac{1}{0} \stackrel{?}{=} \frac{-1}{0}$$

$$\text{not true—division by 0 is undefined}$$

Division by zero is undefined, so the only possibility, 2, is *not* a solution. The solution set is the empty set. ∎

In Example 2 we multiplied both sides of the rational equation $2 - \dfrac{1}{x-2} = \dfrac{x-3}{x-2}$ by the LCD and simplified both sides. The result is the equation $(x-2) \cdot 2 - 1 = x - 3$, which has solution 2. However, 2 is *not* a solution of the original rational equation. We say that 2 is an *extraneous solution*.

To see where we introduced the extraneous solution 2, note that 2 does not satisfy the equation

$$(x-2) \cdot 2 - (x-2) \cdot \frac{1}{x-2} = (x-2) \cdot \frac{x-3}{x-2}$$

but 2 does satisfy the next equation

$$(x-2) \cdot 2 - 1 = x - 3$$

Before Simplifying Both Sides

$$(x-2) \cdot 2 - (x-2) \cdot \frac{1}{x-2} = (x-2) \cdot \frac{x-3}{x-2}$$

$$(2-2) \cdot 2 - (2-2) \cdot \frac{1}{2-2} \stackrel{?}{=} (2-2) \cdot \frac{2-3}{2-2}$$

$$0 \cdot 2 - 0 \cdot \frac{1}{0} \stackrel{?}{=} 0 \cdot \frac{-1}{0}$$

not true—division by 0 is undefined

After Simplifying Both Sides

$$(x-2) \cdot 2 - 1 = x - 3$$

$$(2-2) \cdot 2 - 1 \stackrel{?}{=} 2 - 3$$

$$0 \cdot 2 - 1 \stackrel{?}{=} -1$$

$$-1 \stackrel{?}{=} -1$$

true

Since multiplying both sides of a rational equation by the LCD and then simplifying both sides may introduce extraneous solutions, we must always check that any result satisfies the original equation.

To solve a rational equation, we factor denominators of fractions to help us determine the LCD and later to help us reduce fractions.

Example 3   Solving a Rational Equation

Solve $\dfrac{x}{x+2} - \dfrac{7}{5-x} = \dfrac{14}{x^2 - 3x - 10}$.

Solution

$$\frac{x}{x+2} - \frac{7}{5-x} = \frac{14}{x^2 - 3x - 10}$$

$$\frac{x}{x+2} - \frac{7}{5-x} = \frac{14}{(x-5)(x+2)} \qquad \text{Factor the denominator.}$$

$$\frac{x}{x+2} - \frac{7}{-(x-5)} = \frac{14}{(x-5)(x+2)} \qquad a - b = -(b-a)$$

$$\frac{x}{x+2} + \frac{7}{(x-5)} = \frac{14}{(x-5)(x+2)} \qquad \frac{a}{-b} = -\frac{a}{b}$$

$$\left. (x-5)(x+2)\left(\frac{x}{x+2} + \frac{7}{x-5}\right) = (x-5)(x+2) \right\} \quad \begin{array}{l}\text{Multiply both sides}\\ \text{by the LCD.}\end{array}$$
$$\cdot \frac{14}{(x-5)(x+2)}$$

On the left side, we use the distributive law. On the right side, we simplify.

$$(x-5)(x+2) \cdot \frac{x}{x+2} + (x-5)(x+2) \cdot \frac{7}{x-5} = 14$$

$$(x-5) \cdot x + (x+2) \cdot 7 = 14 \qquad \text{Simplify.}$$

$$x^2 - 5x + 7x + 14 = 14 \qquad \text{Distributive law}$$

$$x^2 + 2x + 14 = 14$$

$$x^2 + 2x = 0$$

$$x(x+2) = 0 \qquad \text{Factor.}$$

$$x = 0 \quad \text{or} \quad x + 2 = 0 \qquad \text{Zero factor property}$$

$$x = 0 \quad \text{or} \quad x = -2$$

We check whether each result satisfies the original equation:

Check $x = 0$	**Check $x = -2$**

$$\frac{x}{x+2} - \frac{7}{5-x} = \frac{14}{x^2 - 3x - 10} \qquad \qquad \frac{x}{x+2} - \frac{7}{5-x} = \frac{14}{x^2 - 3x - 10}$$

$$\frac{0}{0+2} - \frac{7}{5-0} \overset{?}{=} \frac{14}{0^2 - 3(0) - 10} \qquad \frac{-2}{-2+2} - \frac{7}{5-(-2)} \overset{?}{=} \frac{14}{(-2)^2 - 3(-2) - 10}$$

$$0 - \frac{7}{5} \overset{?}{=} \frac{14}{-10} \qquad\qquad\qquad \frac{-2}{0} - \frac{7}{7} \overset{?}{=} \frac{14}{0}$$

$$-\frac{7}{5} \overset{?}{=} -\frac{7}{5} \qquad\qquad \text{not true—division by 0 is undefined}$$

true

The only solution is 0. Substituting -2 for x in the original equation leads to division by zero, so -2 is an extraneous solution.

We use a graphing calculator table to verify that 0 is the only solution and that -2 is an extraneous solution (see Fig. 16).

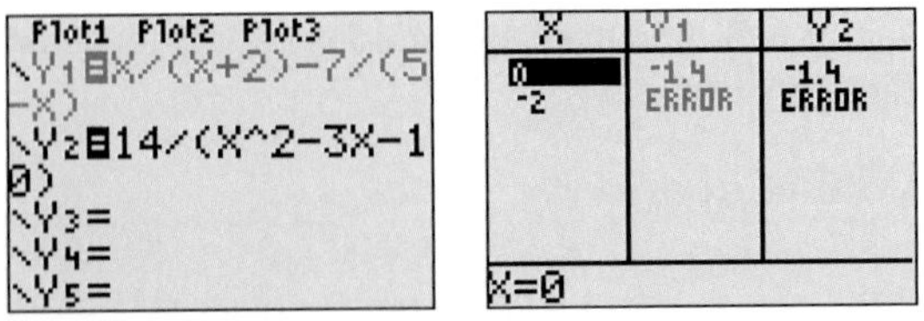

Figure 16 Verify the work

In general, multiplying both sides of a rational equation by the LCD will result in a simpler equation.

Solving a Rational Equation

To solve a rational equation:

1. Factor, if possible, any denominators.
2. Find the LCD of all the fractions.
3. Multiply both sides of the equation by the LCD, which gives a simpler equation to solve.
4. Solve the simpler equation.
5. Check that each result satisfies the original equation.

Example 4   Finding an Input Value of a Rational Function

Let $f(x) = \dfrac{x+1}{x-3} - \dfrac{x-2}{x+3}$. Find x when $f(x) = 1$.

Solution

We substitute 1 for $f(x)$ in the equation $f(x) = \dfrac{x+1}{x-3} - \dfrac{x-2}{x+3}$ and solve for x.

$$1 = \frac{x+1}{x-3} - \frac{x-2}{x+3}$$

$$(x-3)(x+3) \cdot 1 = (x-3)(x+3)\left(\frac{x+1}{x-3} - \frac{x-2}{x+3}\right) \qquad \text{Multiply both sides by the LCD.}$$

$$\left.\begin{array}{r} (x-3)(x+3) = (x-3)(x+3) \cdot \dfrac{x+1}{x-3} \\[2ex] -\,(x-3)(x+3) \cdot \dfrac{x-2}{x+3} \end{array}\right\} \quad \text{Distributive law}$$

$$(x-3)(x+3) = (x+3)(x+1) - (x-3)(x-2) \qquad \text{Simplify.}$$

$$x^2 - 9 = x^2 + 4x + 3 - (x^2 - 5x + 6) \qquad \text{Expand.}$$

$$x^2 - 9 = x^2 + 4x + 3 - x^2 + 5x - 6$$

$$x^2 - 9 = 9x - 3$$

$$x^2 - 9x - 6 = 0$$

$$x = \frac{-(-9) \pm \sqrt{(-9)^2 - 4(1)(-6)}}{2(1)} \qquad \text{Quadratic formula}$$

$$x = \frac{9 \pm \sqrt{105}}{2}$$

We check our work by entering the function $y = \dfrac{x+1}{x-3} - \dfrac{x-2}{x+3}$ in a graphing calculator and checking that for both $\dfrac{9 - \sqrt{105}}{2}$ and $\dfrac{9 + \sqrt{105}}{2}$, the output is 1 (see Fig. 17).

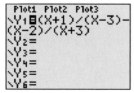 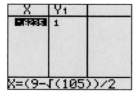

Figure 17 Check that for both $\dfrac{9 - \sqrt{105}}{2}$ and $\dfrac{9 + \sqrt{105}}{2}$, the output is 1

So, if $f(x) = 1$, then $x = \dfrac{9 - \sqrt{105}}{2}$ or $x = \dfrac{9 + \sqrt{105}}{2}$.   ∎

It is important to know the difference between solving an equation and simplifying an expression. In solving an equation, our objective is to find all *numbers* that satisfy the equation. When simplifying an expression, our objective is to find a simpler, yet equivalent, *expression*.

To solve a rational equation, we multiply both sides of the equation by the LCD. To simplify an expression, we do *not* multiply it by the LCD (unless the LCD is 1)—the only multiplication permissible is multiplication by 1, usually in the form $\dfrac{P}{P}$, where P is a nonzero polynomial.

Here we compare solving a rational equation with simplifying a rational expression:

Solving the Equation $1 = \dfrac{3}{x}$ *Simplifying the Expression* $1 + \dfrac{3}{x}$

$$1 = \frac{3}{x}$$

$$x \cdot 1 = x \cdot \frac{3}{x} \qquad \text{Multiply both sides by the LCD, } x.$$

$$x = 3 \qquad \text{The result is a number.}$$

$$1 + \frac{3}{x} = 1 \cdot \frac{x}{x} + \frac{3}{x} \qquad \begin{array}{l}\text{Multiply the 1 term} \\ \text{by } \dfrac{x}{x}.\end{array}$$

$$= \frac{x + 3}{x} \qquad \begin{array}{l}\text{The result is an} \\ \text{expression.}\end{array}$$

Note

We check that 3 is a solution of the equation $1 = \dfrac{3}{x}$: $1 \stackrel{?}{=} \dfrac{3}{3}$, which is true.

For the equation $1 = \dfrac{3}{x}$, the solution is the *number* 3. We simplify the expression $1 + \dfrac{3}{x}$ by writing it as the *expression* $\dfrac{x + 3}{x}$.

EXPLORATION
Simplifying versus solving

1. Two students tried to solve $4 = \dfrac{5}{x} + \dfrac{3}{x}$. Did one, both, or neither of these students solve the equation correctly? Explain.

Student A

$$4 = \frac{5}{x} + \frac{3}{x}$$

$$4x = x\left(\frac{5}{x} + \frac{3}{x}\right)$$

$$4x = x \cdot \frac{5}{x} + x \cdot \frac{3}{x}$$

$$4x = 5 + 3$$

$$4x = 8$$

$$x = 2$$

Student B

$$4 = \frac{5}{x} + \frac{3}{x}$$

$$= \frac{5}{x} + \frac{3}{x} - 4$$

$$= \frac{8}{x} - 4$$

$$= \frac{8}{x} - 4 \cdot \frac{x}{x}$$

$$= \frac{8}{x} - \frac{4x}{x}$$

$$= \frac{-4x + 8}{x}$$

2. Three students tried to simplify $4 + \dfrac{5}{x} + \dfrac{3}{x}$. Which students, if any, simplified the expression correctly? Explain.

Student C

$$4 + \frac{5}{x} + \frac{3}{x} = x\left(4 + \frac{5}{x} + \frac{3}{x}\right)$$

$$= 4x + x \cdot \frac{5}{x} + x \cdot \frac{3}{x}$$

$$= 4x + 5 + 3$$

$$= 4x + 8$$

Student D

$$4 + \frac{5}{x} + \frac{3}{x} = 4 \cdot \frac{x}{x} + \frac{5}{x} + \frac{3}{x}$$

$$= \frac{4x}{x} + \frac{5}{x} + \frac{3}{x}$$

$$= \frac{4x + 8}{x}$$

Student E

$$4 + \frac{5}{x} + \frac{3}{x} = 0$$

$$x\left(4 + \frac{5}{x} + \frac{3}{x}\right) = x \cdot 0$$

$$4x + 5 + 3 = 0$$

$$4x = -8$$

$$x = -2$$

3. a. What is the difference in your goals in solving a rational equation versus simplifying a rational expression?

b. Explain how their difference relates to the techniques you use to solve the equation versus simplify the expression.

Tips on Succeeding in This Course

If you have spent a good amount of time trying to solve an exercise in an assignment, but still can't solve the exercise, consider going on to the next exercise in the assignment. You may find that the next exercise involves a different concept or involves a more familiar situation. You may even find that after completing the rest of the assignment you are able to complete the exercise(s) you skipped. One explanation for this is that you may have learned or remembered some concept in a later exercise that relates to the exercise with which you were struggling.

Key Points
OF THIS SECTION

A solution of an equation is a *number*, whereas the result of simplifying an expression is an *expression*.

Solution versus simplified expression

To solve a rational equation, we multiply both sides of the equation by the LCD. To simplify an expression, we do *not* multiply it by the LCD (unless the LCD is 1)—the only multiplication permissible is multiplication by 1, usually in the form $\frac{P}{P}$, where P is a nonzero polynomial.

Solving versus simplifying

To solve a rational equation:

Solving a rational equation

1. Factor, if possible, the denominators.
2. Find the LCD of all fractions.
3. Multiply both sides by the LCD.
4. Solve the resulting simpler equation.
5. Check that each result satisfies the original equation.

HOMEWORK 8.5

 SSM Tutor Center MathPro 4 MathPro 4 MathPro 5 MathPro 5   CDs/Videos   www

Solve the rational equation.

1. $\dfrac{7}{x} = \dfrac{2}{x} + 1$

2. $\dfrac{8}{x} - 3 = \dfrac{5}{x} + 6$

3. $\dfrac{2x - 3}{x - 8} = \dfrac{x + 4}{x - 8}$

4. $\dfrac{x + 5}{x + 1} = \dfrac{8}{x + 1}$

5. $\dfrac{x - 2}{x - 7} = \dfrac{5}{x - 7}$

6. $3 - \dfrac{1}{x - 5} = \dfrac{x - 6}{x - 5}$

7. $\dfrac{5}{x} + \dfrac{1}{3} = \dfrac{7}{x}$

8. $\dfrac{5}{2} - \dfrac{x - 5}{x} = 1$

9. $\dfrac{2}{x - 1} = \dfrac{3}{x + 1}$

10. $\dfrac{5}{x - 7} = \dfrac{-4}{x + 1}$

11. $\dfrac{3}{x+1} + \dfrac{2}{5} = 1$

12. $\dfrac{2}{3} = \dfrac{7}{x-5} + 2$

13. $\dfrac{1}{x-2} + \dfrac{1}{x+2} = \dfrac{4}{x^2-4}$

14. $\dfrac{60}{x} - \dfrac{60}{x-5} = \dfrac{2}{x}$

15. $2 + \dfrac{4}{x-2} = \dfrac{8}{x^2-2x}$

16. $\dfrac{3}{x} = \dfrac{5}{x+1} - \dfrac{x}{x^2+x}$

17. $\dfrac{48}{x^2-2x-15} + \dfrac{6}{x+3} = \dfrac{7}{x-5}$

18. $\dfrac{7}{x^2+x-20} + \dfrac{2}{x-4} = \dfrac{4}{x+5}$

19. $\dfrac{2}{x+5} + \dfrac{1}{x-5} = \dfrac{16}{x^2-25}$

20. $\dfrac{x}{x-2} + \dfrac{x}{x^2-4} = \dfrac{x+3}{x+2}$

21. $\dfrac{x}{x-5} + \dfrac{2}{x-6} = \dfrac{2}{x^2-11x+30}$

22. $-1 + \dfrac{2x}{x+3} = \dfrac{-4}{x+4}$

23. $3 + \dfrac{1}{x} = \dfrac{10}{x^2}$

24. $\dfrac{5}{x} = \dfrac{3}{x^2} - 2$

25. $2 - \dfrac{2}{x^2} = 5 + \dfrac{1}{x} + \dfrac{3}{x^2}$

26. $\dfrac{7}{x^2} + 1 = \dfrac{4}{x}$

27. $\dfrac{x}{x-3} = 4 - \dfrac{3}{3-x}$

28. $4 - \dfrac{x}{x-5} = \dfrac{1}{5-x}$

29. $\dfrac{x+4}{x^2+4x-21} = \dfrac{x-2}{3-x}$

30. $\dfrac{x-6}{4-x^2} = \dfrac{x+5}{x-2}$

31. $\dfrac{x}{x+1} + \dfrac{1}{2} = \dfrac{-2}{x+2}$

32. $\dfrac{3}{2} - \dfrac{1}{x-4} = \dfrac{-2}{2x-8}$

33. $\dfrac{4}{x-5} + \dfrac{5}{x-2} = \dfrac{x+6}{3x-6}$

34. $\dfrac{5}{x^2-3x+2} - \dfrac{1}{x-2} = \dfrac{1}{3x-3}$

35. $\dfrac{x+2}{x^2-x} - \dfrac{6}{x^2-1} = 0$

36. $\dfrac{x-1}{x^2-7x+12} = \dfrac{x}{x-3} - \dfrac{x+2}{3-x}$

37. $\dfrac{x+3}{x^2+2x-24} + \dfrac{5}{4-x} = \dfrac{x+2}{x+6}$

38. $\dfrac{x+1}{2x+6} = \dfrac{x}{x-3} - \dfrac{x-1}{x^2-9}$

39. $\dfrac{x+1}{x-1} + \dfrac{x-1}{x+1} = \dfrac{1-x}{x^2-1}$

40. $\dfrac{2x+7}{x+6} - \dfrac{3x+1}{x+2} = \dfrac{x-2}{x^2+8x+12}$

41. $\dfrac{x+3}{x^2-x-2} + \dfrac{x+6}{x^2+3x+2} = \dfrac{2x-1}{x^2-4}$

42. $\dfrac{2x}{3x+3} - \dfrac{x+2}{6x+6} = \dfrac{x-6}{8x+8} + \dfrac{5}{12}$

43. $\dfrac{5}{2x+1} - \dfrac{x}{3x-2} = \dfrac{x^2-4}{6x^2-x-2}$

44. $\dfrac{x+2}{4x+1} = \dfrac{3}{2x-5} + \dfrac{x^2-3x+2}{8x^2-18x-5}$

For the given function, find x when y is equal to the indicated value.

45. $f(x) = \dfrac{3}{x-5}, \ y = 4$

46. $g(x) = \dfrac{7}{x+4}, \ y = 3$

47. $h(x) = \dfrac{2x-1}{x^2-2x+5}, \ y = 2$

48. $k(x) = \dfrac{3x-5}{x^2-4}, \ y = 1$

49. $f(x) = \dfrac{5}{x-1} + \dfrac{3}{x+1}, \ y = -1$

50. $g(x) = \dfrac{2}{x+1} - \dfrac{4}{x-2}, \ y = -1$

51. $h(x) = \dfrac{x-2}{x+4} - \dfrac{x+5}{x-3}, \ y = 0$

52. $k(x) = \dfrac{x+1}{x+2} - \dfrac{x+2}{x+1}, \ y = 0$

Find the x-intercept(s) of the function.

53. $f(x) = \dfrac{x-1}{x-5} - \dfrac{x+2}{x+3}$

54. $g(x) = \dfrac{x+4}{x-2} + \dfrac{x-3}{x+6}$

55. A student tried to simplify the expression $\dfrac{3}{x+1} + \dfrac{3}{x-1}$. Did the student simplify the expression correctly? If not, describe the error that was made.

$$\dfrac{3}{x+1} + \dfrac{3}{x-1}$$
$$= (x-1)(x+1)\left(\dfrac{3}{x+1} + \dfrac{3}{x-1}\right)$$
$$= (x-1)(x+1) \cdot \dfrac{3}{x+1} + (x-1)(x+1) \cdot \dfrac{3}{x-1}$$
$$= 3(x-1) + 3(x+1)$$
$$= 3x - 3 + 3x + 3$$
$$= 6x$$

56. A student tried to solve a rational equation. The student's result is $\dfrac{x-2}{x^2-4x+1}$. What would you tell the student?

Solve or simplify, whichever is appropriate.

57. $\dfrac{6}{x} - \dfrac{4}{x} = 1$

58. $\dfrac{x+2}{x+1} = \dfrac{1}{x+1} + 2$

59. $\dfrac{6}{x} - \dfrac{4}{x} - 1$

60. $\dfrac{x+2}{x+1} - \dfrac{1}{x+1} + 2$

61. $\dfrac{5}{x} + \dfrac{4}{x+1} - \dfrac{3}{x}$

62. $\dfrac{2}{x+3} + \dfrac{x}{x-3} + \dfrac{10}{x^2-9}$

63. $\dfrac{5}{x} + \dfrac{4}{x+1} = \dfrac{3}{x}$

64. $\dfrac{2}{x+3} + \dfrac{x}{x-3} = \dfrac{-10}{x^2-9}$

65. $\dfrac{x+2}{x^2-5x+6} - \dfrac{x+1}{x^2-4} = \dfrac{4}{x^2-x-6}$

66. $\dfrac{3}{x^2-4x-12} - \dfrac{x-3}{x^2-7x+6}$

67. $\dfrac{x+2}{x^2-5x+6} - \dfrac{x+1}{x^2-4} + \dfrac{4}{x^2-x-6}$

68. $\dfrac{3}{x^2 - 4x - 12} - \dfrac{x - 3}{x^2 - 7x + 6} = \dfrac{x}{x^2 + x - 2}$

69. Let $L = f(d)$ represent the loudness of the sound of a stereo speaker (in decibels) at a distance of d feet from the speaker.

a. Sketch a qualitative graph that describes the relationship between d and L. Is f an increasing or decreasing function for $d > 0$? What does this mean in terms of the sound level?

b. A model that describes the situation is

$$f(d) = \frac{k}{d^2}$$

where k is a constant. Given that the sound level is 90 decibels at a distance of 5 feet from the speaker, find the constant k. [**Hint:** Substitute the given values for d and L in the equation and solve for k.]

c. Substitute your value for the constant k from part b into the equation $f(d) = \dfrac{k}{d^2}$.

d. Use ZStandard followed by ZOOM OUT to draw a graph of f. Compare your calculator's graph with your qualitative graph in part a.

e. Estimate the sound level at a distance of 8 feet from the speaker.

f. At what distance is the sound level 70 decibels?

70. Let $f(d)$ represent the intensity (in watts per square meter) of a television signal at a distance d kilometers from the transmitter. A model that describes the situation is

$$f(d) = \frac{k}{d^2}$$

where k is a constant.

a. Given that the intensity of the television signal is 32 watts per square meter at a distance of 2 kilometers, find the constant k.

b. Substitute your value of the constant k from part a into the equation $f(d) = \dfrac{k}{d^2}$.

c. Find $f(3)$. What does your result mean in terms of the situation?

d. Use a graphing calculator table to find $f(4)$, $f(5)$, $f(6)$, and $f(7)$.

e. Is f an increasing or decreasing function for $d > 0$? What does this mean in terms of the intensity of the signal?

f. For what distances is the intensity *at least* 4 watts per square meter? Explain.

71. Let $f(x) = \dfrac{x + a}{x + b}$. Find values for a and b so that $f(0) = 2$ and $f(1) = \dfrac{5}{2}$.

72. Let $f(x) = \dfrac{x + a}{x + b}$. Find values for the constants a and b so that $f(1) = -2$ and $f(3) = -8$.

73. Describe how to solve a rational equation.

8.6 MODELING WITH RATIONAL FUNCTIONS

Objectives

▸ Use a rational function to model the mean of a quantity.

▸ Use a rational function to model the time it takes for an object to travel a given distance at a constant speed.

In this section we use a rational function to model a true-to-life situation. To begin, suppose that four students go on a road trip during their spring break and the total cost for gas is $20. Consider the following possibilities:

A. Each person pays $5 for gas.
B. The amounts contributed for gas are $4, $4, $6, and $6.
C. One student pays $20 and the other three students ride for free.

We compute the *mean* amount of money spent per student for gas by dividing the total spent by the number of students:

$$\text{mean amount per student} = \frac{\text{total amount spent for gas}}{\text{number of students}}$$

$$= \frac{20 \text{ dollars}}{4 \text{ students}} = 5 \text{ dollars per student}$$

So, the mean amount of money spent per student is $5. We also say that the *average* amount of money spent per student is $5.

For scenarios A, B, and C we make the following observations:

A. The mean gives the (exact) per-person amount if all students pay an equal amount.
B. The mean gives a reasonable estimate of the per-person amount if the students pay nearly the same amount.
C. The mean gives a poor estimate of the per-person amount if the students pay very different amounts.

Computing the Mean

If a quantity Q is divided into n parts, the mean amount M of quantity per part is the quotient

$$M = \frac{Q}{n}$$

For another example, if a student makes 21 phone calls in 7 days, the mean number of calls he makes per day is $\frac{21}{7} = 3$. The mean is a fairly good estimate of the number of calls on a given day, provided that the student made about the same number of calls each day.

Note

A *DAT tape* is a *digital audio tape*. Music is recorded onto a DAT tape using 0's and 1's. Computer software programs can then *equalize* (*EQ*) the music to adjust the frequencies and improve the quality of the sound.

Example 1   Modeling with a Rational Function

The underground band Melted Zipper wants to make a CD of its original songs and sell the CD at its gigs. It costs about $1000 to record the music onto a DAT tape, $100 to rearrange and EQ the music, $350 for artwork for the cover and inside leaflet, and $350 to set up production. In addition, it will cost $2.50 for each CD manufactured.

1. It is up to the band to decide how many CDs will be manufactured. What is the total cost of making 300 CDs?
2. Let $C(n)$ represent the total cost (in dollars) of making n CDs. Find an equation for C.
3. Let $P(n)$ represent the price the band should set for each CD (in dollars) so it breaks even by making and selling n CDs. Find an equation for P.
4. Find $P(300)$. Describe the result in terms of CDs.
5. Find n when $P(n) = 10$. Describe the result in terms of CDs.
6. Describe the values of $P(n)$ for large values of n.

Melted Zipper
members:

Melted
Zipper

Jay   Jim   Steve

Solution

1. First we compute the total *fixed costs*, the costs that do not depend on how many CDs are manufactured:

$$1000 + 100 + 350 + 350 = 1800 \text{ dollars}$$

The band must also pay $2.50 per CD manufactured. If the band manufactures 300 CDs, this cost, called the *variable cost*, is $2.50(300) = 750$ dollars.

To find the total cost we add the fixed costs and the variable cost:

$$2.50(300) + 1800 = 2550 \text{ dollars}$$

2. The total cost is equal to $2.50 times the number of CDs, plus the fixed cost of $1800:

$$C(n) = 2.50n + 1800$$

3. If the band makes and sells n CDs, it can break even by selling the CD for the amount found by dividing the total cost into n parts. This amount is the mean cost per CD:

$$P(n) = \text{ mean cost per CD}$$

$$= \frac{\text{total cost}}{\text{number of CDs manufactured}}$$

$$= \frac{2.50n + 1800}{n}$$

So, our equation for P is $P(n) = \dfrac{2.50n + 1800}{n}$.

4. $P(300) = \dfrac{2.50(300) + 1800}{300} = 8.50$

If the band makes and sells 300 CDs, it must sell the CD for $8.50 in order to break even.

5. We substitute 10 for $P(n)$ and solve for n.

$$10 = \frac{2.50n + 1800}{n}$$

$$n \cdot 10 = n \cdot \frac{2.50n + 1800}{n} \qquad \text{Multiply both sides by the LCD.}$$

$$10n = 2.50n + 1800 \qquad \text{Simplify.}$$

$$7.5n = 1800$$

$$n = 240$$

If the band can sell the CDs it makes for $10, then it needs to make 240 CDs to break even.

6. First, we use graphing calculator tables to display input-output pairs for P (see Figs. 18, 19, and 20).

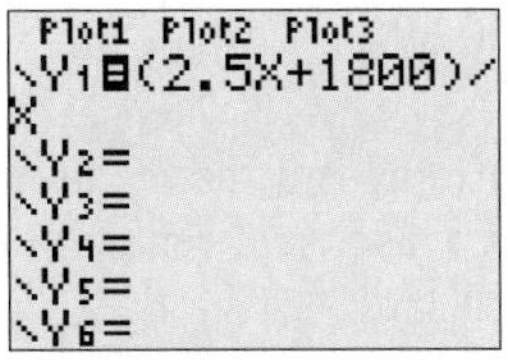

Figure 18 Enter the function

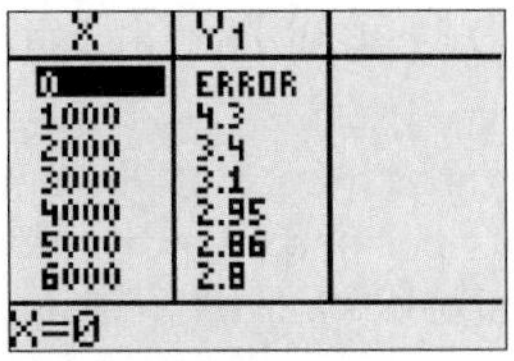

Figure 19 Inputs increasing by 1000

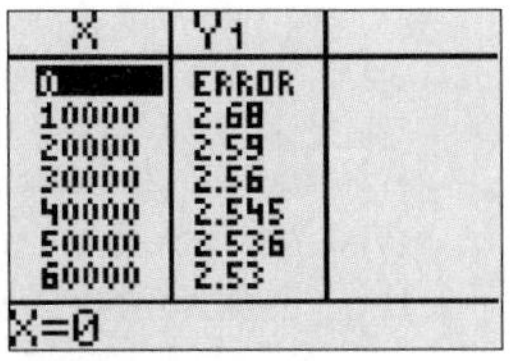

Figure 20 Inputs increasing by 10,000

From the tables, we see that as n increases, the mean cost (price) decreases (in order to break even). This happens because, as the number of disks manufactured increases, the more the fixed cost is spread out, so the smaller the fixed cost that each sale has to cover. In fact, if we continue to scroll down a table for larger and larger inputs n, the outputs $P(n)$ approach 2.50, the variable cost in dollars per CD.

We can also study a graphing calculator graph to observe that the break-even CD price approaches \$2.50 for large manufacturing runs (see Figs. 21, 22, and 23). It appears that for large inputs, the height of the graph of P gets close to 2.50.

Figure 21 Enter the functions; Y_2 is the variable cost per disk

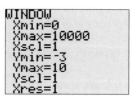

Figure 22 Set up the window to allow for large n

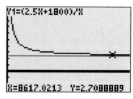

Figure 23 Graph the functions

This means that if the band can sell tens of thousands of CDs, they could price them at a few cents above the \$2.50 cost and still break even, which is probably unrealistic. ∎

Example 2 Modeling with a Rational Function

In Example 3 in Section 7.6, we modeled both video sales and video rentals (see Table 3).

	Video Rentals	Video Sales
Table 3 Video Rentals and Video Sales		
Year	(billions of dollars)	(billions of dollars)
1991	8.4	3.6
1992	9.1	4.0
1994	9.5	5.5
1996	9.3	7.3
2000	8.25	10.8

We found that the numbers of *millions* of dollars in revenues from video rentals, $r(t)$, and video sales, $s(t)$, respectively, during the year that is t years since 1990 can be modeled well by the system

$$r(t) = -54t^2 + 560t + 8030$$
$$s(t) = 19t^2 + 600t + 2860$$

1. Let $V(t)$ represent the total revenues from video rentals and video sales (in millions of dollars) during the year that is t years since 1990. Find an equation for V.
2. The U.S. population (in millions) $P(t)$ can be modeled by the function

$$P(t) = 0.0067t^2 + 2.56t + 250.39$$

where t is the number of years since 1990. Let $M(t)$ represent the mean annual cost per person (in dollars per person) for videos (both sales and rentals) during the year that is t years since 1990. Find an equation for M.

3. Predict the mean annual cost per person to watch videos during the year 2003.
4. When will the mean annual cost per person be \$75?

Solution

1. The total annual video cost, $V(t)$, is the sum of the revenues from video rentals and video sales:

$$V(t) = r(t) + s(t)$$
$$= (-54t^2 + 560t + 8030) + (19t^2 + 600t + 2860)$$
$$= -35t^2 + 1160t + 10{,}890$$

2. The mean annual cost per person is equal to the total annual cost divided by the U.S. population.

$$M(t) = \frac{V(t)}{P(t)} = \frac{-35t^2 + 1160t + 10{,}890}{0.0067t^2 + 2.56t + 250.39}$$

3. To make a prediction for 2003, we find $M(13)$:

$$M(13) = \frac{-35(13)^2 + 1160(13) + 10{,}890}{0.0067(13)^2 + 2.56(13) + 250.39} \approx 70.42$$

According to our model, the mean annual cost per person for videos in 2003 will be approximately $70.42.

4. To find when the mean annual cost will be $75, we substitute 75 for $M(t)$ and solve for t:

$$75 = \frac{-35t^2 + 1160t + 10{,}890}{0.0067t^2 + 2.56t + 250.39}$$

$$(0.0067t^2 + 2.56t + 250.39) \cdot 75 = (0.0067t^2 + 2.56t + 250.39)$$
$$\cdot \frac{-35t^2 + 1160t + 10{,}890}{0.0067t^2 + 2.56t + 250.39}$$

$$0.5025t^2 + 192t + 18{,}779.25 = -35t^2 + 1160t + 10{,}890$$

$$35.5025t^2 - 968t + 7889.25 = 0$$

Now, we find the discriminant $b^2 - 4ac$ of the quadratic formula:

$$b^2 - 4ac = (-968)^2 - 4(35.5025)(7889.25) = -183{,}328.39$$

Since the discriminant is negative, there are no real number solutions. So, *according to our model,* the mean annual cost per person will never reach $75. In fact, we can use "maximum" on a graphing calculator to determine that the maximum value of M is 70.50 for values of t between 0 and 50 (see Fig. 24).

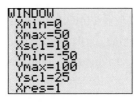

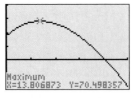

Figure 24 Find the maximum value

Note

The maximum point in Fig. 24 is not the vertex of a parabola; it is the "highest" point on the graph of the rational function.

Fig. 24 also suggests that model breakdown likely occurs in the not-too-distant future, as it is unlikely that the mean annual cost will decrease very much, and it is impossible for it to be negative. ■

We now discuss how to model the distance traveled by an object moving at a constant speed. If a car is driven at 50 mph for 2 hours, it will travel 50 miles in the first hour and 50 miles in the second hour for a total distance of $50 \cdot 2 = 100$ miles. If a car is driven at 60 miles per hour for 3 hours, it will travel $60 \cdot 3 = 180$ miles.

In general, the (constant) speed of the car multiplied by the amount of time the car is in motion gives the distance traveled.

Distance-Speed-Time Relationship

If an object is moving at a constant speed s for an amount of time t, then the distance traveled d is equal to the product

$$d = st$$

and the time t is found from the rational expression

$$t = \frac{d}{s}$$

Note

If the units of s and t are mph and hours, respectively, then the unit for d is miles.

Example 3   Computing a Driving Time

A person plans to drive 55 mph on a 80-mile trip. Compute the driving time.

Solution

Since the person is traveling at a constant rate, we use the equation

$$t = \frac{d}{s}$$

We substitute 80 for d and 55 for s in the equation.

$$t = \frac{80}{55} \approx 1.45$$

So, the driving time will be about 1.5 hours. ■

Example 4   Modeling with a Rational Function

A student from Seattle Central Community College plans to drive from Seattle, Washington to Eugene, Oregon. The speed limit is 70 mph in Washington and 65 mph in Oregon. The trip involves 164 miles of travel in Washington, followed by 121 miles in Oregon.

1. First, suppose that the student plans to drive at the speed limits. Compute the driving time.
2. Now drop the assumption that the student plans to drive at the speed limits. Let $T(a)$ represent the driving time (in hours) at a mph above the speed limits. Find an equation for T.
3. Find $T(0)$. Compare this result with the result in problem 1.
4. If the student drives 5 mph over the speed limits, compute the driving time.
5. If the student wants the driving time to be 4 hours, how much over the speed limits does she need to drive?

Solution

1. Since the student drives at a constant rate in Washington, we can use the equation $t = \frac{d}{s}$ to compute the time (in hours) spent driving in Washington:

$$t = \frac{\text{distance in Washington}}{\text{speed in Washington}}$$

$$= \frac{164}{70}$$

We can also compute the time (in hours) spent driving in Oregon:

$$t = \frac{\text{distance in Oregon}}{\text{speed in Oregon}}$$

$$= \frac{121}{65}$$

The total driving time is the sum of our two computed times. The entire time behind the wheel will be

$$\frac{164}{70} + \frac{121}{65} \approx 4.2 \text{ hours}$$

2. If the student drives, say, 5 mph over the speed limits, then she will drive at $70 + 5 = 75$ mph in Washington and $65 + 5 = 70$ mph in Oregon. If the student drives a miles per hour over the speed limits, she will drive at $70 + a$ mph in Washington and $65 + a$ mph in Oregon. We use these expressions for speeds to find an equation for T:

$$T(a) = \frac{\text{distance in Washington}}{\text{speed in Washington}} + \frac{\text{distance in Oregon}}{\text{speed in Oregon}}$$

$$= \frac{164}{70 + a} + \frac{121}{65 + a}$$

or $T(a) = \dfrac{164}{a + 70} + \dfrac{121}{a + 65}$.

3. $T(0) = \dfrac{164}{0 + 70} + \dfrac{121}{0 + 65} \approx 4.2$

The driving time will be about 4.2 hours if the student drives at the speed limits. We found the same result in problem 1.

4. If the student drives 5 mph over the speed limits, then $a = 5$.

$$T(5) = \frac{164}{5 + 70} + \frac{121}{5 + 65} \approx 3.9$$

The driving time will be about 3.9 hours.

5. If the trip is to take 4 hours, then $T(a) = 4$:

$$4 = \frac{164}{a + 70} + \frac{121}{a + 65}$$

$$(a + 65)(a + 70) \cdot 4 = (a + 65)(a + 70) \cdot \left(\frac{164}{a + 70} + \frac{121}{a + 65} \right)$$

$$(a^2 + 135a + 4550) \cdot 4 = (a + 65)(a + 70) \cdot \frac{164}{a + 70} + (a + 65)(a + 70) \cdot \frac{121}{a + 65}$$

$$4a^2 + 540a + 18{,}200 = (a + 65) \cdot 164 + (a + 70) \cdot 121$$

$$4a^2 + 540a + 18{,}200 = 164a + 10{,}660 + 121a + 8470$$

$$4a^2 + 540a + 18{,}200 = 285a + 19{,}130$$

$$4a^2 + 255a - 930 = 0$$

$$a = \frac{-255 \pm \sqrt{255^2 - 4(4)(-930)}}{2(4)} \qquad \text{Quadratic formula}$$

$$a \approx -67.2 \quad \text{or} \quad a \approx 3.5$$

The value $a = -67.2$ represents driving under the speed limits by 67.2 mph—clearly model breakdown has occurred. So, the student needs to drive 3.5 mph over the speed limits for the driving time to be 4 hours.   ■

EXPLORATION
Looking ahead: Inverse variation

Suppose that you intend to drive a car 100 miles. Let $f(s)$ represent the time (in hours) it will take you to drive 100 miles if you drive the car at a constant speed of s miles per hour.

1. Find an equation for f.

2. Find $f(50)$, $f(55)$, $f(60)$, and $f(70)$. What do your results mean in terms of the situation?

3. Consider completing the 100-mile trip several times, each time at a greater constant speed than the last. What happens to the travel time as speed gets extremely large? Use a graphing calculator table and graph to verify your answer. (Recall that when a function behaves like this, we say that the horizontal axis is a horizontal asymptote for the function.)

4. Consider completing the 100-mile trip several times, each time at a lower constant speed than the last. What happens to the travel time as speed gets extremely close to 0? Use a graphing calculator table and graph to verify your answer.

5. Does the function f have a vertical asymptote? If so, what is it? How does your answer relate to problem 4?

Key Points
OF THIS SECTION

Mean

If a quantity Q is divided into n parts, the mean amount M of quantity per part is equal to the quotient

$$M = \frac{Q}{n}$$

Distance-speed-time relationship

If an object is moving at a constant speed s for an amount of time t, then the distance traveled d is found from the product

$$d = st$$

and the time t is found from the rational expression

$$t = \frac{d}{s}$$

HOMEWORK 8.6

 SSM
Tutor Center

MathPro 4
MathPro 4

MathPro 5
MathPro 5

CDs/Videos

www

1. The ski club at a community college plans to spend $1250 to charter a bus for a ski trip. This cost will be split evenly among the students who sign up for the trip. Each student will also pay $350 for food, lodging, and ski lift tickets.

 a. Let $C(n)$ represent the total cost (in dollars) for n students going on the trip. Find an equation for C.

 b. Let $M(n)$ represent the mean cost per student (in dollars per student) if n students go on the trip. Find an equation for M.

 c. What is the mean cost per student if 30 students go on the trip?

 d. The ski trip will be cancelled unless the mean cost per student is $400 or less. What is the minimum number of students needed to go on the trip?

2. The Borough of Manhattan is one of the most densely populated regions in the United States. This borough is made

up of 792,000,000 square feet of land. Let $f(P)$ represent the number of square feet of Manhattan each Manhattanite would own if Manhattan were divided equally among its P residents.

a. Find an equation for f.

b. Use your formula for f to estimate the amount of land per Manhattanite if 1.5 million people live in Manhattan.

c. If the land in the United States was divided equally among the residents of the United States, each resident would own about 437,000 square feet of land. How many people could live in Manhattan if each resident owned 437,000 square feet of Manhattan? Compare your answer with how many people do live in Manhattan.

d. In your opinion, what is the smallest land-to-person ratio that would still allow people to live enjoyable, healthy lives? Use your estimate and your formula for f to find the maximum number of people that should live in Manhattan.

3. For a high school 5-year reunion, graduates rent out a restaurant for dinner and dancing. The restaurant charges a flat fee of $500 for a band plus $50 per person for food and two drinks.

a. Let $T(n)$ represent the total cost (in dollars) for n people to attend the reunion. Find an equation for T.

b. Let $M(n)$ represent the mean cost per person (in dollars per person) if n people attend the reunion. Find an equation for M.

c. Find $M(270)$. Describe the result in terms of the reunion.

d. Find n when $M(n) = 60$. Describe the result in terms of the reunion.

e. Complete Table 4 by using a graphing calculator table.

Table 4 Mean Cost per Person for the Reunion

Number of Guests n	Mean Cost per Person $M(n)$
100	
200	
300	
400	
500	

f. Describe $M(n)$ for large n. Explain why your observation makes sense in terms of the restaurant fees.

4. A student agrees to throw an end-of-semester party, provided that each guest who attends shares equally in the expenses with the student. A four-person local band will play for $200 and free drinks and snacks. The student estimates that the mean consumption cost of drinks and snacks will be $3 per person.

a. Suppose that n people (including the host and band members) attend the party. Let $C(n)$ represent the total cost (in dollars) of the party. Find an equation for C.

b. Let $P(n)$ represent the equal share of expenses (in dollars) that each guest and host contributes. Find an equation for P. [**Hint**: Recall that the band members get free drinks and snacks.]

c. If the host and each guest is willing to pay at most $5 per person, how many guests must attend the party to cover expenses?

d. What do the values of $P(n)$ get close to as the values of n get very large? Describe this result in terms of the party.

e. If the host and each guest are willing to pay only $2, how many guests must attend the party to cover expenses? Explain why this makes sense.

5. It costs a car manufacturer an average of $90,000 per day to pay for its leasing costs, equipment maintenance, salaries, electricity, and marketing to produce a certain type of car. It also costs the manufacturer an average of $7000 per car to pay for materials, invoices, and shipping.

a. Suppose that the car manufacturer produces and sells n cars per day. Let $C(n)$ represent the total daily cost (in dollars). Find a formula for $C(n)$.

b. Suppose that the car manufacturer produces and sells n cars per day. Let $B(n)$ represent how much (in dollars) the manufacturer should charge for each car so it would break even by selling n cars. Find a formula for B.

c. Suppose that the car manufacturer produces and sells n cars per day. Let $P(n)$ represent how much (in dollars) the manufacturer should charge for each car so it would make a profit of $2000 per car. Find a formula for $P(n)$. [**Hint**: Build on your equation for B to find an equation for P.]

d. Find $P(40)$. What does your result mean in terms of cars?

e. What do the values of $P(n)$ get close to as the values of n get very large? Interpret this result in terms of the cost of the cars.

6. In Example 1 we found the rational function

$$P = f(n) = \frac{2.50n + 1800}{n},$$

where $f(n)$ models how the band Melted Zipper should price its CD (in dollars) so it breaks even by making and selling n CDs. [**Note**: We use different notation than in Example 1. Here, we use f for the name of the function and P for the name of the dependent variable.]

a. If Melted Zipper sets the price of the CD at $7, how many CDs need to be made and sold for the band to break even?

b. Find an equation for f^{-1}. [**Hint**: Perform steps similar to the steps you did in part a, but do not substitute a value for P. After a couple of steps, factor out n on one side of the equation.]

c. Find $f^{-1}(7)$. Compare this result with your result in part a.

7. Book sales (including textbooks) at college stores have increased during the past two decades (see Table 5).

Table 5 Book Sales at College Stores

Year	Book Sales (millions of dollars)
1985	2309
1988	2904
1990	3403
1995	4311
1998	5122
2000	5699

Source: *Book Industry Study Group, Inc.*

a. Let $B(t)$ represent book sales at college stores (in millions of dollars) during the year that is t years since 1970. Perform the first three steps of the modeling process to find an equation for B.

b. In exercise 25 of Homework 3.1, the enrollments (in millions) at U.S. colleges $W(t)$ and $M(t)$ for women and men, respectively, are modeled by the system

$$W(t) = 0.14t + 4.52$$
$$M(t) = 0.083t + 4.60$$

where t represents the number of years since 1970 (see Table 6). Let $E(t)$ represent total college enrollment (in millions) at t years since 1970. Find an equation for E.

Table 6 College Enrollments

| | Enrollment (millions) | |
Year	Women	Men
1980	6.0	5.4
1985	6.6	5.9
1990	7.4	6.2
1995	8.0	6.7
1998	8.6	6.9

c. Let $A(t)$ represent the mean amount of money spent on books per college student (in dollars per student) during the year that is t years since 1970. Find an equation for M.

d. Predict the mean amount of money that will be spent per student in 2008.

e. During which year will the mean amount of money spent per student equal $430?

8. Fuel consumption (in millions of gallons) by vehicles in the United States is listed for various years in Table 7.

Table 7 Vehicle Fuel Consumption

Year	Amount of Fuel (millions of gallons)
1975	109,000
1980	115,000
1985	121,300
1990	130,800
1995	143,800
2000	160,700

Source: *Federal Highway Administration*

a. Let $F(t)$ represent the total amount of fuel (in millions of gallons) used during the year that is t years since 1970. Find an equation for f.

b. The U.S. population (in millions) $P(t)$ can be modeled by the function

$$P(t) = 0.0067t^2 + 2.30t + 201.87$$

where t is the number of years since 1970. Let $M(t)$ represent the mean amount of fuel used per person (in gallons per person) during the year that is t years since 1970. Find an equation for M.

c. Find $M(37)$. Describe your result in terms of fuel.

d. Find t when $M(t) = 700$. Describe your result in terms of fuel.

e. Draw a graph of M using the WINDOW settings displayed in Fig. 25. Describe what the graph tells you in terms of per-person fuel consumption.

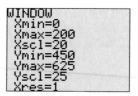

Figure 25 Per-person fuel consumption

f. Is $M(t)$ an underestimate or an overestimate of the mean amount of fuel used per *driver* in the United States? Explain.

9. A person drives at 60 mph for 85 miles. Compute the driving time.

10. A person drives at a constant speed for 100 miles in 1.7 hours. At what speed does the person drive?

11. A student drives from Chicago, Illinois to St. Louis, Missouri, a distance of 295 miles. The speed limit is 65 mph.

a. First suppose that the student plans to drive at the speed limit. Compute the driving time.

b. Now suppose that the student plans to drive at a mph above the speed limit. Let $T(a)$ represent the driving time (in hours) at a mph above the speed limit. Find an equation for T.

c. Find $T(5)$. What does your result mean in terms of the trip?

d. Find a when $T(a) = 4$. What does your result mean in terms of driving from Chicago to St. Louis?

12. In Example 4 we found the equation

$$T(a) = \frac{164}{a + 70} + \frac{121}{a + 65}$$

where $T(a)$ represents the driving time (in hours) for a student to drive nonstop from Seattle to Eugene at a mph above the speed limits.

a. Perform the addition on the right side of the equation for T.

b. Use your result in part a to find the driving time if the student drives 10 mph over the speed limits.

13. A student plans to drive from Grand Valley State University in Allendale, Michigan to Indianapolis, Indiana. The trip involves 114 miles in Michigan, followed by 144 miles in Indiana. The speed limit is 70 mph in Michigan and 65 mph in Indiana.

a. First suppose that the student plans to drive at the speed limits. Compute the driving time.

b. Now suppose that the student plans to drive at a mph above the speed limits. Let $T(a)$ represent the driving time (in hours) at a mph above the speed limits. Find an equation for T.

c. Find $T(0)$. Compare this result with your result from part a.

d. Find $T(10)$. What does your result mean in terms of the trip?

e. Find $T(0) - T(10)$. What does your result mean in terms of traveling from Allendale to Indianapolis?

14. A student plans to drive from Oklahoma City, Oklahoma to Little Rock, Arkansas. The trip involves 183 miles in Oklahoma, followed by 161 miles in Arkansas. The speed limit is 75 mph in Oklahoma and 70 mph in Arkansas.

a. Let $T(a)$ represent the driving time (in hours) at a mph above the speed limits. Find an equation for T.

b. If the student drives 5 mph over the speed limits, compute the driving time.

c. By how much would the student have to exceed the speed limits for the driving time to be 4 hours? Verify your answer using a graphing calculator table.

15. A student plans to drive from Albuquerque Technical Vocational Institute in Albuquerque, New Mexico to visit friends in Dallas, Texas. The trip involves 253 miles in New Mexico,

followed by 410 miles in Texas. The speed limit is 75 mph in New Mexico and 70 mph in Texas.

a. Let $T(a)$ represent the driving time (in hours) if the student drives at a mph above the speed limits. Find an equation for T.

b. By how much would the student have to exceed the speed limits for the driving time to be 8 hours? Verify your answer using a graphing calculator table.

8.7 VARIATION

Objectives

▸ Know the meaning of *direct variation* and *inverse variation*.

▸ For direct variation and inverse variation, know how a change in the independent variable impacts the value of the dependent variable.

▸ Find direct variation equations and inverse variation equations.

▸ Use direct variation models and inverse variation models to make estimates.

In this section we will work with special types of linear functions, rational functions, and other functions.

Example 1   Using a Direct Variation Function to Model Data

A golf ball is dropped from various heights and the bounce height is recorded each time (see Table 8).

Table 8 Drop and Bounce Heights of a Golf Ball

Drop Height (inches)	Bounce Height (inches)	Drop Height (inches)	Bounce Height (inches)
6	4.8	36	31.0
12	10.0	42	37.5
18	15.0	48	44.5
24	20.3	54	47.3
30	26.4	60	52.0

Source: *Data collected by the author*

Let B represent the bounce height (in inches) of the golf ball after being dropped from an initial height of d inches. Find an equation of a function that models the situation well.

Solution

We begin by drawing a scattergram of the data (see Fig. 26). It appears that the variables d and B are approximately linearly related. The linear regression equation is

$$B = 0.90d - 0.77$$

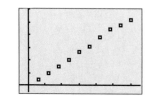

Figure 26 A scattergram of the data

Although this model fits the data very well (see Fig. 27), there is a problem with this model. Note that the model estimates that very small drop heights will have *negative* bounce heights (see Fig. 28). Model breakdown has occurred. Also, as the drop heights get closer to 0 inches, the bounce heights should get close to 0 inches. However, the model estimates that as the drop heights get close to 0 inches, the bounce heights will get close to −0.77 inches.

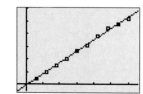

Figure 27 Regression line fits the data well

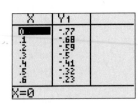

Figure 28 Regression line estimates negative bounce heights

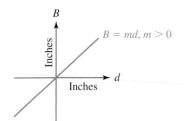

Figure 29 The graph of $B = md$ with $m > 0$

Note that it would be better to find a model of the form $B = md$ with $m > 0$ (see Fig. 29). Such a model estimates a *positive* bounce height for any drop height, including small ones. It also estimates that as drop heights get close to 0 inches, the bounce heights will get close to 0 inches.

To find a slope m for the model $B = md$, we substitute the coordinates of the data point (30, 26.4) into the equation $B = md$.

$$26.4 = m(30)$$

$$m = \frac{26.4}{30}$$

$$m \approx 0.88$$

The equation is $B = 0.88d$. The model fits the data quite well (see Fig. 30). ▪

In Example 1 we found the model $B = 0.88d$. This equation is an example of *direct variation*.

DEFINITION   *Direct Variation*

If $y = kx$ for some positive constant k, we say that y *varies directly as x*, or that y *is directly proportional to x*. We call k the *variation constant* or the *constant of proportionality*.

Note that a direct variation equation $y = kx$ describes a linear function with slope k and y-intercept $(0, 0)$. In Fig. 31 we sketch three such functions with $k = \frac{1}{2}$, $k = 1$, and $k = 2$.

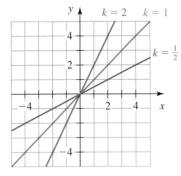

Figure 31 Graphs of $y = kx$ with $k = \frac{1}{2}$, $k = 1$, and $k = 2$

Note that if $k > 0$, then the function $y = kx$ is an increasing line. So, if the value of x increases, the value of y increases.

Changes in Values of Variables for Direct Variation

Assume that y varies directly as x.

- If the value of x increases, then the value of y increases.
- If the value of x decreases, then the value of y decreases.

Note

If we imagine the line that contains the origin (0, 0) and the data point (30, 26.4), it appears that the line might come close to the other data points. If this didn't turn out to be the case, we could try another data point.

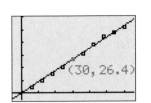

Figure 30 Checking how well $B = 0.88d$ fits the data

For example, consider the golf ball model $B = 0.88d$. As drop heights increase, so do the bounce heights. As drop heights decrease, so do the bounce heights.

In Example 1 we used the function $B = 0.88d$ to model bounce heights for a golf ball. In exercise 44, you will show that the function $B = 0.79d$ is a reasonable model for bounce heights of a racquetball. Since B varies directly as d for a golf ball and a racquetball, it is a reasonable conjecture that B varies directly as d for another type of ball similar in construction such as another golf ball, another racquetball, or perhaps even a tennis ball.

Example 2   Finding a Direct Variation Model

Assume that the bounce height B (in inches) of a tennis ball varies directly as the drop height d (in inches). The bounce height of the tennis ball is 20 inches when dropped from an initial height of 30 inches.

1. Find an equation for B and d.
2. Estimate the bounce height if the drop height is 50 inches.

Solution

1. The equation is of the form $B = kd$. We substitute 30 for d and 20 for B to find the constant k:

$$20 = k(30)$$

$$k = \frac{20}{30}$$

$$k \approx 0.67$$

The equation is $B = 0.67d$.

2. We substitute 50 for d in the equation $B = 0.67d$.

$$B = 0.67(50)$$

$$B = 33.5$$

The bounce height will be 33.5 inches according to the model.   ■

In Example 2 we did not run an experiment and perform the usual modeling steps to find the tennis ball model $B = 0.67d$. Instead, we assumed that the model was of the form $B = kd$ because that is the form of our models for the golf ball and the racquetball. The tennis ball model may or may not be accurate. The only way to know for sure is to run an experiment.

Next, we take a closer look at the significance of $k = 0.88$ for the golf ball model $B = 0.88d$. This equation $B = 0.88d$ tells us that the bounce height is equal to 88% of the drop height. Here we list the values of k and their meanings for the three balls we have investigated:

Type of Ball	Model	Value of k	Height of Bounce
Golf ball	$B = 0.88d$	0.88	88% of drop height
Racquetball	$B = 0.79d$	0.79	79% of drop height
Tennis ball	$B = 0.67d$	0.67	67% of drop height

It is the value of k that takes into account how "bouncy" the ball is.

Example 3   Finding a Direct Variation Equation

Find an equation for x and y if y varies directly as x and $y = 7$ when $x = 3$.

Solution

The equation is of the form $y = kx$. We find the constant k by substituting 3 for x and 7 for y:

$$y = kx$$
$$7 = k(3)$$
$$k = \frac{7}{3}$$

The equation is $y = \frac{7}{3}x$. ■

So far we have described functions where the dependent variable varies directly as the independent *variable*. Here, for positive constant k, we describe functions where the dependent variable varies directly as an *expression*:

$$y = kx^2 \qquad \text{y varies directly as x^2.}$$
$$y = k\sqrt{x} \qquad \text{y varies directly as $\sqrt{x}$.}$$
$$y = k\log(x) \qquad \text{y varies directly as $\log(x)$.}$$

So, we use "varies directly" to mean that the dependent variable is equal to a constant times an expression in terms of the independent variable.

We will now discuss another type of relationship between two variables.

DEFINITION *Inverse Variation*

If $y = \dfrac{k}{x}$ for some positive constant k, we say that y *varies inversely as x*, or that y *is inversely proportional to x*. We call k the *variation constant* or the *constant of proportionality*.

Note that an inverse variation equation $y = \dfrac{k}{x}$ describes a special type of rational function. In Fig. 32 we sketch three such functions with $k = 1$, $k = 2$, and $k = 4$.

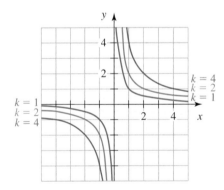

Figure 32 Graphs of $y = \dfrac{k}{x}$ with $k = 1$, $k = 2$, and $k = 4$

Note that if $k > 0$, then the function $y = \dfrac{k}{x}$ is a decreasing function for $x > 0$. This means that if the value of x increases, the value of y decreases.

Changes in Values of Variables for Inverse Variation

Assume that y varies inversely as x. For $x > 0$:

• If the value of x increases, then the value of y decreases.
• If the value of x decreases, then the value of y increases.

Example 4 Finding an Inverse Variation Equation

Find an equation for x and y if y varies inversely as x and $y = 2$ when $x = 6$.

Solution

The equation is of the form $y = \dfrac{k}{x}$. We find the constant k by substituting 6 for x and 2 for y in the equation $y = \dfrac{k}{x}$.

$$2 = \frac{k}{6}$$
$$k = 12$$

The equation is $y = \dfrac{12}{x}$. ■

Example 5   Finding an Inverse Variation Model

If you have ever squeezed a sealed syringe filled with air, you know that it gets harder and harder to squeeze it further. Some volumes in cubic centimeters (cm^3) and corresponding pressures in atmospheres (atm) for a sealed syringe are given in Table 9.

Table 9 Volumes and Pressures in a Syringe

Volume (cm^3)	Pressure (atm)	Volume (cm^3)	Pressure (atm)
3	2.23	12	0.60
4	1.76	13	0.56
5	1.46	14	0.52
6	1.23	15	0.48
7	1.05	16	0.44
8	0.93	17	0.42
9	0.83	18	0.39
10	0.74	19	0.37
11	0.67	20	0.35

Source: *Data collected by the author*

Let P represent the pressure (in atm) in the syringe at volume V (in cm^3).

1. Find an equation for V and P.
2. Estimate at what volume the pressure will be 5 atm.

Solution

1. We draw a scattergram of the data in Fig. 33. Since the points suggest a curve that "bends," we will not use a linear function to model the data. To decide between using an exponential function, a quadratic function, or an inverse variation function, we find equations for each function. To find the constant k for the inverse variation equation, we substitute the data point $(12, 0.60)$ in the equation $P = \dfrac{k}{V}$.

$$0.6 = \frac{k}{12}$$
$$k = 7.2$$

The variation equation is $P = \dfrac{7.2}{V}$. Next, we use a graphing calculator to find the exponential regression equation and the quadratic regression equation:

$P = 2.32(0.90)^V$	Exponential regression equation
$P = 0.0086V^2 - 0.29V + 2.77$	Quadratic regression equation
$P = \dfrac{7.2}{V}$	Inverse variation equation

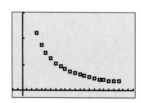

Figure 33 Scattergram of the data

Note

If we imagine an inverse variation curve that contains the data point $(12, 0.60)$, it appears that the curve might come close to the other data points. If this didn't turn out to be the case, we could try another data point.

Next, we see how well each model fits the data (see Figs. 34, 35, and 36).

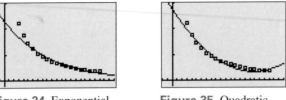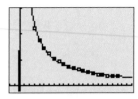

Figure 34 Exponential regression model

Figure 35 Quadratic regression model

Figure 36 Inverse variation model

It appears that the inverse variation model $P = \dfrac{7.2}{V}$ fits the data better than the other two models.

2. We substitute 5 for P in the equation $P = \dfrac{7.2}{V}$ and solve for V.

$$5 = \frac{7.2}{V}$$

$$5V = 7.2$$

$$V = 1.44$$

The pressure is 5 atm when the volume is 1.44 cubic centimeters. ■

Note that as the volume in the syringe increases, the pressure in the syringe decreases. As the volume in the syringe decreases, the pressure in the syringe increases.

It turns out that the equation $P = \dfrac{k}{V}$ is an excellent model for many situations, as long as the temperature and number of molecules in the container are constant. The equation is called **Boyle's law.** The constant k takes into account the temperature and the number of molecules.

Next, for positive constant k, we list some examples of functions in which the dependent variable varies inversely as an *expression*:

$$y = \frac{k}{x^3} \qquad y \text{ varies inversely as } x^3.$$

$$y = \frac{k}{x-5} \qquad y \text{ varies inversely as } x-5.$$

$$y = \frac{k}{2^x} \qquad y \text{ varies inversely as } 2^x.$$

So, we use "varies inversely" to mean that the dependent variable is equal to a positive constant divided by an expression in terms of the independent variable.

Example 6   Finding an Inverse Variation Model

The weight of an object varies inversely as the square of its distance from the center of the Earth. An astronaut weighs 180 pounds at sea level (about 4 thousand miles from the center of the Earth). How much does the astronaut weigh when 2 thousand miles above the surface of the Earth?

Solution

We let w represent the weight (in pounds) of the astronaut when d thousand miles from the center of the Earth. Our desired model is of the form

$$w = \frac{k}{d^2}$$

We substitute 4 for d and 180 for w and solve for k:

$$180 = \frac{k}{4^2}$$

$$k = 2880$$

So, the model is $w = \dfrac{2880}{d^2}$. When the astronaut is 2 thousand miles above the surface of the Earth, we see from Fig. 37 that the astronaut is $4 + 2 = 6$ thousand miles from the center of the Earth. To find w we substitute 6 for d in the equation $w = \dfrac{2880}{d^2}$.

$$w = \frac{2880}{6^2} = 80$$

The astronaut weighs 80 pounds when 2 thousand miles above the surface of the Earth. ■

In Example 7 we will discuss another way to find a variation constant for a direct variation model.

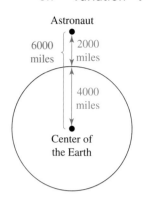

Figure 37   The astronaut is 6000 miles from the center of the Earth

Example 7   Using Ratios to Find a Direct Variation Constant

Use ratios of bounce heights and drop heights to find a golf ball model for the data shown in Example 1.

Solution

From our work in Example 1, we know that we want to find an equation of the form $B = kd$. To find k, we begin by solving the equation $B = kd$ for k:

$$k = \frac{B}{d}$$

Next, we use the first two columns of Table 10 to calculate the ratios $\dfrac{B}{d}$ for the third column.

Table 10 Ratios of Drop and Bounce Heights of a Golf Ball

d	B	$\dfrac{B}{d}$
6	4.8	0.80
12	10.0	0.83
18	15.0	0.83
24	20.3	0.85
30	26.4	0.88
36	31.0	0.86
42	37.5	0.89
48	44.5	0.93
54	47.3	0.88
60	52.0	0.87

According to our model $B = kd$, the ratios $\dfrac{B}{d}$ should be equal to a constant k. The reason that the ratios vary is most likely due to errors in measuring the drop and bounce heights. Other reasons might be that there are imperfections in the golf ball or the surface of the floor.

Next, we find the average of the ratios $\dfrac{B}{d}$ in the third column:

$$\frac{0.80 + 0.83 + 0.83 + 0.85 + 0.88 + 0.86 + 0.89 + 0.93 + 0.88 + 0.87}{10} \approx 0.86$$

Finally, we substitute 0.86 for k in the equation $B = kd$:

$$B = 0.86d$$

Note that the model fits the data reasonably well (see Fig. 38). However, it appears that the model we found in Example 1 fits the data slightly better (see Fig. 39).

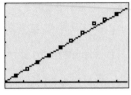

Figure 38 The model
$B = 0.86d$

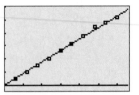

Figure 39 The model
$B = 0.88d$

So, now we have two ways to find a direct variation model. The method shown in Example 1 requires few calculations, but a good selection of an ordered pair is required to find a good value for k. The method shown in Example 7 requires more calculations, but it is not necessary to select an ordered pair to find k.

Example 8 Using Products to Find an Inverse Variation Constant

Use products of volumes and pressures to find a syringe model for the data shown in Example 5.

Solution

From our work in Example 5, we know that we want to find an equation of the form $P = \dfrac{k}{V}$. To find k, we begin by solving the equation $P = \dfrac{k}{V}$ for k:

$$k = VP$$

Next, we use the values of V and P in the first two columns of Table 11 to calculate the products VP for the third column. We also use the values of V and P in the fourth and fifth columns to find the products for the sixth column.

Table 11 Products of Volumes and Pressures in a Syringe					
V	P	VP	V	P	VP
3	2.23	6.69	12	0.60	7.20
4	1.76	7.04	13	0.56	7.28
5	1.46	7.30	14	0.52	7.28
6	1.23	7.38	15	0.48	7.20
7	1.05	7.35	16	0.44	7.04
8	0.93	7.44	17	0.42	7.14
9	0.83	7.47	18	0.39	7.02
10	0.74	7.40	19	0.37	7.03
11	0.67	7.37	20	0.35	7.00

According to our model $P = \dfrac{k}{V}$, the products VP should be equal to a constant k.

The reason that the products vary is most likely due to errors in measuring the volumes and pressures. Another reason might be that there are imperfections in the syringe.

Next, we find the average of the products PV listed in the third and sixth columns:

$$\frac{6.69 + 7.04 + 7.30 + 7.38 + 7.35 + \cdots + 7.00}{18} \approx 7.20$$

Finally, we substitute 7.20 for k in the equation $P = \dfrac{k}{V}$:

$$P = \frac{7.20}{V}$$

Note that this is the same equation that we found in Example 5.

So, we now have two ways to find an inverse variation model. The method shown in Example 5 requires few calculations, but a good selection of an ordered pair must be made to find a good value for k. The method shown in Example 8 requires more calculations, but it is not necessary to select an ordered pair to find k.

EXPLORATION
Looking ahead: Simplifying radical expressions

Recall that $\sqrt[n]{a} = a^{1/n}$. So,

$$\sqrt[5]{a^3} = (a^3)^{1/5} = a^{3 \cdot \frac{1}{5}} = a^{3/5}$$

We say that $\sqrt[5]{a^3}$ is in *radical form* and the expression $a^{3/5}$ is in *exponential form*.

1. Write $\sqrt[7]{a^4}$ in exponential form.

2. Write $\sqrt[9]{a^2}$ in exponential form.

3. Write $\sqrt{a^7}$ in exponential form. [**Hint:** $\sqrt{a}$ is shorthand for $\sqrt[2]{a}$.]

4. Write $\sqrt[n]{a^m}$ in exponential form.

5. Write each of the following in exponential form and then simplify.

   a. $\sqrt{a^2}, \sqrt{a^4}, \sqrt{a^6}, \sqrt{a^8}$

   b. $\sqrt[3]{a^3}, \sqrt[3]{a^6}, \sqrt[3]{a^9}, \sqrt[3]{a^{12}}$

6. Simplify $\sqrt{x^7}$. [**Hint:** $x^7 = x^6 \cdot x$]

7. Simplify $\sqrt{x^{13}}$.

Tips on Succeeding in This Course

Don't wait until the last minute to begin studying for your final exams. Take a look at your finals schedule and decide how you will allocate your time to prepare for each final.

It is important that you are well rested during your finals so you can fully concentrate during your exams. Plan also to do some fun activities that involve exercise, as this is a great way to neutralize stress.

Key Points
OF THIS SECTION

If $y = kx$ for some positive constant k, we say that y *varies directly as x*, or that y *is directly proportional to x*.	**Direct variation**
Assume that y varies directly as x. • If the value of x increases, then the value of y increases. • If the value of x decreases, then the value of y decreases.	**Changes in values for direct variation**
If $y = \dfrac{k}{x}$ for some positive constant k, we say that y *varies inversely as x*, or that y *is inversely proportional to x*.	**Inverse variation**
Assume that y varies inversely as x. For $x > 0$: • If the value of x increases, then the value of y decreases. • If the value of x decreases, then the value of y increases.	**Changes in values for inverse variation**

HOMEWORK 8.7

MathPro 4
MathPro 4

MathPro 5
MathPro 5

SSM
Tutor Center

CDs/Videos

www

Translate each sentence into an equation.

1. w varies directly as t.
2. p varies inversely as s.
3. y varies inversely as the square root of t.
4. z varies directly as u squared.
5. B varies inversely as t cubed.
6. w varies directly as $x + 5$.

Translate the equation with positive constant k by using the phrase "varies directly" or "varies inversely" in a sentence.

7. $w = \dfrac{k}{t}$ 8. $y = ku$ 9. $A = kx^3$

10. $z = \dfrac{k}{\sqrt{t}}$ 11. $y = \dfrac{k}{2^u}$ 12. $y = k\log_3(x)$

Find an equation that meets the given conditions.

13. y varies directly as x, and $y = 6$ when $x = 2$.
14. y varies directly as x, and $y = 12$ when $x = 3$.
15. y varies inversely as x, and $y = 3$ when $x = 5$.
16. y varies inversely as x, and $y = 1$ when $x = 6$.
17. y varies directly as x, and $y = 5.4$ when $x = 2.7$.
18. y varies directly as x, and $y = 6.25$ when $x = 2.27$.
19. p varies inversely as t, and $p = 3.9$ when $t = 5.8$.
20. w varies inversely as u, and $w = 1.35$ when $u = 6.82$.
21. A varies directly as r^2, and $A = 9\pi$ when $r = 3$.
22. s varies directly as $\sqrt{A}$, and $s = 5$ when $A = 4$.
23. H varies directly as $\log_2(t)$, and $H = 12$ when $t = 8$.
24. B varies directly as 3^u, and $B = 35$ when $u = 2$.
25. w varies inversely as $\sqrt[3]{t}$, and $w = 7$ when $t = 8$.
26. C varies inversely as u^4, and $C = 6$ when $u = 2$.
27. A varies inversely as 2^r, and $A = \dfrac{7}{8}$ when $r = 3$.
28. p varies inversely as $\log(t)$, and $p = 3$ when $t = 10$.

29. For a Montgomery County resident, the cost of tuition at Montgomery College varies directly as the number of credit hours a student takes. For academic year 2002–2003, the cost for 15 credit hours is $1185. What is the cost for 12 credit hours?

30. For a Jackson County resident, the cost of tuition at Jackson Community College varies directly as the number of billing contact hours a student takes. The cost for 13 billing contact hours is $793. What is the cost for 15 billing contact hours?

31. When a stone is tied to a string and whirled in a circle with constant speed, the tension in the string varies inversely as the radius of the circle. If the radius is 60 centimeters, the tension is 80 newtons. Find the tension if the radius is 50 centimeters.

32. The current flowing in an electrical circuit at a constant potential varies inversely as the resistance of the circuit. If the current is 25 amperes when the resistance is 4 ohms, what is the current when the resistance is 5 ohms?

33. The distance that an object falls varies directly as the square of the time in motion. If an object falls for 3 seconds, it will fall 144.9 feet. To estimate the height of a cliff, a person drops a stone at the edge of the cliff and measures how long it takes for the stone to reach the base of the cliff. If it takes 3.4 seconds for the stone to reach the base, what is the height of the cliff?

34. A car is traveling at speed s (in mph) on a dry asphalt road and the brakes are suddenly applied. The stopping distance d (in feet) varies directly as the square of the speed s. If a car traveling at 60 mph can stop in 120 feet, what is the stopping distance for a car traveling at 70 mph?

35. A radiation machine can be used to treat a tumor. The intensity of radiation varies inversely as the square of the distance from the machine. If the intensity is 90 milliroentgens per hour (mr/hr) at 2.5 meters, at what distance is the intensity 45 mr/hr?

36. The volume of a sphere varies directly as the cube of its radius. A sphere with a 3-foot radius has a volume of 36π cubic feet. How much air is required to fill a beach ball with radius 1.6 feet?

37. The intensity (in watts per square meter (W/m²)) $I = f(d)$ of a television signal varies inversely as the square of the distance d (in kilometers) from the transmitter. The intensity of a television signal is 30 W/m² at a distance of 2.6 km.

   a. Find an equation for f.

   b. Find $f(1)$, $f(2)$, $f(3)$, and $f(4)$. What do your results mean in terms of the situation?

   c. Use your graphing calculator to sketch a graph of f. What happens to the value of I as the value of d increases, for $d > 0$? What does this mean in terms of the situation?

   d. What can you say about the value of $f(d)$ for an extremely large value of d? Explain. What does this mean in terms of the situation?

38. The force F (in pounds) required to push a couch across a floor varies directly as the weight w (in pounds) of the couch.

   a. A person can push a 120-pound couch across a wood floor by pushing with a force of 50 pounds. Find an equation that describes the relationship between w and F.

   b. How much force is required to push a 150-pound couch across a wood floor?

   c. How would an equation of a model for a *carpeted* floor compare with the model you found in part a? In particular, how would the variation constants compare?

39. The weight (in pounds) $w = f(d)$ of an object varies inversely as the square of its distance (in thousands of miles) d from the center of the Earth.

   a. An astronaut weighs 200 pounds at sea level (about 4 thousand miles from the center of the Earth). Find an equation for f.

   b. How much would the astronaut weigh when 1 thousand miles above the surface of the Earth?

   c. At what distance from the center of the Earth would the astronaut weigh 1 pound?

   d. Use f to estimate how much the astronaut would weigh when on the surface of the moon. The moon is a mean distance of about 239 thousand miles from the Earth. Has model breakdown occurred? Explain.

e. Without finding an equation, discuss how an equation of a model for a 190-pound astronaut would compare with the equation you found in part a. In particular, discuss how the variation constants would compare.

40. The distance d (in miles) that a student travels in a car varies directly as the travel time t (in hours). The student travels 93 miles in 1.5 hours.

   a. Is the student traveling at a constant rate? Explain.

   b. Find an equation that describes the relationship between time and distance.

   c. How far will the student travel in 2 hours?

41. The time T (in seconds) it takes to hear thunder after seeing a lightning strike varies directly as the distance d (in feet) from the lightning strike, assuming that the temperature does not vary much during the storm. In a certain storm it takes 3 seconds to hear the thunder of a lightning strike 3313 feet away.

   a. Find an equation for the relationship between d and T for the storm.

   b. If it takes 4 seconds for a person to hear the thunder from a lightning strike, how far away was the strike?

   c. What does the constant k represent in terms of this situation? Explain. [**Hint**: Recall that slope is the rate of change of the *dependent* variable with respect to the *independent* variable.]

   d. There is a rule of thumb that the number of seconds it takes for a person to hear thunder after seeing a lightning strike is equal to the number of miles that the person is from the strike. Is this rule of thumb a good approximation? If yes, explain. If no, find a better rule of thumb. [**Hint**: There are 5280 feet in 1 mile.]

42. The force F (in pounds) you must exert on a wrench handle to loosen a bolt on a bike varies inversely as the length L (in inches) of the handle. A force of 40 pounds is needed when the handle is 6 inches long.

   a. Find an equation for this relationship between handle length and force.

   b. If an 8-inch long wrench is used to loosen the bolt, how much force would need to be applied to the handle?

   c. Is it easier to use a wrench with a short or long handle? Explain in terms of the equation you found in part a.

43. As you move away from an object such as a car garage, it appears to decrease in height. The author ran an experiment to describe this relationship. To begin, he stood 10 feet away from his garage, held a yardstick one foot away, and measured the image of his garage. The image of the garage had height 16 inches. For this exercise, we call the height of such an image the *apparent height* of the garage. Apparent heights of the garage were collected at various distances from the garage (see Table 12).

   Let $a = f(d)$ be the apparent height (in inches) of the garage when d feet away from the garage.

   a. Use your graphing calculator to draw a scattergram of the data.

   b. Find an equation for f.

   c. Note that your model f is a decreasing function for $d > 0$. Explain why this makes sense in terms of the situation.

Table 12 Apparent Heights of a Car Garage

Distance Away from Garage (feet)	Apparent Height (inches)
10	16.0
20	7.3
30	4.8
40	3.8
50	3.0
60	2.5
70	2.0

Source: *Data collected by the author*

   d. Find the apparent height of the garage from a distance of 100 feet.

   e. Estimate the actual height of the garage. [**Hint**: Think about how the apparent heights were recorded.]

44. A racquetball is dropped from various heights and the bounce height is recorded each time (see Table 13).

Table 13 Drop and Bounce Heights of a Racquetball

Drop Height (inches)	Bounce Height (inches)
6	5.0
12	9.3
18	15.0
24	19.6
30	24.0
36	27.6
42	32.8
48	38.0

Source: *Data collected by the author*

Let $B = f(d)$ represent the bounce height (in inches) of the racquetball after being dropped from an initial height of d inches.

   a. Use your graphing calculator to draw a scattergram of the data.

   b. Find an equation for f.

   c. What is the slope of f? What does it mean in terms of the situation?

   d. What is the B-intercept of f? What does it mean in terms of the situation?

45. When a guitarist picks the second-thickest string on a six-string guitar (the "open 'A' string"), the string vibrates at a frequency of 110 hertz, which means it vibrates 110 times a second. By using a finger to firmly press the "A" string against the fret board, the guitarist shortens the effective length of the string and as a result the frequency increases (and so does the pitch of the note). See Fig. 40.

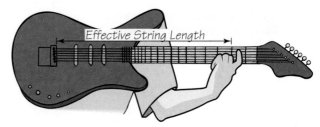

Effective String Length

Figure 40 Exercise 45

We use some of the letters of the alphabet, sometimes in conjunction with the "sharp" symbol #, to refer to these notes. The frequencies for the open "A" string and the next twelve notes are listed in Table 14.

Table 14 Effective Lengths and Frequency of the "A" String

Note	Effective Length of the "A" String (inches)	Frequency (in hertz)
A	25.50	110.0
A#	24.07	116.6
B	22.72	123.5
C	21.44	130.8
C#	20.24	138.6
D	19.10	146.9
D#	18.03	155.6
E	17.02	164.8
F	16.06	174.6
F#	15.16	185.0
G	14.31	196.0
G#	13.51	207.7
A	12.75	220.0

Source: *Math and Music* by *Garland and Kahn, string lengths measured from the author's Fender Squire® guitar*

a. Let F represent the frequency (in hertz) of the "A" string when the effective length is L inches. Draw a scattergram of the data.

b. Find an equation of a reasonable model. Does your model fit the data well?

c. By using a sentence that includes either the phrase "varies directly" or "varies inversely," describe how the variables F and L are related.

d. When the "A" string is vibrating, what is the frequency if the effective length is 7.58 inches?

e. If you halve any effective length of the "A" string, the frequency will double. Use your equation from part b to show that this is true. [**Hint**: First, substitute a for L in your model's equation. Then, substitute $\frac{1}{2}a$ for L in your model's equation.]

46. A pizza with diameter 12 inches ("12-inch pizza") with three toppings weighs 36 ounces at Gabriana's Pizza in San Bruno, California. The weight of a three-toppings pizza varies directly as the square of the diameter of the pizza.

a. Let $W(d)$ represent the weight (in ounces) of a three-toppings pizza with diameter d inches. Find an equation for W.

b. Table 15 lists the prices for various sizes of three-toppings pizzas at Gabriana's Pizza. Let $P(d)$ represent the price (in dollars) of a three-toppings pizza with diameter d inches. Find an equation for P.

Table 15 Prices of Pizzas

Diameter (inches)	Price (dollars)
12	12.20
14	14.80
16	17.40
18	20.00

Source: *Gabriana's Pizza*

c. Let $C(d)$ represent the cost per ounce (in dollars per ounce) of a three-toppings pizza with diameter d inches. Find an equation for C.

d. Use your graphing calculator to sketch a graph of C. Use the "maximum" choice on your calculator to find the highest point of the curve. What does this point represent in terms of the situation?

e. Is the graph of C increasing or decreasing for $d > 6$? What does this mean in terms of the situation?

47. The word *illuminate* means to supply or brighten with light. Obviously, the illumination from a light bulb decreases as you walk away from it. Let I represent the illumination in milliwatts per square centimeter (mW/cm^2) at d centimeters from a 25-watt light bulb. Some values of d and I generated from an experiment are shown in Table 16.

Table 16 Illumination from a Light Bulb

d	I	$d^2 I$
70	0.845	
80	0.677	
90	0.546	
100	0.435	
110	0.349	
120	0.293	
130	0.260	
140	0.214	

Source: *Data collected by the author*

a. Other experiments have shown that the illumination from a light source varies inversely as the square of the distance. Write an equation in terms of d, I, and the variation constant k.

b. Solve the equation that you found in part a for k.

c. Use the first two columns of Table 16 to complete the third column. What is a reasonable value for k? How did you find this value?

d. Substitute your value for k into your equation from part a to find an equation of a model.

e. Draw the graph of your model and a scattergram of the data in the same viewing window. Does your model fit the data well?

f. Estimate the illumination at 160 centimeters from the light bulb.

g. How would an equation of a model for a 50-watt bulb compare with the model you found in part d? In particular, how would the variation constants compare?

48. Do you know what a *pendulum* is? A pendulum can be made by tying one end of some thread to a weight—say, a washer—and attaching the other end of the thread to a surface so that the weight is suspended and can swing freely. The **period** of the pendulum is the amount of time it takes for the weight to swing forward and backward once. Let T represent the period (in seconds) of a pendulum where L is the length (in centimeters) of the thread. Some values of L and T generated from an experiment are shown in Table 17.

a. Other experiments have shown that for a pendulum, the period varies directly as the square root of the length of the thread. Write an equation in terms of L, T, and the variation constant k.

b. Solve the equation that you found in part a for k.

dollars). A graph of f is sketched in Fig. 41.

Table 17 Periods of a Pendulum		
L	T	$\dfrac{T}{\sqrt{L}}$
5.0	0.50	
10.0	0.63	
15.0	0.88	
20.0	1.00	
25.0	1.13	
32.5	1.25	
45.0	1.50	
60.0	1.75	
85.0	2.00	
110.0	2.25	

Source: *Data collected by the author*

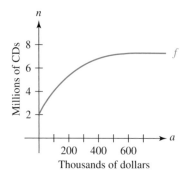

Figure 41 Spending on advertising and CD sales

c. Use the first two columns of Table 17 to complete the third column. What is a reasonable value for k? How did you find this value?

d. Substitute your value for k into your equation from part a.

e. Draw the graph of your model and a scattergram of the data in the same viewing window. Does your model fit the data well?

f. Estimate the period if the length of the thread is 130 inches.

Write a variation equation for the given variables. Also, state the value of the variation constant k.

49. Let $f(L)$ be the area (in square inches) of a rectangle with width 5 inches and length L (in inches).

50. Let $f(s)$ represent the area (in square meters) of a square with side length s (in meters).

51. A group of n people win a total of $2 million from a state lottery drawing. Let $f(n)$ represent each person's share (in millions of dollars).

52. Let $f(s)$ represent the time (in hours) it takes a person to drive 100 miles at a constant speed s (in mph).

53. Let $f(r)$ represent the circumference (in inches) of a circle with radius r (in inches).

54. Let $f(r)$ represent the area (in square feet) of a circle with radius r (in feet).

55. Suppose that the latest Eminem CD is about to be released. Let $n = f(a)$ be the number of CDs (in millions) that will be sold where a is the advertising budget (in thousands of

a. Is f an increasing function? Explain.

b. Does the number of CDs sold vary directly as the amount of money spent on advertising? Explain.

c. Many people say that one quantity varies directly as another quantity when what they mean is that one quantity will increase if the other quantity increases. Compare the meanings of these two statements.

56. For a direct variation equation $y = kx$, we know that as the value of x increases, the value of y increases. When defining the direct variation equation $y = kx$, we state that the constant k must be positive. If k is negative for $y = kx$, what happens to the value of y as the value of x increases?

Decide whether the statement is true or false. Explain.

57. The height of a person varies directly as the person's age.

58. The area of a circle varies directly as the radius of the circle.

59. A person pours a cup of hot coffee. The temperature of the coffee varies inversely as the time since it was poured.

60. A baseball is hit straight up. The height of the baseball varies inversely as the time since it was hit.

61. Assume that y varies directly as x and that the variation constant is k. Does it follow that x varies directly as y? If yes, what is the variation constant? If no, explain. [**Hint:** Solve the equation $y = kx$ for x.]

62. Assume that y varies inversely as x and that the variation constant is k. Does it follow that x varies inversely as y? If yes, what is the variation constant? If no, explain.

Taking it to the lab

For each lab assignment, consult with your instructor on whether to organize your responses as a numbered list or to write them in a paragraph.

LIGHT INTENSITY LAB

In exercise 47 of Homework 8.7, you may have modeled data about the illumination from a light bulb. Such illumination can be measured in milliwatts per square centimeter (mW/cm^2). In this experiment you will study the relationship between the illumination I (mW/cm^2) of a 25-watt light bulb at a distance d (cm).

Materials

You will need the following materials:

1. A 25-watt light bulb
2. Metric tape measure or meter stick
3. Light sensor probe (a Texas Instruments CBL unit and light sensor probe used in conjunction with a TI-82, TI-83, TI-85, or TI-86 graphing calculator works well.)

Preparation

It is best to perform this experiment at night in a room with all other lights turned off and the shades and curtains closed. Place the meter stick so you can measure horizontal distances from the center of the light bulb. Draw any shades, shut all doors, and turn off all lights in the room, except for the 25-watt bulb.

Recording of Data

Record the intensity of the light at 70 cm, 80 cm, 90 cm, and every 10 centimeters thereafter until you reach 140 cm.

Analyzing the Data

1. Your light intensity readings should be adjusted for any small amount of light in the room due to sources other than the 25-watt light bulb. If you are using the CBL unit, there is an initial step that will allow you to input the light intensity of the room with the bulb off. The unit will subtract this value from your light intensity readings. If you don't use a CBL unit, you may have to subtract this small amount of "background" light intensity from your light intensity readings.
2. Complete the second column of Table 18.
3. Create a scattergram of your light data.
4. The illumination from a light bulb varies inversely as the distance squared. Write an equation in terms of d, I, and k.
5. Solve your equation from step 4 for k. Use the first two columns of Table 18 to complete the third column. What is a reasonable value for k? How did you find this value?
6. Substitute your value for k in your equation from step 4. Let $f(d) = I$. Write your equation using the $f(d)$ notation.
7. Use your graphing calculator to draw a graph of f and your scattergram in the same viewing window. Does the function f model the illumination data well?
8. Use the equation for f to estimate the illumination from the bulb at a distance of 150 centimeters. Verify your result graphically.
9. Use the equation for f to estimate the distance at which the illumination from the bulb is 0.1 mW/cm^2. Verify your result graphically.
10. For what values of d is there model breakdown?
11. Explain why it makes sense in terms of illumination that f is a decreasing function for positive values of d.

Table 18 Illuminations from a Light Bulb

d	I	$d^2 I$
70		
80		
90		
100		
110		
120		
130		
140		

BOYLE'S LAW LAB

In Examples 5 and 8 in Section 8.7, we modeled the volume and pressure in a sealed syringe. If you have ever squeezed a sealed syringe filled with air, you know that it gets harder and harder to squeeze it further. That's because as the volume of air in the syringe gets smaller, the pressure of the air inside the syringe increases. In this experiment you will study the relationship between the pressure P in atmospheres (atm) of the air inside a syringe that has been squeezed to a volume of V cubic centimeters (cm^3).

Materials

In order to do this lab you will need the following materials:

1. A Texas Instruments CBL unit
2. A TI-82, TI-83, TI-85, or TI-86 graphing calculator with graph link cable
3. A syringe apparatus (made by the Vernier Company)
4. A CBL-DIN adapter (made by the Vernier Company; Vernier offers the syringe apparatus and CBL-DIN adapter as a package deal.)

Preparation

Follow the instructions that come with the syringe apparatus to set up the equipment. An initial "at rest" volume setting of around 8 cubic centimeters works well. You will have to enter a pressure-gathering program into your CBL unit. (The code for the program is included in the syringe/CBL-DIN package.)

Recording of Data

Record the pressures of the air at volumes 3 cc, 4 cc, 5 cc, and so on until you reach 20 cc. Just get one pressure-reading with each run of the pressure-gathering program.

Analyzing the Data

1. Complete the second and fifth columns of Table 19.

Table 19 Pressures and Volumes in a Syringe

V	P	PV	V	P	PV
3			12		
4			13		
5			14		
6			15		
7			16		
8			17		
9			18		
10			19		
11			20		

2. Create a scattergram of the syringe data.
3. The pressure of air inside a syringe varies inversely as the volume of that air. Write an equation in terms of V, P and k.
4. Solve your equation from step 3 for k. Use the first two columns of Table 19 to complete the third column. Use the fourth and fifth columns to complete the sixth column. What is a reasonable value for k? How did you find this value?
5. Substitute your value for k in your equation from step 3. Let $f(V) = P$. Write your equation using the $f(V)$ notation.
6. Use your graphing calculator to draw a graph of f and your scattergram in the same viewing window. Does the function f model the syringe data well?
7. Use the equation for f to estimate the air pressure in the syringe if the volume is 2 cubic centimeters.

8. Use the equation for f to estimate the volume in the syringe at which air pressure would be 4 atm.

9. Use the equation for f to predict what the air pressure would be if the volume of air in the syringe were squeezed to zero cubic centimeters. Do you think this could be done in reality? Explain.

ROAD TRIP LAB

In this lab you will plan a car trip on which you will drive at least 3000 miles. Let $T(a)$ represent the total amount of time (in hours) behind the wheel if you drive at a mph over all of the speed limits. ($T(a)$ does not include time spent for pit stops, scenic lookouts, or other stops along the way.)

In addition to using a map, you may find it helpful to find distances between cities by logging onto www.mapquest.com. Speed limits can be found at www.hwysafety.org/safety_facts/state_laws/speed_limit_laws.htm.

1. Sketch a map indicating highways and major landmarks along your route. Also list speed limits and related distances for your route.

2. Find an equation for T.

3. What value of the variable a describes your usual driving speed in relation to highway speed limits? Use this value for a and your equation for T to estimate the number of hours that you will be behind the wheel for the trip.

4. Assume that your driving time has a mean of 5 hours per day so that you can sightsee and relax. For how many days will your trip last?

5. How much faster than usual would you have to drive so that you would be behind the wheel for 5 fewer hours than the time you estimated in problem 3?

CHAPTER SUMMARY

Key Points
OF THIS CHAPTER

Throughout these key points, assume that A, B, C, and D are polynomials and that each polynomial appearing in a denominator is nonzero.

Domain of a rational function

The domain of a rational function is the set of real numbers except those that would lead to division by zero.

Asymptote and domain

If a rational function g has a vertical asymptote $x = k$, then k is not in the domain of g. If k is not in the domain of a rational function h, then $x = k$ may or may not be a vertical asymptote of h.

Reducing a rational expression

To reduce a rational expression:
1. Factor the numerator and denominator.
2. Use the property $\frac{AB}{AC} = \frac{B}{C}$, where A is the GCF of the numerator and the denominator.

Multiplying two rational expressions

To multiply rational expressions $\frac{A}{B}$ and $\frac{C}{D}$:
1. Factor the numerators and denominators.
2. Multiply using the property $\frac{A}{B} \cdot \frac{C}{D} = \frac{AC}{BD}$.
3. Simplify.

To divide rational expressions $\dfrac{A}{B}$ and $\dfrac{C}{D}$, where C is nonzero:

1. Rewrite the division as a multiplication using $\dfrac{A}{B} \div \dfrac{C}{D} = \dfrac{A}{B} \cdot \dfrac{D}{C}$.

2. Factor the numerators and denominators.

3. Multiply using the property $\dfrac{A}{B} \cdot \dfrac{D}{C} = \dfrac{AD}{BC}$.

4. Simplify.

Dividing two rational expressions

To add or subtract two rational expressions with different denominators:

1. Factor, if possible, the denominators of the expressions. Identify the LCD.

2. Use the property $\dfrac{a}{a} = 1$ when $a \neq 0$ to write equivalent expressions having the LCD.

3. Add the expressions using the property $\dfrac{A}{B} + \dfrac{C}{B} = \dfrac{A+C}{B}$. Subtract the expressions using the property $\dfrac{A}{B} - \dfrac{C}{B} = \dfrac{A-C}{B}$.

4. Simplify.

Adding or subtracting two rational expressions

To use method 1 to simplify a complex rational expression, first put it in a form where the numerator is a single fraction and the denominator is a single fraction. Then apply the following steps:

$$\dfrac{\dfrac{A}{B}}{\dfrac{C}{D}} = \dfrac{A}{B} \div \dfrac{C}{D} = \dfrac{A}{B} \cdot \dfrac{D}{C} = \dfrac{AD}{BC}$$

Note that C must be nonzero as well as B and D being nonzero.

Simplifying a complex rational expression by method 1

To use method 2 to simplify a complex rational expression, first find the LCD of all fractions appearing in the numerator and denominator. Then multiply by 1 in the form $\dfrac{\text{LCD}}{\text{LCD}}$.

Simplifying a complex rational expression by method 2

To solve a rational equation:

Solving a rational equation

1. Factor, if possible, the denominators.

2. Find the LCD of all fractions.

3. Multiply both sides by the LCD.

4. Solve the resulting simpler equation.

5. Check that each result satisfies the original equation.

A solution of an equation is a *number*, whereas the result of simplifying an expression is an *expression*.

Solution versus simplified expression

When solving an equation, we can multiply both sides of the equation by *any* nonzero expression (including the LCD); however, when simplifying an expression, the only multiplication we can introduce is multiplication by 1.

Solving versus simplifying

If a quantity Q is divided into n parts, the mean amount M of quantity per part is the quotient

Mean

$$M = \dfrac{Q}{n}$$

If an object is moving at a constant speed s for an amount of time t, then the distance traveled d is equal to the product

Distance-speed-time relationship

$$d = st$$

and the time t is found from the rational expression

$$t = \frac{d}{s}$$

Direct variation	If $y = kx$ for some positive constant k, we say that y *varies directly as* x, or that y *is directly proportional to* x.
Changes in values of variables for direct variation	Assume that y varies directly as x. • If the value of x increases, then the value of y increases. • If the value of x decreases, then the value of y decreases.
Inverse variation	If $y = \dfrac{k}{x}$ for some positive constant k, we say that y *varies inversely as* x, or that y *is inversely proportional to* x.
Changes in values of variables for inverse variation	Assume that y varies inversely as x. For $x > 0$: • If the value of x increases, then the value of y decreases. • If the value of x decreases, then the value of y increases.

CHAPTER 8 REVIEW EXERCISES

Find the domain of each rational function.

1. $f(x) = \dfrac{5}{4x^2 - 49}$

2. $f(x) = \dfrac{x^2 - 4}{12x^2 + 13x - 35}$

3. $f(x) = \dfrac{x}{5}$

4. $f(x) = \dfrac{3x + 7}{(x-2)^2 - (x-1)^2}$

Simplify the right side of the equation.

5. $f(x) = \dfrac{3x - 12}{x^2 - 6x + 8}$

6. $f(x) = \dfrac{12 - 6x}{5x - 10}$

7. $f(x) = \dfrac{x^2 - 16}{2x^3 - 16x^2 + 32x}$

8. $f(x) = \dfrac{x + 2}{9x^3 + 18x^2 - x - 2}$

Perform the indicated operation(s).

9. $\dfrac{3x + 6}{2x - 4} \cdot \dfrac{5x - 10}{6x + 12}$

10. $\dfrac{9x - 27}{3x - 33} \cdot \dfrac{x - 11}{6 - 2x}$

11. $\dfrac{4x^2 + 13x + 10}{2x - 6} \cdot \dfrac{3 - x}{x^2 + 7x + 10}$

12. $\dfrac{x^2 - 49}{9 - x^2} \cdot \dfrac{2x^3 + 8x^2 - 42x}{5x - 35}$

13. $\dfrac{x^2 - 4}{x^2 + 3x + 2} \div \dfrac{4x^2 - 24x + 32}{x^2 - 5x + 4}$

14. $\dfrac{4 - x}{4x} \div \dfrac{16 - x^2}{16x^2}$

15. $\dfrac{x^2 - x - 6}{x^2 + 8x + 16} \div \dfrac{x^2 - 5x + 6}{3x^2 + 6x - 24}$

16. $x^2 - 4 \div \dfrac{3x^2 + 6x}{3x^3 - 12x^2 + 12x}$

17. $\dfrac{8x^3 + 4x^2 - 18x - 9}{x^2 - 6x + 9} \div \dfrac{4x^2 + 8x + 3}{x^2 - 9}$

18. $\dfrac{x^2}{x + 1} + \dfrac{3x + 4}{x + 1}$

19. $\dfrac{x}{x^2 + 4x + 4} + \dfrac{2}{3x^2 + 2x - 8}$

20. $\dfrac{x}{x^2 - 5x + 6} + \dfrac{3}{3 - x}$

21. $\dfrac{x - 1}{x^2 - 4} + \dfrac{x + 3}{x^2 - 4x + 4}$

22. $\dfrac{3}{4x - 12} - \dfrac{x}{x^2 - 2x - 3}$

23. $\dfrac{x + 1}{x^2 - 36} - \dfrac{2}{6 - x}$

24. $\dfrac{x + 1}{x^2 - 25} - \dfrac{x - 4}{2x^2 - 14x + 20}$

25. $\dfrac{3x + 2}{16x^2 - 9} - \dfrac{2x - 6}{20x + 15}$

26. $\dfrac{x}{2x^3 - 3x^2 - 5x} + \dfrac{2}{x^3 - x}$

27. $\dfrac{2}{x - 5} - \left(\dfrac{x^2 + 5x + 6}{3x^2 - 75} \div \dfrac{x^2 + 2x}{3x + 15} \right)$

28. $\dfrac{2x - 8}{3x + 4} \cdot \left(\dfrac{x - 2}{x + 1} - \dfrac{x + 3}{x - 4} \right)$

Let $f(x) = \dfrac{x^2 - x - 2}{x^2 + 5x + 6}$ and $g(x) = \dfrac{x + 3}{x + 2}$.

29. Find $f(x)g(x)$.

30. Find $f(x) \div g(x)$.

31. Find $f(x) + g(x)$.

32. Find $f(x) - g(x)$.

33. A student tries to reduce the expression $\dfrac{4x^2 + 1}{6x}$. Show that the work is incorrect.

$$\frac{4x^2 + 1}{6x} = \frac{2x + 1}{3}$$

34. A student tries to find the difference $\dfrac{6x}{x - 9} - \dfrac{2x + 3}{x - 9}$. Describe the error and then perform the subtraction correctly.

$$\frac{6x}{x - 9} - \frac{2x + 3}{x - 9} = \frac{6x - 2x + 3}{x - 9}$$

$$= \frac{4x + 3}{x - 9}$$

Simplify each complex rational expression.

35. $\dfrac{\dfrac{x - 2}{x^2 - 9}}{\dfrac{x^2 - 4}{x + 3}}$

36. $\dfrac{1 + \dfrac{3}{x}}{1 - \dfrac{6}{x}}$

37. $\dfrac{\dfrac{4}{3x^4} - \dfrac{2}{6x^2}}{\dfrac{1}{2x} + \dfrac{1}{4}}$

38. $\dfrac{1 - \dfrac{1}{x + 1}}{1 + \dfrac{1}{x - 1}}$

Solve each equation.

39. $\dfrac{5}{2x} - \dfrac{1}{6} = \dfrac{x}{4x}$

40. $\dfrac{7x + 1}{x^2 - 9} - \dfrac{5}{x - 3} = \dfrac{10}{x + 3}$

41. $\dfrac{1}{x + 5} - \dfrac{2}{x - 2} = \dfrac{-14}{x^2 + 3x - 10}$

42. $\dfrac{x}{x + 2} + \dfrac{3}{x + 4} = \dfrac{14}{x^2 + 6x + 8}$

43. $\dfrac{5}{x} - 3 = \dfrac{4}{x^2}$

44. $\dfrac{x - 3}{2x^2 - 7x - 4} - \dfrac{5}{2x^2 + 3x + 1} = \dfrac{x - 1}{x^2 - 3x - 4}$

Solve or simplify, whichever is appropriate.

45. $\dfrac{3}{2 - x} - \dfrac{7}{x - 2} - 4 = 0$

46. $\dfrac{3}{x^2 - 25} + \dfrac{1}{x^2 - x - 30}$

47. $\dfrac{3}{2 - x} - \dfrac{7}{x - 2} - 4$

48. $\dfrac{3}{x^2 - 25} + \dfrac{1}{x^2 - x - 30} = \dfrac{2}{x^2 - 11x + 30}$

49. Find the x-intercept(s) of $f(x) = \dfrac{x - 7}{x + 1} - \dfrac{x + 3}{x - 4}$.

Translate the equation with positive constant k by using the phrase "varies directly" or "varies inversely" in a sentence.

50. $H = ku^2$

51. $w = \dfrac{k}{\log(t)}$

Find an equation that meets the given conditions.

52. y varies directly as $\sqrt{x}$, and $y = 2$ when $x = 49$.

53. B varies inversely as r^3, and $B = 9$ when $r = 2$.

54. The number of inches of water varies directly as the number of inches of snow. If 20 inches of snow will melt to 2.24 inches of water, to how many inches of water will 37 inches of snow melt?

55. Let m represent the mass (in grams) of a ball bearing with radius r (in cm). Table 20 shows values of r and m for ball bearings of various sizes.

Table 20 Radii and Masses of Ball Bearings		
r	m	$\dfrac{m}{r^3}$
1.0	17.1	
1.2	29.4	
1.4	46.7	
1.6	69.6	
1.8	99.1	
2.0	135.9	

  a. The mass of a ball bearing is directly proportional to the radius cubed. Write an equation in terms of r, m, and k.

  b. Solve the equation that you found in part a for k.

  c. Use the first two columns of Table 20 to complete the third column. What is a reasonable value for k? How did you find this value?

  d. Substitute your value for k into your equation from part a to find an equation of a model.

  e. Draw the graph of your model and a scattergram of the data in the same viewing window. Does your model fit the data well?

  f. What is the mass of a ball bearing with radius 2.3 cm?

56. A hotel offers a one-day rental of a conference room (capacity 120) for a flat fee of $600 plus a per-person charge of $40 for lunch.

  a. Let $C(n)$ represent the total cost (in dollars) of renting the room if n people use the room for one day. Find an equation for C.

  b. Let $M(n)$ represent the mean cost per person (in dollars per person) if n people use the room for one day. Find an equation for M.

  c. Find $M(270)$. Describe your result in terms of the situation.

  d. Find n when $M(n) = 50$. Describe your result in terms of the situation.

57. Until 1997, the percentage of New York prison inmates who are nonviolent felons had been on the rise in New York prisons. To change this trend, New York Governor George E. Pataki's administration started implementing in 1998 a strategy where more nonviolent offenders are granted parole and more violent offenders are denied parole. The number of nonviolent felons and the total number of felons for various years are listed in Table 21.

Year	Number of Nonviolent Felons (thousands)	Total Number of Felons (thousands)
1982	8.3	28.3
1985	10.0	35.0
1988	18.3	45.0
1991	27.5	58.3
1994	32.5	67.5
1997	34.2	70.0

Source: *New York State Department of Correctional Services*

a. Let $N(t)$ represent the number of nonviolent prisoners (in thousands) and $T(t)$ represent the total number of prisoners (in thousands) at t years since 1980. Find formulas for $N(t)$ and $T(t)$.

b. Let $P(t)$ represent the percentage of prisoners who are nonviolent at t years since 1980. Which of the following expressions equals $P(t)$?

$$\frac{T(t)}{N(t)} \quad \frac{N(t)}{T(t)} \quad \frac{T(t)}{N(t)} \cdot 100 \quad \frac{N(t)}{T(t)} \cdot 100 \quad N(t)T(t)$$

c. Substitute the formulas for $N(t)$ and $T(t)$ into your expression from part b to get a formula for $P(t)$ in terms of t.

d. Find $P(10)$. What does your result mean in terms of prisoners?

e. Use a graphing calculator to draw a graph of P, for $t \geq 0$. For $t \geq 0$, is P an increasing or decreasing function? What does this mean in terms of the prisoners?

f. According to the model, P, when would half of the prisoners be nonviolent prisoners?

g. Explain why there may be model breakdown for 1998 and thereafter. [**Hint:** Reread the opening paragraph.]

58. A student plans to drive 75 miles on an undivided highway and another 40 miles on a divided highway. The speed limit for the undivided highway is 50 mph and the speed limit for the divided highway is 65 mph.

a. Let $T(a)$ represent the driving time (in hours) if the student drives at a mph above the speed limits. Find a formula for $T(a)$.

b. Find $T(5)$. What does your result mean in terms of the length of the trip?

c. By how much over the speed limits would the student need to drive for the driving time to be 2 hours? Use a graphing calculator table to verify your result.

CHAPTER 8 TEST

Find the domain of each function.

1. $f(x) = \dfrac{5}{6x^2 + 11x - 10}$

2. $g(x) = \dfrac{2}{72 - 2x^2}$

3. $h(x) = \dfrac{x}{5}$

4. Give examples of three functions each of whose domains are the set of real numbers except -3 and 7.

Simplify the rational expression.

5. $\dfrac{6 - 3x}{x^2 - 5x + 6}$

6. $\dfrac{9x^2 - 1}{18x^3 - 12x^2 + 2x}$

Perform the indicated operation.

7. $\dfrac{5x^4}{3x^2 + 6x + 12} \cdot \dfrac{x^3 - 8}{15x^7}$

8. $\dfrac{x^2 - 25}{2x^3 - 18x^2 + 28x} \div \dfrac{5 - x}{8x - 16}$

9. $\dfrac{5x + 12}{-2x^2 - 8x} - \dfrac{2x + 1}{x^2 + 2x - 8}$

10. $\dfrac{x + 2}{x^2 - 9} + \dfrac{3}{x^2 + 11x + 24}$

11. Perform the indicated operations:

$$\dfrac{3}{x^2 - 2x} \div \left(\dfrac{x}{5x - 10} - \dfrac{x - 1}{x^2 - 4} \right)$$

12. Let $f(x) = \dfrac{x + 1}{x - 5}$ and $g(x) = \dfrac{x - 2}{x + 4}$. Find $f(x) - g(x)$.

13. Simplify $\dfrac{5 + \dfrac{2}{x}}{3 - \dfrac{4}{x - 1}}$.

Solve the equation.

14. $\dfrac{1}{x - 1} - \dfrac{1}{x + 1} = \dfrac{3x}{x^2 - 1}$

15. $\dfrac{1}{x + 3} - \dfrac{x}{x^2 - 9} = \dfrac{2}{3 - x}$

16. Let $f(x) = \dfrac{2}{x - 4} + \dfrac{3}{x + 1}$. Find a when $f(a) = 5$.

Let $f(x) = \dfrac{(x - 5)(x + 2)}{(x - 1)(x + 3)}$.

17. Find $f(-2)$.   18. Find $f(1)$.

19. Find x when $y = 0$.

Find an equation that meets the given conditions.

20. W varies directly as t^2, and $W = 3$ when $t = 7$.

21. y varies inversely as $\sqrt{x}$, and $y = 8$ when $x = 25$.

22. It costs a bike manufacturer about \$200 per bike for the materials to manufacture a line of mountain bikes. It also costs \$10,000 each month for the lease, electricity, salaries, and so on.

a. Let $C(n)$ represent the total monthly cost (in dollars) if n bikes are manufactured that month. Find a formula for C.

b. Let $B(n)$ represent how the manufacturer should price each bike (in dollars) so that the company breaks even by making and selling *n* bikes in a month. Find a formula for *B*.

c. Let $P(n)$ represent how the manufacturer should price each bike (in dollars) so that there is a profit of $150 per bike by making and selling *n* bikes in a month. Find a formula for *P*. Simplify your formula, if possible.

d. Find $P(100)$. What does your result mean in terms of profit from selling bikes?

23. Suppose that a student plans to drive from San Francisco, California to Salt Lake City, Utah. The trip involves 400 miles of travel in California followed by 920 miles in Nevada and Utah. The speed limit is 70 mph in California and 75 mph in Nevada and Utah.

a. Let $T(a)$ represent the driving time (in hours) if the student drives at *a* mph above the speed limits. Find a formula for $T(a)$.

b. Find $T(5)$. What does your result mean in terms of the length of the trip?

c. Find *a* when $T(a) = 17$. What does your result mean in terms of the length of the trip?

24. The frequency $g(L) = F$ (in hertz) of a tuning fork varies inversely as the square of the length *L* (in cm) of the prongs. The frequency is 50 hertz when the length of the prongs is 8 cm.

a. Find an equation for *g*.

b. Find the frequency if the length of the prongs is 6 cm.

c. What is the length of the prongs if the frequency is 200 hertz?

d. Is the graph of *g* increasing or decreasing for $L > 0$? What does this mean in terms of the situation?

Radical Functions

Many of life's failures are people who did not realize how close they were to success when they gave up.
—*Thomas Edison, 1847–1931*

9.1 SIMPLIFYING RADICAL EXPRESSIONS

Objectives

▸ Convert expressions in radical form to exponential form and vice versa.
▸ Know the power property and product property for radicals.
▸ Simplify radical expressions.

In this chapter we work with expressions of the form $\sqrt[n]{x}$, where n is a counting number greater than 1. We say that n is the *index* and the symbol $\sqrt{}$ is the *radical sign*. The notation for square roots, $\sqrt{x}$, is shorthand for $\sqrt[2]{x}$.

For any index n, the expression under the radical sign is called the *radicand*. The radical sign together with the radicand is called a *radical*. An expression written with a radical sign is called a *radical expression*. Here are some examples of radical expressions:

$$5\sqrt{x+1} + 2\sqrt{x+1}, \qquad 2\sqrt[3]{x}(\sqrt[4]{x} + 5), \qquad \frac{5 + \sqrt{7}}{2 - 3\sqrt{7}}$$

Recall from Section 4.2 that for nonnegative a, $\sqrt[n]{a}$ is the nonnegative number whose nth power is a. If a is negative and n is odd, then $\sqrt[n]{a}$ is the (negative) number whose nth power is a. Recall also that both $\sqrt[n]{a}$ and $a^{1/n}$ represent the principal nth root of a.

Example 1 Evaluating Radicals

Evaluate each radical.

1. $\sqrt[4]{81}$ **2.** $\sqrt[3]{-8}$

Solution

1. $\sqrt[4]{81} = 3$ since $3^4 = 81$. **2.** $\sqrt[3]{-8} = -2$ since $(-2)^3 = -8$. ■

Recall from Section 4.2 that if $a^{1/n}$ is defined, then

$$a^{m/n} = \left(a^{1/n}\right)^m = (a^m)^{1/n}$$

or, we can write

$$a^{m/n} = (\sqrt[n]{a})^m = \sqrt[n]{a^m}$$

For example, $x^{3/5} = (\sqrt[5]{x})^3 = \sqrt[5]{x^3}$. We say that the expressions $(\sqrt[n]{x})^m$ and $\sqrt[n]{x^m}$ are in *radical form* and the expression $x^{m/n}$ is in *exponential form*.

Example 2   Exponential and Radical Forms

If the expression is in exponential form, write it in radical form. If it is in radical form, write it in exponential form.

1. $x^{3/7}$ **2.** $\sqrt{x^5}$ **3.** $(3x+1)^{4/5}$ **4.** $\sqrt[8]{(2x+5)^3}$

Solution

1. $x^{3/7} = \sqrt[7]{x^3}$

2. $\sqrt{x^5}$ is shorthand for $\sqrt[2]{x^5}$. So, $\sqrt{x^5} = x^{5/2}$.

3. $(3x+1)^{4/5} = \sqrt[5]{(3x+1)^4}$

4. $\sqrt[8]{(2x+5)^3} = (2x+5)^{3/8}$ ∎

It will be helpful, at times, to write a radical expression as an exponential expression so that we can apply exponential properties to simplify the expression.

Example 3   Simplifying Radical Expressions

Simplify the expression.

1. $\sqrt[3]{x^3}$ **2.** $\sqrt[3]{x^{12}}$ **3.** $\sqrt[3]{x^{21}}$

Solution

1. $\sqrt[3]{x^3} = x^{3/3} = x^1 = x$ **2.** $\sqrt[3]{x^{12}} = x^{12/3} = x^4$

3. $\sqrt[3]{x^{21}} = x^{21/3} = x^7$ ∎

From problems 1–3 of Example 3, we see that we can eliminate the radical sign for expressions of the form $\sqrt[3]{x^k}$ if k is a multiple of 3. If $\sqrt[n]{x}$ is defined, we can eliminate the radical sign for $\sqrt[n]{x^k}$ if k is a multiple of the index n. In this case, we say x^k is a *perfect nth power*.

For example, perfect 5th powers are x^5, x^{10}, x^{15}, x^{20}, …. Furthermore, we can find the 5th root of these expressions. For example,

$$\sqrt[5]{x^{15}} = x^{15/5} = x^3$$

In Section 7.1 we studied the product property for square roots:

$$\sqrt{ab} = \sqrt{a}\sqrt{b}$$

where $a \geq 0$ and $b \geq 0$. The more general product property for radicals

$$\sqrt[n]{ab} = \sqrt[n]{a}\sqrt[n]{b}$$

is true for any counting number index n greater than 1, where $\sqrt[n]{a}$ and $\sqrt[n]{b}$ are defined. Here is the proof:

$$\sqrt[n]{ab} = (ab)^{1/n} \qquad \text{Write in exponential form.}$$
$$= a^{1/n}b^{1/n} \qquad (ab)^m = a^m b^m$$
$$= \sqrt[n]{a}\sqrt[n]{b} \qquad \text{Write in radical form.}$$

Product Property for Radicals

If $\sqrt[n]{a}$ and $\sqrt[n]{b}$ are defined, then

$$\sqrt[n]{ab} = \sqrt[n]{a}\sqrt[n]{b}$$

We can say the nth root of a product is equal to the product of the nth roots.

We use the product property to simplify some radical expressions with index n. If possible, we write the radicand as a product of an expression and one or more perfect nth powers. Then, we apply the product property.

For example, consider $\sqrt{x^7}$, where $x \geq 0$. The radicand x^7 is not a perfect 2nd power (perfect square), but we can write $x^7 = x^6 \cdot x$ and x^6 *is* a perfect square:

$$\sqrt{x^7} = \sqrt{x^6 \cdot x} \qquad x^6 \text{ is a perfect square.}$$
$$= \sqrt{x^6}\sqrt{x} \qquad \sqrt[n]{ab} = \sqrt[n]{a}\,\sqrt[n]{b}$$
$$= x^3\sqrt{x}$$

Example 4   Simplifying Radical Expressions

Simplify the expressions. Assume that all variables are nonnegative.

1. $\sqrt{18}$ 2. $\sqrt{25x^6}$ 3. $\sqrt{x^{13}}$

4. $\sqrt{8x^{11}}$ 5. $\sqrt{48x^4y^{13}}$ 6. $\sqrt{(5x+3)^9}$

Solution

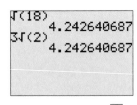

Figure 1 Computing $\sqrt{18}$ and $3\sqrt{2}$

1. $\sqrt{18} = \sqrt{9 \cdot 2}$ 9 is a perfect square.
   $\qquad = \sqrt{9}\sqrt{2}$ $\sqrt[n]{ab} = \sqrt[n]{a}\,\sqrt[n]{b}$
   $\qquad = 3\sqrt{2}$

We use a calculator to confirm that $\sqrt{18}$ and $3\sqrt{2}$ are equal (see Fig. 1).

2. $\sqrt{25x^6} = \sqrt{25}\sqrt{x^6}$ $\sqrt[n]{ab} = \sqrt[n]{a}\,\sqrt[n]{b}$
   $\qquad = 5x^3$

We verify our work by entering the functions $y = \sqrt{25x^6}$ and $y = 5x^3$ in a graphing calculator and comparing graphs (see Fig. 2).

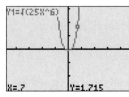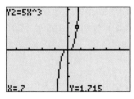

Figure 2 Compare the graphs for $x \geq 0$

Recall that we assumed that $x \geq 0$. Note that for these values of x, the graphs appear to be the same. We will investigate working with radical expressions where variables represent negative numbers at the end of this section.

3. $\sqrt{x^{13}} = \sqrt{x^{12} \cdot x}$ x^{12} is a perfect square.
   $\qquad = \sqrt{x^{12}}\sqrt{x}$ $\sqrt[n]{ab} = \sqrt[n]{a}\,\sqrt[n]{b}$
   $\qquad = x^6\sqrt{x}$

4. $\sqrt{8x^{11}} = \sqrt{4 \cdot 2 \cdot x^{10} \cdot x}$ 4 and x^{10} are perfect squares.
   $\qquad = \sqrt{4 \cdot x^{10} \cdot 2x}$ Rearrange factors.
   $\qquad = \sqrt{4}\sqrt{x^{10}}\sqrt{2x}$ $\sqrt[n]{abc} = \sqrt[n]{a}\,\sqrt[n]{b}\,\sqrt[n]{c}$
   $\qquad = 2x^5\sqrt{2x}$

Note

You will show that
$\sqrt[n]{abc} = \sqrt[n]{a}\,\sqrt[n]{b}\,\sqrt[n]{c}$ in exercise 79.

5. $\sqrt{48x^4y^{13}} = \sqrt{16 \cdot 3 \cdot x^4 \cdot y^{12} \cdot y}$ 16, x^4, and y^{12} are perfect squares.
   $\qquad = \sqrt{16 \cdot x^4 \cdot y^{12} \cdot 3y}$
   $\qquad = \sqrt{16}\sqrt{x^4}\sqrt{y^{12}}\sqrt{3y}$
   $\qquad = 4x^2y^6\sqrt{3y}$

6. $\sqrt{(5x+3)^9} = \sqrt{(5x+3)^8(5x+3)}$ $\qquad$ $(5x+3)^8$ is a perfect square.

$\qquad\qquad\quad = \sqrt{(5x+3)^8}\sqrt{5x+3}$

$\qquad\qquad\quad = (5x+3)^4\sqrt{5x+3}$ $\qquad\qquad$ ■

In Example 4 we simplified square root expressions. In Example 5 we simplify radical expressions with index n greater than 2.

Example 5 Simplifying Radical Expressions

Simplify the expressions. Assume that all variables are nonnegative.

1. $\sqrt[3]{40x^{17}}$ $\qquad$ **2.** $\sqrt[4]{80x^{20}y^{15}}$ $\qquad$ **3.** $\sqrt[5]{(2x+7)^{34}}$

Solution

1. $\sqrt[3]{40x^{17}} = \sqrt[3]{8 \cdot 5 \cdot x^{15} \cdot x^2}$ $\qquad$ 8 and x^{15} are perfect 3rd powers (*perfect cubes*).

$\qquad\qquad\quad = \sqrt[3]{8 \cdot x^{15} \cdot 5x^2}$ $\qquad$ Rearrange factors.

$\qquad\qquad\quad = \sqrt[3]{8}\sqrt[3]{x^{15}}\sqrt[3]{5x^2}$ $\qquad$ $\sqrt[n]{abc} = \sqrt[n]{a}\sqrt[n]{b}\sqrt[n]{c}$

$\qquad\qquad\quad = 2x^5\sqrt[3]{5x^2}$

We verify our work by comparing graphing calculator tables for $y = \sqrt[3]{40x^{17}}$ and $y = 2x^5\sqrt[3]{5x^2}$ (see Fig. 3).

Figure 3 Compare tables for $y = \sqrt[3]{40x^{17}}$ and $y = 2x^5\sqrt[3]{5x^2}$

Graphing Calculator

Instructions for $\sqrt[3]{x}$: Press $\boxed{\text{MATH}}$

4. Then press $\boxed{\text{X, T, }\Theta,n}\boxed{)}$.

2. $\sqrt[4]{80x^{20}y^{15}} = \sqrt[4]{16 \cdot 5 \cdot x^{20} \cdot y^{12} \cdot y^3}$ $\qquad$ 16, x^{20}, and y^{12} are perfect 4th powers.

$\qquad\qquad\quad = \sqrt[4]{16 \cdot x^{20} \cdot y^{12} \cdot 5y^3}$ $\qquad$ Rearrange factors.

$\qquad\qquad\quad = \sqrt[4]{16}\sqrt[4]{x^{20}}\sqrt[4]{y^{12}}\sqrt[4]{5y^3}$

$\qquad\qquad\quad = 2x^5y^3\sqrt[4]{5y^3}$

Graphing Calculator

Instructions for $\sqrt[4]{x}$: Press 4 $\boxed{\text{MATH}}$ 5. Then press $\boxed{\text{X, T, }\Theta,n}$.

3. $\sqrt[5]{(2x+7)^{34}} = \sqrt[5]{(2x+7)^{30}(2x+7)^4}$ $\qquad$ $(2x+7)^{30}$ is a perfect 5th power.

$\qquad\qquad\quad = \sqrt[5]{(2x+7)^{30}}\sqrt[5]{(2x+7)^4}$

$\qquad\qquad\quad = (2x+7)^6\sqrt[5]{(2x+7)^4}$ $\qquad\qquad$ ■

Consider the radical expression $\sqrt[8]{x^6}$, $x \geq 0$. Although the radicand x^6 does not have factors that are perfect 8th powers, we can write the radical with a smaller index:

$$\sqrt[8]{x^6} = x^{6/8} = x^{3/4} = \sqrt[4]{x^3}$$

Now consider the expression $\sqrt[n]{x^m} = x^{m/n}$, where $\sqrt[n]{x}$ is defined. Note that we can write the radical expression $\sqrt[n]{x^m}$ with a smaller index if we can reduce the fraction $\dfrac{m}{n}$.

Simplifying a radical expression includes writing the result with as small an index as possible.

Example 6 Simplifying Radical Expressions

Simplify the expression. Assume that $x \geq 0$.

1. $\sqrt[12]{(3x+7)^8}$ $\qquad$ **2.** $\sqrt[10]{4}$ $\qquad$ **3.** $\sqrt[4]{81x^6}$

Solution

1. $\sqrt[12]{(3x+7)^8} = (3x+7)^{8/12}$ Write in exponential form.

                   $= (3x+7)^{2/3}$ Reduce the exponent.

                   $= \sqrt[3]{(3x+7)^2}$ Write in radical form.

2. We note that $4 = 2^2$.

     $\sqrt[10]{4} = \sqrt[10]{2^2}$

         $= 2^{2/10}$ Write in exponential form.

         $= 2^{1/5}$ Reduce the exponent.

         $= \sqrt[5]{2}$ Write in radical form.

3. $\sqrt[4]{81x^6} = \sqrt[4]{81x^4x^2}$ 81 and x^4 are perfect 4th powers.

          $= \sqrt[4]{81}\sqrt[4]{x^4}\sqrt[4]{x^2}$

          $= 3x\sqrt[4]{x^2}$

          $= 3x \cdot x^{2/4}$ Write in exponential form.

          $= 3x \cdot x^{1/2}$ Reduce the exponent.

          $= 3x\sqrt{x}$ Write in radical form.

We verify our work by comparing graphing calculator tables for $y = \sqrt[4]{81x^6}$ and $y = 3x\sqrt{x}$ for $x \geq 0$ (see Fig. 4).

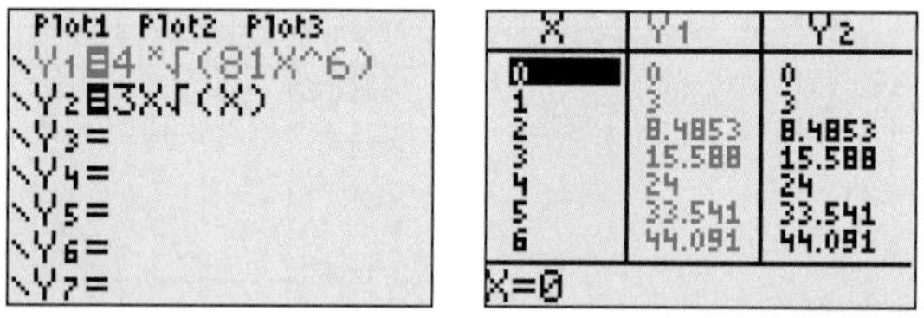

Figure 4 Verify the work

Simplifying a Radical Expression

To simplify a radical expression with index n:

1. Find perfect nth power factors of the radicand.

2. Apply the product property for radicals.

3. Find the nth root of each perfect nth power.

4. Write the radical with as small an index as possible.

Here we consider the expression $\sqrt[n]{x^n}$ where x is negative. First, we compare $\sqrt{x^2}$ for $x = -4$ and $x = 4$:

$$\sqrt{(-4)^2} = \sqrt{16} = 4$$

$$\sqrt{4^2} = \sqrt{16} = 4$$

So when -4 or 4 is substituted for x in the expression $\sqrt{x^2}$, we get the same result, namely 4. This is precisely what happens if we substitute -4 or 4 in the absolute value expression $|x|$:

$$|-4| = 4$$

$$|4| = 4$$

Note

To review absolute value, see Section A.3.

This suggests that $\sqrt{x^2} = |x|$. It turns out that $\sqrt[n]{x^n} = |x|$ for even index n.

Here are some examples where n is odd:

$$\sqrt[3]{4^3} = \sqrt[3]{64} = 4 \qquad\qquad \sqrt[5]{2^5} = \sqrt[5]{32} = 2$$

$$\sqrt[3]{(-4)^3} = \sqrt[3]{-64} = -4 \qquad \sqrt[5]{(-2)^5} = \sqrt[5]{-32} = -2$$

These examples suggest that $\sqrt[n]{x^n} = x$ if n is odd.

Power Property for Radicals

Let n be a counting number greater than 1.

- If n is even, then $\sqrt[n]{x^n} = |x|$.
- If n is odd, then $\sqrt[n]{x^n} = x$.

For example, $\sqrt[7]{(-3)^7} = -3$ and $\sqrt[8]{(-5)^8} = |-5| = 5$.

EXPLORATION
Index property for radicals

Assume that $x \geq 0$.

1. Write the expression with as small an index as possible. [**Hint**: First write the expression in exponential form.]

   a. $\sqrt[8]{x^6}$ b. $\sqrt[6]{x^4}$ c. $\sqrt[16]{x^{10}}$

   d. $\sqrt[9]{x^3}$ e. $\sqrt[22]{x^4}$ f. $\sqrt[30]{x^{25}}$

2. Let k, m, and n be counting numbers, where m and n have no common factors. Write the expression $\sqrt[kn]{x^{km}}$ with as small an index as possible. Include in your work the exponential form of the expression $\sqrt[kn]{x^{km}}$.

3. Describe what your result from problem 2 tells you about simplifying a radical expression. Use this observation to simplify $\sqrt[20]{x^{16}}$ in one step.

EXPLORATION
Looking ahead: Combining like radicals

1. Use factoring to explain why $2x + 3x = 5x$.

2. Use factoring to simplify the expression $7\sqrt{x} - 4\sqrt{x}$. Then discuss how you can simplify $7\sqrt{x} - 4\sqrt{x}$ in one step.

3. A student tries to simplify $5\sqrt{x} + 2$.

   $$5\sqrt{x} + 2 = 7\sqrt{x}$$

   Is the work correct? If yes, use factoring to show why. If no, use graphing calculator tables to show that the work is incorrect.

4. Simplify each expression, if possible. If you cannot simplify it further, explain why not.

   a. $2\sqrt[3]{x} + 4\sqrt[3]{x}$ b. $4\sqrt[3]{x} + 5\sqrt{x}$ c. $3\sqrt{x} + 7\sqrt{x+1}$

5. Summarize your findings from this exploration. In particular, discuss when you can further simplify the sum or difference of two radical expressions.

Tips on Succeeding in This Course

Do you have trouble memorizing definitions and properties? If so, try writing a word or phrase on one side of a 3-by-5 card and putting its definition or property statement and how it can be applied on the other side. For example, you could write "product property for radicals" on one side of a card and "$\sqrt[n]{ab} = \sqrt[n]{a}\,\sqrt[n]{b}$, where n is a counting number and $\sqrt[n]{a}$ and $\sqrt[n]{b}$ are defined" on the other side. You could also describe the meaning of the property and how you can apply it, in your own words.

(Continued)

Once you have completed a card for each definition and property, you can shuffle the cards and quiz yourself until you are confident that you know the definitions and properties and how to apply them. It is a good idea to quiz yourself again later to make sure that you have retained the information.

In addition to memorizing definitions and properties, it is important that you continue to strive to understand their meanings and how to apply them.

Key Points OF THIS SECTION

Product property	If $\sqrt[n]{a}$ and $\sqrt[n]{b}$ are defined, then $\sqrt[n]{ab} = \sqrt[n]{a}\,\sqrt[n]{b}$.
Power property	Let n be a counting number greater than 1. • If n is even, then $\sqrt[n]{x^n} = |x|$. • If n is odd, then $\sqrt[n]{x^n} = x$.
Simplifying a radical expression	To simplify a radical expression with index n: **1.** Find perfect nth power factors of the radicand. **2.** Apply the product property for radicals. **3.** Find the nth root of each perfect nth power. **4.** Write the radical with as small an index as possible.

HOMEWORK 9.1

SSM
Tutor Center

MathPro 4
MathPro 4

MathPro 5
MathPro 5

CDs/Videos

www

If an expression is in exponential form, write it in radical form. If it is in radical form, write it in exponential form.

1. $\sqrt[5]{x^2}$

2. $x^{3/8}$

3. $\sqrt[4]{x^3}$

4. $\sqrt[9]{x^5}$

5. $\sqrt{x}$

6. $\sqrt[3]{x}$

7. $(2x + 9)^{3/7}$

8. $(5x + 1)^{5/6}$

9. $\sqrt[7]{(3x + 2)^4}$

10. $\sqrt[6]{8x + 3}$

11. $(7x + 4)^{1/2}$

12. $(6x + 1)^{3/4}$

Simplify each expression. Assume that each variable is non-negative.

13. $\sqrt{49}$

14. $\sqrt{81}$

15. $\sqrt{50}$

16. $\sqrt{20}$

17. $\sqrt{x^2}$

18. $\sqrt{x^4}$

19. $\sqrt{x^8}$

20. $\sqrt{x^{18}}$

21. $\sqrt{36x^6}$

22. $\sqrt{4x^4}$

23. $\sqrt{5x^2}$

24. $\sqrt{7x^{10}}$

25. $\sqrt{x^9}$

26. $\sqrt{x^{15}}$

27. $\sqrt{24x^5}$

28. $\sqrt{12x^{13}}$

29. $\sqrt{80x^3 y^8}$

30. $\sqrt{27x^{10}y^7}$

31. $\sqrt{200x^3 y^5}$

32. $\sqrt{75x^{15}y^9}$

33. $\sqrt{(2x + 5)^8}$

34. $\sqrt{(3x + 4)^2}$

35. $\sqrt{(6x + 3)^5}$

36. $\sqrt{(7x + 1)^{13}}$

37. $\sqrt[3]{27}$

38. $\sqrt[5]{32}$

39. $\sqrt[6]{x^6}$

40. $\sqrt[9]{x^9}$

41. $\sqrt[3]{8x^3}$

42. $\sqrt[4]{16x^4}$

43. $\sqrt[5]{32x^5}$

44. $\sqrt[3]{27x^3}$

45. $\sqrt[4]{81x^{12}}$

46. $\sqrt[3]{27x^{15}}$

47. $\sqrt[6]{x^{17}}$

48. $\sqrt[9]{x^{73}}$

49. $\sqrt[3]{125x^{17}}$

50. $\sqrt[4]{81x^{21}}$

51. $\sqrt[5]{64x^{40}y^7}$

52. $\sqrt[4]{32x^{19}y^{13}}$

53. $\sqrt[5]{(6xy)^5}$

54. $\sqrt[7]{(4x^2)^7}$

55. $\sqrt[4]{(3x + 6)^4}$

56. $\sqrt[8]{(5x + 2)^8}$

57. $\sqrt[5]{(4x + 7)^{20}}$

58. $\sqrt[3]{(3x + 5)^{12}}$

59. $\sqrt[4]{(x + 7)^{24}}$

60. $\sqrt[7]{(x + 3)^{21}}$

61. $\sqrt[6]{(2x + 9)^{31}}$

62. $\sqrt[5]{(4x + 5)^{43}}$

Simplify each expression. Remember to write your result with as small an index n as possible. Assume that $x \geq 0$.

63. $\sqrt[8]{x^6}$

64. $\sqrt[6]{x^3}$

65. $\sqrt[6]{x^4}$

66. $\sqrt[9]{x^6}$

67. $\sqrt[12]{(2x + 7)^{10}}$

68. $\sqrt[21]{(3x + 5)^{14}}$

69. $\sqrt[6]{x^{14}}$

70. $\sqrt[8]{x^{22}}$

71. $\sqrt[6]{27}$

72. $\sqrt[8]{25}$

73. $\sqrt[10]{16x^8}$

74. $\sqrt[12]{125x^9}$

75. The distance d (in miles) to the horizon at an altitude h feet above sea level is given by the equation

$$d = \sqrt{\frac{3h}{2}}$$

a. Simplify the right side of the horizon-distance equation. Use a graphing calculator table or graph to verify your result.

b. The Sears Tower in Chicago is a 1450-foot-tall skyscraper, the third tallest habitable building in the world. What would be the distance to the horizon from the top of the skyscraper? Assume that the base of the building is at sea level.

c. If an airplane flies over the skyscraper at an altitude of 30,000 feet, what is the distance to the horizon from the airplane?

76. If a car is traveling on a dry asphalt road at a speed of s miles per hour, the car will skid about d feet upon sudden braking. The relationship between d and s is described by the model

$$s = \sqrt{30d}$$

a. A motorist who was involved in an accident claims that she was driving at the posted speed limit of 50 miles per hour. A police officer measures the motorist's car skid marks to be 120 feet long. Assuming that the motorist suddenly applied the brakes, estimate the speed at which the motorist was traveling before braking.

Honestly officer, this guy came out of nowhere and hit me!

b. If the motorist first lightly applied the brakes and then forcefully applied the brakes after a few seconds, explain why your result in part b is an underestimate of the speed at which the motorist was traveling before braking.

77. A student says that $\sqrt{x^{16}}$ is equal to x^4, since $\sqrt{16} = 4$. What would you tell this student?

78. Is the statement $\sqrt[n]{a+b} = \sqrt[n]{a} + \sqrt[n]{b}$ true or false, for all values of a and b where $\sqrt[n]{a}$ and $\sqrt[n]{b}$ are defined? [**Hint:** Substitute values for a, b, and n in the two expressions and compare the results.]

79. Show that the following statement is true:

$$\sqrt[n]{abc} = \sqrt[n]{a}\,\sqrt[n]{b}\,\sqrt[n]{c}$$

where $\sqrt[n]{a}$, $\sqrt[n]{b}$, and $\sqrt[n]{c}$ are defined. [**Hint:** Use the product property twice.]

80. Write the expression $\sqrt[n]{\sqrt[n]{x}}$ with one radical sign. [**Hint:** Write the expression in exponential form.]

81. a. Use a graphing calculator to draw a graph of the given function. What do you notice about the graph? [**Graphing Calculator:** Instructions for $\sqrt[7]{x^7}$: Press 7 [MATH] 5. Then press [X,T,Θ,n] [∧] 7.]

  i. $y = \sqrt[3]{x^3}$
  ii. $y = \sqrt[5]{x^5}$
  iii. $y = \sqrt[7]{x^7}$

b. Compare your graphs in part a with the graph of $y = x$. Explain how your observation relates to the power property.

c. Use a graphing calculator to draw a graph of the given function. What do you notice about the graph?

  i. $y = \sqrt{x^2}$
  ii. $y = \sqrt[4]{x^4}$
  iii. $y = \sqrt[6]{x^6}$

d. Compare your graphs in part c with the graph of $y = |x|$. Explain how your observation relates to the power property. [**Graphing Calculator:** The absolute value choice "abs(" is located in the "NUM" menu. ("NUM" is in the "MATH" menu.)]

82. In this exercise you will explore another version of the power property. Simplify the expression.

  a. $\left(\sqrt{x}\right)^2$ **b.** $\left(\sqrt[3]{x}\right)^3$ **c.** $\left(\sqrt[n]{x}\right)^n$

83. Describe how to simplify a radical expression.

9.2 ADDING, SUBTRACTING, AND MULTIPLYING RADICAL EXPRESSIONS

Objectives

▸ Add, subtract, and multiply radical expressions.

▸ Simplify radical expressions.

In this section we add, subtract, and multiply radical expressions. We add radical expressions by using the distributive law:

$$2\sqrt[3]{x} + 5\sqrt[3]{x} = (2+5)\sqrt[3]{x} \qquad \text{Distributive law}$$
$$= 7\sqrt[3]{x}$$

We say that the radical expressions $2\sqrt[3]{x}$ and $5\sqrt[3]{x}$ are *like radicals* since they have the same index *and* the same radicand. When we add or subtract like radicals, we say that we *combine like radicals*.

Example 1   Combining Like Radicals

Combine like radicals.

1. $3\sqrt{x} + 6\sqrt{x}$ **2.** $4\sqrt[5]{3x} - 2\sqrt[5]{3x}$ **3.** $4\sqrt[3]{x} + 5\sqrt[6]{x}$ **4.** $3\sqrt[4]{x} - 2\sqrt[4]{x+1}$

Solution

1. $3\sqrt{x} + 6\sqrt{x} = (3+6)\sqrt{x}$ Distributive law

   $\qquad\qquad = 9\sqrt{x}$

We verify our work by comparing graphing calculator graphs for $y = 3\sqrt{x} + 6\sqrt{x}$ and $y = 9\sqrt{x}$ for $x \geq 0$ (see Fig. 5).

 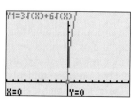 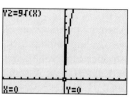

Figure 5 Compare graphs for $y = 3\sqrt{x} + 6\sqrt{x}$ and $y = 9\sqrt{x}$

2. $4\sqrt[5]{3x} - 2\sqrt[5]{3x} = (4-2)\sqrt[5]{3x}$ Distributive law

   $\qquad\qquad\quad = 2\sqrt[5]{3x}$

3. Since the radicals $4\sqrt[3]{x}$ and $5\sqrt[6]{x}$ have different indexes, we can not use the distributive law. $4\sqrt[3]{x} + 5\sqrt[6]{x}$ is already in simplified form.

4. Since the radicals $3\sqrt[4]{x}$ and $2\sqrt[4]{x+1}$ have different radicands, we can not use the distributive law. $3\sqrt[4]{x} - 2\sqrt[4]{x+1}$ is already in simplified form. ∎

Example 2   Performing Operations with Radical Expressions

Perform the indicated operations.

1. $3\sqrt[4]{x} + 4\sqrt{x} + 2\sqrt[4]{x} + 7\sqrt{x}$ 2. $3(5\sqrt[3]{x+1} - 2) - 4\sqrt[3]{x+1}$

Solution

1. $3\sqrt[4]{x} + 4\sqrt{x} + 2\sqrt[4]{x} + 7\sqrt{x}$

   $= (4\sqrt{x} + 7\sqrt{x}) + (3\sqrt[4]{x} + 2\sqrt[4]{x})$ Group the like radicals.

   $= (4+7)\sqrt{x} + (3+2)\sqrt[4]{x}$ Distributive law

   $= 11\sqrt{x} + 5\sqrt[4]{x}$

2. $3(5\sqrt[3]{x+1} - 2) - 4\sqrt[3]{x+1}$

   $= 3 \cdot 5\sqrt[3]{x+1} - 3 \cdot 2 - 4\sqrt[3]{x+1}$ Distributive law

   $= 15\sqrt[3]{x+1} - 4\sqrt[3]{x+1} - 6$ Group the like radicals.

   $= (15-4)\sqrt[3]{x+1} - 6$ Distributive law

   $= 11\sqrt[3]{x+1} - 6$ ∎

Sometimes simplifying radicals will allow us to combine like radicals.

Example 3   Adding or Subtracting Radical Expressions

Perform the indicated operations.

1. $\sqrt{8x} + 3\sqrt{2x}$ 2. $2\sqrt[3]{16x^4} - 4x\sqrt[3]{54x}$

Solution

1. $\sqrt{8x} + 3\sqrt{2x} = \sqrt{4 \cdot 2x} + 3\sqrt{2x}$ 4 is a perfect square.

   $= \sqrt{4}\sqrt{2x} + 3\sqrt{2x}$ $\sqrt[n]{ab} = \sqrt[n]{a}\sqrt[n]{b}$

   $= 2\sqrt{2x} + 3\sqrt{2x}$

   $= (2+3)\sqrt{2x}$ Distributive law

   $= 5\sqrt{2x}$

2. $2\sqrt[3]{16x^4} - 4x\sqrt[3]{54x}$

$\quad = 2\sqrt[3]{8 \cdot x^3 \cdot 2x} - 4x\sqrt[3]{27 \cdot 2x}$ $\qquad$ 8, x^3, and 27 are perfect cubes.

$\quad = 2\sqrt[3]{8}\sqrt[3]{x^3}\sqrt[3]{2x} - 4x\sqrt[3]{27}\sqrt[3]{2x}$ $\qquad$ $\sqrt[n]{ab} = \sqrt[n]{a}\sqrt[n]{b}$

$\quad = 2 \cdot 2 \cdot x \cdot \sqrt[3]{2x} - 4x \cdot 3 \cdot \sqrt[3]{2x}$

$\quad = 4x\sqrt[3]{2x} - 12x\sqrt[3]{2x}$

$\quad = (4x - 12x)\sqrt[3]{2x}$ $\qquad$ Distributive law

$\quad = -8x\sqrt[3]{2x}$ $\qquad$ ■

Next we multiply radical expressions. We will use the product property

$$\sqrt[n]{ab} = \sqrt[n]{a}\sqrt[n]{b}, \qquad \text{where } \sqrt[n]{a} \text{ and } \sqrt[n]{b} \text{ are defined}$$

Here, we multiply $5\sqrt{2x}$ and $4\sqrt{3}$ and simplify the result:

$$5\sqrt{2x} \cdot 4\sqrt{3} = 5 \cdot 4 \cdot \sqrt{2x} \cdot \sqrt{3} \qquad \text{Rearrange factors.}$$
$$= 5 \cdot 4 \cdot \sqrt{2x \cdot 3} \qquad \text{Product property}$$
$$= 20\sqrt{6x}$$

It is good practice to check whether the product of rational expressions can be simplified.

Example 4   Multiplying Radical Expressions

Find the indicated product.

1. $2\sqrt{6x} \cdot 5\sqrt{2x}$ $\qquad$ **2.** $3\sqrt{5x}(4\sqrt{x} - \sqrt{5})$

Solution

1. $2\sqrt{6x} \cdot 5\sqrt{2x} = 2 \cdot 5\sqrt{6x} \cdot \sqrt{2x}$ $\qquad$ Rearrange factors.

$\qquad\qquad\qquad = 2 \cdot 5 \cdot \sqrt{6x \cdot 2x}$ $\qquad$ Product property

$\qquad\qquad\qquad = 10 \cdot \sqrt{12x^2}$

$\qquad\qquad\qquad = 10 \cdot \sqrt{4 \cdot x^2 \cdot 3}$

$\qquad\qquad\qquad = 10 \cdot 2x\sqrt{3}$

$\qquad\qquad\qquad = 20x\sqrt{3}$

We verify our work by comparing graphing calculator tables for $y = 2\sqrt{6x} \cdot 5\sqrt{2x}$ and $y = 20x\sqrt{3}$ for $x \geq 0$ (see Fig. 6).

Figure 6 Compare tables for $y = 2\sqrt{6x} \cdot 5\sqrt{2x}$ and $y = 20x\sqrt{3}$

2. $3\sqrt{5x}(4\sqrt{x} - \sqrt{5}) = 3\sqrt{5x} \cdot 4\sqrt{x} - 3\sqrt{5x} \cdot \sqrt{5}$ $\qquad$ Distributive law

$\qquad\qquad\qquad\qquad = 3 \cdot 4 \cdot \sqrt{5x}\sqrt{x} - 3\sqrt{5x}\sqrt{5}$

$\qquad\qquad\qquad\qquad = 12\sqrt{5x \cdot x} - 3\sqrt{5x \cdot 5}$ $\qquad$ Product property

$\qquad\qquad\qquad\qquad = 12\sqrt{x^2 \cdot 5} - 3\sqrt{25x}$

$\qquad\qquad\qquad\qquad = 12(x\sqrt{5}) - 3(5\sqrt{x})$ $\qquad$ Simplify radicals.

$\qquad\qquad\qquad\qquad = 12x\sqrt{5} - 15\sqrt{x}$ $\qquad$ ■

If $\sqrt[n]{x}$ is defined, then note that

$$(\sqrt[n]{x})^n = x^{n/n} = x^1 = x$$

This property is helpful for simplifying products.

Another Version of the Power Property for Radicals

If $\sqrt[n]{x}$ is defined, then

$$(\sqrt[n]{x})^n = x$$

In particular, we have $(\sqrt{x})^2 = x$, if $x \geq 0$.

Example 5   Expanding Radical Expressions

Find the indicated product.

1. $(2\sqrt{x} - 7)(3\sqrt{x} + 4)$ **2.** $(x - \sqrt{3})^2$ **3.** $(2\sqrt{3x} + 5)(2\sqrt{3x} - 5)$

Solution

1. Multiply each term of the first factor by each term of the second factor and combine like radicals.

$$(2\sqrt{x} - 7)(3\sqrt{x} + 4) = 2\sqrt{x} \cdot 3\sqrt{x} + 2\sqrt{x} \cdot 4 - 7 \cdot 3\sqrt{x} - 7 \cdot 4$$
$$= 6\sqrt{x^2} + 8\sqrt{x} - 21\sqrt{x} - 28$$
$$= 6x - 13\sqrt{x} - 28$$

We can verify our result by comparing tables for $y = (2\sqrt{x} - 7)(3\sqrt{x} + 4)$ and $y = 6x - 13\sqrt{x} - 28$ for $x \geq 0$ (see Fig. 7).

Figure 7 Compare tables for $y = (2\sqrt{x} - 7)(3\sqrt{x} + 4)$ and $y = 6x - 13\sqrt{x} - 28$

2. We expand $(x - \sqrt{3})^2$ using the square of a difference formula.

$$(x - \sqrt{3})^2 = x^2 - 2(x)(\sqrt{3}) + (\sqrt{3})^2 \qquad (a - b)^2 = a^2 - 2ab + b^2$$
$$= x^2 - 2x\sqrt{3} + 3$$

3. We expand $(2\sqrt{3x} + 5)(2\sqrt{3x} - 5)$ using the difference of two squares formula.

$$(2\sqrt{3x} + 5)(2\sqrt{3x} - 5) = (2\sqrt{3x})^2 - 5^2 \qquad (a + b)(a - b) = a^2 - b^2$$
$$= 2^2(\sqrt{3x})^2 - 5^2 \qquad (ab)^2 = a^2b^2$$
$$= 4(3x) - 25$$
$$= 12x - 25$$

We can verify our result by comparing tables for $y = (2\sqrt{3x} + 5)(2\sqrt{3x} - 5)$ and $y = 12x - 25$ for $x \geq 0$. ∎

Note

In general, $(\sqrt{a} + \sqrt{b})^2 \neq a + b$. For example, $(\sqrt{1} + \sqrt{1})^2 \neq 1 + 1$. Because $4 \neq 2$.

Recall that the square of a sum formula states that $(a + b)^2 = a^2 + 2ab + b^2$. In general, $(a + b)^2$ is not equal to $a^2 + b^2$. So, in general, $(\sqrt{a} + \sqrt{b})^2$ is not equal to $a + b$.

Rather, for $a \geq 0$ and $b \geq 0$ we have:

$$(\sqrt{a} + \sqrt{b})^2 = (\sqrt{a})^2 + 2\sqrt{a}\sqrt{b} + (\sqrt{b})^2 = a + 2\sqrt{ab} + b$$

In Example 6 we find the product of two radical expressions with index $n = 4$.

Note

Also, we have $(\sqrt{a} - \sqrt{b})^2 = a - 2\sqrt{ab} + b$.

Example 6   Multiplying Radical Expressions

Find the product $(\sqrt[4]{x^3} + 5)(\sqrt[4]{x^3} - 6)$.

Solution

$$
\begin{aligned}
(\sqrt[4]{x^3} + 5)(\sqrt[4]{x^3} - 6) &= \sqrt[4]{x^3}\sqrt[4]{x^3} - \sqrt[4]{x^3} \cdot 6 + 5 \cdot \sqrt[4]{x^3} - 5 \cdot 6 \\
&= \sqrt[4]{x^3 \cdot x^3} - 6\sqrt[4]{x^3} + 5\sqrt[4]{x^3} - 30 \\
&= \sqrt[4]{x^6} - \sqrt[4]{x^3} - 30 \\
&= x\sqrt[4]{x^2} - \sqrt[4]{x^3} - 30 \qquad \sqrt[4]{x^6} = \sqrt[4]{x^4 \cdot x^2} = x\sqrt[4]{x^2} \\
&= x\sqrt{x} - \sqrt[4]{x^3} - 30 \qquad \sqrt[4]{x^2} = x^{2/4} = x^{1/2} = \sqrt{x}
\end{aligned}
$$

We cannot combine the radicals $\sqrt{x}$ and $\sqrt[4]{x^3}$ since the indexes (and the radicands) are different. So, we are done. ■

If the indexes of two radicals are equal, we find the product of the radicals by using the product property. If the indexes are different and the radicands are equal, we find the product by writing the radicals in exponential form. This way, we can apply properties of exponents. Here, we find the product $\sqrt[3]{x} \cdot \sqrt[4]{x}$.

$$
\begin{aligned}
\sqrt[3]{x} \cdot \sqrt[4]{x} &= x^{\frac{1}{3}} \cdot x^{\frac{1}{4}} && \text{Write in exponential form.} \\
&= x^{\frac{1}{3} + \frac{1}{4}} && a^m a^n = a^{m+n} \\
&= x^{\frac{4}{12} + \frac{3}{12}} && \text{Get a common denominator.} \\
&= x^{\frac{7}{12}} \\
&= \sqrt[12]{x^7} && \text{Write in radical form.}
\end{aligned}
$$

Example 7   Expanding Radical Expressions

Perform the indicated operations. Assume that $x \geq 0$.

1. $2\sqrt{x}(\sqrt[3]{x} - 5)$ **2.** $(\sqrt[3]{x} + 3\sqrt[5]{x^2})^2$

Solution

1.
$$
\begin{aligned}
2\sqrt{x}(\sqrt[3]{x} - 5) &= 2\sqrt{x}\sqrt[3]{x} - 2\sqrt{x} \cdot 5 && \text{Distributive law} \\
&= 2x^{\frac{1}{2}}x^{\frac{1}{3}} - 10\sqrt{x} && \text{Write in exponential form.} \\
&= 2x^{\frac{1}{2} + \frac{1}{3}} - 10\sqrt{x} && a^m a^n = a^{m+n} \\
&= 2x^{\frac{3}{6} + \frac{2}{6}} - 10\sqrt{x} && \text{Get a common denominator.} \\
&= 2x^{\frac{5}{6}} - 10\sqrt{x} \\
&= 2\sqrt[6]{x^5} - 10\sqrt{x} && \text{Write in radical form.}
\end{aligned}
$$

We can verify our result by comparing tables for $y = 2\sqrt{x}(\sqrt[3]{x} - 5)$ and $y = 2\sqrt[6]{x^5} - 10\sqrt{x}$ for $x \geq 0$ (see Fig. 8).

Figure 8 Verify the work

2. $(\sqrt[3]{x} + 3\sqrt[5]{x^2})^2$

$\quad = (\sqrt[3]{x})^2 + 2(\sqrt[3]{x})(3\sqrt[5]{x^2}) + (3\sqrt[5]{x^2})^2 \qquad (a+b)^2 = a^2 + 2ab + b^2$

$\quad = (\sqrt[3]{x})^2 + 6x^{\frac{1}{3}}x^{\frac{2}{5}} + 3^2(\sqrt[5]{x^2})^2 \qquad$ Write in exponential form.

$\quad = \sqrt[3]{x^2} + 6x^{\frac{1}{3}+\frac{2}{5}} + 9\sqrt[5]{(x^2)^2} \qquad (\sqrt[n]{x})^m = \sqrt[n]{x^m}$

$\quad = \sqrt[3]{x^2} + 6x^{\frac{5}{15}+\frac{6}{15}} + 9\sqrt[5]{x^4}$

$\quad = \sqrt[3]{x^2} + 6x^{\frac{11}{15}} + 9\sqrt[5]{x^4}$

$\quad = \sqrt[3]{x^2} + 6\sqrt[15]{x^{11}} + 9\sqrt[5]{x^4} \qquad$ Write in radical form. ■

Simplifying a Radical Expression

To simplify a radical expression:
1. Perform any indicated multiplications.
2. Combine like radicals.
3. Write a radicand as a product of an expression and one or more perfect powers. Then apply the product property for radicals.
4. Write a radical with as small an index as possible.

Depending on the radical expression, we may need to perform these steps in a different order or we may need to return to a step once again at a later stage in the process of simplifying the expression. We will discuss more ways to simplify radical expressions in Section 9.3.

EXPLORATION
Looking ahead: Rationalizing the denominator

In Section 7.1 you "rationalized the denominator" of fractions of the form $\dfrac{1}{\sqrt{a}}$ by finding an equivalent expression that does not have a radical in any denominator. Here, you will explore how to rationalize the denominator of a fraction with a denominator that is a sum or a difference involving radicals.

1. Perform the indicated multiplication.
   a. $(x - \sqrt{2})(x + \sqrt{2})$ b. $(x + \sqrt{5})(x - \sqrt{5})$
   c. $(\sqrt{x} - 4)(\sqrt{x} + 4)$ d. $(\sqrt{x} + 3)(\sqrt{x} - 3)$

2. What patterns do you notice from your work in part 1?

3. Rationalize the denominator of $\dfrac{1}{\sqrt{x} - 7}$ by performing the multiplication

$$\frac{1}{\sqrt{x} - 7} \cdot \frac{\sqrt{x} + 7}{\sqrt{x} + 7}$$

   Use graphing calculator tables to verify your work.

4. Rationalize the denominator of the expression $\dfrac{1}{\sqrt{x} + 5}$.

5. Describe how to rationalize the denominator of a radical expression.

Key Points
OF THIS SECTION

Another version of the power property   If n is a counting number and $\sqrt[n]{x}$ is defined, then $(\sqrt[n]{x})^n = x$.

To add or subtract like radicals, we use the distributive law.

Adding or subtracting like radicals

To multiply radicals that have the same index, we use the product property.

Multiplying radicals with the same index

To multiply two radicals with different indexes and equal radicands, or to find a smaller index for a radical:

Multiplying radicals with different indexes

- Write the radical(s) in exponential form.
- Simplify the exponential expression.
- Write the simplified exponential expression in radical form.

To simplify a radical expression:

Simplifying radical expressions

- Perform any indicated multiplications.
- Combine like radicals.
- Write a radicand as a product of an expression and one or more nth perfect powers. Then apply the product property.
- Write a radical with as small an index as possible.

HOMEWORK 9.2

   SSM Tutor Center   MathPro 4   MathPro 5   CDs/Videos   www

Simplify each expression. Use graphing calculator tables to compare your result and the original expression. Assume that $x \geq 0$.

1. $4\sqrt{x} + 5\sqrt{x}$
2. $8\sqrt{x} - 4\sqrt{x}$

3. $2.3\sqrt{x} - 4.8\sqrt{x}$
4. $7.1\sqrt{x} + 1.7\sqrt{x}$

5. $3\sqrt{5x} + 2\sqrt{3x} - 6\sqrt{3x} + 7\sqrt{5x}$

6. $10\sqrt{2x+1} - 9 - 2\sqrt{2x+1} + 1$

7. $2\sqrt{x} + 5\sqrt[3]{x} - 5\sqrt{x}$

8. $8 - \sqrt[3]{x} + 2 + \sqrt{x} - 4\sqrt[3]{x}$

9. $6\sqrt[3]{x-1} - 3\sqrt[3]{x-1} - 2\sqrt{x-1}$

10. $4\sqrt[7]{3x+1} + 3\sqrt[7]{3x+1}$

11. $5\sqrt{x} + 3\sqrt{x} + 4\sqrt[3]{x} + 2\sqrt[3]{x}$

12. $2\sqrt[5]{x} - 7\sqrt[4]{x} - 5\sqrt[5]{x} - 4\sqrt[4]{x}$

13. $3.7\sqrt[4]{x} - 1.1\sqrt[4]{x} - 4.2\sqrt[6]{x} + 4.2\sqrt[6]{x}$

14. $4.1\sqrt{x} - 2.9\sqrt[3]{x} - 5.8\sqrt[3]{x} + 2.3\sqrt{x}$

15. $3(7 - \sqrt{x} + 2) - (\sqrt{x} + 2)$

16. $4(3\sqrt{x} - 8)$
17. $-5(2\sqrt{x} + 4)$

18. $-2(5\sqrt{x} + 1) - 3\sqrt{x}$
19. $7(\sqrt[3]{x} + 1) - 7(\sqrt[3]{x} - 1)$

20. $13 - 2(5\sqrt[4]{x} + 2)$
21. $\sqrt{25x} + \sqrt{4x}$

22. $\sqrt{8x} - \sqrt{18x}$
23. $3\sqrt{20x} + 2\sqrt{45x}$

24. $\sqrt{27x^7} + 2x^2\sqrt{12x^3}$
25. $5\sqrt{x^3} - x\sqrt{49x}$

26. $\sqrt{9x} - \sqrt{49x}$
27. $3\sqrt{81x^2} - 2\sqrt{100x^2}$

28. $2\sqrt{36x^2} + 5\sqrt{16x^2}$
29. $\sqrt{12x^3} + x\sqrt{75x}$

30. $\sqrt{20x^5} - x\sqrt{45x^3}$
31. $\sqrt[3]{27x^5} - x\sqrt[3]{8x^2}$

32. $\sqrt[3]{54x^3} + x\sqrt[3]{16}$
33. $\sqrt[4]{16x^{11}} - 3x\sqrt[4]{x^7}$

34. $7x^2\sqrt[5]{x^3} + 3\sqrt[5]{32x^{13}}$

Simplify each expression. Verify your result by comparing graphing calculator tables or graphs. Assume that $x \geq 0$.

35. $3\sqrt{x} \cdot 2\sqrt{x}$
36. $-5\sqrt{x} \cdot 4\sqrt{x}$

37. $-2\sqrt{5x} \cdot 4\sqrt{3x}$
38. $-3\sqrt{10x} \cdot 2\sqrt{5x}$

39. $2\sqrt{7x}(\sqrt{7x} + \sqrt{2x})$
40. $4\sqrt{2x}(\sqrt{8} - 3\sqrt{x})$

41. $(6 - 2\sqrt{x})(5\sqrt{x} - 4)$
42. $(3\sqrt{x} + 7)(2\sqrt{x} + 5)$

43. $(2\sqrt{x} + 1)(\sqrt{x} - 4)$
44. $(10\sqrt{x} + 3)(\sqrt{x} - 9)$

45. $(1 - \sqrt{x})(1 + \sqrt{x})$
46. $(2 + 3\sqrt{x})(2 - 3\sqrt{x})$

47. $(\sqrt{x} + 5)(\sqrt{x} - 5)$
48. $(4\sqrt{x} + 3)(4\sqrt{x} - 3)$

49. $(5 + 6\sqrt{x})^2$
50. $(3\sqrt{x} + 2)^2$

51. $(7\sqrt{x} - 1)^2$
52. $(2\sqrt{x} - 3)^2$

53. $(4\sqrt{x} + 5)^2$
54. $(2\sqrt{x} - 4)^2$

55. $(\sqrt{x} + 1)^2$
56. $(\sqrt{x} - 1)^2$

57. $\sqrt{x}\sqrt[5]{x}$
58. $\sqrt[4]{x}\sqrt[6]{x}$

59. $\sqrt[5]{x^4}\sqrt[5]{x^3}$
60. $\sqrt[4]{3x^2}\sqrt[4]{3x^2}$

61. $-5\sqrt{x}(\sqrt[4]{2x} - 4)$
62. $-4\sqrt[3]{x}(\sqrt{x} + 3)$

63. $(\sqrt[3]{x} + 1)^2$
64. $(\sqrt[4]{x} + 5)^2$

65. $(\sqrt[3]{x} - 2)^2$
66. $(\sqrt[5]{x^2} + 3)^2$

67. $(\sqrt[4]{x} + \sqrt[3]{x})^2$
68. $(2\sqrt[5]{x} - \sqrt{x})^2$

69. $(2\sqrt{x} - 6)(3\sqrt[3]{x} + 1)$
70. $(4\sqrt[3]{x^2} + 1)(5\sqrt[4]{x^2} + 2)$

71. $(3\sqrt[4]{x} + 5)(3\sqrt[4]{x} - 5)$
72. $(2\sqrt[5]{x} + 1)(3\sqrt[5]{x} - 2)$

73. The time $f(d)$ (in seconds) that it takes for an object to fall d feet can be modeled by the equation

$$f(d) = \sqrt{\frac{2d}{g}}$$

where g is the constant 32.2 feet per second squared.

a. Find $f(100)$. What does your result mean in terms of a falling object?

b. On December 28, 2002, sky divers jumped off the 1483-foot-tall Petronas Towers (in Malaysia), the tallest habitable buildings in the world. Estimate how long a sky diver was in free fall by finding the time it would take to fall 1483 feet with a closed parachute. Ideally, the parachute does open. In this case, is your estimate an underestimate or an overestimate? Explain.

c. Is f an increasing or decreasing function? Explain why this makes sense in terms of falling objects.

74. The time it takes for a planet to make one revolution around the sun is called the planet's *period*. The period $f(d)$ (in years) of a planet whose average distance from the sun is d million kilometers is modeled by the equation

$$f(d) = 0.0005443\sqrt{d^3}$$

a. What is the period of Pluto, whose average distance from the sun is 5913 million kilometers?

b. Use a graphing calculator table to find the average distance that the Earth is from the sun.

c. Suppose that in the future we colonize Mars, whose average distance from the sun is 228 million kilometers. What is the period of Mars? If a person is 20 years old in "Earth years," how old is the person in "Mars years?"

*Write the expression as a single radical. [**Hint:** Write the expression in exponential form.]*

75. $\dfrac{\sqrt{x}}{\sqrt[3]{x}}$

76. $\sqrt[3]{\sqrt{x}}$

77. Find two radical expressions whose sum is $9\sqrt{x} + 7$ and whose difference is $\sqrt{x} + 3$.

78. Find two radical expressions whose sum is $6\sqrt{x} + 2\sqrt[4]{x} + 2$ and whose difference is $2\sqrt{x} - 8$.

79. a. Write the expression $\sqrt[3]{x}\sqrt[4]{x}$ as a single radical.

b. Write the expression $\sqrt[k]{x}\sqrt[n]{x}$ as a single radical. [**Hint:** Perform steps similar to your work in part a.]

c. Use your result from part b to find the product $\sqrt[3]{x}\sqrt[4]{x}$. Compare your result to your result from part a.

d. Use your result from part b to find the product $\sqrt[5]{x}\sqrt[7]{x}$.

80. We cannot factor $x^2 - 3$ over the integers. We *can* factor $x^2 - 3$ over the real numbers:

$$x^2 - 3 = (x - \sqrt{3})(x + \sqrt{3})$$

a. Factor $x^2 - 2$ over the real numbers.

b. Factor $x^2 - 5$ over the real numbers.

c. Simplify $\dfrac{x^2 - 2}{x - \sqrt{2}}$.

d. Simplify $\dfrac{x^2 - 7}{x + \sqrt{7}}$.

81. Describe how to multiply two radical expressions. Include a discussion of various types of formulas, laws, and techniques you can use to find such products.

9.3 RATIONALIZING DENOMINATORS AND SIMPLIFYING QUOTIENTS OF RADICAL EXPRESSIONS

Objectives

‣ Rationalize the denominator of a radical expression.

‣ Simplify a radical expression.

In Section 9.2 we discussed how to add, subtract, and multiply radical expressions. In this section we discuss how to simplify quotients of radical expressions.

A simplified radical expression does not have radicals in any denominator. Recall from Section 7.1, we rationalize the denominator of $\dfrac{2}{\sqrt{5}}$ by multiplying by $1 = \dfrac{\sqrt{5}}{\sqrt{5}}$:

$$\frac{2}{\sqrt{5}} = \frac{2}{\sqrt{5}} \cdot \frac{\sqrt{5}}{\sqrt{5}}$$

$$= \frac{2\sqrt{5}}{5}$$

Example 1   Rationalizing a Denominator

Simplify the expression $\dfrac{4}{5\sqrt{3x}}$.

Solution

Since $\sqrt{3x} \cdot \sqrt{3x} = 3x$ when $x \geq 0$, we rationalize the denominator of $\dfrac{4}{5\sqrt{3x}}$ by multiplying by $\dfrac{\sqrt{3x}}{\sqrt{3x}}$.

$$\frac{4}{5\sqrt{3x}} = \frac{4}{5\sqrt{3x}} \cdot \frac{\sqrt{3x}}{\sqrt{3x}}$$

$$= \frac{4\sqrt{3x}}{5(\sqrt{3x})^2}$$

$$= \frac{4\sqrt{3x}}{5(3x)}$$

$$= \frac{4\sqrt{3x}}{15x}$$

We verify our work by comparing graphing calculator graphs for $y = \dfrac{4}{5\sqrt{3x}}$ and $y = \dfrac{4\sqrt{3x}}{15x}$ for $x > 0$ (see Fig. 9).

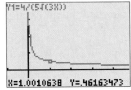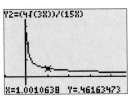

Figure 9 Compare tables for $y = \dfrac{4}{5\sqrt{3x}}$ and $y = \dfrac{4\sqrt{3x}}{15x}$

We can also rationalize the denominator for indexes different from 2. With any index n, our intermediate goal is the same: To write the denominator so that the radicand of its radical is a perfect nth power.

Example 2   Rationalizing Denominators

Simplify the expression.

1. $\dfrac{1}{\sqrt[3]{x}}$ **2.** $\dfrac{3}{\sqrt[5]{8x^2}}$

Solution

1. For the radicand to be a perfect cube, x needs to be multiplied by $x \cdot x = x^2$. So, we multiply $\dfrac{1}{\sqrt[3]{x}}$ by $\dfrac{\sqrt[3]{x^2}}{\sqrt[3]{x^2}}$.

$$\frac{1}{\sqrt[3]{x}} = \frac{1}{\sqrt[3]{x}} \cdot \frac{\sqrt[3]{x^2}}{\sqrt[3]{x^2}} \qquad \text{Rationalize the denominator.}$$

$$= \frac{\sqrt[3]{x^2}}{\sqrt[3]{x^3}}$$

$$= \frac{\sqrt[3]{x^2}}{x} \qquad \sqrt[3]{x^3} = x$$

2. To become a perfect 5th power, the radicand $8x^2 = 2 \cdot 2 \cdot 2 \cdot x \cdot x$ needs to be multiplied by $2 \cdot 2 \cdot x \cdot x \cdot x = 4x^3$.

$$\frac{3}{\sqrt[5]{8x^2}} = \frac{3}{\sqrt[5]{8x^2}} \cdot \frac{\sqrt[5]{4x^3}}{\sqrt[5]{4x^3}} \qquad \text{Rationalize the denominator.}$$

$$= \frac{3\sqrt[5]{4x^3}}{\sqrt[5]{8x^2 \cdot 4x^3}}$$

$$= \frac{3\sqrt[5]{4x^3}}{\sqrt[5]{32x^5}}$$

$$= \frac{3\sqrt[5]{4x^3}}{\sqrt[5]{32}\sqrt[5]{x^5}}$$

$$= \frac{3\sqrt[5]{4x^3}}{2x} \qquad \sqrt[5]{32} = 2, \ \sqrt[5]{x^5} = x \qquad ■$$

To rationalize the denominator of a radical expression of the form $\dfrac{1}{\sqrt[n]{x^m}}$, we multiply the expression by a fraction of the form $\dfrac{\sqrt[n]{x^k}}{\sqrt[n]{x^k}}$ so that the radical in the denominator has a perfect nth power radicand.

A simplified radical expression also does not have any fraction under a radical sign. Recall that in Section 7.1, we used the quotient property for square roots

$$\sqrt{\frac{a}{b}} = \frac{\sqrt{a}}{\sqrt{b}}, \qquad \text{where } a \geq 0 \text{ and } b > 0$$

to simplify expressions such as $\sqrt{\dfrac{2}{9}}$:

$$\sqrt{\frac{2}{9}} = \frac{\sqrt{2}}{\sqrt{9}}$$

$$= \frac{\sqrt{2}}{3}$$

Next, we describe the quotient property for any index n.

Note

You will prove the quotient property for radicals in exercise 68.

Quotient Property for Radicals

If $\sqrt[n]{a}$ and $\sqrt[n]{b}$ are defined, and b is nonzero, then

$$\sqrt[n]{\frac{a}{b}} = \frac{\sqrt[n]{a}}{\sqrt[n]{b}}$$

We can say the nth root of a quotient is equal to the quotient of the nth roots.

We use the quotient property to simplify radical expressions.

Example 3 Simplifying Radical Expressions

Simplify the expression.

1. $\sqrt{\dfrac{5}{x}}$

2. $\sqrt[3]{\dfrac{7}{2x^2}}$

Solution

1. $\sqrt{\dfrac{5}{x}} = \dfrac{\sqrt{5}}{\sqrt{x}}$ Quotient property

   $= \dfrac{\sqrt{5}}{\sqrt{x}} \cdot \dfrac{\sqrt{x}}{\sqrt{x}}$ Rationalize the denominator.

   $= \dfrac{\sqrt{5x}}{x}$

2. $\sqrt[3]{\dfrac{7}{2x^2}} = \dfrac{\sqrt[3]{7}}{\sqrt[3]{2x^2}}$ Quotient property

   $= \dfrac{\sqrt[3]{7}}{\sqrt[3]{2x^2}} \cdot \dfrac{\sqrt[3]{4x}}{\sqrt[3]{4x}}$ To become a perfect cube, $2x^2 = 2 \cdot x \cdot x$ needs to be multiplied by $2 \cdot 2 \cdot x = 4x$.

   $= \dfrac{\sqrt[3]{28x}}{\sqrt[3]{8x^3}}$

   $= \dfrac{\sqrt[3]{28x}}{2x}$ ■

Consider rationalizing the denominator of $\dfrac{3}{\sqrt{x} - 2}$. If we multiply the denominator $\sqrt{x} - 2$ by $\sqrt{x} + 2$, we can apply the difference of two squares formula to get:

$$(\sqrt{x} - 2)(\sqrt{x} + 2) = (\sqrt{x})^2 - 2^2 = x - 4, \quad \text{(where } x \geq 0\text{)}$$

Note that the result does not contain any radicals. So, we rationalize the denominator of $\dfrac{3}{\sqrt{x} - 2}$ by multiplying $\dfrac{3}{\sqrt{x} - 2}$ by $\dfrac{\sqrt{x} + 2}{\sqrt{x} + 2}$:

$$\dfrac{3}{\sqrt{x} - 2} = \dfrac{3}{\sqrt{x} - 2} \cdot \dfrac{\sqrt{x} + 2}{\sqrt{x} + 2} \qquad \text{Rationalize the denominator.}$$

$$= \dfrac{3(\sqrt{x} + 2)}{x - 4} \qquad (a - b)(a + b) = a^2 - b^2$$

Since the denominator does not contain any radicals, we have successfully rationalized the denominator.

We say that $\sqrt{x} + 2$ is the *conjugate* of $\sqrt{x} - 2$ and vice versa. Here we list expressions, their conjugates and the products of the expressions and their conjugates.

Expression	*Conjugate*	*Product*
$3\sqrt{x} + 7$	$3\sqrt{x} - 7$	$9x - 49$
$\sqrt{x} - 4$	$\sqrt{x} + 4$	$x - 16$
$2\sqrt{x} + 3\sqrt{5}$	$2\sqrt{x} - 3\sqrt{5}$	$4x - 45$

More generally, $a\sqrt{b} + c\sqrt{d}$ and $a\sqrt{b} - c\sqrt{d}$ are conjugates of each other. Note that the product of a square root expression and its conjugate does not contain any square roots.

Example 4   Rationalizing a Denominator

Simplify the expression $\dfrac{5}{\sqrt{x} + 3}$.

Solution

$$\frac{5}{\sqrt{x}+3} = \frac{5}{\sqrt{x}+3} \cdot \frac{\sqrt{x}-3}{\sqrt{x}-3}$$

The conjugate of $\sqrt{x}+3$ is $\sqrt{x}-3$.

$$= \frac{5\sqrt{x}-15}{x-9}$$

We verify our result by comparing graphing calculator graphs for $y = \dfrac{5}{\sqrt{x}+3}$ and $y = \dfrac{5\sqrt{x}-15}{x-9}$ for $x \geq 0$ and $x \neq 9$ (see Fig. 10).

Note

The graph of $y = \dfrac{5\sqrt{x}-15}{x-9}$ does not contain a point with x-coordinate 9, since a value of 9 for x leads to division by zero.

 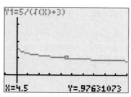 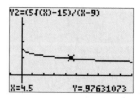

Figure 10 Compare graphs for $x > 0$ and $x \neq 9$

Rationalizing a Denominator

To rationalize the denominator of a square root expression that has a denominator that is a sum or a difference involving radicals:

1. Determine the conjugate of the denominator.
2. Multiply the original fraction by the fraction $\dfrac{\text{conjugate}}{\text{conjugate}}$.
3. Find the product of the denominators using the difference of two squares formula.

Example 5 Rationalizing a Denominator

Simplify the expression $\dfrac{\sqrt{x}+4}{3\sqrt{x}-\sqrt{2}}$.

Solution

$$\frac{\sqrt{x}+4}{3\sqrt{x}-\sqrt{2}} = \frac{\sqrt{x}+4}{3\sqrt{x}-\sqrt{2}} \cdot \frac{3\sqrt{x}+\sqrt{2}}{3\sqrt{x}+\sqrt{2}}$$

The conjugate of $3\sqrt{x}-\sqrt{2}$ is $3\sqrt{x}+\sqrt{2}$.

$$= \frac{(\sqrt{x}+4)(3\sqrt{x}+\sqrt{2})}{(3\sqrt{x})^2 - (\sqrt{2})^2}$$

$(a-b)(a+b) = a^2 - b^2$

$$= \frac{3\sqrt{x}\sqrt{x} + \sqrt{x}\sqrt{2} + 4 \cdot 3\sqrt{x} + 4\sqrt{2}}{3^2(\sqrt{x})^2 - (\sqrt{2})^2}$$

$$= \frac{3x + \sqrt{2x} + 12\sqrt{x} + 4\sqrt{2}}{9x - 2}$$

Table 1 Values of the Function $y = \sqrt{x}$

x	y
0	
1	
4	
9	
16	
25	

EXPLORATION
Looking ahead: Sketching graphs of square root functions

1. Find values of the function $y = \sqrt{x}$ by completing Table 1. Then use pencil and paper to plot the ordered pairs and sketch a curve that contains the points.
2. Verify your sketch in problem 1 by using a graphing calculator to draw a graph of $y = \sqrt{x}$.

3. Use your graphing calculator to compare graphs of $y = 0.5\sqrt{x}, y = \sqrt{x},$ $y = 2\sqrt{x},$ and $y = -2\sqrt{x}$. Describe the effect that a has on the graph of $y = a\sqrt{x}, a \neq 0$.

4. Use your graphing calculator to compare graphs of $y = \sqrt{x-2}, y = \sqrt{x},$ and $y = \sqrt{x+4}$. Describe the effect that h has on the graph of $y = \sqrt{x-h}$.

5. Use your graphing calculator to compare graphs of $y = \sqrt{x} - 2, y = \sqrt{x},$ and $y = \sqrt{x} + 4$. Describe the effect that k has on the graph of $y = \sqrt{x} + k$.

6. Graph

$$y = \sqrt{x}$$

$$y = 0.5\sqrt{x}$$

$$y = 0.5\sqrt{x+3}$$

$$y = 0.5\sqrt{x+3} - 2$$

in order and explain how these graphs relate to your observations in problems 3, 4, and 5.

7. Sketch the graph of $y = 2\sqrt{x-3} + 1$. Use a graphing calculator to verify your sketch.

8. Describe how $a, h,$ and k affect the graph of $f(x) = a\sqrt{x-h} + k$ with $a \neq 0$. Compare their effects for this function with their effects for the quadratic function $g(x) = a(x-h)^2 + k$.

Tips on Succeeding in This Course

To study for the final exam, consider retaking your quizzes and exams. These quizzes and exams can reveal your weak areas. If you have difficulty with a certain concept, you can refer to homework exercises that address this concept. It is also wise to reflect on *why* you are having such difficulty, rather than to just do more homework problems that address the concept.

Key Points
OF THIS SECTION

To rationalize the denominator of a radical expression of the form $\dfrac{1}{\sqrt[n]{x^m}}$, multiply the expression by a fraction of the form $\dfrac{\sqrt[n]{x^k}}{\sqrt[n]{x^k}}$ so that the radical in the denominator has a perfect nth power radicand.	**Rationalizing the denominator of** $\dfrac{1}{\sqrt[n]{x^m}}$
If $\sqrt[n]{a}$ and $\sqrt[n]{b}$ are defined and b is non zero, then $\sqrt[n]{\dfrac{a}{b}} = \dfrac{\sqrt[n]{a}}{\sqrt[n]{b}}$.	**Quotient property**
The radical expressions $a\sqrt{b} + c\sqrt{d}$ and $a\sqrt{b} - c\sqrt{d}$ are *conjugates* of each other.	**Conjugates**
To rationalize the denominator of a square root expression whose denominator is a sum or a difference:	**Rationalizing the denominator where the denominator is a sum or difference**

1. Determine the conjugate of the denominator.

2. Multiply the original fraction by the fraction $\dfrac{\text{conjugate}}{\text{conjugate}}$.

3. Find the product of the denominators using the difference of two squares formula.

HOMEWORK 9.3

MathPro 4
MathPro 4

MathPro 5
MathPro 5

SSM
Tutor Center

CDs/Videos

www

Simplify the expression. Use a graphing calculator to verify your result. Assume that $x \geq 0$.

1. $\dfrac{2}{\sqrt{3}}$

2. $\dfrac{5}{\sqrt{7}}$

3. $\dfrac{8}{\sqrt{x}}$

4. $\dfrac{2}{\sqrt{x}}$

5. $\dfrac{3}{\sqrt{5x}}$

6. $\dfrac{2}{\sqrt{7x}}$

7. $\dfrac{\sqrt{5}}{2\sqrt{x}}$

8. $\dfrac{\sqrt{3}}{7\sqrt{x}}$

9. $\dfrac{4}{3\sqrt{2x}}$

10. $\dfrac{7}{6\sqrt{3x}}$

11. $\sqrt{\dfrac{4}{x}}$

12. $\sqrt{\dfrac{25}{x}}$

13. $\sqrt{\dfrac{7}{2}}$

14. $\sqrt{\dfrac{5}{3}}$

15. $\sqrt{\dfrac{2}{x}}$

16. $\sqrt{\dfrac{11}{x}}$

17. $\sqrt{\dfrac{x}{3}}$

18. $\sqrt{\dfrac{x}{5}}$

19. $\dfrac{3}{\sqrt{x-4}}$

20. $\dfrac{5}{\sqrt{2x+1}}$

21. $\dfrac{2}{\sqrt[3]{5}}$

22. $\dfrac{5}{\sqrt[3]{2}}$

23. $\dfrac{5}{\sqrt[3]{16}}$

24. $\dfrac{1}{\sqrt[3]{25}}$

25. $\dfrac{4}{5\sqrt[3]{x}}$

26. $\dfrac{7}{4\sqrt[3]{x^2}}$

27. $\dfrac{6}{\sqrt[3]{2x^2}}$

28. $\dfrac{1}{\sqrt[3]{9x}}$

29. $\dfrac{7}{\sqrt[4]{4x^3}}$

30. $\dfrac{2}{\sqrt[5]{16x^2}}$

31. $\dfrac{\sqrt[3]{x}}{\sqrt{x}}$

32. $\dfrac{\sqrt[5]{x}}{\sqrt[4]{2x}}$

33. $\sqrt[5]{\dfrac{2}{x^3}}$

34. $\sqrt[3]{\dfrac{4}{x^2}}$

35. $\sqrt[4]{\dfrac{4}{9x^2}}$

36. $\sqrt[3]{\dfrac{7}{25x}}$

37. $\sqrt[5]{\dfrac{3}{4x^4}}$

38. $\sqrt[6]{\dfrac{5}{8x^2}}$

Simplify each expression. Use a graphing calculator to verify your result.

39. $\dfrac{1}{5+\sqrt{3}}$

40. $\dfrac{1}{1-\sqrt{5}}$

41. $\dfrac{2}{4+\sqrt{7}}$

42. $\dfrac{4}{2+\sqrt{x}}$

43. $\dfrac{1}{\sqrt{x}-7}$

44. $\dfrac{6}{5\sqrt{x}+2}$

45. $\dfrac{\sqrt{x}}{\sqrt{x}-1}$

46. $\dfrac{\sqrt{x}}{\sqrt{x}+1}$

47. $\dfrac{3\sqrt{x}}{4\sqrt{x}-5}$

48. $\dfrac{4\sqrt{x}}{2\sqrt{x}+6}$

49. $\dfrac{2\sqrt{x}}{3\sqrt{x}-7}$

50. $\dfrac{\sqrt{x}+3}{\sqrt{x}+4}$

51. $\dfrac{\sqrt{x}-5}{\sqrt{x}+5}$

52. $\dfrac{\sqrt{x}+9}{\sqrt{x}-9}$

53. $\dfrac{\sqrt{x}+3}{4-\sqrt{x}}$

54. $\dfrac{\sqrt{x}-2}{8-\sqrt{x}}$

55. $\dfrac{2\sqrt{x}+5}{3\sqrt{x}-1}$

56. $\dfrac{4\sqrt{x}-3}{2\sqrt{x}-5}$

57. $\dfrac{6\sqrt{x}+\sqrt{5}}{3\sqrt{x}-\sqrt{7}}$

58. $\dfrac{8\sqrt{x}-\sqrt{3}}{4\sqrt{x}-\sqrt{2}}$

59. Two students both tried to rationalize the denominator of the expression $\dfrac{2}{\sqrt{x}}$. Did one, both, or neither of these students correctly rationalize the denominator? If any errors were made, describe them and where they occurred.

Student 1's work

$$\dfrac{2}{\sqrt{x}} = \dfrac{2}{\sqrt{x}} \cdot \dfrac{\sqrt{x}}{\sqrt{x}}$$

$$= \dfrac{2\sqrt{x}}{\sqrt{x}\sqrt{x}}$$

$$= \dfrac{2\sqrt{x}}{x}$$

Student 2's work

$$\dfrac{2}{\sqrt{x}} = \left(\dfrac{2}{\sqrt{x}}\right)^2$$

$$= \dfrac{2^2}{(\sqrt{x})^2}$$

$$= \dfrac{4}{x}$$

60. Two students tried to rationalize the denominator of $\dfrac{3}{\sqrt{x^3}}$. Did one, both, or neither of the students correctly rationalize the denominator? Explain.

Student 1's work

$$\dfrac{3}{\sqrt{x^3}} = \dfrac{3}{\sqrt{x^3}} \cdot \dfrac{\sqrt{x^3}}{\sqrt{x^3}}$$

$$= \dfrac{3\sqrt{x^3}}{\sqrt{x^3}\sqrt{x^3}}$$

$$= \dfrac{3(x\sqrt{x})}{x^3}$$

$$= \dfrac{3\sqrt{x}}{x^2}$$

Student 2's work

$$\dfrac{3}{\sqrt{x^3}} = \dfrac{3}{x\sqrt{x}}$$

$$= \dfrac{3}{x\sqrt{x}} \cdot \dfrac{\sqrt{x}}{\sqrt{x}}$$

$$= \dfrac{3\sqrt{x}}{x \cdot x}$$

$$= \dfrac{3\sqrt{x}}{x^2}$$

61. A student tried to rationalize the denominator of $\dfrac{5}{\sqrt[3]{x}}$. Describe the error and show how to rationalize the denominator.

$$\dfrac{5}{\sqrt[3]{x}} = \dfrac{5}{\sqrt[3]{x}} \cdot \dfrac{\sqrt[3]{x}}{\sqrt[3]{x}}$$

$$= \dfrac{5\sqrt[3]{x}}{x}$$

62. Two students both tried to rationalize the denominator of the expression $\dfrac{4}{2+\sqrt{x}}$. Did one, both, or neither of these students correctly rationalize the denominator? If any errors were made, describe them and where they occurred.

Student 1's work

$$\frac{4}{2+\sqrt{x}} = \frac{4}{2+\sqrt{x}} \cdot \frac{\sqrt{x}}{\sqrt{x}}$$

$$= \frac{4\sqrt{x}}{2+\sqrt{x}\sqrt{x}}$$

$$= \frac{4\sqrt{x}}{2+x}$$

Student 2's work

$$\frac{4}{2+\sqrt{x}} = \frac{4}{2+\sqrt{x}} \cdot \frac{2-\sqrt{x}}{2-\sqrt{x}}$$

$$= \frac{8-4\sqrt{x}}{2^2-(\sqrt{x})^2}$$

$$= \frac{8-4\sqrt{x}}{4-x}$$

We rationalize the numerator of a radical expression by finding an equivalent expression whose numerator does not contain any radicals. Rationalize the numerator of the given expression.

63. $\dfrac{\sqrt{x}}{3}$

64. $\dfrac{\sqrt{x}}{\sqrt{2}}$

65. $\dfrac{\sqrt{x+2}-\sqrt{x}}{2}$

66. $\dfrac{\sqrt{x+3}-\sqrt{x}}{3}$

67. Simplify the expression.

$$\frac{\dfrac{1}{\sqrt{x}} - \dfrac{3}{x}}{\dfrac{2}{\sqrt{x}} + \dfrac{1}{x}}$$

68. Prove the quotient property for radicals:

$$\sqrt[n]{\frac{a}{b}} = \frac{\sqrt[n]{a}}{\sqrt[n]{b}}$$

where $\sqrt[n]{a}$ and $\sqrt[n]{b}$ are defined, and b is nonzero.

69. Find and simplify the exact solution of $x\sqrt{2} + 3\sqrt{5} = 9\sqrt{5}$.

70. Find the exact solution of $x^2\sqrt{2} + x\sqrt{17} + \sqrt{2} = 0$. Simplify your result. [**Hint**: Use the quadratic formula.]

71. Describe how to rationalize the denominator of a radical expression.

9.4 SKETCHING GRAPHS AND COMBINING SQUARE ROOT FUNCTIONS

I love math because I think that it is like magic. Every time I solve a problem I am always surprised and amazed at what I can do. —*Ana J., student*

Objectives

▸ Know the meaning of *square root function*.

▸ Know the graphical significance of a, h, and k for a function of the form $y = a\sqrt{x-h} + k, a \neq 0$.

▸ Sketch graphs of square root functions.

▸ Add, subtract, multiply, and divide square root functions.

In this section we sketch the graph of a *square root function*. Here are some examples of square root functions:

$$f(x) = \sqrt{x}, \qquad g(x) = 3\sqrt{x-4} + 5, \qquad h(x) = -2\sqrt{x+7}$$

Example 1 Graphing a Square Root Function

Sketch the graph of $f(x) = \sqrt{x}$.

Solution

We list some input-output pairs in Table 2. We choose perfect square inputs because it is easiest to find their outputs mentally. Since the radicand of $\sqrt{x}$ must be nonnegative, we cannot choose any negative numbers as inputs. Then we sketch the graph of f (see Fig. 11).

Table 2 Input-Output Pairs of $f(x) = \sqrt{x}$	
x	**y**
0	0
1	1
4	2
9	3
16	4

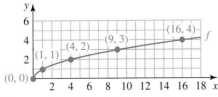

Figure 11 Graph of $f(x) = \sqrt{x}$

We use a graphing calculator to verify our graph (see Fig. 12).

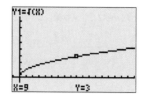

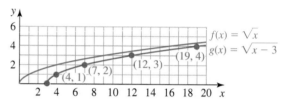

Figure 12 Verify the graph of $f(x) = \sqrt{x}$

Example 2   Horizontal Translation

Compare the graph of $g(x) = \sqrt{x - 3}$ to the graph of $f(x) = \sqrt{x}$.

Solution

We list input-output pairs for g in Table 3. We choose inputs that lead to easily found outputs. Then, we sketch graphs of g and f (see Fig. 13).

Table 3 Input-Output Pairs of $g(x) = \sqrt{x-3}$	
x	y
3	0
4	1
7	2
12	3
19	4

Figure 13 Graphs of $g(x) = \sqrt{x - 3}$ and $f(x) = \sqrt{x}$

We see that the graph of $g(x) = \sqrt{x - 3}$ is the translation of the graph of $f(x) = \sqrt{x}$ three units to the right. To see why this makes sense, we solve each equation for x:

$$f(x) = \sqrt{x} \qquad\qquad g(x) = \sqrt{x - 3}$$
$$y = \sqrt{x} \qquad\qquad y = \sqrt{x - 3}$$
$$\sqrt{x} = y \qquad\qquad \sqrt{x - 3} = y$$
$$(\sqrt{x})^2 = y^2 \qquad\qquad (\sqrt{x - 3})^2 = y^2$$
$$x = y^2 \qquad\qquad x - 3 = y^2$$
$$x = y^2 + 3$$

For each positive value of y, the input value of x for g is 3 more than the input value of x for f. Therefore, the graph of $g(x) = \sqrt{x - 3}$ lies 3 units to the right of the graph of $f(x) = \sqrt{x}$.

In Example 2 we found that to graph $g(x) = \sqrt{x - 3}$, we translate the graph of $f(x) = \sqrt{x}$ by 3 units to the right. Note that this is the same pattern we observed with quadratic functions. To graph the quadratic function $g(x) = (x - 3)^2$, we translate the graph of $f(x) = x^2$ by 3 units to the right.

In fact, the values of a, h, and k have similar effects on a square root function $g(x) = a\sqrt{x - h} + k, a \neq 0$ as they do on a quadratic function $Q(x) = a(x - h)^2 + k$. Here we summarize the roles of h and k:

Function	*To graph the function $g(x) = \sqrt{x - h} + k$,*
$g(x) = \sqrt{x - 2}$	translate the graph $y = \sqrt{x}$ by 2 units to the right.
$g(x) = \sqrt{x + 2}$	translate the graph $y = \sqrt{x}$ by 2 units to the left.
$g(x) = \sqrt{x} - 2$	translate the graph $y = \sqrt{x}$ down 2 units.
$g(x) = \sqrt{x} + 2$	translate the graph $y = \sqrt{x}$ up 2 units.

The graphs of $f(x) = -a\sqrt{x}$ and $g(x) = a\sqrt{x}$ are reflections of each other across the x-axis (see Fig. 14).

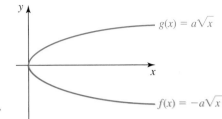

Figure 14 Typical graphs of $f(x) = -a\sqrt{x}$ and $g(x) = a\sqrt{x}$, where $a > 0$

If $a > 0$, then $g(x) = a\sqrt{x - h} + k$ is an increasing function and (h, k) is the minimum point (see Fig. 15). If $a < 0$, then g is a decreasing function and (h, k) is the maximum point (see Fig. 16).

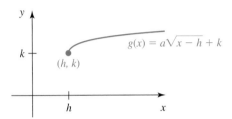

Figure 15 If $a > 0$, then g is increasing with minimum point (h, k)

Figure 16 If $a < 0$, then g is decreasing with maximum point (h, k)

If a is a large number, then the graph of $g(x) = a\sqrt{x - h} + k$ rises much more quickly than the graph of $y = \sqrt{x - h} + k$, and if a is a positive number near zero, the graph rises much more slowly.

Example 3   Graphing a Square Root Function

Sketch the graph of $g(x) = -2\sqrt{x + 4} - 1$. Also, find the domain and range of g.

Solution

First we sketch the graph of $y = -2\sqrt{x}$ in Fig. 17. Next, we translate the graph 4 units to the left and down 1 unit.

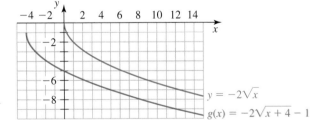

Figure 17 Graphs of $g(x) = -2\sqrt{x + 4} - 1$ and $y = -2\sqrt{x}$

We check that the solutions of g listed in Table 4 lie on our sketched square root function. Also, we use a graphing calculator to verify our sketch (see Fig. 18).

Figure 18 Verify the graph of $g(x) = -2\sqrt{x + 4} - 1$

```
WINDOW
Xmin=-5
Xmax=15
Xscl=1
Ymin=-10
Ymax=1
Yscl=1
Xres=1
```

Table 4 Input-Output Pairs of $g(x) =$ $-2\sqrt{x+4} - 1$	
x	y
-4	-1
-3	-3
0	-5
5	-7
12	-9

From the graph of g, we see that the values of x are greater than or equal to -4. So the domain is the set of numbers x where $x \geq -4$.

We can also find the domain by using the fact that the radicand $x + 4$ must be nonnegative:

$$x + 4 \geq 0$$
$$x + 4 - 4 \geq 0 - 4$$
$$x \geq -4$$

From the graph of g, we see that the maximum point is $(-4, -1)$, so the values of y are less than or equal to -1. So the range is the set of numbers y where $y \leq -1$. ∎

In Example 4 we add, subtract, multiply, and divide two square root functions to form new functions.

Example 4 Performing Operations with Square Root Functions

Let $f(x) = 2\sqrt{x} - 3$ and $g(x) = 5\sqrt{x} + 4$. Find an equation for h.

1. $h(x) = f(x) + g(x)$ 　　　　**2.** $h(x) = f(x) - g(x)$

3. $h(x) = f(x)g(x)$ 　　　　　**4.** $h(x) = \dfrac{f(x)}{g(x)}$

Solution

1. $h(x) = f(x) + g(x) = (2\sqrt{x} - 3) + (5\sqrt{x} + 4)$
$$= 7\sqrt{x} + 1$$

2. $h(x) = f(x) - g(x) = (2\sqrt{x} - 3) - (5\sqrt{x} + 4)$
$$= 2\sqrt{x} - 3 - 5\sqrt{x} - 4$$
$$= -3\sqrt{x} - 7$$

3. $h(x) = f(x)g(x) = (2\sqrt{x} - 3)(5\sqrt{x} + 4)$
$$= 2\sqrt{x} \cdot 5\sqrt{x} + 2\sqrt{x} \cdot 4 - 3 \cdot 5\sqrt{x} - 12$$
$$= 10x + 8\sqrt{x} - 15\sqrt{x} - 12$$
$$= 10x - 7\sqrt{x} - 12$$

4. $h(x) = \dfrac{f(x)}{g(x)} = \dfrac{2\sqrt{x} - 3}{5\sqrt{x} + 4}$

$$= \frac{2\sqrt{x} - 3}{5\sqrt{x} + 4} \cdot \frac{5\sqrt{x} - 4}{5\sqrt{x} - 4} \qquad \text{Rationalize the denominator.}$$

$$= \frac{2\sqrt{x} \cdot 5\sqrt{x} - 2\sqrt{x} \cdot 4 - 3 \cdot 5\sqrt{x} + 12}{(5\sqrt{x})^2 - 4^2}$$

$$= \frac{10x - 8\sqrt{x} - 15\sqrt{x} + 12}{5^2(\sqrt{x})^2 - 4^2}$$

$$= \frac{10x - 23\sqrt{x} + 12}{25x - 16}$$

We verify our work by comparing graphing calculator tables for $y = \dfrac{2\sqrt{x} - 3}{5\sqrt{x} + 4}$ and $y = \dfrac{10x - 23\sqrt{x} + 12}{25x - 16}$ for $x > 0$ and $x \neq \dfrac{16}{25}$ (see Fig. 19).

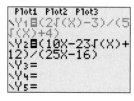

Figure 19 Verify the work

EXPLORATION
Translating and reflecting the absolute value function

Note

To review absolute value, see Section A.3.

1. Complete Table 5 for the absolute value function $y = |x|$. Recall that $|-2| = 2$, $|0| = 0$, and $|3| = 3$.

2. Sketch a graph of $y = |x|$. Use your graphing calculator to verify your graph.

3. Translate and/or reflect the graph of $y = |x|$ to sketch the graph of the given function. Use your graphing calculator to verify your sketch.

   a. $y = |x| - 2$ **b.** $y = |x - 4|$

   c. $y = -|x + 3|$ **d.** $y = -|x - 2| + 5$

4. Sketch the graph of the given function.

   a. $y = 2|x|$ **b.** $y = -3|x|$

5. Describe the graphical significance of a, h, and k for a function of the form $y = a|x - h| + k$, where $a \neq 0$.

Table 5 Values of the Function $y=\lvert x\rvert$	
x	y
-3	
-2	
-1	
1	
2	
3	

Graphing Calculator

To enter $|x|$:

Press [MATH] [▷] **1** [X,T,Θ,n] [ )].

EXPLORATION
Looking ahead: Finding an equation using trial and error

1. In this problem you will use your graphing calculator to graph a radical curve like the one in Fig. 20. Use a function of the form $f(x) = a\sqrt{x} + b$, where a and b are real number constants. What is the equation for f that works? You should be able to determine the value for b by inspecting the graph. You can find the value of a through trial-and-error graphing.

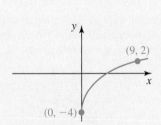

Figure 20 Problem 1

Table 6 Percentages of Americans Who are High School Graduates	
Year	Percent
1980	66.5
1985	73.9
1990	77.6
1995	81.7
2000	84.1

Source: *U.S. Census Bureau*

2. The percentages of Americans 25 years old and over who are high school graduates for various years are listed in Table 6.

    Let $f(t)$ represent the percentage of Americans who are high school graduates at t years since 1980. Find an equation of the form $f(t) = a\sqrt{t} + b$ that models the high school graduate data well. Find the equation through trial-and-error graphing.

Key Points
OF THIS SECTION

For a square root function $f(x) = a\sqrt{x - h} + k$ with $a \neq 0$:

Properties for a square root function

- The graph of f can be found by translating the graph of $y = a\sqrt{x}$ horizontally by $|h|$ units (right if $h > 0$, left if $h < 0$) and vertically by $|k|$ units (up if $k > 0$, down if $k < 0$).

- If $a > 0$, then f is an increasing function and (h, k) is a minimum point.

- If $a < 0$, then f is a decreasing function and (h, k) is a maximum point.
- If a is a large number, then the graph of $f(x) = a\sqrt{x - h} + k$ rises much more quickly than the graph of $y = \sqrt{x - h} + k$, and if a is a positive number near zero, the graph rises much more slowly.
- We add, subtract, multiply, or divide two square root functions to form a new function.

HOMEWORK 9.4

 SSM Tutor Center   MathPro 4 MathPro 4 MathPro 5 MathPro 5   CDs/Videos www www

Use pencil and paper to sketch the graph of the function. Use your graphing calculator to verify your sketch.

1. $y = 2\sqrt{x}$

2. $y = -3\sqrt{x}$

3. $y = -\sqrt{x}$

4. $y = -\dfrac{1}{2}\sqrt{x}$

5. $y = \sqrt{x} + 3$

6. $y = \sqrt{x} - 5$

7. $y = 2\sqrt{x} - 5$

8. $y = 3\sqrt{x} + 2$

9. $y = -3\sqrt{x} + 4$

10. $y = -2\sqrt{x} + 1$

11. $y = \sqrt{x - 2}$

12. $y = \sqrt{x + 5}$

13. $y = -\sqrt{x + 2}$

14. $y = -2\sqrt{x - 5}$

15. $y = \dfrac{1}{2}\sqrt{x - 4}$

16. $y = \dfrac{1}{4}\sqrt{x + 1}$

17. $y = \sqrt{x + 3} + 2$

18. $y = \sqrt{x - 1} - 3$

19. $y = -2\sqrt{x + 3} - 4$

20. $y = -3\sqrt{x - 2} + 1$

21. $y = 4\sqrt{x - 1} + 3$

22. $y = 2\sqrt{x + 6} - 2$

23. $\sqrt{x} + y = 4$

24. $2\sqrt{x} - y = 3$

25. $2y - 6\sqrt{x} = 8$

26. $y - 2\sqrt{x} + 9 = 0$

Use pencil and paper to sketch the graph of the equation. Use your graphing calculator to verify your sketch. Also, find the domain and range of the function.

27. $y = -2\sqrt{x}$

28. $y = \sqrt{x - 6}$

29. $y = \sqrt{x} + 2$

30. $y = \sqrt{x} - 1$

31. $y = -2\sqrt{x + 5} + 4$

32. $y = 3\sqrt{x - 3} - 6$

33. Four functions of the form $y = a\sqrt{x - h} + k$ have been sketched in Fig. 21. Describe the signs of the constants a, h, and k for each of these functions.

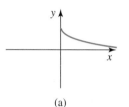
(a)

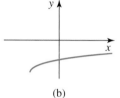
(b)

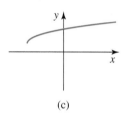
(c)

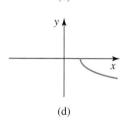
(d)

Figure 21 Exercise 33

34. In this exercise you will use your graphing calculator to graph a family of "square root curves" similar to the one in Fig. 22. List the equations of your square root curves.

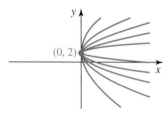

Figure 22 A family of square root curves—exercise 34

35. In this exercise you will use your graphing calculator to graph a family of square root curves similar to the one in Fig. 23. List the equations of your square root curves.

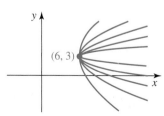

Figure 23 Another family of square root curves—exercise 35

36. For each part, copy the graph of $f(x) = a\sqrt{x-h}+k$ as shown in Fig. 24. On the same coordinate system, use the graph of f to sketch the graph of the given function. Be sure to label each graph.

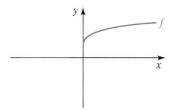

Figure 24 Exercise 36

  a. Sketch the graph of f and a graph of $h(x) = a\sqrt{x-h}+2k$.

  b. Sketch the graph of f and a graph of $l(x) = a\sqrt{x-2h}+k$.

  c. Sketch the graph of f and a graph of $k(x) = -a\sqrt{x-h}+k$.

  d. Sketch the graph of f and a graph of $g(x) = (2a)\sqrt{x+h}-k$.

37. For what values of a, h, and k for the square root function $f(x) = a\sqrt{x-h}+k$ with $a \neq 0$ is there a point on the graph of f that is higher than all other points on the graph? (Recall that this point is called the maximum point.) For what values of a, h, and k does f have a minimum point? Describe the maximum or minimum point in terms of a, h, and k.

38. Solve the system.

$$y = 2\sqrt{x-3}+1$$
$$y = -2\sqrt{x-3}+1$$

Evaluate the function $f(x) = 7\sqrt{x} - 3$ *at each indicated value of x. Assume that* $c \geq 0$.

39. $f(4)$ **40.** $f(0)$ **41.** $f(9c)$ **42.** $f(4c)$

Let $f(x) = \sqrt{x}+3$ *and* $g(x) = \sqrt{x}-5$. *Find an equation for h.*

43. $h(x) = f(x)+g(x)$ **44.** $h(x) = f(x)-g(x)$

45. $h(x) = g(x)-f(x)$ **46.** $h(x) = f(x)g(x)$

47. $h(x) = \dfrac{f(x)}{g(x)}$ **48.** $h(x) = \dfrac{g(x)}{f(x)}$

Let $f(x) = 5\sqrt{x} - 9$ *and* $g(x) = 4\sqrt{x} + 1$. *Find an equation for h.*

49. $h(x) = f(x)+g(x)$ **50.** $h(x) = f(x)-g(x)$

51. $h(x) = g(x)-f(x)$ **52.** $h(x) = f(x)g(x)$

53. $h(x) = \dfrac{f(x)}{g(x)}$ **54.** $h(x) = \dfrac{g(x)}{f(x)}$

Let $f(x) = 2\sqrt{x} - 3\sqrt{5}$ *and* $g(x) = 2\sqrt{x} + 3\sqrt{5}$. *Find an equation for h.*

55. $h(x) = f(x)+g(x)$ **56.** $h(x) = f(x)-g(x)$

57. $h(x) = g(x)-f(x)$ **58.** $h(x) = f(x)g(x)$

59. $h(x) = \dfrac{f(x)}{g(x)}$ **60.** $h(x) = \dfrac{g(x)}{f(x)}$

Let $f(x) = \sqrt{x+1} - 2$ *and* $g(x) = \sqrt{x+1}+2$. *Find an equation for h.*

61. $h(x) = f(x)+g(x)$ **62.** $h(x) = f(x)-g(x)$

63. $h(x) = g(x)-f(x)$ **64.** $h(x) = f(x)g(x)$

65. $h(x) = \dfrac{f(x)}{g(x)}$ **66.** $h(x) = \dfrac{g(x)}{f(x)}$

67. In the Third International Mathematics and Science Study, American high school seniors scored last in physics and second to last in advanced math. The good news is that U.S. seniors' scores in standard testing on science have improved since 1982 (see Table 7).

Table 7 U.S. High School Seniors' Test Scores on Science

Year	Average Test Scores
1982	280
1986	287
1990	288
1992	293
1994	293
1996	295
1998	296

Source: *National Assessment of Educational Progress*

  a. Let T represent the average test score for seniors at t years since 1982. A *square root model* for the data is
$$T = S(t) = 3.9\sqrt{t} + 280$$
A quadratic model is
$$T = Q(t) = -0.027t^2 + 1.41t + 280.43$$
If scores continue to improve well into the future, which function best models the situation? Explain.

  b. Use S to predict the average test score in 2002.

  c. Use S to predict the average test score in 2005.

68. Since 1995, Airbus has narrowed the gap with Boeing in receiving orders to manufacture airplanes (see Table 8).

Table 8 Orders for Airplanes

| Year | Firm Orders | |
	Airbus	Boeing
1995	116	484
1996	326	726
1997	458	579
1998	558	653

Source: New York Times

  a. Let n represent the number of firm orders for Airbus airplanes at t years since 1995. A *square root model* for the data is
$$n = S(t) = 248\sqrt{t} + 116$$
A quadratic model is
$$n = Q(t) = -27.5t^2 + 228.3t + 118.3$$
Compare $S(2)$ with $Q(2)$, $S(3)$ with $Q(3)$, and $S(5)$ with $Q(5)$. Explain in terms of graphing calculator graphs of S and Q why the values for $S(5)$ and $Q(5)$ are so different.

b. If the number of orders continues to increase well into the future, which function best models the situation? Explain.

c. Use S to predict the number of firm orders for Airbus airplanes in 2004.

For exercises 69–76 refer to Fig. 25.

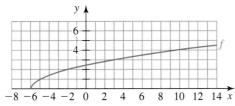

Figure 25 Exercises 69–76

69. Estimate $f(-6)$.

70. Estimate $f(-2)$.

71. Estimate $f(0)$.

72. Estimate $f(5)$.

73. Estimate x when $f(x) = 0$.

74. Estimate x when $f(x) = 2$.

75. Estimate x when $f(x) = 3$.

76. Estimate x when $f(x) = 4$.

77. Use ZStandard followed by ZSquare to draw the graphs of $f(x) = 2\sqrt{x+3} + 2$ and $g(x) = -2\sqrt{x+3} + 2$ on the same coordinate system. Consider the combined graph as the graph of a single relation. What do you notice about the graph? Is the relation described by this graph a function? Explain.

78. Find two functions that have the same domain and range as the function $f(x) = \sqrt{x+7} - 2$.

79. Describe how to sketch the graph of the function $g(x) = a\sqrt{x-h} + k$, $a \neq 0$, given the graph of $f(x) = \sqrt{x}$.

9.5 SOLVING SQUARE ROOT EQUATIONS

Objectives

▸ Solve square root equations.

▸ Find the x-intercepts of square root functions.

In Sections 9.1–9.3 we simplified radical expressions. In this section we solve *square root equations*. Here are some examples of square root equations:

$$\sqrt{x} = 5, \qquad 4\sqrt{7x - 3} = 8, \qquad \sqrt{x+3} - \sqrt{2x - 4} = 6$$

Consider the square root equation

$$\sqrt{x} = 3$$

We solve the equation by squaring both sides:

$$(\sqrt{x})^2 = 3^2$$
$$x = 9$$

So, the solution is 9. This checks out, because $\sqrt{9} = 3$.

Note

Although we can see that the solution of $\sqrt{x} = 3$ is 9 without squaring both sides, the technique shown will be very helpful when solving more complicated equations.

Example 1 Solving a Square Root Equation

Solve the equation $2\sqrt{x} + 5 = 13$.

Solution

First, we isolate $\sqrt{x}$ on one side of the equation:

$$2\sqrt{x} + 5 = 13$$
$$2\sqrt{x} = 8 \qquad \text{Subtract 5 from both sides.}$$
$$\sqrt{x} = 4 \qquad \text{Divide both sides by 2.}$$
$$(\sqrt{x})^2 = 4^2 \qquad \text{Square both sides.}$$
$$x = 16$$

We check that 16 satisfies the original equation.

$$2\sqrt{x} + 5 = 13$$
$$2\sqrt{16} + 5 \stackrel{?}{=} 13$$
$$2 \cdot 4 + 5 \stackrel{?}{=} 13$$
$$13 \stackrel{?}{=} 13$$
$$\text{true}$$

So the solution is 16.

If a square root equation contains the variable in exactly one square root, we isolate the square root on one side of the equation and then square both sides. If the equation contains *like* square roots, we combine the like square roots, isolate the square root containing the variable on one side of the equation, and then square both sides.

Example 2   Solving a Square Root Equation

Solve the equation $x = \sqrt{x - 1} + 3$.

Solution

$$x = \sqrt{x - 1} + 3$$

$x - 3 = \sqrt{x - 1}$	Isolate the radical.
$(x - 3)^2 = (\sqrt{x - 1})^2$	Square both sides.
$x^2 - 6x + 9 = x - 1$	$(a - b)^2 = a^2 - 2ab + b^2$
$x^2 - 7x + 10 = 0$	
$(x - 2)(x - 5) = 0$	Factor the left side.
$x - 2 = 0 \quad$ or $\quad x - 5 = 0$	Zero factor property
$x = 2 \quad$ or $\quad x = 5$	

Note

Recall that $(x - 3)^2$ is equal to $x^2 - 6x + 9$, *not* $x^2 + 9$.

We check that 2 and 5 satisfy the original equation.

Check $x = 2$	**Check** $x = 5$
$x = \sqrt{x - 1} + 3$	$x = \sqrt{x - 1} + 3$
$2 \overset{?}{=} \sqrt{2 - 1} + 3$	$5 \overset{?}{=} \sqrt{5 - 1} + 3$
$2 \overset{?}{=} 4$	$5 \overset{?}{=} 5$
false	true

Since 2 does not satisfy the original equation, it is an extraneous solution. Therefore, the only solution is 5.

To verify that 2 is an extraneous solution and that 5 is a solution, we compare graphing calculator tables for $y = x$ and $y = \sqrt{x - 1} + 3$, the left and right sides, respectively, of the original equation (see Fig. 26).

Figure 26 Verify that 2 is an extraneous solution and 5 is a solution

In Example 2 we introduced the extraneous solution 2 by squaring both sides of the equation $x - 3 = \sqrt{x - 1}$. Here, we show that although 2 does *not* satisfy the equation $x - 3 = \sqrt{x - 1}$, it *does* satisfy the equation that results from squaring both sides:

Before Squaring Both Sides	***After Squaring Both Sides***
$x - 3 = \sqrt{x - 1}$	$(x - 3)^2 = (\sqrt{x - 1})^2$
$2 - 3 \overset{?}{=} \sqrt{2 - 1}$	$(2 - 3)^2 \overset{?}{=} (\sqrt{2 - 1})^2$
$-1 \overset{?}{=} 1$	$(-1)^2 \overset{?}{=} 1^2$
false	$1 \overset{?}{=} 1$
	true

Since squaring both sides of an equation may introduce extraneous solutions, it is an essential part of the solving process to check that each proposed solution satisfies the original equation.

If a square root equation contains two or more unlike square roots, we isolate one of the square roots and then square both sides of the equation. After squaring both sides, the equation will have at least one less square root. We repeat this process of isolating and squaring until no square roots remain.

Example 3   Solving a Square Root Equation

Solve $2\sqrt{x-3} - \sqrt{x} = 0$.

Solution

First, we isolate $2\sqrt{x-3}$ and then square both sides of the equation.

$$2\sqrt{x-3} - \sqrt{x} = 0$$
$$2\sqrt{x-3} = \sqrt{x} \qquad \text{Isolate } 2\sqrt{x-3}.$$
$$(2\sqrt{x-3})^2 = (\sqrt{x})^2 \qquad \text{Square both sides.}$$
$$2^2(\sqrt{x-3})^2 = (\sqrt{x})^2 \qquad (ab)^2 = a^2b^2$$
$$4(x-3) = x$$
$$4x - 12 = x$$
$$3x = 12$$
$$x = 4$$

To verify that 4 is a solution, we first enter the function $y = 2\sqrt{x-3} - \sqrt{x}$. Then, we check that for the input 4, the output is 0 (see Fig. 27).

 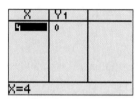

Figure 27 Verify that 4 is a solution

In Example 3, after squaring both sides of $2\sqrt{x-3} = \sqrt{x}$, we simplified $(2\sqrt{x-3})^2$:

$$(2\sqrt{x-3})^2 = 2^2(\sqrt{x-3})^2 = 4(x-3) \qquad \text{Correct}$$

Beware! It is a common student error to omit squaring the coefficient 2 and write:

$$(2\sqrt{x-3})^2 = 2(x-3) \qquad \text{Incorrect}$$

Recall that when simplifying $(ab)^2$, both a and b are squared:

$$(ab)^2 = a^2b^2$$

Example 4   Solving a Square Root Equation

Solve the equation $\sqrt{x-5} - \sqrt{x} = -1$.

Solution

First, we isolate the radical $\sqrt{x-5}$ and then square both sides of the equation.

$$\sqrt{x-5} - \sqrt{x} = -1$$
$$\sqrt{x-5} = \sqrt{x} - 1 \qquad \text{Isolate the radical } \sqrt{x-5}.$$
$$(\sqrt{x-5})^2 = (\sqrt{x} - 1)^2 \qquad \text{Square both sides.}$$
$$x - 5 = (\sqrt{x})^2 - 2\sqrt{x} \cdot 1 + 1^2 \qquad (a-b)^2 = a^2 - 2ab + b^2$$
$$x - 5 = x - 2\sqrt{x} + 1$$

Next, we isolate the radical $\sqrt{x}$ and square both sides.

$$2\sqrt{x} = x + 1 - x + 5$$
$$2\sqrt{x} = 6$$
$$\sqrt{x} = 3$$
$$(\sqrt{x})^2 = 3^2$$
$$x = 9$$

Next, we check that 9 satisfies the original equation.

$$\sqrt{x-5} - \sqrt{x} = -1$$
$$\sqrt{9-5} - \sqrt{9} \stackrel{?}{=} -1$$
$$2 - 3 \stackrel{?}{=} -1$$
$$-1 \stackrel{?}{=} -1$$
$$\text{true}$$

So the solution is 9. ∎

Example 5   Solving a Square Root Equation

Solve $\sqrt{2x + 1} - \sqrt{x + 2} = 1$.

Solution

$$\sqrt{2x + 1} - \sqrt{x + 2} = 1$$

$$\sqrt{2x + 1} = \sqrt{x + 2} + 1 \qquad\qquad \text{Isolate the radical } \sqrt{2x + 1}.$$

$$(\sqrt{2x + 1})^2 = (\sqrt{x + 2} + 1)^2 \qquad\qquad \text{Square both sides.}$$

$$2x + 1 = (\sqrt{x + 2})^2 + 2\sqrt{x + 2} \cdot 1 + 1^2$$

$$2x + 1 = x + 2 + 2\sqrt{x + 2} + 1$$

$$x - 2 = 2\sqrt{x + 2} \qquad\qquad \text{Isolate the radical } 2\sqrt{x + 2}.$$

$$(x - 2)^2 = (2\sqrt{x + 2})^2 \qquad\qquad \text{Square both sides.}$$

$$x^2 - 4x + 4 = 4(x + 2)$$

$$x^2 - 4x + 4 = 4x + 8$$

$$x^2 - 8x - 4 = 0 \qquad\qquad \text{A quadratic equation.}$$

$$x = \frac{-(-8) \pm \sqrt{(-8)^2 - 4(1)(-4)}}{2(1)} \qquad\qquad \text{Use the quadratic formula.}$$

$$= \frac{8 \pm \sqrt{80}}{2}$$

$$= \frac{8 \pm 4\sqrt{5}}{2}$$

$$= 4 \pm 2\sqrt{5}$$

So $x = 4 - 2\sqrt{5} \approx -0.47$ or $x = 4 + 2\sqrt{5} \approx 8.47$. We check that the approximations for the solutions approximately satisfy the original equation $\sqrt{2x + 1} - \sqrt{x + 2} = 1$:

Check $x \approx -0.47$

$$\sqrt{2(-0.47) + 1} - \sqrt{-0.47 + 2} \approx -0.9920 \qquad\qquad \text{not close to 1}$$

Check $x \approx 8.47$

$$\sqrt{2(8.47) + 1} - \sqrt{8.47 + 2} \approx 0.9998 \qquad\qquad \text{close to 1}$$

So, the only solution is $4 + 2\sqrt{5}$.

Now, we store $4 - 2\sqrt{5}$ for the variable "X" in a graphing calculator and perform a similar but more precise check. Then, we do the same for $4 + 2\sqrt{5}$ (see Fig. 28).

Figure 28 Verify that $4 - 2\sqrt{5}$ is an extraneous solution and $4 + 2\sqrt{5}$ is a solution

In Example 5 we expanded the right side of the equation

$$(\sqrt{2x + 1})^2 = (\sqrt{x + 2} + 1)^2$$

using the square of a sum formula $(a + b)^2 = a^2 + 2ab + b^2$:

$$(\sqrt{x + 2} + 1)^2 = (\sqrt{x + 2})^2 + 2\sqrt{x + 2} \cdot 1 + 1^2 \qquad \text{Correct}$$

Remember that, in general, $(a + b)^2$ is *not* equal to $a^2 + b^2$, so it is incorrect to say

$$(\sqrt{x + 2} + 1)^2 = (\sqrt{x + 2})^2 + 1^2 \qquad \text{Incorrect}$$

In Example 6 we solve a square root equation to help ourselves find the x-intercepts of a function.

Example 6 Find the *x*-intercept of a Square Root Function

Find the x-intercepts of $f(x) = \sqrt{3x - 4} - \sqrt{x + 2}$.

Solution

We substitute 0 for $f(x)$ and solve for x.

$$\sqrt{3x - 4} - \sqrt{x + 2} = 0$$
$$\sqrt{3x - 4} = \sqrt{x + 2} \qquad \text{Isolate at least one radical.}$$
$$(\sqrt{3x - 4})^2 = (\sqrt{x + 2})^2 \qquad \text{Square both sides.}$$
$$3x - 4 = x + 2$$
$$2x = 6$$
$$x = 3$$

We check that 3 satisfies the equation $\sqrt{3x - 4} - \sqrt{x + 2} = 0$.

$$\sqrt{3x - 4} - \sqrt{x + 2} = 0$$
$$\sqrt{3(3) - 4} - \sqrt{3 + 2} \stackrel{?}{=} 0$$
$$\sqrt{5} - \sqrt{5} \stackrel{?}{=} 0$$
$$0 \stackrel{?}{=} 0$$
$$\text{true}$$

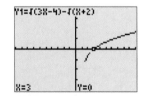

Figure 29 Verify that $(3, 0)$ is an x-intercept

Since 3 is the only number that satisfies the equation $\sqrt{3x - 4} - \sqrt{x + 2} = 0$, we can conclude that $(3, 0)$ is the only x-intercept of $f(x) = \sqrt{3x - 4} - \sqrt{x + 2}$.

We use a graphing calculator to verify that $(3, 0)$ is an x-intercept (see Fig. 29).

EXPLORATION
Extraneous solutions

1. Solve the equation $\sqrt{3x - 2} + 2 = x$. Carefully record each step of your work.

2. In problem 1 you found that 1 is an extraneous solution and 6 is the only solution. Now, substitute 1 for x in each equation for the steps you recorded in problem 1. Which of the equations are satisfied by 1?

3. What does it mean to say that 1 is an extraneous solution? Why do we sometimes get extraneous solutions when solving square root equations but not when solving linear, exponential, or quadratic equations?

Key Points
OF THIS SECTION

If a square root equation contains the variable in exactly one square root, we isolate the square root on one side of the equation and then square both sides.	**An equation with a variable in exactly one square root**
If a square root equation contains *like* square roots, we combine the like square roots, isolate the square root containing the variable on one side of the equation, and then square both sides.	**An equation with like square roots**
If a square root equation contains two or more unlike square roots, we isolate one of the square roots and then square both sides of the equation. After squaring both sides, the equation will have at least one less square root. We repeat this process of isolating and squaring until no square roots remain.	**An equation with unlike square roots**
When solving a square root equation, always check results for extraneous solutions.	**Check for extraneous solutions**

HOMEWORK 9.5

 SSM Tutor Center   MathPro 4   MathPro 5   CDs/Videos   www

Solve each square root equation.

1. $\sqrt{x} = 5$
2. $\sqrt{x} = 8$
3. $\sqrt{x} = -2$
4. $\sqrt{x} = -7$
5. $3\sqrt{x} - 1 = 5$
6. $5\sqrt{x} + 2 = 37$
7. $4 - 5\sqrt{x} = 2\sqrt{x} - 10$
8. $12\sqrt{x} + 13 - 10\sqrt{x} = 15\sqrt{x}$
9. $3\sqrt{7x} - 24 = -9\sqrt{7x}$
10. $8 - 4\sqrt{2x} = 3\sqrt{2x} - 6$
11. $\sqrt{x - 1} = 2$
12. $\sqrt{13 - 2x} = 3$
13. $\sqrt{5x - 7} = -8$
14. $\sqrt{15 + x} + 8 = 2$
15. $10\sqrt{6x + 3} = 100$
16. $10 - 8\sqrt{2x - 1} = 14$
17. $\sqrt{3x + 1} = \sqrt{2x + 6}$
18. $\sqrt{10x - 3} = \sqrt{6x + 2}$
19. $2\sqrt{1 - x} - \sqrt{5} = 0$
20. $2\sqrt{x - 1} - \sqrt{3x - 1} = 0$
21. $-4.91\sqrt{3.18x - 7.14} = -2.19$
22. $-7.9 = 5.6 - 3.7\sqrt{4.4 - 9.8x}$

23. $\sqrt{3x + 3} = x - 5$
24. $\sqrt{x + 10} = x - 2$
25. $\sqrt{12x + 13} + 2 = 3x$
26. $\sqrt{x + 2} - x = 2$
27. $\sqrt{3x + 4} - x = 3$
28. $1 + \sqrt{2x + 5} = 2x$
29. $\sqrt{x^2 - 5x + 1} = x - 3$
30. $\sqrt{x^2 + 2x} = x + 5$
31. $2 + \sqrt{x} = \sqrt{x + 12}$
32. $3 - 2\sqrt{x} + \sqrt{9 - x} = 0$
33. $\sqrt{x} - 1 = \sqrt{5 - x}$
34. $3 = \sqrt{6 + x} + \sqrt{x}$
35. $\sqrt{x} - \sqrt{2x} = -1$
36. $\sqrt{x} = \sqrt{3x} - 2$
37. $\sqrt{x + 1} = -\sqrt{x} - 7$
38. $\sqrt{x + 3} = \sqrt{x + 2} + 1$
39. $\sqrt{x - 3} + \sqrt{x + 5} = 4$
40. $\sqrt{x + 6} - \sqrt{x - 2} = 2$
41. $\sqrt{x - 4} = \sqrt{x + 6} + 2$
42. $\sqrt{x - 1} = 1 - \sqrt{x + 1}$
43. $\sqrt{2x - 1} + \sqrt{3x - 2} = 2$
44. $\sqrt{3x - 1} - \sqrt{4x + 1} = -1$
45. $\sqrt{\sqrt{x} - 2} = 3$
46. $\sqrt{\sqrt{x} + 1} = 1$
47. $\dfrac{1}{\sqrt{x}} + \sqrt{x} = 2$
48. $3 - \sqrt{x} = \dfrac{2}{\sqrt{x}}$

49. $\dfrac{1}{\sqrt{x+2}} = 3 - \sqrt{x+2}$ **50.** $\dfrac{1}{\sqrt{x-5}} = 2 - \sqrt{x-5}$

Solve or simplify, whichever is appropriate.

51. $3\sqrt{x} + 4 - 7\sqrt{x} + 1$

52. $2\sqrt{x+1} - 2 + 5\sqrt{x+1} - 9 = 3$

53. $3\sqrt{x} + 4 - 7\sqrt{x} + 1 = -7$

54. $2\sqrt{x+1} - 2 + 5\sqrt{x+1} - 9$

55. $(\sqrt{x} + 3)(\sqrt{x} + 1) = 3$

56. $(\sqrt{x} - 2)(\sqrt{x} - 3)$

57. $(\sqrt{x} + 3)(\sqrt{x} + 1)$

58. $(\sqrt{x} - 2)(\sqrt{x} - 3) = 2$

Find the x-intercept(s) of the function. Use a graphing calculator to verify your result(s).

59. $f(x) = 5\sqrt{x} - 7$

60. $g(x) = 2\sqrt{x} - 6$

61. $h(x) = 3\sqrt{-3x+4} - 15$

62. $k(x) = 2\sqrt{4x+1} - 22$

63. $f(x) = \sqrt{3x-2} - \sqrt{x+8}$

64. $g(x) = \sqrt{2x-5} + \sqrt{4x+1}$

65. $h(x) = 2\sqrt{x+4} + 3\sqrt{x-5}$

66. $k(x) = \sqrt{5x} - \sqrt{x+1}$

Let $f(x) = 3\sqrt{x} - 7$.

67. Find x when $f(x) = -1$.

68. Find x when $f(x) = -8$.

Let $f(x) = -2\sqrt{x-4} + 5$.

69. Find x when $f(x) = 7$.

70. Find x when $f(x) = 1$.

71. In exercise 67 in Section 9.4, you may have worked with the model $S(t) = 3.9\sqrt{t} + 280$, where $S(t)$ represents the average high school seniors' scores in standard testing on science at t years since 1982 (see Table 9).

Table 9 U.S. Seniors' Scores in Science

Year	Average Test Scores
1982	280
1986	287
1990	288
1992	293
1994	293
1996	295
1998	296

  a. Although test scores have been improving since 1982, the test score in 1970 was 305. How would you assess student performance since 1970?

  b. Use S to predict when the average score will return to the 1970 average of 305.

  c. Use S to predict when the average score will reach 500, the highest possible score.

72. In exercise 68 in Section 9.4, you may have worked with the model $S(t) = 248\sqrt{t} + 116$, where $S(t)$ represents the number of firm orders for Airbus airplanes at t years since 1995 (see Table 10).

Table 10 Numbers of Firm Orders

	Firm Orders	
Year	Airbus	Boeing
1995	116	484
1996	326	726
1997	458	579
1998	558	653

  a. Use S to predict when the number of firm orders for Airbus airplanes will reach Boeing's 1998 level of 653 airplanes.

  b. Use S to predict when the number of firm orders for Airbus airplanes will reach Boeing's 1996 level of 726 airplanes.

73. A student tries to solve the equation $5\sqrt{x+3} = 4$. Did the student solve the equation correctly? Explain.

$$5\sqrt{x+3} = 4$$
$$5(x+3) = 16$$
$$5x + 15 = 16$$
$$5x = 1$$
$$x = \frac{1}{5}$$

74. A student tries to solve the equation $\sqrt{x} + 1 = \sqrt{2x+1}$. The student believes that the work is correct because the check shows that 0 is a solution. What would you tell the student?

$$\sqrt{x} + 1 = \sqrt{2x+1}$$
$$(\sqrt{x})^2 + 1^2 = (\sqrt{2x+1})^2$$
$$x + 1 = 2x + 1$$
$$-x = 0$$
$$x = 0$$

  Check $x = 0$
$$\sqrt{x} + 1 = \sqrt{2x+1}$$
$$\sqrt{0} + 1 \stackrel{?}{=} \sqrt{2(0)+1}$$
$$1 \stackrel{?}{=} 1$$
  true

75. Solve the system. Then verify your solution graphically.
$$y = 3\sqrt{x} - 4$$
$$y = -2\sqrt{x} + 6$$

76. Create a system of two square root equations that has $(4, 5)$ as its only solution. Verify your system graphically.

77. The first of the following statements is true, yet the last statement is false. Describe the error(s).

$$2x - x = x$$
$$(2x)^2 - x^2 = x^2$$
$$4x^2 - x^2 = x^2$$
$$3x^2 = x^2$$
$$3 = 1$$

78. Describe how to solve square root equations that contain one square root. Also describe how to solve square root equations that contain two or more unlike square roots.

9.6 MODELING WITH SQUARE ROOT FUNCTIONS

Objectives

▸ Find an equation of a square root curve that contains two given points.

▸ Find an equation of a square root model.

▸ Use a square root model to make estimates and predictions.

In this section we use a square root function of the form

$$f(t) = a\sqrt{t} + b, \quad a \neq 0$$

to model data. We begin by finding an equation for a square root curve that contains two given points.

Example 1   Finding an Equation of a Square Root Function

Find the equation of a square root curve that contains the points $(2, 3)$ and $(5, 7)$.

Solution

We substitute the points $(2, 3)$ and $(5, 7)$ into the equation $f(x) = a\sqrt{x} + b$.

$$3 = a\sqrt{2} + b$$

$$7 = a\sqrt{5} + b$$

We then calculate approximate values for $\sqrt{2}$ and $\sqrt{5}$:

$$1.41a + b = 3 \tag{1}$$

$$2.24a + b = 7 \tag{2}$$

Next, we multiply both sides of equation (1) by -1.

$$-1.41a - b = -3 \tag{3}$$

Then, we eliminate b by adding the left sides of equations (2) and (3) and the right sides of equations (2) and (3).

$$0.83a = 4$$

$$a = \frac{4}{0.83}$$

$$a \approx 4.82$$

So, our equation has the form $f(x) = 4.82\sqrt{x} + b$. To find b, we substitute the point $(2, 3)$ into the equation $f(t) = 4.82\sqrt{t} + b$.

$$3 = 4.82\sqrt{2} + b$$

$$b = 3 - 4.82\sqrt{2}$$

$$b \approx -3.82$$

So, the equation is $f(t) = 4.82\sqrt{t} - 3.82$.

We use a graphing calculator to verify that the graph of f approximately contains the two given points (see Fig. 30). ∎

To find an equation of a square root curve that contains two given points, we substitute the points into the equation $f(x) = a\sqrt{x} + b$ which gives a system of two equations. We find values for a and b by solving the system using elimination or substitution.

In Example 2 we find an equation of a square root model.

Note

The graph approximately contains the points since we used approximate values for $\sqrt{2}$ and $\sqrt{5}$ and we rounded the values for a and b to the nearest hundredth.

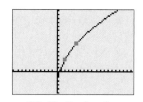

Figure 30 Verify that the graph of $f(x) = 4.82\sqrt{t} - 3.82$ approximately contains the points $(2, 3)$ and $(5, 7)$

Example 2 Finding an Equation of a Square Root Model

The values of money lent to students by direct student loans and guaranteed student loans for various years are listed in Table 11. Let $f(t)$ represent the value of student loans (in billions of dollars) made during the year that is t years since 1993. Find a formula for $f(t)$.

Solution

First, we use a graphing calculator to draw a scattergram of the data (see Fig. 31).

Table 11 Student Loan Totals	
Year	New Student Loans (billions of dollars)
1993	12.0
1994	18.0
1995	22.0
1996	24.0
1997	25.5

Source: *U.S. Department of Education*

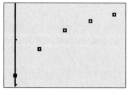

Figure 31 Student loan scattergram

It appears that the data might be modeled well by a square root function or a quadratic function. To decide which model to use, we find an equation for each type of function and then compare the fit of each model.

For the square root model, we use a function of the form

$$f(t) = a\sqrt{t} + b$$

We use the data points $(0, 12)$ and $(4, 25.5)$ to find values for a and b. To find b, we substitute the data point $(0, 12)$ in the equation $f(t) = a\sqrt{t} + b$:

$$12 = a\sqrt{0} + b$$
$$12 = b$$

So, the equation is of the form

$$f(t) = a\sqrt{t} + 12$$

To find a, we substitute the data point $(4, 25.5)$ into the equation $f(t) = a\sqrt{t} + 12$.

$$25.5 = a\sqrt{4} + 12$$
$$25.5 = 2a + 12$$
$$13.5 = 2a$$
$$a = 6.75$$

So, the square root function is $f(t) = 6.75\sqrt{t} + 12$.

The quadratic regression equation is:

$$f(t) = -0.79t^2 + 6.44t + 12.13$$

Next, we use a graphing calculator to compare the fits of the two models (see Figs. 32 and 33).

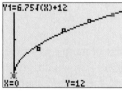

Figure 32 Square root model

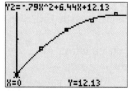

Figure 33 Quadratic model

The fit appears to be fairly good with each model. Finally, we ZOOM OUT and compare the graphs again (see Figs. 34 and 35).

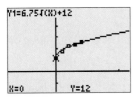

Figure 34 Square root model

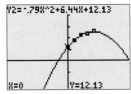

Figure 35 Quadratic model

The square root model predicts that the values for student loans for each year will continue to slowly increase, whereas the quadratic model predicts that the values will significantly decrease and, in fact, reach zero dollars (in the year 2003—as we can show, with work). Taking into account inflation, the square root model will probably prove to be the better of the two models. ■

Note

The quadratic model predicts that after 2003, the total loan values will be negative. Model breakdown has occurred for these predictions.

In Example 2 we found the equation of a square root model $f(t) = a\sqrt{t} + b$ without having to solve a system of equations. We found b by substituting the point $(0, 12)$ into the equation $f(t) = a\sqrt{t} + b$:

$$12 = a\sqrt{0} + b$$
$$b = 12$$

Then we found a by substituting another data point into the equation $f(t) = a\sqrt{t} + 12$.

In general, if one of the two points we choose to use is of the form $(0, k)$, we can find a square root equation without having to solve a system of equations.

Example 3   Making Estimates and Predictions

In Example 2 we found the equation $L = f(t) = 6.75\sqrt{t} + 12$, where $f(t)$ represents the value of student loans (in billions of dollars) during the year that is t years since 1993.

1. Find the L-intercept. What does it represent in terms of the value of student loans?
2. Predict the value of student loans for 2003.
3. Predict the year during which student loans will total $40 billion.

Solution
1. To find the L-intercept, we find $f(0)$:

$$f(0) = 6.75\sqrt{0} + 12 = 12$$

The L-intercept is $(0, 12)$. This means that the value of student loans in 1993 was $12.0 billion.

2. We represent 2003 by $t = 10$. We predict the value of student loans in 2003 by finding $f(10)$:

$$f(10) = 6.75\sqrt{10} + 12 \approx 33.35$$

According to the model, the value of student loans in 2003 will be approximately $33.4 billion.

3. To find the year during which the value of student loans will be $40 billion, we substitute 40 for $f(t)$ and solve for t:

$$6.75\sqrt{t} + 12 = 40$$
$$6.75\sqrt{t} = 28$$
$$\sqrt{t} = \frac{28}{6.75}$$
$$(\sqrt{t})^2 = \left(\frac{28}{6.75}\right)^2$$
$$t \approx 17.21$$

We check that 17.21 approximately satisfies the original equation:

$$6.75\sqrt{17.21} + 12 \approx 40.002$$

Since $1993 + 17 = 2010$, the model predicts that in 2010 the value of student loans will be approximately \$40 billion. ∎

In Example 3 we found that for $L = 6.75\sqrt{t} + 12$, the L-intercept is $(0, 12)$. Note that 12 is the constant term of the right side of the equation $L = 6.75\sqrt{t} + 12$.

More generally, substituting 0 for t in the model $y = a\sqrt{t} + b$ gives

$$y = a\sqrt{0} + b$$
$$y = b$$

So, the y-intercept of $y = a\sqrt{t} + b$ is $(0, b)$. See Fig. 36.

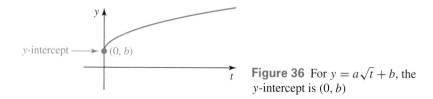

Figure 36 For $y = a\sqrt{t} + b$, the y-intercept is $(0, b)$

We close this section with a reminder about the four-step modeling process.

Four-Step Modeling Process

1. Create a scattergram of the data. Decide whether any of a line, an exponential curve, a parabola, or a square root curve comes close to the points. If a square root curve seems suitable, choose two points to use to derive the equation of your square root model.

2. Find an equation for your function.

3. Verify that your equation has a graph that comes close to the points in the scattergram. If it doesn't, check for calculation errors or try using different points to find the equation. An alternative is to reconsider your choice of model in step 1.

4. Use your equation of the model to draw conclusions, make estimates, and/or make predictions.

Tips on Succeeding in This Course

In preparing for a final examination, it is helpful to consider how all the concepts you have learned are interconnected. Recall that one way to help yourself do this is to make a mind map. (See the "Tips on Succeeding in This Course" in Section 1.4.) A portion of a mind map that describes this course is illustrated in Fig. 37. Obviously, many more "concept rectangles" could be added to it. You could make one mind map showing an overview of the course and then several mind maps for the components of the overview mind map.

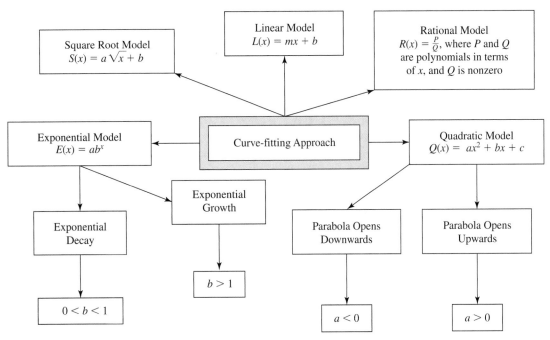

Figure 37 A portion of a mind map describing the course

EXPLORATION

Looking ahead: Arithmetic sequences

A math tutor is willing to tutor individual students or groups of students. The tutor charges $25 per hour for one student plus $8 per hour for each additional student.

1. Let $f(n)$ represent the amount of money (in dollars) the tutor will charge per hour if n students want to be tutored. Find an equation for f.

2. Evaluate f at each given value of n. Explain what each result means in terms of the situation.

   a. $f(3.8)$ **b.** $f(-2)$ **c.** $f(0)$

3. Based on your results in problem 2, determine a domain for the *model* f. [**Hint**: Which inputs make sense in terms of the situation?]

4. Based on your domain for f, find the range of f. List the values of the range of f in this order: $f(1), f(2), f(3), f(4), \ldots$ What do you notice about these numbers?

5. Sketch a graph of f, but only plot points whose n-coordinates are in the domain. [**Hint**: Your graph will look like a scattergram.]

OF THIS SECTION

Use two points to find a square root model of the form $y = a\sqrt{t} + b, a \neq 0$.

Finding an equation for a square root model

- If one of the two points is the y-intercept $(0, k)$, then the equation has the form $y = a\sqrt{t} + k$. To find a, substitute the other point into this equation and solve for a.

- If neither of the two points is the y-intercept, substitute the two points into the equation $y = a\sqrt{t} + b$ to obtain two equations. Then find values for a and b by solving the system of two equations.

HOMEWORK 9.6

SSM
Tutor Center

MathPro 4
MathPro 4

MathPro 5
MathPro 5

CDs/Videos

www
www

Find an equation of a square root curve that contains the given points.

1. (0, 3) and (4, 5)

2. (0, 5) and (4, 3)

3. (0, 2) and (9, 6)

4. (0, 3) and (1, 2)

5. (0, 2) and (3, 5)

6. (0, 1) and (2, 3)

7. (1, 2) and (4, 3)

8. (4, 5) and (9, 8)

9. (2, 4) and (3, 5)

10. (5, 2) and (7, 4)

11. (2, 6) and (5, 4)

12. (3, 8) and (6, 5)

Consider the scattergram of data and the graph of a model $f(t) = a\sqrt{t} + b$ *in the indicated figure. Sketch the graph of a square root model that better describes the data and then explain how you would adjust the values of a and b of the original model in order to better describe the data.*

13. See Fig. 38.

14. See Fig. 39.

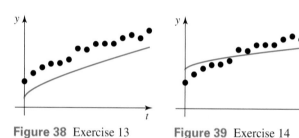

Figure 38 Exercise 13   **Figure 39** Exercise 14

15. As the number of births out of wedlock has increased and society's attitude toward single-parent households headed by fathers has changed, the number of such households has been increasing since 1976. Also rising is the percentage of such fathers who have never been married. (See Table 12.)

Table 12 Never-Married Single-Parent Fathers

Year	Percent of Single-Parent Fathers Who Have Never Been Married
1982	10.0
1984	17.0
1986	20.0
1988	21.0
1990	24.5
1992	27.0
1994	29.0
1996	29.9
1998	35.0
2000	34.0

Source: *National Center for Children in Poverty*

Let $p = f(t)$ represent the percentage of single-parent fathers who have never been married and let t represent the number of years since 1982.

a. Find an equation for f.

b. Find $f(26)$. What does your result mean in terms of single-parent fathers?

c. Find t when $f(t) = 39$. What does your result mean in terms of single-parent fathers?

d. Find the p-intercept. What does it mean in terms of single-parent fathers?

16. The median heights (in inches) for girls of various ages (in months) in the United States are listed in Table 13.

Table 13 Girls' Heights

Age (months)	Median Height (inches)
0	19.0
6	25.2
12	28.4
18	30.9
24	32.9
36	37.5
48	40.2
60	43.1

Source: The Portable Pediatrician for Parents
by Laura Walther Nathanson, M.D., FAAP

Let $h = f(t)$ represent the median height (in inches) of girls who are t months of age.

a. Find an equation for f.

b. Use f to estimate the median height of girls who are 6 years old. [**Hint**: Use the correct units.]

c. Use the equation for f to estimate the age at which the median height of girls is 4 feet.

d. Find the h-intercept. What does it mean in terms of median heights of girls?

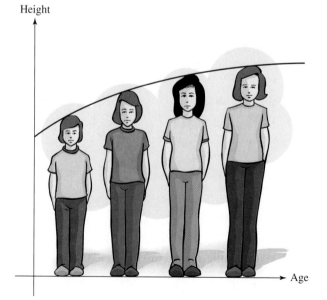

e. Use f to estimate the age at which the median height of girls is 10 feet. [**Hint:** Remember, if you think model breakdown occurs, say so, say where, and explain why.]

17. Some students ran an experiment to explore the relationship between the time it takes a baseball to fall to the ground when dropped from various heights (see Table 14). Let $T = S(h)$ represent the time (in seconds) it takes a baseball to fall to the ground when dropped from h feet above the ground.

Table 14 Drop Heights and Falling Times of Baseball*

Drop Height (feet)	Falling Time (seconds)
0.00	0.00
3.28	0.53
13.10	0.96
26.30	1.40
39.40	1.70
52.50	1.94

Source: *Dan R., Tim M., Lorraine L., Conrad H., and Maurisa M. collected the data during Spring Semester, 1994*

a. Find a square root function for S.

b. A linear model for the baseball drop data is given by $L(h) = 0.034h + 0.327$. A quadratic model is given by $Q(h) = -0.00058h^2 + 0.064h + 0.165$.

   i. Describe how well each of the functions S, L, and Q fits the data points in the scattergram.

   ii. Find $S(0)$, $L(0)$, and $Q(0)$. Which function models the baseball situation the best near $h = 0$? Explain.

   iii. ZOOM OUT and decide which of the functions could not possibly be a good model for the baseball situation when the drop heights are very large. Explain.

   iv. Based on your responses to parts i–iii, determine which of S, L, or Q best models the baseball drop situation. Explain.

   v. An equation that is often used to model drop times is $T = \sqrt{\dfrac{2h}{g}}$, where g is the constant 32.2 feet per second squared. Find an approximate form $T = a\sqrt{h}$ of this equation so that you can compare the model to your square root model S.

c. Suppose that you wish to estimate the height of a sheer cliff. While standing, you drop a stone from the cliff at shoulder level, and it takes 3 seconds to reach the foot of the cliff. Use your model to estimate the height of the cliff.

d. Use your model to estimate how long it would take for a baseball to reach the ground if it were to be dropped from the top of New York City's Empire State Building, which is 1250 feet tall.

18. Suppose that all soup cans are 10 centimeters tall. Then, if the radius of the base of the can is relatively large, the amount of soup in the can will be relatively large. How large should a soup can manufacturer make the radius of the base so the can will hold a certain amount of soup? Possible volumes of a 10-centimeter tall soup can and the corresponding radii of the can's base are listed in Table 15.

Table 15 Radii of the Bases of Soup Cans†

Volume (cubic centimeters)	Radius (centimeters)
0	0
200	2.52
400	3.57
600	4.37
800	5.05

Let $f(V)$ represent the radius (in centimeters) of a 10-centimeter tall soup can that has volume V cubic centimeters.

a. Find an equation for f.

b. A can of Campbell's® tomato soup is 10 centimeters tall and has a base radius of 3.3 centimeters. What is the volume of the can?

c. A soup can is 10 centimeters tall and has a volume of 384.85 cubic centimeters. What is the radius of the base of the can?

d. Is f an increasing function, a decreasing function, or neither? What does your answer mean in terms of the radius of a 10-centimeter tall soup can?

19. The percentages of convicts released from a state prison who are arrested for a new crime after being out of prison for various numbers of years are shown in Table 16.

Table 16 Released Convicts Who Have Been Arrested for a New Crime

Number of Years Since Release	Percent
0	0
0.5	30
1	44
2	59
3	68

Source: *U.S. Department of Justice*

---

*Although a ball cannot be dropped from height zero feet, the data point $(0, 0)$ is included because drop heights near zero will correspond to falling times near zero.

†Although a can cannot have radius zero, the data point $(0, 0)$ is included because a can with volume near zero will have a radius near zero.

Let $S(t)$ represent the percentage of convicts released from a state prison who are arrested for a new crime after being out of prison for t years.

a. Find a square root function for S.

b. A quadratic model for the data is given by

$$Q(t) = -9.05t^2 + 48.09t + 3.47$$

i. Describe how well each of the functions S and Q fits the data points in the scattergram.

ii. Explain in terms of the situation why the model should be increasing for $t \geq 0$. Are either of the functions S or Q increasing for $t \geq 0$?

c. Use S to estimate the percentage of released convicts who have been arrested for a new crime after being out of prison for four years.

d. Use S to estimate after how many years since being released, all convicts will have been arrested for a new crime.

20. In 1970, 5.4% of Americans lived alone. In 1990, this percentage had nearly doubled to 9.2%. This increase is due in part to greater acceptance by society of those who live alone, an increase in the divorce rate, as well as people waiting longer to get married (see Table 17).

Table 17 People Living Alone

Year	Number of People (in millions)
1970	10.9
1980	18.3
1985	20.6
1990	23.0
1995	24.7
2000	26.7

Source: *U.S. Bureau of the Census*

Let $n = S(t)$ represent the number (in millions) of people who live alone at t years since 1970.

a. Find a square root function for S.

b. A quadratic model for the living alone data is given by $Q(t) = -0.0087t^2 + 0.78t + 11.03$.

i. Describe how well each of the functions S and Q fits the data points in the scattergram.

ii. ZOOM OUT and decide which of the functions will most likely best model the living alone situation in the future. Explain.

iii. Which function best models the living alone situation for years before 1970? Explain.

c. Use S to predict the number of people who will live alone in 2008.

d. Use S to predict when 1 out of every 10 people will live alone. Use the 2002 U.S. population of 287 million.

21. As the world population continues to grow, birth control has become a critical issue for the well-being of everyone. Although some of the population boom is due to couples who don't use contraceptives (even if their family size is large), there are many couples who have babies in spite of using contraceptives.

In a study of births among 1444 fertile couples, the birth order and claims by the parents that conception happened in spite of contraception were recorded (see Table 18). The percentage, for example, of first births that happened in spite of contraception, according to the parents, was 37.6%.

Table 18 Percentages of Births "in Spite of Contraception"

Birth Order	Percentage of Conceptions "in Spite of Contraception"
1	37.6
2	54.3
3	66.5
4	73.0
5	84.1
6	81.2

Source: Social and Psychological Factors Affecting Fertility *by Whelpton and Kiser*

Let $p = f(n)$ represent the percentage of births of the nth-born child that happened in spite of contraception at the time of conception, as claimed by the parents.

a. Perform the first three steps of the modeling process to find an equation for f. [**Hint:** Use the points $(2, 54.3)$ and $(4, 73.0)$.]

b. Find $f(7)$. What does your result mean in terms of contraception?

c. Find n when $f(n) = 100$. What does your result mean in terms of contraception?

d. Note that f is an increasing function. What does this mean in terms of contraception? Form a theory to explain why this happens.

22. A study of children who were adopted after being in temporary foster care explored the relationship between the age of the children when separated from their foster parents and the percentage of children showing problems immediately after the separation. Some data are listed in Table 19.

Table 19 Adopted Children Separated from Foster Parents

Age When Separated from Foster Parents	Middle of Age Group	Percent Showing Problems
<3	1.5	4
3–4	3.5	40
4–5	4.5	70
6–8	7.0	90
9	9.0	100

Source: The Immediate Impact of Separation: Reactions of Infants to a Change in Mother Figures *by Yarrow and Goodwin*

Let $p = f(t)$ represent the percentage of children showing problems immediately after separation if they are separated at age t years.

a. Find an equation for f.

b. Is f an increasing function, decreasing function, or neither? What does this mean in terms of adopted children?

c. Use f to estimate the percentage of children who show problems if separated at age 10. Is this a reasonable estimate? If so, explain why. If not, explain why not and suggest a better value.

d. Sketch a qualitative graph that describes the relationship between t and p for *all* ages of children.

23. Do children from large households tend to try to take more control or less control of their environment than do children from small households? A study was performed at Yale University to address this question. In this experiment, a child would receive a marble if she hit a key when a light was a certain color. There were two components in this experiment. In one component, the subject could trade a marble for a candy selected by the experimenter. In the other component, the subject could trade a marble for a self-selected candy. The experiment lasted for 360 seconds and would automatically switch from one component to the other every 30 seconds. However, the subject could also switch the component by pressing a switching key. The household sizes of the children and the corresponding amounts of time spent in each component are listed in Table 20.

Let $f(n)$ represent the average time (in seconds) spent in the experimenter-selected candy component by a child living with n people.

a. Perform the first three steps of the modeling process. [**Hint**: Use the points (2, 85.3) and (6, 160).]

Table 20 Average Times Spent in Each Component

Number of People Living with Child	Self-selected Candy	Experimenter-selected Candy
2	274.7	85.3
3	248.8	111.2
4	217.4	142.6
5	234.8	125.2
6	200.0	160.0
7	176.6	183.4
8	180.2	179.8

Source: *Journal of Experimental Social Psychology*

b. Is f an increasing function, decreasing function, or neither? What does your answer suggest in terms of how the size of a child's household affects the extent to which the child tries to take control?

Taking it to the lab

PENDULUM LAB

In exercise 48 of Homework 8.7, you may have modeled data about a pendulum. You can construct a pendulum by tying one end of some thread to a washer and attaching the other end of the thread to a surface so the washer is suspended and can swing freely.

Let the washer swing forward and backward. Start a timer just when the washer begins to swing forward. The **period** of the pendulum is the amount of time it takes for the washer to swing forward *and* backward once.

The period of a pendulum depends on the length of the thread. In this lab you will discover the relationship between the period of a pendulum and the length of its thread.

Materials

You will need the following materials:

1. Thread
2. Scissors
3. A timing device
4. Tape
5. A washer or some other small object with high density
6. A meter stick

Preparation

Knot one end of the thread to the washer. Tape the other end of the thread to a surface so the washer is suspended and can swing freely. The distance from the middle of the washer to the tape should be at least 110 centimeters.

Note

You will introduce some error in estimating the period by not starting and stopping your timing device at the right moments. By timing 4 "cycles" and then dividing by 4, the ratio of the error to the period will be one-fourth the ratio should you time just one cycle.

Recording of Data

Record the distance from the middle of the washer to the tape. Then time how long it takes for the washer to swing back and forth four times. Divide this time by 4 to find the period of the pendulum. Repeat this several times. Discard the times that are very different from most of the times and average the remaining times.

Repeat this procedure for various lengths of thread. When the thread is quite short, time how long it takes for the washer to swing back and forth eight times and then divide this time by 8 to find the period.

Analyzing the Data

Consult with your instructor as to how you should organize your responses to the following questions. Your instructor may want you to number your responses and to list them in order. On the other hand, your instructor may want you to present your responses in paragraph form without numbering your responses.

1. Display your data in a table.
2. Let $f(L)$ represent the period (in seconds) of the pendulum where L is the length (in centimeters) of the thread. Create a scattergram of your pendulum data.
3. Perform the first three steps of the modeling process.
4. What are the intercept(s) of f? What do these intercept(s) mean in terms of the pendulum?
5. Is f an increasing function, decreasing function, or neither? What does this mean in terms of the pendulum?
6. Use your formula for f to predict the period of the pendulum if its length is 150 centimeters.
7. How long is the pendulum if its period is 0.1 second?
8. Suppose that a big chunk of concrete attached to some thin, strong cable is suspended over Wacker Drive from the skydeck of the Sears Tower in Chicago. The skydeck is 1353 feet above Wacker Drive. Assume that the concrete chunk almost touches the street when at rest. Estimate the period of this pendulum. (Note: f can be used to model the Sears Tower pendulum, even though a concrete chunk weighs a lot more than a washer.)
9. What is the length of a pendulum that has a period of one minute?

Note

What *is* crucial is that the weight of the concrete should weigh substantially more than the thin cable, just as the washer weighs substantially more than the thread.

CHAPTER SUMMARY

Key Points
OF THIS CHAPTER

Power property	Let n be a counting number greater than 1. • If n is even, then $\sqrt[n]{x^n} = \lvert x \rvert$. • If n is odd, then $\sqrt[n]{x^n} = x$.
Another version of the power property	If $\sqrt[n]{x}$ is defined, then $(\sqrt[n]{x})^n = x$.
Product and quotient properties	Assume that $\sqrt[n]{a}$ and $\sqrt[n]{b}$ are defined. • Product property: $\sqrt[n]{ab} = \sqrt[n]{a}\,\sqrt[n]{b}$. • Quotient property: If $b \neq 0$, then $\sqrt[n]{\dfrac{a}{b}} = \dfrac{\sqrt[n]{a}}{\sqrt[n]{b}}$.
Adding or subtracting like radicals	To add or subtract like radicals, we use the distributive law.
Multiplying radicals with the same index	To multiply radicals that have the same index, we use the product property.

To multiply two radicals with different indexes and equal radicands, or to find a smaller index for a radical:

1. Write the radical(s) in exponential form.
2. Simplify the exponential expression.
3. Write the simplified exponential expression in radical form.

To simplify a radical expression:

1. Perform any indicated multiplications.
2. Combine like radicals.
3. Write a radicand as a product of an expression and one or more nth perfect powers. Then apply the product property.
4. Write a radical with as small an index as possible.

To rationalize the denominator of a radical expression of the form $\dfrac{1}{\sqrt[n]{x^m}}$, multiply the expression by a fraction of the form $\dfrac{\sqrt[n]{x^k}}{\sqrt[n]{x^k}}$ so that the radical in the denominator has a perfect nth power radicand.

The radical expressions $a\sqrt{b} + c\sqrt{d}$ and $a\sqrt{b} - c\sqrt{d}$ are *conjugates* of each other.

To rationalize the denominator of a square root expression whose denominator is a sum or a difference:

1. Determine the conjugate of the denominator.
2. Multiply the original fraction by the fraction $\dfrac{\text{conjugate}}{\text{conjugate}}$.
3. Find the product of the denominators using the difference of two squares formula.

For a square root function $f(x) = a\sqrt{x - h} + k$ with $a \neq 0$:

- The graph of f can be found by translating the graph of $y = a\sqrt{x}$ horizontally by $|h|$ units (right if $h > 0$, left if $h < 0$) and vertically by $|k|$ units (up if $k > 0$, down if $k < 0$).
- If $a > 0$, then f is an increasing function and (h, k) is a minimum point.
- If $a < 0$, then f is an decreasing function and (h, k) is a maximum point.
- If a is a large number, then the graph of $f(x) = a\sqrt{x - h} + k$ rises much more quickly than the graph of $y = \sqrt{x - h} + k$, and if a is a positive number near zero, the graph rises much more slowly.

If a square root equation contains the variable in exactly one square root, we isolate the square root on one side of the equation and then square both sides.

If a square root equation contains *like* square roots, we combine the like square roots, isolate the square root containing the variable on one side of the equation, and then square both sides.

If a square root equation contains two or more unlike square roots, we isolate one of the square roots and then square both sides of the equation. After squaring both sides, the equation will have at least one less square root. We repeat this process of isolating and squaring until no square roots remain.

When solving a square root equation, always check results for extraneous solutions.

Use two points to find a square root model of the form $y = a\sqrt{t} + b$, $a \neq 0$.

- If one of the two points is the y-intercept $(0, k)$, then the equation has the form $y = a\sqrt{t} + k$. To find a, substitute the other point into this equation and solve for a.
- If neither of the two points is the y-intercept, substitute the two points into the equation $y = a\sqrt{t} + b$ to obtain two equations. Then find values for a and b by solving the system of two equations.

CHAPTER 9 REVIEW EXERCISES

If an expression is in exponential form, then write it in radical form. If it is in radical form, then write it in exponential form.

1. $x^{3/7}$

2. $\sqrt{x}$

3. $(2x+1)^{2/9}$

4. $\sqrt[5]{(3x+4)^7}$

Simplify each expression. Assume that $x \geq 0$. Use a graphing calculator to verify your result.

5. $\sqrt{16x}$

6. $\sqrt{8x^6}$

7. $\sqrt{3x^7}$

8. $\sqrt[8]{x^6}$

9. $\sqrt[3]{24x^{10}}$

10. $\sqrt[5]{(6x+11)^{27}}$

11. $2\sqrt{x} - 5\sqrt{x} + 3\sqrt{x}$

12. $5\sqrt[3]{x} - 2\sqrt{x} + 7\sqrt[3]{x} + 4\sqrt{x}$

13. $5(4\sqrt{x} - \sqrt[3]{x}) - 2\sqrt[3]{x} + 8\sqrt{x}$

14. $3\sqrt{x}(\sqrt{x} - 7)$

15. $(4\sqrt{x} - 3)(2\sqrt{x} + 1)$

16. $2(3\sqrt{x} + 8)(4\sqrt{x} + 2)$

17. $(\sqrt{x} + 1)(\sqrt{x} - 1)$

18. $(3\sqrt{x} + 4)(3\sqrt{x} - 4)$

19. $(2\sqrt[3]{x} + 5)^2$

20. $(5\sqrt{x} - 2)^2$

21. $\sqrt[4]{x}\sqrt[5]{x}$

22. $\sqrt[3]{\sqrt[6]{x}}$

23. $\dfrac{\sqrt[4]{x}}{\sqrt[6]{x}}$

24. $\dfrac{x}{\sqrt{2}}$

25. $\sqrt{\dfrac{3}{x}}$

26. $\dfrac{5}{\sqrt[3]{x}}$

27. $\sqrt[5]{\dfrac{7}{27x^2}}$

28. $\dfrac{5}{3 + \sqrt{x}}$

29. $\dfrac{2}{2 - 3\sqrt{x}}$

30. $\dfrac{5\sqrt{x} - 4}{2\sqrt{x} + 3}$

Let $f(x) = 3\sqrt{x} + 5$ and $g(x) = 2 - 4\sqrt{x}$. Find an equation for h.

31. $h(x) = f(x) + g(x)$

32. $h(x) = f(x) - g(x)$

33. $h(x) = f(x)g(x)$

34. $h(x) = \dfrac{f(x)}{g(x)}$

Solve the equation.

35. $3\sqrt{x} + 4 = 13$

36. $\sqrt{2x+1} + 4 = 7$

37. $\sqrt{4x+5} = x$

38. $\sqrt{2x-4} = x - 2$

39. $\sqrt{x} + 6 = x$

40. $\sqrt{13x+4} = \sqrt{5x-20}$

41. $\sqrt{2x-1} = 1 + \sqrt{x+3}$

42. $\sqrt{x+2} + \sqrt{x+9} = 7$

Use pencil and paper to sketch the graph of the function. Use a graphing calculator to verify your sketch.

43. $y = -2\sqrt{x}$

44. $y = 3\sqrt{x} + 1$

45. $y = -\sqrt{x-5} + 3$

46. $y = 2\sqrt{x+4} - 1$

Find the x-intercept(s) of the function. Use a graphing calculator to verify your result(s).

47. $f(x) = \sqrt{4x-7} - \sqrt{2x+1}$

48. $g(x) = -3\sqrt{x+2} + 9$

49. Consider the scattergram of data and the graph of the model $f(x) = a\sqrt{x} + b$ in Fig. 40. Sketch the graph of a square root model that fits the data better and then explain how you would adjust the values of a and b of the original model in order to better describe the data.

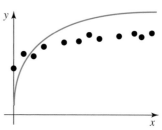

Figure 40 Exercise 49

50. Find an equation for the square root function whose graph is sketched in Fig. 41. Verify your equation using your graphing calculator.

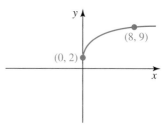

Figure 41 Exercise 50

Find an equation of a square root curve that contains the given points.

51. $(0, 3)$ and $(4, 8)$

52. $(0, 7)$ and $(9, 3)$

53. $(2, 5)$ and $(3, 6)$

54. $(3, 7)$ and $(5, 4)$

55. The most popular video game console in the United States is the Sony PlayStation™. The numbers of households in the United States that have this console for various years are listed in Table 21.

Table 21 Numbers of Households That Have a Sony PlayStation	
Year	Number of Households That Have a Sony PlayStation (millions)
1998	16.0
1999	21.9
2000	25.0
2001	27.4
2002	28.8

Source: *International Development Group*

Let $n = f(t)$ represent the number of households (in millions) that have a Sony PlayStation at t years since 1998.

a. Find an equation for f.

b. Find the n-intercept of f. What does it mean in terms of the situation?

c. Find t when $f(t) = 36$. What does your result mean in terms of the situation?

d. Use f to predict the number of households that will have Sony PlayStations in 2003. Compare your prediction with International Development Group's prediction of 29.4 million households.

CHAPTER 9 TEST

Simplify each expression. Assume that $x \geq 0$.

1. $\sqrt{32x^9}$

2. $\sqrt[3]{64x^{22}}$

3. $\sqrt[4]{(2x+8)^{27}}$

4. $\dfrac{4\sqrt[3]{x}}{6\sqrt[5]{x}}$

5. $\dfrac{\sqrt{x}+1}{2\sqrt{x}-3}$

6. $3\sqrt{12x} + 2\sqrt[4]{x} - 8\sqrt{27x} - 5\sqrt[4]{x}$

7. $3\sqrt{x}(6\sqrt{x}-5)$

8. $(2+4\sqrt{x})(3-5\sqrt{x})$

9. $(4+3\sqrt{x})(4-3\sqrt{x})$

10. $(4\sqrt[5]{x}-3)^2$

11. Show that the following statement is true:

$$\frac{\sqrt[n]{x}}{\sqrt[k]{x}} = \sqrt[kn]{x^{k-n}}$$

where n and k are counting numbers greater than 1 and $x > 0$.

Let $f(x) = 7 - 3\sqrt{x}$ and $g(x) = 4 + 5\sqrt{x}$. Find an equation for $h(x)$.

12. $h(x) = f(x) + g(x)$

13. $h(x) = f(x) - g(x)$

14. $h(x) = f(x)g(x)$

15. $h(x) = \dfrac{f(x)}{g(x)}$

Let $f(x) = 6 - 4\sqrt{x+1}$.

16. Find $f(8)$.

17. Find a value of x so that $f(x) = -2$.

18. Sketch the graph of the function $y = -2\sqrt{x+3} + 1$. Show work to demonstrate that you can sketch the graph without using a graphing calculator.

19. Let $f(x) = a\sqrt{x-h} + k$, $a \neq 0$.

a. What must be true of a, h, and k for f to have an x-intercept? [**Hint**: Think graphically.]

b. Now assume that f has an x-intercept. Find the x-intercept in terms of a, h, and k.

Solve each equation.

20. $2\sqrt{x} + 3 = 13$

21. $3\sqrt{5x-4} = 27$

22. $3 - 2\sqrt{x} + \sqrt{9-x} = 0$

23. Find the x-intercept(s) of the function

$$f(x) = 3\sqrt{2x-4} - 2\sqrt{2x+1}.$$

24. Consider the scattergram of the data and the graph of the model $y = a\sqrt{x} + b$ sketched in Fig. 42. Sketch the graph of a square root model that fits the data better and then explain how you would adjust the values of a and b of the original model in order to better describe the data.

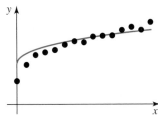

Figure 42 Exercise 24

25. Find an equation of a square root curve that contains the points $(2, 4)$ and $(5, 6)$.

26. The median heights for boys of various ages to five years in the United States are listed in Table 22.

Table 22 Boys' Median Heights	
Age (months)	Height (inches)
0	20.5
6	27.0
12	30.8
18	32.9
24	35.0
36	37.5
48	40.8
60	43.4

Source: The Portable Pediatrician for Parents *by Laura Walther Nathanson, M.D., FAAP*

Let $h = f(t)$ represent the median height (in inches) of boys who are t months of age.

a. Find an equation for f.

b. Estimate the median height of boys who are 6 years old.

c. Estimate the age at which the median height of boys is 3 feet.

d. Find the h-intercept. What does it mean in terms of the median height of boys?

Modeling with Sequences and Series

Here, where we reach the sphere of mathematics, we are among processes which seem to some the most inhuman of all human activities and the most remote from poetry. Yet it is here that the artist has the fullest scope of his imagination.

—*Havelock Ellis, The Dance of Life, 1923*

10.1 ARITHMETIC SEQUENCES

Objectives

▸ Know the meaning of *arithmetic sequence*.

▸ Find a formula of an arithmetic sequence.

▸ Make estimates and predictions using an arithmetic sequence.

In earlier chapters we worked with linear and exponential functions. In this chapter we reexamine these functions from a different perspective. First, we take another look at linear functions.

To begin, suppose that a math tutor charges $23 per hour for one student plus $6 for each additional student. The tutor will take up to 10 students. We list the total charges (in dollars) for 1, 2, 3, ..., 9, and 10 students in order:

$$23, 29, 35, 41, 47, 53, 59, 65, 71, 77$$

We call the list of numbers a *sequence*.

DEFINITION Sequence

Any ordered list of numbers is called a **sequence**. Each number is a **term** of the sequence.

A sequence that has a last term, such as the math tutor sequence, is called a **finite sequence**. A sequence that does not have a last term is called an **infinite sequence**. For example, the sequence of odd numbers

$$1, 3, 5, 7, 9, \ldots$$

is an infinite sequence.

For the math tutor sequence, note that the difference between any term and the preceding term is 6:

$$29 - 23 = 6, 35 - 29 = 6, 41 - 35 = 6, \ldots, 77 - 71 = 6$$

We call 6 the *common difference* of the sequence and we say that the sequence is *arithmetic*.

DEFINITION *Arithmetic Sequence*

If the difference between any term of a sequence and the preceding term is a constant d for every such pair of terms, the sequence is an **arithmetic sequence**. We call the constant d the **common difference**.

Example 1   Identifying Arithmetic Sequences

Determine whether each sequence is arithmetic. If it is, find the common difference d.

1. $2, 6, 10, 14, 18, \ldots$ **2.** $80, 77, 74, 71, 68, \ldots$

3. $3, 6, 12, 24, 48, \ldots$

Solution

1. The sequence has a common difference of 4:

$$6 - 2 = 4, 10 - 6 = 4, 14 - 10 = 4, 18 - 14 = 4, \ldots$$

The sequence is arithmetic.

2. The sequence has a common difference of -3:

$$77 - 80 = -3, 74 - 77 = -3, 71 - 74 = -3, 68 - 71 = -3, \ldots$$

The sequence is arithmetic.

3. We can see from the first two differences that the sequence does not have a common difference:

$$6 - 3 = 3, 12 - 6 = 6$$

The sequence is not arithmetic.   ■

We use the notation $a_1, a_2, a_3, \ldots$ to denote the terms of a sequence. For the math tutor sequence we write:

$$a_1 = 23, a_2 = 29, a_3 = 35, \ldots, a_{10} = 77$$

where a_n represents the charge (in dollars per hour) for n students. We call the term's position in the list the *term number*. For example, the term number of $a_3 = 35$ is 3.

Next, we find a formula that describes the terms of an arithmetic sequence. Since the math tutor sequence has a common difference of 6, we add 6 to the first term 23 to find the second term, we add 6 two times to 23 to find the third term, we add 6 three times to 23 to find the fourth term, and so on:

$$23 + 6 = 29, 23 + 6 + 6 = 35, 23 + 6 + 6 + 6 = 41, \ldots$$

In general, for an arithmetic sequence with common difference d we have the terms

$$a_1, a_1 + d, a_1 + d + d, a_1 + d + d + d, \ldots$$

or by simplifying we have

$$a_1, a_1 + d, a_1 + 2d, a_1 + 3d, \ldots$$

There is a pattern that we can use to find a formula for any term a_n of the sequence:

$a_1 = a_1$

$a_2 = a_1 + d$ Add d once to the first term to get the second term.

$a_3 = a_1 + 2d$ Add d twice to the first term to get the third term.

$a_4 = a_1 + 3d$ Add d three times to the first term to get the fourth term.

$$\vdots$$

$a_n = a_1 + (n-1)d$ Add d a total of $(n-1)$ times to the first term to get the nth term.

Formula for the nth Term of an Arithmetic Sequence

If an arithmetic sequence $a_1, a_2, a_3, \ldots, a_n, \ldots$ has common difference d, then

$$a_n = a_1 + (n-1)d$$

In words, the nth term is equal to the first term plus $n - 1$ times the common difference.

Example 2   Finding a Formula

Find a formula that describes the terms of the math tutor sequence.

Solution

To find a formula for the math tutor sequence 23, 29, 35, 41, 47, ..., 77, we substitute $a_1 = 23$ and $d = 6$ in the formula $a_n = a_1 + (n-1)d$:

$$a_n = 23 + (n-1)(6) \qquad \blacksquare$$

In Example 2 we found the math tutor formula $a_n = 23 + (n-1)(6)$. Here we write the equation in the form $a_n = mn + b$:

$$a_n = 23 + (n-1)(6)$$
$$= 23 + 6n - 6$$
$$= 6n + 17$$

The equation $a_n = 6n + 17$ describes a linear function where the inputs are only the numbers of students 1, 2, 3, ..., 10 and the outputs are only the dollar charges 23, 29, 35, ..., 77. For example:

$$a_1 = 6(1) + 17 = 23$$
$$a_2 = 6(2) + 17 = 29$$
$$a_3 = 6(3) + 17 = 35$$
$$a_{10} = 6(10) + 17 = 77$$

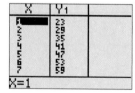

Figure 1 Verify the formula $a_n = 6n + 17$

We can verify our formula by entering $y = 6x + 17$ in a graphing calculator and checking that the inputs 1, 2, 3, ..., 10 give the outputs 23, 29, 35, ..., 77 (see Fig. 1).

We can also sketch a graph of the math tutor sequence. The graph consists of the ten input-output pairs (1, 23), (2, 29), (3, 35) ... (10, 77). For $a_n = 6n + 17$, the coefficient of the independent variable n is 6, so the points lie on an increasing line with slope 6 (see Fig. 2).

For an arithmetic sequence, if we know values for three of the four variables a_1, a_n, n, and d, we can find the value of the fourth variable by using the formula $a_n = a_1 + (n-1)d$.

As the formula $a_n = a_1 + (n-1)d$ is valid only for arithmetic sequences, we first check that a sequence is arithmetic before using the formula.

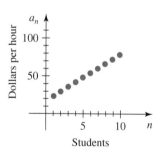

Figure 2 Math tutor sequence

Example 3 Finding a Term

Find the 25th term of the sequence 13, 20, 27, 34, 41, ...

Solution

The sequence has a common difference ($d = 7$), so the sequence is arithmetic. We substitute $a_1 = 13$, $n = 25$, $d = 7$ in the formula $a_n = a_1 + (n - 1)d$.

$$\begin{aligned} a_{25} &= 13 + (25 - 1)(7) \\ &= 13 + 24(7) \\ &= 181 \end{aligned}$$

So, the 25th term is $a_{25} = 181$. To verify our work, we enter $y = 13 + (x - 1)(7)$ in a graphing calculator and check that the first five terms of the sequence are 13, 20, 27, 34, and 41, and that the 25th term is 181 (see Fig. 3).

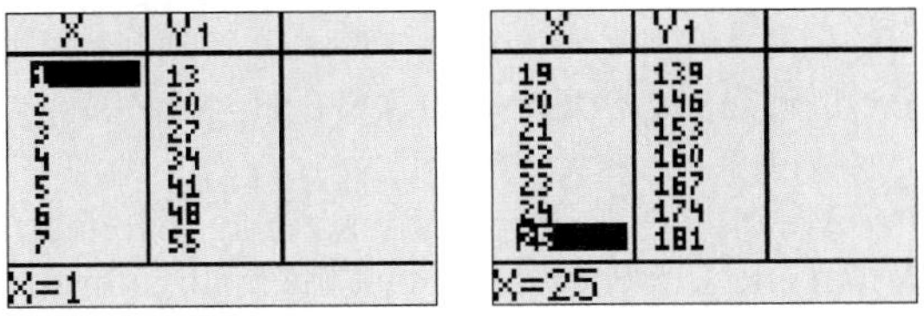

Figure 3 Verify the formula for the first five terms and the 25th term

■

Example 4 Finding a Term Number

The number 23 is a term in the sequence 155, 151, 147, 143, 139, ... What is its term number?

Solution

The sequence has a common difference ($d = -4$), so the sequence is arithmetic. We substitute $a_1 = 155$, $d = -4$, $a_n = 23$ in the formula $a_n = a_1 + (n - 1)d$ and solve for n.

$$\begin{aligned} 23 &= 155 + (n - 1)(-4) \\ 23 &= 155 - 4n + 4 \\ 4n &= 136 \\ n &= 34 \end{aligned}$$

So, 23 is the 34th term. In symbols, $a_{34} = 23$. We can use a graphing calculator table to verify our work. ■

Example 5 Modeling with an Arithmetic Sequence

A person's salary is $25,000 for the first year and will increase by $750 each year. Let a_n represent the salary (in dollars) for the nth year.

1. Find a formula for a_n.
2. What will be the salary for the 32nd year?
3. In what year will the salary be $40,000?

Solution

1. The salary sequence has a common difference of 750, so the sequence is arithmetic. We substitute $a_1 = 25{,}000$ and $d = 750$ in the formula $a_n = a_1 + (n - 1)d$ to find a formula for the sequence:

$$\begin{aligned} a_n &= 25{,}000 + (n - 1)(750) \\ &= 25{,}000 + 750n - 750 \\ &= 750n + 24{,}250 \end{aligned}$$

The formula is $a_n = 750n + 24{,}250$. We can verify our work using a graphing calculator table by checking that $a_1 = 25{,}000$ and that the common difference is 750.

2. To find the salary for the 32nd year, we substitute $n = 32$ in the formula $a_n = 750n + 24{,}250$.

$$a_{32} = 750(32) + 24{,}250 = 48{,}250$$

The salary will be \$48,250 during the 32nd year.

3. To determine when the salary will be \$40,000, we substitute $a_n = 40{,}000$ in the formula $a_n = 750n + 24{,}250$ and solve for n.

$$40{,}000 = 750n + 24{,}250$$
$$15{,}750 = 750n$$
$$n = \frac{15{,}750}{750}$$
$$n = 21$$

The salary will be \$40,000 during the 21st year. ∎

We close this section by noticing a connection between a linear function and an arithmetic sequence. Note that the math tutor formula $a_n = 6n + 17$ describes a linear function with slope equal to 6, which is the common difference of the sequence. If we let $f(n)$ represent the tutor's charge (in dollars) for n students, then

$$f(n) = 6n + 17$$

and the sequence 23, 29, 35, 41, 47, …, 77 is given by

$$f(1), \ f(2), \ f(3), \ \ldots, \ f(10)$$

Connection between a Linear Function and an Arithmetic Sequence

If f is a linear function, then

$$f(1), \ f(2), \ f(3), \ \ldots$$

is an arithmetic sequence with common difference equal to the slope of f.

So, for the linear function $g(x) = -5x + 105$, the sequence $g(1), g(2), g(3), \ldots$ is an arithmetic sequence with common difference -5, which is the slope of g. By computing the outputs $g(1), g(2), g(3), \ldots$, we see that the sequence is 100, 95, 90, 85, 80, ….

Recall that the slope addition property for linear functions states that if the inputs of a linear function increase by 1, the outputs change by the slope. If we think in terms of the slope addition property, it makes sense that the slope of a linear function and the common difference of the corresponding sequence are equal.

EXPLORATION

Connection between combining linear functions and forming new sequences

1. Let $f(x) = 3x + 5$. Use a graphing calculator table to find the values of the sequence $f(1), \ f(2), \ f(3), \ f(4), \ f(5)$. Determine whether the sequence is arithmetic. Explain. If the sequence is arithmetic, find the common difference and compare it to the slope of f.

2. For $f(x) = 3x + 5$, find the difference $f(x + 1) - f(x)$. Compare your result with your result in problem 1.

3. Let $f(x) = 2x + 1$ and $g(x) = 3x - 4$. For each of the following definitions of h, decide whether the sequence $h(1), h(2), h(3), h(4), h(5)$ is arithmetic. Explain.

   a. $h(x) = f(x) + g(x)$

   b. $h(x) = f(x) - g(x)$

   c. $h(x) = f(x)g(x)$

4. Assume that the sequence $a_1, a_2, a_3, \ldots$ is arithmetic with common difference d_a and the sequence $b_1, b_2, b_3, \ldots$ is arithmetic with common difference d_b. Decide whether the sequence $a_1 + b_1, a_2 + b_2, a_3 + b_3, \ldots$ is arithmetic. If so, what is the common difference?

Tips on Succeeding in This Course

At various points during the term, it is helpful to think about how different topics you have studied share certain characteristics. For example, you should think about the connection discussed in this section between a linear function and an arithmetic sequence. If you can make the connection, then it is easier to learn about arithmetic sequences since you have already learned a lot about linear functions. Also, your work with arithmetic sequences will reinforce what you know about linear functions.

Key Points OF THIS SECTION

If the difference between any term of a sequence and the preceding term is equal to a constant d for every such pair of terms, we call the sequence an **arithmetic sequence** and we call the constant d the **common difference**.

Arithmetic sequence

If an arithmetic sequence $a_1, a_2, a_3, \ldots, a_n, \ldots$ has common difference d, then $a_n = a_1 + (n-1)d$.

Formula for an arithmetic sequence

As the formula $a_n = a_1 + (n-1)d$ is valid only for arithmetic sequences, we must first check that a sequence is arithmetic before using the formula.

Check that a sequence is arithmetic

If f is a linear function, then $f(1), f(2), f(3), \ldots$ is an arithmetic sequence with common difference equal to the slope of f.

Connection between a linear function and an arithmetic sequence

HOMEWORK 10.1

 SSM Tutor Center

 MathPro 4 MathPro 5

 CDs/Videos

 www

Check whether the sequence is arithmetic. If the sequence is arithmetic, find the common difference d.

1. $7, 2, -5, -8, -11, \ldots$

2. $1, 5, 7, 11, 13, \ldots$

3. $40, 38, 36, 34, 32, \ldots$

4. $-20, -13, -6, 1, 8, \ldots$

5. $3, 11, 19, 27, 35, \ldots$

6. $-2, -5, -8, -11, -14, \ldots$

7. $4, 44, 444, 4444, 44{,}444, \ldots$

8. $1, 1, 2, 2, 3, 3, \ldots$

Find a formula, using a_n notation, for the sequence. Use a graphing calculator table to verify your result.

9. $5, 11, 17, 23, 29, \ldots$

10. $7, 11, 15, 19, 23, \ldots$

11. $-4, -15, -26, -37, -48, \ldots$

12. $-3, -7, -11, -15, -19, \ldots$

13. 100, 94, 88, 82, 76, . . .

14. 72, 69, 66, 63, 60, . . .

15. 1, 3, 5, 7, 9, . . .

16. 1, 2, 3, 4, 5, . . .

Find the indicated term for the sequence. Verify your result using a graphing calculator table.

17. the 37th term of 5, 8, 11, 14, 17, . . .

18. the 52nd term of 4, 19, 34, 49, 64, . . .

19. the 45th term of 200, 191, 182, 173, 164, . . .

20. the 21st term of 83, 79, 75, 71, 67, . . .

21. a_{96} for 4.1, 5.7, 7.3, 8.9, 10.5, . . .

22. a_{31} for 23.8, 21.5, 19.2, 16.9, 14.6, . . .

23. a_{64} for 8, 1, −6, −13, −20, . . .

24. a_{400} for 1, 2, 3, 4, 5, . . .

Find the term number n of the last term of the finite sequence. Verify your result using a graphing calculator table.

25. 3, 8, 13, 18, 23, . . . , 533

26. 4, 10, 16, 22, 28, . . . , 1426

27. 7, 15, 23, 31, 39, . . . , 695

28. 10, 19, 28, 37, 46, . . . , 415

29. −27, −19, −11, −3, 5, . . . , 2469

30. −11, −5, 1, 7, 13, . . . , 409

31. 29, 25, 21, 17, 13, . . . , −14,251

32. −27, −39, −51, −63, −75, . . . , −999

33. If $a_7 = 24$ and $a_{13} = 66$ are terms of an arithmetic sequence, find a_{40}.

34. If $a_4 = 7$ and $a_{11} = -14$ are terms of an arithmetic sequence, find a_{98}.

35. If $a_{41} = 500$ and $a_{81} = 500$ are terms of an arithmetic sequence, find a_{990}.

36. If $a_{12} = 12$ and $a_{78} = 78$ are terms of an arithmetic sequence, find a_{103}.

37. Let $f(x) = 4x - 2$. Is the sequence $f(1), f(2), f(3), \ldots$ arithmetic? Explain.

38. Let $f(x) = -5x + 1$. Is the sequence $f(1), f(2), f(3), \ldots$ arithmetic? Explain.

39. Let $f(x) = x^2$. Is the sequence $f(1), f(2), f(3), \ldots$ arithmetic? Explain.

40. Let $f(x) = \sqrt{x}$. Is the sequence $f(1), f(2), f(3), \ldots$ arithmetic? Explain.

41. A student tries to find a_{54} for the sequence 2, 7, 11, 16, 20, 25, 29, 34, 38, ... Is the student's work correct? Explain.

$$a_{54} = 2 + (54 - 1)(5)$$
$$= 2 + 53(5)$$
$$= 267$$

42. An arithmetic sequence is described by $a_n = 3n + 7$. A student concludes that the first term is 7 and the common difference is 3. What would you tell the student?

43. Find an equation for a function f so that $f(1), f(2), f(3), f(4), f(5), \ldots$ is the sequence 8, 17, 26, 35, 44, ...

44. Find an equation for a function g so that $g(1), g(2), g(3), g(4), g(5), \ldots$ is the sequence 75, 65, 55, 45, 35, ...

45. A person's starting salary is $27,500. Each year the person receives an $800 raise.

a. Let a_n represent the person's salary (in dollars) for the nth year. Find a formula for a_n.

b. What will be the salary for the 22nd year?

c. In what year will the salary first be above $50,000?

46. A person's starting salary is $30,700. At the end of each of the first 9 years, there is a $950 raise. After that, there will be an $1150 raise at the end of each year. What will be the salary for the 17th year?

47. A math instructor at a community college estimates that it takes an average of 10 minutes per student to grade students' quizzes and tests each week. This instructor also spends a total of 35 hours per week facilitating classroom activities, holding office hours, planning for classes, and attending committee meetings. Let a_n represent the number of hours that this instructor works per week, when n students are enrolled in her courses.

a. Find a formula for a_n.

b. Find the values of a_1, a_2, a_3, and a_4. What do these four terms represent in terms of the instructor?

c. If the instructor has 130 students, how many hours does she work per week?

d. What is the greatest number of students the instructor can have without having to work over 60 hours per week?

48. A full bottle of household glass cleaner holds 22 fluid ounces. It takes about 500 squeezes of the bottle's trigger to use all the cleaner.

a. Let a_n represent the number of ounces of cleaner that remains in the bottle after the trigger has been squeezed n times. Find a formula for a_n.

b. Find a_1, a_2, a_3, and a_4. What do these four terms represent in terms of the liquid?

c. Assume that it takes about 7 squeezes of the trigger to clean one side of a 4-by-3-foot window. How much liquid would remain in the bottle after cleaning both sides of the 4-by-3-foot windows in an apartment complex that has 32 such windows? The bottle is filled at the start of the cleaning.

49. The underground rock band Little Muddy spends $50 to send postcards announcing its latest one-night gig at The Rathskeller ("The Rat") in Boston. There are three bands playing that night. Each band gets 30% of the money collected from a $6 per person cover charge. Let a_n represent Little Muddy's profit (in dollars) if n people pay the cover charge.

a. Find a formula for a_n in terms of n.

b. If the band's profit is $256.00, how many people paid the cover charge?

c. The maximum capacity at The Rat is 200 people. Assume that 18 people are on the guest list (they get in free), 11 people are in the 3 bands (they get in free), and 6 people are working for the club (they too get in free). What is the greatest profit that Little Muddy can earn, assuming no one leaves the club until closing time?

d. Sketch a graph of Little Muddy's profit sequence.

50. In 1996, a new main library was built in San Francisco. Inside the library, there is a five-story-high glass sculpture that looks like a star constellation from a distance. This constellation is made up of white lights. However, on approach, one can see that each light is actually an illuminated disk displaying the name of an author etched into the glass. This sculpture was created by Nayland Blake. Although the sculpture originally had 160 names, Blake left room for 200 more names, 5 to be added each year. The names will be determined by an annual contest organized by the library staff.

Let a_n represent the number of names in the sculpture in the nth year, where 1996 is the first year. So $a_1 = 160$, $a_2 = 165$, $a_3 = 170$, and so on.

a. Find a formula for a_n.

b. Use your formula for a_n to predict the number of names there will be in the 33rd year (2028).

c. Sketch a graph of the library sculpture sequence.

51. Postage for a first-class letter depends on its weight. In November 2002, the cost for a letter weighing 1 ounce or less was $0.37. The postage increased by $0.23 for each additional ounce through 13 total ounces.

a. Let a_n represent first-class postage (in dollars) for a letter that weighs n ounces. Find a formula for a_n.

b. Use your formula for a_n to find the postage for a letter that weighs 13 ounces.

c. Actual postage for a 1-pound large envelope sent by priority mail (there is no first-class service for this weight) is $3.85. Is this a better deal than the postage that would be required by your formula for a_n? (There are 16 ounces in 1 pound.)

d. Actual first-class postage from San Francisco to Chicago for a 5-pound package is $9.43. According to your formula for a_n, how much would the postage be?

52. Prescription acne medicine sales have greatly increased in recent years (see Table 1).

Table 1 Prescription Acne Medicine Sales	
Year	Sales (billions of dollars)
1996	0.54
1997	0.63
1998	0.80
1999	0.96
2000	1.11

Source: *IMS Health, A.C. Nielsen*

Let $f(t)$ represent the prescription acne medicine sales (in billions of dollars) in the year that is t years since 1996.

a. Use your graphing calculator to draw a scattergram of the data.

b. Find a linear equation for f.

c. Use a graphing calculator table to find the values of the sequence $f(0)$, $f(1)$, $f(2)$, $f(3)$, $f(4)$, $f(5)$. What do these values mean in terms of the situation?

d. Use f to predict the prescription acne medicine sales in 2008.

53. Describe an arithmetic sequence. Also, if the first few terms of an arithmetic sequence are given, explain how to find:

- A term with a known term number.
- The term number of a known term.

10.2 GEOMETRIC SEQUENCES

Objectives

▸ Know the meaning of *geometric sequence*.

▸ Find a formula of a geometric sequence.

▸ Make estimates and predictions using a geometric sequence.

In Section 10.1 we worked with arithmetic sequences. In this section we study another type of sequence. Consider the sequence

$$3, 6, 12, 24, 48, \ldots$$

Note that the *ratio* of any term to its preceding term is 2:

$$\frac{6}{3} = 2, \frac{12}{6} = 2, \frac{24}{12} = 2, \frac{48}{24} = 2, \ldots$$

We call 2 the *common ratio* and we call the sequence a *geometric sequence*.

DEFINITION *Geometric Sequence*

If the ratio between any term of a sequence and the preceding term is equal to a constant r for every such pair of terms, then the sequence is a **geometric sequence**. We call the constant r the **common ratio**.

Example 1 Identifying Geometric Sequences

Determine whether each sequence is geometric. If so, find the common ratio r.

1. $2, 6, 18, 54, 162, \ldots$ **2.** $4, 8, 24, 48, 240, \ldots$ **3.** $32, 16, 8, 4, 2, \ldots$

Solution

1. The sequence has a common ratio of 3:

$$\frac{6}{2} = 3, \frac{18}{6} = 3, \frac{54}{18} = 3, \frac{162}{54} = 3, \ldots$$

The sequence is geometric.

2. We can see from the first two ratios that the sequence does not have a common ratio:

$$\frac{8}{4} = 2, \frac{24}{8} = 3$$

The sequence is not geometric.

3. The sequence has a common ratio of $\frac{1}{2}$:

$$\frac{16}{32} = \frac{1}{2}, \frac{8}{16} = \frac{1}{2}, \frac{4}{8} = \frac{1}{2}, \frac{2}{4} = \frac{1}{2}, \ldots$$

The sequence is geometric. ∎

Next, we find a general formula for geometric sequences. Earlier, we determined that the geometric sequence $3, 6, 12, 24, 48, \ldots$ has common ratio 2. This means that we multiply the first term 3 by 2 to find the second term, we multiply 3 by 2 two times to find the third term, we multiply 3 by 2 three times to find the fourth term, and so on.

$$3 \cdot 2 = 6, 3 \cdot 2 \cdot 2 = 12, 3 \cdot 2 \cdot 2 \cdot 2 = 24, \ldots$$

In general, for a geometric sequence with common ratio r we have the terms

$$a_1, \; a_1 \cdot r, \; a_1 \cdot r \cdot r, \; a_1 \cdot r \cdot r \cdot r, \ldots$$

or by using exponents, we have

$$a_1, a_1 r, a_1 r^2, a_1 r^3, \ldots$$

There is a pattern that we can use to find a formula for any term a_n of the geometric sequence.

$a_1 = a_1$

$a_2 = a_1 r$ Multiply a_1 by r once to get the second term.

$a_3 = a_1 r^2$ Multiply a_1 by r two times to get the third term.

$a_4 = a_1 r^3$ Multiply a_1 by r three times to get the fourth term.

$\vdots$

$a_n = a_1 r^{n-1}$ Multiply a_1 by r a total of $(n-1)$ times to get the nth term.

Formula for the nth Term of a Geometric Sequence

If a sequence has common ratio r, then

$$a_n = a_1 r^{n-1}$$

In words, the nth term is equal to the first term times the $(n-1)$th power of r.

Example 2 Finding a Formula

Find a formula for the sequence 12, 36, 108, 324, 972, . . .

Solution

The sequence has a common ratio ($r = 3$), so the sequence is geometric. We substitute $a_1 = 12$ and $r = 3$ into the formula $a_n = a_1 r^{n-1}$ to get a formula for the sequence:

$$a_n = 12(3)^{n-1}$$

We verify our formula by entering $y = 12(3)^{x-1}$ in a graphing calculator and checking that the first five terms are 12, 36, 108, 324, and 972 (see Fig. 4). ∎

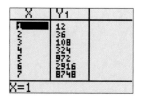

Figure 4 Verify the formula $a_n = 12(3)^{n-1}$

Here we write the formula $a_n = 12(3)^{n-1}$ in the form $a_n = ab^n$:

$$
\begin{aligned}
a_n &= 12(3)^{n-1} \\
&= 12(3)^n(3)^{-1} \qquad b^{m+n} = b^m b^n \\
&= \frac{12(3)^n}{3} \qquad\qquad b^{-n} = \frac{1}{b^n} \\
&= 4(3)^n
\end{aligned}
$$

The formula $a_n = 4(3)^n$ describes an exponential function where the inputs are only the counting numbers 1, 2, 3, . . . and the outputs are only the terms of the sequence 12, 36, 108, 324, 972, ...

In Fig. 5 we sketch a graph of the first five terms of the sequence 12, 36, 108, 324, 972, ... For $a_n = 4(3)^n$, the base 3 is greater than 1, so the points lie on an *increasing* exponential curve.

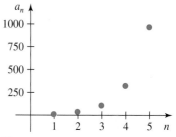

Figure 5 The first five terms of the sequence 12, 36, 108, 324, 972, . . .

For a geometric sequence, if we know values for any three of the four variables a_1, a_n, n, and r, we can find the value of the fourth variable using the formula $a_n = a_1 r^{n-1}$.

Example 3 Finding a Term

Find the 12th term of the sequence 160, 80, 40, 20, 10, ...

Solution

The sequence has a common ratio $\left(r = \dfrac{1}{2} \right)$, so the sequence is geometric. We substitute $a_1 = 160$, $n = 12$, and $r = \dfrac{1}{2}$ in the formula $a_n = a_1 r^{n-1}$.

$$
\begin{aligned}
a_{12} &= 160 \left(\frac{1}{2} \right)^{12-1} \\
&= 160 \left(\frac{1}{2} \right)^{11} \\
&= 160 \left(\frac{1}{2^{11}} \right) \qquad \left(\frac{a}{b} \right)^n = \frac{a^n}{b^n} \\
&= \frac{160}{2^{11}} \\
&= 0.078125
\end{aligned}
$$

We enter $y = 160 \left(\dfrac{1}{2} \right)^{x-1}$ in a graphing calculator and check that the first five terms are 160, 80, 40, 20, and 10 and that the 12th term is 0.078125 (see Fig. 6).

X	Y1	
1	160	
2	80	
3	40	
4	20	
5	10	
6	5	
7	2.5	
X=1		

X	Y1	
6	5	
7	2.5	
8	1.25	
9	.625	
10	.3125	
11	.15625	
12	.078125	
Y1=.078125		

Figure 6 Check the first five terms and the 12th term

Note

The entry across from $x = 12$ has been rounded. However, the bottom entry of the table displays the exact value $a_{12} = 0.078125$.

∎

When working with a sequence, we first determine whether the sequence is arithmetic, geometric, or something else. An arithmetic sequence has a common difference, whereas a geometric sequence has a common ratio.

Example 4 Finding a Term Number

The number 1,572,864 is a term in the sequence 6, 24, 96, 384, 1536, ... What is its term number?

Solution

The sequence has a common ratio ($r = 4$), so the sequence is geometric. We substitute $a_1 = 6$, $a_n = 1,572,864$, and $r = 4$ in the formula $a_n = a_1 r^{n-1}$ and solve for n.

$$1,572,864 = 6(4)^{n-1}$$

$$262,144 = 4^{n-1} \qquad \text{Divide both sides by 6.}$$

$$\log(262,144) = \log(4^{n-1}) \qquad \text{Take the log of both sides.}$$

$$\log(262,144) = (n-1)\log(4) \qquad \text{Power Property}$$

$$\frac{\log(262,144)}{\log(4)} = n - 1 \qquad \text{Divide each side by } \log(4).$$

$$n = \frac{\log(262,144)}{\log(4)} + 1$$

$$n = 10$$

So, 1,572,864 is the 10th term. We can use a graphing calculator table to verify our work. ■

Example 5 Modeling with a Geometric Sequence

A person's salary is $25,000 for the first year. It will increase by 3% each year. Let a_n represent the salary (in dollars) for the nth year.

1. Find a formula for a_n.
2. Predict the salary for the 32nd year.
3. Compare the result from problem 2 with the result in Section 10.1, Example 5, problem 2 where we assumed that the salary increases by a constant $750 each year.

Solution

1. The salary for the second year is equal to 103% of $25,000, or $25,000(1.03) = 25,750$ dollars. Each year, the salary is equal to 1.03 times the salary for the preceding year. So, a_n is a geometric sequence with common ratio 1.03. We substitute $a_1 = 25,000$ and $r = 1.03$ in the formula $a_n = a_1 r^{n-1}$:

$$a_n = 25,000(1.03)^{n-1}$$

2. To find the salary for the 32nd year, we substitute $n = 32$ in the formula $a_n = 25,000(1.03)^{n-1}$.

$$a_{32} = 25,000(1.03)^{32-1} \approx 62,502.01$$

The salary will be $62,502.01 for the 32nd year.

3. First, we note that 3% of $25,000 is $750, so the first raise is the same in either scenario. In Example 5 in Section 10.1, we found that if the salary increases by $750 each year, the salary for the 32nd year will be $48,250, which is considerably less than the salary of $62,502 from receiving 3% raises each year.

To verify our comparisons, we enter the constant-raise formula $a_n = 750n + 24{,}250$ and the percentage-raise formula $a_n = 25{,}000(1.03)^{n-1}$ in a graphing calculator and compare tables (see Fig. 7).

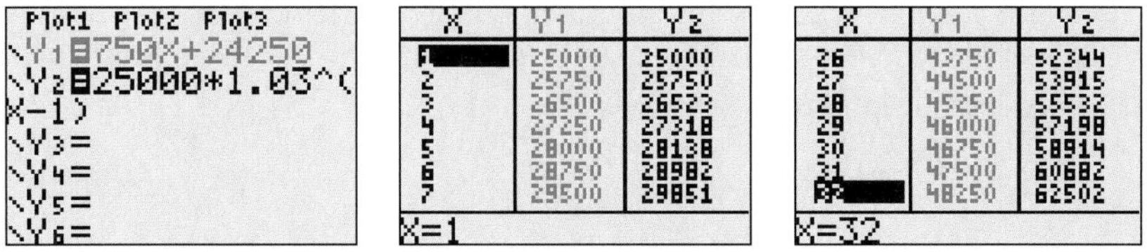

Plot1 Plot2 Plot3		
\Y₁■750X+24250		
\Y₂■25000*1.03^(		
X-1)		
\Y₃=		
\Y₄=		
\Y₅=		
\Y₆=		

X	Y₁	Y₂
1	25000	25000
2	25750	25750
3	26500	26523
4	27250	27318
5	28000	28138
6	28750	28982
7	29500	29851

X=1

X	Y₁	Y₂
26	43750	52344
27	44500	53915
28	45250	55532
29	46000	57198
30	46750	58914
31	47500	60682
32	48250	62502

X=32

Figure 7 Verify the comparisons of the two scenarios

Now, we write the geometric sequence's salary formula $a_n = 25{,}000(1.03)^{n-1}$ in the form $a_n = ab^n$:

$$a_n = 25{,}000(1.03)^{n-1}$$

$$= 25{,}000(1.03)^n(1.03)^{-1} \qquad b^{m+n} = b^m b^n$$

$$= \frac{25{,}000(1.03)^n}{1.03} \qquad b^{-n} = \frac{1}{b^n}$$

$$= \frac{25{,}000}{1.03}(1.03)^n$$

Note that the base of the exponential function $a_n = \dfrac{25{,}000}{1.03}(1.03)^n$ is equal to 1.03, which is also the common ratio of the geometric sequence. If we let $f(n)$ represent the person's salary (in dollars) for the nth year, then

$$f(n) = \frac{25{,}000}{1.03}(1.03)^n$$

and the salary geometric sequence is given by

$$f(1), f(2), f(3), \ldots$$

Connection between an Exponential Function and a Geometric Sequence

For an exponential function $f(x) = ab^x$, the sequence

$$f(1), f(2), f(3), \ldots$$

is a geometric sequence with common ratio equal to the base b of f.

For example, if g is the exponential function $g(x) = 7(5)^x$, then the sequence $g(1)$, $g(2), g(3), \ldots$ is a geometric sequence with common ratio 5, which is also the base of g.

Recall that the base multiplier property for exponential functions states that if the value of the independent variable is increased by 1, then the value of the dependent variable is multiplied by the base of the exponential function. If we think in terms of the base multiplier property for exponential functions, it makes sense that the base of an exponential function is equal to the common ratio of the corresponding geometric sequence.

EXPLORATION
Connection between combining exponential functions and forming new sequences

1. Let $f(x) = 3(2)^x$. Use a graphing calculator table to find the values of the sequence $f(1)$, $f(2)$, $f(3)$, $f(4)$, $f(5)$. Determine whether the sequence is geometric. Explain. If the sequence is geometric, find the common ratio and compare it to the base of f.

2. For $f(x) = 3(2)^x$, find the ratio $\dfrac{f(x+1)}{f(x)}$. Compare your result with your result in problem 1.

3. Let $f(x) = 4(2)^x$ and $g(x) = 2(3)^x$. For each of the following definitions of h, decide whether the sequence $h(1), h(2), h(3), h(4), h(5)$ is geometric. Explain.

   a. $h(x) = f(x) + g(x)$

   b. $h(x) = f(x)g(x)$

   c. $h(x) = \dfrac{f(x)}{g(x)}$

4. Assume that the sequence $a_1, a_2, a_3, \ldots$ is geometric with common ratio r_a and the sequence $b_1, b_2, b_3, \ldots$ is geometric with common ratio r_b. Is the sequence $a_1b_1, a_2b_2, a_3b_3, \ldots$ geometric? If so, what is the common ratio?

Key Points

OF THIS SECTION

Geometric sequence	If the ratio between any term of a sequence and the preceding term is equal to a constant r for every such pair of terms, then we call the sequence a **geometric sequence** and we call the constant r the **common ratio**.
Formula for a geometric sequence	For a geometric sequence with common ratio r, $a_n = a_1 r^{n-1}$.
Check that a sequence is geometric	When working with a sequence, we first check whether the sequence is arithmetic, geometric, or neither. An arithmetic sequence has a common difference, whereas a geometric sequence has a common ratio.
Connection between an exponential function and a geometric sequence	For an exponential function $f(x) = ab^x$, the sequence $f(1), f(2), f(3), \ldots$ is a geometric sequence with common ratio equal to the base b of f.

HOMEWORK 10.2

 SSM Tutor Center   MathPro 4   MathPro 5 CDs/Videos www

Check whether the sequence is arithmetic, geometric, or neither. If the sequence is geometric, find the common ratio r. If the sequence is arithmetic, find the common difference d.

1. 4, 28, 196, 1372, 9604, ...

2. 62, 57, 54, 42, 39, ...

3. 13, 6, −1, −8, −15, ...

4. 75, 72, 69, 66, 63, ...

5. 3, 4, 6, 9, 13, ...

6. 3, 7, 11, 15, 19, ...

7. 200, 40, 8, $\dfrac{8}{5}$, $\dfrac{8}{25}$, ...

8. 0.08, 0.8, 8, 80, 800, ...

Find a formula for the sequence. Use a_n notation. Use a graphing calculator table to verify your formula.

9. 3, 6, 12, 24, 48, ...

10. 162, 54, 18, 6, 2, ...

11. 800, 200, 50, 12.5, 3.125, ...

12. 4, 20, 100, 500, 2500, ...

13. 100, 50, 25, 12.5, 6.25, ...

14. 5, 15, 45, 135, 405, ...

15. 1, 4 , 16, 64, 256, ...

16. 32, 16, 8, 4, 2, ...

17. 14, 19, 24, 29, 34, ...

18. 57, 49, 41, 33, 25, ...

Find the indicated term for the sequence. Use a graphing calculator table to verify your result.

19. the 34th term of 4, 20, 100, 500, 2500, ...

20. the 103rd term of 2, 8, 32, 128, 512, ...

21. the 27th term of 80, 40, 20, 10, 5, ...

22. the 15th term of 36, 12, 4, $\dfrac{4}{3}$, $\dfrac{4}{9}$, ...

23. a_{23} for 8, 16, 32, 64, 128, ...

24. a_{99} for 17, 12, 7, 2, −3, ...

25. a_{96} for 9.5, 12.9, 16.3, 19.7, 23.1, ...

26. a_{19} for 1, $\dfrac{3}{2}$, $\dfrac{9}{4}$, $\dfrac{27}{8}$, $\dfrac{81}{16}$, ...

Find the term number n of the last term of the finite sequence. Verify your result using a graphing calculator.

27. 240, 120, 60, 30, 15, . . . , 0.46875

28. 80, 20, 5, 1.25, 0.3125, . . . , 0.01953125

29. 0.00224, 0.0112, 0.056, 0.28, 0.14, . . . , 109,375

30. 192, 96, 48, 24, 12, . . . , 0.046875

31. 4, 7, 10, 13, 16, . . . , 367

32. 88, 81, 74, 67, 60, . . . , −801

33. The number 3,407,872 is a term in the sequence 13, 26, 52, 104, 208, ... Which term is it? Verify your answer with the help of a graphing calculator table.

34. The number 2,470,629 is a term in the sequence 3, 21, 147, 1029, 7203, ... Which term is it? Verify your answer with the help of a graphing calculator table.

35. The number 28,697,814 is a term in the sequence 2, 6, 18, 54, 162, ... Which term is it? Verify your answer with the help of a graphing calculator table.

36. Is 9,238,946 a term in the sequence 13, 26, 52, 104, 208, . . . ? Explain.

37. Let $f(x) = 2(5)^x$. Is the sequence $f(1)$, $f(2)$, $f(3)$, . . . arithmetic, geometric, or neither? Explain.

38. Let $f(x) = 8\left(\dfrac{1}{2}\right)^x$. Is the sequence $f(1)$, $f(2)$, $f(3)$, . . . arithmetic, geometric, or neither? Explain.

39. Let $f(x) = 7x - 3$. Is the sequence $f(1)$, $f(2)$, $f(3)$, . . . arithmetic, geometric, or neither? Explain.

40. Let $f(x) = 3(x)^2$. Is the sequence $f(1)$, $f(2)$, $f(3)$, . . . arithmetic, geometric, or neither? Explain.

41. Find an equation for a function f so that $f(1)$, $f(2)$, $f(3)$, $f(4)$, $f(5)$, . . . is the sequence 8, 24, 72, 216, 648, ...

42. Find an equation for a function g so that $g(1)$, $g(2)$, $g(3)$, $g(4)$, $g(5)$, . . . is the sequence 48, 24, 12, 6, 3, ...

43. A student tries to find a_{17} for the sequence 2, 6, 10, 14, 18, ... Is the student's work correct? Explain.

$$a_n = a_1 r^{n-1}$$
$$a_{17} = 2(3)^{17-1}$$
$$= 2(3)^{16}$$
$$= 86{,}093{,}442$$

44. A geometric sequence is described by the formula $a_n = 4(6)^n$. A student concludes that the first term is 4 and the common ratio is 6. What would you tell the student?

45. Assume that a person's salary is $27,000 for the first year and that the salary increases by 4% each year.

a. Let a_n represent the salary (in dollars) for the nth year. Find a formula for a_n.

b. What will be the person's salary for the 10th year?

c. In what year will the salary first be above $50,000?

46. A person's salary is $24,000 for the first year.

a. If the salary increases by $960 each year, calculate the salary for the 2nd year and for the 30th year.

b. If the salary increases by 4% each year, calculate the salary for the 2nd year and for the 30th year.

c. Compare your results for parts a and b. Explain why the salaries are the same for the 2nd year and different for the 30th year.

47. A person's ancestors one generation back are that person's natural parents. A person's ancestors two generations back are that person's natural grandparents. Let a_n represent the number of ancestors that a person has in the nth generation back.

Family Tree

a. List the first five terms of the sequence a_n.

b. Find a formula for a_n.

c. Use your formula for a_n to find the number of ancestors that a person has in the 8th generation back.

d. Use your formula for a_n to find the number of ancestors that a person has in the 35th generation back. Explain why model breakdown has occurred. Describe any assumptions that you made in finding the formula for a_n.

48. A rubber ball is dropped. The height of the ball is measured from the floor to the bottom of the ball. The maximum height of the ball after one bounce is 4 feet. The maximum height of the ball after the second bounce is 70% of 4 feet, or 2.8 feet. This pattern continues, that is, the maximum height after each bounce is 70% of the maximum height after the preceding bounce. Let a_n represent the maximum height (in feet) of the ball after the nth bounce.

a. Find a formula for a_n.

b. Predict the maximum height of the ball after the 5th bounce.

c. For which bounce does the ball reach at least half a foot for the last time?

d. Sketch a graph of the bouncing ball sequence.

e. Does model breakdown occur? Explain.

49. Suppose that a rumor is spreading on campus that students will not have to take any final exams this semester in any of their courses. Suppose that 5 students hear the rumor on the first day. Assume that each person who hears the rumor tells the rumor, approximately 24 hours later, to exactly 3 students who have not yet heard the rumor. Let a_n represent the number of students who hear the rumor on the nth day.

a. Find a formula for a_n.

b. How many students will hear the rumor on the 5th day?

c. Use your formula for a_n to predict the number of students who will hear the rumor on the 11th day. Has model breakdown occurred? Explain.

d. In order to find the formula for a_n, you made some assumptions about the way in which the rumor would spread. Describe each assumption and discuss whether you think each assumption is reasonable.

50. Although having a cellular phone can be expensive, the appeal of their convenience outweighs their cost for a growing number of consumers. The numbers of subscribers to cellular telephone services in the United States for various years are listed in Table 2.

Table 2 Cellular Telephone Subscribers	
Year	Subscribers (millions)
1990	5.3
1992	11.0
1994	24.1
1996	44.0
1998	69.2
2000	109.5
2001	128.4

Source: *Cellular Telecommunications Industry Association*

Let $f(t)$ represent the number of subscribers (in millions) at t years since 1990.

a. Find an exponential equation for f.

b. Use a graphing calculator table to find the values of the sequence $f(0), f(1), f(2), f(3), f(4)$. What do these values mean in terms of the situation?

c. Use f to predict when every American will have a cell phone. Assume that the U.S. population is 287 million.

51. Describe a geometric sequence. Also, if the first few terms of a geometric sequence are given, explain how to find:

• A term with a known term number.

• The term number of a known term.

10.3 ARITHMETIC SERIES

Objectives

▸ Know the meaning of *arithmetic series*.

▸ Evaluate the sum of an arithmetic series.

▸ Make estimates and predictions using arithmetic series.

So far in this chapter, we have worked with arithmetic sequences and geometric sequences. In the next two sections, we discuss topics that are related to these sequences.

Suppose that a person's salary is $23,000 during the first year and it increases by $2000 each year. Here, we describe the salaries (in thousands of dollars) for the first 32 years using an arithmetic sequence:

$$23, 25, 27, \ldots, 81, 83, 85$$

To find the *total* earnings (in thousands of dollars) during the first 32 years, we find the sum

$$23 + 25 + 27 + \cdots + 81 + 83 + 85$$

We call this sum an *arithmetic series*.

DEFINITION Arithmetic Series

If the sequence $a_1, a_2, a_3, \ldots, a_n$ is an arithmetic sequence, then the sum $a_1 + a_2 + a_3 + \cdots + a_n$ is an **arithmetic series**. Each a_i in the series is a **term** of the series.

For the series $23 + 25 + 27 + \cdots + 81 + 83 + 85$, the number 27 is the *third term*. Likewise, a_3 is the third term of the series $a_1 + a_2 + a_3 + \cdots + a_n$. We use the notation S_n to represent the sum of the first n terms of an arithmetic sequence:

$$S_n = a_1 + a_2 + a_3 + \cdots + a_n$$

Next, we find the total earnings (in thousands of dollars) for 32 years, S_{32}, where

$$S_{32} = 23 + 25 + 27 + \cdots + 81 + 83 + 85$$

We find the sum in a way that suggests a general formula for S_n for any arithmetic series.

To find the sum, we write the equation for S_{32} twice, the second time with the terms in reverse order. Then we add the left sides and add the right sides of the two equations.

$$
\begin{array}{rcccccccc}
S_{32} & = & 23 & + & 25 & + & 27 & + \cdots + & 81 & + & 83 & + & 85 \\
+ \; S_{32} & = & 85 & + & 83 & + & 81 & + \cdots + & 27 & + & 25 & + & 23 \\
\hline
2S_{32} & = & 108 & + & 108 & + & 108 & + \cdots + & 108 & + & 108 & + & 108
\end{array}
$$

On the right side of the equation, the number 108 appears 32 times, so the sum equals $32(108)$.

$$2S_{32} = 32(108) \tag{1}$$

$$S_{32} = \frac{32(108)}{2} \qquad \text{Divide both sides by 2.} \tag{2}$$

$$S_{32} = 1728 \tag{3}$$

The total earnings for 32 years will be $1,728,000.

Note that 108 is the sum of the first and last terms of the series:

$$108 = 23 + 85$$

So, we can write equation (2) as

$$S_{32} = \frac{32(23 + 85)}{2}$$

Our process and result suggest that we can find S_n for any arithmetic series by first multiplying the sum of the first and last terms $a_1 + a_n$ by the number of terms n, and then dividing the product by 2.

Sum of an Arithmetic Series Formula

If $S_n = a_1 + a_2 + a_3 + \cdots + a_n$ is an arithmetic series, then

$$S_n = \frac{n(a_1 + a_n)}{2}$$

We derive the formula $S_n = \dfrac{n(a_1 + a_n)}{2}$ at the end of this section.

Example 1   Evaluating Sums

1. Evaluate S_{50}, where $S_{50} = 3 + 7 + 11 + 15 + 19 + \cdots + 199$.
2. Evaluate S_{80}, where $S_{80} = 60 + 53 + 46 + 39 + 32 + \cdots + (-493)$.

Solution

1. The sequence 3, 7, 11, 15, 19, ..., 199 is arithmetic with common difference $d = 4$. We find S_{50} by substituting $n = 50$, $a_1 = 3$, and $a_n = 199$ in the equation $S_n = \dfrac{n(a_1 + a_n)}{2}$.

$$S_{50} = \frac{50(3 + 199)}{2} = 5050$$

So, $S_{50} = 5050$.

2. The sequence 60, 53, 46, 39, 32, ..., −493 is arithmetic with common difference $d = -7$. We find S_{80} by substituting $n = 80$, $a_1 = 60$, and $a_n = -493$ in the equation $S_n = \dfrac{n(a_1 + a_n)}{2}$.

$$S_{80} = \frac{80(60 + (-493))}{2} = -17{,}320$$

So, $S_{80} = -17{,}320$. ◾

To evaluate the sum $S_n = a_1 + a_2 + a_3 + \cdots + a_n$, we first check that the sequence $a_1, a_2, a_3, \ldots, a_n$ is arithmetic before using the formula

$$S_n = \frac{n(a_1 + a_n)}{2}$$

In some problems, it may be necessary to use the formula $a_n = a_1 + (n-1)d$ to determine a_1, a_n, or n, so that we can use the formula $S_n = \dfrac{n(a_1 + a_n)}{2}$ to evaluate S_n.

Example 2   Evaluating a Sum

Evaluate S_{43}, where $S_{43} = 150 + 147 + 144 + 141 + 138 + \cdots + a_{43}$.

Solution

The sequence 150, 147, 144, 141, 138, ..., a_{43} is arithmetic with common difference $d = -3$. Although we know that $a_1 = 150$ and $n = 43$, we must first find a_{43} before we can use the formula $S_n = \dfrac{n(a_1 + a_n)}{2}$.

We find a_{43} by substituting $a_1 = 150$, $n = 43$, and $d = -3$ in the equation $a_n = a_1 + (n-1)d$:

$$a_{43} = 150 + (43 - 1)(-3) = 24$$

Next, we substitute $n = 43$, $a_1 = 150$, and $a_n = 24$ in the equation $S_n = \dfrac{n(a_1 + a_n)}{2}$.

$$S_{43} = \frac{43(150 + 24)}{2} = 3741$$

So, $S_{43} = 3741$. ◾

Example 3   Evaluating a Sum

Evaluate the sum $2 + 8 + 14 + 20 + 26 + \cdots + 338$.

Solution

The sequence 2, 8, 14, 20, 26, ..., 338 is arithmetic with common difference $d = 6$. Although we know that $a_1 = 2$ and $a_n = 338$, we must first find n before we can use the formula $S_n = \dfrac{n(a_1 + a_n)}{2}$.

We find n by substituting $a_1 = 2$, $a_n = 338$, and $d = 6$ in the equation $a_n = a_1 + (n-1)d$:

$$338 = 2 + (n-1)6$$
$$336 = (n-1)6 \qquad \text{Subtract 2 from both sides.}$$
$$56 = n - 1 \qquad \text{Divide both sides by 6.}$$
$$n = 57$$

Now, we can find S_n by substituting $n = 57$, $a_1 = 2$, and $a_n = 338$ into the equation $S_n = \dfrac{n(a_1 + a_n)}{2}$:

$$S_{57} = \frac{57(2 + 338)}{2} = 9690$$

So, $S_{57} = 9690$. ◾

Example 4   Modeling with an Arithmetic Series

A person's salary is \$30,000 for the first year and it increases by \$1200 each year. Find the person's total earnings for the first 25 years.

Solution

Let a_n represent the person's salary (in thousands of dollars) for the nth year. The salary sequence $a_1, a_2, a_3, \ldots, a_n$ is arithmetic with common difference 1200. First, we find the salary for the 25th year by substituting $a_1 = 30{,}000$, $n = 25$, and $d = 1200$ in the equation $a_n = a_1 + (n - 1)d$.

$$a_{32} = 30{,}000 + (25 - 1)(1200) = 58{,}800$$

Next we find S_{25} by substituting $n = 25$, $a_1 = 30{,}000$, and $a_n = 58{,}800$ in the equation $S_n = \dfrac{n(a_1 + a_n)}{2}$.

$$S_{25} = \frac{25(30{,}000 + 58{,}800)}{2} = 1{,}110{,}000$$

The total earnings for 25 years will be \$1,110,000. ■

We close this section by deriving the formula $S_n = \dfrac{n(a_1 + a_n)}{2}$. To begin, consider an arithmetic series:

$$S_n = a_1 + a_2 + a_3 + \cdots + a_n$$

Since the series is arithmetic, each term of the series is found by adding d (the common difference) to the preceding term. This means that a_2 is found by adding d to a_1, so $a_2 = a_1 + d$. Also, a_3 is found by adding d twice to a_1, so $a_3 = a_1 + 2d$. The pattern continues, so the series can be expressed as:

$$S_n = a_1 + (a_1 + d) + (a_1 + 2d) + \cdots + a_n$$

If we list backwards from a_n, it follows that the term before a_n can be found by subtracting d, so the term before a_n is $a_n - d$. Likewise, the term before that is $a_n - 2d$. This means that the series can be expressed as:

$$S_n = a_1 + (a_1 + d) + (a_1 + 2d) + \cdots + (a_n - 2d) + (a_n - d) + a_n$$

Now, just as we did at the start of the section, we write the equation for S_n twice, the second time with the terms in reverse order. Then we add the left sides and add the right sides of the two equations.

$$S_n = a_1 + (a_1 + d) + (a_1 + 2d) + \cdots + (a_n - 2d) + (a_n - d) + a_n$$
$$S_n = a_n + (a_n - d) + (a_n - 2d) + \cdots + (a_1 + 2d) + (a_1 + d) + a_1$$

$$2S_n = (a_1 + a_n) + (a_1 + a_n) + (a_1 + a_n) + \cdots + (a_1 + a_n) + (a_1 + a_n) + (a_1 + a_n)$$

Note that the expression $a_1 + a_n$ appears n times on the right side of the last equation. Thus, the sum on the right side is $n(a_1 + a_n)$. We have:

$$2S_n = n(a_1 + a_n)$$
$$S_n = \frac{n(a_1 + a_n)}{2}$$

This formula is what we set out to derive, the sum of an arithmetic series formula.

EXPLORATION
Other ways to evaluate the sum of an arithmetic series

A person's salary is \$26,000 for the first year and it increases by \$1500 each year.

1. Find the person's total earnings for the first 19 years.
2. Find the person's earnings for the 19th year. Find the average of the salaries for the 1st year and the 19th year. Multiply the average by 19.

3. Find the person's earnings for the 10th year (the "middle" year). Multiply the result by 19.

4. Explain why it makes sense that your results in problems 1, 2, and 3 are equal.

5. Your work for problems 1–3 suggests three ways to compute the person's total earnings for the first 19 years, an odd number of years. Which of these methods will give correct results for calculating the total earnings for an *even* number of years (such as 40 years)? Explain.

Tips on Succeeding in This Course

To prepare for the final exam, it may be helpful to form a study group. The group could list important concepts of the course and then discuss the meaning of these concepts and how to apply them. You could also list important techniques learned in the course and then practice these techniques. It is good practice to set aside some solo study time after the group study session to make sure you can do the mathematics without the help of other members of the study group.

Key Points

OF THIS SECTION

Arithmetic series	If the sequence $a_1, a_2, a_3, \ldots, a_n$ is an arithmetic sequence, then the sum $a_1 + a_2 + a_3 + \cdots + a_n$ is an **arithmetic series**. Each a_i in the series is a **term** of the series.
Formula for an arithmetic series	If $S_n = a_1 + a_2 + a_3 + \cdots + a_n$ is an arithmetic series, then $S_n = \dfrac{n(a_1 + a_n)}{2}$.
Check that a series is arithmetic	To evaluate S_n for a series, we first check that the sequence is arithmetic in order to use the formula $S_n = \dfrac{n(a_1 + a_n)}{2}$.
Using a combination of formulas	When evaluating S_n for an arithmetic series, we sometimes must first use the formula $a_n = a_1 + (n-1)d$ to find n, a_1, or a_n in order to use the formula $S_n = \dfrac{n(a_1 + a_n)}{2}$.

HOMEWORK 10.3

 SSM
Tutor Center

MathPro 4
MathPro 4

MathPro 5
MathPro 5

 CDs/Videos

 www

Evaluate the sum of the arithmetic series with the given values for a_1, a_n, and n.

1. $a_1 = 2$, $a_n = 447$, and $n = 90$

2. $a_1 = 7$, $a_n = 187$, and $n = 61$

3. $a_1 = 13$, $a_n = 548$, and $n = 108$

4. $a_1 = 38$, $a_n = 605$, and $n = 82$

5. $a_1 = 37$, $a_n = -1099$, and $n = 72$

6. $a_1 = -208$, $a_n = 386$, and $n = 67$

Evaluate the sum of the series.

7. $S_{74} = 5 + 13 + 21 + 29 + 37 + \cdots + 589$

8. $S_{59} = 14 + 17 + 20 + 23 + 26 + \cdots + 188$

9. $S_{101} = 93 + 89 + 85 + 81 + 77 + \cdots + (-307)$

10. $S_{45} = 131 + 129 + 127 + 125 + 123 + \cdots + 43$

11. $S_{117} = 4 + 4 + 4 + 4 + 4 + \cdots + 4$

12. $S_{46} = 7 + 14 + 21 + 28 + 35 + \cdots + 322$

13. $S_{125} = 3 + 13 + 23 + 33 + 43 + \cdots + a_{125}$

14. $S_{125} = 4 + 14 + 24 + 34 + 44 + \cdots + a_{125}$

15. $S_{81} = 8 + 19 + 30 + 41 + 52 + \cdots + a_{81}$

16. $S_{87} = 11 + 17 + 23 + 29 + 35 + \cdots + a_{87}$

17. $S_{152} = 15 + 28 + 41 + 54 + 67 + \cdots + a_{152}$

18. $S_{48} = -37 + (-30) + (-23) + (-16) + (-9) + \cdots + a_{48}$

19. $S_{100} = -40 - 37 - 34 - 31 - 28 - \cdots - a_{100}$

20. $14 + 19 + 24 + 29 + 34 + \cdots + 1389$

21. $19 + 25 + 31 + 37 + 43 + \cdots + 247$

22. $14 + 26 + 38 + 50 + 62 + \cdots + 794$

23. $900 + 892 + 884 + 876 + 868 + \cdots + (-900)$

24. $207 + 203 + 199 + 195 + 191 + \cdots + 3$

25. $4 + 7 + 10 + 13 + 16 + \cdots + 334 + 337 + 340$

26. $1 + 3 + 5 + 7 + 9 + \cdots + 10{,}001$

27. $1 + 2 + 3 + 4 + 5 + \cdots + 9{,}999 + 10{,}000$

28. $2 + 3 + 4 + 5 + 6 + \cdots + 10{,}000 + 10{,}001$

*In each situation that follows, determine whether the sum of an arithmetic series S_n is positive or negative. Explain why your response makes sense. [**Hint:** Try experimenting with specific values of a_1, d, and n that meet the stated conditions. Then explain why your answer makes sense for any values that meet the stated conditions.]*

29. $a_1 > 0, d > 0$, and n is any counting number

30. $a_1 < 0, d < 0$, and n is any counting number

31. $a_1 = -20, d = 8$, and n is a very large counting number

32. $a_1 = 10, d = -4$, and n is a very large counting number

33. If $f(x) = 7x - 1$, is the series $f(1) + f(2) + f(3) + f(4) + \cdots + f(100)$ arithmetic? Explain.

34. If $g(x) = 4(3)^x$, is the series $g(1) + g(2) + g(3) + \cdots + g(50)$ arithmetic? Explain.

35. A first-year salary is $28,500. Each year there is a raise of $1100.

   a. Calculate the salary for the 28th year.

   b. Find the total earnings for 28 years of work.

36. A first-year salary is $35,100. Each year there is a raise of $1400.

   a. What will be the salary for the 30th year of work?

   b. What will be the total earnings for 30 years of work?

37. Two companies have made you job offers. Company A offers a first-year salary of $35,000 with a $700 raise at the end of each year. Company B offers a first-year salary of $27,000 with a $1500 raise at the end of each year. At which company would your total earnings for 20 years be greater? By how much?

38. An auditorium has 30 rows of seats. There are 20 seats in the front row, 24 seats in the second row, 28 seats in the third row, and so on. In other words, each row has four more seats than the row in front of it.

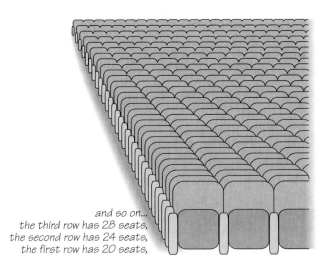

and so on...
the third row has 28 seats,
the second row has 24 seats,
the first row has 20 seats,

   a. How many seats are in the back row?

   b. How many seats are in the auditorium?

39. An auditorium has 50 rows of seats. There are 16 seats in the front row, 18 seats in the second row, 20 seats in the third row,

and so on. In other words, each row has two more seats than the row in front of it.

   a. How many seats are in the auditorium?

   b. If a ticket costs $20 for a seat in the first 10 rows and $15 for a seat in the remaining rows, what is the revenue for a sellout performance? The "revenue" is the total amount of money paid for tickets.

   c. Suppose that 2900 people buy tickets for one performance. Describe all possibilities for the revenue from the performance.

40. In exercise 52 of Homework 10.1, you found a function close to $f(t) = 0.15t + 0.37$, where $f(t)$ represents the prescription acne medicine sales (in billions of dollars) in the year that is t years since 1995 (see Table 3).

Table 3 Prescription Acne Medicine Sales

Year	Sales (billions of dollars)
1996	0.54
1997	0.63
1998	0.80
1999	0.96
2000	1.11

   a. Find $f(1)$. What does your result mean in terms of the situation?

   b. Find $f(15)$. What does your result mean in terms of the situation?

   c. Find $f(1) + f(2) + f(3) + \cdots + f(15)$. What does your result mean in terms of the situation?

41. It is not surprising that sales at food and drink places have increased over the years as the population continues to increase (see Table 4).

Table 4 Sales at Food and Drink Places

Year	Sales (billions of dollars)
1992	203.4
1994	225.6
1996	242.9
1998	272.6
2000	306.1

Source: *U.S. Census Bureau*

Let $f(t)$ represent sales (in billions of dollars) at food and drink places at t years since 1990.

   a. Create a scattergram for the sales data.

   b. Find a linear equation for f.

   c. Find $f(0)$. What does your result mean in terms of sales?

   d. Find $f(19)$. What does your result mean in terms of sales?

   e. Find $f(0) + f(1) + f(2) + \cdots + f(19)$. What does your result mean in terms of sales? [**Hint:** Think carefully about the value of n.]

   f. Many restaurant owners report to the Internal Revenue Service (IRS) that their wait staffs receive between 8% and 10% of each check as tip income, while in reality most wait staff receive between 10% and 20% of each check as tip income. Wait staff are expected to report to the IRS the portion of their tip income unreported by their restaurant owner, yet most do not. Assuming that 60% of food and drink sales in the United States occur in restaurants that have wait staff (most fast-food restaurants don't), predict the total amount of unreported income by wait staff from 1990 through 2009.

42. A first-year salary is $24,800. Each year there is a raise of $1200.

a. What will be the total amount of money earned in 26 years?

b. What will be the total amount of money earned from raises in 26 years?

c. What is the mean amount of money earned per year for the 26 years? For which of the 26 years will this mean be greater than the actual amount of money earned? For

which years will this mean be less than the actual amount of money earned?

d. Assume that for each of the 26 years, *taxable income* is equal to salary minus $4250. Also assume that the federal income tax rate is 15.016% on the first $25,000 of taxable income and 17.04% on the remaining taxable income. Estimate the total amount paid in federal income tax for the 26 years. (Our simplified tax rules are based on the IRS Code for 1998.)*

43. Describe an arithmetic series. Also, explain how to evaluate the sum of an arithmetic series $S_n = a_1 + a_2 + a_3 + \cdots + a_n$, if you know a_1, a_n, and the common difference d of the arithmetic sequence $a_1, a_2, a_3, \ldots, a_n$.

10.4 GEOMETRIC SERIES

Objectives

▸ Know the meaning of *geometric series*.

▸ Evaluate the sum of a geometric series.

▸ Make estimates and predictions using geometric series.

In Section 10.3 we worked with arithmetic series. In this section we discuss another type of series.

Consider the geometric sequence

$$3, 6, 12, 24, 48, \ldots, 1536$$

We call the sum

$$3 + 6 + 12 + 24 + 48 + \cdots + 1536$$

a *geometric series*.

DEFINITION Geometric Series

If the sequence $a_1, a_2, a_3, \ldots, a_n$ is a geometric sequence, then the sum $a_1 + a_2 + a_3 + \cdots + a_n$ is a **geometric series**. Each a_i in the series is a **term** of the series.

We now derive a general formula for the sum of a geometric series $S_n = a_1 + a_2 + a_3 + \cdots + a_n$. In Section 10.2 we described the terms of a geometric sequence $a_1, a_2, a_3, \ldots, a_n$ in terms of a_1 and r:

$$a_1 = a_1$$
$$a_2 = a_1 r$$
$$a_3 = a_1 r^2$$
$$a_4 = a_1 r^3$$
$$\vdots$$
$$a_n = a_1 r^{n-1}$$

In each case, the exponent of r is one less than the term number. So, the series $S_n = a_1 + a_2 + a_3 + \cdots + a_n$ can be expressed as

$$S_n = a_1 + a_1 r + a_1 r^2 + \cdots + a_1 r^{n-1}$$

If we work backwards from $a_1 r^{n-1}$, it follows that the term before $a_1 r^{n-1}$ will have an exponent for r that is one less than $n - 1$. So the term before $a_1 r^{n-1}$ is $a_1 r^{n-2}$.

---

*The actual tax rates change at $25,000, $30,000, and then every additional $10,000.

Similarly, the term before that is $a_1 r^{n-3}$. This means that the series can be expressed as:

$$S_n = a_1 + a_1 r + a_1 r^2 + \cdots + a_1 r^{n-3} + a_1 r^{n-2} + a_1 r^{n-1} \qquad (4)$$

We multiply both sides of this equation by r to obtain:

$$r S_n = a_1 r + a_1 r^2 + a_1 r^3 + \cdots + a_1 r^{n-2} + a_1 r^{n-1} + a_1 r^n \qquad (5)$$

We now multiply both sides of equation (5) by -1 and then add the left sides and add the right sides of this new equation and equation (4).

$$S_n = a_1 + a_1 r + a_1 r^2 + \cdots + a_1 r^{n-2} + a_1 r^{n-1}$$
$$-r S_n = -a_1 r - a_1 r^2 - \cdots - a_1 r^{n-2} - a_1 r^{n-1} - a_1 r^n$$
$$\overline{S_n - r S_n = a_1 + 0 + 0 + \cdots + 0 + 0 - a_1 r^n}$$

Note that the last equation can be simplified:

$$S_n - r S_n = a_1 - a_1 r^n$$

We now factor out S_n from the left side and a_1 from the right side:

$$S_n(1 - r) = a_1(1 - r^n)$$

$$S_n = \frac{a_1(1 - r^n)}{1 - r}, \quad r \neq 1 \qquad \text{Divide both sides by } 1 - r.$$

Sum of a Geometric Series Formula

If $S_n = a_1 + a_2 + a_3 + \cdots + a_n$ is a geometric series with common ratio $r \neq 1$, then

$$S_n = \frac{a_1(1 - r^n)}{1 - r}.$$

Example 1   Evaluating a Sum

Evaluate the sum of the series.

1. $S_{15} = 4 + 12 + 36 + 108 + 324 + \cdots + a_{15}$
2. $S_{13} = 486 + 162 + 54 + 18 + 6 + \cdots + a_{13}$

Solution

1. The sequence 4, 12, 36, 108, 324, ..., a_{15} is geometric with common ratio $r = 3$.

   We substitute $a_1 = 4$, $r = 3$, and $n = 15$ in the equation $S_n = \dfrac{a_1(1 - r^n)}{1 - r}$:

   $$S_{15} = \frac{4(1 - 3^{15})}{1 - 3} = 28{,}697{,}812$$

   So, $S_{15} = 28{,}697{,}812$.

2. The sequence 486, 162, 54, 18, 6, ..., a_{13} is geometric with common ratio $r = \dfrac{1}{3}$.

   We substitute $a_1 = 486$, $r = \dfrac{1}{3}$, and $n = 13$ in the equation $S_n = \dfrac{a_1(1 - r^n)}{1 - r}$:

   $$S_{13} = \frac{486\left(1 - \left(\dfrac{1}{3}\right)^{13}\right)}{1 - \dfrac{1}{3}} \approx 728.9995428$$

   So, $S_{13} \approx 728.9995428$. ∎

To find the sum $a_1 + a_2 + a_3 + \cdots + a_n$, we first determine whether the series is arithmetic, geometric, or neither. If the series is arithmetic, we use $S_n = \dfrac{n(a_1 + a_n)}{2}$. If the series is geometric, we use $S_n = \dfrac{a_1(1 - r^n)}{1 - r}$.

For a geometric series, we sometimes use $a_n = a_1 r^{n-1}$ to find $a_1, r,$ or n and then find S_n by using $S_n = \dfrac{a_1(1 - r^n)}{1 - r}$.

Example 2 Evaluating a Sum

Evaluate the sum of the series $24{,}576 + 12{,}288 + 6144 + 3072 + 1536 + \cdots + 3$.

Solution

The sequence $24{,}576, 12{,}288, 6144, 3072, 1536, \ldots, 3$ is geometric with common ratio $r = \dfrac{1}{2}$. First, we find the term number n of the last term 3 and then we find S_n. To find n, we substitute $a_1 = 24{,}576$, $a_n = 3$, and $r = \dfrac{1}{2}$ in the equation $a_n = a_1 r^{n-1}$ and solve for n:

$$3 = 24{,}576 \left(\frac{1}{2}\right)^{n-1}$$

$$\frac{3}{24{,}576} = \left(\frac{1}{2}\right)^{n-1} \qquad \text{Divide both sides by 24,576.}$$

$$\log\left(\frac{3}{24{,}576}\right) = \log\left(\frac{1}{2}\right)^{n-1} \qquad \text{Take the log of both sides.}$$

$$\log\left(\frac{3}{24{,}576}\right) = (n - 1)\log\left(\frac{1}{2}\right) \qquad \text{Power property}$$

$$\frac{\log\left(\dfrac{3}{24{,}576}\right)}{\log\left(\dfrac{1}{2}\right)} = n - 1 \qquad \text{Divide both sides by } \log\left(\frac{1}{2}\right).$$

$$\frac{\log\left(\dfrac{3}{24{,}576}\right)}{\log\left(\dfrac{1}{2}\right)} + 1 = n$$

$$n = 14$$

Next, we substitute $a_1 = 24{,}576$, $r = \dfrac{1}{2}$, and $n = 14$ in the equation $S_n = \dfrac{a_1(1 - r^n)}{1 - r}$:

$$S_{14} = \frac{24{,}576\left(1 - \left(\dfrac{1}{2}\right)^{14}\right)}{1 - \dfrac{1}{2}} = 49{,}149$$

So, $S_{14} = 49{,}149$. ∎

Example 3 Modeling with a Geometric Series

A person's salary is \$30,000 for the first year and it increases by 4% at the end of each year.

1. Calculate the person's total earnings for the first 25 years.
2. Compare the result from problem 1 with the result from Example 4 in Section 10.3, where we assumed that the person's salary increases by a constant \$1200 each year.

Solution

1. Let a_n represent the person's salary (in thousands of dollars) during the nth year. Since the salary for each year is 104% of the salary for the previous year, the sequence $a_1, a_2, a_3, \ldots, a_n$ is geometric with common ratio 1.04. To find the

total earnings, we substitute $a_1 = 30{,}000$, $r = 1.04$, and $n = 25$ in the equation
$S_n = \dfrac{a_1(1 - r^n)}{1 - r}$:

$$S_{25} = \frac{30{,}000(1 - 1.04^{25})}{1 - 1.04} \approx 1{,}249{,}377.25$$

So, the total earnings will be $1,249,377.25.

2. First, note that 4% of $30,000 is $1200, so the first raise is the same in either scenario. In Example 4 in Section 10.3, we found that if the person receives constant raises of $1200, the total earnings will be $1,110,000 in 25 years, which is about $140,000 less than the total earnings of $1,249,377.25 from earning 4% raises each year. In some parts of the country, this would amount to what you would pay for a small home with a 25-year mortgage. ■

EXPLORATION
Stacks of pennies on a chessboard

A chessboard (or checkerboard) has 64 squares. Suppose that you have won the lottery and may choose between payment plan A or B. By plan A you will receive $50 million. By plan B, you will receive a chessboard with 1 penny on the first square, 2 pennies stacked on the second square, 4 pennies stacked on the third square, 8 pennies stacked on the fourth square, 16 pennies stacked on the fifth square, and so on, where each square has twice as many pennies as the previous square.

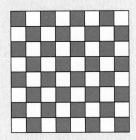

1. What is the total number of pennies paid under plan B? How much are they worth in dollars?

2. By which plan would you receive more money?

3. Measure the height (in inches) of a stack of 10 pennies. Then divide the result by 10 to get an estimate of the height (in inches) of 1 penny.

4. Assuming that they do not fall over, what would be the height (in inches) of the stack of pennies on the 64th square? Convert your result using units of miles. [**Hint**: There are 5280 feet in a mile.] Would the stack of pennies on the 64th square reach the moon? [**Hint**: Consult a dictionary.]

OF THIS SECTION

If the sequence $a_1, a_2, a_3, \ldots, a_n$ is a geometric sequence, then the sum $a_1 + a_2 + a_3 + \cdots + a_n$ is a **geometric series**.	**Geometric series**
If $S_n = a_1 + a_2 + a_3 + \cdots + a_n$ is a geometric series with common ratio r, then $S_n = \dfrac{a_1(1 - r^n)}{1 - r}$.	**Formula for a geometric series**

Check that a series is geometric	To evaluate the sum $S_n = a_1 + a_2 + a_3 + \cdots + a_n$, we first check whether the series is arithmetic, geometric, or neither. If the series is geometric, we use $S_n = \dfrac{a_1(1 - r^n)}{1 - r}$.
Using a combination of formulas	For a geometric series $a_1 + a_2 + a_3 + \cdots + a_n$, it sometimes is necessary to use $a_n = a_1 r^{n-1}$ to find a_1, r, or n in order to then find S_n by using $S_n = \dfrac{a_1(1 - r^n)}{1 - r}$.

HOMEWORK 10.4

SSM
Tutor Center

 MathPro 4
MathPro 4

 MathPro 5
MathPro 5

CDs/Videos

www
www

Evaluate the sum of the geometric series with the given values for a_1, r, and n.

1. $a_1 = 5$, $r = 2$, and $n = 27$

2. $a_1 = 6$, $r = 3$, and $n = 19$

3. $a_1 = 6$, $r = 1.3$, and $n = 30$

4. $a_1 = 10$, $r = 1.5$, and $n = 33$

5. $a_1 = 13$, $r = 0.8$, and $n = 27$

6. $a_1 = 9$, $r = 0.2$, and $n = 20$

7. $a_1 = 2.3$, $r = 0.9$, and $n = 50$

8. $a_1 = 4$, $r = 0.6$, and $n = 38$

Find the sum of the series.

9. $2 + 10 + 50 + 250 + 1250 + \cdots + a_{13}$

10. $1 + 2 + 4 + 8 + 16 + \cdots + a_{28}$

11. $600 + 180 + 54 + 16.2 + 4.86 + \cdots + a_{11}$

12. $625 + 500 + 400 + 320 + 256 + \cdots + a_{12}$

13. $3 + 2 + \dfrac{4}{3} + \dfrac{8}{9} + \dfrac{16}{27} + \cdots + a_{10}$

14. $6 + 0.6 + 0.06 + 0.006 + 0.0006 + \cdots + a_8$

15. $1 + 4 + 16 + 64 + 256 + \cdots + 67,108,864$

16. $7 + 21 + 63 + 189 + 567 + \cdots + 33,480,783$

17. $5 + 6 + 7.2 + 8.64 + 10.368 + \cdots + 21.4990848$

18. $800 + 1120 + 1568 + 2195.2 + 3073.28 + \cdots + 11,806.312448$

19. $10,000 + 5,000 + 2,500 + 1,250 + 625 + \cdots + 4.8828125$

20. $3 + 6 + 12 + 24 + 48 + \cdots + 196,608$

21. $S_{100} = 1 + 1 + 1 + 1 + 1 + \cdots + 1$

22. $3 + 30 + 300 + 3000 + 30,000 + \cdots + 3,000,000,000,000$

23. $3 + 9 + 15 + 21 + 27 + \cdots + 351$

24. $351 + 347 + 343 + 339 + 335 + \cdots + 103$

25. $30 + 5 + \dfrac{5}{6} + \dfrac{5}{36} + \dfrac{5}{216} + \cdots + \dfrac{5}{362,797,056}$

26. $80 + 40 + 20 + 10 + 5 + \cdots + \dfrac{5}{1024}$

*Let S_n represent the sum of a geometric series. Determine whether S_n is positive or negative in each of the following situations. Explain. [**Hint:** Try experimenting with specific values of a_1, r, and n that meet the conditions stated. Then explain why your response makes sense for any values that meet the conditions.]*

27. $a_1 > 0$, $r > 0$, and n is a counting number

28. $a_1 < 0$, $r > 0$, and n is a counting number

29. If $f(x) = 7 - x$, is the series $f(1) + f(2) + f(3) + \cdots + f(30)$ arithmetic, geometric, or neither? Explain.

30. If $f(x) = 2(4)^x$, is the series $f(1) + f(2) + f(3) + \cdots + f(70)$ arithmetic, geometric, or neither? Explain.

31. A person's starting salary is $23,500. Each year the salary increases by 4%. What will be the person's total earnings after 20 years of work?

32. A person's first-year salary is $32,000. The salary increases by 3% each year. What will be the person's total earnings after 30 years of work?

33. Two companies make you job offers. Company A offers a first-year salary of $26,000 and a 5% raise at the end of each year. Company B offers a first-year salary of $31,000 and 3% raises at the end of each year. At which company would your total earnings for 30 years be greater? By how much?

34. Two companies are bidding against each other to hire you. Company A offers a first-year salary of $25,000, a 4% raise at the end of each year, and a $500 bonus at the end of each year. (The 4% raise is based on the salary and not the bonus.) Company B offers a first-year salary of $30,000 and 3% raises at the end of each year. At which company would your total earnings for 26 years be greater? By how much?

35. In exercise 47 of Homework 10.2, you found the number (actually the greatest possible number) of ancestors a person has in the nth generation back. Find the total number of ancestors a person has through 10 generations back.

36. Suppose that a rumor is spreading in the United States that chlorine in swimming pools causes skin cancer. Suppose that 4 people hear the rumor on the first day. Assume that each person who hears the rumor tells the rumor to exactly 5 people who have not yet heard the rumor. Also assume that each person tells their 5 people approximately 24 hours after hearing the rumor.

    a. How many people will have heard the rumor after 10 days?

    b. After how many days will everyone in the United States have heard the rumor? Use the 2002 U.S. population of 287 million people.

    c. In order to model the spread of the chlorine-causes-cancer rumor, you made some assumptions about the way in which the rumor would spread. Describe each assumption and discuss whether you think each assumption is reasonable or not.

37. An entrepreneur writes letters to eight people (the first round of letters), explaining that she has found a way for herself and many other people to get rich. On each letter, the entrepreneur has written her name and address. She asks each of the eight people to send her $5 and to add their name and address below hers so each letter will now have two names on it. The entrepreneur also instructs each of the eight people to send the list of two names with the instructions off to eight more people (the second round). Then all these people should send the entrepreneur $5, add their names and addresses to the list, send the list of three names off to eight more people (the third round), and so on. Each person who receives a letter is instructed to send $5 to the person whose name is at the top of the list. When there are 10 people on the list, then the next person should send $5 to the person whose name is at the top of the list, scratch that name off the list, and add his name to the bottom of the list. The instructions include a warning that something terrible will happen to those people who do not send the money as well as the eight letters. (These letters are called *chain letters*. They are illegal.)

Assume that the letters of a round are received at approximately the same time and no one receives more than one letter.

a. In which round would the entrepreneur's name be taken off the list? How much money could the entrepreneur receive?

b. By which round would everyone in the world (about 6 billion people) have received a letter?

c. How many people will receive money from the chain letters? How much will they receive? [**Hint**: With 6 billion people in the world, there would be only $30 billion to go around.]

38. Suppose you win a contest and choose between two award plans. If you choose award plan A, you will receive $100,000 per day for 30 days. If you choose award plan B, you will receive one cent the first day, two cents the second day, four cents the third day, and so on (each day you receive twice as much as you did on the preceding day) for 30 days. Which plan would you choose? Explain your reasoning.

39. Prior to cassettes, popular recordings were sold on *eight-track* tape cartridges. In 1980, 89.5 million eight-track cartridges

were sold. After that year, sales dropped off sharply due to consumers' preference for cassettes over eight-tracks. In fact, sound recordings are no longer made on eight-tracks. The numbers of eight-track cartridges (in millions) sold for various years are listed in Table 5.

Table 5 Eight-track Cartridge Sales

Year	Sales (millions)
1980	89.5
1981	32.0
1982	20.0
1983	10.0
1984	5.0
1985	1.5

Source: *Recording Industry Association of America*

a. Let $f(t)$ represent the number of eight track cartridges (in millions) sold at t years since 1980. Find an exponential equation for f.

b. Find $f(0)$. What does your result mean in terms of eight-track cartridges?

c. Use the formula for the sum of a geometric series to find $f(0) + f(1) + f(2) + f(3) + f(4) + f(5)$. [**Hint**: Think carefully about the value of n.]

d. Compare your result from part c to the actual total.

e. Use the formula for the sum of a geometric series to predict the total number of eight-track cartridges that would have been sold from 1980 through 2008. Explain why this total is not much more than your total found in part c.

40. Describe a geometric series. Also, explain how to evaluate the sum of a geometric series $S_n = a_1 + a_2 + a_3 + \cdots + a_n$ if you know a_1, a_n, and the common ratio r of the geometric sequence $a_1, a_2, a_3, \ldots, a_n$.

Taking it to the lab

For each lab assignment, consult with your instructor on whether to organize your responses as a numbered list or to write them in a paragraph.

BOUNCING BALL LAB

In this lab you will analyze the heights reached by a bouncing ball. The heights represent the distance from the floor to the bottom of the ball.

Check with your instructor whether you should collect your own data or use the data listed in Table 6. In this table, $f(n)$ represents the maximum height (in centimeters) reached by a ball after n bounces.

Table 6 Maximum Heights of a Bouncing Ball

n	$f(n)$
1	72.930
2	48.673
3	33.745
5	18.379
6	11.301
7	10.805

Source: *Data collected by the author*

Materials

If you are going to perform your own experiment, you will need the following materials:

1. a rubber ball

2. a Texas Instruments CBL unit

3. a TI-82 or TI-83 graphing calculator

4. a Vernier motion detector

5. (optional) a Texas Instruments CBR unit (to be used in place of the CBL unit and motion detector)

If you don't have items 2, 4, and/or 5, another option is to use a video camera to tape the bouncing ball. You can make a background indicating heights in large print so you can estimate the heights by watching the video recording and using "pause" on your video player.

Preparation

First find a level surface for bouncing the ball. For the experimenting, it is ideal for the ball to bounce almost straight up and down at least 7 times. Attach the motion detector to a fixed object so the motion detector is above where the ball will bounce, facing directly downward (see Fig. 8). If you have a low ceiling, you can tape the motion detector to it.

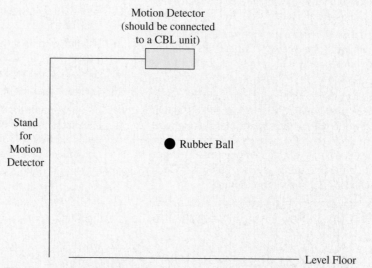

Figure 8 Equipment setup

Recording of Data

Use the CBL unit and motion detector to measure the maximum height (in centimeters) of the ball after each bounce. It is ideal to measure the heights reached by the ball for at least 7 bounces.

Analyzing the Data

1. Let $f(n)$ represent the maximum height (in centimeters) of the ball after n bounces. Display your data in a table or use the data in Table 6.

2. Recall that we have defined the ball's "height" to be the distance from the floor to the bottom of the ball. If you measured the distance from the floor to the top of the ball, adjust your data accordingly. (The data in Table 6 have already been adjusted.)

3. Find a quadratic and an exponential equation for f. Which model fits the data better? Which model is likely to make better predictions? Decide which equation you will use for f for the rest of this lab.

4. Is the sequence $f(1), f(2), f(3), f(4), \ldots, f(7)$ arithmetic, geometric, or neither? Explain.

5. If you performed your own experiment, you may skip this question. If you are using the data from Table 6, notice that no information is given for the height between the

4th and 5th bounces. That's because the motion detector malfunctioned. Use your equation for f to estimate the height reached after the 4th bounce.

6. Use your equation for f to estimate the height of the ball after the 8th bounce.
7. After which bounces will the ball reach a height of at least one foot?
8. Use your equation for f to estimate the height reached by the ball after the 30th bounce.
9. If your function f is an exponential function, then describe how the base of your function's equation relates to the situation.

Note

Remember, if you think model breakdown occurs, say so, say where, and explain why.

STACKED CUPS LAB

In this lab you will compare the height of a stack of plastic cups (one placed inside the next) to the number of cups in the stack.

Check with your instructor whether you should collect your own data or use the data listed in Table 7.

Materials

If you are going to perform your own experiment, you will need the following materials:

1. Some plastic cups that can be stacked
2. A ruler

Table 7 Heights of Stacks of Cups

Number of Cups in Stack	Height of Stack (centimeters)
1	12.00
2	14.75
3	17.50
4	20.25
5	23.00

Source: *Data collected by the author*

Recording of Data

Measure (in centimeters) the height of one cup. Then measure the heights of stacks of two, three, four, and five cups.

Analyzing the Data

1. Display your data in a table, or use the data in Table 7.
2. Let $h = f(n)$ represent the height (in centimeters) of n cups. Find an equation for f.
3. Use f to estimate the height of a stack of 70 cups.
4. Find the h-intercept of f. What does the intercept mean in terms of the cups?
5. Find the slope of f. What does the slope mean in terms of the stacked cups?
6. Sketch a graph of your model. Sketch only the portion for which your model makes reasonably good estimates.
7. Suppose that $f(n) = 0.3n + 20$ for a different set of cups.
   a. What does this equation tell you about the cups?
   b. What is the arithmetic sequence that corresponds to f?

Note

Remember, if you think model breakdown occurs, say so, say where, and explain why.

RETIREMENT LAB

In this lab, you will create a savings plan for your retirement and calculate how much money you will have at retirement.

Analyzing the Data

1. State the amount of money you have saved so far. (This can be a fictitious amount.)
2. State a constant amount of money you plan to save each month or year.
3. Describe how you will invest the money until you retire and what interest rate you will earn.
4. Calculate the amount of money you will have at retirement.
5. Describe how you will invest your savings after you retire and what interest rate you will earn. Calculate how much money you will have to live on each year after retirement.

CHAPTER SUMMARY

Key Points
OF THIS CHAPTER

Arithmetic sequence	If the difference between any term of a sequence and the preceding term is equal to a constant d for every such pair of terms, we call the sequence an **arithmetic sequence** and we call the constant d the **common difference**.
Formulas for arithmetic sequences and series	If $a_1, a_2, a_3, \ldots a_n$ is an arithmetic sequence with common difference d, then: • $a_n = a_1 + (n-1)d$ • $S_n = a_1 + a_2 + a_3 + \cdots + a_n$ is an arithmetic series and $S_n = \dfrac{n(a_1 + a_n)}{2}$.
Connection between a linear function and an arithmetic sequence	For a linear function $f(x) = mx + b$, the sequence $f(1), f(2), f(3), \ldots$ is an arithmetic sequence with common difference equal to the slope m of f.
Geometric sequence	If the ratio between any term of a sequence and the preceding term is equal to a constant r for every such pair of terms, then we call the sequence a **geometric sequence** and we call the constant r the **common ratio**.
Formulas for geometric sequences and series	If $a_1, a_2, a_3, \ldots a_n$ is a geometric sequence with common ratio r, then: • $a_n = a_1 r^{n-1}$ • $S_n = a_1 + a_2 + a_3 + \cdots + a_n$ is a geometric series and $S_n = \dfrac{a_1(1 - r^n)}{1 - r}$.
Connection between an exponential function and a geometric sequence	For an exponential function $f(x) = ab^x$, the sequence $f(1), f(2), f(3), \ldots$ is a geometric sequence with common ratio equal to the base b of f.
Determine the type of sequence or series	When working with a sequence (or series), we first determine whether the sequence (or series) is arithmetic, geometric, or neither in order to use the appropriate formula.

CHAPTER 10 REVIEW EXERCISES

Determine whether each of the following is an arithmetic sequence, arithmetic series, geometric sequence, geometric series, or none of these types of sequences or series.

1. $160, 40, 10, 2.5, 0.625, \ldots$

2. $13 + 24 + 35 + 46 + 57 + \cdots$

3. $101, 95, 89, 83, 77, \ldots$

4. $9 + 18 + 36 + 72 + 144 + \cdots$

5. $7 + \dfrac{7}{5} + \dfrac{7}{25} + \dfrac{7}{125} + \dfrac{7}{625} + \cdots$

6. $3, 4, 6, 9, 13, \ldots$

Find a formula, using a_n notation, for the sequence.

7. $2, 6, 18, 54, 162, \ldots$

8. $25, 28, 31, 34, 37, \ldots$

9. $9, 4, -1, -6, -11, \ldots$

10. $200, 100, 50, 25, 12.5, \ldots$

11. $3.2, 5.9, 8.6, 11.3, 14, \ldots$

12. $800, 560, 392, 274.4, 192.08, \ldots$

Find the indicated term of the sequence.

13. the 47th term of $6, 12, 24, 48, 96, \ldots$

14. the 9th term of $768, 192, 48, 12, 3, \ldots$

15. the 98th term of $87, 84, 81, 78, 75, \ldots$

16. the 87th term of $2.3, 4.9, 7.5, 10.1, 12.7, \ldots$

Find the term number of the last term in the finite sequence.

17. $7, 11, 15, 19, 23, \ldots, 2023$

18. $501, 493, 485, 477, 469, \ldots, -107$

19. The number $470{,}715{,}894{,}135$ is a term in the sequence $5, 15, 45, 135, 405, \ldots$. What is its term number?

20. If $a_5 = 52$ and $a_9 = 36$ are terms from an arithmetic sequence, find a_{69}.

21. Find the sum of the first 43 terms of an arithmetic series with $a_1 = 52$ and $a_{43} = -200$.

22. Find the sum of the first 22 terms of a geometric series with $a_1 = 4$, $r = 1.7$, and $n = 22$.

Evaluate the sum of the series.

23. $3 + 6 + 12 + 24 + 48 + \cdots + 1{,}610{,}612{,}736$

24. $30 + 36 + 42 + 48 + 54 + \cdots + 1200$

25. $11 + 7 + 3 - 1 - 5 - \cdots - 1197$

26. $531,441 + 177,147 + 59,049 + 19,683 + 6,561 + \cdots + 1$

27. If $f(x) = 4(5)^x$, is $f(1) + f(2) + f(3) + \cdots + f(80)$ an arithmetic sequence, arithmetic series, geometric sequence, or geometric series? Explain.

28. If $f(x) = -9x + 40$, is $f(1), f(2), f(3), \ldots, f(80)$ an arithmetic sequence, arithmetic series, geometric sequence, or geometric series? Explain.

29. Two companies have made you job offers. Company A offers a first-year salary of $28,000 with a 4% raise at the end of each year. Company B offers a first-year salary of $34,000 with a constant raise of $1500 each year.

   a. What would be the salary for the 25th year at company A? at company B?

   b. What would be the total earnings for 25 years of work at company A? at company B?

   c. Explain how it is possible for the salary for the 25th year to be greater at company A than at company B, yet the total earnings for 25 years to be greater at company B than at company A.

30. The number of deaths at traffic signals in the United States was on the rise during the 1990s. Since the leading cause of such deaths is drivers who run red lights, dozens of cities have begun using cameras to catch these violators. Many cities first use the cameras to issue warnings and then, several months later, use the cameras to issue fines of about $100. The numbers of deaths at intersections for various years are listed in Table 8.

Table 8 Deaths at Intersections

Year	Deaths (thousands)
1992	6.2
1993	6.3
1994	6.7
1995	6.9
1996	7.3
1997	7.8
1998	8.0

Source: *Federal Highway Administration*

   a. Let $f(t)$ represent the number of deaths (in thousands) at intersections during the year that is t years since 1990. Assuming that f is a linear function, find an equation for f.

   b. Use f to predict the number of deaths at intersections in 2009.

   c. Estimate the total number of deaths at intersections from 1990 through 2009.

   d. If more cities begin using cameras to give fines for running red lights, do you think that your results to parts b and c will be underestimates or overestimates? Explain.

CHAPTER 10 TEST

Determine whether each of the following is an arithmetic sequence, arithmetic series, geometric sequence, geometric series, or none of these types of sequences or series.

1. $3, 6, 12, 24, 48, \ldots$

2. $20, 19, 17, 14, 10, \ldots$

3. $7 + 35 + 175 + 875 + 4375 + \cdots$

4. $69 + 61 + 53 + 45 + 37 + \cdots$

Find a formula for the sequence.

5. $31, 25, 19, 13, 7, \ldots$

6. $6, 24, 96, 384, 1536, \ldots$

7. Find the 87th term of the sequence $4, 7, 10, 13, 16, \ldots$.

8. Find the 31st term of the sequence $100, 50, 25, 12.5, 6.25, \ldots$.

Find the term number of the last term of the finite sequence.

9. $-27, -23, -19, -15, -11, \ldots, 1789$

10. $200, 220, 242, 266.2, 292.82, \ldots, 428.717762$

Evaluate the sum of the series.

11. $27 + 9 + 3 + 1 + \dfrac{1}{3} + \cdots + a_{20}$

12. $4 + 8 + 16 + 32 + 64 + \cdots + 2{,}147{,}483{,}648$

13. $50 + 46 + 42 + 38 + 34 + \cdots + (-78)$

14. $19 + 33 + 47 + 61 + 75 + \cdots + a_{400}$

15. Evaluate the sum of the series. [**Hint**: Begin by writing the series as a sum of two series.]

$$(7 + 2) + (7 \cdot 2 + 2^2) + (7 \cdot 3 + 2^3) + (7 \cdot 4 + 2^4)$$
$$+ (7 \cdot 5 + 2^5) + \cdots + (7 \cdot 20 + 2^{20})$$

16. Let $f(x) = 3x^2 + 1$. Is $f(1) + f(2) + f(3) + \cdots + f(100)$ an arithmetic sequence, arithmetic series, geometric sequence, geometric series, or none of these types of sequences or series? Explain.

17. Let S_n equal the sum of an arithmetic series. Determine whether S_n is positive or negative if $a_1 = 10$, $d = -3$, and n is a very large counting number. Explain.

18. As the popularity of jet skis and other "wet bikes" has increased, so have the number of fatalities. Some data are listed in Table 9.

Table 9 Numbers of Fatalities From Wet Bikes

Year	Number of Fatalities
1987	5
1989	20
1992	34
1994	56
1998	78

Source: *Boating Accident Statistics, www.commanderbob.com/cbstats.html. Accessed 4/7/02.*

a. Let $f(t)$ represent the number of fatalities from wet bikes at t years since 1980. Find an equation for f.

b. Find the values of the sequence $f(7)$, $f(8)$, $f(9)$, $f(10)$, $f(11)$. What do these values mean in terms of the situation?

c. Use your model to predict the total number of wet-bike fatalities from 1987 to 2007.

19. Assume that a person's salary is $32,000 for the first year and that the salary increases by 3% each year.

a. Let a_n represent the person's salary (in thousands of dollars) for the nth year. Find a formula for a_n.

b. When will the salary first be above $40,000?

c. What will the salary be for the 25th year?

d. What will be the total earnings during the first 25 years?

Cumulative Review of Chapters 1–10

Solve.

1. $4x^2 - 49 = 0$

2. $5e^x = 98$

3. $2b^7 - 3 = 51$

4. $\log_3(4x - 7) = 4$

5. $6x^2 + 13x = 5$

6. $(x + 3)(x - 4) = 5$

7. $4\ln(2x^3) - \ln(8x^6) = 9$

8. $6(3)^x - 5 = 52$

9. $3(2x - 5) + 4 = (x - 3)^2$

10. $\log_b(81) = 4$

11. $\dfrac{1}{x^2 - x - 6} - \dfrac{x}{x + 2} = \dfrac{x - 2}{x - 3}$

12. $5(3x - 2)^2 + 7 = 17$

13. $20 - 4x = 7(2x + 9)$

14. $\sqrt{x + 1} - \sqrt{2x - 5} = 1$

15. $\log_6(3x) + \log_6(x - 1) = 1$

16. Solve $3x^2 - 5x + 1 = 0$ by completing the square.

Solve each system.

17. $2x + 4y = 0$
$5x + 3y = 7$

18. $y = 3x + 9$
$4x + 2y = -2$

19. Solve the inequality $5 - 2(3x - 5) + 1 \geq 2 - 4x$. Describe the solution set with inequality notation, interval notation, and a graph.

Simplify.

20. $(3b^{-2}c^{-3})^4(6b^{-5}c^2)^2$

21. $\dfrac{8b^{1/2}c^{-4/3}}{10b^{3/4}c^{-7/3}}$

22. $\left(\dfrac{14b^3c^5}{21b^{-9}c^{-4}}\right)^2$

23. $ab^{-1} - a^{-1}b$

24. $3\sqrt{8x^3} - 2x\sqrt{18x}$

25. $\sqrt{12x^7}$

26. $\sqrt[4]{(5x - 7)^{21}}$

27. $\sqrt[3]{\dfrac{4}{x}}$

28. $\dfrac{5\sqrt{x} - 4}{3\sqrt{x} + 2}$

Simplify. Write your result as a single logarithm with a coefficient of 1.

29. $2\ln(3x^4) + 3\ln(5x^9)$

30. $4\log_b(2x^5) - 5\log_b(2x^2)$

Expand.

31. $(3x - 5)^2$

32. $(3\sqrt{x} - 4)(2\sqrt{x} + 7)$

33. $-2x(x^2 + 1)(x^3 + 5)$

34. $(4x - 1)(x^2 - 2x - 3)$

Perform the indicated operation.

35. $\dfrac{x^2 - 9}{x^2 - 9x + 20} \div \dfrac{5x - 15}{x^2 - 4x - 5}$

36. $\dfrac{3x}{x^2 - 10x + 25} - \dfrac{x + 2}{x^2 - 7x + 10}$

37. $\dfrac{4x - x^2}{6x^2 + 10x - 4} \cdot \dfrac{7 - 21x}{x^2 - 8x + 16}$

38. $\dfrac{1}{x^2 + 12x + 27} + \dfrac{x + 2}{x^3 + x^2 - 9x - 9}$

39. Simplify $\dfrac{\dfrac{x + 2}{x^2 - 64}}{\dfrac{x^2 + 4x + 4}{3x + 24}}$.

40. Write $f(x) = -3(x + 3)^2 - 7$ in standard form.

Factor.

41. $4x^3 - 8x^2 - 25x + 50$

42. $2x^3 - 4x^2 - 30x$

43. $6x^2 + 2x - 20$

44. $100x^2 - 1$

For exercises 45–51, refer to Fig. 9.

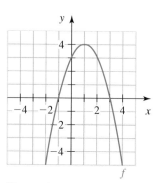

Figure 9 Exercises 45–51

45. Find $f(2)$.

46. Find x when $f(x) = 3$.

47. Find x when $f(x) = 4$.

48. Find x when $f(x) = 5$.

49. Find the y-intercept of f.

50. Find the x-intercepts of f.

51. Find an equation for f.

Sketch the graph.

52. $y = 5(2)^x$

53. $y = -3(x - 4)^2 + 3$

54. $y = -\dfrac{3}{5}x + 4$

55. $3(2x - y - 2) = y - x + 2$

56. $y = 2\sqrt{x + 5} - 4$

57. $y = 15\left(\dfrac{1}{3}\right)^x$

58. $y = 2x^2 + 5x - 1$

59. $2x(x - 3) + y = 5(x + 1)$

60. Find an equation of the line that contains the points $(-3, 2)$ and $(2, -5)$.

61. Find an equation of an exponential curve that contains the points $(3, 95)$ and $(6, 12)$.

62. Find an equation of a parabola that contains the points $(2, 1)$, $(3, 6)$, and $(4, 15)$.

63. Find an equation of a square root curve that contains the points $(2, 5)$ and $(6, 17)$.

64. Let f be the linear function, g be the exponential function, and h be a quadratic function whose graphs contain the points $(0, 2)$ and $(1, 4)$.
  a. Find possible equations for f, g, and h.
  b. Use your graphing calculator to draw the graphs of f, g, and h in the same viewing window.

Find the logarithm.

65. $\log_3(81)$ **66.** $\log_b(\sqrt{b})$

67. $\log(0.0001)$ **68.** $\log_3(8)$

Find the inverse of the function.

69. $g(x) = \log_2(x)$

70. $f(x) = -4x - 7$

71. Find the domain of $f(x) = \dfrac{x - 3}{x^2 - 2x - 35}$.

72. Find a formula of the sequence $96, 48, 24, 12, 6, \ldots$.

73. Find the 10th term of the sequence $2, 6, 18, 54, 162, \ldots$.

74. Find the term number of the last term of the finite sequence $-86, -82, -78, -74, -70, \ldots, 170$.

Find the sum of the series.

75. $98{,}304 + 49{,}152 + 24{,}576 + 12{,}288 + 6{,}144 + \cdots + 3$

76. $11 + 14 + 17 + 20 + 23 + \cdots + 182$

77. Table 10 lists sales of books and total recreational expenditures in the United States for various years.

Table 10 Sales of Books and Total Recreational Expenditures

	Books		All Forms of Recreation
Year	Sales (in billions of dollars)	Year	Sales (in billions of dollars)
1982	9.9	1987	223.7
1985	12.6	1990	284.9
1990	19.0	1993	340.1
1995	25.2	1996	429.6
2000	32.1	1999	534.9

Source: *Book Industry Study Group and U.S. Bureau of Economic Analysis*

  a. Let $B(t)$ represent total book sales (in billions of dollars) for the year that is t years since 1980. Perform the first three steps of the modeling process to find an equation for B.

  b. Let $R(t)$ represent total recreational expenditures (in billions of dollars) for the year that is t years since 1980. Find a quadratic equation for R.

  c. Let $P(t)$ represent the percentage of total recreational expenditures that consist of book sales for the year that is t years since 1980. Find an equation for P. [**Hint:** Use your equations for B and R to build an equation for P.]

  d. Explain in terms of the situation how it is possible for P to be a decreasing function for $t \geq 10$ when B is an increasing function.

  e. Predict when 3.8% of recreational expenditures will consist of book sales.

78. In 2000, the two most populous nations were China and India with populations of 1.261 billion and 1.014 billion, respectively. India is expected to pass China as the most populous nation within the next 50 years. The population of India in 1980 was 690 million.

  a. First, assume that India's population is growing linearly. Let $L(t)$ represent India's population (in millions) at t years since 1980. Find an equation for L.

  b. Now, assume that India's population is growing exponentially. Let $E(t)$ represent India's population (in millions) at t years since 1980. Find an equation for E.

  c. Find $L(70)$ and $E(70)$. What do these results mean in terms of India's population?

  d. Find $E(70) - L(70)$. What does this result mean in terms of India's population? To get an idea of the size of your result, compare it with 393, a prediction of the United State's population (in millions) in 2050 (Source: U.S. Census Bureau).

79. The percentage of passenger vehicles sold in the United States that are light trucks has more than quadrupled since 1979 (see Table 11).

  Let $f(t)$ represent the percentage of passenger vehicles sold that are light trucks for the year that is t years since 1979.

  a. Perform the first three steps of the modeling process to find an equation for f.

  b. Estimate the percentage of passenger vehicles sold in 2002 that were light trucks.

c. Predict when 49% of passenger vehicles sold will be light trucks.

Table 11 Percentages of Vehicles Sold that Are Light Trucks

Year	Percent
1979	9.4
1984	24.0
1989	30.2
1994	40.6
1998	44.8
2000	46.0

Source: *U.S. Department of Transportation*

80. The numbers of men studying to be Roman Catholic priests in graduate level programs in the United States for various years are listed in Table 12.

Let $f(t)$ represent the number of men (in thousands) studying to be priests in a graduate program at t years since 1960.

a. Perform the first three steps of the modeling process to find an equation for f.

Table 12 Men Studying to Be Roman Catholic Priests

Year	Number in Graduate Programs (thousands)
1965	8.3
1970	6.6
1975	5.3
1980	4.1
1985	4.0
1990	3.6
1995	3.2
1998	3.3
2001	3.5

Source: *CARA—Center for Applied Research in the Apostolate, http://cara.georgetown.edu/bulletin/index.htm. Accessed 4/7/02.*

b. Find $f(49)$. What does your result mean in terms of the situation?

c. Find t when $f(t) = 5$. What does your result mean in terms of the situation?

d. For what values of t is f decreasing? For what values of t is f increasing? What do your responses mean in terms of the situation?

Additional Topics

Mathematics abounds in bright ideas. No matter how long and hard one pursues her, mathematics never seems to run out of exciting surprises. And by no means are these gems to be found only in difficult work at an advanced level. All kinds of simple notions are full of ingenuity.

—*Ross Honsberger,* Mathematical Morsels, *Washington DC: Mathematical Association of America, 1978, p. vii*

11.1 ABSOLUTE VALUE EQUATIONS AND INEQUALITIES

Objectives

▸ Know the graphical meaning of absolute value.
▸ Solve absolute value equations and absolute value inequalities.

In this section we work with absolute value. Recall that $|-6| = 6$ and $|6| = 6$. We can think of the absolute value of a number as the *distance* that the number is from 0 on the number line (see Fig. 1).

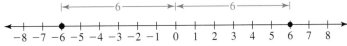

$$\overset{\longleftarrow \!\!\!\rule[0.5ex]{2.2cm}{0.4pt}\!\!\! 6 \!\!\!\rule[0.5ex]{2.2cm}{0.4pt}\!\!\! \longrightarrow\!\!|\!\!\longleftarrow\!\!\!\rule[0.5ex]{2.2cm}{0.4pt}\!\!\! 6 \!\!\!\rule[0.5ex]{2.2cm}{0.4pt}\!\!\!\longrightarrow}{}$$

$$-8 \; -7 \; -6 \; -5 \; -4 \; -3 \; -2 \; -1 \quad 0 \quad 1 \quad 2 \quad 3 \quad 4 \quad 5 \quad 6 \quad 7 \quad 8$$

Figure 1 The numbers -6 and 6 are both a distance of 6 from 0

So $|-6| = 6$ means that -6 is a distance of 6 units from 0 and $|6| = 6$ means that 6 is a distance of 6 units from 0.

Absolute Value
The absolute value of a number is the distance that the number is from 0 on the number line.

To solve an *absolute value equation* such as $|x| = 4$, we must determine all numbers that are a distance of 4 units from 0. There are two such numbers, -4 and 4. So the statements $|x| = 4$ and $x = \pm 4$ are equivalent statements.

An Equivalent Form of an Absolute Value Equation
For an expression A and nonnegative constant k, the equation $$|A| = k$$ is equivalent to the statement $A = -k$ or $A = k$.

Example 1   Solving Absolute Value Equations

Solve the equation.

1. $|2x + 1| = 11$

2. $3|x| + 4 = 12$

3. $|4x + 12| = 0$

4. $|5x - 8| = -4$

Solution

1. For $|2x + 1| = 11$, the expression $2x + 1$ represents numbers that are a distance of 11 from 0, which are the numbers -11 and 11.

$$
\begin{array}{lcl}
2x + 1 = -11 & \text{or} & 2x + 1 = 11 \\
2x = -12 & \text{or} & 2x = 10 \\
x = -6 & \text{or} & x = 5
\end{array}
$$

We check that both -6 and 5 satisfy the original equation.

Check x = -6	**Check x = 5**
$\|2x + 1\| = 11$	$\|2x + 1\| = 11$
$\|2(-6) + 1\| \stackrel{?}{=} 11$	$\|2(5) + 1\| \stackrel{?}{=} 11$
$\|-11\| \stackrel{?}{=} 11$	$\|11\| \stackrel{?}{=} 11$
$11 \stackrel{?}{=} 11$	$11 \stackrel{?}{=} 11$
true	true

2. We begin solving $3|x| + 4 = 12$ by isolating the absolute value.

$$
\begin{aligned}
3|x| + 4 &= 12 \\
3|x| &= 8 \\
|x| &= \frac{8}{3} \\
x &= \pm\frac{8}{3}
\end{aligned}
$$

We can check that $-\dfrac{8}{3}$ and $\dfrac{8}{3}$ satisfy the original equation.

3. For $|4x + 12| = 0$, the expression $4x + 12$ represents the one number that is a distance of 0 units from 0. This is the number 0.

$$
\begin{aligned}
4x + 12 &= 0 \\
4x &= -12 \\
x &= -3
\end{aligned}
$$

4. Since $|5x - 8|$ is nonnegative, the solution set for $|5x - 8| = -4$ is the empty set. ∎

We now turn our attention to solving *absolute value inequalities* such as $|x| < 3$. To solve $|x| < 3$, we find all numbers whose distance from 0 is less than 3 units. So the solutions of $|x| < 3$ are all the numbers between -3 and 3 (see Fig. 2).

Figure 2 Graph of numbers whose distance from 0 is less than 3 units

We can describe the numbers between -3 and 3 using the inequality $-3 < x < 3$. The inequality $-3 < x < 3$ describes each number that is *both* greater than -3 *and* less than 3.

In Fig. 3 we describe some typical solution sets for absolute value inequalities in words, graphs, and notations.

In Words	Inequality Notation	Graph	Interval Notation
Numbers between 1 and 3	$1 < x < 3$	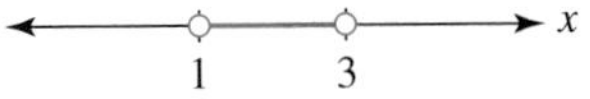	$(1, 3)$
Numbers between 1 and 3, inclusive	$1 \leq x \leq 3$		$[1, 3]$
Numbers less than 1 or greater than 3	$x < 1 \text{ or } x > 3$	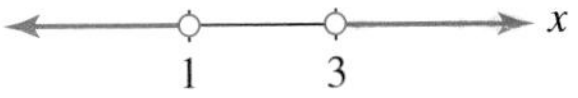	$(-\infty, 1) \cup (3, \infty)$
Numbers less than or equal to 1 or greater than or equal to 3	$x \leq 1 \text{ or } x \geq 3$		$(-\infty, 1] \cup [3, \infty)$

Figure 3 Words, graphs, and notation for inequalities

If A and B are sets, then $A \cup B$, the *union of A and B,* is the set of members of A together with members of B. So $(-\infty, 1) \cup (3, \infty)$ is the set of numbers less than 1 together with numbers greater than 3.

We refer to a set of numbers such as $(1, 3)$ as an *open interval*. We call a set of numbers such as $[1, 3]$ a *closed interval*.

Example 2   Solving Absolute Value Inequalities

Solve the inequality and graph the solution set.

1. $|x| > 2$ **2.** $|2x - 3| \leq 9$

3. $-3|x + 1| + 4 < -2$ **4.** $|7x - 10| < -2$

5. $|5x - 8| > -1$

Solution

1. For $|x| > 2$, the solutions are numbers each of whose distance from 0 is greater than 2 units. These are all the numbers that are either less than -2 *or* greater than 2 (see Fig. 4).

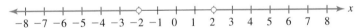

Figure 4 Graph of numbers whose distance from 0 is greater than 2 units

So, the solution set is the set of numbers x where $x < -2$ or $x > 2$, or in interval notation, $(-\infty, -2) \cup (2, \infty)$.

2. For $|2x - 3| \leq 9$, the expression $2x - 3$ represents numbers each of whose distance from 0 is less than or equal to 9. Such a number is between -9 and 9, inclusive:

$$-9 \leq 2x - 3 \leq 9$$

This means that $2x - 3$ represents numbers that are *both* greater than or equal to -9 *and* less than or equal to 9:

$$2x - 3 \geq -9 \quad \text{and} \quad 2x - 3 \leq 9$$
$$2x \geq -6 \quad \text{and} \quad 2x \leq 12$$
$$x \geq -3 \quad \text{and} \quad x \leq 6$$

So, the solution set is the set of numbers x where $-3 \leq x \leq 6$, or in interval notation, $[-3, 6]$. We graph the solution set in Fig. 5.

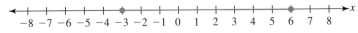

Figure 5 Graph of $-3 \leq x \leq 6$

Instead of writing $-9 \le 2x - 3 \le 9$ as two inequalities and then solving, we can apply the same steps to $-9 \le 2x - 3 \le 9$:

$$-9 \le 2x - 3 \le 9$$
$$-9 + 3 \le 2x - 3 + 3 \le 9 + 3$$
$$-6 \le 2x \le 12$$
$$-3 \le x \le 6$$

3. To solve $-3|x + 1| + 4 < -2$, we begin by isolating the absolute value:

$$-3|x + 1| + 4 < -2$$
$$-3|x + 1| < -6$$
$$\frac{-3|x + 1|}{-3} > \frac{-6}{-3} \qquad \text{Reverse the direction of the inequality.}$$
$$|x + 1| > 2$$

So $x + 1$ represents numbers whose distance is more than 2 units from 0. These numbers are less than -2 or greater than 2:

$$x + 1 < -2 \quad \text{or} \quad x + 1 > 2$$
$$x < -3 \quad \text{or} \quad x > 1$$

Note

The solution set is *not* the set of numbers *x* where $1 < x < -3$, which would say that *x* is both greater than 1 *and* less than -3. There is no such number.

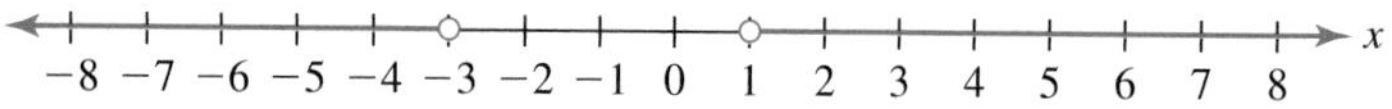

The solution set is the set of numbers x where $x < -3$ or $x > 1$, or in interval notation, $(-\infty, -3) \cup (1, \infty)$. We graph the solution set in Fig. 6.

Figure 6 Graph of $x < -3$ or $x > 1$

4. Since $|7x - 10|$ is nonnegative, the inequality $|7x - 10| < -2$ has an empty set solution.

5. Since $|5x - 8|$ is nonnegative for *any* real number x, the solution set of $|5x - 8| > -1$ is the set of all real numbers, or in interval notation, $(-\infty, \infty)$. ∎

Key Points

OF THIS SECTION

Absolute value	The absolute value of a number is the distance that the number is from 0 on the number line.
Equivalent statements	For an expression A and nonnegative constant k, the equation $|A| = k$ is equivalent to the statement $A = -k$ or $A = k$.
Solving absolute value equations or inequalities	To solve absolute value equations or absolute value inequalities, first isolate the absolute value and then think of absolute value as the distance that a number is from 0.

EXPLORATION

Graphical meaning of |a − b|

In this exploration you will explore the graphical meaning of $|a - b|$.

1. Plot the points 1 and 6 on a number line. What is the distance between 1 and 6? Compare your result with each of $|1 - 6|$ and $|6 - 1|$.

2. Plot the points -2 and 3 on a number line. What is the distance between -2 and 3? Compare your result with each of $|(-2) - (3)|$ and $|(3) - (-2)|$.

3. Find the distance between -7 and -3 and compare your result with each of $|(-7) - (-3)|$ and $|(-3) - (-7)|$.

4. Describe the graphical meaning for $|a - b|$.

5. Solve the equation. Then find the distance between 5 and each solution. Explain why your result makes sense if you think in terms of the graphical meaning of $|x - 5|$.

   a. $|x - 5| = 1$

   b. $|x - 5| = 2$

   c. $|x - 5| = 3$

6. Solve the equation or inequality and graph the solutions. Explain why your result makes sense if you think in terms of the graphical meaning of $|x - 4|$.

   a. $|x - 4| = 3$ [**Hint**: For the graph, plot the two solutions.]

   b. $|x - 4| < 3$

   c. $|x - 4| > 3$

HOMEWORK 11.1

 SSM
Tutor Center

 MathPro 4

 MathPro 5

CDs/Videos   www

Solve the equation.

1. $|x| = 7$

2. $|x| = 4$

3. $|x| = -3$

4. $|x| = -1$

5. $5|x| - 3 = 15$

6. $-7|x| + 6 = 4$

7. $-4|x| - 11 = 2$

8. $2|x| + 7 = 1$

9. $|x + 2| = 5$

10. $|x - 3| = 8$

11. $|x - 5| = 0$

12. $-|x + 1| = 0$

13. $|3x - 1| = 11$

14. $|6x + 4| = 7$

15. $|5 - 2x| = 9$

16. $|4 - 7x| = 3$

17. $|2x + 9| = -6$

18. $|5x - 1| = -3$

19. $2|x + 5| = 8$

20. $-3|x - 4| = -15$

21. $|3x - 9| = 0$

22. $-5|4x - 8| = 0$

23. $-6|2x - 7| + 4 = -2$

24. $3|6x + 5| - 2 = 7$

25. $\left| \dfrac{1}{2}x - \dfrac{5}{3} \right| = \dfrac{7}{6}$

26. $\left| \dfrac{3}{4}x + \dfrac{7}{2} \right| = \dfrac{1}{3}$

Find approximate solution(s) for the equation.

27. $4.7|x| - 3.9 = 8.8$

28. $1.9|x| + 4.1 = 12.8$

29. $3.7|2.1x + 5.8| - 9.7 = 10.2$

30. $-5.4|3.6x - 2.1| + 9.4 = 2.8$

Solve the inequality. Describe the solution set with inequality notation, interval notation, and a graph.

31. $|x| < 4$

32. $|x| > 1$

33. $|x| \geq 3$

34. $|x| \leq 2$

35. $|x| < -3$

36. $|x| > -3$

37. $2|x| - 5 > 3$

38. $-3|x| + 11 \geq -1$

39. $2 - 5|x| \leq -8$

40. $10 - 4|x| < -2$

41. $|x - 6| \geq 7$

42. $|x + 1| < 3$

43. $|2x + 5| \leq 15$

44. $|3x - 4| > 25$

45. $|7x + 15| > -4$

46. $|2x - 1| < -2$

47. $2|x - 3| + 1 \geq 17$

48. $-4|x + 5| + 8 \geq 4$

49. $-3|5 - 2x| + 1 \leq -8$

50. $4|3 - x| - 5 > 25$

51. $5|2x + 3| + 7 < 2$

52. $-3|5x - 1| - 2 \leq 7$

53. $\left| \dfrac{2x}{5} + \dfrac{3}{2} \right| \leq \dfrac{9}{20}$

54. $\left| \dfrac{5x}{3} - \dfrac{1}{4} \right| > \dfrac{7}{12}$

Find approximate solutions for the inequality. Describe your result with inequality notation, interval notation, and a graph.

55. $2.9|x| - 5.6 > 13.9$

56. $-3.8|x| + 4.4 \geq -10.1$

57. $-6.2|3.5x - 1.3| \geq -14.5$

58. $2.8|4.9x - 3.1| < 10.4$

59. Is the statement $|a + b| = |a| + |b|$ for all real numbers a and b true or false? Explain. [**Hint**: Try substituting a positive number for a and a negative number for b.]

60. A student tries to solve the inequality $|x + 2| < 7$. What would you tell the student?

$$|x + 2| < 7$$
$$x + 2 < 7$$
$$x < 5$$

61. a. Solve $|2x + 1| = 11$.

   b. Solve $|2x + 1| < 11$.

   c. Solve $|2x + 1| > 11$.

   d. Graph the solutions in parts a, b, and c on the same number line. Use three colors to identify the different solutions. Make some observations about the solutions. Explain these observations.

62. List three numbers that satisfy the inequality $-2|3x - 4| < -4$.

63. Assume that $a \neq 0$ and $b \neq 0$ and that the equation $a|bx + c| + d = k$ has at least one solution for x. Solve the equation for x.

64. What must be true of the constants a, b, c, d, and k if the equation $a|bx + c| + d = k$ has exactly one solution?

65. A student tries to solve the inequality $|x+3| < 10$. Did the student solve the inequality correctly? Explain.

$$|x+3| < 10$$
$$x+3 < -10 \quad \text{or} \quad x+3 < 10$$
$$x < -13 \quad \text{or} \quad x < 7$$

66. Explain how to solve an inequality of the form $a|bx+c|+d < k$, where $a \neq 0$ and $b \neq 0$.

Section 11.1 Quiz

Solve the equation.

1. $3|x| - 4 = 11$

2. $|2x - 3| = 7$

3. $-2|3x + 5| + 9 = 1$

4. $5|6x - 5| = 15$

5. $|2x - 5| = 0$

6. $|7x + 1| = -3$

Solve the inequality. Describe the solution set with inequality notation, interval notation, and a graph.

7. $|x| \leq 6$

8. $|x| \geq 4$

9. $|4x - 8| > 12$

10. $|3x + 1| < 5$

11. $-2|x + 5| + 10 \geq 4$

12. $7|3x - 2| \leq 42$

13. $|x - 5| < -7$

14. $3|2x - 9| > -1$

15. Is the statement $|a - b| = |a| - |b|$ for all real numbers a and b true or false? Explain.

16. Solve the inequality $|3x - 6| \leq 0$.

11.2 INEQUALITIES IN TWO VARIABLES

Objectives

▸ Graph a linear inequality in two variables.

▸ Solve a system of linear inequalities in two variables.

In this section we work with *linear inequalities in two variables*. Here are some examples:

$$y < 3x - 5, \qquad 2x + 4y \geq 9, \qquad 5(y - 3) + x \leq 0$$

We say that an ordered pair (a, b) *satisfies* an inequality in two variables, such as $y < 3x - 5$, if the inequality becomes a true statement after substituting a for x and b for y. We call such an ordered pair a *solution* of the inequality and we call the set of all solutions the *solution set*. We describe the solution set by graphing the solutions.

In the first part of this section we discuss how to graph such inequalities. For example, to graph $y > \frac{1}{2}x + 1$, we begin by sketching a graph of $y = \frac{1}{2}x + 1$ (see Fig. 7).

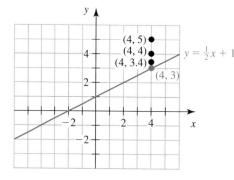

Figure 7 Graph of $y = \frac{1}{2}x + 1$

To investigate solving $y > \frac{1}{2}x + 1$, we choose a value of x, say 4, and find several solutions with x-coordinate 4. For $y = \frac{1}{2}x + 1$, if $x = 4$, then $y = \frac{1}{2}(4) + 1 = 3$. So the point $(4, 3)$ is on the graph of $y = \frac{1}{2}x + 1$.

For $y > \frac{1}{2}x + 1$, if $x = 4$, then we have

$$y > \frac{1}{2}(4) + 1$$
$$y > 3$$

So, if $x = 4$, some possible values for y are $y = 3.4$, $y = 4$, and $y = 5$. Note that the points $(4, 3.4)$, $(4, 4)$, $(4, 5)$ lie *above* the point $(4, 3)$, which lies on the line $y = \frac{1}{2}x + 1$ (see Fig. 7 once again).

We could choose other values of x beside 4 and go through a similar argument. These investigations would suggest that the solutions of $y > \frac{1}{2}x + 1$ lie *above* the graph of $y = \frac{1}{2}x + 1$. This is, in fact, true. In Fig. 8 we shade the region that contains all of the solutions of $y > \frac{1}{2}x + 1$. We call this region the *solution region* of the inequality.

We dash the line $y = \frac{1}{2}x + 1$ to indicate that its points are *not* solutions of $y > \frac{1}{2}x + 1$. For example, the point $(4, 3)$ on the line $y = \frac{1}{2}x + 1$ does *not* satisfy the inequality $y > \frac{1}{2}x + 1$:

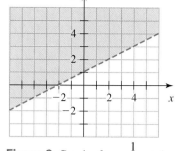

Figure 8 Graph of $y > \frac{1}{2}x + 1$

$$y > \frac{1}{2}x + 1$$

$$3 \overset{?}{>} \frac{1}{2}(4) + 1$$

$$3 \overset{?}{>} 3$$

false

We can draw a graph of $y > \frac{1}{2}x + 1$ using a graphing calculator (see Fig. 9), but we have to imagine that the border $y = \frac{1}{2}x + 1$ is drawn with a dashed line.

Figure 9 Graph of $y > \frac{1}{2}x + 1$ where we imagine that the border is drawn with a dashed line

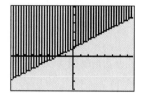

Graphing Calculator

To shade above a line, press $\boxed{\text{Y=}}$ and then press $\boxed{\triangleleft}$ twice. Next press $\boxed{\text{ENTER}}$ as many times as necessary for the triangle shown in Fig. 9 to appear.

To graph $y \geq \frac{1}{2}x + 1$, we use a *solid* line along the border $y = \frac{1}{2}x + 1$ to indicate that the points on the line $y = \frac{1}{2}x + 1$ *are* solutions of $y \geq \frac{1}{2}x + 1$ (see Fig. 10).

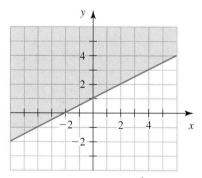

Figure 10 Graph of $y \geq \frac{1}{2}x + 1$

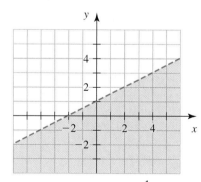

Figure 11 Graph of $y < \frac{1}{2}x + 1$

For the graph of $y < \frac{1}{2}x + 1$, the solutions lie *below* the line $y = \frac{1}{2}x + 1$, so we shade the region *below* the line $y = \frac{1}{2}x + 1$ and use a dashed line for the border $y = \frac{1}{2}x + 1$ (see Fig. 11). For the graph of $y \leq \frac{1}{2}x + 1$, we would use a solid line for the border $y = \frac{1}{2}x + 1$.

Graph of an Inequality in Two Variables

- The graph of an inequality of the form $y > mx + b$ is the region above the line $y = mx + b$. The graph of an inequality of the form $y < mx + b$ is the region below the line $y = mx + b$. The line $y = mx + b$ is not part of the graph. We show this with a dashed line.
- The graph of an inequality of the form $y \geq mx + b$ is the region above the line $y = mx + b$ as well as the line $y = mx + b$. The graph of an inequality of the form $y \leq mx + b$ is the region below the line $y = mx + b$ as well as the line $y = mx + b$.

To sketch a graph of an inequality in two variables, we begin by isolating y on one side of the inequality.

Example 1   Sketching the Graph of an Inequality

Sketch the graph of $-2x - 3y > 6$.

Solution

First, we isolate y.

$$-2x - 3y > 6$$

$$-3y > 2x + 6 \qquad \text{Add } 2x \text{ to both sides.}$$

$$\frac{-3y}{-3} < \frac{2x}{-3} + \frac{6}{-3} \qquad \text{Reverse the direction of the inequality.}$$

$$y < -\frac{2}{3}x - 2$$

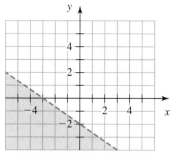

Figure 12 Graph of $-2x - 3y > 6$

The graph of $y < -\frac{2}{3}x - 2$ is the region below the line $y = -\frac{2}{3}x - 2$. We show that $y = -\frac{2}{3}x - 2$ is not part of the graph by using a dashed line (see Fig. 12).

As a check, we choose a point in our solution region, say $(-3, -1)$, and see whether it satisfies the inequality:

$$-2x - 3y > 6$$

$$-2(-3) - 3(-1) \overset{?}{>} 6$$

$$9 \overset{?}{>} 6$$

true

To check further, we could choose several other points in the solution region and check that each point satisfies the inequality.

We could also choose several points that are not in the solution region and check that each point does not satisfy the inequality. For example, note that $(0, 0)$ is not in the solution region and that this point does not satisfy the inequality $-2x - 3y > 6$:

$$-2x - 3y > 6$$

$$-2(0) - 3(0) \overset{?}{>} 6$$

$$0 \overset{?}{>} 6$$

false ■

Two or more inequalities form a *system of inequalities*. Here is an example of such a system:

$$y \geq -2x + 1$$

$$y < \frac{1}{2}x - 3$$

An ordered pair (a, b) is a *solution* of a system of inequalities if it satisfies all the inequalities in the system. The set of all solutions is called the *solution set* of the system. In this text, we show a solution set graphically. A solution of the system

$$y \geq -2x + 1$$
$$y < \frac{1}{2}x - 3$$

is a point that lies both in the solution region for $y \geq -2x + 1$ (see Fig. 13, blue region) and the solution region for $y < \frac{1}{2}x - 3$ (see Fig. 13, red region). So, the solution set of the system is the intersection of the two solution regions (the region shaded by a cross hatch of blue lines and red lines). We graph the solution set of the system in Fig. 14.

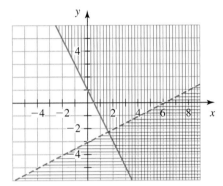

Figure 13 The solution region of $y \geq -2x + 1$ and the solution region of $y < \frac{1}{2}x - 3$

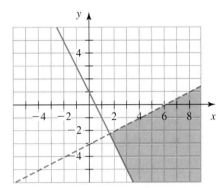

Figure 14 Graph of the solution set of the system

We use a graphing calculator to verify our work (see Fig. 15), where we imagine that the border $y = \frac{1}{2}x - 3$ is drawn with a dashed line. The solution set is the region shaded both by horizontal lines (in red) and vertical lines (in blue).

Figure 15 Verify our work, where we imagine that the border $y = \frac{1}{2}x - 3$ is drawn with a dashed line

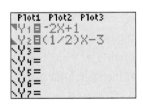

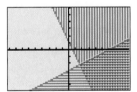

In general, the solution set of a system of inequalities is the region that is the intersection of all the solution regions of the inequalities in the system.

Example 2 Graphing the Solution Set of a System

Graph the solution set of the system.

$$y \geq 2x - 3$$
$$y \leq -x + 5$$
$$x \geq 0$$
$$y \geq 0$$

Solution

In Fig. 16 we use arrows to indicate the solution regions for each of the four inequalities. The solution set of the system is the intersection of the four solution regions, which is shown in Fig. 17.

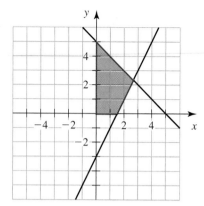

Figure 16 The solution regions of $y \geq 2x - 3$, $y \leq -x + 5$, $x \geq 0$, and $y \geq 0$

Figure 17 Graph of the solution set of the system

EXPLORATION

Meaning of solution of a system of inequalities in two variables

The graphs of $y = ax + b$ and $y = cx + d$ are sketched in Fig. 18.

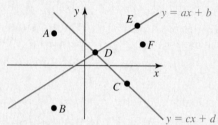

Figure 18 Graphs of $y = ax + b$ and $y = cx + d$

1. For each part, decide which of the points A, B, C, D, E, or F:
   a. satisfy the inequality $y < ax + b$
   b. satisfy the inequality $y \geq cx + d$
   c. are solutions of the system of inequalities

   $$y < ax + b$$
   $$y \geq cx + d$$

2. Write a system of inequalities in terms of a, b, c, d, x, and y so that the points B, C, and D are solutions and the points A, E, and F are not solutions.

3. Write a system of inequalities in terms of a, b, c, d, x, and y so that the points B, and C are solutions and the points A, D, E, and F are not solutions.

Key Points
OF THIS SECTION

The graph of $y > mx + b$ or $y < mx + b$

The graph of an inequality of the form $y > mx + b$ is the region above the line $y = mx + b$. The graph of an inequality of the form $y < mx + b$ is the region below the line $y = mx + b$. For either inequality, we use a dashed line to show that $y = mx + b$ is not part of the graph.

The graph of an inequality of the form $y \geq mx + b$ is the region above the line $y = mx + b$ as well as the line $y = mx + b$. The graph of an inequality of the form $y \leq mx + b$ is the region below the line $y = mx + b$ as well as the line $y = mx + b$.

The graph of $y \geq mx + b$ or $y \leq mx + b$

The solution region of a system of inequalities is the intersection of all of the solution regions of the inequalities in the system.

System of inequalities

HOMEWORK 11.2

 SSM
Tutor Center

 MathPro 4
 MathPro 4

 MathPro 5
MathPro 5

CDs/Videos

www

Graph the inequality.

1. $y \geq 2x - 4$

2. $y > x + 1$

3. $y < -\dfrac{1}{2}x + 3$

4. $y \leq -2x + 6$

5. $y \leq \dfrac{2}{3}x - 5$

6. $y \geq -3x + 4$

7. $y > x$

8. $y \leq 2x$

9. $2x + 5y < 10$

10. $3x - 2y \geq 8$

11. $4x - 6y - 6 \geq 0$

12. $2y - x + 1 > 0$

13. $3(x - 2) + y \leq -2$

14. $y > -2(x - 1) + 3$

15. $y \leq 2$

16. $y < -3$

17. $y > -5$

18. $y \geq 1$

Graph the solution set of the system of inequalities.

19. $y \geq \dfrac{1}{3}x - 2$
$y > -x + 3$

20. $y > \dfrac{1}{2}x + 3$
$y \leq -2x + 5$

21. $y \leq x - 4$
$y \geq -3x$

22. $y < 2x - 3$
$y < -2x + 1$

23. $y \leq -3x + 9$
$y \geq 2x - 3$
$x \geq 0$
$y \geq 0$

24. $y \leq \dfrac{1}{3}x + 4$
$y \geq 2x - 5$
$x \geq 0$
$y \geq 0$

25. $y < -x + 5$
$y \leq x + 5$
$y > \dfrac{1}{2}x + 1$

26. $y > -x - 4$
$y \geq 2x + 6$
$y \leq \dfrac{1}{2}x + 6$

27. $y \leq -3$
$y \geq -5$

28. $y \leq 2$
$y > -1$

29. $2x - 4y \leq 8$
$3x + 5y \leq 10$

30. $x - 2y > 6$
$x + 3y \leq 3$

31. $y \geq \dfrac{1}{2}(x - 4) - 1$
$y < -2(x - 1) + 3$

32. $2 - y < 3(x + 2)$
$y - 7 \leq 2(x - 3)$

33. A student believes that the graph of $2x - 3y < 6$ is the region below the line $2x - 3y = 6$. What would you tell the student?

34. Find three ordered pairs that are solutions of the inequality $y \leq -x - 2$. Also, find three ordered pairs that are not solutions.

35. Give an example of an inequality in terms of x and y for which $(3, 4)$ is a solution and $(4, 3)$ is not a solution.

36. Give an example of an inequality in terms of x and y where $(3, 3)$, $(-3, 3)$, and $(-3, -3)$ are solutions and $(3, -3)$ is not a solution.

37. Graph the solution set of the system
$$y \geq 2x + 1$$
$$y \leq 2x + 1$$

38. Describe the solution set of the system
$$y > 2x + 1$$
$$y < 2x + 1$$

39. Explain how to solve a system of inequalities in two variables.

Section 11.2 Quiz

Graph the inequality.

1. $y \leq 2x - 6$

2. $y \geq -3x + 9$

3. $4x - 2y > 8$

4. $3y - 5x < 15$

5. $-2(y + 3) + 4x \geq -8$

6. $y < -2$

Graph the solution set of the system of inequalities.

7. $y \leq -x + 4$
$y > 2x - 3$

8. $y \leq \dfrac{2}{5}x + 1$
$y < -\dfrac{1}{4}x + 2$

9. $3x - 4y \geq 12$
$6y - 2x \leq 12$

10. $x - y > 3$
$x + y < 5$
$x > 0$
$y > 0$

11. Find three ordered pairs that are solutions of the inequality $2x - 5y > 10$. Also, find three ordered pairs that are not solutions.

12. Give an example of an inequality where $(-2, -5)$, $(-2, 5)$, and $(2, -5)$ are solutions and $(2, 5)$ is not a solution.

11.3 FACTORING SUMS AND DIFFERENCES OF CUBES

Objective

▸ Factor sums of cubes and differences of cubes.

In earlier chapters we factored polynomials. In this section we focus on factoring special types of polynomials that are sums of cubes or differences of cubes. Here are some examples of such polynomials:

$$x^3 + 8, \qquad x^3 - 27, \qquad 8x^3 + 125, \qquad 64x^3 - 1$$

To see how to factor the sum of two cubes, we begin by expanding the expression $(x + a)(x^2 - ax + a^2)$:

$$(x + a)(x^2 - ax + a^2) = x \cdot x^2 - x \cdot ax + x \cdot a^2 + a \cdot x^2 - a \cdot ax + a \cdot a^2$$
$$= x^3 - ax^2 + a^2x + ax^2 - a^2x + a^3$$
$$= x^3 + a^3$$

So, $(x + a)(x^2 - ax + a^2) = x^3 + a^3$. Note that the right side of the equation, $x^3 + a^3$, is a sum of two cubes.

Sum of Two Cubes Formula

$$x^3 + a^3 = (x + a)(x^2 - ax + a^2)$$

We can use this formula to factor any polynomial that is a sum of two cubes.

Example 1 Factoring a Sum of Two Cubes

Factor $x^3 + 8$.

Solution

$$x^3 + 8 = x^3 + 2^3$$
$$= (x + 2)(x^2 - 2x + 2^2)$$
$$= (x + 2)(x^2 - 2x + 4)$$

We verify our work by using a graphing calculator to compare graphs of $y = x^3 + 8$ and $y = (x + 2)(x^2 - 2x + 4)$. See Fig. 19.

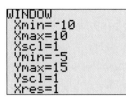  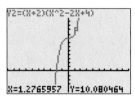

Figure 19 Verifying that $x^3 + 8 = (x + 2)(x^2 - 2x + 4)$

We can also factor a difference of two cubes. To see how to factor such a polynomial, we begin by expanding the expression $(x - a)(x^2 + ax + a^2)$:

$$(x - a)(x^2 + ax + a^2) = x \cdot x^2 + x \cdot ax + x \cdot a^2 - a \cdot x^2 - a \cdot ax - a \cdot a^2$$
$$= x^3 + ax^2 + a^2x - ax^2 - a^2x - a^3$$
$$= x^3 - a^3$$

So, $(x - a)(x^2 + ax + a^2) = x^3 - a^3$. Note that the right side of the equation, $x^3 - a^3$, is a difference of two cubes.

Difference of Two Cubes Formula

$$x^3 - a^3 = (x - a)(x^2 + ax + a^2)$$

Example 2   Factoring a Difference of Two Cubes

Factor $x^3 - 27$.

Solution

$$x^3 - 27 = x^3 - 3^3$$
$$= (x - 3)(x^2 + 3x + 3^2)$$
$$= (x - 3)(x^2 + 3x + 9)$$

We use graphing calculator tables to verify our work (see Fig. 20).

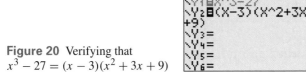

Figure 20 Verifying that
$x^3 - 27 = (x - 3)(x^2 + 3x + 9)$

Example 3   Factoring Sums or Differences of Cubes

Factor.

1. $64x^3 + 125$ **2.** $8x^3 - 1$

Solution

1. $64x^3 + 125 = (4x)^3 + 5^3$
$$= (4x + 5)((4x)^2 - 4x(5) + 5^2)$$
$$= (4x + 5)(16x^2 - 20x + 25)$$

2. $8x^3 - 1 = (2x)^3 - 1^3$
$$= (2x - 1)((2x)^2 + 2x(1) + 1^2)$$
$$= (2x - 1)(4x^2 + 2x + 1)$$

Recall that if the terms of a polynomial have a GCF not equal to one, we factor out the GCF first and then factor the result further, if possible.

Example 4   Factoring a Polynomial

Factor $5x^4 - 40x$.

Solution

To factor $5x^4 - 40x$, we first factor out the GCF $5x$:

$$5x^4 - 40x = 5x(x^3 - 8) \qquad \text{Factor out } 5x.$$
$$= 5x(x - 2)(x^2 + 2x + 4) \qquad x^3 - a^3 = (x - a)(x^2 + ax + a^2)$$

We verify our work using graphing calculator tables (see Fig. 21).

Figure 21 Verifying that
$5x^4 - 40x = 5x(x - 2)(x^2 + 2x + 4)$

In Section 6.4 we discussed a five-step factoring strategy. Here is a similar strategy which includes factoring the sum or difference of two cubes.

Five-Step Factoring Strategy

Here is a five-step method for factoring many polynomials. Steps 2–4 can be applied to the entire polynomial or to a factor of the polynomial.

1. If the GCF is not 1, factor it out.
2. For a binomial, try using the difference of two squares formula, the sum of two cubes formula, or the difference of two cubes formula.
3. For a trinomial $ax^2 + bx + c$:
   a. If $a = 1$, then try to find two integers whose product is c and whose sum is b.
   b. If $a \neq 1$, then try to factor using the trial and error method.
4. For an expression with four terms, try factoring by grouping.
5. Continue applying steps 2–4 until the polynomial is completely factored.

Example 5   Factoring a Polynomial

Factor $x^4 - 2x^2 + 1000x - 2000$.

Solution

Since $x^4 - 2x^2 + 1000x - 2000$ has four terms, we try to factor by grouping.

$$x^4 - 2x^3 + 1000x - 2000$$

$$\left. \begin{array}{l} = x^3(x - 2) + 1000(x - 2) \\ = (x - 2)(x^3 + 1000) \end{array} \right\} \qquad \text{Factor by grouping.}$$

$$= (x - 2)(x + 10)(x^2 - 10x + 100) \qquad x^3 + a^3 = (x + a)(x^2 - ax + a^2) \quad \blacksquare$$

 EXPLORATION

Finding a factor of a difference of two powers

1. Expand each product.
   a. $(x - a)(x + a)$
   b. $(x - a)(x^2 + ax + a)$
   c. $(x - a)(x^3 + ax^2 + a^2x + a^3)$
   d. $(x - a)(x^4 + ax^3 + a^2x^2 + a^3x + a^4)$
   e. $(x - a)(x^5 + ax^4 + a^2x^3 + a^3x^2 + a^4x + a^5)$
2. Write $x^6 - a^6$ as a product of $x - a$ and another polynomial.
3. Write $x^5 - 32$ as a product of $x - 2$ and another polynomial.

 Key Points

OF THIS SECTION

Sum of two cubes	$x^3 + a^3 = (x + a)(x^2 - ax + a^2)$
Difference of two cubes	$x^3 - a^3 = (x - a)(x^2 + ax + a^2)$

HOMEWORK 11.3

 MathPro 4
MathPro 4

MathPro 5
MathPro 5

SSM
Tutor Center

CDs/Videos

www
www

Factor. Verify your result using graphing calculator tables or graphs.

1. $x^3 + 27$

2. $x^3 - 8$

3. $x^3 - 64$

4. $x^3 + 64$

5. $x^3 - 1000$

6. $x^3 + 1000$

7. $x^3 + 125$

8. $x^3 - 125$

9. $x^3 + 1$

10. $x^3 - 1$

11. $64x^3 - 125$

12. $8x^3 + 1$

13. $27x^3 + 64$

14. $27x^3 - 64$

15. $1000x^3 - 1$

16. $1000x^3 + 1$

17. $8x^3 - 27$

18. $8x^3 + 27$

19. $125x^3 + 8$

20. $125x^3 - 8$

21. $4x^3 - 32$

22. $2x^3 + 54$

23. $7x^3 - 7000$

24. $5x^3 + 5$

25. $10x^3 + 640$

26. $8x^3 - 8$

27. $x^4 + 27x$

28. $x^4 - 64x$

29. $3x^4 - 24x$

30. $2x^4 + 250x$

31. $32x^4 - 4x$

32. $54x^4 - 16x$

Factor. Verify your result using graphing calculator tables or graphs.

33. $2x^3 - 50x$

34. $x^2 - 3x - 28$

35. $2x^3 - 16$

36. $8x^2 + 10x - 3$

37. $x^2 - 12x + 36$

38. $54x^4 - 2x$

39. $x^4 + x^3 + x + 1$

40. $8x^3 - 72x$

41. $6x^2 + 11x - 10$

42. $x^4 - 2x^3 + 8x - 16$

43. $2x^3 - 20x^2 + 32x$

44. $125x^3 + 27$

45. $8x^4 + 32x^3 - x - 4$

46. $4x^3 - 100x$

47. $x^3 - x^2 - 30x$

48. $x^4 - 2x^3 + 27x - 54$

49. $125 - 64x^3$

50. $4x - 4x^4$

51. A student tries to factor $x^3 - 8$. Did the student factor the polynomial correctly? Explain.

$$x^3 - 8 = (x - 2)(x^2 + 4x + 4)$$

52. A student tries to factor $x^3 + 27$. Did the student factor the polynomial correctly? Explain.

$$x^3 + 27 = (x + 3)(x^2 + 3x + 9)$$

53. Is the following statement true? Explain.

$$x^3 + a^3 = (x + a)(x^2 - 2ax + a^2)$$

54. Is the following statement true? Explain.

$$(x + a)^3 = x^3 + a^3$$

55. Describe how to factor:
- a sum of two cubes.
- a difference of two cubes.

Section 11.3 Quiz

Factor.

1. $x^3 - 216$

2. $x^3 + 216$

3. $64x^3 + 1$

4. $27x^3 - 125$

5. $16x^4 + 54x$

6. $375x^4 - 24x$

7. $5x - 5x^4$

8. $56 - 7x^3$

9. $x^4 + 2x^3 - 8x - 16$

10. $x^4 - 5x^3 + x - 5$

11.4 COMPLEX NUMBERS

Objectives

▸ Know the meaning of *complex number, imaginary number,* and *pure imaginary number.*

▸ Perform operations of complex numbers.

▸ Solve quadratic equations with complex number solutions.

So far in this course, we have worked with real numbers. In this section we discuss numbers that are not real numbers.

To start, consider the equation

$$x^2 = -1$$

We define $\sqrt{-1}$ to be a number whose square is -1. We represent this number more simply as i.

DEFINITION *Imaginary Unit i*

$$i = \sqrt{-1} \quad \text{and} \quad i^2 = -1$$

Next, we define the square root of any negative number.

Here, we write i in front of the radical sign so that no one mistakenly sees it as being under the radical sign.

DEFINITION *Square Root of a Negative Number*

If k is a positive real number, then

$$\sqrt{-k} = i\sqrt{k}$$

If b is a nonzero real number, we call an expression of the form bi a **pure imaginary number**.

Example 1 Writing Numbers in *bi* Form

Write the number in bi form, where b is a real number. Simplify the result.

1. $\sqrt{-49}$ 2. $-\sqrt{-12}$

Solution

1. $\sqrt{-49} = i\sqrt{49} = 7i$
2. $-\sqrt{-12} = -i\sqrt{12} = -i\sqrt{4 \cdot 3} = -2i\sqrt{3}$ ■

We can combine real numbers a and b with the imaginary unit i to form a *complex number* of the form

$$a + bi$$

DEFINITION *Complex Number*

A **complex number** is a number of the form

$$a + bi$$

where a and b are real numbers.

Here are some examples of complex numbers:

$$2 + 5i, \qquad 3 - 4i, \qquad 5 + 0i = 5, \qquad 0 - 7i = -7i$$

Since $a = a + 0i$ and $bi = 0 + bi$, we see that all real numbers and all pure imaginary numbers are complex numbers.

A complex number that is not a real number is called an *imaginary number*.

DEFINITION *Imaginary Number*

An **imaginary number** is a number $a + bi$ where a and b are real numbers and $b \neq 0$.

Since $i = \sqrt{-1}$, we add, subtract, multiply, and divide complex numbers in much the same way as we do radical expressions. For example,

$$(2 + 3\sqrt{7}) + (1 + 5\sqrt{7}) = 3 + 8\sqrt{7} \text{ and}$$
$$(2 + 3i) + (1 + 5i) = 3 + 8i$$

If a radicand is negative, we first write the radical in terms of i before performing any operations:

$$\sqrt{-1}\sqrt{-1} = i \cdot i = i^2 = -1 \qquad \text{Correct}$$
$$\sqrt{-1}\sqrt{-1} = \sqrt{-1 \cdot -1} = \sqrt{1} = 1 \qquad \text{Incorrect}$$

So, there is no product property $\sqrt{a}\sqrt{b} = \sqrt{ab}$ when a and b are both negative. To find $\sqrt{-2}\sqrt{-3}$, we first write each radical in terms of i and then find the product:

$$\sqrt{-2}\sqrt{-3} = i\sqrt{2} \cdot i\sqrt{3} = i^2\sqrt{6} = -\sqrt{6}$$

When we combine two complex numbers using an operation, we write the result in the form $a + bi$, where a and b are in lowest terms.

Example 2   Performing Operations with Complex Numbers

Perform the indicated operation. Simplify the result.

1. $(5 + 9i) + (3 - 2i)$

2. $(3 - \sqrt{-36}) - 5(2 - \sqrt{-16})$

3. $4i \cdot 6i$

4. $\sqrt{-4}\sqrt{-9}$

5. $(2 + 5i)(3 - 7i)$

Solution

1. $(5 + 9i) + (3 - 2i) = 5 + 3 + 9i - 2i$ Rearrange the terms.

$= 8 + 7i$

2. $(3 - \sqrt{-36}) - 5(2 - \sqrt{-16}) = (3 - i\sqrt{36}) - 5(2 - i\sqrt{16})$ Write in terms of i.

$= (3 - 6i) - 5(2 - 4i)$

$= 3 - 6i - 10 + 20i$ Distributive law

$= 3 - 10 - 6i + 20i$ Rearrange the terms.

$= -7 + 14i$

3. $4i \cdot 6i = 24i^2$

$= 24(-1)$ $i^2 = -1$

$= -24$

4. $\sqrt{-4}\sqrt{-9} = i\sqrt{4} \cdot i\sqrt{9}$ Write radicals in terms of i.

$= 2i \cdot 3i$

$= 6i^2$

$= -6$

5. $(2 + 5i)(3 - 7i) = 2 \cdot 3 - 2 \cdot 7i + 5i \cdot 3 - 5i \cdot 7i$

$= 6 - 14i + 15i - 35i^2$

$= 6 + i - 35(-1)$

$= 6 + i + 35$

$= 41 + i$

We use a graphing calculator to verify our work (see Fig. 22).   ■

Graphing Calculator

The imaginary unit i is listed in the "CATALOG" menu.

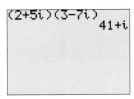

Figure 22 Verify that $(2 + 5i)(3 - 7i) = 41 + i$

The *conjugate* of a complex number $a + bi$ is $a - bi$ and vice versa. Just as we use the conjugate of a denominator containing a radical to remove the radical from the denominator (we rationalize the denominator), we can use the conjugate of a denominator containing i to remove i from the denominator.

Here, we simplify the expression $\dfrac{5}{2+3i}$:

$$\frac{5}{2+3i} = \frac{5}{2+3i} \cdot \frac{2-3i}{2-3i} \qquad \text{The conjugate of } 2+3i \text{ is } 2-3i.$$

$$= \frac{10-15i}{4-9i^2} \qquad \text{Distributive law; difference of two squares formula}$$

$$= \frac{10-15i}{4-9(-1)} \qquad i^2 = -1$$

$$= \frac{10-15i}{13}$$

$$= \frac{10}{13} - \frac{15}{13}i \qquad \text{Write in } a+bi \text{ form.}$$

To simplify the quotient of two complex numbers, we multiply the quotient by $\dfrac{\text{conjugate of the denominator}}{\text{conjugate of the denominator}}$.

Example 3 Simplifying the Quotient of Two Complex Numbers

Simplify $\dfrac{2+4i}{3-5i}$.

Solution

$$\frac{2+4i}{3-5i} = \frac{2+4i}{3-5i} \cdot \frac{3+5i}{3+5i} \qquad \text{The conjugate of } 3-5i \text{ is } 3+5i.$$

$$= \frac{6+10i+12i+20i^2}{9-25i^2}$$

$$= \frac{6+22i+20(-1)}{9-25(-1)}$$

$$= \frac{-14+22i}{34}$$

$$= -\frac{14}{34} + \frac{22}{34}i \qquad \text{Write in } a+bi \text{ form.}$$

$$= -\frac{7}{17} + \frac{11}{17}i \qquad \text{Reduce.}$$

We use a graphing calculator to verify our work (see Fig. 23). ∎

Graphing Calculator

Here, we set the float to 2, so the numbers in the result are rounded to the nearest hundredth.

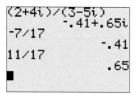

Figure 23 Verify that $\dfrac{2+4i}{3-5i} \approx -0.41 + 0.65i$

Recall that we can find solutions of the equation $ax^2 + bx + c = 0$, $a \neq 0$, by using the quadratic formula:

$$x = \frac{-b \pm \sqrt{b^2 - 4ac}}{2a}$$

Now we can go a step further and find the complex number solutions as well. They occur when the discriminant $b^2 - 4ac$ is negative.

Example 4 Solving a Quadratic Equation

Solve $4x^2 + 2x + 3 = 0$.

Solution

By the quadratic formula, we have:

$$x = \frac{-2 \pm \sqrt{2^2 - 4(4)(3)}}{2(4)}$$

$$= \frac{-2 \pm \sqrt{-44}}{8}$$

$$= \frac{-2 \pm i\sqrt{44}}{8}$$

$$= \frac{-2 \pm 2i\sqrt{11}}{8}$$

$$= -\frac{2}{8} \pm \frac{2i\sqrt{11}}{8}$$

$$= -\frac{1}{4} \pm \frac{\sqrt{11}}{4}i$$

So, the equation has two complex number solutions. We can use a graphing calculator to verify each solution by setting x equal to a solution and then checking that $4x^2 + 2x + 3$ is indeed 0 (see Fig. 24).

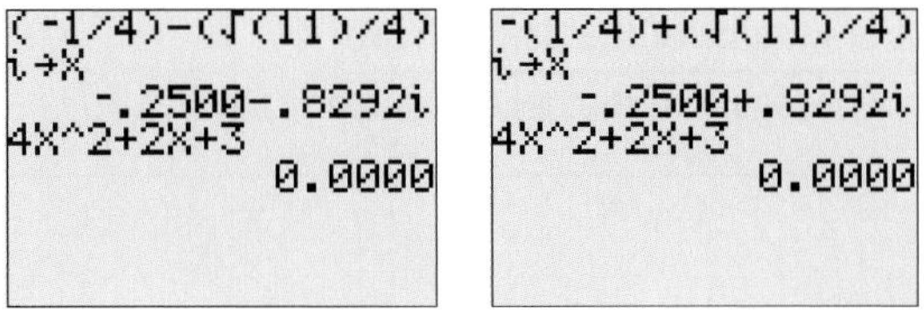

Figure 24 Verifying that $-\frac{1}{4} - \frac{\sqrt{11}}{4}i$ and $-\frac{1}{4} + \frac{\sqrt{11}}{4}i$ are solutions

```
(-1/4)-(√(11)/4)
i→X
         -.2500-.8292i
4X^2+2X+3
          0.0000
```

```
-(1/4)+(√(11)/4)
i→X
         -.2500+.8292i
4X^2+2X+3
          0.0000
```

We can describe the types of solutions of a quadratic equation of the form $ax^2 + bx + c = 0$, where a, b, and c are real, by considering the value of the discriminant $b^2 - 4ac$ of the quadratic formula

$$x = \frac{-b \pm \sqrt{b^2 - 4ac}}{2a}$$

Discriminant $b^2 - 4ac$	Number and Type of Solutions
positive	2 real numbers
zero	1 real number
negative	2 imaginary numbers

EXPLORATION
Finding powers of i

1. Find the indicated power of i. Verify your work using a graphing calculator.

   a. i^2

   b. i^3 [**Hint**: $i^3 = i^2 \cdot i$]

   c. i^4 [**Hint**: $i^4 = i^3 \cdot i$]

   d. i^5

2. Continue finding powers of i such as $i^6, i^7, i^8, \ldots$ until you see a pattern in your results. Describe the pattern.

3. Find the indicated power of i. Verify your result using a graphing calculator.

    a. i^{23} **b.** i^{41} **c.** i^{102} **d.** i^{400}

Key Points

OF THIS SECTION

Meaning of i	$i = \sqrt{-1}$ and $i^2 = -1$
Imaginary number	If a is a real number and b is a nonzero real number, then $a + bi$ is an *imaginary number*.
Pure imaginary number	If b is a nonzero real number, then bi is a *pure imaginary number*.
Simplifying $\sqrt{-k}$	If k is a positive real number, then $\sqrt{-k} = i\sqrt{k}$.
Complex number	A *complex number* is a number of the form $a + bi$, where a and b are real numbers.
First write radicals in terms of i	If a radicand is a negative real number, we first write the radical in terms of i before performing any operations.
Conjugates	$a + bi$ and $a - bi$ are conjugates of each other.
Simplifying the quotient of two complex numbers	To simplify the quotient of two complex numbers, multiply the quotient by $\dfrac{\text{conjugate of the denominator}}{\text{conjugate of the denominator}}$.

HOMEWORK 11.4

  SSM Tutor Center   MathPro 4   MathPro 5   CDs/Videos   www

Write the imaginary number in bi form. Simplify the result.

1. $\sqrt{-36}$ **2.** $\sqrt{-4}$ **3.** $-\sqrt{-25}$

4. $-\sqrt{-49}$ **5.** $\sqrt{-13}$ **6.** $\sqrt{-3}$

7. $\sqrt{-20}$ **8.** $\sqrt{-32}$

Write the complex number in a + bi form. Simplify the result.

9. $\dfrac{4 - \sqrt{-24}}{8}$

10. $\dfrac{6 + \sqrt{-45}}{12}$

11. $\dfrac{20 + \sqrt{-50}}{10}$

12. $\dfrac{12 - \sqrt{-32}}{8}$

Perform the indicated operation. Write your result in a + bi form.

13. $\sqrt{-9} + \sqrt{-9}$ **14.** $5\sqrt{-36} - \sqrt{-36}$

15. $4 + 2\sqrt{-25} - 1 - 8\sqrt{-4}$ **16.** $3 - 6\sqrt{-9} - 7 + \sqrt{-16}$

17. $\sqrt{-4}\sqrt{-25}$ **18.** $\sqrt{-49}\sqrt{-16}$

19. $\sqrt{-3}\sqrt{-5}$ **20.** $\sqrt{-2}\sqrt{-6}$

Perform the indicated operations. Write your result in a + bi form. Verify your result using a graphing calculator.

21. $(4 - 7i) + (3 + 10i)$ **22.** $(15 + 2i) + (6 - 17i)$

23. $(6 - 5i) - (2 + 13i)$ **24.** $(9 + 4i) - (8 - 2i)$

25. $2(3 - 5i) - 4(2 - 6i)$ **26.** $3(7 + i) - 5(6 + 4i)$

27. $2i \cdot 9i$ **28.** $15i \cdot 3i$

29. $-10i(-5i)$ **30.** $-7i \cdot 4i$

31. $5i(3 - 2i)$ **32.** $6i(1 + 3i)$

33. $20 - 3i(2 - 7i)$ **34.** $2 + 4i(3 - 8i)$

35. $(2 + 5i)(3 + 4i)$ **36.** $(7 + 3i)(10 + 2i)$

37. $(3 - 6i)(5 + 2i)$ **38.** $(4 - 3i)(2 - 7i)$

39. $(5 + 4i)(5 - 4i)$ **40.** $(8 - 3i)(8 + 3i)$

41. $(1 + i)(1 - i)$ **42.** $(10 - 2i)(10 + 2i)$

43. $(2 + 3i)^2$ **44.** $(5 - 2i)^2$

45. $(4 - i)^2$ **46.** $(1 + i)^2$

47. $\dfrac{3}{2 + 5i}$ **48.** $\dfrac{4}{7 - 3i}$

49. $\dfrac{7}{6 + 3i}$ **50.** $\dfrac{1}{3 - i}$

51. $\dfrac{2 - 3i}{7 + i}$ **52.** $\dfrac{5 + 8i}{3 - 2i}$

53. $\dfrac{3 + 4i}{3 - 4i}$ **54.** $\dfrac{4 - 7i}{4 + 7i}$

55. $\dfrac{3 + 5i}{2 + 9i}$ **56.** $\dfrac{5 - i}{1 + i}$

Solve.

57. $x^2 + 2x + 3 = 0$

58. $x^2 - x + 1 = 0$

59. $x^2 - 2x + 5 = 0$

60. $x^2 + x + 4 = 0$

61. $2x^2 + x + 1 = 0$

62. $3x^2 - 2x + 3 = 0$

63. $5x^2 - 4x = -1$

64. $4x^2 - x = -2$

65. $2x(2x - 1) = -3$

66. $3(x^2 + 1) = x$

67. $(x + 1)(x + 2) = x$

68. $(x - 3)(2x + 1) = -10$

69. $x(3x - 2) = 2 + 2(x - 3)$

70. $5 - 2(x - 4) = -2x(x - 2)$

71. $x^2 = -9$

72. $-x^2 = 7$

73. $-3x^2 = 5$

74. $2x^2 = -8$

75. $(5x + 3)^2 = -7$

76. $(4x - 9)^2 = -3$

77. $-2(3x - 1)^2 + 4 = 12$

78. $3(2x + 5)^2 + 8 = 5$

79. Two students try to find the product $\sqrt{-2}\sqrt{-8}$. Did one, both, or neither student find the product correctly? Explain.

Student 1's work	**Student 2's work**
$\sqrt{-2}\sqrt{-8} = \sqrt{-2(-8)}$	$\sqrt{-2}\sqrt{-8} = i\sqrt{2} \cdot i\sqrt{8}$
$= \sqrt{16}$	$= i^2\sqrt{16}$
$= 4$	$= -4$

80. The square of a real number is a nonnegative real number. What can you say about the square of a pure imaginary number? Explain.

81. Find nonzero real number values for a, b, c, and d so that the sum $(a + bi) + (c + di)$:

a. is an imaginary number (not pure).

b. is a real number.

c. is a pure imaginary number.

82. Find nonzero real number values for a, b, c, and d so that the product $(a + bi)(c + di)$:

a. is an imaginary number (not pure).

b. is a real number.

c. is a pure imaginary number.

83. For the quadratic equation $ax^2 + bx + c = 0$, explain what the value of the discriminant tells you about the number and type of solutions.

Section 11.4 Quiz

Perform the indicated operation. Write the complex number in $a + bi$ form.

1. $3 + 4\sqrt{-36} - 2 - 8\sqrt{-36}$

2. $\sqrt{-2}\sqrt{-7}$

3. $(6 - 2i) + (3 - 4i)$

4. $-4i \cdot 3i$

5. $(5 - 3i)(7 + i)$

6. $(4 - 3i)^2$

7. $(8 + 5i)(8 - 5i)$

8. $\dfrac{2}{4 - i}$

9. $\dfrac{3 + 2i}{5 - 4i}$

10. $\dfrac{1 - i}{1 + i}$

Solve.

11. $x^2 - 2x + 3 = 0$

12. $-3x^2 + x - 4 = 0$

13. $(x + 4)(x + 5) = x$

14. $-3(2x - 5)^2 + 1 = 13$

15. For the equation $ax^2 + bx - a = 0$, each solution is a real number. Explain.

16. True or false? A complex number times a pure imaginary number must be an imaginary number. Explain.

11.5 INTEREST AND REVENUE APPLICATIONS

Objective

▸ Model and analyze situations that involve interest or revenue.

In this section we will model situations that involve money. We begin by considering interest from investments.

Money deposited in an account such as a savings account, CD, or mutual fund is called the **principal**. A person invests money in hopes of later getting back the principal plus additional money called the **interest** (see Fig. 25). The **annual interest rate** is the percentage of the principal that equals the interest earned in one year. So, if $100 is invested and the interest earned in one year is $5, then the annual interest rate is 5%.

Principal   Time passes

Figure 25 Invest the principal, time passes, get back the principal plus interest

Example 1 Interest from an Investment

How much interest will a person earn from investing $2500 in an account at 6% annual interest for one year?

Solution

We find 6% of 2500:

$$0.06(2500) = 150$$

The person will earn $150 in interest. ∎

    With many forms of investment, you run the risk that you will not earn interest. In fact, you may lose some or all of your principal. Accounts such as savings accounts and CDs are extremely safe investments, but they pay low interest rates. Accounts such as mutual funds often pay higher interest rates, but they are riskier investments.

    When investing money, it is wise to invest in a variety of accounts, including some lower-risk accounts and some higher-risk accounts. This way, you might earn large amounts of interest from the higher-risk accounts, but have a safety net of principal and interest from your lower-risk accounts.

    For many investments such as mutual funds, the interest rate changes over time. We can estimate the annual interest rate for such investments by calculating the average annual interest rate over the past, say, three years. For accounts with interest rates that change, the text gives this type of average and we will assume that all stated annual interest rates do turn out to be accurate in the coming year.

Example 2   Interest Application

A person plans to invest a total of $6000 in a Gabelli ABC mutual fund that has a 3-year average annual interest of 6% and in a Presidential Bank Internet CD account at 2.25% annual interest.* Let x and y represent the money (in dollars) invested in the mutual fund and the CD, respectively.

1. Let $I = f(x)$ represent the total interest (in dollars) earned from investing the $6000 for one year. Find an equation for f.

2. Use a graphing calculator to draw a graph of f for $0 \leq x \leq 6000$. Is f increasing, decreasing, or neither? What does this mean in terms of the situation?

3. Use a graphing calculator to create a table of values of f. Explain how such a table could help the person decide how much money to invest in each account.

4. How much money should be invested in each account to earn $300 in one year?

Solution

1. The interest earned from investing x dollars in a 6% annual interest account is $0.06x$ and the interest earned from investing y dollars in a 2.25% annual interest account is $0.0225y$ dollars. We add the two interest earnings to find the total interest earned:

$$I = 0.06x + 0.0225y$$

So far, we have described I in terms of x and y. Next, we describe I in terms of just x. To start, recall that the person plans to invest $6000:

$$x + y = 6000$$

Then, we isolate y:

$$y = 6000 - x$$

Next, we substitute $6000 - x$ for y in the equation $I = 0.06x + 0.0225y$:

$$I = 0.06x + 0.0225(6000 - x) \qquad \text{Substitute } 6000 - x \text{ for } y.$$
$$= 0.06x + 135 - 0.0225x \qquad \text{Distributive law}$$
$$= 0.0375x + 135 \qquad \text{Combine like terms.}$$

So, our equation for f is

$$f(x) = 0.0375x + 135$$

---

*Presidential Bank—Deposit Rates, www.presidential.com/DEP_RATE.HTM. Quicken.com—Mutual Fund Finder, www.quicken.com/investments/mutualfunds/finder/. Accessed 10/16/02.

2. We draw a graph of f in Fig. 26. The function f is an increasing linear function with slope 0.0375. This means that if one more dollar is invested at 6% (and one less dollar is invested at 2.25%), the total interest will increase by 3.75 cents.

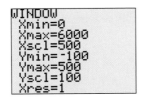 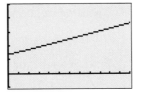

Figure 26 The revenue model

3. We create a table in Fig. 27. The person may know that she wants to invest some of the $6000 in the safe Presidential CD account and the rest in the riskier Gabelli mutual fund, but she may not be clear exactly how much risk she is willing to take. By seeing some possible interest earnings, the person may get a clearer idea of how much money she is willing to invest in each account.

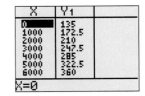

Figure 27 Table of values of f

4. We substitute 300 for I in the equation $I = 0.0375x + 135$ and solve for x.

$$300 = 0.0375x + 135$$
$$-0.0375x = -165$$
$$x = 4400$$

The person should invest $4400 in the Presidential mutual fund account and $6000 - 4400 = 1600$ dollars in the Gabelli CD account. ∎

In Example 2 we worked with the equation $I = 0.06x + 0.0225y$ which describes the *interest* from the investments and the equation $x + y = 6000$ which describes the *principal* of the investments. Keeping in mind what type of quantity you are describing will be essential when finding equations in this section.

The equation $x + y = 6000$ limits the possible pairs of values of x and y. We call this equation a *constraint equation*. We used this equation and substitution to help us describe revenue in terms of just x (rather than x and y). In this section we will use constraint equations to help us describe a quantity with an expression in terms of just one variable.

In the next example we will model money collected from selling tickets at a rock concert. We refer to the money collected as the *revenue* from selling the tickets.

Example 3   Revenue Application

A 10,000-seat amphitheater will sell tickets at $45 and $65 for a Foo Fighters concert. Let x and y represent the number of tickets that will sell for $45 and $65, respectively. Assume that the show will sell out.

1. Let $R = f(x)$ represent the total revenue (in dollars) from selling the $45 and $65 tickets. Find an equation for f.

2. Use a graphing calculator to sketch a graph of f for $0 \leq x \leq 10{,}000$. Is f increasing, decreasing, or neither? What does this mean in terms of the situation?

3. Find $f(8500)$. What does the result mean in terms of the situation?

4. Find $f(11{,}000)$. What does the result mean in terms of the situation?

5. The total cost of the production is $350,000. How many of each ticket must be sold to make a profit of $150,000?

Solution

1. The price of a ticket times the number of those tickets sold gives the revenue from selling those tickets. So, the revenue from selling x tickets at $45 is $45x$ and the revenue from selling y tickets at $65 is $65y$. We add the two revenues to find the total revenue:

$$R = 45x + 65y$$

So far, we have described R in terms of x and y. Next, we describe R in terms of just x. To begin, recall that the total number of tickets sold for the sell-out performance is 10,000:

$$x + y = 10,000$$

Then, we isolate y in this constraint equation:

$$y = 10,000 - x$$

Next, we substitute $10,000 - x$ for y in the equation $R = 45x + 65y$:

$$R = 45x + 65(10,000 - x) \qquad \text{Substitute } 10,000 - x \text{ for } y.$$
$$= 45x + 650,000 - 65x \qquad \text{Distributive law}$$
$$= -20x + 650,000 \qquad \text{Combine like terms.}$$

So, an equation for f is

$$f(x) = -20x + 650,000$$

2. We draw a sketch of f in Fig. 28. The function f is a decreasing linear function with slope -20. This means that if one more ticket is sold for \$45 (and one less ticket is sold for \$65), the revenue will decrease by \$20.

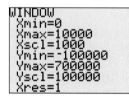 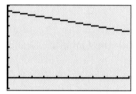

Figure 28 The revenue model

3. $f(8500) = -20(8500) + 650,000 = 480,000$. This means that if 8500 tickets sell for \$45 (and 1500 tickets sell for \$65), the total revenue will be \$480,000.

4. $f(11,000) = -20(11,000) + 650,000 = 430,000$. This means that if 11,000 tickets sell for \$45, the total revenue will be \$430,000. Since there are only 10,000 seats, model breakdown has occurred.

5. To make a profit of \$150,000, the revenue would need to be $350,000 + 150,000 = 500,000$ dollars. We substitute 500,000 for R in the equation $R = -20x + 650,000$ and solve for x:

$$500,000 = -20x + 650,000$$
$$20x = 150,000$$
$$x = 7500$$

So, 7500 \$45 tickets and $10,000 - 7500 = 2500$ \$65 tickets would need to be sold for the profit to be \$150,000. ∎

In Example 3 the equation $R = 45x + 65y$ describes the *value* of tickets and the equation $x + y = 10,000$ describes the *number* of tickets. We used the constraint equation $x + y = 10,000$ and substitution to help us describe revenue in terms of just x (rather than x and y).

In Example 3 we made use of the concept that the number of \$45 tickets times the price of these tickets is equal to the revenue from selling these tickets. In general, the number of units sold times the (constant) price (in dollars per unit) of the units gives the revenue (in dollars).

Revenue Relationship

If n units are sold at price p dollars per unit, the total revenue (in dollars) is equal to the product

$$R = np$$

Example 4   Maximizing Revenue

The average number of programs sold per day at a baseball stadium depends on the price of the program. Over the course of a season, the stadium varies the price of a program and computes the average number of programs sold at each price (see Table 1).

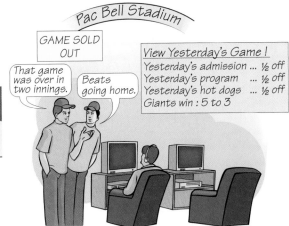

Table 1 Average Numbers of Programs Sold

Price (in dollars)	Average Number of Programs Sold per Day
4	17,390
5	15,789
6	13,112
7	11,421
8	9273

1. Let n represent the average number of programs sold per day when the price of a program is p dollars. Find an equation for p and n.
2. Let $R = f(p)$ represent the daily revenue (in dollars) from selling programs at p dollars per program. Find an equation for f.
3. Use a graphing calculator to draw a graph of f. It is a common student error to think that the revenue can always be increased by raising the price. Explain why this is incorrect.
4. What are the p-intercepts of f? What do they mean in terms of the situation?
5. What should be the price so that the daily revenue is maximized? What is the maximum daily revenue?

Solution

1. First, we draw a scattergram of the data (see Fig. 29). The data appear to be approximately linearly related. The linear regression equation is

$$n = -2060.2p + 25{,}758.2$$

The model appears to fit the data well (see Fig. 30).

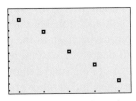

Figure 29 Number of programs scattergram

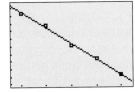

Figure 30 Verify the number of programs model

2. We substitute $-2060.2p + 25{,}758.2$ for n in the revenue equation $R = np$

$R = (-2060.2p + 25{,}758.2)p$ Substitute $-2060.2p + 25{,}758.2$ for n.

$= -2060.2p^2 + 25{,}758.2p$ Distributive law

So, an equation for f is

$$f(p) = -2060.2p^2 + 25{,}758.2p$$

3. We draw a graph of f in Fig. 31. Note that to the right of the vertex, the function f is decreasing. So, if the price is relatively high and it is increased, the revenue will

decrease. That's because when we raise the price, the number of programs sold decreases by so much that the revenue decreases.

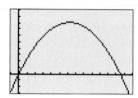

Figure 31 The revenue model

4. We substitute 0 for $f(p)$ in the equation $f(p) = -2060.2p^2 + 25,758.2p$ and solve for p:

$$0 = -2060.2p^2 + 25,758.2p$$

$$0 = p(-2060.2p + 25,758.2)$$

$$p = 0 \quad \text{or} \quad -2060.2p + 25,758.2 = 0$$

$$p = 0 \quad \text{or} \quad p = \frac{-25,758.2}{-2060.2}$$

$$p = 0 \quad \text{or} \quad p \approx 12.50$$

So, the p-intercepts are approximately $(0, 0)$ and $(12.50, 0)$. This means that the revenue will be zero if the programs are free or if they are sold for $12.50. The revenue will be zero if the price is $12.50 because no will buy a program at that price, according to our demand equation.

5. To find the maximum point, we first note that the p-intercepts $(0, 0)$ and $(12.50, 0)$ are symmetric points. We find the p-coordinate of the vertex by averaging the p-coordinates of the p-intercepts:

$$p\text{-coordinate of the vertex} = \frac{0 + 12.50}{2} \approx 6.25,$$

Next, we find $f(6.25)$.

$$f(6.25) = -2060.2(6.25)^2 + 25,758.2(6.25) \approx 80,512.19$$

So, the vertex is approximately $(6.25, 80,512.19)$. This means that the maximum revenue of $80,512.19 can be achieved if the price of the programs is $6.25. We can use "maximum" on a graphing calculator to verify our work (see Fig. 32). ∎

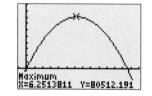

Figure 32 Verify that the vertex is approximately $(6.25, 80,512.19)$

In Example 4 we worked with the revenue equation $R = np$ and the constraint equation $n = -2060.2p + 25,758.2$. The constraint equation describes a decreasing function. It makes sense that the function is decreasing because as the price increases, the average number of programs sold per day will decrease. We call an equation that describes the relationship between price and number of units sold a **demand equation**. For some exercises, a demand equation will be provided and in other exercises you will have to model some given data to find the demand equation.

EXPLORATION
Elasticity of demand

The demand equation for a digital television model is $n = -30p + 22,500$, where n is the number of units sold per month at price p dollars.

1. Let $R = f(p)$ represent the total monthly revenue (in dollars) from selling the televisions at price p dollars per television. Find an equation for R.

2. Use a graphing calculator to draw a graph of f.

3. Use "maximum" on a graphing calculator to find the maximum point of f. What does it mean in terms of the situation?

4. Suppose that the current price of the television is less than $375. If the price is increased (but not beyond $375), how will that impact the revenue? Explain. We say that the demand is *inelastic* when changes in price and revenue are related in this way. [**Hint**: Think in terms of the graph of f.]

5. Now suppose that the current price of the television is more than $375. If the price is increased, how will that impact the revenue? Explain. We say that the demand is *elastic* when changes in price and revenue are related in this way.

Key Points

OF THIS SECTION

Money deposited in an account is called the *principal*. The money earned is called the *interest*.	**Principal, interest**
It is essential that you keep in mind the quantities that you are describing when finding equations such as those in this section.	**Finding an equation**
We use the constraint equation and substitution to help us describe a quantity with an expression in just one variable.	**Using a constraint equation**
If n units are sold at price p dollars per unit, the total revenue (in dollars) is equal to the product $R = np$.	**Revenue relationship**
An equation that describes the relationship between the price per unit and the number of units sold is called a *demand equation*.	**Demand equation**

HOMEWORK 11.5

SSM Tutor Center   MathPro 4   MathPro 5   CDs/Videos   www

1. A person plans to invest a total of $10,000 in a Charter One Bank CD at 2.87% annual interest and in a Dodge & Cox Balanced mutual fund that has a 3-year average annual interest of 8.10%. Let x and y represent the money (in dollars) invested in the CD and mutual fund, respectively.*

   a. Let $I = f(x)$ represent the total interest (in dollars) earned from investing the $10,000 for one year. Find an equation for f.

   b. Use a graphing calculator to draw a graph of f for $0 \le x \le 10{,}000$. Is f increasing, decreasing, or neither? What does this mean in terms of the situation?

   c. How much money should be invested in each account to earn a total of $400 in one year?

2. A person plans to invest a total of $5000 in a Citizens Bank CD at 2% annual interest and in a Calamos Market Neutral A mutual fund that has a 3-year average annual interest of 9.45%. Let x and y represent the money (in dollars) invested in the CD and the mutual fund, respectively.

   a. Let $I = f(x)$ represent the total interest (in dollars) earned from investing the $5000 for one year. Find an equation for f.

   b. Use a graphing calculator to draw a graph of f for $0 \le x \le 5000$. Is f increasing, decreasing, or neither? What does this mean in terms of the situation?

   c. How much money should be invested in each account to earn a total of $350 in one year?

3. A person plans to invest a total of $9000 in an ING Direct CD at 2.5% annual interest and in a Thompson Plumb Balanced mutal fund that has a 3-year average annual interest of 9.45%. Let x and y represent the money (in dollars) invested in the CD and the mutual fund, respectively.

   a. Let $I = f(x)$ represent the total interest (in dollars) earned from investing the $9000 for one year. Find an equation for f.

   b. Find $f(500)$. What does your result mean in terms of the situation?

---

*All interest rates in this homework section were obtained at various Internet sites, including Quicken.com—Mutual Fund Finder, www.quicken.com/investments/mutualfunds/finder/ and Best Savings Deals Nationwide for 1 Yr CD, www.bankrate.com/brm/rate/high_ratehome.asp?web=brm&prodtype=high&sort=2&product=3&Submit1=GO. Accessed 10/16/02.

c. Find the value of x when $f(x) = 500$. What does your result mean in terms of the situation?

d. Find $f(10,000)$. What does your result mean in terms of the situation?

4. A person plans to invest a total of $12,000 in a Savings Bank of Manchester CD at 1.92% annual interest and in a Vanguard Wellesley Income mutual fund that has a 3-year average annual interest of 8%. Let x and y represent the money (in dollars) invested in the CD and the mutual fund, respectively.

a. Let $I = f(x)$ represent the total interest (in dollars) earned from investing the $12,000 for one year. Find an equation for f.

b. Find $f(800)$. What does your result mean in terms of the situation?

c. Find the value of x when $f(x) = 800$. What does your result mean in terms of the situation?

d. Find $f(15,000)$. What does your result mean in terms of the situation?

5. A person plans to invest a total of $8000 in a LaSalle Bank CD at 1.5% annual interest and in a Bridgeway Aggressive Investors 1 mutual fund that has a 3-year average annual interest of 11.6%. Let x and y represent the money (in dollars) invested in the CD and the mutual fund, respectively.

a. Let $I = f(x)$ represent the total interest (in dollars) earned from investing the $8000 for one year. Find an equation for f.

b. A minimum principal of $2500 is required for the LaSalle Bank CD account. Describe the various possible total interest earnings from investing the $8000 in the accounts for one year.

c. How much of the $8000 should be invested in each account so that the total interest earned in one year is $400?

6. A person plans to invest a total of $7000 in a Security Bank USA CD at 2.55% annual interest and in an Artisan Mid Cap mutual fund that has a 3-year average annual interest of 8%. Let x and y represent the money (in dollars) invested in the CD and the mutual fund, respectively.

a. Let $I = f(x)$ represent the total interest (in dollars) earned from investing the $7000 for one year. Find an equation for f.

b. A minimum principal of $500 is required for the Security Bank USA CD account. Describe the various possible total interest earnings from investing the $7000 in the accounts for one year.

c. How much of the $7000 should be invested in each account so that the total interest earned in one year is $300?

7. A person plans to invest a total of $6000 in a Nexity Bank CD at 2.85% annual interest and in an FMI Focus mutual fund that has a 3-year average annual interest of 9%. Let x and y represent the money (in dollars) invested in the CD and the mutual fund, respectively.

a. Let $I = f(x)$ represent the total interest (in dollars) earned from investing the $6000 for one year. Find an equation for f.

b. Find the I-intercept of f. What does it mean in terms of the situation?

c. Find the x-intercept of f. What does it mean in terms of the situation?

d. What is the slope of f? What does it mean in terms of the situation?

8. A person plans to invest a total of $14,000 in a North Middlesex Savings Bank CD at 2.27% annual interest and in a Calamos Growth A mutual fund that has a 3-year average annual interest of 15%. Let x and y represent the money (in dollars) invested in the CD and the mutual fund, respectively.

a. Let $I = f(x)$ represent the total interest (in dollars) earned from investing the $14,000 for one year. Find an equation for f.

b. Find the I-intercept of f. What does it mean in terms of the situation?

c. Find the x-intercept of f. What does it mean in terms of the situation?

d. What is the slope of f? What does it mean in terms of the situation?

9. A 20,000-seat amphitheater will sell tickets at $50 and $75 for a blink-182 concert. Let x and y represent the number of tickets that will sell for $50 and $75, respectively. Assume that the show will sell out.

a. Let $R = f(x)$ represent the total revenue (in dollars) from selling the $50 and $75 tickets. Find an equation for f.

b. Use a graphing calculator to draw a graph of f for $0 \le x \le 20,000$. Is f increasing, decreasing, or neither? What does this mean in terms of the situation?

c. Find $f(16,000)$. What does the result mean in terms of the situation?

d. The total cost of the production is $475,000. How many of each ticket must be sold to make a profit of $600,000?

10. An 8000-seat amphitheater will sell tickets at $30 and $55 for a Garbage concert. Let x and y represent the number of tickets that will sell for $30 and $55, respectively. Assume that the show will sell out.

a. Let $R = f(x)$ represent the total revenue (in dollars) from selling the $30 and $55 tickets. Find an equation for f.

b. Use a graphing calculator to draw a graph of f for $0 \le x \le 8000$. Is f increasing, decreasing, or neither? What does this mean in terms of the situation?

c. Find $f(6500)$. What does the result mean in terms of the situation?

d. How many of each ticket must be sold for the revenue to be $250,000?

11. A 12,000-seat amphitheater will sell tickets at $45 and $70 for a Cake concert. Let x and y represent the number of tickets that will sell for $45 and $70, respectively. Assume that the show will sell out.

a. Let $R = f(x)$ represent the total revenue (in dollars) from selling the $45 and $70 tickets. Find an equation for f.

b. Use a graphing calculator table to find the values $f(0)$, $f(2000)$, $f(4000)$, ..., $f(12,000)$. What do these values mean in terms of the situation?

c. Describe the various possible total revenues from selling the tickets.

d. How many of each ticket must be sold for the revenue to be $602,500?

12. A 25,000-seat amphitheater will sell tickets at $55 and $85 for a U2 concert. Let x and y represent the number of tickets that will sell for $55 and $85, respectively. Assume that the show will sell out.

a. Let $R = f(x)$ represent the total revenue (in dollars) from selling the $55 and $85 tickets. Find an equation for f.

b. Use a graphing calculator table to find the values $f(0)$, $f(5000)$, $f(10,000)$, ..., $f(25,000)$. What do these values mean in terms of the situation?

c. Describe the various possible total revenues from selling the tickets.

d. How many of each ticket must be sold for the revenue to be $1,510,000?

13. For flights between Los Angeles and San Francisco, United Airlines® usually uses a 737 airplane which has 8 first-class seats and 126 coach seats. The average price of round-trip first-class tickets is $242 more than the average price of round-trip coach tickets. Assume that a round-trip flight is sold out. Let x and y represent the prices (in dollars) of coach and first-class tickets, respectively.

a. Let $R = f(x)$ represent the total revenue (in dollars) from selling the tickets. Find an equation for f.

b. Find the slope of f. What does it represent in terms of the situation?

c. If United Airlines wants the total revenue from the tickets for the round trip to be $14,130, what should be the average selling prices of both types of tickets?

14. An orchestra will perform in an auditorium with 1400 main-level seats and 300 balcony seats. Tickets for the balcony seats will be priced at $12 less than the main-level seats. Assume that there is a sell-out performance. Let x and y represent the prices (in dollars) for the main-level seats and the balcony seats, respectively.

a. Let $R = f(x)$ represent the total revenue (in dollars) from ticket sales. Find an equation for f.

b. Find $f(55)$. What does your result mean in terms of the situation?

c. Find the value of x when $f(x) = 101,800$. What does your result mean in terms of the situation?

15. A board of trustees of a college is considering opening a satellite college. Based on the main college statistics, researchers estimate that there will be three times as many full-time students as part-time students and that full-time students will

take an average of 14 units and part-time students will take an average of 3 units each semester. The charge for tuition will be $13 per unit. How many of each type of student would need to attend the college for the total revenue to be $900,000 each semester, the minimum amount needed for the college to be a success financially?

16. A college may be opened in the future. Full-time instructors will teach 15 units each semester. Researchers estimate that adjunct (part-time) instructors teach an average of 6 units each semester and that there is student demand for a total of 3300 units of courses. There will be enough office space for 250 instructors. If the college wants to keep costs low by hiring as many adjunct instructors as possible, yet have office space for all instructors, how many full-time instructors and adjunct instructors should be hired?

17. The demand equation for an electric guitar sales model is $n = -2.25p + 3162.5$, where n is the number of guitars sold per month when the price of a guitar is p dollars.

a. Let $R = f(p)$ represent the total monthly revenue (in dollars) from selling the guitars at p dollars per guitar. Find an equation for f.

b. Find the p-intercepts of f. What do they mean in terms of the situation?

c. What should be the price so that the monthly revenue is maximized? What is that maximum monthly revenue?

18. The demand equation for a washing machine sales model is $n = -14p + 12,600$, where n is the number of washing machines sold per month at p dollars per machine.

a. Let $R = f(p)$ represent the total monthly revenue (in dollars) from selling the washing machines at p dollars per machine. Find an equation for f.

b. Find the p-intercepts of f. What do they mean in terms of the situation?

c. What should be the price so that the monthly revenue is maximized? What is that maximum monthly revenue?

19. A record company decides to perform some research on one of their classic rock band's CDs that has had consistent sales for the past few years. The company varies the price of the CD and computes the average number of CDs sold per month at each price (see Table 2).

Table 2 Average Numbers of CDs Sold per Month	
Price (in dollars)	Average Number of CDs Sold per Month
15.99	3695
16.99	3542
17.99	3389
18.99	3292
19.99	3139

a. Let n represent the average number of CDs sold per month when the price of a CD is p dollars. Find an equation for p and n.

b. Let $R = f(p)$ represent the monthly revenue (in dollars) from selling the CDs at p dollars per CD. Find an equation for f.

c. Find the p-intercepts of f. What do they mean in terms of the situation?

d. What should be the price so that the monthly revenue is maximized? What is that maximum monthly revenue?

20. A company varies the price of one of their bicycle models and computes the average number of these bicycles sold per month at each price (see Table 3).

Table 3 Average Numbers of Bicycles Sold per Month

Price (in dollars)	Average Number of Bicycles Sold per Month
600	2600
650	2475
700	2281
750	2150
800	1972

a. Let n represent the average number of bicycles sold per month when the price of a bicycle is p dollars. Find an equation for p and n.

b. Let $R = f(p)$ represent the monthly revenue (in dollars) from selling the bicycles at p dollars per bicycle. Find an equation for f.

c. What are the p-intercepts of f? What do they mean in terms of the situation?

d. What should be the price so that the monthly revenue is maximized? What is that maximum monthly revenue?

21. A company varies the price of one of their computer printer models and computes the average number of these printers sold per month at each price (see Table 4).

Table 4 Average Numbers of Printers Sold per Month

Price (in dollars)	Average Number of Printers Sold per Month
600	1186
650	1105
700	978
750	855
800	770

a. Let n represent the average number of printers sold per month when the price of a printer is p dollars. Find an equation for p and n.

b. For the demand equation you found in part a, which variable is the dependent variable? Explain.

c. Let $R = f(p)$ represent the monthly revenue (in dollars) from selling the printers at p dollars per printer. Find an equation for f.

d. For what price(s) is the monthly revenue equal to $240,000?

e. What should be the price so that the monthly revenue is maximized? What is that maximum monthly revenue?

22. A health club varies its initiation fee for new members and computes the average number of people who join the club per month at each price (see Table 5).

Table 5 Average Numbers of People Who Join

Initiation Fee (in dollars)	Average Number of People Who Join per Month
50	22
100	19
150	13
200	9
250	6

a. Let n represent the average number of people who join the club when the initiation fee for a new member is p dollars. Find an equation for p and n.

b. What is the slope of the demand equation you found in part a? What does it mean in terms of the situation?

c. Let $R = f(p)$ represent the monthly revenue (in dollars) from initiation fees when the fee is p dollars per new member. Find an equation for f.

d. What is the slope of f? What does it mean in terms of the situation?

e. What should be the initiation fee so that the monthly revenue from the fee is a maximum? What is that maximum monthly revenue?

Section 11.5 Quiz

1. A person plans to invest a total of $8000 in an IndyMac Bank CD at 2.9% annual interest and in a Meridian Value mutual fund that has a 3-year average annual interest of 12.7%. Let x and y represent the money (in dollars) invested in the CD and mutual fund, respectively.

a. Let $I = f(x)$ represent the total interest (in dollars) earned from investing the $8000 for one year. Find an equation for f.

b. Find $f(500)$. What does your result mean in terms of the situation?

c. Find the value of x where $f(x) = 500$. What does your result mean in terms of the situation?

2. A 16,000-seat amphitheater will sell tickets at $30 and $50 for a Weezer concert. Let x and y represent the number of tickets that will sell for $30 and $50, respectively. Assume that the show will sell out.

a. Let $R = f(x)$ represent the total revenue (in dollars) from selling the $30 and $50 tickets. Find an equation for f.

b. Find the slope of f. What does it mean in terms of the situation?

c. How many of each ticket must be sold for the revenue to be $500,000?

3. The demand equation for a DVD player sales model is $n = -120p + 65{,}000$, where n is the average number of DVD players sold per month at p dollars per player.

a. Let $R = f(p)$ represent the total monthly revenue (in dollars) from selling the DVD players at p dollars per player. Find an equation for f.

b. Find the p-intercepts of f. What do they mean in terms of the situation?

c. What should be the price so that the monthly revenue is maximized? What is that maximum monthly revenue?

11.6 PYTHAGOREAN THEOREM, DISTANCE FORMULA, AND CIRCLES

I'm very well acquainted too with matters mathematical,
I understand equations, both the simple and quadratical,
About the binomial theorem I'm teeming with a lot of news
With many cheerful facts about the square of the hypotenuse.
—Gilbert & Sullivan, *The Pirates of Penzance*

Objectives

▸ Find the length of a side of a right triangle using the Pythagorean theorem.

▸ Find the distance between two points using the distance formula.

▸ Find or graph the equation of a circle.

If one angle of a triangle measures $90°$, the triangle is a **right triangle** (see Fig. 33).
We call the side opposite the right angle (the longest side) the **hypotenuse** and we call the other two sides the **legs**. The Pythagorean theorem describes a relationship between the lengths of the legs and hypotenuse of a right triangle.

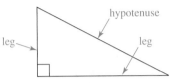

Figure 33 A right triangle

Pythagorean Theorem

If a and b are the lengths of the legs of a right triangle and c is the length of the hypotenuse, then

$$a^2 + b^2 = c^2$$

In words, the sum of the squares of the lengths of the legs is equal to the square of the length of the hypotenuse (see Fig. 34).

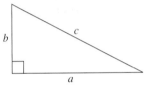

Figure 34 The Pythagorean theorem: $a^2 + b^2 = c^2$

If we know the lengths of two of the three sides of a right triangle, we can use the Pythagorean theorem to find the length of the third side.

Example 1   Finding the Length of a Side of a Right Triangle

The lengths of two sides of a right triangle are given. Find the length of the third side.

1.   **2.**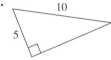

Solution

1. Since the lengths of the legs are given, we are to find the length of the hypotenuse. We substitute $a = 3$ and $b = 4$ into the equation $c^2 = a^2 + b^2$ and solve for c.

$$c^2 = a^2 + b^2$$
$$c^2 = 3^2 + 4^2$$
$$c^2 = 9 + 16$$
$$c^2 = 25$$
$$c = 5 \qquad \text{\textit{c} is nonnegative.}$$

The length of the hypotenuse is 5 units.

2. The length of the hypotenuse is 10 and the length of one of the legs is 5. We substitute $a = 5$ and $c = 10$ into the equation $a^2 + b^2 = c^2$ and solve for b.

$$5^2 + b^2 = 10^2$$

$$25 + b^2 = 100$$

$$b^2 = 75$$

$$b = \sqrt{75} \qquad b \text{ is nonnegative.}$$

$$b = 5\sqrt{3}$$

$$b \approx 8.66$$

The length of the other leg is about 8.66 units. ■

We can use the Pythagorean theorem to find a formula for the distance between two points in the coordinate system. Let $P(x_1, y_1)$ and $Q(x_2, y_2)$ represent two points, where $x_2 > x_1$ and $y_2 > y_1$ (see Fig. 35).

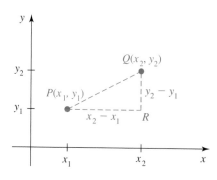

Figure 35 Find the distance between points P and Q

Note that $\triangle PQR$ is a right triangle with hypotenuse $\overline{PQ}$. Also, $PR = x_2 - x_1$ and $QR = y_2 - y_1$. We find the distance between points P and Q by using the Pythagorean theorem.

$$c^2 = a^2 + b^2$$

$$(PQ)^2 = (PR)^2 + (QR)^2$$

$$(PQ)^2 = (x_2 - x_1)^2 + (y_2 - y_1)^2$$

$$PQ = \sqrt{(x_2 - x_1)^2 + (y_2 - y_1)^2} \qquad PQ \text{ is nonnegative.}$$

Although we assumed that $x_2 > x_1$ and $y_2 > y_1$ to find the *distance formula*, it can be shown that the formula gives the correct distance between *any* two points.

Distance Formula

The distance d between points (x_1, y_1) and (x_2, y_2) is

$$d = \sqrt{(x_2 - x_1)^2 + (y_2 - y_1)^2}$$

Example 2   Finding the Distance Between Two Points

Find the distance between $(-2, 5)$ and $(3, -1)$.

Solution

We substitute $x_1 = -2$, $y_1 = 5$, $x_2 = 3$, and $y_2 = -1$ into the distance formula.

$$d = \sqrt{(3 - (-2))^2 + (-1 - 5)^2}$$

$$= \sqrt{5^2 + (-6)^2}$$

$$= \sqrt{61}$$
$$\approx 7.81$$

The distance between $(-2, 5)$ and $(3, -1)$ is about 7.81 units.

We can use the distance formula to find an equation whose graph is a circle. To see how, we first state the definition of a circle in terms of its *center* and its *radius*.

DEFINITION *Circle*

A **circle** with **center** point C and **radius** r, where $r > 0$, is the set of all points in a plane that are r units from the point C in that plane. See Fig. 36.

Figure 36 Circle with center C and radius r

Now we find an equation for a circle with center $C(h, k)$ and radius r (see Fig. 37). If $P(x, y)$ is a point on the circle, then the distance between $P(x, y)$ and $C(h, k)$ is the radius r:

$$\sqrt{(x - h)^2 + (y - k)^2} = r \qquad \text{Distance formula}$$

Squaring both sides of the equation gives:

$$(x - h)^2 + (y - k)^2 = r^2$$

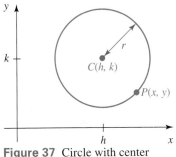

Figure 37 Circle with center $C(h, k)$ and radius r

Equation for a Circle

If a circle has center $C(h, k)$ and radius r, then an equation for the circle is
$$(x - h)^2 + (y - k)^2 = r^2$$

Also, the graph of an equation of the above form with $r > 0$ is a circle with center (h, k) and radius r.

Example 3 Finding an Equation for a Circle

Find an equation for the circle with center C and radius r.

1. $C(0, 0)$ and $r = 3$ **2.** $C(2, -5)$ and $r = \sqrt{13}$

Solution

1. We substitute $h = 0$, $k = 0$, and $r = 3$ into the equation $(x - h)^2 + (y - k)^2 = r^2$.

$$(x - 0)^2 + (y - 0)^2 = 3^2$$
$$x^2 + y^2 = 9$$

2. We substitute $h = 2$, $k = -5$, and $r = \sqrt{13}$ into the equation $(x - h)^2 + (y - k)^2 = r^2$.

$$(x - 2)^2 + (y - (-5))^2 = (\sqrt{13})^2$$
$$(x - 2)^2 + (y + 5)^2 = 13$$

The equation for a circle centered at the origin $(0, 0)$ with radius r is

$$(x - 0)^2 + (y - 0)^2 = r^2$$
$$x^2 + y^2 = r^2$$

Equation for a Circle Centered at the Origin

If a circle has center $(0, 0)$ and radius r, then an equation for the circle is

$$x^2 + y^2 = r^2$$

Example 4   Finding the Center and Radius of a Circle

Determine the center and radius of the circle. Also, sketch the circle.

1. $x^2 + y^2 = 19$ 2. $(x + 5)^2 + (y - 3)^2 = 4$

Solution

1. The equation has the form $x^2 + y^2 = r^2$. The circle is centered at the origin $(0, 0)$ and

$$r^2 = 19$$
$$r = \sqrt{19} \qquad r \text{ is positive.}$$

So, the radius is $\sqrt{19} \approx 4.36$. We sketch the circle in Fig. 38.

2. We write $(x + 5)^2 + (y - 3)^2 = 4$ in the form $(x - h)^2 + (y - k)^2 = r^2$:

$$(x - (-5))^2 + (y - 3)^2 = 2^2$$

So, $h = -5$, $k = 3$, and $r = 2$. So, the circle has center $(-5, 3)$ and radius 2. We sketch the circle in Fig. 39.

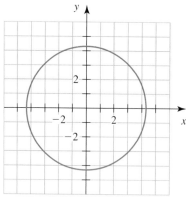

Figure 38 Graph of $x^2 + y^2 = 19$

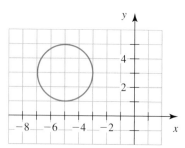

Figure 39 Graph of $(x + 5)^2 + (y - 3)^2 = 4$

To use a graphing calculator to draw the circle $(x + 5)^2 + (y - 3)^2 = 4$, we first isolate y:

$$(x + 5)^2 + (y - 3)^2 = 4$$
$$(y - 3)^2 = 4 - (x + 5)^2$$
$$y - 3 = \pm\sqrt{4 - (x + 5)^2}$$
$$y = 3 \pm \sqrt{4 - (x + 5)^2}$$

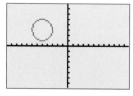

Figure 40 Use ZStandard, then ZInteger, and then ZOOM IN

Then we enter $y = 3 - \sqrt{4 - (x + 5)^2}$ and $y = 3 + \sqrt{4 - (x + 5)^2}$ and draw the graphs of both functions on the same coordinate system (see Fig. 40).

EXPLORATION
Pythagorean theorem and its converse

For this exploration, you will need a ruler, scissors, and paper. It is also helpful to have a tool for drawing right angles such as a protractor or graph paper, although a corner of a piece of paper will suffice. For each triangle, assume that c represents the length of (one of) the *longest* side(s) and a and b represent the lengths of the other sides.

1. Sketch three right triangles of different sizes. Measure the sides and show that for each right triangle, $a^2 + b^2 = c^2$.

2. Now sketch three triangles of different sizes that are *not* right triangles. For these triangles, check whether $a^2 + b^2 = c^2$.

3. Sketch a triangle that has an angle close, but not equal, to $90°$. Check whether $a^2 + b^2 \approx c^2$. If you cannot show this for your triangle, repeat this problem with a triangle that has an angle even closer to $90°$.

4. Note that if $a = 3$, $b = 5$, and $c = \sqrt{34}$, then $a^2 + b^2 = c^2$. Cut three thin strips of paper that are about 3, 5, and $\sqrt{34}$ inches in length. Form a triangle with the three strips of paper. Is the triangle a right triangle?

5. Find values for a, b, and c, other than the ones in problem 4, so that $a^2 + b^2 = c^2$. Then repeat problem 4 using your values.

6. Find three more values of a, b, and c so that $a^2 + b^2 = c^2$. Then repeat problem 4 using your values.

7. Summarize at least three concepts addressed in this exploration.

Key Points
OF THIS SECTION

Pythagorean theorem: If a and b are the lengths of the legs of a right triangle and c is the length of the hypotenuse, then $a^2 + b^2 = c^2$.	**Pythagorean theorem**
Distance formula: The distance d between the points (x_1, y_1) and (x_2, y_2) is given by the formula $d = \sqrt{(x_2 - x_1)^2 + (y_2 - y_1)^2}$.	**Distance formula**
If a circle has center $C(h, k)$ and radius r, then an equation for the circle is $(x - h)^2 + (y - k)^2 = r^2$. Also, the graph of an equation of this form with $r > 0$ is a circle with center (h, k) and radius r.	**Equation for a circle**
A circle centered at the origin $(0, 0)$ with radius r, has $x^2 + y^2 = r^2$ as its equation.	**Circle centered at the origin**

HOMEWORK 11.6

SSM
Tutor Center

MathPro 4
MathPro 4

MathPro 5
MathPro 5

CDs/Videos

www
www

Let a and b represent the lengths of the legs of a right triangle and let c represent the length of the hypotenuse. Values for two of the three lengths are given. Find the third length.

1. $a = 5$ and $b = 12$
2. $a = 6$ and $b = 8$
3. $a = 4$ and $b = 5$
4. $a = 2$ and $b = 7$
5. $a = 3$ and $c = 8$
6. $b = 2$ and $c = 9$
7. $b = 5$ and $c = 7$
8. $a = 4$ and $c = 10$
9. $a = \sqrt{2}$ and $b = \sqrt{5}$
10. $a = \sqrt{3}$ and $c = \sqrt{11}$

The lengths of two sides of a right triangle are given. Find the length of the third side.

11. See Fig. 41.
12. See Fig. 42.

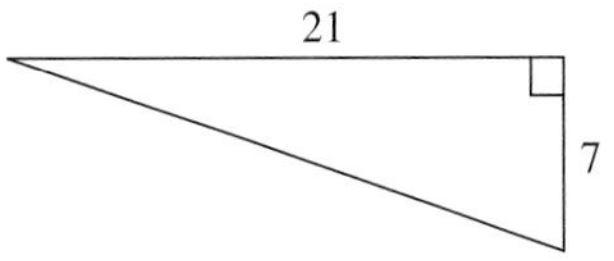

Figure 41 Exercise 11

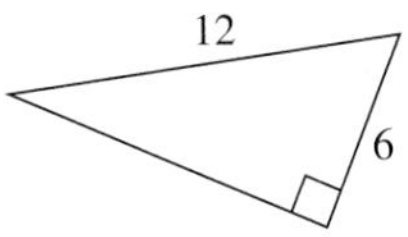

Figure 42 Exercise 12

Find the distance between the two given points. Round your result to the nearest hundredth.

13. $(2, 5)$ and $(6, 7)$

14. $(3, 1)$ and $(7, 4)$

15. $(-3, 5)$ and $(4, 2)$

16. $(1, -5)$ and $(3, 1)$

17. $(-4, -1)$ and $(-1, -5)$

18. $(-7, -4)$ and $(-2, -3)$

19. $(2.1, 8.9)$ and $(5.6, 1.7)$

20. $(3.2, 7.1)$ and $(6.6, 8.4)$

21. $(-2.18, -5.74)$ and $(3.44, 6.29)$

22. $(-6.41, 1.12)$ and $(2.89, -3.55)$

Find an equation for the circle with the given center C and radius r.

23. $C(0, 0)$ and $r = 5$

24. $C(0, 0)$ and $r = 10$

25. $C(0, 0)$ and $r = 6.7$

26. $C(0, 0)$ and $r = 2.3$

27. $C(5, 3)$ and $r = 2$

28. $C(4, 7)$ and $r = 5$

29. $C(-2, 1)$ and $r = 4$

30. $C(3, -4)$ and $r = 6$

31. $C(-7, -3)$ and $r = \sqrt{3}$

32. $C(-6, -1)$ and $r = \sqrt{2}$

Find the radius and center of the circle. Sketch the circle.

33. $x^2 + y^2 = 25$

34. $x^2 + y^2 = 9$

35. $x^2 + y^2 = 8$

36. $x^2 + y^2 = 17$

37. $(x - 3)^2 + (y - 5)^2 = 16$

38. $(x - 2)^2 + (y - 4)^2 = 4$

39. $(x + 6)^2 + (y - 1)^2 = 7$

40. $(x - 5)^2 + (y + 2)^2 = 3$

41. $(x + 3)^2 + (y + 2)^2 = 1$

42. $(x + 1)^2 + (y + 1)^2 = 1$

43. The bottom of a ladder is placed 5 feet from the base of a building. How long must the ladder be so that its other end will reach a window that is 20 feet above the ground?

44. Salt Lake City, Utah, is about 498 miles almost directly south of Helena, Montana, and is about 1030 miles almost directly west of Omaha, Nebraska. What is the approximate distance between Omaha and Helena?

45. Los Angeles is about 465 miles almost directly south of Reno, Nevada, and is almost directly west of Albuquerque, New Mexico. The distance between Albuquerque and Reno is about 964 miles. If some students plan to go on a road trip from Los Angeles to Reno to Albuquerque to Los Angeles, what would be the total distance for the trip?

46. A person drives 30 miles south from home and then 20 miles west to get to the nearest airport. How much shorter would the trip be if it was possible to drive along a straight line from home to the airport?

47. Find an equation for the circle in Fig. 43.

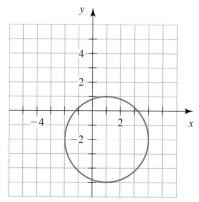

Figure 43 Exercise 47

48. Find an equation for the circle in Fig. 44.

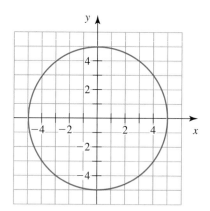

Figure 44 Exercise 48

49. A circle with center $(3, 2)$ contains the point $(5, 6)$. Find an equation for the circle.

50. A circle with center $(-4, 3)$ contains the point $(2, -1)$. Find an equation for the circle.

51. Find the equations for two circles that both contain the point $(5, 3)$. Also, sketch the graph of the two circles on the same coordinate system.

52. Show with a sketch that there are many circles that contain $(2, 1)$ and $(4, 6)$. Which of these circles has the smallest radius? What is the equation for this circle?

53. Give the coordinates of five points that are each a distance of 4 units from the point $(3, 2)$.

54. Give the coordinates of five points that are each a distance of 1 unit from the point $(-4, 1)$.

55. Is the relation $x^2 + y^2 = 49$ a function? Explain.

56. Is the relation $(x - 4)^2 + (y + 2)^2 = 16$ a function? Explain.

57. a. For $y = \sqrt{25 - x^2}$, explain why $y \geq 0$ for real number values of y.

b. Use pencil and paper to sketch a graph of the function $y = \sqrt{25 - x^2}$. [**Hint:** Square both sides of the equation.] Use a graphing calculator to verify your graph.

58. a. For $y = \sqrt{9 - (x - 5)^2}$, explain why $y \geq 0$ for real number values of y.

b. Use pencil and paper to sketch a graph of the function $y = \sqrt{9 - (x - 5)^2}$. Use a graphing calculator to verify your graph.

59. If the lengths of the legs of a right triangle are equal, we call the triangle an **isosceles right triangle**.

a. Sketch an example of an isosceles right triangle.

b. Show that the length of the hypotenuse of an isosceles right triangle is $\sqrt{2}$ times the length of either leg of the triangle. [**Hint:** Let $a = k$, $b = k$ and apply the Pythagorean theorem.]

c. If the length of a leg of an isosceles right triangle is 3, what is the length of the hypotenuse?

d. If the length of the hypotenuse of an isosceles right triangle is 5, what is the length of each leg?

60. Explain how to sketch the graph of an equation of the form $(x - h)^2 + (y - k)^2 = a$, where $a > 0$.

Section 11.6 Quiz

1. The length of a leg of a right triangle is 4 inches and the length of the hypotenuse is 8 inches. Find the length of the other leg.

2. The size of a rectangular television screen is usually described in terms of the length of its diagonal. If a 19-inch television screen has width 16 inches, what is the height of the screen?

Find the distance between the two given points.

3. $(-2, -5)$ and $(3, -1)$ **4.** $(-3, 2)$ and $(4, 2)$

Find an equation for the circle with given center C and radius r.

5. $C(-3, 2)$ and $r = 6$ **6.** $C(0, 0)$ and $r = 2.8$

Find the center and radius of the circle. Also, sketch the circle.

7. $x^2 + y^2 = 12$

8. $(x + 4)^2 + (y - 3)^2 = 25$

9. Find an equation for a circle with center $(2, -1)$ that contains the point $(4, 7)$.

10. Find equations for two circles that both contain the point $(0, 0)$. Also, sketch the two circles on the same coordinate system.

11.7 ELLIPSES AND HYPERBOLAS

Objective

▸ Sketch graphs of ellipses and hyperbolas.

In this text we work with four types of curves that are cross sections of cones: circles, *ellipses*, parabolas, and *hyperbolas* (see Fig. 45).

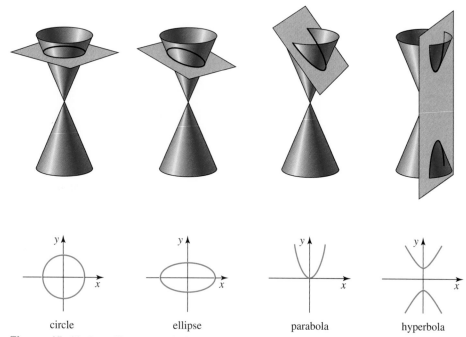

circle   ellipse   parabola   hyperbola

Figure 45 Circles, ellipses, parabolas, and hyperbolas are cross sections of cones

In Chapters 6 and 7 we worked with parabolas and in Section 11.6 we studied circles. In this section we first discuss ellipses and then we discuss hyperbolas.

To begin our study of ellipses, we sketch a graph of

$$\frac{x^2}{25} + \frac{y^2}{9} = 1$$

First, we find the y-intercepts by substituting 0 for x and solving for y.

$$\frac{0^2}{25} + \frac{y^2}{9} = 1$$

$$\frac{y^2}{9} = 1$$

$$y^2 = 9$$

$$y = \pm 3$$

So, the y-intercepts are $(0, -3)$ and $(0, 3)$.

Then, we find the x-intercepts by substituting 0 for y and solving for x.

$$\frac{x^2}{25} + \frac{0^2}{9} = 1$$

$$\frac{x^2}{25} = 1$$

$$x^2 = 25$$

$$x = \pm 5$$

So, the x-intercepts are $(-5, 0)$ and $(5, 0)$.

All of the points on the graph of the relation have x-coordinates between -5 and 5, inclusive. To see why, we isolate the term $\frac{y^2}{9}$ in the equation $\frac{x^2}{25} + \frac{y^2}{9} = 1$

$$\frac{y^2}{9} = 1 - \frac{x^2}{25}$$

Note that $\frac{y^2}{9}$ is nonnegative, so

$$1 - \frac{x^2}{25} \geq 0$$

$$1 \geq \frac{x^2}{25}$$

$$25 \geq x^2$$

$$x^2 \leq 25$$

$x^2 \leq 25$ implies that x is between -5 and 5, inclusive. This is what we set out to show.

Next, we find points on the graph when x is $-4, -3, -2, \ldots, 4$ (see Table 6) and plot these points as well as the intercepts (see Fig. 46).

Table 6 Solutions of $\frac{x^2}{25} + \frac{y^2}{9} = 1$	
x	y
-4	± 1.8
-3	± 2.4
-2	± 2.75
-1	± 2.94
0	± 3
1	± 2.94
2	± 2.75
3	± 2.4
4	± 1.8

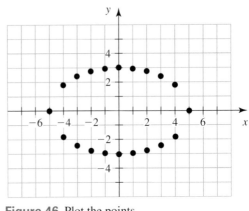

Figure 46 Plot the points

Finally, we sketch a curve through the points we've plotted (see Fig. 47).

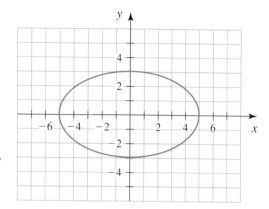

Figure 47 Graph of
$$\frac{x^2}{25} + \frac{y^2}{9} = 1$$

The graph is an ellipse.

Ellipse

An equation that can be put into the form

$$\frac{x^2}{a^2} + \frac{y^2}{b^2} = 1, \qquad a > 0, \ b > 0$$

has an **ellipse** as its graph.

Next, we discuss an easy way to sketch the graph of an equation in the form $\frac{x^2}{a^2} + \frac{y^2}{b^2} = 1$. We begin by finding the x-intercepts:

$$\frac{x^2}{a^2} + \frac{0^2}{b^2} = 1$$

$$\frac{x^2}{a^2} = 1$$

$$x^2 = a^2$$

$$x = \pm\sqrt{a^2}$$

$$x = \pm a$$

So the x-intercepts are $(-a, 0)$ and $(a, 0)$.

By similar steps, we can show that the y-intercepts are $(0, -b)$ and $(0, b)$.

Intercepts of an Ellipse

The ellipse described by

$$\frac{x^2}{a^2} + \frac{y^2}{b^2} = 1$$

has x-intercepts $(-a, 0)$ and $(a, 0)$ and y-intercepts $(0, -b)$ and $(0, b)$. See Fig. 48.

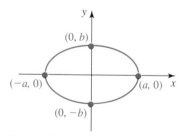

Figure 48 Intercepts of the ellipse $\dfrac{x^2}{a^2} + \dfrac{y^2}{b^2} = 1$

Example 1 Sketching the Graph of an Ellipse

Sketch the graph of $9x^2 + 4y^2 = 36$.

Solution

First, we divide both sides of the equation $9x^2 + 4y^2 = 36$ by 36 so that the right side of the equation is 1:

$$\frac{9x^2}{36} + \frac{4y^2}{36} = \frac{36}{36}$$

$$\frac{x^2}{4} + \frac{y^2}{9} = 1$$

The equation is of the form $\dfrac{x^2}{a^2} + \dfrac{y^2}{b^2} = 1$ with $a^2 = 4$ and $b^2 = 9$. Since $a^2 = 4$, we have $a = 2$. So, the x-intercepts are $(-2, 0)$ and $(2, 0)$. Since $b^2 = 9$, we have $b = 3$. So, the y-intercepts are $(0, -3)$ and $(0, 3)$. We plot the intercepts and then sketch an ellipse that contains them (see Fig. 49).

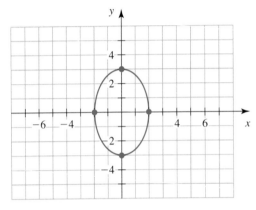

Figure 49 Graph of $9x^2 + 4y^2 = 36$

To use a graphing calculator to draw the ellipse $9x^2 + 4y^2 = 36$, we begin by isolating y:

$$9x^2 + 4y^2 = 36$$

$$4y^2 = 36 - 9x^2$$

$$y^2 = \frac{36 - 9x^2}{4}$$

$$y = \pm\sqrt{\frac{36 - 9x^2}{4}}$$

$$y = \pm\sqrt{\frac{9(4 - x^2)}{4}}$$

$$y = \pm\frac{3}{2}\sqrt{4 - x^2}$$

Then, we enter the functions $y = \dfrac{3}{2}\sqrt{4 - x^2}$ and $y = -\dfrac{3}{2}\sqrt{4 - x^2}$ (or $Y_2 = -Y_1$) and graph both functions on the same coordinate system (see Fig. 50).

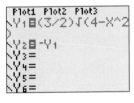
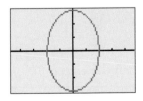

Figure 50 The graph of $9x^2 + 4y^2 = 36$ using ZDecimal

We now turn our attention to sketching graphs of hyperbolas. The general equation for a hyperbola is similar to the general equation for an ellipse, except that the left side of the equation is a difference.

Hyperbola

- An equation that can be put into the form

$$\frac{x^2}{a^2} - \frac{y^2}{b^2} = 1, \qquad a > 0, b > 0$$

has a **hyperbola** as its graph (see Fig. 51). The x-intercepts are $(-a, 0)$ and $(a, 0)$. There are no y-intercepts.

- An equation that can be put into the form

$$\frac{y^2}{b^2} - \frac{x^2}{a^2} = 1, \qquad a > 0, b > 0$$

has a **hyperbola** as its graph (see Fig. 52). The y-intercepts are $(0, -b)$ and $(0, b)$. There are no x-intercepts.

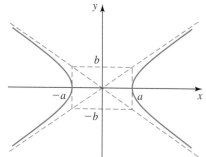

Figure 51 Graph of $\dfrac{x^2}{a^2} - \dfrac{y^2}{b^2} = 1$

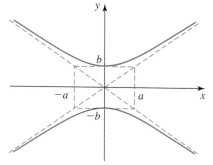

Figure 52 Graph of $\dfrac{y^2}{b^2} - \dfrac{x^2}{a^2} = 1$

Each *pair* of curves in Figs. 51 and 52 is a hyperbola. Each curve is a *branch*. The dashed figures are *not* parts of the hyperbola. They are simply tools to guide us in sketching the branches. Note that the dashed rectangles are centered at the origin and stretch a units either direction horizontally and b units either direction vertically. Through their opposite corners we draw the dashed lines that are *inclined asymptotes*. Note that the branches of the hyperbolas approach the inclined asymptotes as $|x|$ gets large.

To sketch a hyperbola based on its equation:

1. Sketch a dashed rectangle whose sides are parallel to the axes and contain the points $(-a, 0)$, $(a, 0)$, $(0, -b)$, and $(0, b)$.
2. Sketch two dashed lines (the inclined asymptotes) that contain the diagonals of the rectangle.
3. Plot the intercepts of the hyperbola.
4. Sketch the branches to contain the intercepts and get closer to the asymptotes as $|x|$ gets large.

Example 2 Sketching the Graph of a Hyperbola

Sketch the graph of $\dfrac{y^2}{4} - \dfrac{x^2}{9} = 1$.

Solution

Since the equation is of the form $\dfrac{y^2}{b^2} - \dfrac{x^2}{a^2} = 1$, we have $a^2 = 9$ and $b^2 = 4$. So $a = 3$ and $b = 2$. We sketch a dashed rectangle that contains the points $(-3, 0)$, $(3, 0)$, $(0, -2)$, and $(0, 2)$, and then we sketch the inclined asymptotes (see Fig. 53).

Note

Recall that we assume that a and b are positive.

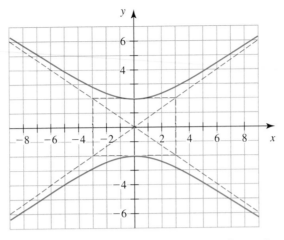

Figure 53 Graph of
$$\frac{y^2}{4} - \frac{x^2}{9} = 1$$

Since the equation is of the form $\dfrac{y^2}{b^2} - \dfrac{x^2}{a^2} = 1$ (with $\dfrac{y^2}{b^2}$ first), the graph has y-intercepts at $(0, -2)$ and $(0, 2)$ and there are no x-intercepts. We sketch the branches to contain the y-intercepts and approach the inclined asymptotes for large $|x|$. ∎

Example 3   Sketching the Graph of a Hyperbola

Sketch the graph of $4x^2 - 25y^2 = 100$.

Solution

We divide both sides of the equation by 100 so that the right side of the equation is equal to one:

$$4x^2 - 25y^2 = 100$$

$$\frac{4x^2}{100} - \frac{25y^2}{100} = \frac{100}{100}$$

$$\frac{x^2}{25} - \frac{y^2}{4} = 1$$

The equation is in $\dfrac{x^2}{a^2} - \dfrac{y^2}{b^2} = 1$ form where $a^2 = 25$ and $b^2 = 4$. So, $a = 5$ and $b = 2$. We sketch the dashed rectangle that contains $(-5, 0)$, $(5, 0)$, $(0, -2)$, and $(0, 2)$, and then sketch the inclined asymptotes (see Fig. 54).

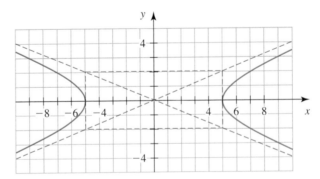

Figure 54 Graph of
$4x^2 - 25y^2 = 100$

Since the equation is in $\dfrac{x^2}{a^2} - \dfrac{y^2}{b^2} = 1$ form (with $\dfrac{x^2}{a^2}$ first), the graph has x-intercepts at $(-5, 0)$ and $(5, 0)$ and there are no y-intercepts. We sketch the branches to contain the intercepts and approach the inclined asymptotes for large $|x|$. ∎

To use a graphing calculator to draw a graph of $4x^2 - 25y^2 = 100$, we begin by isolating y:

$$4x^2 - 25y^2 = 100$$

$$4x^2 - 100 = 25y^2$$

$$\frac{4x^2 - 100}{25} = y^2$$

$$y = \pm\sqrt{\frac{4x^2 - 100}{25}}$$

$$y = \pm\sqrt{\frac{4(x^2 - 25)}{25}}$$

$$y = \pm\frac{2}{5}\sqrt{x^2 - 25}$$

Then, we enter the functions $y = \frac{2}{5}\sqrt{x^2 - 25}$ and $y = -\frac{2}{5}\sqrt{x^2 - 25}$ (or $Y_2 = -Y_1$) and draw the graphs on the same coordinate system (see Fig. 55).

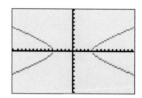

Figure 55 The graph of $4x^2 - 25y^2 = 100$ using ZStandard followed by ZSquare

EXPLORATION
Graphical significance of a and b for ellipses and hyperbolas

1. Sketch two ellipses that:
   a. Intersect in four points.
   b. Intersect in two points.
   c. Intersect in no points.
2. Write and graph equations to correspond to each of your sketches in problem 1.
3. Sketch an ellipse and a hyperbola that:
   a. Intersect in four points.
   b. Intersect in two points.
   c. Intersect in no points.
4. Write and graph equations to correspond to each of your sketches in problem 3.
5. Sketch two hyperbolas that:
   a. Intersect in four points.
   b. Intersect in two points.
   c. Have no intersection points.
6. Write and graph equations to correspond to each of your sketches in problem 5.

Key Points
OF THIS SECTION

Throughout these key points, assume that a and b are positive.

An equation that can be put into the form $\dfrac{x^2}{a^2} + \dfrac{y^2}{b^2} = 1$ has an ellipse as its graph.   **Equation of an ellipse**

The x-intercepts are $(-a, 0)$ and $(a, 0)$. The y-intercepts are $(0, -b)$ and $(0, b)$.

An equation that can be put into the form $\dfrac{x^2}{a^2} - \dfrac{y^2}{b^2} = 1$ has a hyperbola as its graph.   **Equation of a hyperbola with x-intercepts**

The x-intercepts are $(-a, 0)$ and $(a, 0)$. There are no y-intercepts.

An equation that can be put into the form $\dfrac{y^2}{b^2} - \dfrac{x^2}{a^2} = 1$ has a hyperbola as its graph.   **Equation of a hyperbola with y-intercepts**

The y-intercepts are $(0, -b)$ and $(0, b)$. There are no x-intercepts.

HOMEWORK 11.7

SSM
Tutor Center

MathPro 4
MathPro 4

MathPro 5
MathPro 5

CDs/Videos

www

Sketch the ellipse.

1. $\dfrac{x^2}{36} + \dfrac{y^2}{9} = 1$

2. $\dfrac{x^2}{49} + \dfrac{y^2}{16} = 1$

3. $\dfrac{x^2}{4} + \dfrac{y^2}{36} = 1$

4. $\dfrac{x^2}{9} + \dfrac{y^2}{64} = 1$

5. $\dfrac{x^2}{100} + \dfrac{y^2}{16} = 1$

6. $\dfrac{x^2}{81} + \dfrac{y^2}{25} = 1$

7. $25x^2 + 4y^2 = 100$

8. $4x^2 + 16y^2 = 64$

9. $9x^2 + 100y^2 = 900$

10. $16x^2 + 25y^2 = 400$

11. $x^2 + y^2 = 36$

12. $2x^2 + 2y^2 = 50$

13. $x^2 + 25y^2 = 25$

14. $64x^2 + y^2 = 64$

15. $5x^2 + 16y^2 = 80$

16. $22x^2 + 4y^2 = 88$

17. $16x^2 + 25y^2 = 1$

18. $36x^2 + 9y^2 = 4$

Sketch the hyperbola.

19. $\dfrac{x^2}{16} - \dfrac{y^2}{4} = 1$

20. $\dfrac{y^2}{25} - \dfrac{x^2}{9} = 1$

21. $\dfrac{y^2}{16} - \dfrac{x^2}{25} = 1$

22. $\dfrac{x^2}{49} - \dfrac{y^2}{9} = 1$

23. $\dfrac{x^2}{25} - \dfrac{y^2}{81} = 1$

24. $\dfrac{x^2}{64} - \dfrac{y^2}{9} = 1$

25. $16x^2 - 4y^2 = 64$

26. $25x^2 - 16y^2 = 400$

27. $x^2 - 9y^2 = 9$

28. $y^2 - 4x^2 = 4$

29. $y^2 - x^2 = 4$

30. $4x^2 - 4y^2 = 36$

31. $16y^2 - x^2 = 16$

32. $25x^2 - y^2 = 25$

33. $25x^2 - 7y^2 = 175$

34. $30x^2 - 9y^2 = 270$

Sketch the graph.

35. $\dfrac{x^2}{64} + \dfrac{y^2}{4} = 1$

36. $x^2 + y^2 = 49$

37. $x^2 - y^2 = 1$

38. $\dfrac{x^2}{16} + \dfrac{y^2}{16} = 1$

39. $81x^2 + 49y^2 = 3969$

40. $36y^2 - 4x^2 = 44$

41. $x^2 + y^2 = 1$

42. $\dfrac{x^2}{49} + \dfrac{y^2}{100} = 1$

43. $9y^2 - 4x^2 = 144$

44. $25x^2 + 25y^2 = 100$

45. $\dfrac{x^2}{25} - \dfrac{y^2}{25} = 1$

46. $4x^2 + 16y^2 = 256$

47. $x^2 + y^2 = 16$

48. $\dfrac{x^2}{9} - \dfrac{y^2}{25} = 1$

49. $4x^2 + 4y^2 = 64$

50. $\dfrac{x^2}{16} + \dfrac{y^2}{16} = 1$

51. $9x^2 - 9y^2 = 81$

52. $\dfrac{x^2}{36} + \dfrac{y^2}{36} = 1$

53. a. Sketch a graph of the equation $\dfrac{x^2}{c} + \dfrac{y^2}{d} = 1$ for the given values of c and d.

   i. $c = 4$ and $d = 16$.

   ii. $c = 4$ and $d = -16$.

   iii. $c = -4$ and $d = 16$.

   iv. $c = 4$ and $d = 4$.

b. Discuss, in terms of the values of the constants c and d, whether the graph of the equation $\dfrac{x^2}{c} + \dfrac{y^2}{d} = 1$ is a circle, an ellipse, a hyperbola with x-intercepts, or a hyperbola with y-intercepts.

54. Sketch graphs of the following equations on the same coordinate system.

  a. $\dfrac{x^2}{36} + \dfrac{y^2}{9} = 1$ **b.** $\dfrac{x^2}{36} - \dfrac{y^2}{9} = 1$ **c.** $\dfrac{y^2}{9} - \dfrac{x^2}{36} = 1$

55. Is the relation $\dfrac{x^2}{4} + \dfrac{y^2}{81} = 1$ a function? Explain.

56. Is the relation $\dfrac{x^2}{9} - \dfrac{y^2}{81} = 1$ a function? Explain.

57. a. For $y = \dfrac{5}{2}\sqrt{4 - x^2}$, explain why $y \geq 0$ for real number values of y.

b. Use pencil and paper to sketch a graph of the function $y = \dfrac{5}{2}\sqrt{4 - x^2}$. [**Hint**: Square both sides of the equation.] Use a graphing calculator to verify your graph.

58. a. For $y = \dfrac{2}{3}\sqrt{x^2 - 9}$, explain why $y \geq 0$ for real number values of y.

b. Use pencil and paper to sketch a graph of the function $y = \dfrac{2}{3}\sqrt{x^2 - 9}$. Use a graphing calculator to verify your graph.

59. Find equations for five ellipses that do not intersect each other. Sketch the five ellipses on the same coordinate system.

60. Show that the y-intercepts of $\dfrac{x^2}{a^2} + \dfrac{y^2}{b^2} = 1$ are $(0, -b)$ and $(0, b)$.

61. Is the graph of the equation $x^2 + y^2 = r^2$ with $r > 0$ a circle, an ellipse, both, or neither? Explain. [**Hint**: Is it possible to write the equation in the form $\dfrac{x^2}{a^2} + \dfrac{y^2}{b^2} = 1$?]

62. Describe how to graph an equation of the form $\dfrac{x^2}{a^2} - \dfrac{y^2}{b^2} = 1$.

Section 11.7 Quiz

Sketch the graph.

1. $\dfrac{x^2}{9} + \dfrac{y^2}{25} = 1$

2. $\dfrac{y^2}{49} - \dfrac{x^2}{9} = 1$

3. $4x^2 - y^2 = 16$

4. $16x^2 + 3y^2 = 48$

5. $\dfrac{x^2}{8} - \dfrac{y^2}{3} = 1$

6. $4y^2 - 4x^2 = 16$

7. $\dfrac{x^2}{5} + \dfrac{y^2}{14} = 1$

8. $x^2 + 9y^2 = 81$

9. Is the relation $\dfrac{x^2}{9} - \dfrac{y^2}{4} = 1$ a function? Explain.

10. Find equations for three ellipses that all contain the points $(0, 3)$ and $(0, -3)$.

11.8 SOLVING NONLINEAR SYSTEMS OF EQUATIONS

Objectives

▸ Solve nonlinear systems by graphing.
▸ Solve nonlinear systems by substitution or elimination.

In this section we solve *nonlinear systems*. Here is an example of such a system:

$$x^2 + y^2 = 4$$
$$y = x^2$$

Note that the graph of $x^2 + y^2 = 4$ is a circle and the graph of $y = x^2$ is a parabola. A **nonlinear system of equations** is a system of equations where *at least* one of the equations is not linear.

When solving nonlinear systems in this text, we find only solutions that have real number coordinates.

Just as with linear systems, we can solve nonlinear systems by graphing. Recall that for a system of two equations, the solution(s) of the system is (are) the intersection point(s) of the graphs of the two equations.

We can also solve some nonlinear systems by substitution or elimination.

Example 1   Solving a System

Solve the system

$$y = x^2 - 5$$
$$y = -x + 1$$

by graphing and also by substitution.

Solution

The graph of $y = x^2 - 5$ is a parabola and the graph of $y = -x + 1$ is a line. We sketch graphs of both equations on the same coordinate system (see Fig. 56).

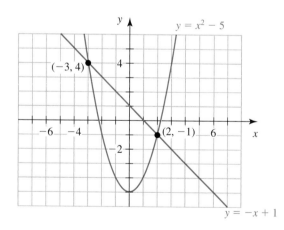

Figure 56 Graphs of $y = x^2 - 5$ and $y = -x + 1$

The graphs appear to intersect at $(-3, 4)$ and $(2, -1)$. The two intersection points are the solutions of the system.

Next, we solve the system

$$y = x^2 - 5$$
$$y = -x + 1$$

using substitution. To begin, we substitute $x^2 - 5$ for y in the equation $y = -x + 1$ and solve for x.

$$x^2 - 5 = -x + 1$$
$$x^2 + x - 6 = 0$$

$$(x + 3)(x - 2) = 0$$
$$x + 3 = 0 \quad \text{or} \quad x - 2 = 0$$
$$x = -3 \quad \text{or} \quad x = 2$$

Next, we substitute $x = -3$ and $x = 2$ into the equation $y = -x + 1$ to find the corresponding values of y.

$$y = -(-3) + 1 \qquad y = -2 + 1$$
$$= 4 \qquad\qquad = -1$$

So $(-3, 4)$ and $(2, -1)$ are the solutions of the system.

We can check that both $(-3, 4)$ and $(2, -1)$ satisfy both equations in the original system. However, the fact that we have solved the system in two different ways is itself a check of our work. ∎

We could also solve the nonlinear system in Example 1 by using "intersect" on a graphing calculator (see Fig. 57).

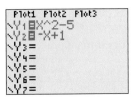 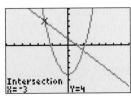 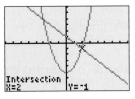

Figure 57 Verifying that $(-3, 4)$ and $(2, -1)$ are solutions

Example 2 Solving a System

Solve the system

$$x^2 + y^2 = 9$$
$$9x^2 + 4y^2 = 36$$

by graphing and also by elimination.

Solution

The graph of the equation $x^2 + y^2 = 9$ is a circle and the graph of $9x^2 + 4y^2 = 36$ is an ellipse. We sketch the graphs on the same coordinate system (see Fig. 58).

Note

We sketched the ellipse $9x^2 + 4y^2 = 36$ in Example 1 in Section 11.7.

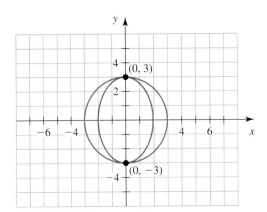

Figure 58 Graphs of $x^2 + y^2 = 9$ and $9x^2 + 4y^2 = 36$

The intersection points appear to be $(0, -3)$ and $(0, 3)$. These points are the solutions of the system.

Now, we solve the system

$$x^2 + y^2 = 9$$
$$9x^2 + 4y^2 = 36$$

using elimination. First, we multiply both sides of the equation $x^2 + y^2 = 9$ by -4.

$$-4x^2 - 4y^2 = -36$$
$$9x^2 + 4y^2 = 36$$

Next, we add the left sides of the two equations and the right sides of the two equations and solve for x.

$$5x^2 = 0$$
$$x^2 = 0$$
$$x = 0$$

Then, we substitute 0 for x in the equation $x^2 + y^2 = 9$ and solve for y.

$$0^2 + y^2 = 9$$
$$y^2 = 9$$
$$y = \pm 3$$

The solutions are $(0, -3)$ and $(0, 3)$. We found the same result when we solved the system by graphing. ■

If the equations of a nonlinear system are fairly easy to graph, the best solution process may be to first graph the equations. Then use the graphs to get an idea of the number of solutions and the values of the coordinates of the solution(s). Finally, find the exact solution(s) by substitution or elimination.

Example 3   Solving a System

Solve the system

$$x^2 + y^2 = 25$$
$$-4x^2 + 9y^2 = 36$$

by graphing and also by elimination. Round coordinates of any solutions to the nearest hundredth.

Solution

The graph of $x^2 + y^2 = 25$ is a circle and the graph of $-4x^2 + 9y^2 = 36$ is a hyperbola. We sketch both graphs on the same coordinate system (see Fig. 59).

Note

If we divide both sides of $-4x^2 + 9y^2 = 36$ by 36, we have $-\dfrac{x^2}{9} + \dfrac{y^2}{4} = 1$. We sketched the graph of this equation in Example 2 in Section 11.7.

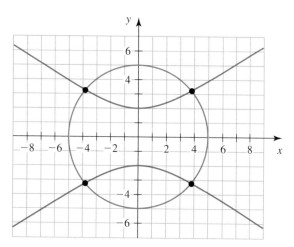

Figure 59 Graphs of $x^2 + y^2 = 25$ and $-4x^2 + 9y^2 = 36$

The graphs suggest that there are four solutions: $(-3.8, -3.2)$, $(-3.8, 3.2)$, $(3.8, -3.2)$, and $(3.8, 3.2)$.

Now, we solve the system

$$x^2 + y^2 = 25$$
$$-4x^2 + 9y^2 = 36$$

using elimination. First, we multiply both sides of the equation $x^2 + y^2 = 25$ by 4.

$$4x^2 + 4y^2 = 100$$
$$-4x^2 + 9y^2 = 36$$

Next, we add the left sides of the two equations and add the right sides of the two equations and solve for y.

$$13y^2 = 136$$
$$y^2 = \frac{136}{13}$$

$$y = \pm\sqrt{\frac{136}{13}}$$
$$y \approx \pm 3.234$$

Then, we substitute -3.234 and 3.234 for y in the equation $x^2 + y^2 = 25$ and solve for x.

$$x^2 + (-3.234)^2 = 25 \qquad\qquad x^2 + 3.234^2 = 25$$
$$x^2 + 3.234^2 = 25 \qquad\qquad x^2 = 25 - 3.234^2$$
$$x^2 = 25 - 3.234^2 \qquad\qquad x = \pm\sqrt{25 - 3.234^2}$$
$$x = \pm\sqrt{25 - 3.234^2} \qquad\qquad x \approx \pm 3.81$$
$$x \approx \pm 3.81$$

So, the four approximate solutions are $(-3.81, -3.23)$, $(3.81, -3.23)$, $(-3.81, 3.23)$, and $(3.81, 3.23)$, which agrees with what we found graphically. ■

EXPLORATION
Using graphs to find the number of solutions

For each problem, think graphically. It is not necessary to solve the systems in problems 2 and 4. Assume that a, b, and r are positive constants.

1. If $r < a$ and $r < b$, explain why there are no solutions with real number coordinates for the following system.

$$x^2 + y^2 = r^2$$
$$\frac{x^2}{a^2} + \frac{y^2}{b^2} = 1$$

2. If $a < r < b$, explain why the following system has four solutions.

$$x^2 + y^2 = r^2$$
$$\frac{x^2}{a^2} + \frac{y^2}{b^2} = 1$$

3. If $a < r < b$, explain why there are no solutions with real number coordinates for the following system.

$$x^2 + y^2 = r^2$$

$$\frac{y^2}{b^2} - \frac{x^2}{a^2} = 1$$

4. If $a < r < b$, explain why the following system has four solutions.

$$x^2 + y^2 = r^2$$

$$\frac{x^2}{a^2} - \frac{y^2}{b^2} = 1$$

Key Points
OF THIS SECTION

For a nonlinear system of two equations, the intersection point(s) of the graphs of the equations is (are) the solution(s) of the system.	**Solutions of a nonlinear system**
If it is reasonable to do so, first graph a nonlinear system to determine the number of solutions and to find approximate coordinates of the solutions. Then solve the system by substitution or elimination.	**First graph a nonlinear system**

HOMEWORK 11.8

  SSM Tutor Center   **MathPro 4** MathPro 4   **MathPro 5** MathPro 5   CDs/Videos   www

Solve the system by graphing. Also, solve the system by substitution or elimination.

1. $x^2 + y^2 = 25$
  $4x^2 + 25y^2 = 100$

2. $x^2 + y^2 = 4$
  $4x^2 + 16y^2 = 64$

3. $y = x^2 + 1$
  $y = -x + 3$

4. $y = x^2 - 3$
  $y = 2x$

5. $y = x^2 - 2$
  $y = -x^2 + 6$

6. $y = 2x^2 - 8$
  $y = x^2 + 1$

7. $x^2 + y^2 = 49$
  $x^2 + y^2 = 16$

8. $4x^2 + 9y^2 = 36$
  $9x^2 + 25y^2 = 225$

9. $x^2 + y^2 = 25$
  $y = -x - 1$

10. $x^2 + y^2 = 36$
  $y = x + 6$

11. $y^2 - x^2 = 16$
  $y + x^2 = 4$

12. $y^2 - 4x^2 = 4$
  $4x^2 + y^2 = 4$

13. $25x^2 - 9y^2 = 225$
  $4x^2 + 9y^2 = 36$

14. $y^2 - x^2 = 9$
  $4x^2 + 100y^2 = 400$

15. $9x^2 + y^2 = 9$
  $y = 3x + 3$

16. $x^2 - y^2 = 16$
  $y = -x + 1$

17. $4x^2 + 9y^2 = 36$
  $16x^2 + 25y^2 = 225$

18. $y = -3x^2 + 7$
  $y = -x^2 + 3$

19. $y = \sqrt{x} - 3$
  $y = -x - 1$

20. $y = \sqrt{x} + 1$
  $y = -\sqrt{x} + 5$

21. $y = 2x^2 - 5$
  $y = x^2 - 2$

22. $4x^2 + 9y^2 = 36$
  $x^2 + y = -2$

Solve the system by graphing. Also, solve the system using substitution or elimination. Round the coordinates of your solution(s) to the nearest hundredth.

23. $25y^2 - 4x^2 = 100$
  $9x^2 + y^2 = 9$

24. $16x^2 - 4y^2 = 64$
  $x^2 + 16y^2 = 16$

25. $25x^2 + 9y^2 = 225$
  $x^2 + y^2 = 16$

26. $36x^2 + 4x^2 = 144$
  $x^2 + y^2 = 9$

Solve the system using substitution or elimination. Check that each result satisfies both equations.

27. $9x^2 + y^2 = 85$
  $2x^2 - 3y^2 = 6$

28. $4y^2 + x^2 = 25$
  $y = -x + 5$

29. $y = 2x^2 - 5x - 11$
  $y = x^2 - 3x + 4$

30. $x^2 + 9y^2 = 13$
  $y = x - 1$

31. $y = x^2 - 3x + 2$

$y = 2x - 4$

32. $x^2 - 6y = 34$

$x^2 + y^2 = 25$

Solve the system of three equations by graphing. Check that each result satisfies each equation.

33. $x^2 + y^2 = 25$

$4x^2 - 25y^2 = 100$

$4x^2 + 25y^2 = 100$

34. $x^2 + y^2 = 1$

$9x^2 + y^2 = 9$

$y = x + 1$

35. Create a nonlinear system of two equations whose solutions are $(-4, 0)$ and $(4, 0)$.

36. Create a nonlinear system of two equations whose solutions are $(0, -3)$ and $(0, 3)$.

37. Find values for c and d so that $(1, 4)$ is a solution of the system

$$2x^2 + cy^2 = 82$$
$$y = x^2 + dx + 5$$

38. Find values for c and d so that $(2, -5)$ is a solution of the system

$$y = cx^2 - 4x^2 - 5$$
$$y = dx - 13$$

39. In your own words, describe a linear system and a nonlinear system. Also, compare the numbers of possible solutions for a linear system with the numbers of possible solutions for a nonlinear system.

Section 11.8 Quiz

Solve the system.

1. $9x^2 + y^2 = 81$

$x^2 + y^2 = 9$

2. $y = x^2 - 2$

$y = -2x + 1$

3. $y = x^2 + 3$

$y = x^2 - 6x + 9$

4. $25x^2 - 4y^2 = 100$

$9x^2 + y^2 = 9$

5. Use graphing to solve the system of three equations.

$$x^2 - y^2 = 16$$
$$x^2 + y^2 = 16$$
$$y = (x + 4)^2$$

6. Create a nonlinear system of two equations whose solution is $(0, 5)$.

Reviewing Prerequisite Material

In this appendix we review skills that you will need for this text. Review these skills before you begin Section 1.2. The answers to exercises in this appendix are located toward the end of the Answers to Odd-Numbered Exercises.

A.1 PLOTTING POINTS

In this section we review how to plot points. To plot the point $(6, 4)$, we draw a dot that is directly in line with 6 on the x-axis and 4 on the y-axis (see Fig. 1). The points $(-4, 2)$, $(-5, -3)$, and $(3, -4)$ have also been plotted in Fig. 1.

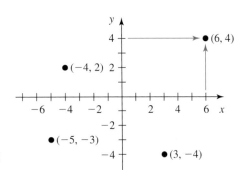

Figure 1 Plotting the points $(6, 4)$, $(-4, 2)$, $(-5, -3)$, and $(3, -4)$

1. Plot the points $(2, 4)$, $(-3, 1)$, $(-4, -3)$, and $(4, -2)$.

A.2 IDENTIFYING REAL NUMBERS

Recall that a *real number* is a number on the real number line. For example, -2.3781, $\frac{3}{7}$, and $\sqrt{4}$ are real numbers. The number $\sqrt{5}$ is a real number because there is such a number whose square is 5. We can use a calculator to find a decimal approximation of this number: $\sqrt{5} \approx 2.2361$. Decimal notation for $\sqrt{5}$ neither terminates nor repeats.

The expression $\frac{3}{0}$ is undefined because we can't divide by zero. Also, the expression $\sqrt{-4}$ is *not* a real number because no real number squared is equal to -4 (the square of any real number is a nonnegative number).

Decide whether the expression is a real number.

1. $\frac{2}{5}$ **2.** $\frac{4}{0}$ **3.** $\sqrt{-9}$ **4.** -3

5. $\sqrt{9}$ **6.** 4.21529 **7.** $\frac{0}{5}$ **8.** $\sqrt{7}$

Recall that the *counting numbers* is the set $\{1, 2, 3, \ldots\}$. The *integers* is the set $\{0, \pm 1, \pm 2, \pm 3, \ldots\}$. The *rational numbers* is the set of numbers that can be written in the form $\frac{n}{m}$ where n is an integer and m is a counting number.

Note that every counting number is an integer, every integer is a rational number, and every rational number is a real number. The set of all real numbers that are *not* rational numbers is the set of *irrational numbers*.

A.3 ABSOLUTE VALUE

The absolute value of a number a, written $|a|$, is the distance that the number is from 0 on the number line. For example, $|-5| = 5$ because -5 is a distance of 5 units from 0 (see Fig. 2). Also, $|5| = 5$ because 5 is a distance of 5 units from 0 (see Fig. 2).

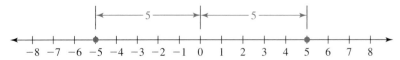

Figure 2 The numbers -5 and 5 are both a distance of 5 from 0.

Here are two more examples:

$$\left|\frac{2}{3}\right| = \frac{2}{3}$$

$$|-2.89| = 2.89$$

Compute.

1. $|-3|$ **2.** $|4.69|$ **3.** $|0|$ **4.** $|-\pi|$

A.4 PERFORMING ARITHMETIC WITH SIGNED NUMBERS

In this section you will review how to add, subtract, multiply, and divide signed numbers.

1. Do you recall how to multiply and divide signed numbers? Here are some examples to refresh your memory.

$$2(-5) = -10$$

$$-2(-5) = 10$$

$$\frac{-10}{5} = -2$$

$$\frac{-10}{-5} = 2$$

In general:

- The product or quotient of two numbers with the same sign is positive.
- The product or quotient of two numbers with different signs is negative.

Compute.

a. $-3(7)$ **b.** $-9(-4)$ **c.** $5(-6)$ **d.** $-8(-2)$

e. $-(-4)$ **f.** $-(-(-9))$ **g.** $\dfrac{8}{-2}$ **h.** $\dfrac{-6}{-2}$

    [**Hint:** $-(-4) = (-1)(-4)$]

i. $-\dfrac{-12}{4}$ **j.** $\dfrac{9}{-3}$

2. Do you recall how to add signed numbers? Here are some examples.

$$2 + 5 = 7$$

$$-2 + (-5) = -7$$

$$-2 + 5 = 3$$

$$2 + (-5) = -3$$

When adding signed numbers, it helps to think in terms of the number line. To find $-2 + (-5)$, imagine moving 2 units to the left of 0, then 5 more units to the left. Figure 3 illustrates that $-2 + (-5) = -7$.

Figure 3 Illustration of $-2 + (-5) = -7$

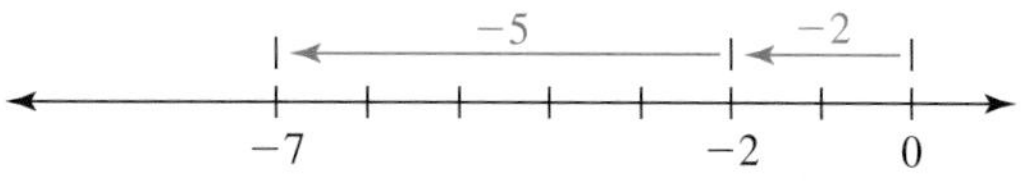

To find $2 + (-5)$, imagine moving 2 units to the right of 0, then 5 units to the left of 2. Figure 4 illustrates that $2 + (-5) = -3$.

Figure 4 Illustration of $2 + (-5) = -3$

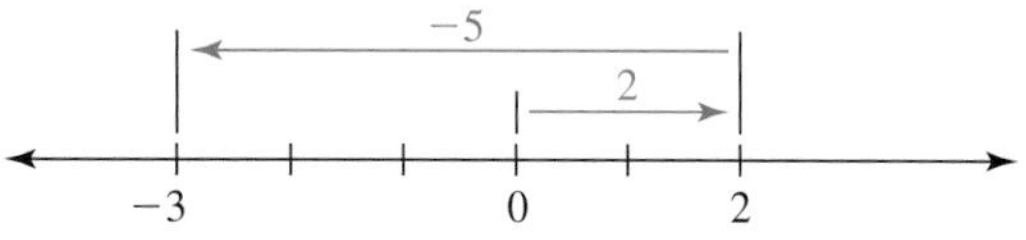

Compute.

a. $-3 + (-5)$ **b.** $9 + (-3)$ **c.** $2 + (-8)$ **d.** $-7 + 3$

e. $-1 + (-8)$ **f.** $-17 + 15$ **g.** $-4 + (-6)$ **h.** $-2 + (-3)$

3. We also need to discuss subtraction of signed numbers. Here are some examples.

$$10 - 2 = 10 + (-2) = 8$$
$$10 - (-2) = 10 + [-(-2)] = 10 + 2 = 12$$

In general, subtracting a number is the same as adding the opposite of that number. In symbols, we write $x - y = x + (-y)$.

Compute.

a. $3 - 7$ **b.** $-2 - 8$ **c.** $5 - (-3)$ **d.** $-9 - (-4)$

e. $3 - 8$ **f.** $-2 - 4$ **g.** $1 - (-1)$ **h.** $-10 - (-6)$

A.5 CONSTANTS, VARIABLES, EXPRESSIONS, AND EQUATIONS

A **constant** is a symbol used to represent a specific number. So, 2, 0, $\frac{1}{2}$, and π are constants. A **variable** is a symbol used to represent a quantity that changes.

An **expression** consists of constants, variables, and operation symbols. Here are some examples of expressions:

$$2\pi x + 3y, \qquad ax^2 + bx + c, \qquad x^2 - 4$$

In an **equation**, there are expressions on both sides of the equality symbol, $=$. Here are some examples of equations:

$$2\pi x + 3y = k, \qquad y = ax^2 + bx + c, \qquad x^2 - 4 = 0$$

We say an equation is an *equation in one variable* if its expressions use exactly one variable. If an equation in one variable becomes a true statement when a number is substituted for the variable, we call the number a **solution** of the equation. The **solution set** of an equation is the set of all solutions of the equation. We **solve** an equation by finding its solution set.

Identify each of the following as an expression or an equation.

1. $y = mx + b$ **2.** $x^3 - 8$ **3.** $2x - 5\pi + 1$ **4.** $3x^2 - 5x + 4 = 8$

A.6 USING DISTRIBUTIVE LAWS

Recall the distributive laws:

- $a(b + c) = ab + ac$
- $a(b - c) = ab - ac$
- $(b + c)a = ba + ca$
- $(b - c)a = ba - ca$

For example, $2(x - 3) = 2x - 2(3) = 2x - 6$.

Apply the distributive law for the expression.

1. $2(x + 4)$ **2.** $3(x - 5)$ **3.** $-5(x + 6)$ **4.** $-3(x - 8)$

5. $4(3x - 2)$ **6.** $5(2x + 7)$ **7.** $(x - 3)(4)$ **8.** $(x - 4)(-7)$

9. $2.8(x + 4.1)$ **10.** $-5.2(x + 3.9)$

A.7 COMBINING LIKE TERMS

Recall that we can simplify the expression $3x + 5x$ by adding the coefficients of the two terms: $3x + 5x = 8x$. We can use the distributive law to see why this "works":

$$3x + 5x = (3 + 5)x = 8x$$

We can simplify the expression $2x^2 + 3x + 5x^2 + 2x + 8$ by adding the coefficients of terms that have the same power of x: $2x^2 + 3x + 5x^2 + 2x + 8 = 7x^2 + 5x + 8$. We can use the distributive law twice to see why this "works":

$$2x^2 + 3x + 5x^2 + 2x + 8 = 2x^2 + 5x^2 + 3x + 2x + 8$$
$$= (2 + 5)x^2 + (3 + 2)x + 8$$
$$= 7x^2 + 5x + 8$$

Simplify.

1. $4x + 3x$ **2.** $7x - 5x$

3. $3x^2 + 2x + 7x^2 + 4x$ **4.** $6x^2 - 3x^2 + 9x + 2x + 5 - 3$

5. $8x^2 - 4x + 1 - x^2 + x - 5$ **6.** $x^2 + x + 1 + x^2 - x - 1$

7. $4(x - 1) + 2x + 3$ **8.** $-5(x - 2) + 3x^2 - x + 4 - 7x^2$

9. $5.4x^2 - 2.1x + 3.9 - 1.3x^2 + 4.4x + 1.6$

10. $-2.4(x - 6.2) + 3.9(x + 5.1)$

A.8 REVIEWING ORDER OF OPERATIONS

In this section we review order of operations.

- First, perform operations within parentheses.
- Second, perform exponentiations.
- Third, perform multiplications and divisions, going from left to right.
- Fourth, perform additions and subtractions, going from left to right.

For example, here's how to compute $5 - 8 \div (7 - 5)^2 + 2 \cdot 3$:

$5 - 8 \div (7 - 5)^2 + 2 \cdot 3 = 5 - 8 \div (2)^2 + 2 \cdot 3$	Perform the subtraction within the parentheses.
$= 5 - 8 \div 4 + 2 \cdot 3$	Perform the exponentiation.
$= 5 - 2 + 2 \cdot 3$	Perform the division, since it is to the left of the multiplication.
$= 5 - 2 + 6$	Perform the multiplication.
$= 3 + 6$	Perform the subtraction, since it is left of the addition.
$= 9$	

Compute.

1. $3 + 5(2)$

2. $2(8) - 4$

3. $2 + 10 \div 5$

4. $14 \div 7 - 1$

5. $2(3 - 1) + 4(2)$

6. $20 \div 5 \cdot 2 - 4$

7. $(9 - 5)(3 + 2) \div 5 + 2$

8. $4 - (8 - 3 \cdot 2) - 1$

9. $4(3)^2$

10. $5(2)^3$

11. -3^2 [**Hint**: $-3^2 = (-1)(3)^2$]

12. $(-3)^2$

13. $5 - 4^2 + 3$

14. $2^3 - 10 \div 5 + 1$

15. $6 - (3 - 1)^3 + 8$

16. $10 - 4(2)^3 + 5$

17. $5.4 + 7.1 \cdot 2.4 - 1.8$

18. $-2.9(4.1 - 9.9) - 24.7 \div 6.4$

A.9 SOLVING LINEAR EQUATIONS IN ONE VARIABLE

In this section you will review how to solve linear equations in one variable. For example, here's how to solve the equation $5(x - 2) = 3x + 4$:

$$5(x - 2) = 3x + 4$$

$$5x - 10 = 3x + 4 \qquad \text{Distributive law}$$

$$5x - 10 - 3x = 3x + 4 - 3x \qquad \text{Subtract } 3x \text{ from both sides.}$$

$$2x - 10 = 4 \qquad \text{Combine like terms.}$$

$$2x - 10 + 10 = 4 + 10 \qquad \text{Add 10 to both sides.}$$

$$2x + 0 = 14 \qquad \text{Simplify.}$$

$$2x = 14$$

$$\frac{2x}{2} = \frac{14}{2} \qquad \text{Divide both sides by 2.}$$

$$x = 7$$

We check that the original equation becomes a true statement if 7 is substituted for x.

$$5(x - 2) = 3x + 4$$

$$5(7 - 2) \stackrel{?}{=} 3(7) + 4$$

$$5(5) \stackrel{?}{=} 21 + 4$$

$$25 \stackrel{?}{=} 25$$

$$\text{true}$$

So, the solution of the equation is 7.

If an equation contains fractions, it is often helpful to multiply both sides of the equation by the *least common denominator* (*LCD*) of the fractions. To illustrate, we solve $\frac{5}{6}x - \frac{1}{2} = \frac{9}{4}$. To find the LCD, we first "prime factor" the denominators:

$$6 = 2 \cdot 3$$

$$2 = 2$$

$$4 = 2 \cdot 2$$

Next, we count the most times that each prime appears in any one factorization. The most that 2 appears is twice. The most that 3 appears is once. Then we find the product of that many 2's and 3's. So, the LCD is $2 \cdot 2 \cdot 3 = 12$.

Now we multiply both sides of the equation $\frac{5}{6}x - \frac{1}{2} = \frac{9}{4}$ by 12.

$$\frac{5}{6}x - \frac{1}{2} = \frac{9}{4}$$

$$12\left(\frac{5}{6}x - \frac{1}{2}\right) = 12 \cdot \frac{9}{4} \qquad \text{Multiply both sides by the LCD 12.}$$

$$12 \cdot \frac{5}{6}x - 12 \cdot \frac{1}{2} = 12 \cdot \frac{9}{4} \qquad \text{Distributive law}$$

$$10x - 6 = 27 \qquad \text{Simplify.}$$

$$10x = 33 \qquad \text{Add 6 to both sides.}$$

$$x = \frac{33}{10} \qquad \text{Divide both sides by 10.}$$

Since $\frac{33}{10} = 3.3$, we can use a calculator to check that the original equation becomes a true statement if 3.3 is substituted for x.

Solve.

1. $4x = 12$ **2.** $x - 2 = 10$

3. $-5x + 3 = 13$ **4.** $-9x - 4 = 5$

5. $2(x + 1) = 7$ **6.** $3(x - 4) = x + 8$

7. $x - 3 = -5(x - 9)$ **8.** $6(x - 2) + 3x = x - 4 + 3x$

9. $2.17x - 6.88 = 3.32x + 9.19$ **10.** $3.4(x - 1.2) = 8.8x + 4.7$

11. $\frac{2}{3}x + \frac{1}{4} = \frac{5}{12}$ **12.** $\frac{x-3}{5} - \frac{3}{10} = \frac{1}{2}$

A.10 SOLVING LITERAL EQUATIONS IN TWO VARIABLES

In this section you will review how to solve *literal equations in two variables*. For example, here's how to solve the equation $5x + 3y = 15$ for y:

$$5x + 3y = 15$$

$$5x + 3y - 5x = 15 - 5x \qquad \text{Subtract } 5x \text{ from both sides.}$$

$$3y + 0 = 15 - 5x$$

$$3y = -5x + 15 \qquad \text{Rearrange the terms on the right side.}$$

$$\frac{3y}{3} = \frac{-5x + 15}{3} \qquad \text{Divide both sides by 3.}$$

$$y = \frac{-5x}{3} + \frac{15}{3} \qquad \text{Separate the fraction.}$$

$$y = -\frac{5}{3}x + 5$$

For another example, we solve $-\frac{1}{4}x + \frac{7}{10}y = \frac{3}{5}$ for y. To find the LCD, we first prime factor the denominators:

$$4 = 2 \cdot 2$$

$$10 = 5 \cdot 2$$

$$5 = 5$$

Next, we count the most times that each prime appears in any one factorization. The most that 2 appears is twice. The most that 5 appears is once. Then we find the product of that many 2's and 5's. So, the LCD is $2 \cdot 2 \cdot 5 = 20$.

Next, we multiply both sides of $-\dfrac{1}{4}x + \dfrac{7}{10}y = \dfrac{3}{5}$ by the LCD 20.

$$-\frac{1}{4}x + \frac{7}{10}y = \frac{3}{5}$$

$$20\left(-\frac{1}{4}x + \frac{7}{10}y\right) = 20 \cdot \frac{3}{5} \qquad \text{Multiply both sides by the LCD 20.}$$

$$20\left(-\frac{1}{4}x\right) + 20\left(\frac{7}{10}y\right) = 20 \cdot \frac{3}{5} \qquad \text{Distributive law}$$

$$-5x + 14y = 12 \qquad \text{Simplify both sides.}$$

$$-5x + 14y + 5x = 12 + 5x \qquad \text{Add } 5x \text{ to both sides.}$$

$$14y = 5x + 12$$

$$y = \frac{5}{14}x + \frac{12}{14} \qquad \text{Divide both sides by 14.}$$

$$y = \frac{5}{14}x + \frac{6}{7} \qquad \text{Reduce } \frac{12}{14}.$$

1. Solve $2x + y = 8$ for y.

2. Solve $3x - y = 18$ for y.

3. Solve $5w + 2t = 10$ for w.

4. Solve $3(w - 2) - t = 1$ for w.

5. Solve $p = 2.5d - 8.1$ for d.

6. Solve $p = -1.6d + 10.2$ for d.

7. Solve $-4x + 3y = 2x + 9$ for y.

8. Solve $-4x + 3y = 2x + 9$ for x.

9. Solve $3a + 4b = 2(a + 5)$ for b.

10. Solve $7a - 3b + a = 0$ for a.

11. Solve $\dfrac{1}{2}x - \dfrac{3}{4}y = \dfrac{5}{8}$ for y.

12. Solve $-\dfrac{3}{4}x - \dfrac{2}{3}y = \dfrac{1}{4}$ for y.

A.11 EQUIVALENT EXPRESSIONS AND EQUIVALENT EQUATIONS

By the distributive law, we have $2(x + 3) = 2x + 6$. We say that the expressions $2(x + 3)$ and $2x + 6$ are **equivalent expressions** since the expressions attain equal values for *any* number substituted for x. In Table 1, we show that the expressions $2(x + 3)$ and $2x + 6$ attain equal values when 0, 1, 2, 3, and 4 are substituted for x.

Table 1 Substituting Values for x in $2(x + 3)$ and $2x + 6$

x	$2(x + 3)$	$2x + 6$
0	$2(0 + 3) = 6$	$2(0) + 6 = 6$
1	$2(1 + 3) = 8$	$2(1) + 6 = 8$
2	$2(2 + 3) = 10$	$2(2) + 6 = 10$
3	$2(3 + 3) = 12$	$2(3) + 6 = 12$
4	$2(4 + 3) = 14$	$2(4) + 6 = 14$

For another example, consider the expression $2x + 4x + 3$. By combining like terms, we have $2x + 4x + 3 = 6x + 3$. The expressions $2x + 4x + 3$ and $6x + 3$ are equivalent expressions.

We *simplify* an expression such as $2(x - 5) + 4x + 1$ by applying the distributive law to remove parentheses, combining like terms, and/or performing as many operations with numbers as possible. The result is a *simplified expression* that is equivalent to the original expression.

Now we turn our attention to *equivalent equations*. To solve the equation $x + 3 = 7$, we write:

$$x + 3 = 7$$
$$x + 3 - 3 = 7 - 3$$
$$x = 4$$

Each of the equations $x + 3 = 7$, $x + 3 - 3 = 7 - 3$, and $x = 4$ has 4 as its only solution. So the three equations have the same solution set. Equations that have the same solution set are called **equivalent equations**.

It is important to know the difference between simplifying an expression and solving an equation. To simplify an *expression*, we find a simpler, equivalent *expression*. To solve an *equation*, we find *number(s)* that satisfy the equation.

Determine whether the pair are two equivalent expressions, two equivalent equations, or neither. Explain.

1. $5(x - 4)$ and $5x - 20$

2. $x + 8 = 0$ and $x = -8$

3. $4x - 3x + 8$ and $-12x^2 + 8$

4. $3(x + 1) + 7$ and $3x + 8$

5. $4(x + 2)$ and $4x + 8 = 0$

6. $-3(2x - 5)$ and $-6x + 15$

7. $3x + 1 = 16$ and $3x = 15$

8. $2(x - 3) + 5 = 25$ and $2x = 23$

9. $-3(x - 4) = -18$ and $x = 2$

10. $3x^2 + 5x - 2$ and $2x^2 + x - 2 + 5x$

Using a TI-82 or TI-83 Graphing Calculator

The more you experiment with your calculator, the more comfortable and efficient you will become in using it.

There are several types of errors that a TI graphing calculator can catch you making and when this happens, it will display an error message. When this occurs, refer to Section B.26 for some explanations of the more common error messages and how to fix these types of mistakes. When you make an error, this does not hurt your calculator. In fact, you can't hurt your calculator no matter what order you press its keys. So the more you experiment with your calculator, the better.

Most of the key combinations for a TI-82 and a TI-83 are the same. When there is a difference, the instructions for a TI-82 will be put in parenthesis in red type.

To access a command written in yellow above a key, you will first need to press 2nd and then the key. In this case the instructions will use brackets for the key. For example, "Press 2nd [OFF]" means to press 2nd and then press ON (because "OFF" is written in yellow above the ON key).

B.1 TURNING YOUR GRAPHING CALCULATOR ON AND OFF

To turn your graphing calculator on, press ON . To turn it off, press 2nd . Then press [OFF].

Graphing Calculator

When told to "press" a key, this means to press and release the key.

B.2 MAKING YOUR SCREEN LIGHTER OR DARKER

To make the screen darker, press 2nd . Then hold the △ key down for a while.

To make the screen lighter, press 2nd . Then hold the ▽ key down for a while.

B.3 ENTERING A FUNCTION

Suppose that you want to enter the function $y = 2x + 1$.

1. Press Y= .

2. If necessary, press CLEAR to erase a previously entered function.

3. To enter the function $y = 2x + 1$, press **2** X,T,Θ,*n* + **1**. Your calculator screen should look like the one displayed in Fig. 1.

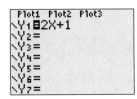

Figure 1 Entering a function

4. If you want to enter another function, press ENTER . Then type in the next expression. If you don't want to enter a function, there is no need to press ENTER .

5. You can use the △ or ▽ keys to get from one function to another.

B.4 GRAPHING A FUNCTION

Suppose that you want to graph the function $y = 2x + 1$.

1. See Section B.3 to enter the function $y = 2x + 1$.

2. Press ZOOM **6**. This will draw a graph of $y = 2x + 1$ between the values of -10 and 10 for both x and y.

3. See Section B.6 if you want to ZOOM IN or ZOOM OUT to get another part of the graph to appear on the calculator screen. Or see Section B.7 to manually change the window format and then press GRAPH .

B.5 TRACING A FUNCTION WITHOUT A SCATTERGRAM

Suppose that you want to find the coordinates of some points that lie on the graph of $y = 2x + 1$.

1. See Section B.4 to graph $y = 2x + 1$.

2. Press TRACE .

3. Notice the flashing "X" on the curve. The coordinates for this point should be listed at the bottom of the screen. If you don't see the flashing "X," press ENTER and your calculator will adjust the viewing window so you can see it.

4. To find coordinates of points on your curve off to the right, press ▷ .

5. To find coordinates of points on your curve off to the left, press ◁ .

6. The y-coordinate of a point can be found by entering the x-coordinate. For example, to find the y-coordinate of the point with x-coordinate 3, press **3** ENTER . Your calculator screen should look like the one displayed in Fig. 2. This feature works for values of x between X_{min} and X_{max}, inclusive (see Section B.7).

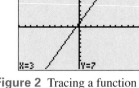

Figure 2 Tracing a function

7. If more than one function has been graphed, press ▽ to trace the second function. Continue pressing ▽ to trace the third function, and so on. Press △ to return to the previous function. Notice that the equation of the function as it is defined in the Y= mode is listed in the upper left corner of the screen.

(TI-82: Notice that the number of the function as it is defined in the Y= mode is listed in the upper right corner of the screen. The feature described in step 6 is not available on the TI-82.)

B.6 ZOOMING

The ZOOM menu has several features. These features allow you to adjust the viewing window. Some of the features adjust which values of x are used when tracing.

- **ZOOM IN** magnifies the graph around the cursor location. The following instructions are for ZOOMing IN on the graph of $y = 2x + 1$.

  1. See Section B.4 to graph $y = 2x + 1$.

  2. Press ZOOM **2**.

  3. Use ◁ , ▷ , △ , and ▽ to position the cursor on a portion of the line that you want to ZOOM IN on.

  4. To ZOOM IN press ENTER .

  5. To ZOOM IN on the graph again you have two options:

     a. To ZOOM IN at the same point, press ENTER .
     b. To ZOOM IN at a new point, move the cursor to the new point, and press ENTER .

  6. To return to your original graph, press ZOOM **6**. Or, you can ZOOM OUT (see the next instructions) the same number of times you ZOOMed IN.

Graphing Calculator

If you lose sight of the line, you can always press TRACE ENTER .

- **ZOOM OUT** does the reverse of ZOOM IN, it allows you to see *more* of a graph. To ZOOM OUT, follow the above instructions, but press $\boxed{\text{ZOOM}}$ **3** instead of $\boxed{\text{ZOOM}}$ **2** in step 2.

- **ZStandard** will change your viewing screen so that x and y will both go from -10 to 10. To use ZStandard, press $\boxed{\text{ZOOM}}$ **6**.

- **ZDecimal** allows you to trace using the numbers 0, ± 0.1, ± 0.2, ± 0.3, ... for x. ZDecimal will change your viewing screen so that x will go from -4.7 to 4.7 and y will go from -3.1 to 3.1. To use ZDecimal, press $\boxed{\text{ZOOM}}$ **4**.

- **ZInteger** allows you to trace using the numbers 0, ± 1, ± 2, ± 3, ... for x. ZInteger can be used for any viewing window, although it will change the view. To use ZInteger, press $\boxed{\text{ZOOM}}$ **8** $\boxed{\text{ENTER}}$.

- **ZSquare** will change your viewing window so that if you measured a change of one unit on both axes with a ruler, the two distances would be equal. To use ZSquare, press $\boxed{\text{ZOOM}}$ **5**.

- **ZoomStat** will change your viewing window so that you can see a scattergram of points that you have entered in the statistics editor. To use ZoomStat, press $\boxed{\text{ZOOM}}$ **9**.

- **ZoomFit** will adjust the dimensions of the y-axis so that as much of a curve as possible is displayed. The dimensions of the x-axis will remain unchanged. To use ZoomFit press $\boxed{\text{ZOOM}}$ **0**.
  (**ZoomFit is not available on the TI-82.**)

Graphing Calculator

When ZOOMing OUT, you will only return to the exact original graph if you did not move the cursor while ZOOMing IN.

B.7 SETTING THE WINDOW FORMAT

Suppose that you would like to graph the function $y = 2x + 1$ between the values of -2 and 3 for x and between the values of -5 and 7 for y.

1. See Section B.3 to enter the function $y = 2x + 1$.

2. Press $\boxed{\text{WINDOW}}$. You will now change the WINDOW settings so that it looks like the one displayed in Fig. 3.

3. Press $\boxed{\text{(-)}}$ **2** $\boxed{\text{ENTER}}$.
   (**TI-82: Press** $\boxed{\triangledown}$ $\boxed{\text{(-)}}$ **2** $\boxed{\text{ENTER}}$.)

4. Press **3** $\boxed{\text{ENTER}}$.

5. Press **1** $\boxed{\text{ENTER}}$ to set the scaling for the x-axis to be increments of 1.

6. Press $\boxed{\text{(-)}}$ **5** $\boxed{\text{ENTER}}$.

7. Press **7** $\boxed{\text{ENTER}}$.

8. Press **1** $\boxed{\text{ENTER}}$ to set the scaling for the y-axis to be increments of 1.

9. Press $\boxed{\text{GRAPH}}$ to view the graph of $y = 2x + 1$. Your calculator screen should look like the graph drawn in Fig. 4.

Graphing Calculator

If you press $\boxed{\text{ZOOM}}$ **6**, $\boxed{\text{ZOOM}}$ **9**, or ZOOM IN or ZOOM OUT, your window settings will change accordingly.

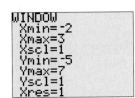

Figure 3 Window setup

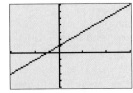

Figure 4 Graph of $y = 2x + 1$

B.8 PLOTTING POINTS FOR A SCATTERGRAM

Suppose that you want to create a scattergram for the data displayed in Table 1.

1. To create a table, press $\boxed{\text{STAT}}$ **1**.

Table 1 Creating a Scattergram

x	y
2	4
3	7
4	10
5	11

Graphing Calculator

When clearing a *list,* make sure you press CLEAR rather than DEL . If you press DEL , the column will vanish. If you ever do this by mistake, press STAT 5 ENTER to get back the missing column.

2. To clear list L_1, press ◁ as many times as necessary to get to column L_1. Then press △ once to get to the top of column L_1. Then press CLEAR ENTER .

3. To clear list L_2, press ▷ to move the cursor to column L_2. Then press △ CLEAR ENTER .

4. To return to the first entry position of list L_1, press ◁ .

5. To enter the data in the first column, press **2** ENTER **3** ENTER **4** ENTER **5** ENTER to enter the elements of L_1. (If you make a mistake you can delete an entry by pressing DEL and you can insert an entry by pressing 2nd [DEL].)

6. Press ▷ to move to the first entry position of list L_2.

7. Press **4** ENTER **7** ENTER **10** ENTER **11** ENTER to enter the elements of L_2.

8. Press 2nd [STAT PLOT].

9. Press **1** to select "Plot1."

10. Press ENTER to turn Plot1 on.

11. Press ▽ ENTER to choose the scattergram mode.

12. Press ▽ so that the cursor is at "Xlist." Then press 2nd [L_1].
(**TI-82: Press** ▽ **twice so that the cursor is on one of six choices for the "Xlist."**
Then press ◁ **as many times as necessary to select** L_1. **Then press** ENTER .)

13. Press ▽ so that the cursor is at "Ylist." Then press 2nd [L_2].
(**TI-82: Press** ▽ **once so that the cursor is on one of six choices for the "Ylist."**
Then press ▷ **or** ◁ **as many times as necessary to select** L_2. **Then press** ENTER .)

14. For the points plotted on the scattergram, you can use squares, plus signs, or dots (these three symbols are called "Marks"). Press ▽ so that the cursor is on one of the three choices for the "Mark." Then press ▷ and/or ◁ to select a marking. Then press ENTER .

15. Press ZOOM **9**. Your calculator screen should look like the one displayed in Fig. 5.

Graphing Calculator

If Plot1 is "off," your points are saved in columns L_1 and L_2, but they will not be plotted.

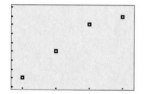

Figure 5 Creating a scattergram

B.9 TRACING A SCATTERGRAM

Suppose that you want to see the coordinates of a point in a scattergram.

1. See Section B.8 to draw a scattergram.

2. Press TRACE .

3. Notice the flashing "X" on one of the points of the scattergram. The coordinates for this point should be listed at the bottom of the screen.

4. To find the coordinates of the next point of the scattergram that lies to the right, press ▷ .

5. To find the coordinates of the next point of the scattergram that lies to the left, press ◁ .

B.10 GRAPHING FUNCTIONS WITH A SCATTERGRAM

Suppose that you want to graph the function $y = 2x + 1$ with a scattergram of the table of values listed in Section B.8.

1. Follow the instructions in Section B.3 to enter the function $y = 2x + 1$.

2. Follow the instructions in Section B.8 to draw the scattergram. (The graph of the function will also be drawn since you turned the function on.) Your calculator screen should look like the one displayed in Fig. 6.

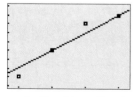

Figure 6 Graphing a function and a scattergram

B.11 TRACING A FUNCTION WITH A SCATTERGRAM

1. See Section B.10 to draw a function with a scattergram.

2. Press $\boxed{\text{TRACE}}$ to trace points that are part of the scattergram. Press $\boxed{\text{TRACE}}$ $\boxed{\triangledown}$ to trace points that lie on the curve. If other functions are graphed, continue pressing $\boxed{\triangledown}$ to trace the second function, and so on. Press $\boxed{\triangle}$ to begin to return to the scattergram. Notice that the label "P1:L_1, L_2" is in the upper left corner of the screen when Plot1's points are being traced, and that the equation of the function as it is defined in the $\boxed{\text{Y=}}$ mode is listed in the upper left corner of the screen when it is being traced.

(**TI-82: Notice that the label "P1" is in the upper right corner of the screen when Plot1's points are being traced, and that the number of the function as it is defined in the $\boxed{\text{Y=}}$ mode is listed in the upper right corner of the screen when it is being traced.**)

Graphing Calculator

Recall that if you do not see the flashing "X," press $\boxed{\text{ENTER}}$ and your calculator will adjust the WINDOW settings so that you can see it.

B.12 TURNING A PLOTTER "ON" OR "OFF"

To change the on/off status of the plotter, follow the instructions below.

1. Press $\boxed{\text{Y=}}$.

2. Press $\boxed{\triangle}$. A flashing rectangle will be on "Plot1."

3. Press $\boxed{\triangleright}$ if necessary to move the flashing rectangle to the plotter you wish to turn "on" or "off."

4. Press $\boxed{\text{ENTER}}$ to turn your plotter "on" or "off." The plotter is "on" if the plotter icon is highlighted.

(**TI-82:**

1. **Press $\boxed{\text{2nd}}$ [STAT PLOT].**

2. **Press $\boxed{\triangledown}$ if necessary to highlight the plotter you wish to turn on or off. There are three plotters to choose from. Then press $\boxed{\text{ENTER}}$.**

3. **Press $\boxed{\triangleright}$ $\boxed{\text{ENTER}}$ to turn the plotter off. Press $\boxed{\triangleleft}$ $\boxed{\text{ENTER}}$ to turn the plotter on.**)

B.13 CREATING A TABLE

Suppose that you would like to create a table of ordered pairs for the function $y = 2x+1$, where the values of x are 3, 4, 5, ... The table is displayed in Fig. 7.

1. Enter the function $y = 2x + 1$ for Y_1. See Section B.3 to enter the function $y = 2x + 1$.

2. Press $\boxed{\text{2nd}}$ [TBLSET].

3. Press 3 $\boxed{\text{ENTER}}$. This tells your calculator that the first x value in your table should be 3.

4. Press 1 $\boxed{\text{ENTER}}$. This tells your calculator that the x values in your table should increase by 1.

5. Press $\boxed{\text{ENTER}}$ $\boxed{\triangledown}$ $\boxed{\text{ENTER}}$ to highlight "Auto" for both "Indpnt" and "Depend." Your screen should now look like the one displayed in Fig. 8.

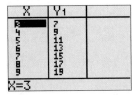

Figure 7 Table of ordered pairs for $y = 2x + 1$

Figure 8 Table setup

6. Press $\boxed{\text{2nd}}$ [TABLE].

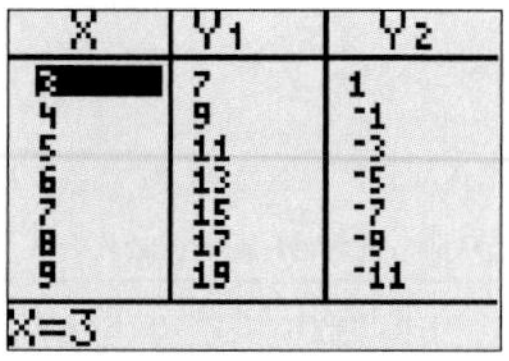
X Y1 Y2
3 7 1
4 9 -1
5 11 -3
6 13 -5
7 15 -7
8 17 -9
9 19 -11
X=3

Figure 9 Table for two (functions)

B.14 CREATING A TABLE FOR TWO FUNCTIONS

Suppose that you would like to create a table of ordered pairs for the functions $y = 2x+1$ and $y = -2x + 7$, where the values of x are 3, 4, 5, ... The table is displayed in Fig. 9.

1. Enter the function $y = 2x + 1$ for Y_1 and enter the function $y = -2x + 7$ for Y_2. See Section B.3 to enter these functions.
2. Follow steps 2–5 in Section B.13.

B.15 USING "ASK" IN A TABLE

Suppose that you would like to complete Table 2 for $y = 2x + 1$. One way to do this is to use the "Ask" option in the Table Setup mode.

1. Enter the function $y = 2x + 1$ for Y_1. See Section B.3 to enter the function $y = 2x + 1$.
2. Press [2nd] [TBLSET].
3. Press [ENTER] twice, then press [▷], then press [ENTER]. The "Ask" choice for "Indpnt" should now be highlighted. Make sure that the "Auto" choice for "Depend" is highlighted.
4. Press [2nd] [TABLE].
5. Press **2** [ENTER] **2.9** [ENTER] **5.354** [ENTER] **7** [ENTER] **100** [ENTER]. Your calculator screen should now look like the one displayed in Fig. 10.

Table 2 Using "Ask" in a Table with $y = 2x + 1$

x	y
2	
2.9	
5.354	
7	
100	

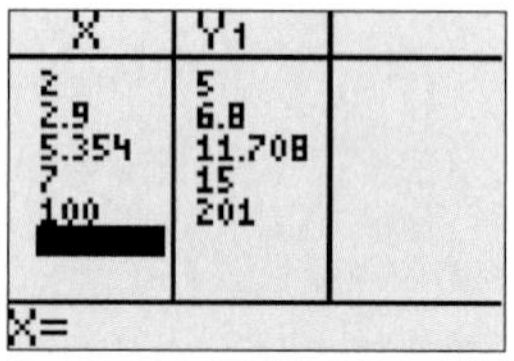
X Y1
2 5
2.9 6.8
5.354 11.708
7 15
100 201
X=

Figure 10 Using "Ask" for a table with $y = 2x + 1$

B.16 FINDING A REGRESSION CURVE FOR SOME DATA

Table 3 Finding a Regression Curve for Data

x	y
2	4
3	7
4	10
5	11

Suppose that you want to find a regression curve for the data displayed in Table 3.

1. See Section B.8 to create a scattergram of the data in Table 3. Enter your data in the first two columns (L_1 and L_2) of the STAT list editor.
2. Clear your Home Screen by pressing [2nd] [QUIT] [CLEAR].
3. Press [STAT].
4. To choose the CALC menu, press [▷]. Your screen should look like the one displayed in Fig. 11 (see Figs. 11, 12, and 13).

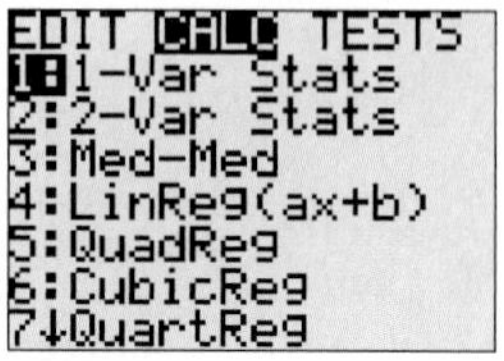
EDIT CALC TESTS
1:1-Var Stats
2:2-Var Stats
3:Med-Med
4:LinReg(ax+b)
5:QuadReg
6:CubicReg
7↓QuartReg

Figure 11 CALC menu

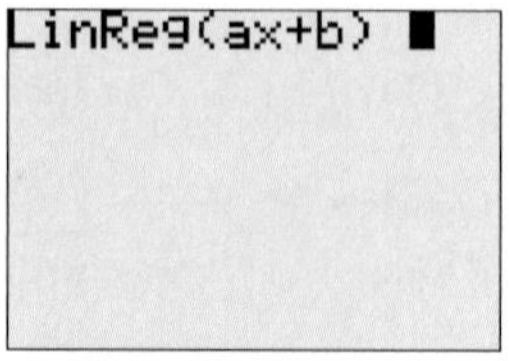
LinReg(ax+b)

Figure 12 About to find the equation

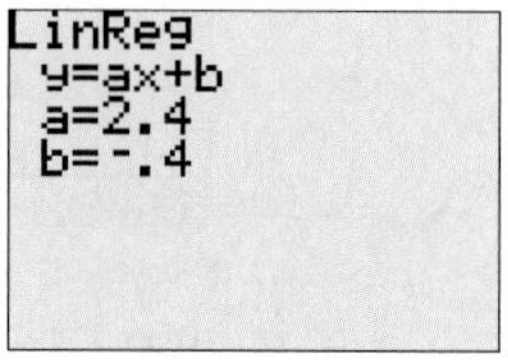
LinReg
y=ax+b
a=2.4
b=-.4

Figure 13 The equation

5. To choose Linear Regression, press **4**. Your screen should now look like the one displayed in Fig. 12. (Note that you can choose Quadratic Regression by pressing **5**, Exponential Regression by pressing **0**, and Power Regression by pressing ALPHA [A].)

(**TI-82: Choose Linear Regression by pressing 5, Quadratic Regression by pressing 6, Exponential Regression by pressing** ALPHA **[A], and Power Regression by pressing** ALPHA **[B].**)

6. Press ENTER . Your screen should now look like the one displayed in Fig. 13. This means that the equation of the regression line is $y = 2.4x - 0.4$.

In order to draw a graph of the regression line, you may either enter the function manually (see Section B.3) or use the command

$$\text{LinReg(ax + b) } L_1, L_2, Y_1$$

which saves the equation to Y_1. Here are the keystrokes:

1. Follow the earlier instructions to get "LinReg(ax + b)" on your screen.

2. Press 2nd [L_1] , 2nd [L_2] , .

3. Press VARS ▷ **1** ENTER . Your screen should look like the one displayed in Fig. 14.

4. Press ENTER . Your screen should now look like the one displayed in Fig. 13. In addition, if you press Y= , your screen will look like the one displayed in Fig. 15.

> **Graphing Calculator**
>
> You can perform regression on columns other than L_1 and L_2 by listing the two columns separated by commas after the "LinReg(ax + b)" command on the Home screen. For example, "LinReg(ax + b) L_4, L_6" will perform a linear regression on columns 4 and 6 of the STAT list editor.

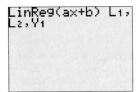

Figure 14 About to save the equation to Y_1

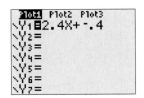

Figure 15 The equation is saved in Y_1

(**TI-82: To save the regression equation to** Y_1**, do these steps:**

1. **Press** Y= **.**

2. **Move the cursor to where you would like to enter the formula for the regression line.**

3. **Press** CLEAR **.**

4. **Press** VARS **.**

5. **To select the Statistics menu, press 5.**

6. **To select the EQ menu, press** ▷ **twice.**

7. **Select "RegEQ" by pressing 7. Your screen should look like the one displayed in Fig. 15.**)

B.17 PLOTTING POINTS FOR TWO SCATTERGRAMS

Suppose that you want to draw two scattergrams on the same calculator screen using different markings for the points.

Follow the instructions in Section B.8 in order to create a scattergram for the data values in Table 4.

These data are stored in columns L_1 and L_2. These points are plotted by the plotter called "Plot1."

You will now create a scattergram for the data values in Table 5.

Table 4 Creating the First of Two Scattergrams	
x	y
2	4
3	7
4	10
5	11

Table 5 Creating the Second of Two Scattergrams	
x	y
2	11
2	9
3	6
5	4

You will store these data in columns L_3 and L_4. These points will be plotted by the plotter called "Plot2." To do this, perform the following instructions.

Graphing Calculator

Make sure you press CLEAR rather than DEL. If you press DEL, the column will vanish. If you ever do this by mistake, press STAT 5 ENTER to get back the missing column.

1. To create a table, press STAT 1.
2. To clear list L_3, press ▷ and/or ◁ to move the cursor to column L_3. Then press △ CLEAR ENTER.
3. To clear list L_4, press ▷ to move the cursor to column L_4. Then press △ CLEAR ENTER.
4. To return to the first entry position of list L_3, press ◁.
5. Press **2** ENTER **2** ENTER **3** ENTER **5** ENTER to enter the elements of L_3.
6. Press ▷ to move to the first entry position of list L_4.
7. Press **11** ENTER **9** ENTER **6** ENTER **4** ENTER to enter the elements of L_4.
8. Press 2nd [STAT PLOT].
9. Press **2** to select "Plot2."
10. Press ENTER to turn Plot2 on.
11. Press ▽ ENTER to choose the scattergram mode.
12. Press ▽ so that the cursor is at "Xlist." Then press 2nd [L_3].
   (**TI-82: Press** ▽ **twice so that the cursor is on one of six choices for the "Xlist." Then press** ▷ **or** ◁ **as many times as necessary to select** L_3**. Then press** ENTER **.**)
13. Press ENTER so that the cursor is at "Ylist." Then press 2nd [L_4].
   (**TI-82: Press** ▽ **once so that the cursor is on one of six choices for the "Ylist." Then press** ▷ **or** ◁ **as many times as necessary to select** L_4**. Then press** ENTER **.**)
14. Press ▽ once so that the cursor is on one of the three choices for the "Mark." Then press ▷ and/or ◁ to select a different marking than what you used for the first scattergram. Then press ENTER.

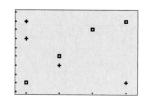

Figure 16 Creating two scattergrams

15. Press ZOOM **9** to obtain the two scattergrams with different markings. Your calculator screen should look like the one displayed in Fig. 16.

B.18 FINDING THE INTERSECTION POINT(S) OF TWO CURVES

Suppose that you would like to find the intersection point of the lines $y = 2x + 1$ and $y = -2x + 7$.

1. Enter the function $y = 2x + 1$ for Y_1 and enter the function $y = -2x + 7$ for Y_2. See Section B.3 to enter these functions.
2. By ZOOMing IN or OUT or by changing the WINDOW settings, draw a graph of both curves so that you can see an intersection point. For our example, press ZOOM **6**.

3. Press $\boxed{\text{2nd}}$ [CALC]. You will see a menu of choices as in Fig. 17.

4. Press **5** to select "intersect."

5. You will now see a flashing cursor on your first curve. If there is more than one intersection point on your display screen, move the cursor by pressing $\boxed{\triangleright}$ or $\boxed{\triangleleft}$ so that it is closer to the intersection point you want to find. Your screen should look something like the one displayed in Fig. 18.

6. Press $\boxed{\text{ENTER}}$; the cursor will now be on the second curve. Press $\boxed{\text{ENTER}}$ again; your screen will now display "Guess?" Press $\boxed{\text{ENTER}}$ once more. Your calculator screen should look like the one displayed in Fig. 19. The intersection point is $(1.5, 4)$.

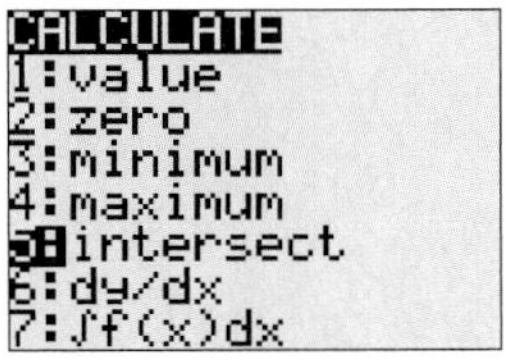

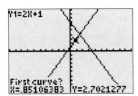
Figure 18 Put cursor near the intersection point

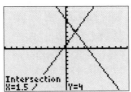
Figure 19 Location of intersection point

Figure 17 Menu of choices

B.19 FINDING THE MINIMUM OR MAXIMUM OF A CURVE

Suppose that you want to the find the minimum point of the function $y = x^2 - 3x + 1$ (which is the vertex of the quadratic function).

1. See Section B.3 to enter the function $y = x^2 - 3x + 1$.

2. See Section B.6 to use ZDecimal to draw a graph of the function.

3. Press $\boxed{\text{2nd}}$ [CALC].

4. Press **3** to choose the "minimum" option.

5. Move the flashing cursor to the left of the minimum point and press $\boxed{\text{ENTER}}$. Or, type a number between Xmin and the x-coordinate of the vertex and then press $\boxed{\text{ENTER}}$. **(TI-82: You must use the cursor.)**

6. Move the flashing cursor to the right of the minimum point and press $\boxed{\text{ENTER}}$. Or, type a number between the x-coordinate of the vertex and Xmax and then press $\boxed{\text{ENTER}}$. **(TI-82: You must use the cursor.)**

7. Press $\boxed{\text{ENTER}}$.

8. Note that your calculator displays coordinates of the minimum point of about $(1.50, -1.25)$ (see Fig. 20).

You can find the maximum point of a function in a similar fashion, but use the "maximum" choice, rather than "minimum."

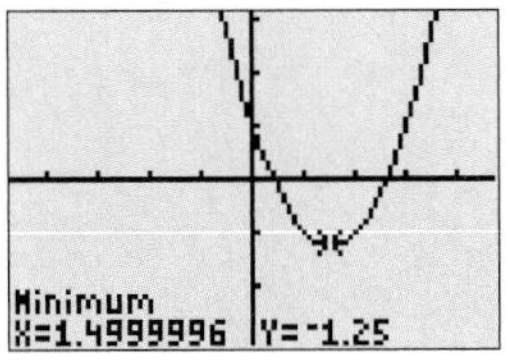

Graphing Calculator

You can use any window settings to find a maximum or minimum point. It's usually best for the point to be visible on the screen.

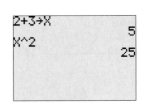
Figure 20 Finding the minimum point of $y = x^2 - 3x + 1$

B.20 STORING A VALUE

It is possible to store a number as x and then perform operations with x. For example, to find $(2 + 3)^2$:

1. Press $2 \boxed{+} 3 \boxed{\text{STO} \triangleright} \boxed{\text{X,T,}\Theta\text{,}n} \boxed{\text{ENTER}}$.

2. Press $\boxed{\text{X,T,}\Theta\text{,}n} \boxed{\wedge} 2 \boxed{\text{ENTER}}$. Your screen should now look like the one displayed in Fig. 21.

Figure 21 Computing $(2 + 3)^2 = 25$

B.21 GRAPHING SEQUENCES

Suppose that you want to graph the sequence $a_n = 2n + 1$ with domain $1, 2, 3, \ldots, 9$.

1. Press $\boxed{\text{MODE}}$.

2. Move the cursor to "Seq" and press ENTER. Also move the cursor to "Dot" and press ENTER. Your calculator screen should look like the one displayed Fig. 22.

3. Press Y=.

4. Set n Min equal to 1. Then set $u(n)$ equal to $2n + 1$ by pressing **2** X,T,Θ,n + **1**. Your screen should look like the one displayed in Fig. 23.
   (**TI-82: Set Un equal to $2n + 1$ by pressing 2** 2nd **9** + **1**.)

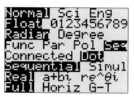

Figure 22 Getting in Seq and Dot mode

Figure 23 Enter $u(n) = 2n + 1$

5. Press WINDOW and set nMin equal to 1, nMax equal to 9, PlotStart equal to 1, and PlotStep equal to 1. (See Fig. 24.)
   (**TI-82: Press WINDOW and set UnStart equal to 3, VnStart equal to 1, nStart equal to 1, nMin equal to 1, and nMax equal to 9.**)

6. Then press ZOOM **0** to graph the sequence, followed by ZOOM **3** ENTER to ZOOM OUT so that you can see the axes. (See Fig. 25.)
   (**TI-82: Press** ZOOM **8.**)

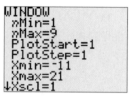

Figure 24 WINDOW setup

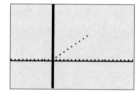

Figure 25 Graph of $a_n = 2n + 1$

When done graphing sequences, highlight "Func" and "Connected" in the MODE menu.

B.22 TURNING A FUNCTION "ON" OR "OFF"

A function will only be graphed if its equality sign is highlighted (the function is then "on"). Up to ten functions can be graphed at one time. To change the on/off status of a function:

1. Press Y=.

2. Move the cursor to the function whose status you want to change.

3. Use ◁ to place the cursor over the "=" sign of the function.

4. Press ENTER to change the status.

B.23 FINDING COORDINATES OF POINTS

1. Press GRAPH to get into graphing mode.

2. Press ▷ to get a cursor to appear on your calculator's screen. [If you cannot see it, it is probably on one or both of the axes. If it is on one of the axes, you should still be able to see a small flashing dot.] Notice that the coordinates of the point where the cursor is currently positioned are at the bottom of the screen.

3. Use ◁, ▷, △, and ▽ to move the cursor left, right, up, and down.

B.24 GRAPHING FUNCTIONS WITH AXES "TURNED OFF"

Suppose that you would like to draw a graph of $y = 0$. The axes will "cover up" the graph of $y = 0$. The following instructions are for graphing without the axes appearing on your screen.

1. Enter the function $y = 0$ for Y_1. See Section B.3 to enter the function $y = 0$.
2. Press 2nd [FORMAT]. You are now at the FORMAT menu.
   (TI-82: Press WINDOW ▷.)
3. Press ▽ three times, then press ▷, then press ENTER. "AxesOff" should now be highlighted.
   (TI-82: Press ▽ four times, then press ▷, then press ENTER. "AxesOff" should now be highlighted.)
4. Using ZDecimal, your calculator screen should look like the one displayed in Fig. 26.

Figure 26 Graph of $y = 0$ with axes "turned off"

You can turn the axes back on by highlighting "AxesOn" in the FORMAT menu.

B.25 ENTERING A FUNCTION USING Y_n REFERENCES

Suppose that you would like to enter the function $y = \dfrac{x+1}{x-3} \div \dfrac{x-2}{x+5}$. One way to enter lengthy formulas is to use Y_n references.

1. Follow the instructions in Section B.3 to enter $Y_1 = \dfrac{x+1}{x-3}$ and $Y_2 = \dfrac{x-2}{x+5}$.
2. Follow the instructions in Section B.22 to turn both functions "off."
3. Move the flashing cursor to the right of "$Y_3 =$."
4. Press VARS, ▷, ENTER.
5. Move the cursor to "$1:Y_1$" and press ENTER. "Y_1" should now appear to the right of "$Y_3 =$" in the Y= window.
6. Press ÷.
7. Press VARS, ▷, ENTER.
8. Move the cursor to "$2:Y_2$" and press ENTER. "Y_1/Y_2" should now appear to the right of "$Y_3 =$" in the Y= window.
9. You can now draw a graph or make a table for the function Y_3.

B.26 RESPONDING TO ERROR MESSAGES

There are several types of errors that a TI graphing calculator can catch you making, and when this happens, it will display an error message. When you make an error, this does not hurt your calculator. In fact, you can't hurt your calculator no matter what order you press its keys. So the more you experiment with your calculator, the better.

1. The "Syntax" error is displayed in Fig. 27.
   This means that you have misplaced one or more parentheses, operations, commas, or functions. Your calculator will find this type of error if you choose "Goto" by pressing ▽ followed by ENTER. Your error will be highlighted by a flashing black rectangle.
   (TI-82: Choose "Goto" by pressing ENTER.)

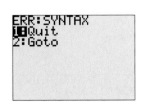

Figure 27 Syntax error message

The most common error is made by pressing (-) when you should press −, or vice versa. Here's the meaning of each symbol:

- Press the (-) key when you want take the opposite of a number or when working with negative numbers. So to compute $-5(-2)$, press (-) 5 ((-) 2).

- Press the $\boxed{-}$ key when you want to subtract two numbers. So to compute $5 - 2$, press $5 \boxed{-} 2$.

2. The "Invalid" error is displayed in Fig. 28. This means that you have tried to enter an inappropriate number, expression, or command. The most common error is to try to enter a number that is not between Xmin and Xmax, inclusive, when using commands such as $\boxed{\text{TRACE}}$, "minimum," or "maximum."

3. The "Invalid dimension" error is displayed in Fig. 29.

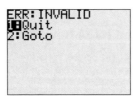

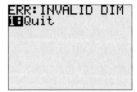

Figure 28 Invalid error message

Figure 29 Invalid dimension error message

This means that you have your plotter turned on (See Fig. 30), but you do not have any data points entered in the STAT list editor (See Fig. 31). In this case, first press $\boxed{\text{ENTER}}$ to exit the error message display, and then either turn the plotter off or enter some data in the STAT list editor.
(TI-82: The error message reads "ERR: STAT PLOT.")

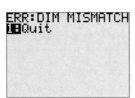

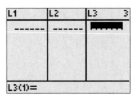

Figure 30 Plotter is on

Figure 31 STAT list editor's columns are empty

4. The "Dimension mismatch" error is displayed in Fig. 32. This means one of two things:

- One possibility is that for the two columns you are trying to plot with, there are more numbers in one column than the other in the STAT list editor (see Fig. 33). In this case, first press $\boxed{\text{ENTER}}$ to exit the error message display, and then correct your mistake so that the two columns have the same length.

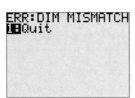

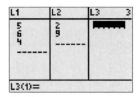

Figure 32 Dimension mismatch error message

Figure 33 Columns of unequal length in STAT list editor

- A second possibility is that for the two columns you are trying to plot with, there are more numbers in one column than the other in the STAT list editor, but you didn't notice the difference in length because you deleted one or both of the columns by mistake. The missing column(s) can be found by pressing $\boxed{\text{STAT}}$ 5 $\boxed{\text{ENTER}}$.
(TI-82: It's impossible to delete the STAT list editor's columns.)

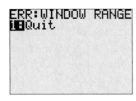

Figure 34 Window range error message

5. The "Window range" error is displayed in Fig. 34. This means one of two things:

- One possibility is that you made an error in setting up your WINDOW. This usually means that you entered a larger number for Xmin than for Xmax or that you entered a larger number for Ymin than Ymax. In this case, first press $\boxed{\text{ENTER}}$ to exit the error message display, and then correct your WINDOW settings accordingly (see Section B.7).

- Another possibility is that you pressed $\boxed{\text{ZOOM}}$ **9** when there was only one data pair entered in the STAT list editor (The command ZoomStat only works if you have two or more pairs of data in the STAT list editor.) In this case, first press $\boxed{\text{ENTER}}$ to exit the error message display, and then either add more points to the STAT list editor or avoid pressing $\boxed{\text{ZOOM}}$ **9** and set up your WINDOW settings manually (see Section B.7).

6. The "No sign change" error is displayed in Fig. 35. This means one of two things:

- One possibility is that you are trying to locate a point that does not appear on your screen. For example, you may be trying to find an intersection point of two curves that does not appear on your screen. Or you may be trying to find a zero of a function that does not appear on your screen. In this case, press $\boxed{\text{ENTER}}$ and change your WINDOW settings so that the point you are trying to locate is on your screen.

- Another possibility is that you are trying to locate a point that does not exist. For example, you may be trying to find an intersection point for two parallel lines. Or you may be trying to find a zero of a function that does not have any. In this case, press $\boxed{\text{ENTER}}$ and stop looking for the point that doesn't exist!

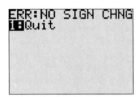

Figure 35 No sign change error message

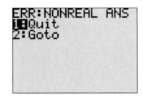

Figure 36 Nonreal answer error message

7. The "Nonreal answer" error is displayed in Fig. 36. This means that your computation did not yield a real number. For example, $\sqrt{-4}$ is not a real number. Your calculator will locate this computation if you choose "Goto" by pressing $\boxed{\triangledown}$ followed by $\boxed{\text{ENTER}}$.

(TI-82: The error message reads "ERR: DOMAIN." Choose "Goto" by pressing $\boxed{\text{ENTER}}$.)

8. The "Domain" error is displayed in Fig. 37. This means that you have asked your calculator to find a power regression equation for some data which has zero or a negative number as one of its values for the independent variable.

Figure 37 Domain error message

Figure 38 Divide by zero error message

9. The "Divide by zero" error is displayed in Fig. 38. This means that you asked your calculator to perform a calculation that involves a division by zero. For example, $3 \div (5 - 5)$ will yield an error message.

Your calculator will locate this computation if you choose "Goto" by pressing $\boxed{\triangledown}$ followed by $\boxed{\text{ENTER}}$.

(TI-82: Choose "Goto" by pressing $\boxed{\text{ENTER}}$.)

Comparing TI-82/83/85/86 Calculator Commands

Although this text has referred to commands of the TI-82/83 graphing calculators, most of these commands also exist on the TI-85/86 graphing calculators. In fact, in most cases, the names of the commands are similar (see Table 1).

Table 1 Comparing Command Names

TI-82/83 Command	TI-85/86 Command	Description of Command
ZDecimal	ZDECM	Allows tracing by 0.1 increments in x
ZInteger	ZINT	Allows tracing by unit increments in x
ZoomFit*	ZFIT	Adjusts the dimensions of the y-axis so that as much of a curve as possible is displayed
ZoomStat	ZDATA†	Displays all data points
ZSquare	ZSQR	Displays a graph with equal-size pixels on the x-axis and y-axis
ZStandard	ZSTD	Displays a graph for x and y both between -10 and 10
ZOOM IN	ZIN	Magnifies the graph around the cursor
ZOOM OUT	ZOUT	Displays more of the graph around the cursor
TRACE	TRACE	Gives coordinates of points on a curve
intersect	ISECT	Finds intersection points of two curves
LinReg(ax + b)	LinR	Finds a linear regression curve
QuadReg	P2Reg	Finds a quadratic regression curve
ExpReg	ExpR	Finds an exponential regression curve
PwrReg	PwrR	Finds a power regression curve

*ZoomFit is not available on the TI-82

†ZDATA is not available on the TI-85

Recall that instructions for using a TI-82/83/85/86 graphing calculator can be found at the site www.prenhall.com/divisions/esm/app/calc_v2/.

Answers To Odd-Numbered Exercises

Answers to most discussion exercises and to exercises where answers may vary have been omitted.

Chapter 1

Exercise Set 1.1

1. a) (d) **b)** (c) **c)** (a) **d)** (b)

3.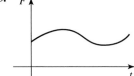

5.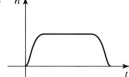

7.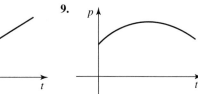

9.

11.

13.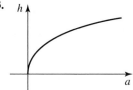

15.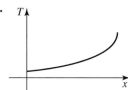

17.

19.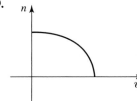

21.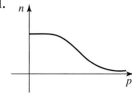

23.

Exercise Set 1.2

1.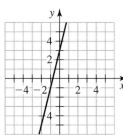

3.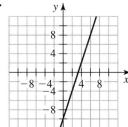

5.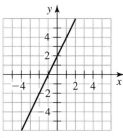

7.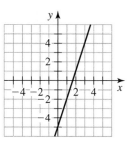

9.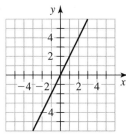

11.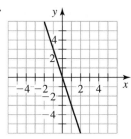

13.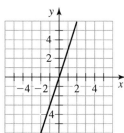

15.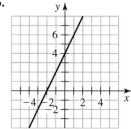

17.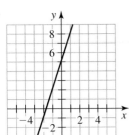

19.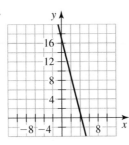

21.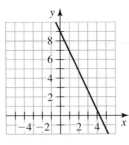

23.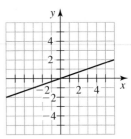

25.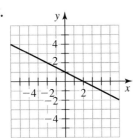

27. a)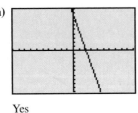

Yes

b)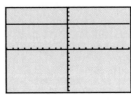

Yes

29. a) i.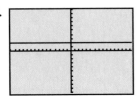

ii.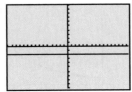

iii.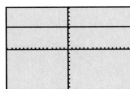

b) The graph is a horizontal line.

31.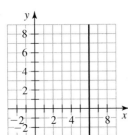

33.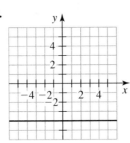

35. $y = 0$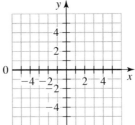

37.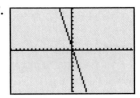

39.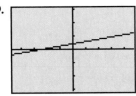

41. x-intercept: $(-5, 0)$; y-intercept: $(0, 10)$ **51.** 3

43. x-intercept: $(6, 0)$; y-intercept: $(0, 4)$ **53.** -2

45. x-intercept: $(0, 0)$; y-intercept: $(0, 0)$ **55.** 6

47. No x-intercept; y-intercept: $(0, 3)$ **57.** -1

59. a)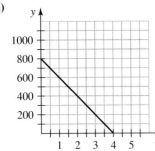

b) $(0, 800)$; The balloon was at height 800 feet when hot air was first released from the balloon.

c) $(4, 0)$; The balloon reached the ground after 4 minutes of hot air being released from the balloon.

61. B, C, F

63. $(1, 3)$

65. $b = -9$

Exercise Set 1.3 **1.** 2; increasing **3.** -2; decreasing **5.** 4; increasing **7.** $\frac{3}{5}$; increasing **9.** 1; increasing **11.** 1; increasing **13.** -1.04; decreasing **15.** 1.28; increasing **17.** 0; horizontal **19.** Undefined slope; vertical **21.** (a) Positive; (b) Negative; (c) Zero; (d) Undefined slope **23.** $-\frac{2}{5}$ **25.** Parallel **27.** Neither **29.** Perpendicular **31.** Neither **33.** Perpendicular

35. Answers may vary.

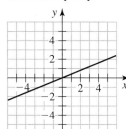

37. Answers may vary.

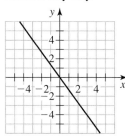

39. Answers may vary.

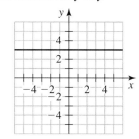

41. The line with slope 3 is steeper.

43. Road A is steeper.

45. Ski run A is steeper.

47. Answers may vary.

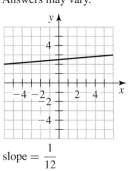

49. Answers may vary.

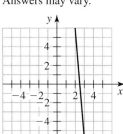

51. Answers may vary.

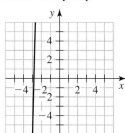

53. Answers may vary.

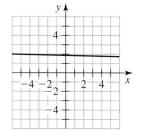

slope $= \dfrac{1}{12}$

slope $= -12$

55. (1, 4), (4, 13), (5, 16); Answers may vary.

57. a) i.

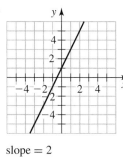

ii.

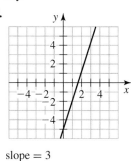

iii.

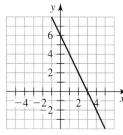

slope $= 2$

slope $= 3$

slope $= -2$

b) The slope is equal to the coefficient of x.

59. a)

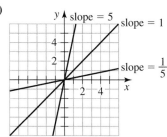

b)

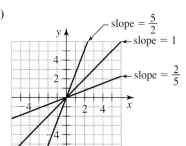

c)

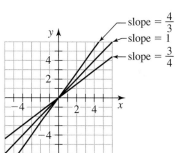

e)

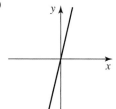

61. Yes; Yes

Exercise Set 1.4

1. Slope: 6; y-intercept: (0, 1)   **3.** Slope: −2; y-intercept: (0, 7)   **5.** Slope: $\frac{5}{4}$; y-intercept: (0, −2)   **7.** Slope: $-\frac{3}{7}$; y-intercept: (0, 2)

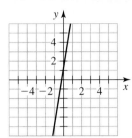
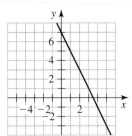
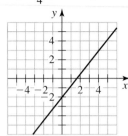
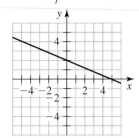

9. Slope: $-\frac{5}{3}$; y-intercept: (0, −1)   **11.** Slope: −1; y-intercept: (0, 5)   **13.** Slope: $\frac{7}{2}$; y-intercept: (0, 5)

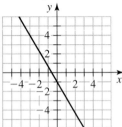
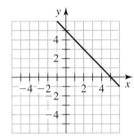
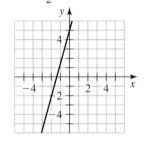

15. Slope: $\frac{1}{2}$; y-intercept: $\left(0, -\frac{3}{2}\right)$ **17.** Slope: $\frac{2}{3}$; y-intercept: (0, −1)   **19.** Slope: $\frac{7}{2}$; y-intercept: (0, −2)

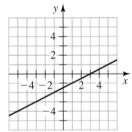
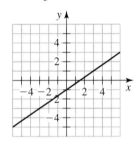
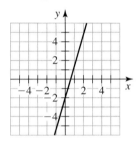

21. Slope: 4; y-intercept: (0, 0)   **23.** Slope: −1.5; y-intercept: (0, 3)   **25.** Slope: 1; y-intercept: (0, 0)

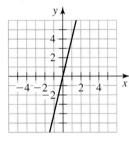
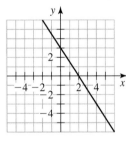
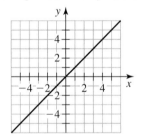

27. Slope: 0; y-intercept: (0, 4)   **29.** Slope: 0; y-intercept: (0, −2)

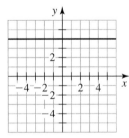
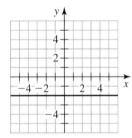

31. There is a line that comes close to every point of Set 1. There is a line with slope −0.3 that passes through every point of Set 2. No line comes close to every point of Set 3. There is a line with slope −10 that passes through every point of Set 4.

33.

x	y		x	y		x	y		x	y
1	12		23	69		1	−7		30	5.0
2	15		24	61		2	−2		31	4.4
3	18		25	53		3	3		32	3.8
4	21		26	45		4	8		33	3.2
5	24		27	37		5	13		34	2.6
6	27		28	29		6	18		35	2.0

35. (a) $m < 0$ and $b > 0$; (b) $m > 0$ and $b < 0$; (c) $m = 0$ and $b < 0$; (d) $m < 0$ and $b = 0$ **37.** Parallel **39.** Neither **41.** Parallel

43. Perpendicular **45.** Perpendicular **47.** Perpendicular

49. a) The values of y in the second column are 18, 15, 12, 9, 6, 3, 0.

  b) The gas is decreasing at a rate of 3 gallons per hour, which is the slope of $y = -3x + 18$. **c)** 20 miles per gallon

51. a) The values of y in the second column are 26, 28, 30, 32, 34.

  b) The person's salary is increasing at a rate of $2000 per year, which corresponds to the slope of $y = 2x + 26$ since y is in thousands of dollars.

53. a)
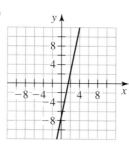
b)
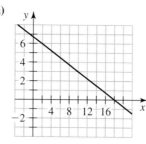
c)
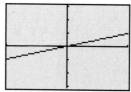

  d) The graph in part b appears steeper than the graph in part a. The graph in part c appears less steep than the graph in part a.

  e) No; No; The line will always appear to be increasing and it will always lie in quadrants one and three, but it may appear to have any steepness within these constraints.

55. $y = 2x - 3$ **57. a)**

  b) $y = 5x - 7$

59. a)
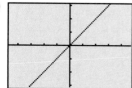

  b) $y = -\dfrac{3}{8}x + \dfrac{27}{4}$

61. a) The slope of each line is undefined.

  b) The slope is undefined.

Exercise Set 1.5

1. $y = 3x - 13$ **3.** $y = -2x - 3$ **5.** $y = 1.6x + 0.44$ **7.** $y = \dfrac{3}{5}x - 5$ or $y = 0.6x - 5$

9. $y = -\dfrac{1}{6}x - \dfrac{8}{3}$ or $y = -0.17x - 2.67$ **11.** $y = -\dfrac{5}{2}x - \dfrac{23}{2}$ or $y = -2.5x - 11.5$ **13.** $y = 2$ **15.** $x = 3$ **17.** $y = x + 1$

19. $y = -2x + 2$ **21.** $y = -2x - 22$ **23.** $y = x$ **25.** $y = -0.74x + 7.67$ **27.** $y = 1.09x - 2.5$ **29.** $y = \dfrac{4}{5}x - \dfrac{3}{5}$ or $y = 0.8x - 0.6$

31. $y = \dfrac{5}{2}x + \dfrac{11}{2}$ or $y = 2.5x + 5.5$ **33.** $y = -\dfrac{7}{6}x - \dfrac{8}{3}$ or $y = -1.17x - 2.67$ **35.** $y = 5$ **37.** $x = -3$ **39.** $y = 3x - 7$

41. $y = -2x + 2$ **43.** $y = \dfrac{1}{2}x - 1$ or $y = 0.5x - 1$ **45.** $y = \dfrac{3}{4}x + \dfrac{7}{4}$ or $y = 0.75x + 1.75$ **47.** $y = \dfrac{1}{6}x - \dfrac{3}{2}$ or $y = 0.17x - 1.5$

49. $y = 3$ **51.** $x = -5$ **53.** $y = -\dfrac{1}{2}x + \dfrac{19}{2}$ or $y = -0.5x + 9.5$ **55.** $y = \dfrac{1}{3}x + \dfrac{22}{3}$ **57.** $y = \dfrac{5}{2}x + 2$ or $y = 2.5x + 2$

59. $y = -\dfrac{5}{4}x + \dfrac{31}{2}$ or $y = -1.25x + 15.5$ **61.** $y = -\dfrac{1}{2}x - \dfrac{5}{2}$ or $y = -0.5x - 2.5$ **63.** $y = 3$ **65.** $x = 2$

67. $y = -2x + 19$ **69.** $y = \dfrac{1}{3}x - \dfrac{2}{3}$ **71.** $y = -\dfrac{3}{2}x + \dfrac{13}{2}$ or $y = -1.5x + 6.5$ **73. a)** It is possible. **b)** It is possible.

c) It is not possible. **d)** It is possible. **75.** $y = -2x + 7$

Exercise Set 1.6

1. Relation 2, Relation 3 **3.** No **5.** Yes **7.** The graphs a and c are graphs of functions. The graphs b and d are not graphs of functions. **9.** The relation is a function. **11.** The relation is a function. **13.** The relation is a function. **15.** The relation is not a function. **17.** The relation is a function. **19.** Yes **21.** No **27.** Yes **29.** No **31.** Yes **33.** No

Chapter 1 Review

1.
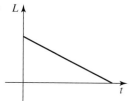

2.
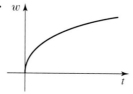

3.
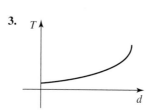

4.
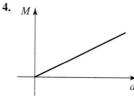

5.

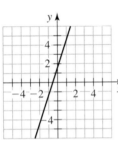

7. -4 **8.** -0.67 **9.** -4 **10.** -1 **11.** -1 **12.** $-\dfrac{3}{5}$

13. $\dfrac{1}{2}$ **14.** 3.81 **15.** 0 **16.** Undefined slope

17.

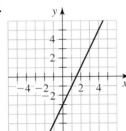

18.

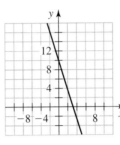

19.

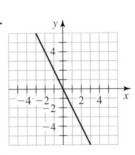

20.

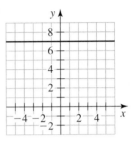

21.

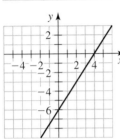

22.

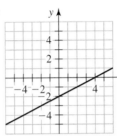

23.

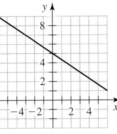

24.

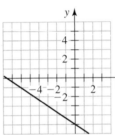

25.

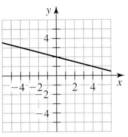

26.

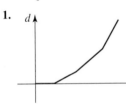

27. Neither **28.** Perpendicular **29.** $y = -4x - 5$ **30.** $y = -\dfrac{2}{3}x - \dfrac{2}{3}$ or $y = -0.67x - 0.67$ **31.** $y = -3x + 17$

32. $y = 2x - 5$ **33.** $y = -0.72x + 5.72$ **34.** $y = \dfrac{8}{5}x + \dfrac{14}{5}$ or $y = 1.6x + 2.8$ **35.** $y = -\dfrac{4}{3}x + \dfrac{2}{3}$ or $y = -1.33x + 0.67$

36. $x = 3$ **37.** $y = -2x$; $y = 2x + 2$; $y = -3$ **38.** The values of y in the second column are $20, 16, 12, 8, 4, 0$. **39.** $y = 0.5x + 3$

40. x-intercept: $\left(\dfrac{17}{3}, 0\right)$; y-intercept: $\left(0, -\dfrac{17}{5}\right)$ **41. a)** B, C, F **b)** C, E **c)** C **d)** A, D **42.** $y = -\dfrac{3}{2}x + \dfrac{7}{2}$ or $y = -1.5x + 3.5$

43. $y = 3x + 11$ **44.** $y = 0$ **45.** Relation 1, Relation 3 **47.** Yes **48.** Yes **49.** No **50.** Yes **51.** Yes **52.** No

Chapter 1 Test

1.

3. Equation for line 1: $y = -\dfrac{5}{2}x + 10$, equation for line 2: $y = \dfrac{2}{7}x + 2$, equation for line 3: $x = -3$

4. a) $k > m$ **b)** $b > c$ **5.** Ski run A is steeper.

7.

x	y
4	25
5	29
6	33
7	37
8	41
9	45

8.

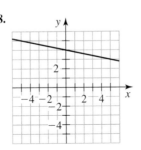

9.

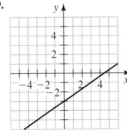

10. $-\dfrac{5}{4}$

11. $(-1, 10), (8, 4), (11, 2)$; Answers may vary.

12. $y = -3x - 1$

13. $y = -\dfrac{12}{5}x - \dfrac{1}{5}$ or $y = -2.4x - 0.2$

14. No. $y = -2x + 1$

15. $y = -\dfrac{5}{3}x + \dfrac{17}{3}$ or $y = -1.67x + 5.67$

16. x-intercept: $\left(\dfrac{3}{2}, 0\right)$; y-intercept: $(0, -3)$

18.

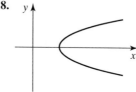

The graph of a relation that is not a
function is sketched. Answers may vary.

19. No

20. Yes

Chapter 2

Exercise Set 2.1

1. a–b)

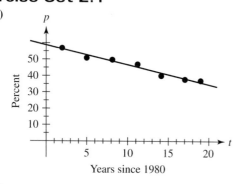

Years since 1980

c) 44%   **d)** 1985

3. a–b)

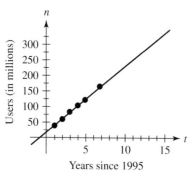

Years since 1980

c) 4000 collisions; interpolation

d) 2009; extrapolation

e) $(0, 8.5)$; In 1980 there were 8500 collisions.

f) $(32.5, 0)$; There will be no collisions in 2013.
  Model breakdown has likely occurred.

5. a–b)

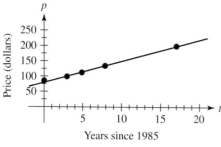

Years since 1995

c) $(0, 18.8)$; In 1995 there were 18.8 million users.

d) 21 million users per year

e) 2008; Model breakdown has likely occurred.

7. a–b)

Years since 1985

c) $177.69; $27.69   **d)** $191.50; $31.50

f)

Years since 1985

9. a–b)

Months since January 1, 1998

c)

Months since January 1, 1998

The stock will be worth $152 on January 1, 2001. **d)** $109

11. a) (7.5, 0); The difference in percent of sales was 0% in 1998, according to the model.

 b) (0, 5.1); The difference in percent of sales was 5.1% in 1990, according to the model.

Exercise Set 2.2

1. Increase b **3.** $y = 2.5x - 2.7$; Answers may vary. **5.** Student B

7. a)

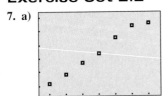

b) $p = 0.81t - 45.86$; Answers may vary. **c)**

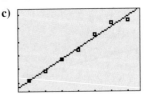

9. a)

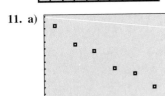

b) No **c)** $T = 5.85n + 81.18$; Answers may vary.

11. a)

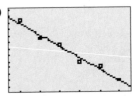

b) $p = -3.63t + 117.68$; Answers may vary. **c)**

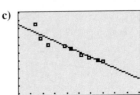

13. a)

b) $r = -0.27t + 70.45$; Answers may vary. **c)**

15. a)

b) Yes; 42.62 seconds; 2003 **c)** Yes; After 2003

17. a)

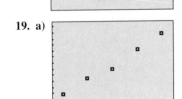

b) $p = -0.28t + 3.32$; Answers may vary. **c)**

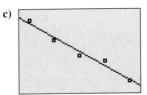

d) $p = -0.28t + 3.32$

19. a)

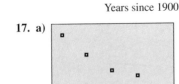

b) $p = 2.48x - 23.64$; Answers may vary. **c)**

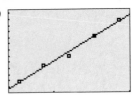

Exercise Set 2.3

1. 5   **3.** $\frac{19}{3}$ **5.** $-\frac{79}{2}$ **7.** -2 **9.** 11.47   **11.** -1 **13.** -25 **15.** 8   **17.** 9.8   **19.** -3 **21.** -8

23. $-4a - 1$ **25.** $-4a - 5$ **27.** $-4a - 4h - 1$ **29.** $\frac{1}{3}$ **31.** 3.97   **33.** $\frac{13}{6}$ **35.** $\frac{a+5}{2}$ **37.** 4   **39.** 1   **41.** 1.2   **43.** 6   **45.** -3

47. 4.5   **49.** 4   **51.** 1 and 3   **53.** x-intercept: $\left(\frac{8}{5}, 0\right)$; y-intercept: (0, −8)   **55.** x-intercept: (0, 0); y-intercept: (0, 0)

57. No x-intercepts; y-intercept: (0, 5)   **59.** x-intercept: (−2, 0); y-intercept: (0, −4.2)   **61.** x-intercept: (6, 0); y-intercept: (0, −3)

63. 2   **65.** $\frac{5}{4}$ **67.** 4.35   **69.** -17 **71.** $\frac{11}{4}$ **73.** $-\frac{3}{2}$ **75.** Empty set solution   **77.** Both students solved the equation correctly.

79. a) $f(t) = 0.81t - 45.86$ **b)** 41.62; In 2008, 41.6% of births will be out of wedlock.   **c)** 109.70; In 2010, 43% of births will be out of wedlock.   **d)** 2080; Model breakdown has likely occurred.   **e)** 32.7; 0.3%   **81. a)** $f(n) = 5.85n + 81.18$ **b)** The T-intercept: (0, 81.18). It took the student 81 seconds to complete the task when sober.   **c)** The model is an increasing function. The amount of time needed to complete the task increased as more screwdrivers were consumed.   **d)** 81.18, 87.03, 92.88, 98.73, 104.58, 110.43, 116.28, 122.13, 127.98, 133.83, 139.68; The estimate $f(4)$ has the least error and the estimate $f(10)$ has the most error.

83. a)

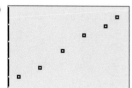

b) $f(t) = 24.07t + 423.92$ **c)** 689,000,000; 67,000,000   **d)** $8,375,000,000

85. a) $f(c) = 1.8c + 32$ **b)** 77°F   **c)** $4.\overline{4}$°C

87. a) $f(t) = 0.29t + 3.36$ **b)** 5970 bald eagle pairs   **c)** 7.71; There will be 7710 bald eagles in 2005.   **d)** 22.90; There will be 10,000 bald eagles in 2013.

89. a) $f(x) = 2.48x - 23.64$ **b)** 50 points   **c)** less than 10 points   **d)** 30 students   **e)** 83 students

91. a) $f(t) = -160t + 640$

b)

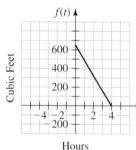

Hours

c) The domain is the set of real numbers t that are between 0 and 4, inclusive. The range is the set of real numbers $f(t)$ that are between 0 and 640, inclusive.

93. Input: 3; Output: 5

Exercise Set 2.4

1. a) 70; The car is moving at 70 miles per hour.   **b)** $f(t) = 70t$

3. a)

t	n
0	$1.7 + 0(2.1)$
1	$1.7 + 1(2.1)$
2	$1.7 + 2(2.1)$
3	$1.7 + 3(2.1)$
4	$1.7 + 4(2.1)$
5	$1.7 + 5(2.1)$
t	$1.7 + t(2.1)$

b) $n = 2.1t + 1.7$ **d)** 2.1; The number of households that pay their bills online is increasing by 2.1 million households per year.   **5. a)** 2.5; AOL charges $2.50 per hour.   **b)** $f(t) = 2.5t + 4.95$ **d)** The flat fee plan is a better deal if a customer is online for more than 7.58 hours per month.   **7. a)** 15.75; The charge is $15.75 per textbook.   **b)** $f(n) = 15.75n + 1224.60$ **c)** 5 textbooks   **9. a)** −0.05; The car uses 0.05 gallon of gas per mile.   **b)** $g(x) = -0.05x + 15.3$ **c)** (306, 0); The car can go 306 miles on a full tank of gasoline.   **d)** The domain is the set of numbers x between 0 and 306, inclusive.   **e)** 286 miles

11. 1.70; The average salaries increase by $1,700 per year.   **13.** −0.27; The record times decrease by 0.27 second per year.   **15.** Yes; 0.99; The charge for a call increases by $0.99 per minute.   **17. a)** $f(t) = -0.34t + 70.00$ **b)** −0.34; The percentage of world population living in rural areas decreases by 0.34% per year.   **c)** 2156; Model breakdown has likely occurred.   **19. a)** $f(t) = 160.86t + 2186.24$ **b)** $g(t) = 926.74t + 9956.10$ **c)** 160.86; 926.74; Public tuition is increasing by $160.86 per year and private tuition is increasing by $926.74 per year.   **d)** Public: $20,648.60; private: $108,403.00   **21. a)**

t	0	1	2	3	4	5
$f(t)$	0	500	1000	1500	1900	2300

b)

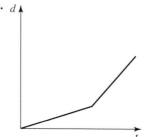

Hours

c) No   **23.**   **25.**

Chapter 2 Review
1. $\dfrac{13}{2}$ 2. $-\dfrac{2}{9}$ 3. -3.09 4. $\dfrac{21}{8}$ 5. -6 6. 9 7. 19 8. -25.45 9. 11 10. $-5a+4$
11. 3 12. $-\dfrac{3}{2}$ 13. $-\dfrac{9}{2}$ 14. -0.595 15. $x=-\dfrac{7}{6}$ 16. $\dfrac{a-3}{2}$ 17. -2 18. $-\dfrac{1}{2}$ 19. 0 20. 1 21. -7 22. -1 23. -5
24. 7 25. 1 26. 4 27. 4 28. 1 29. x-intercept: $\left(\dfrac{3}{7},0\right)$; y-intercept: $(0,3)$ 30. x-intercept: $(-1.84,0)$; y-intercept: $(0,-5.7)$

31. No x-intercept; y-intercept: $(0,4)$ 32. x-intercept: $\left(\dfrac{7}{2},0\right)$; y-intercept: $(0,2)$ 33. Lower y-intercept, increase slope

34. a) $f(t)=-1.8t+13$ b) -1.8; The amount of gasoline in the gas tank decreases at the rate of 1.8 gallons per hour. c) $(0,13)$; The original amount of gasoline in the gas tank is 13 gallons. d) $(7.22,0)$; The gas tank will be empty after 7.22 hours. e) The domain is the set of numbers t between 0 and 7.22, inclusive. The range is the set of numbers A between 0 and 13, inclusive. 35. a) 1309; The income increases by $1309 per year. b) $g(t)=1309t+50{,}500$ c) 66,208; The income will be $66,208 in 2007. d) 14.90; The income will be $70,000 in 2010. 36. a) $f(t)=0.55t-25.13$ b) 34.3% c) $(45.69,0)$; Not one Californian approved of the split in 1945, according to the model. d) 2037; We are not confident in this prediction. 37. a) $g(t)=0.80t+24.44$ b) 104.44; In 2000, 104% of CEOs will have at least a bachelor's degree. Model breakdown has occurred. c) 94.45; In 1994, all CEOs had at least a bachelor's degree. d) Model breakdown definitely occurs before 1869 and after 1994.
e)

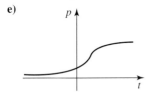

Chapter 2 Test
1. 1 2. 0 3. -1 4. -2.7 5. -6 6. -3 7. 3 8. 4.5 9. 19 10. $\dfrac{5}{4}$
11. x-intercept: $\left(\dfrac{7}{3},0\right)$; y-intercept: $(0,-7)$ 12. x-intercept: $(0,0)$; y-intercept: $(0,0)$ 13. x-intercept: $(-4,0)$; y-intercept: $(0,10)$
14. x-intercept: $(24,0)$; y-intercept: $(0,-8)$ 15. -1 16. 2
17.

18. a) $f(t)=750t+8320$ b) 750; The college's tuition is increasing at the rate of $750 per year. c) $(0,8320)$; The college's tuition was $8320 in 2000. d) 2016
19. a) $f(A)=-0.74A+164.47$ b) 151 beats per minute c) 63 years old d) $(222.26,0)$; A 222 year-old person's target pulse rate is 0 beats per minute. Model breakdown has occurred. e) $(0,164.47)$; The target pulse rate for a person at birth is 164 beats per minute. Model breakdown has occurred.
20. a) $f(t)=0.39t-24.47$ b) 0.39; The percent of the military that is women is increasing at the rate of 0.39 percent per year. c) 14.53; In the year 2000, about 14.5% of the military were women. d) 319.15; In the year 2219, the military will be 100% women. e) Model breakdown definitely occurs when $t<62.74$ and when $t>319.15$.

Chapter 3

Exercise Set 3.1
1. $(1,4)$ 3. $(-6,6)$ 5. $(2,3)$ 7. $(2,-1)$ 9. $(0,5)$ 11. $(4,2)$ 13. $(1,4)$ 15. $(480,460)$
17. All points on the line $y=2x-1$; dependent system 19. Empty set solution; inconsistent system 21. $(3.33,1.33)$ 23. a) Women's time is 37.65 seconds, men's time is 34.78 seconds; The error for the women's time estimate is 0.275 seconds and the error for the men's time estimate is 0.165 seconds. b) The absolute value of the slope of W is more than the absolute value of the slope of M. Women's times are decreasing at a greater rate than men's times. d) 2098; 19.47 seconds 25. a) $W(t)=0.14t+4.52$; $M(t)=0.083t+4.60$ b) 1971 c) 17.8 million students 27. a) $f(t)=-2.45t+72.82$, $g(t)=3.08t+12.58$ b) 2001; 2000; In 2000 the percentage of voters using punch cards or lever machines was 48.3% and the percentage of voters using optical scan or other modern electronic systems was 43.4%. c) 8.3% 29. $(-1.9,-2.8)$ 31. $(3.5,19.5)$ 33. 0 35. 5 37. -1 41. $(2,5)$

Exercise Set 3.2
1. $(7,2)$ 3. $(-5,-3)$ 5. $(3,-4)$ 7. $(4,-3)$ 9. $(0,0)$ 11. $(2,1)$ 13. $(4,-3)$ 15. $(4,-7)$
17. $(1,-2)$ 19. $(-4,3)$ 21. $(-2,3)$ 23. $(3,-2)$ 25. $(2,1)$ 27. $(5,4)$ 29. $(-2,3)$ 31. Empty set solution; inconsistent system 33. $(1,-2)$ 35. The solution set is the infinite number of points on the line $4x-5y=3$; Dependent system 37. $(3.79,1.72)$
39. $(2,1)$ 41. Empty set solution; inconsistent system 43. The solution set is the infinite number of points on the line $y=\dfrac{1}{2}x+3$; Dependent system 45. $(3,2)$ 47. $(2,5)$ 49. No; $(200,403)$ 51. $(4.7,21.8)$
53. $A(0,0)$, $B(0,3)$, $C(3,9)$, $D(6,8)$, $E(7.2,4.4)$, $F(5,0)$ 55. a) $\left(\dfrac{ce-bf}{ae-bd},\dfrac{af-cd}{ae-bd}\right)$, assuming $ae-bd\neq 0$ b) $\left(\dfrac{14}{11},-\dfrac{4}{11}\right)$

Exercise Set 3.3

1. a) 2098, 19.47 seconds   **3.** 1971   **5.** 2001, 2000   **7. a)** 1997   **b)** The competition heated up because the two newspapers had approximately equal circulations.   **c)** 147 thousand bonus copies   **d)** 688 thousand newspapers **e)** Overestimate   **9. a)** $W(t) = -1.22t + 338.47$; $M(t) = -0.36t + 240.44$ **b)** 2014   **c)** $W(t) = -1.01t + 323.57$; $M(t) = -0.36t + 240.73$; 2027

11. a) $T(t) = -1582t + 12{,}939$; $E(t) = -1024t + 9330$ **b)** 2008; \$2707   **c)**

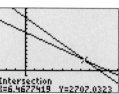

13. a) $A(t) = 670t + 6100$ and $B(t) = 440t + 8500$ **b)** 2010; \$13,091   **c)**

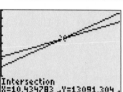

15. a) $J(t) = 72t + 19$; $W(t) = 77t$ **b)** 4 weeks; \$293   **c)**

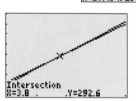

17. 2004

Exercise Set 3.4

1. In Words	**Inequality Notation**	**Graph**	**Interval Notation**
Numbers greater than 3	$x > 3$		$(3, \infty)$
Numbers less than or equal to -4	$x \le -4$		$(-\infty, -4]$
Numbers less than 5	$x < 5$		$(-\infty, 5)$
Numbers greater than or equal to -1	$x \ge -1$		$[-1, \infty)$

3. $x \ge 3$; $[3, \infty)$;   **5.** $x \le -3$; $(-\infty, -3]$;   **7.** $x < 2$; $(-\infty, 2)$;

9. $x < 1$; $(-\infty, 1)$;   **11.** $x \le 8.48$; $(-\infty, 8.48]$;

13. $x < -5$; $(-\infty, -5)$;   **15.** $x \le -2$; $(-\infty, -2]$;

17. $x < \dfrac{5}{3}$; $\left(-\infty, \dfrac{5}{3}\right)$;   **19.** $x > -2.06$; $(-2.06, \infty)$;

21. $x \ge 23$; $[23, \infty)$;   **23.** $x < -6$; $(-\infty, -6)$;

25. $x < 20$; $(-\infty, 20)$;   **27.** $x \ge -\dfrac{22}{9}$; $\left[-\dfrac{22}{9}, \infty\right)$;

29. $x \le -1$; $(-\infty, -1]$;   **31.** For years before 2008   **33. a)** $U(x) = 0.69x + 19.95$; $P(x) = 0.39x + 29.95$ **b)** For miles driven less than $33.\overline{3}$ miles   **35. a)** 5.0 years   **b)** For birth years after 2061   **c) i.** Younger man.   **c) ii.** For birth years 1986 and thereafter   **45.** No.   **47.** No.   **49.** $x < 2.8$

Chapter 3 Review

1. $(4, -5)$ **2.** $(3, 2)$ **3.** $(1, 7)$ **4.** $(-3, 2)$ **5.** The system is dependent. The solution set is the set of points on the line $-2x + 3y = 7$.   **6.** The system is inconsistent. The solution set is the empty set.   **7.** The system is dependent. The solution set is the set of points on the line $-4x - 5y = 3$.   **8.** $(1.29, -2.5)$ **9.** $(0, 0)$ **10.** $(-2, 4)$ **11.** $(2, -3)$ **12.** $(10, 3)$ **13.** $(2, -1)$ **14.** $(5, -1)$ **16.** $(2.6, 21.9), (0.4, 15.1)$ **17.** $a = 19, b = 18$ **18.** $A: (0, 0)$, $B: (0, 4)$, $C: (2, 10)$, $D: (5, 8)$, $E: (6, 4)$, $F: \left(\dfrac{14}{3}, 0\right)$ **19.** The solution set is the empty set.

20. $x \ge -9$; $[-9, \infty)$;   **21.** $x \le 7$; $(-\infty, 7]$;

22. $x < -5$; $(-\infty, -5)$; 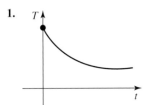 **23.** $x \leq -0.532$; $(-\infty, -0.532]$;

24. $x \leq -\dfrac{7}{16}$; $\left(-\infty, -\dfrac{7}{16}\right]$; **25.** $x > \dfrac{1}{4}$; $\left(\dfrac{1}{4}, \infty\right)$;

26. $x \geq -\dfrac{10}{3}$; $\left[-\dfrac{10}{3}, \infty\right)$; **27.** $x < \dfrac{45}{14}$; $\left(-\infty, \dfrac{45}{14}\right)$; **30.** 1 **31.** -3 **32.** 1

33. -5 **34.** -2 **35.** $x > -2$ **37. a)** $R(x) = 0.22x + 75$; $U(x) = 0.69x + 29.95$ **b)** 95.9 miles **c)** For miles driven over 95.9 miles **38.** 2003 **39. a)** $n(t) = 0.52t + 35.04$; $f(t) = 1.50t + 13.89$ **b)** The slope of n is smaller than the slope of f. North America's petroleum consumption is increasing by 0.52 quadrillion Btus per year; the Far East's petroleum consumption is increasing by 1.50 quadrillion Btus per year. **c)** 2002 **d)** $t > 21.58$; North America's petroleum consumption is less than the Far East's for years after 2002. **40. a)** 1993 **b)** After 1993 **c)** Hollywood

Chapter 3 Test

1. $(1, 2)$ **2.** The system is dependent. The solution is the set of points on the line $2x - 5y = 3$. **3.** The system is inconsistent. Its solution is the empty set. **4.** $(5, -8)$ **5.** $(-2, 4)$ **6.** $x = 5$, $y = 2$; Answers may vary. **7.** $(3, 1)$ **8.** $m = 5$, $b \neq -13$

9. $x \leq -\dfrac{12}{13}$; $\left(-\infty, -\dfrac{12}{13}\right]$; **10.** $x > \dfrac{23}{2}$; $\left(\dfrac{23}{2}, \infty\right)$;

11. $x < -1.03$; $(-\infty, -1.03)$; **12.** $x \geq \dfrac{2}{41}$; $\left[\dfrac{2}{41}, \infty\right)$;

13. 3 **14.** -4 **15.** 2 **16.** $x \leq 2$ **17.** $x > 13$ **18. a)** $-1, 0, 1$; Answers may vary. **b)** 3, 4, 5; Answers may vary.

19. a) College A, by 2320 students **b)** The slopes of A and B are -0.12 and 0.08, respectively. The enrollment at college A is decreasing by 120 students per year and the enrollment at college B is increasing by 80 students per year. **c)** 2012 **d)** Before 2012

20. a) $f(t) = 25.59t + 929.18$; $g(t) = 23.14t + 1685.30$ **b)** 4466 **c)** 2279

Cumulative Review for Chapters 1–3

1. 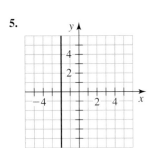 **2.** **3.** 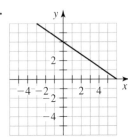 **4.**

5. **6.** $-\dfrac{3}{7}$ **7.** $y = -\dfrac{3}{5}x - \dfrac{9}{5}$ or $y = -0.6x - 1.8$

8. $y = \dfrac{5}{3}x + \dfrac{19}{3}$ or $y = 1.67x + 6.33$

9. $y = -\dfrac{5}{2}x - \dfrac{19}{2}$ or $y = -2.5x - 9.5$

11.

x	$f(x)$	x	$g(x)$	x	$h(x)$	x	$k(x)$
0	97	4	-22	1	23	10	-28
1	84	5	-9	2	14	11	-25
2	71	6	4	3	5	12	-22
3	58	7	17	4	-4	13	-19
4	45	8	30	5	-13	14	-16
5	32	9	43	6	-22	15	-13

12. 32 **13.** 8 **14.** 13 **15.** $\dfrac{22}{3}$ **16.** $-\dfrac{3}{2}$ **17.** $\left(\dfrac{14}{3}, 0\right)$ **18.** $(0, 7)$

19. 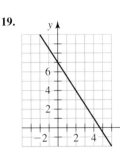 **20.** -3 **21.** -6 **22.** $(0, 1)$ **23.** $-\dfrac{1}{3}$ **24.** $f(x) = \dfrac{2}{3}x + 1$ **25.** -3 **26.** $x \leq -3$ **27.** $-\dfrac{20}{9}$

28. $\dfrac{3}{5}$ **29.** Yes **30.** Yes **32.** $(2, -3)$ **33.** The solution set is the infinite number of points on the line $y = \dfrac{3}{7}x - 2$. **34.** $(3, 2)$ **35.** $x \leq \dfrac{10}{11}$; $\left(-\infty, \dfrac{10}{11}\right]$;

36. $x < 1$; $(-\infty, 1)$; **37.** Decrease m and increase b

38. a) 2 **b)** $x > 2$, or $(2, \infty)$;

c)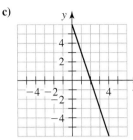

d) $(1, 3)$

40. a) $h(t) = -0.0031t + 0.42$ **b)** 0.09; The grass height will be 0.09 inch in 2005. **c)** 115; The grass height will be $\frac{1}{16}$ inch in 2015. **d)** -0.0031; The grass height decreases by .0031 inch per year. **e)** $(135.48, 0)$; Putting surfaces will have no grass in 2035.

41. a) $f(t) = -12t + 446$ **b)** -12; The number of children hit and killed by motor vehicles declines by 12 children per year. **c)** $(0, 446)$; A total of 446 children were hit and killed by motor vehicles in 1975. **d)** $(37.17, 0)$; No children will be hit and killed by motor vehicles in 2012. **e)** After 2012

42. a) $A(t) = 1.3t + 9.5$; $B(t) = 1.8t + 5.2$

b) 2009; \$20.68 million   **c)** 2009; \$20.68 million

d) $t > 8.6$; Company A's sales will be less than company B's sales after 2009.

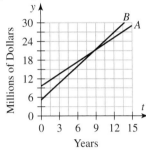

43. a) $W(t) = -0.064t + 27.00$; $M(t) = -0.029t + 22.10$ **b)** $W(107) \approx 20.15$; $M(107) \approx 19.00$ **c)** The absolute value of W's slope is greater than the absolute value of M's slope. Women's record times are decreasing at a greater rate than men's record times.
e) 2040 **f)** $t < 140$; Women's record times are greater than men's record times for years before 2040. **g)** t-intercept for W: $(421.88, 0)$; t-intercept for M: $(762.07, 0)$; According to the models, the women's record time will be 0 seconds in 2322 and the men's record time will be 0 seconds in 2662. Model breakdown has occurred. **h)** Answers may vary.

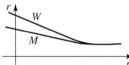

Chapter 4

Exercise Set 4.1

1. $\frac{1}{2}$ **3.** $\frac{1}{9}$ **5.** 16 **7.** $\frac{5}{6}$ **9.** 49 **11.** 1 **13.** 1 **15.** b^{16} **17.** 1 **19.** $-\frac{6}{b}$ **21.** b^8

23. $\frac{32b^{26}}{9c^2}$ **25.** $\frac{1}{b^5}$ **27.** $\frac{2}{b^3}$ **29.** $-\frac{6b}{7}$ **31.** $-\frac{1}{3bc^5}$ **33.** $-\frac{1}{4b^{10}c^{14}}$ **35.** $\frac{3b^5}{c^{13}}$ **37.** $\frac{27b^{18}}{7c^7}$ **39.** $\frac{b^6}{c^{12}}$ **41.** 1 **43.** $\frac{1}{16b^{12}c^{12}}$

45. $\frac{1}{bc}$ **47.** $b + c$ **49.** b^{7n} **51.** b^{5n-4} **53.** Student B **55.** $-2^2, 2(-1), \left(\frac{1}{2}\right)^2, 2^{-1}, \frac{1}{2}, \left(\frac{1}{2}\right)^{-1}, (-2)^2, 2^2$; Ties: $2^{-1} = \frac{1}{2}$, $(-2)^2 = 2^2$ **57. a)** 1 **b)** 0 **59. a)** $\frac{1}{b}$ **b)** b **c)** $\frac{1}{b}$ **d)** b **e)** The expression is equal to b if n is even. The expression is equal to $\frac{1}{b}$ if n is odd.

Exercise Set 4.2

1. 4 **3.** 10 **5.** 7 **7.** 5 **9.** 16 **11.** 27 **13.** 4 **15.** 32 **17.** $\frac{1}{3}$ **19.** $-\frac{1}{6}$ **21.** $\frac{1}{32}$ **23.** $\frac{1}{81}$

25. 2 **27.** 3 **29.** 49 **31.** b^2 **33.** $\frac{1}{b^2}$ **35.** $2b^2$ **37.** $\frac{4}{5b^4c^7}$ **39.** $\frac{b}{c^2}$ **41.** $5bcd$ **43.** $8b^{12}c^5$ **45.** $\frac{b^{3.3}}{c^{2.5}}$ **47.** $\frac{3b^4}{5c^2}$ **49.** $b^{29/35}$

51. $b^{7/12}$ **53.** $\frac{8}{b^{1/5}}$ **55.** $\frac{2b^{1/12}c^{1/4}}{3}$ **57.** 396.5 **59.** 0.239 **61.** 520 **63.** 0.00009113 **65.** $-652,000$ **67.** 900,000

69. -0.08 **71.** 5.426×10^3 **73.** 2.3587×10^4 **75.** 9.8×10^{-4} **77.** 3.46×10^{-5} **79.** -4.2215×10^4 **81.** -8.928×10^{-3}

83. 8.0×10^0 **85.** -1.0×10^5 **87.** 287,000,000 **89.** 0.000000063 **91.** 0.000000000001 **93.** 8.3×10^5 **95.** 4.7×10^{-7}

Exercise Set 4.3

1.
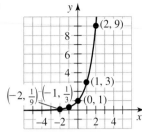

3.
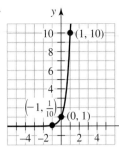

5.
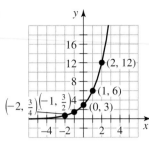

7.
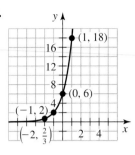

9.
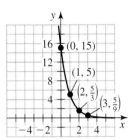

11.
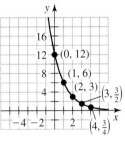

13.
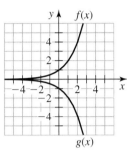

15.
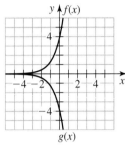

17.
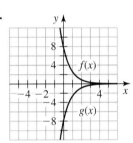

19. (a) $a < 0, b > 1$; (b) $a > 0, b > 1$; (c) $a > 0, 0 < b < 1$; (d) $a < 0, 0 < b < 1$

23. $f(x) = 3(2)^x$

27. f might be linear, g might be exponential, h might be exponential, k is neither linear nor exponential

29.

x	$f(x)$	$g(x)$	$h(x)$	$k(x)$
0	5	160	162	3
1	10	80	54	12
2	20	40	18	48
3	40	20	6	192
4	80	10	2	768

31. 8 **33.** 1 **35.** -2 **37.** 0 **39.** 24 **41.** 96 **43.** 0 **45.** 3

47. a)
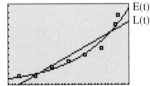

The graph of the exponential model comes closer to the points. The exponential model makes better estimates for years before 1980.

b) The models L and E estimate the cost to be $-\$226,000$ and $\$203,934$, respectively. The exponential model does a better job.

c) $\$16.62$ million

49. No x-intercept; y-intercept: $(0, 1)$

51. No x-intercept; y-intercept: $(0, 3)$

53. a) The y-intercept for f is $(0, 100)$. The y-intercept for g is $(0, 5)$.

b) For f, as the value of x increases by 1, the value of $f(x)$ is multiplied by 2. For g, as the value of x increases by 1, the value of $g(x)$ is multiplied by 3.

c) The values of g will be much greater.

55. 4 **57.** 64 **59.** 1 **61.** 5 **63.** 2 **65.** 1 **67.** 0

69. The entries in the second column are $\frac{1}{8}, \frac{1}{4}, \frac{1}{2}, 1, 2, 4, 8, 16$.

83. The entries in the second column are 0.0000000009; 0.00000095; 0.00098; 1; 1024; 1,050,000; 1,070,000,000.

Exercise Set 4.4

1. $f(x) = 4(2)^x, g(x) = 36\left(\frac{1}{3}\right)^x, h(x) = 5(10)^x, k(x) = 250\left(\frac{1}{5}\right)^x$ **3.** $f(x) = 100\left(\frac{1}{2}\right)^x,$

$g(x) = -50x + 100, h(x) = 4x + 2, k(x) = 2(3)^x$ **5.** ± 4 **7.** ± 3 **9.** 2 **11.** ± 2.11 **13.** 2.28 **15.** ± 0.89 **17.** 2.22

19. ± 3 **21.** 1.74 **23.** b^5 **25.** 2.38 **27.** $\frac{4b^4}{3}$ **29.** ± 0.75 **31.** $y = 4(2)^x$ **33.** $y = 3(2.02)^x$ **35.** $y = 87(0.74)^x$

37. $y = 7.4(0.56)^x$ **39.** $y = 5.5(3.67)^x$ **41.** $y = 39.18(0.85)^x$ **43.** $y = 4\left(\dfrac{1}{2}\right)^x$ **45.** $y = 1.33(3)^x$ **47.** $y = 1.19(1.50)^x$

49. $y = 1170.33(0.88)^x$ **51.** $y = 37.05(0.74)^x$ **53.** $y = 146.91(0.71)^x$ **55.** $y = 0.072(1.57)^x$ **57.** $y = 1.26(1.58)^x$

59. a) $f(x) = 4x + 2$ **b)** $g(x) = 2(3)^x$ **c)**

61. $(0, 6)$

63. a) No   **b)** Yes   **c)** Yes   **d)** Yes

Exercise Set 4.5

1. a) $f(t) = 40(3)^t$ **b)** 2.36 million people   **c)** 574 million people; Yes

3. a) $f(t) = 108\left(\dfrac{1}{2}\right)^t$ **b)** \$13,184   **5. a)** $f(t) = 2(2)^t$ **b)** 512 billion web pages   **c)** 32,323 miles

7. a) $f(t) = 3000(1.08)^t$ **b)** The base is 1.08. The value of the account increases by 8% per year.   **c)** 3000; The original value of the account is \$3000.   **9. a)** $f(t) = 4000(2)^{t/6}$ or $f(t) = 4000(1.1225)^t$ **b)** \$40,317.47   **11. a)** $C(t) = 800(1.03)^t$
b) $S(t) = 24t + 800$ **c)** $C(1) = 824, C(2) = 848.72, S(1) = 824, S(2) = 848$ **d)** $C(20) = 1444.89, S(20) = 1280$

13. a) $f(t) = 57.7\left(\dfrac{1}{2}\right)^{t/40}$ or $f(t) = 57.7(0.9828)^t$ **b)** 43.73; The population will be 43.7 million people in 2018.

15. a) $g(t) = 100\left(\dfrac{1}{2}\right)^{t/5.3}$ or $g(t) = 100(0.8774)^t$ **b)** 2.57%   **c)** 12.26%   **17.** Decrease b

19. a)

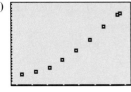

Exponential function

b) $f(t) = 0.33(1.30)^t$ **c)** 1.30; The chip speeds increase by 30% per year.   **d)** It would be 13.8 times quicker.

21. a)

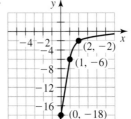

Exponential function

b) $f(t) = 1.197(1.0163)^t$ **c)** 1.0163; The population is growing exponentially at a rate of 1.63% per year.   **d)** 1.197; The population was 1.197 billion people in 1900.   **e)** 7.09; The population will be 7.09 billion people in 2010.

23. a) $f(t) = 1.46(1.48)^t$ **b)** 85,444 stores   **c)** 48% per year
25. a) The entries in the second column of the table are 1.36, 1.36, 1.33, 1.34, 1.33, 1.36, 1.35.   **b)** The ratios are approximately equal
to 1.35.   **c)** Exponential function   **d)** $f(t) = 3.94(1.03)^t$ **e)** The entries in the third column of the table are 1.27, 1.26, 1.25,
1.21.   **f)** No   **g)** 2.07 billion people; 1.78 billion people
27. a) $f(t) = 55.15(1.10)^t$ **c)** i. \$15.3 trillion;   ii. \$51,515;   iii. \$760 trillion;   iv. \$1.93 million

Chapter 4 Review

1. 32   **2.** $\dfrac{48c^3}{b^{12}}$ **3.** $\dfrac{bc^9}{4}$ **4.** $\dfrac{8c^6}{9b^{23}}$ **5.** 0   **6.** $\dfrac{b}{c}$ **7.** 1   **8.** 16   **9.** $\dfrac{1}{8}$ **10.** $\dfrac{1}{b^{5/3}}$ **11.** $\dfrac{2b^{11}}{125c^7}$

12. $2b^2c$ **13.** $\dfrac{b^{1/6}}{c^{7/4}}$ **14.** b^{6n+2} **15.** $b^{n/6}$ **16.** $3^{2x} = (3^2)^x = 9^x$ **17.** 64   **18.** 1   **19.** $\dfrac{1}{64}$ **20.** 4   **21.** $\dfrac{1}{3}$ **22.** -1
23. 44,487,000   **24.** 0.0000385   **25.** 5.4698201×10^7 **26.** -8.97×10^{-3}

27.   **28.**   **29.**   **30.**

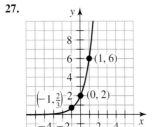
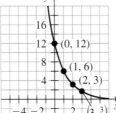

31. 2 **32.** 1.9744 **33.** 1.8447 **34.** ± 2 **35.** ± 1.6125 **36.** $f(x) = -4x + 34$, linear; $g(x) = \dfrac{5}{3}(3)^x$, exponential; h is neither;

$k(x) = 192\left(\dfrac{1}{2}\right)^x$, exponential **37.** 18 **38.** 6 **39.** 1 **40.** 5 **41.** $y = 2(1.08)^x$ **42.** $y = 3.8(2.34)^x$ **43.** $y = 62.11(0.78)^x$

44. $y = 3.07(1.18)^x$ **45.** Increase a and decrease b. **46. a)** $f(t) = 2000(1.07)^t$ **b)** \$2805.10 **47. a)** $g(t) = 185.75(2)^{t/7}$ or

$g(t) = 185.75(1.10)^t$ **b)** The total sales will be \$1,219,000.

48. a) $f(t) = 100\left(\dfrac{1}{2}\right)^{t/5730}$ or $f(t) = 100(.999879)^t$ **b)** 98.8%

49. a)

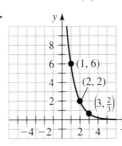

Exponential function

b) $f(t) = 48.46(1.206)^t$ **c)** 48.46; There were 48 stores in 1990. **d)** 1.206; The number of stores is growing exponentially by 20.6% per year. **e)** 1411.35; There will be 1411 stores in 2008.

50. a) $f(t) = 44.32(2.01)^t$ **b)** 1.57 million lawsuits **c)** \$785 billion **d)** No

Chapter 4 Test

1. 4 **2.** $-\dfrac{1}{16}$ **3.** $8b^9c^{24}$ **4.** 1 **5.** $b^{1/6}$ **6.** $\dfrac{5b}{7c^5}$ **7.** $\dfrac{4b^4}{c^{14}}$ **8.** $\dfrac{250b^{14}}{7c^6}$

9. $8^{x/3}2^{x+3} = (2^3)^{x/3}2^{x+3} = 2^x 2^{x+3} = 2^{2x+3} = 2^3(2)^{2x} = 8(2^2)^x = 8(4)^x$

10.

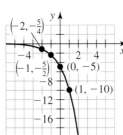

11.

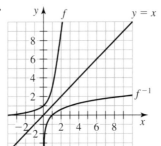

12. $f(x) = 20(0.5)^x$, $k(x) = 30(0.3)^x$, $h(x) = 30(3)^x$, $g(x) = 50(1.2)^x$; Answers may vary. **13.** $f(t) = 160\left(\dfrac{1}{2}\right)^t$

14. ± 1.7248 **15.** $y = 70(0.81)^x$ **16.** $y = 0.91(1.77)^x$ **17.** 6

18. 1 **19.** $f(x) = 6\left(\dfrac{1}{2}\right)^x$

20. a) $f(t) = 40(2)^t$ **b)** 2560; There will be 2560 leaves on the tree 6 weeks after March 1. **c)** 1.80×10^{17}; There will be 1.80×10^{17} leaves on the tree one year later. Model breakdown has occurred.

21. a) $f(t) = 759.79(1.076)^t$ **b)** 7.6% **c)** 759.79; There were 760 multiple births in 1970. **d)** 12,290 multiple births

Chapter 5

Exercise Set 5.1

1. 4 **3.** 6 **5.** 5 **7.**

x	$f^{-1}(x)$
2	6
4	5
6	4
8	3
10	2

9. 6 **11.** 1 **13.**

x	$g^{-1}(x)$
2	1
6	2
18	3
54	4
162	5
486	6

15. The entries in the fourth column of the table are 6, 5, 4, 3, 2, 1.

17.

x	$f^{-1}(x)$
3	0
6	1
12	2
24	3
48	4

19. 24 **21.** 0 **23.**

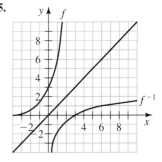

25.

27.
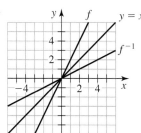

29.
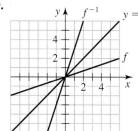

31.
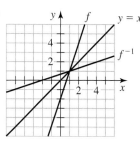

33.
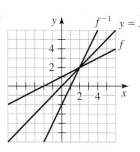

35.
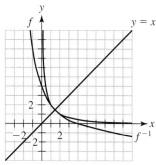

37.
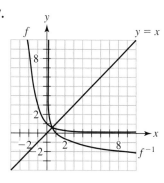

39. a) $f^{-1}(p) = 1.23p + 56.62$ **b)** 35.14; According to the model, in 2000, 35.1% of births were out of wedlock. **c)** 179.62;
According to the inverse model, all births will be out of wedlock by 2080.

41. a)
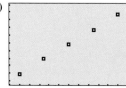
Linear function

b) $f(t) = 0.78t - 52.56$ **c)** $f^{-1}(p) = 1.28p + 67.38$ **d)** 2031 **e)** 2031 **f)** The results
are the same.

43. a)
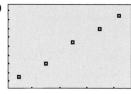
Linear function

b) $f(t) = 0.38t - 20.13$ **c)** $f^{-1}(p) = 2.63p + 52.97$ **d)** 21.29; The price of CDs will be
$21.29 in 2009. **e)** 118.72; The price of CDs will be $25 in 2019.

45. $f^{-1}(x) = x + 3$ **47.** $f^{-1}(x) = x - 8$ **49.** $f^{-1}(x) = \dfrac{x}{4}$ **51.** $f^{-1}(x) = 7x$ **53.** $f^{-1}(x) = \dfrac{1}{4}x - \dfrac{5}{4}$ **55.** $f^{-1}(x) = -\dfrac{1}{6}x - \dfrac{1}{3}$

57. $f^{-1}(x) = 2.5x + 19.75$ **59.** $f^{-1}(x) = -0.014x + 1.36$ **61.** $f^{-1}(x) = \dfrac{3}{7}x - \dfrac{3}{7}$ **63.** $f^{-1}(x) = -\dfrac{6}{5}x - \dfrac{18}{5}$

65. $f^{-1}(x) = \dfrac{5}{6}x + \dfrac{1}{3}$ **67.** $f^{-1}(x) = -\dfrac{1}{6}x + \dfrac{5}{2}$ **69.** $f^{-1}(x) = -\dfrac{1}{8}x - \dfrac{1}{8}$ **71.** $f^{-1}(x) = x$ **75.** No; No

77. a) $f^{-1}(x) = \dfrac{1}{m}x - \dfrac{b}{m}$

Exercise Set 5.2
1. 2 **3.** 3 **5.** 5 **7.** 3 **9.** 2 **11.** 2 **13.** 3 **15.** 6 **17.** −1 **19.** −3 **21.** −1 **23.** 0
25. $\dfrac{1}{2}$ **27.** $\dfrac{1}{3}$ **29.** $\dfrac{1}{2}$ **31.** $\dfrac{1}{4}$ **33.** 2 **35.** 0 **37.** 1 **39.** 2 **41.** $\dfrac{1}{2}$ **43.** 0 **45.** $f^{-1}(x) = \log_3(x)$ **47.** $h^{-1}(x) = \log(x)$
49. $f^{-1}(x) = 5^x$ **51.** $h^{-1}(x) = 10^x$ **53.** 4 **55.** 1 **57.** 1 **59.** 27 **61.** 3 **63.** 0; $\log_3(1)$

65.
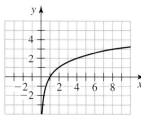

67.
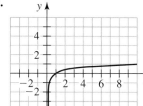

69.
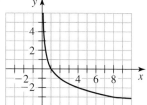

71. $f^{-1}(x) = \log_5(x)$

x	$f^{-1}(x) = \log_5(x)$
1	0
5	1
25	2
125	3
625	4

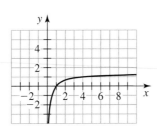

73. a) 2 **b)** 2

75. a) 8.6 **b)** 7.2 **c)** 1.2 **d)** 25

77. The decibel readings for the entries in the second column of the table are 0, 20, 40, 60, 80, 100, and 120.

Exercise Set 5.3

1. $3^5 = 243$ **3.** $10^2 = 100$ **5.** $b^c = a$ **7.** $10^n = m$ **9.** $\log_5(125) = 3$ **11.** $\log(1000) = 3$
13. $\log_y(x) = w$ **15.** $\log(q) = p$ **17.** 16 **19.** $\dfrac{1}{16}$ **21.** 0.01 **23.** 1 **25.** 2 **27.** 3.3019 **29.** 6561 **31.** 7 **33.** 2 **35.** 2
37. 1.2380 **39.** 1.5850 **41.** 2 **43.** 3.1842 **45.** 3.8278 **47.** 4.8738 **49.** −0.2281 **51.** 0.8644 **53.** Empty set solution
55. 3 **57.** 81 **59.** 64 **61.** 7 **63.** Empty set solution **65.** 4 **67.** 1 **69.** 1.1487 **73.** $\dfrac{\log\left(\frac{c}{a}\right)}{\log(b)}$ **75.** $\dfrac{\log\left(\frac{c-d}{a}\right)}{k\log(b)}$ **77.** 256
79. 0.7925 **81.** 3 **83.** 32

Exercise Set 5.4

1. a) $f(t) = 2000(1.05)^t$ **b)** (0, 2000); The original value is $2000. **c)** 8.31 years

3. a) 2031 **b)**

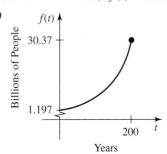

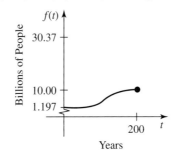

There is model breakdown for the years after 2031.

5. a) 1.46; There was one store in 1980. **b)** No **c)** 1.48; The number of stores is growing exponentially by 48% per year.
d) 2007 **e)** 2026

7. a)

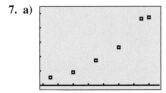

Exponential function

b) $f(t) = 1.60(1.018)^t$ **c)** 9.53; In 1900, the percentage of food expenditures devoted to food away from home was 9.5%. **d)** 231.79; In 2032, all food expenditures will be devoted to food away from home. **e)** 231.79; In 2032, all food expenditures will be devoted to food away from home.

9. a) $E(s) = 0.36(1.0036)^s$; $R(s) = 0.037(1.0049)^s$ **b)** 60.3%; 39.2% **c)** 1373; 1475 **d)** 102 points **e)** (1757.58, 199.19);
Students who score 1758 points have the same chance (199%) of being selected from the early decision and the regular decision systems.
Model breakdown has occurred.

11. a) $f(t) = 30(2)^t$ **b)** 24 days **13. a)** $f(d) = 8\left(\dfrac{1}{2}\right)^{d/5}$ or $f(d) = 8(0.8706)^d$ **b)** 0.29 hour; Yes **c)** 97 decibels

15. 871 years **17.** 3.6 hours **19. a)** $f(t) = 100\left(\dfrac{1}{2}\right)^{t/1600}$ or $f(t) = 100(0.999567)^t$ **b)** 5315 years from now

Exercise Set 5.5

1. $\log_b(2x^2)$ **3.** $\log_b(5x)$ **5.** $\log_b(3x^4)$ **7.** $\log_b(16x^{11})$ **9.** $\log_b\left(\dfrac{1}{2x^3}\right)$ **11.** $\log_b(x^9)$
13. $\log_b\left(\dfrac{9}{x^8}\right)$ **15.** 3 **17.** 4 **19.** 1.9129 **21.** 0.9057 **23.** 1.6818 **25.** $\log_2(x^7)$ **27.** 1.4860 **29.** 1.0596 **31.** $\log_9(16x^{28})$
33. 1.7712 **35.** 3.8553 **37.** 2.3104 **39.** −2.0431 **41.** $\log_7(x)$ **43.** $\log_s(r)$ **45.** All three students did the problem correctly.
47. 1.5937 **49.** −0.4307 **51.** 1.1402 **53.** 1.4531 **55.** $\log_b(b^2)$, $\log_b\left(\dfrac{b^6}{b^4}\right)$, 2, $\log_b(b^6) - \log_b(b^4)$ **59. a)** $\log_2(x^8)$ **b)** 1.8340

Exercise Set 5.6

1. 1.9459 **3.** 4.0037 **5.** −0.2231 **7.** −0.6931 **9.** 4 **11.** 1 **13.** −1 **15.** 3 **17.** e^2
19. 2.9957 **21.** 15.0855 **23.** 2.1313 **25.** 1.8383 **27.** 0.9061 **29.** 1.2777 **31.** 3.2459 **33.** 3.4541 **35.** 1.9458
37. $\ln(12x^2)$ **39.** $\ln(5x)$ **41.** $\ln(72x^{11})$ **43.** $\ln(3x^5)$ **45.** $\ln(6x+3)$ **47.** 4.2661 **49.** 37.1033 **51.** 1.2227 **53.** $\ln(x^5)$

55. 2.2255 **59.** $\dfrac{\ln\left(\dfrac{c}{a}\right)}{b}$ **61.** $3\ln(x)$, $\ln(x^7) - \ln(x^4)$, $\ln(x^3)$ **63. a)** 207°F **b)** 3.66 minutes **c)** 70°F

65. a) 20.91 feet **b)** 20.32; The height of the cable four feet to the left of the right pole is 20.32 feet. **c)** 20 feet.

Chapter 5 Review

1. $f^{-1}(x) = \dfrac{x}{3}$ **2.** $g^{-1}(x) = 2x + \dfrac{7}{4}$ **3.** $h^{-1}(x) = \log_3(x)$ **4.** $f^{-1}(x) = 4^x$
5. $g^{-1}(x) = \log(x)$ **6.** $h^{-1}(x) = 10^x$ **7.** 1 **8.** 4 **9.** 4 **10.** 1

11.
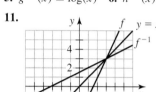

12.
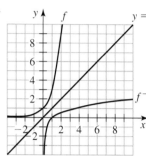

13. 2 **14.** 5 **15.** −2 **16.** −3 **17.** $\dfrac{1}{2}$ **18.** $\dfrac{1}{3}$
19. −3 **20.** 1.7712 **21.** 1.6094 **22.** 1 **23.** 7
24. 0 **25.** $\log_d(k) = t$ **26.** $y^r = w$

27.
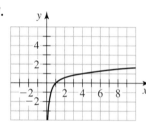

28. 2.3219 **29.** 81 **30.** 2.9312 **31.** 0.4310 **32.** 2.0886 **33.** 7.3891 **34.** 2
35. 164.3368 **36.** 2.8333 **37.** 3.4950 **38.** 2 **39.** 8.4853 **40.** 2.1297 **41.** 1.4609
42. 74.2066 **43.** 16 **44.** 1.2027 **45.** $\log_b(3x)$ **46.** $\log_b(72x^5)$ **47.** $\ln(512x^{24})$

48. $\log_b\left(\dfrac{1}{x^2}\right)$ **49.** $\ln\left(\dfrac{1}{8x^5}\right)$ **50.** $\log_y(w)$ **51.** $\log_b(b^5) - \log_b(b^2)$, $\log_b(b^3)$, 3, $\log_b\left(\dfrac{b^5}{b^2}\right)$
52. a) $f(t) = 8(1.05)^t$ **b)** \$12,410.63 **c)** 14.2 years
53. a) $f(t) = 30(4)^t$ **b)** 30,720; There are 30,720 leaves on the tree at 5 weeks after April 1. **c)** 5.85; There are 100,000 leaves on the tree at 5.85 weeks after April 1.

54. a)
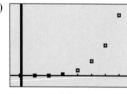
Exponential function

b) $f(t) = 0.083(2.65)^t$ **c)** 2.65; Sales are growing exponentially by approximately 165% per year; In 1994, sales grew 500%. **d)** 2000
55. a) $f(n) = 9.33(1.31)^n$ **b)** 1.31; As each cassette is added to the bag, the length increases by 31%. **c)** 9.33; The initial length of the rubberband is 9.33 inches. **d)** 80.92 inches
e) 10 cassettes

Chapter 5 Test

1. $g^{-1}(x) = \dfrac{x+9}{2}$ **2.** $h^{-1}(x) = \log_4(x)$ **3.** $f^{-1}(x) = 5^x$

4.
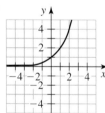

5. 4 **6.** −3 **7.** 1.1833 **8.** −1 **9.** 2 **10.** 0 **11.** 3 **12.** $\dfrac{1}{2}$ **13.** −2 **14.** 0
15. 2.6591 **16.** 2.4150 **17.** 11.0227 **18.** 1.4319 **19.** 1.6330 **20.** 4.0427 **21.** $\log_s(w) = t$
22. $e^p = k$ **23.** $\log_b(5x^4)$ **24.** $\ln\left(\dfrac{x^2}{32}\right)$ **25. a)** $f(t) = 100\left(\dfrac{1}{2}\right)^{t/5730}$ or $f(t) = 100(0.999879)^t$
b) 7575 years **26. a)** $f(t) = 0.37(1.77)^t$ **b)** 0.37; There were 370 arrests in 1990. **c)** 1.77; The number of arrests increases by 77% each year. **d)** 2004

Cumulative Review for Chapters 1–5

1. 0.5087 **2.** 86 **3.** 1.3538 **4.** −1 **5.** 1.2528 **6.** 3.1054
7. ±0.7811 **8.** 0.71 **9.** 1.0782 **10.** (−1, 2) **11.** (2, 4) **12.** (−6, 12) **13.** $x \geq \dfrac{9}{10}$; $\left[\dfrac{9}{10}, \infty\right)$;

14. $\dfrac{1600c^4}{b^{23}}$ **15.** $\dfrac{4b^{5/6}}{3c^{5/4}}$ **16.** $\dfrac{8c^3}{27b^9}$ **17.** $\log_b\left(\dfrac{16x^{24}}{49}\right)$ **18.** $\ln(5184x^{26})$ **19.** $f(x) = 5(3)^x$ **20.** $g(x) = 3x + 25$ **21.** −7
22. 40 **23.** 8 **24.** 0 **25.** (0, 1280)

26.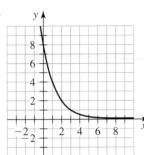

27.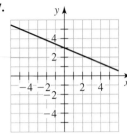

28.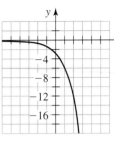

29.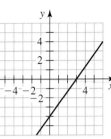

30.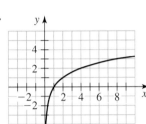

31. $y = -\dfrac{10}{9}x + \dfrac{23}{9}$ or $y = -1.11x + 2.56$ **32.** $y = 8.1(1.25)^x$ **33.** $y = 347.56(0.63)^x$

34. $\dfrac{2}{81}$ **35.** 1.8928 **36.** $(0, 2)$

37.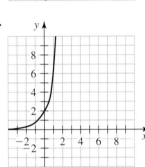

38.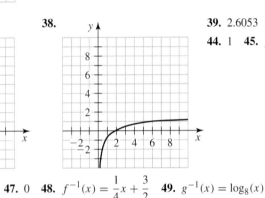

39. 2.6053 **40.** -4 **41.** 1 **42.** $\dfrac{1}{7}$ **43.** 2.0633

44. 1 **45.** $f(x) = 2(2)^x$

46.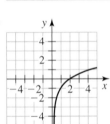

47. 0 **48.** $f^{-1}(x) = \dfrac{1}{4}x + \dfrac{3}{2}$ **49.** $g^{-1}(x) = \log_8(x)$

51. a) The y-intercept for both functions is $(0, 2)$.

b) For f, as the value of x increases by 1, the value of y increases by 3. For g, as the value of x increases by 1, the value of y is multiplied by 3.

c) g

d)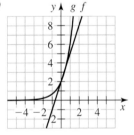

52. a) $f(x) = \dfrac{5}{3}x - \dfrac{11}{3}$ or $f(x) = 1.67x - 3.67$ **b)** $g(x) = 0.81(1.39)^x$ **c)**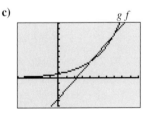

53. a) $f(2) = 6; g(2) = 9$ **b)** $f^{-1}(x) = \dfrac{x}{3}; g^{-1}(x) = \log_3(x)$ **c)** $f^{-1}(81) = 27; g^{-1}(81) = 4$

54. a) $U(x) = 0.69x + 19.95; B(x) = 0.45x + 29.95$ **b)** The slope of U is 0.69. The slope of B is 0.45. U-Haul charges \$0.69 per mile and Budget charges \$0.45 per mile. **c)** 41.67 miles **d)** $x < 41.67$; U-Haul charges less than Budget for mileages under 41.67 miles.

55. a) $f(t) = 20.38(0.79)^t$ **b)** $(0, 20.38)$; There were 20.4 cases per 100,000 people in 1990, according to the model. **c)** 21%; The number of cases decreases by 21% each year. **d)** 2003 **e)** 22.56; There will be 1 case per million people in 2013. **f)** 0.75; 2168 cases

56. a)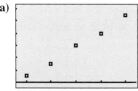

Linear function

b) $f(t) = 0.25t - 1.67$ **c)** 0.25; The number of laser eye surgeries is increasing by 250 thousand surgeries per year.   **d)** (6.68, 0); There were no laser eye surgeries in 1997, according to the model. Model breakdown has occurred.   **e)** (0, −1.67); There were −1.67 million laser eye surgeries in 1990. Model breakdown has occurred.   **f)** 2.83; There will be 2.83 million laser eye surgeries in 2008.   **g)** $f^{-1}(n) = 4n + 6.68$ **h)** 22.68; There will be 4 million laser eye surgeries in 2013.

Chapter 6

Exercise Set 6.1

1.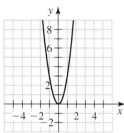

Vertex (0, 0)

3.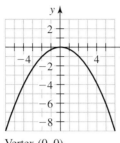

Vertex (0, 0)

5.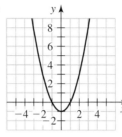

Vertex (0, −1)

7.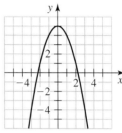

Vertex (0, 5)

9.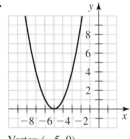

Vertex (−5, 0)

11.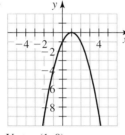

Vertex (1, 0)

13.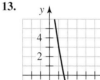

Vertex (4, −6)

15.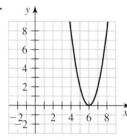

Vertex (6, 0)

17.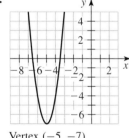

Vertex (−5, −7)

19.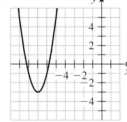

Vertex (−7, −3)

21.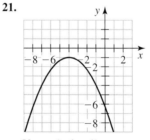

Vertex (−4, −1)

23.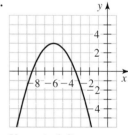

Vertex (−6, 3)

25.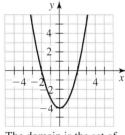

The domain is the set of real numbers. The range is the set of numbers y where $y \geq -4$.

27.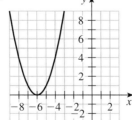

The domain is the set of real numbers. The range is the set of numbers y where $y \geq 0$.

29.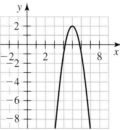

The domain is the set of real numbers. The range is the set of numbers y where $y \leq 2$.

31. (a) $a > 0$, $h < 0$, and $k < 0$; (b) $a < 0$, $h < 0$, and $k > 0$; (c) $a > 0$, $h > 0$, and $k = 0$; (d) $a < 0$, $h = 0$, and $k < 0$

35. $y = 0.625(x - 5)^2 - 6$ **37.** $y = -2.1(x + 7)^2 + 3.71$ **39.** It is possible.   **41.** It is possible.

43. a)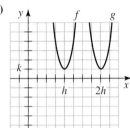

Quadratic function

b)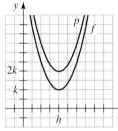

Yes

c) (8.86, 47.49); The percentage of Americans who were pro-choice was at a minimum of 48% in 1999, according to the model.   **d)** 1995, 2003   **e)** 94%

45. (2, 5)   **51. a)** 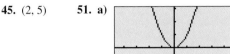   **b)** 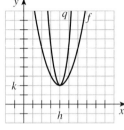   **c)**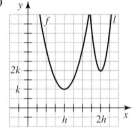

55. a) 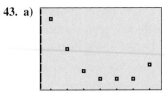   **b)**   **c)**   **d)**

Exercise Set 6.2

1. $x^2 - x$ **3.** $-7x^2 + 14x$ **5.** $-24x^2 - 15x$ **7.** $x(x+4)$ **9.** $-9x(3x-4)$ **11.** $-5x(5x-7)$
13. $-4x(x+2)$ **15.** $x(3.8x + 4.7)$ **17.** $f(x) = x(x-7)$ **19.** $h(x) = 6x(4x-5)$ **21.** $f(x) = -2x(x+4)$ **23.** $x(2.5x - 6.2)$
27. All three students did the problem correctly, although we usually write the result as student 2 and student 3 did.   **29.** $14x^2 - 8x$
31. $x^2 + 6x + 8$ **33.** $x^2 + 3x - 18$ **35.** $x^2 + 16x + 64$ **37.** $x^2 - 14x + 49$ **39.** $x^2 - 25$ **41.** $12x^2 + 23x + 5$
43. $24x^2 - 26x + 5$ **45.** $x^3 + 2x^2 - 4x - 8$ **47.** $x^5 + 3x^3 + 6x^2 + 18$ **49.** $4x^2 + 20x + 25$ **51.** $9x^2 - 4$ **53.** $x^2 - x$
55. $-x^2 + 2x - 1$ **57.** $75x^3 - 30x^2 + 3x$ **59.** $-32x^3 - 80x^2 - 50x$ **61.** $3.78x^2 + 4.92x - 21.28$ **63.** $x^3 + 6x^2 + 7x + 10$
65. $6x^3 - 7x^2 - 11x + 12$ **67.** $x^3 + 64$ **69.** $x^4 + 3x^3 + 7x^2 + 7x + 6$ **71.** $2x^4 - 3x^3 - 3x^2 + 7x - 3$ **73.** $x^3 + 6x^2 + 11x + 6$
75. $\frac{1}{25}x^2 - 4$ **77.** $-20x$ **79.** $2x^2 - 16x + 32$ **81.** $f(x) = x^2 + 12x + 36$ **83.** $h(x) = 2x^2 + 12x + 19$
85. $p(x) = 4x^2 - 40x + 102$ **87.** $g(x) = -4x^2 + 8x - 5$ **89.** $k(x) = 1.5x^2 + 8.4x + 8.06$ **91.** $f(x) = 5x^2 - 10x$; quadratic
93. $h(x) = x^2 - 36$; quadratic   **95.** $p(x) = 24x$; linear   **97.** 4   **99.** -0.56 **101.** $a^2 - 3a$ **103.** $a^2 - a - 2$ **105.** -1 **107.** 1
109. $-2a^2 + 5a - 1$ **111.** $-2a^2 - 7a - 4$

Exercise Set 6.3

1. $(x+3)(x+4)$ **3.** $(x-20)(x-1)$ **5.** Prime   **7.** $3(x-3)(x+2)$ **9.** $-5(x-6)(x+2)$
11. $6(x+3)^2$ **13.** $(x-4)(x+4)$ **15.** $(x-1)(x+1)$ **17.** $-(x-5)(x+5)$ **19.** $(10x-1)(10x+1)$ **21.** $-(x-7)(x+7)$
23. Prime   **25.** $(8x-7)(8x+7)$ **27.** $\left(x - \frac{1}{3}\right)\left(x + \frac{1}{3}\right)$ **29.** $(3x+4)(x+2)$ **31.** $(x-5)(2x-3)$ **33.** $(3x+1)(3x+2)$
35. Prime   **37.** $(3x-4)(3x-1)$ **39.** $10(x-2)(2x+7)$ **41.** $(x+5)(x+6)$ **43.** $(x-2)(x-1)$ **45.** $3x(5x-9)$
47. $(2x+3)(5x+4)$ **49.** Prime   **51.** $(x-20)(x+5)$ **53.** $-3(x^2+25)$ **55.** $(x-6)(x+8)$ **57.** $(x-1)(12x+5)$
59. $7(x^2+2x+3)$ **61.** $(9x-4)(9x+4)$ **63.** Prime   **65.** $(2x-3)(4x+5)$ **67.** $(x-10)(x+10)$ **69.** $(x-7)^2$
71. $(3x-4)(3x-2)$ **73.** $(x+20)(4x-1)$ **75.** $2(x-8)(x-3)$ **77.** $(2x-3)(5x+8)$ **79.** $4(3x+2)^2$ **81.** No   **83.** No

Exercise Set 6.4

1. $3x(x-5)(x+5)$ **3.** $7x(3x-1)(3x+1)$ **5.** $8x^3(x-3)(x+3)$ **7.** $-3x^2(5x-3)(5x+3)$
9. $2x(x+1)(x+3)$ **11.** $4x^2(x-2)(x+5)$ **13.** $-2x(x-5)^2$ **15.** $3x^3(x+3)(2x+5)$ **17.** $10x(2x+5)(4x-3)$
19. $-5x^4(2x+1)(3x-1)$ **21.** $(x+3)(x^2+4)$ **23.** $(x-4)(x^2+3)$ **25.** $(x-5)(3x^2-4)$ **27.** $(x-4)(x+4)(2x+1)$
29. $(x-5)(x-2)(x+2)$ **31.** $(x-3)(2x-3)(2x+3)$ **33.** $(2x-1)(2x+1)(3x+1)$ **35.** $2x(x-5)(x+5)$ **37.** $(x+4)(x^2-7)$
39. $(x-9)(x+9)$ **41.** $(2x-1)(4x-3)$ **43.** $3x(3x-5)$ **45.** $3x^2(x+2)(x+4)$ **47.** $(x+7)^2$ **49.** $-4x^2(x+2)(x+3)$
51. $x(x+1)$ **53.** $-x(x-4)(x+4)$ **55.** $(x+5)(2x-1)(2x+1)$ **57.** $x(x-10)(x+10)$ **59.** $x^3 - 6x^2 - 5x + 15$
61. $5x^2(x-4)(x-3)$ **63.** $2x(2x+3)(3x+2)$ **65.** $4(x-2)^2$ **67.** $-(x+1)^2$ **69.** $-2x^4(3x-4)$ **71.** $x^3 - 4x^2 - 9x + 9$
73. $5x^3(2x-1)(2x+3)$

Exercise Set 6.5

1. $0, 3$ **3.** $0, \dfrac{3}{2}$ **5.** $-5, -3$ **7.** ± 2 **9.** $-\dfrac{6}{7}, 0$ **11.** $0, 2$ **13.** $-5, \dfrac{1}{2}$ **15.** $\dfrac{2}{3}$ **17.** ± 2

19. $\pm \dfrac{5}{4}$ **21.** $-\dfrac{3}{5}, 0$ **23.** $0, \dfrac{10}{3}$ **25.** $\pm \dfrac{7}{3}$ **27.** ± 1 **29.** $\dfrac{8}{3}$ **31.** $-2, 4$ **33.** 6 **35.** $\pm \dfrac{1}{5}$ **37.** $-\dfrac{1}{3}, \dfrac{1}{2}$ **39.** $-4, 6$ **41.** $\pm 2, 0$

43. $\pm 3, \dfrac{1}{2}$ **45.** $0, \dfrac{7}{4}$ **47.** $\pm \dfrac{3}{2}, 5$ **49.** $-3, \pm \dfrac{2}{3}$ **51.** $-\dfrac{2}{3}, 0, \dfrac{1}{2}$ **53.** $-\dfrac{4}{3}, 7$ **55.** $-10, 3$ **57.** $-6, 4$ **59.** $(x + 2)(x + 3)$

61. $-3, -2$ **63.** $2, 5$ **65.** $(x - 5)(x - 2)$ **67.** False **71.** $(4, 0)$ and $(5, 0)$ **73.** $(-4, 0)$ and $(4, 0)$ **75.** $(2, 0), (8, 0)$

77. $\left(-\dfrac{5}{6}, 0\right), \left(\dfrac{5}{6}, 0\right)$ **79.** $\left(-\dfrac{2}{3}, 0\right), \left(\dfrac{5}{4}, 0\right)$ **81.** $(0, 0), (10, 0)$ **83.** $(-6, 0), (0, 0), (7, 0)$ **85.** $(-2, 0), (-1, 0), (1, 0)$

87. a)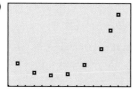

Quadratic function

b)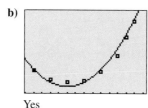

Yes

c) 1981, 2001 **d)** 298 million cases

89. a)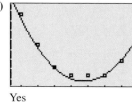

Yes

b) 2004 **c)** 2008

91. 66 **93.** 50 **95.** $-4, 5$ **97.** $0, 1$ **99.** 5 **101.** -1 **103.** -1.7 and 3.7

105. 1 **107.** 19 **109.** 3 **111.** $0, 6$ **113.** 3 **115. a)** $1, 5$

121. $h(x) = x^2 + 9x + 18$; Answers may vary. **123.** Student 1's work is correct.

Exercise Set 6.6

1. 5 **3.** 3 **5.** -2 **7.** $\dfrac{7}{2}$ **9.** 5 **11.** 1 **13.** $(4, 9)$ **15.** $(5, 1)$

17.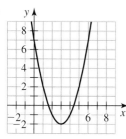

Vertex $(3, -2)$

19.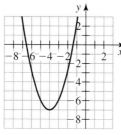

$(-4, -7)$

21.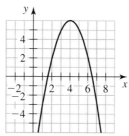

Vertex $(4, 6)$

23.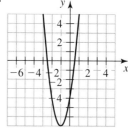

Vertex: $(-1, -7)$

25.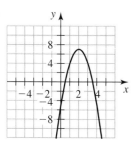

Vertex $(2, 7)$

27.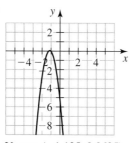

Vertex: $(-1.125, 0.0625)$

29.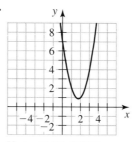

Vertex $(1.75, 0.875)$

31.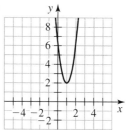

Vertex: $(1, 2)$

33.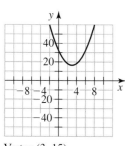

Vertex $(3, 15)$

35.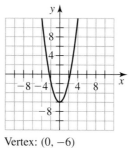

Vertex: $(0, -6)$

37.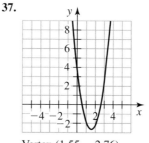

Vertex $(1.55, -2.76)$

39. Vertex: $(-0.88, -6.45)$

41.

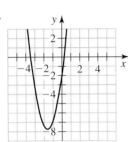

Vertex $(-1.58, -7.76)$

43. 4   **45.** $-\dfrac{5}{2}$ **47.** x-intercepts: $(0, 0)$ and $(2, 0)$;
y-intercept: $(0, 0)$; vertex: $(1, -5)$

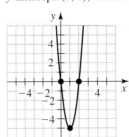

49. x-intercepts: $(0, 0)$ and $(4, 0)$;
y-intercept: $(0, 0)$; vertex: $(2, -4)$

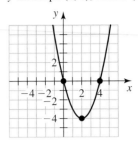

51. x-intercepts: $(4, 0)$ and $(6, 0)$;
y-intercept: $(0, 24)$; vertex: $(5, -1)$

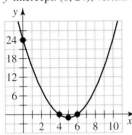

53. x-intercepts: $(1, 0)$ and $(7, 0)$;
y-intercept: $(0, 7)$; vertex: $(4, -9)$

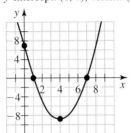

55. x-intercepts: $(-3, 0)$ and $(3, 0)$;
y-intercept: $(0, -9)$; vertex: $(0, -9)$

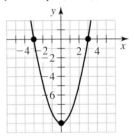

57. x-intercepts: $\left(-\dfrac{3}{2}, 0\right)$ and $(7, 0)$;
y-intercept: $(0, -21)$; vertex: $(2.75, -36.125)$

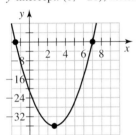

59. a) 3 feet
b) 309.25 feet; 4.375 seconds
c)

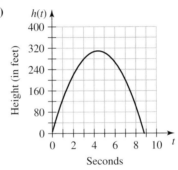

61. a) 1999; $2.16 million
b)

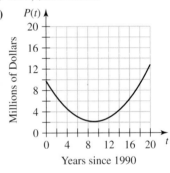

63. 23.0 years; 1958   **65.** 2000; 62.0 thousand cars   **67. a)** -2 **b)** -2 **c)** Yes   **e)** The method in part a can be used.
69. $(3, 2)$ is the vertex of both f and k. The vertex of g is approximately $(2.7, 1.8)$. The vertex of h is approximately $(3.3, 1.7)$.

Chapter 6 Review

1.

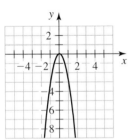

2.

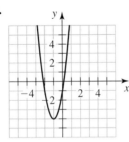

3.

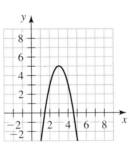

4.

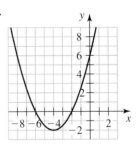

5. $a < 0, h < 0, k > 0$ **6.** $y = -(x - 4)^2 + 4$ **7.** $-3x^2 - 6x$ **8.** $x^2 + 8x + 16$ **9.** $x^2 - 3x - 10$ **10.** $12x^3 - 60x^2 + 75x$
11. $-10x^2 + 13x + 3$ **12.** $-32x^3 + 98x$ **13.** $\dfrac{1}{25}x^2 - \dfrac{1}{64}$ **14.** $x^4 - 10x^2 + 21$ **15.** $x^3 - 3x^2 - 6x + 8$
16. $3x^4 + 7x^3 - 11x^2 + 7x - 2$ **17.** $y = 3x^2 + 9x - 12$; $y = 3(x + 4)(x - 1)$; $y = 3x(x + 3) - 12$; $y = (x + 4)(3x - 3)$
18. $f(x) = -2x^2 - 4x - 5$ **19.** $-4x(x - 3)$ **20.** $(x - 6)(x + 4)$ **21.** $7(x + 3)(x - 3)$ **22.** $\left(x - \dfrac{1}{2}\right)\left(x + \dfrac{1}{2}\right)$
23. $3(x - 5)(x + 3)$ **24.** $x(2.4x - 7.9)$ **25.** $2(x - 4)^2$ **26.** $(2x + 5)(4x - 3)$ **27.** $3(2x - 3)(x - 4)$ **28.** $5x(x - 3)(x + 3)$
29. $3x(x - 7)(x - 2)$ **30.** $(x + 2)(2x - 3)(2x + 3)$ **31.** $-4, 6$ **32.** ± 5 **33.** $\pm\dfrac{1}{9}$ **34.** $\dfrac{1}{3}, 2$ **35.** $0, 10$ **36.** $-5, \dfrac{1}{2}$

37. $-2, 0, \dfrac{5}{2}$ **38.** $-3, \pm 2$ **39.** $1, 20$ **40.** $\dfrac{3}{4}$ **41.** $\left(-\dfrac{1}{2}, 0\right), \left(\dfrac{5}{3}, 0\right)$ **42.** $\left(-\dfrac{7}{8}, 0\right), \left(\dfrac{7}{8}, 0\right)$ **43.** -4 **44.** -4

45. $5a^2 + 19a + 14$ **46.** $-\dfrac{4}{5}, 1$ **47.** $-1, \dfrac{6}{5}$ **48.** $0, \dfrac{1}{5}$ **49.** $\dfrac{7}{2}$

50. *y*-intercept: $(0, 5)$; vertex: $(2, 13)$. **51.** *y*-intercept: $(0, 9)$; vertex: $(0, 9)$ **52.** *y*-intercept: $(0, 0)$; vertex: $(1, -3)$

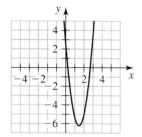 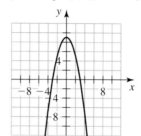 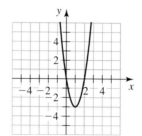

53. *y*-intercept: $(0, 2.1)$; vertex: $(1.43, -6.25)$ **56. a)** 159.25 feet; 3.125 seconds **57.** 20.4 years of age; 1957
b) 6.25 seconds

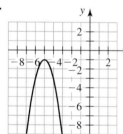 **c)**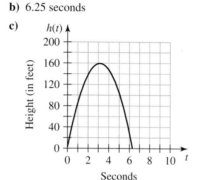

Chapter 6 Test

1. 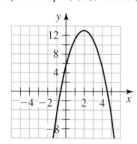 **2.** $a > 0, h > 0, k = 0$ **4.** $-2x^2 - 10x + 28$ **5.** $9x^2 - 49$ **6.** $50x^3 + 160x^2 + 128x$
7. $2x^4 + 3x^3 - x^2 + 7x - 3$ **8.** $-2x^2 - 24x - 61$ **9.** $(x + 2)(8x - 3)$ **10.** $2x(x - 3)^2$
11. $-2, 5$ **12.** $1, \dfrac{11}{2}$ **13.** $\pm\dfrac{4}{5}$ **14.** $-\dfrac{3}{2}, \pm 3$ **15.** $(7, 3)$

16. a) $(-3, 0)$ and $(8, 0)$ **17.** -8 **18.** -11 **19.** $-4, \dfrac{2}{3}$ **26. a)** 1995, 2008
b) $(2.5, -30.25)$ **20.** $-\dfrac{13}{3}, 1$ **21.** 0.75 **22.** -3 **b)** 2002, $5.65 million
c) 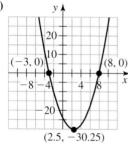 **23.** This is not true for any value of x. **c)**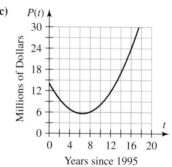
24. 1 **25.** $-3, 5$

Chapter 7

Exercise Set 7.1
1. 4 **3.** $2\sqrt{2}$ **5.** $2\sqrt{3}$ **7.** $\dfrac{2}{3}$ **9.** $\dfrac{\sqrt{6}}{7}$ **11.** $\dfrac{5\sqrt{2}}{2}$ **13.** $\dfrac{3\sqrt{6}}{2}$ **15.** $\dfrac{3\sqrt{2}}{8}$ **17.** $\dfrac{\sqrt{6}}{2}$ **19.** $\dfrac{\sqrt{55}}{10}$
21. ± 6 **23.** $\pm\sqrt{5}$ **25.** $\pm\sqrt{3}$ **27.** $\pm 4\sqrt{2}$ **29.** $\pm 10\sqrt{3}$ **31.** $\pm 4\sqrt{3}$ **33.** No real number solutions **35.** No real number
solutions **37.** ± 3 **39.** $\pm\dfrac{2\sqrt{10}}{5}$ **41.** $\pm\dfrac{\sqrt{42}}{3}$ **43.** $-1, 9$ **45.** $-\dfrac{9}{8}, \dfrac{3}{8}$ **47.** $\dfrac{7}{9}$ **49.** $\dfrac{3 \pm 2\sqrt{21}}{8}$ **51.** No real number solutions

53. 4, 8   **55.** $\dfrac{9 \pm 2\sqrt{3}}{4}$ **57.** $(-\sqrt{17}, 0), (\sqrt{17}, 0)$ **59.** $(0, 0), (4, 0)$ **61.** $(-3, 0), (1, 0)$ **63.** No x-intercepts

65. x-intercepts: $(2, 0)$ and $(8, 0)$;   **67.** x-intercepts: $(-4.88, 0)$ and $(-9.12, 0)$;   **69.** x-intercepts: $(-5.57, 0)$ and $(5.57, 0)$;
vertex: $(5, -9)$ vertex: $(-7, 9)$ vertex: $(0, -31)$

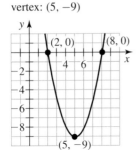
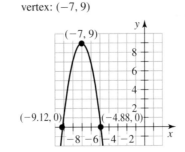
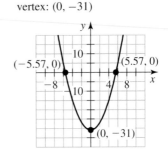

71. No

Exercise Set 7.2

1. $36; (x + 6)^2$ **3.** $49; (x - 7)^2$ **5.** $\dfrac{49}{4}; \left(x - \dfrac{7}{2}\right)^2$ **7.** $\dfrac{9}{4}; \left(x + \dfrac{3}{2}\right)^2$ **9.** $\dfrac{1}{16}; \left(x + \dfrac{1}{4}\right)^2$

11. $\dfrac{4}{25}; \left(x - \dfrac{2}{5}\right)^2$ **13.** $-3 \pm \sqrt{10}$ **15.** $-6 \pm \sqrt{38}$ **17.** $1 \pm \sqrt{11}$ **19.** $9 \pm 2\sqrt{21}$ **21.** $-9 \pm 3\sqrt{10}$ **23.** No real number solutions

25. $\dfrac{7 \pm \sqrt{61}}{2}$ **27.** $\dfrac{-5 \pm \sqrt{41}}{2}$ **29.** No real number solutions   **31.** $\dfrac{5 \pm \sqrt{33}}{4}$ **33.** $\dfrac{4 \pm \sqrt{22}}{2}$ **35.** $\dfrac{-15 \pm 3\sqrt{5}}{10}$ **37.** $\dfrac{-2 \pm \sqrt{19}}{3}$

39. $\dfrac{1 \pm \sqrt{57}}{4}$ **41.** No real number solutions   **43.** $\dfrac{5 \pm 3\sqrt{17}}{16}$ **45.** $\dfrac{5 \pm \sqrt{65}}{4}$ **47.** $3 \pm \sqrt{11}$ **49.** No   **51. a)** No values   **b)** -3

c) $-4, -2$ **53.** $(4 - \sqrt{13}, 0), (4 + \sqrt{13}, 0)$ **55.** No x-intercepts   **57.** $(-5, 0)$

Exercise Set 7.3

1. $\dfrac{-5 \pm \sqrt{41}}{4}$ **3.** $\dfrac{-7 \pm \sqrt{73}}{6}$ **5.** $\dfrac{5 \pm \sqrt{17}}{4}$ **7.** No real number solutions   **9.** $-3, \dfrac{1}{2}$

11. $\pm \dfrac{\sqrt{51}}{3}$ **13.** $-\dfrac{5}{2}, 0$ **15.** $\dfrac{1 \pm \sqrt{13}}{4}$ **17.** No real number solutions   **19.** $\dfrac{5 \pm \sqrt{97}}{6}$ **21.** No real number solutions

23. $-3 \pm 2\sqrt{3}$ **25.** $-0.64, 3.14$ **27.** $-0.52, 3.05$ **29.** $0.020, -8.54$ **31.** ± 5 **33.** 4   **35.** $\pm 2\sqrt{5}$ **37.** $\dfrac{-3 \pm 2\sqrt{5}}{4}$

39. No real number solutions   **41.** $0, \dfrac{5}{9}$ **43.** $-1, \dfrac{8}{5}$ **45.** $0, 9$ **47.** No real number solutions   **49.** -6 **51.** $-\dfrac{5}{4}, 2$

53. $4 \pm \sqrt{10}$ **55.** $-1 \pm 2\sqrt{2}$ **57.** No real number solutions   **59.** $-5, 8$ **61.** $0, 7$ **63.** $3, 7$ **65.** $\dfrac{1 \pm \sqrt{7}}{3}$ **67.** $\dfrac{7 \pm \sqrt{13}}{2}$

69. No real number solutions   **71.** $x^2 - 3x - 10$ **73.** $\dfrac{3 \pm \sqrt{61}}{2}$ **75.** $1, 3$ **77.** $-4x^2 + 16x - 13$

79. a) No such points   **b)** One point   **c)** Two points   **d)**

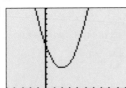

81.

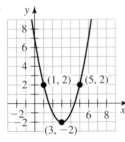

Points of height 2: $(1, 2), (5, 2)$;
Vertex: $(3, -2)$

83. a)

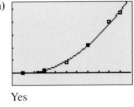

Yes

b) 6796 schools   **c)** 2010

85. a)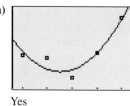

Yes

b) 1406.62; There will be 1407 convictions in 2003. **c)** $t = -3.48$ or $t = 14.54$; There were 2000 convictions in 1987 and there will be 2000 convictions in 2005.

87. No **89.** $(-1.16, 0), (5.16, 0)$ **91.** No x-intercepts **93. a)** $x = -\dfrac{b}{m}$ **b)** -3

95. 9 **97.** ± 8 **99.** 1 **101.** $-4, 5$

Exercise Set 7.4
1. $y = x^2 + 2x + 3$ **3.** $y = x^2 + 3x + 1$ **5.** $y = -3x^2 + 7x + 5$ **7.** $y = 2x^2 - x - 4$
9. $y = x^2 + 2x - 6$ **11.** $y = -2x^2 + 7x + 4$ **13.** $y = 3x^2 - 7x - 4$ **15.** $y = 2x^2 - 8x + 3$ **17.** $y = 2x^2 - 6x + 3$
19. $y = -3x^2 + 8x + 4$ **21.** $y = 3x^2 + x - 1$ **23.** $y = -2x^2 + x + 17$ **25.** $y = x^2$ **27.** $f(x) = x$; linear **29.** $y = 2(x - 5)^2 - 7$
or $y = 2x^2 - 20x + 43$ **31.** $y = 2x^2 - 6x + 4$ **33.** Linear: $y = 2x + 2$; Quadratic: $y = 2x^2 + 2$ (answers may vary);
Exponential: $y = 2(2)^x$ **35.** $y = x^2 - 9x + 22$ **37.** $y = -2x^2 + 20x - 42$

Exercise Set 7.5
1. a) Quadratic function **b)** Linear function **c)** Exponential function **d)** None of the mentioned
types of functions **5.** $f(t) = -0.66t^2 + 21.49t - 95.85$ **7. a)** Linear: $f(t) = 5.09t + 11.75$; Exponential: $f(t) = 16.63(1.14)^t$;
Quadratic: $f(t) = 0.28t^2 + 1.98t + 16.49$ **b)** The exponential model is best. **9.** $f(t) = 0.013t^2 - 1.19t + 28.24$
11. $f(t) = 0.0067t^2 - 0.12t + 6.39$ **13. a)** Quadratic function **b)** $f(d) = -0.000333d^2 + 0.152d + 4$
15. a) $C(t) = 0.90t^2 - 19.94t + 121.38$, $H(t) = 2.48t - 1.30$ **b)** $(8.12, 18.83)$ and $(16.79, 40.35)$; Corona beer and Heineken beer
had equal sales of 18.8 million cases in 1988 and had equal sales of 40.4 million cases in 1997, according to the models.

Exercise Set 7.6
1. a) $(5.33, 0), (27.23, 0)$; No firms performed drug tests in 1985 and 2007. **b)** For years up to and
including 1985 and for years after 2007 **c)** 79%; 1996 **3. b)** 46 year-old drivers; 1 fatal crash per 100 million miles **5. a)** 21.3 feet
b) Yes **7. a)** 53.33; In 2010, 53.3% of households will have cable. Model breakdown has likely occurred. **b)** $t = 2.01$ or $t = 37.72$;
30% of households had cable in 1982 and will have cable in 2018. Model breakdown is likely for the 2018 prediction. **c)** 64.4%; 2000
d) For years before 1976 and for years after 2000 **9.** 1988, 1997 **11. a)** 295.85; The population will be 295.9 million in 2007.
b) $t = 218.49$ or $t = -200.57$; The population was 300 million in 1589 (model breakdown) and will be 300 million in 2008.

c)

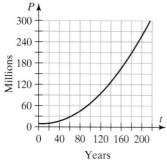

Years

d) $t < 8.96$; For years before 1799 **e)**

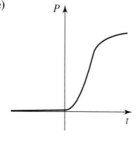

13. a) 2.1 billion people **b)**

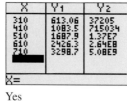

Yes

15. a) $f(t) = -1.30t^2 + 28.99t - 61.29$

b) $(2.36, 0), (19.94, 0)$; There were no schools with Internet access in 1992 and there will be no schools with Internet access in 2010 (model breakdown for the 2010 prediction).

c) The vertex is $(11.15, 100.33)$; The percentage of schools with Internet access will reach a maximum of 100% in 2001.

d) $t < 2.36$ or $t > 19.94$; For years before 1992 and for years after 2001

e)

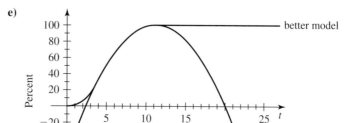

17. a) Linear: $f(t) = 1.86t - 1.61$; Exponential: $f(t) = 1.63(1.26)^t$; Quadratic: $f(t) = 0.25t^2 - 0.92t + 2.62$. The exponential model and the quadratic model fit the data well. **b)** Exponential model **c)** 2026 **d)** 2149

Chapter 7 Review

1. $6\sqrt{2}$ **2.** $\dfrac{\sqrt{15}}{5}$ **3.** $\dfrac{5\sqrt{2}}{7}$ **4.** $\dfrac{7}{10}$ **5.** $\dfrac{3 \pm \sqrt{57}}{12}$ **6.** $\pm\dfrac{\sqrt{35}}{5}$ **7.** $\dfrac{15 \pm \sqrt{15}}{5}$ **8.** $\pm 7\sqrt{2}$

9. $3 \pm 2\sqrt{5}$ **10.** No real number solutions **11.** $0, 2$ **12.** $-4, 6$ **13.** $\pm\dfrac{5}{2}$ **14.** $\dfrac{1}{3}, 2$ **15.** $\pm\dfrac{3}{5}$ **16.** $\dfrac{-5 \pm \sqrt{41}}{4}$

17. $\dfrac{-5 \pm \sqrt{57}}{4}$ **18.** $2 \pm \sqrt{3}$ **19.** $\pm\dfrac{\sqrt{10}}{2}$ **20.** $5 \pm \sqrt{17}$ **21.** $1, 20$ **22.** No real number solutions **23.** No real number solutions

24. $\pm\dfrac{7}{5}$ **25.** $-1.18, 3.07$ **26.** $-3.05, 1.41$ **27.** $-3 \pm \sqrt{13}$ **28.** No real number solutions **29.** $\dfrac{-3 \pm \sqrt{57}}{4}$ **30.** $\dfrac{-5 \pm \sqrt{29}}{2}$

31. $\left(-\dfrac{\sqrt{30}}{3}, 0\right), \left(\dfrac{\sqrt{30}}{3}, 0\right)$ **32.** $\left(\dfrac{-6 - \sqrt{10}}{2}, 0\right), \left(\dfrac{6 + \sqrt{10}}{2}, 0\right)$ **33.** $\left(\dfrac{-1 - \sqrt{7}}{3}, 0\right), \left(\dfrac{-1 + \sqrt{7}}{3}, 0\right)$ **34.** No x-intercepts

35. $\left(\dfrac{3 - \sqrt{41}}{4}, 0\right), \left(\dfrac{3 + \sqrt{41}}{4}, 0\right)$ **36.** $\left(\dfrac{5 - \sqrt{109}}{6}, 0\right), \left(\dfrac{5 + \sqrt{109}}{6}, 0\right)$ **37.** $-2, 4$ **38.** ± 12 **39. a)** No such value **b)** 1

c) $\dfrac{3 \pm \sqrt{3}}{3}$ **40.** $y = 3(x + 4)^2 + 3$ **41.** $y = 2x^2 - 3x + 4$ **42.** $y = -2x^2 + 5x + 1$ **43.** $y = 2x^2 - x + 3$ **44.** $y = -2x^2 + 3x + 5$

45. $y = -3x^2 + 4x + 7$ **46.** $y = 2x^2 + x + 3$ **47.** Linear: $y = -2x + 4$; Exponential: $y = 4\left(\dfrac{1}{2}\right)^x$; Quadratic: $y = -2x^2 + 4$

(answers may vary) **48.** $y = x^2 - 5x + 7$ **49. a)** $f(t) = -0.0123t^2 + 1.85t + 0.64$ **b)** 30.0% **c)** 35 years **d)** 75 years

50. a)

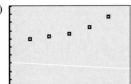

b) $f(t) = 0.037t^2 - 0.47t + 31.84$

c) 65.10; In 2007, 65.1% of the grades at Princeton will be A's.

d) $-37.04, 49.74$; In 1933, all grades were A's. In 2020 all grades will be A's. Model breakdown has occurred.

Chapter 7 Test

1. $4\sqrt{2}$ **2.** $\dfrac{7\sqrt{2}}{2}$ **3.** $\dfrac{2\sqrt{15}}{15}$ **4.** $-2, 5$ **5.** $\pm\dfrac{5\sqrt{6}}{3}$ **6.** No real number solutions

7. $\dfrac{6 \pm \sqrt{6}}{2}$ **8.** $0, 7$ **9.** ± 9 **10.** $-1 \pm \sqrt{22}$ **11.** No real number solutions **12.** $\dfrac{-3 \pm \sqrt{3}}{2}$ **13.** $\dfrac{4 \pm 4\sqrt{2}}{3}$ **14.** $-1.93, 0.34$

15. $4 \pm 3\sqrt{2}$ **16.** $\dfrac{-3 \pm \sqrt{73}}{4}$ **17.** $\left(\dfrac{4 - \sqrt{13}}{3}, 0\right), \left(\dfrac{4 + \sqrt{13}}{3}, 0\right)$

18. $(1.42, 0), (4.58, 0); (3, 5)$;

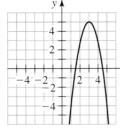

19. ± 1 **20.** $y = x^2 + 2x + 1$ **21.** $y = 2x^2 - 20x + 53$ **22. a)** 0 **b)** 1 **c)** 2

23. a) $f(t) = -1.14t^2 + 25.26t - 81.20$ **b)** 41.20; In 2005, 41.2% of television shows will be at least an hour long. **c)** 2004 **d)** $(3.90, 0), (18.26, 0)$; No television shows were at least an hour long in 1994 and no television shows will be at least an hour long in 2008. **e)** For years before 1994 and for years after 2008

24. 2.5 seconds, 103 feet

Cumulative Review for Chapters 1–7

1. $\pm\dfrac{2\sqrt{30}}{5}$ **2.** $\dfrac{19}{5}$ **3.** 13 **4.** 2.8904 **5.** ± 1.5807

6. $\dfrac{5 \pm \sqrt{73}}{6}$ **7.** 2.4962 **8.** $-\dfrac{5}{2}, \dfrac{4}{3}$ **9.** 2.8394 **10.** $-1, \dfrac{5}{2}$ **11.** -30 **12.** ± 1.4142 **13.** $\dfrac{4 \pm \sqrt{2}}{7}$ **14.** 1.1540

15. $\dfrac{15 \pm 3\sqrt{5}}{5}$ **16.** $\dfrac{-3 \pm \sqrt{57}}{4}$ **17.** $(1, 2)$ **18.** $(2, 5)$ **19.** $(-3, -4)$ **20.** $x < -\dfrac{1}{12}, \left(-\infty, -\dfrac{1}{12}\right)$,

21. $\dfrac{648b^8}{c^{23}}$ **22.** $\dfrac{2}{3b^{1/4}c^{1/5}}$ **23.** $\dfrac{27b^{36}}{64c^6}$ **24.** $\ln(128x^{31})$ **25.** $\log_b(4x^{12})$ **26.** $9x^2 - 24x + 16$ **27.** $25x^2 - 49$

28. $-3x^6 + 15x^4 - 24x^3 + 120x$ **29.** $2x^3 + 5x^2 - 22x + 15$ **30.** $f(x) = -2x^2 + 20x - 47$ **31.** $(9x - 5)(9x + 5)$

32. $x(x - 8)(x - 5)$ **33.** $(2x + 7)(4x - 3)$ **34.** $(x - 3)(x + 3)(x + 4)$ **35.** $f(x) = -3x + 20$ **36.** $g(x) = \dfrac{4}{3}(3)^x$

37. 4 **38.** 108 **39.** 1 **40.** 4 **41.** $(2, 0)$

42. 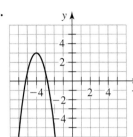 **43.** 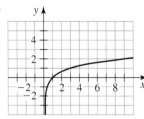 **44.** 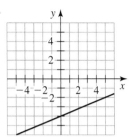 **45.**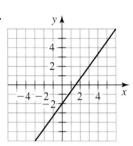

46. 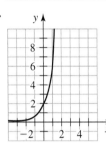 **47.** 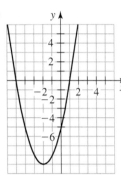 **48.**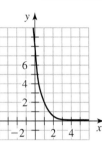

49. 3 **50.** $y = -\dfrac{4}{3}x + \dfrac{10}{3}$ or $y = -1.33x + 3.33$ **51.** $y = 78(0.82)^x$ **52.** $y = 12.77(1.45)^x$ **53.** $y = x^2 + 2x - 4$

54. a) $f(x) = 3x + 3$; $g(x) = 3(2)^x$; $h(x) = 3x^2 + 3$ (answers may vary) **b)**

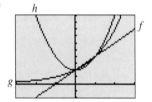

55. 2, 4 **56.** 3 **57.** No such values **58.** $(1, 0), (5, 0)$ **59.** $(0, -5)$

60. 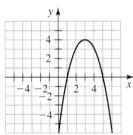 **61.** 4 **62.** 3 **63.** -2 **64.** 0.3155 **65.**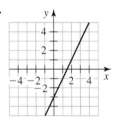

66. $g^{-1}(x) = \log_3(x)$ **67.** $f^{-1}(x) = \dfrac{5}{2}x - \dfrac{5}{2}$ **68. a)** $f(t) = 3.97t + 3.02$ **b)** 74.48; In 2008, 74% of companies will offer stock options to at least half of their employees. **c)** 24.43; All companies will offer stock options to at least half of their employees in 2014. Model breakdown is likely. **d)** $(-0.76, 0)$; No companies offered stock options to at least half of their employees in 1989. Model breakdown has occurred. **e)** $t < -0.76$ or $t > 24.43$ **69. a)** Linear: $f(t) = 0.18t - 0.61$; Exponential: $f(t) = 0.15(1.22)^t$; Quadratic: $f(t) = 0.0216t^2 - 0.19t + 0.88$ **c)** Quadratic model **d)** \$0.46 billion **e)** 22% per year **f)** Quadratic model prediction: 2009, Exponential model prediction: 2008 **70. a)** $h(t) = 2.4t + 10.7, c(t) = t + 53$ **b)** Rate of change at Hartford: 2.4 percent per year, Rate of change in Connecticut: 1 percent per year **c)** $h(18) = 53.9, c(18) = 71$ **d)** 2020

Chapter 8

Exercise Set 8.1

1. The set of real numbers except 0 **3.** The set of real numbers **5.** The set of real numbers except -3 **7.** The set of real numbers except 2 **9.** The set of real numbers except $-\dfrac{1}{2}$ **11.** The set of real numbers except -8 and 4 **13.** The set of real numbers except -2 and 5 **15.** The set of real numbers except ± 3 **17.** The set of real numbers except $\pm\dfrac{5}{2}$ **19.** The set of real numbers **21.** The set of real numbers except $-\dfrac{3}{2}$ and 5 **23.** The set of real numbers **25.** The set of real numbers except $\dfrac{1 \pm \sqrt{22}}{3}$ **27.** The set of real numbers except $\pm\dfrac{3}{2}$, and 2 **29.** The set of real numbers except $\dfrac{8}{5}$

31. The set of real numbers except $\dfrac{-5 \pm \sqrt{33}}{4}$ **33.** $f(x) = \dfrac{1}{x^4}$ **35.** $f(x) = \dfrac{4x^3}{3}$ **37.** $f(x) = \dfrac{4}{5}$ **39.** $f(x) = \dfrac{x+5}{x-9}$

41. $f(x) = \dfrac{5}{x-2}$ **43.** $f(x) = \dfrac{3}{4x-5}$ **45.** $f(x) = \dfrac{x+7}{x-7}$ **47.** $f(x) = \dfrac{x+5}{3(x+1)}$ **49.** $f(x) = \dfrac{x-4}{x+4}$ **51.** $f(x) = -1$

53. $f(x) = -\dfrac{2}{3}$ **55.** $f(x) = \dfrac{-6}{x+3}$ **57.** $f(x) = -\dfrac{x+7}{x+2}$ **59.** $f(x) = -\dfrac{3x+2}{5}$ **61.** $f(x) = \dfrac{3x(x+4)}{x-3}$ **63.** $f(x) = \dfrac{x-2}{3}$

65. a) 8.4; If the person drives 50 mph, the trip will take 8.4 hours.   **b)** 7.64, 7, 6.46, 6   **c)** Decreasing function; For greater speeds, the trip takes less time.   **69.** Members of the domain: $-3, -2, -1, 0, 1, 2, 3$; Members of the range: 7.2, 8.4, 10.08, 12.6, 16.8, 25.2, 50.4
71. The student's work is incorrect.   **73.** 0   **75.** Not defined

Exercise Set 8.2

1. $\dfrac{10}{x^2}$ **3.** $\dfrac{21x^2}{5}$ **5.** $\dfrac{3}{8}$ **7.** $\dfrac{6x^2}{25}$ **9.** $\dfrac{4(x+8)}{3(x+3)}$ **11.** $\dfrac{2(x+9)}{x-3}$ **13.** $\dfrac{3}{x^5}$ **15.** $\dfrac{2(x+7)}{x-3}$

17. $\dfrac{2(x+1)}{x-3}$ **19.** $-\dfrac{8}{3}$ **21.** $\dfrac{2(x-4)(x-1)}{x+6}$ **23.** $\dfrac{x-5}{3(x+1)(x+5)}$ **25.** $-\dfrac{x+4}{(x-4)(x+1)}$ **27.** $\dfrac{2(x-5)(2x+3)}{(x+2)(x+4)}$

29. $\dfrac{8}{(x-6)(3x-4)}$ **31.** $\dfrac{x+4}{2}$ **33.** $-x^2$ **35.** $-\dfrac{(x-4)(x-2)}{8x(x+4)}$ **37.** $-\dfrac{2(x-2)(4x+5)}{9(x^2+4)}$ **39.** $\dfrac{(x+2)(x-3)}{x^6}$

41. $\dfrac{(x+2)(x+3)}{3x}$ **43.** $\dfrac{10x(3x+1)}{x+2}$ **45.** $\dfrac{x-1}{(x-2)(x+5)}$ **47.** $\dfrac{25}{2x^6(x-12)}$ **49.** $-\dfrac{36}{x^4}$ **51.** 1   **53.** $\dfrac{(x-8)^2}{(x-5)^2}$ **55.** $\dfrac{(x+8)^2}{(x+2)^2}$

57. $-\dfrac{(x+1)^2}{(x-7)^2}$ **61. a)** 66.7 million households   **b)** 103.7 million households   **c)** 64.3%   **d)** $p(t) = \dfrac{-7.7t^2 + 406t + 1738}{1.20t + 80.92}$
e) 64.3%   **f)** 63.90; In 1998, 63.9% of households had cable television.

g)

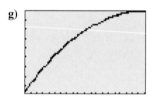 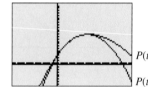

$P(t) = \dfrac{-7.7\,t^2 + 406\,t + 1738}{1.20\,t + 80.92}$

$P(t) = -0.108\,t^2 + 4.29\,t + 21.83$

63. $\dfrac{x+5}{x-5}, \dfrac{x+5}{x-5}$; Answers may vary.

Exercise Set 8.3

1. $\dfrac{5x+1}{x-1}$ **3.** $\dfrac{x-5}{x+7}$ **5.** $\dfrac{2(x-2)}{x^2}$ **7.** $\dfrac{x^2+2}{4x^6}$ **9.** $\dfrac{7x-2}{(x-2)(x+1)}$ **11.** $\dfrac{2(x+9)}{(x-6)(x-1)(x+4)}$

13. $\dfrac{19x+87}{15(x-2)(x+3)}$ **15.** $\dfrac{8x+25}{x(x-5)(x+5)}$ **17.** $\dfrac{5x+1}{(x-4)(x-3)(x+3)}$ **19.** $\dfrac{3x-1}{x+1}$ **21.** $\dfrac{2x}{x+1}$ **23.** $\dfrac{12}{x-6}$ **25.** $\dfrac{1}{2(x+3)}$

27. $\dfrac{4x^2+13x-1}{(2x-7)(2x+7)}$ **29.** $\dfrac{2x^2+2x+5}{(x-1)(x+2)}$ **31.** $\dfrac{2x^2+x+17}{(x-5)(x+4)}$ **33.** $\dfrac{x-6}{2(x-5)}$ **35.** $\dfrac{2x^2+5x-1}{(x-4)(x+1)(x+3)^2}$ **37.** $\dfrac{4}{x-2}$

39. $\dfrac{x^2-15x-21}{(2x+5)^2(3x+1)}$ **41.** $\dfrac{7x^2-11x-1}{(x+2)^2(3x-1)}$ **43.** $\dfrac{x^2+x+8}{6x(x-4)(x+2)}$ **45.** $\dfrac{5}{2(x+2)}$ **47.** $\dfrac{-x+16}{(x+1)(x+5)}$ **49.** $\dfrac{2(2x^2+8x+5)}{(x+2)^2}$

51. $\dfrac{5(x+2)}{3(x+1)}$ **53.** $\dfrac{2x^2-25}{(x-4)(x-3)}$ **55.** $-\dfrac{7}{(x-4)(x-3)}$ **57.** $\dfrac{-x^2+6x-2}{3(x-4)(x+2)}$ **63.** $\dfrac{3x}{x-1}, \dfrac{x}{x-1}$; Answers may vary.

Exercise Set 8.4

1. $\dfrac{2}{3}$ **3.** $\dfrac{x^3}{3}$ **5.** $\dfrac{15}{16x^4}$ **7.** $\dfrac{(x-6)(x+1)}{5}$ **9.** $\dfrac{7(x+7)}{3x(x+2)}$ **11.** $\dfrac{x}{2}$ **13.** $\dfrac{-3x^2+2}{4x+5}$

15. $\dfrac{-5}{3x}$ **17.** $-\dfrac{4x+3}{3x-2}$ **19.** $\dfrac{x}{2x+1}$ **21.** $\dfrac{x-3}{x}$ **23.** $\dfrac{x-9}{x+6}$ **25.** $\dfrac{x^2-4x+2}{x^2-4x-3}$ **27.** $-\dfrac{1}{x(x+3)}$ **29.** $-\dfrac{2(x+1)}{x^2(x+2)^2}$

31. $-\dfrac{(x+3)(x+4)(3x+10)}{x(x+2)}$ **33.** $h(x) = \dfrac{x^2+4}{10}$ **35.** $h(x) = \dfrac{5(x-1)}{4(x-3)}$ **37.** $h(x) = \dfrac{(x-2)(x+1)}{(x-1)(x+2)}$ **39.** $-\dfrac{x+1}{x-1}$

41. x^2-x **45.** $\dfrac{-2(5x-6)}{3x+4}$

Exercise Set 8.5

1. 5   **3.** 7   **5.** Empty set solution   **7.** 6   **9.** 5   **11.** 4   **13.** Empty set solution   **15.** -2
17. Empty set solution   **19.** 7   **21.** -2 **23.** -2 and $\dfrac{5}{3}$ **25.** Empty set solution   **27.** Empty set solution   **29.** $-3 \pm \sqrt{19}$

31. -3 and $-\dfrac{2}{3}$ **33.** 3 and 23 **35.** 2 **37.** Empty set solution **39.** Empty set solution **41.** $\dfrac{5}{8}$ **43.** $\dfrac{7 \pm \sqrt{31}}{3}$ **45.** $\dfrac{23}{4}$

47. No such value **49.** $-4 \pm \sqrt{15}$ **51.** -1 **53.** $\left(-\dfrac{7}{5}, 0\right)$ **55.** No **57.** 2 **59.** $\dfrac{-x + 2}{x}$ **61.** $\dfrac{2(3x + 1)}{x(x + 1)}$

63. $-\dfrac{1}{3}$ **65.** $-\dfrac{15}{2}$ **67.** $\dfrac{10x - 1}{(x - 3)(x - 2)(x + 2)}$

69. a) f is a decreasing function for $d > 0$.
This means that the farther you are from
the speaker, the lower the sound level will be.

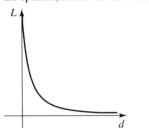

b) 2250

c) $f(d) = \dfrac{2250}{d^2}$

d)

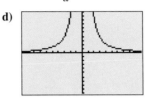

e) 35.16 decibels

f) 5.67 feet

71. $a = -6, b = -3$

Exercise Set 8.6

1. a) $C(n) = 350n + 1250$ **b)** $M(n) = \dfrac{350n + 1250}{n}$ **c)** $\$391.67$ **d)** 25 students

3. a) $T(n) = 50n + 500$ **b)** $M(n) = \dfrac{50n + 500}{n}$ **c)** 51.85; If 270 people go to the restaurant, the mean cost per person is $\$51.85$.

d) 50; If 50 people go to the restaurant, the mean cost per person is $\$60$. **e)** The entries in the second column are

55, 52.5, 51.67, 51.25, and 51. **f)** The values get close to 50. **5. a)** $C(n) = 7000n + 90,000$ **b)** $B(n) = \dfrac{7000n + 90,000}{n}$

c) $P(n) = \dfrac{9000n + 90,000}{n}$ **d)** 11,250; If 40 cars are produced and sold per day, the price should be $\$11,250$ per car for the company

to make a profit of $\$2000$ on each car. **e)** The values get close to 9000. This means that if a very large number of cars are produced

and sold, the price can be set a few cents more than $\$9000$ to insure a profit of $\$2000$ per car. **7. a)** $B(t) = 221.52t - 1063.14$

b) $E(t) = 0.223t + 9.12$ **c)** $A(t) = \dfrac{221.52t - 1063.14}{0.223t + 9.12}$ **d)** $\$418.02$ **e)** 2010 **9.** 1.4 hours **11. a)** 4.5 hours

b) $T(a) = \dfrac{295}{a + 65}$ **c)** 4.21; If the student drives at 70 mph, the driving time will be 4.21 hours. **d)** 8.75; If the student drives

at 73.75 mph, the driving time will be 4 hours. **13. a)** 3.8 hours **b)** $T(a) = \dfrac{114}{a + 70} + \dfrac{144}{a + 65}$ or $T(a) = \dfrac{258a + 17,490}{(a + 65)(a + 70)}$ **c)** 3.8

d) 3.35; This means that if the student drives 10 mph over the speed limits, the driving time will be 3.35 hours. **e)** 0.45; This means that

if the student drives 10 mph over the speed limits, the driving time will be 0.5 hours less than if the student drives at the speed limits.

15. a) $T(a) = \dfrac{253}{a + 75} + \dfrac{410}{a + 70}$ or $T(a) = \dfrac{663a + 48,460}{(a + 75)(a + 70)}$ **b)** The student would have to drive 11.0 mph over the speed limits.

Exercise Set 8.7

1. $w = kt$ **3.** $y = \dfrac{k}{\sqrt{t}}$ **5.** $B = \dfrac{k}{t^3}$ **7.** w varies inversely as t. **9.** A varies directly as x^3.

11. y varies inversely as 2^u. **13.** $y = 3x$ **15.** $y = \dfrac{15}{x}$ **17.** $y = 2x$ **19.** $p = \dfrac{22.62}{t}$ **21.** $A = \pi r^2$ **23.** $H = 4\log_2(t)$

25. $w = \dfrac{14}{\sqrt[3]{t}}$ **27.** $A = \dfrac{7}{2^r}$ **29.** $\$948$ **31.** 96 newtons **33.** 186.1 feet **35.** 3.5 meters

37. a) $f(d) = \dfrac{202.8}{d^2}$

b) 202.8, 50.7, 22.5, 12.7; The values
are the intensities (in W/m^2) at distances
of 1, 2, 3, and 4 km, respectively.

c)

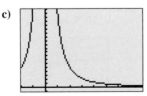

The value of I decreases as the value of
d increases; The intensity of the signal
decreases as the distance increases.

d) The value of $f(d)$ is positive
and nearly zero. If the distance
is extremely large, then the intensity
of the signal will be almost zero.

39. a) $f(d) = \dfrac{3200}{d^2}$ **b)** 128 pounds **c)** 56,569 miles **d)** 0.056 pounds; Yes

41. a) $T = 0.000906d$ **b)** 4415 feet **c)** For each additional foot away from the lightning strike, it takes another 0.000906 seconds to

hear the thunder. **d)** No; The number of seconds divided by 5 is approximately the number of miles that the person is from the strike.

43. a)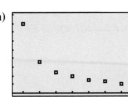

b) $f(d) = \dfrac{148.86}{d}$

c) As distance increases, the apparent height decreases.

d) 1.5 inch

e) 148.9 inches or 12.4 feet

45. a)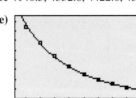

b) $F = \dfrac{2805.21}{L}$

c) F varies inversely as L.

d) 370.1 hertz

47. a) $I = \dfrac{k}{d^2}$ **b)** $k = Id^2$ **c)** The entries in the third column of the table are 4140.5, 4332.8, 4422.6, 4350.0, 4222.9, 4219.2, 4394.0, 4194.4; The average of these entries is about 4284.55. **d)** $I = \dfrac{4284.55}{d^2}$ **e)** 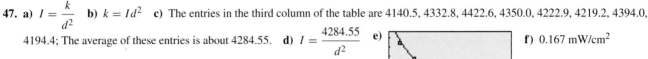 **f)** 0.167 mW/cm^2

49. $f(L) = 5L; 5$ **51.** $f(n) = \dfrac{2}{n}, 2$ **53.** $f(r) = 2\pi r; 2\pi$ **55. a)** Yes **b)** No **57.** False **59.** False **61.** Yes; $\dfrac{1}{k}$

Chapter 8 Review

1. The domain is the set of real numbers except $\pm\dfrac{7}{2}$. **2.** The domain is the set of real numbers except $-\dfrac{7}{3}$ and $\dfrac{5}{4}$. **3.** The domain is the set of real numbers. **4.** The domain is the set of real numbers except $\dfrac{3}{2}$. **5.** $f(x) = \dfrac{3}{x-2}$

6. $f(x) = -\dfrac{6}{5}$ **7.** $f(x) = \dfrac{x+4}{2x(x-4)}$ **8.** $f(x) = \dfrac{1}{(3x+1)(3x-1)}$ **9.** $\dfrac{5}{4}$ **10.** $-\dfrac{3}{2}$ **11.** $-\dfrac{4x+5}{2(x+5)}$ **12.** $-\dfrac{2x(x+7)^2}{5(x+3)}$

13. $\dfrac{x-1}{4(x+1)}$ **14.** $\dfrac{4x}{x+4}$ **15.** $\dfrac{3(x+2)}{x+4}$ **16.** $(x-2)^3$ **17.** $\dfrac{(x+3)(2x-3)}{x-3}$ **18.** $\dfrac{x^2+3x+4}{x+1}$ **19.** $\dfrac{3x^2-2x+4}{(x+2)^2(3x-4)}$

20. $-\dfrac{2}{x-2}$ **21.** $\dfrac{2(x^2+x+4)}{(x-2)^2(x+2)}$ **22.** $-\dfrac{1}{4(x+1)}$ **23.** $\dfrac{3x+13}{(x-6)(x+6)}$ **24.** $\dfrac{x^2-3x+16}{2(x+5)(x-5)(x-2)}$ **25.** $\dfrac{-8x^2+45x-8}{5(4x-3)(4x+3)}$

26. $\dfrac{(x-2)(x+5)}{x(x-1)(x+1)(2x-5)}$ **27.** $\dfrac{x-3}{x(x-5)}$ **28.** $-\dfrac{10(2x-1)}{(x+1)(3x+4)}$ **29.** $f(x)g(x) = \dfrac{(x-2)(x+1)}{(x+2)^2}$

30. $f(x) \div g(x) = \dfrac{(x-2)(x+1)}{(x+3)^2}$ **31.** $f(x) + g(x) = \dfrac{2x^2+5x+7}{(x+2)(x+3)}$ **32.** $f(x) - g(x) = \dfrac{-7x-11}{(x+2)(x+3)}$

34. The correct answer is $\dfrac{4x-3}{x-9}$. **35.** $\dfrac{1}{(x-3)(x+2)}$ **36.** $\dfrac{x+3}{x-6}$ **37.** $-\dfrac{4(x-2)}{3x^3}$ **38.** $\dfrac{x-1}{x+1}$ **39.** 6 **40.** 2 **41.** Empty set solution **42.** $-8, 1$ **43.** Empty set solution **44.** $-3 \pm 3\sqrt{3}$ **45.** $-\dfrac{1}{2}$ **46.** $\dfrac{4x-23}{(x-6)(x-5)(x+5)}$ **47.** $-\dfrac{2(2x+1)}{x-2}$ **48.** $\dfrac{33}{2}$

49. $\left(\dfrac{5}{3}, 0\right)$ **50.** H varies directly as u^2. **51.** w varies inversely as $\log(t)$. **52.** $y = \dfrac{2}{7}\sqrt{x}$ **53.** $B = \dfrac{72}{r^3}$ **54.** 4.14 inches

55. a) $m = kr^3$ **b)** $k = \dfrac{m}{r^3}$ **c)** The entries in the third column of the table are 17.1, 17.01, 17.02, 16.99, 16.99, 16.99; The average of these entries is about 17.02. **d)** $m = 17.02r^3$ **e)** 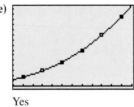 **f)** 207.1 grams

Yes

56. a) $C(n) = 40n + 600$ **b)** $M(n) = \dfrac{40n + 600}{n}$ **c)** 42.22; When 270 people use the room, the mean cost per person is $42.22. Model breakdown has occurred since the room capacity is 120. **d)** 60; When the mean cost per person is $50, 60 people are using the room.

57. a) $N(t) = 1.96t + 3.14, T(t) = 3.04t + 21.79$

b) $P(t) = \dfrac{N(t)}{T(t)} \cdot 100$

c) $P(t) = \dfrac{196t + 314}{3.04t + 21.79}$

d) $P(10) = 43.57$; This means that 43.6% of prisoners were nonviolent in 1990.

e)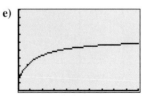

P is an increasing function for $t \geq 0$. This means that the percentage of prisoners that are nonviolent is increasing.

f) According to the model, half of the prisoners were nonviolent in 1998.

58. a) $T(a) = \dfrac{75}{a+50} + \dfrac{40}{a+65}$ **b)** 1.94; When the student drives 5 mph above the speed limits, the driving time is 1.9 hours.

c) 3.1 mph over the speed limits

Chapter 8 Test

1. The domain is the set of real numbers except $-\dfrac{5}{2}$ and $\dfrac{2}{3}$. **2.** The domain is the set of real numbers except -6 and 6. **3.** The domain is the set of real numbers. **5.** $-\dfrac{3}{x-3}$ **6.** $\dfrac{3x+1}{2x(3x-1)}$ **7.** $\dfrac{x-2}{9x^3}$ **8.** $\dfrac{-4(x+5)}{x(x-7)}$

9. $\dfrac{-9x^2-4x+24}{2x(x-2)(x+4)}$ **10.** $\dfrac{x^2+13x+7}{(x+3)(x-3)(x+8)}$ **11.** $\dfrac{15(x+2)}{x(x^2-3x+5)}$ **12.** $\dfrac{6(2x-1)}{(x-5)(x+4)}$ **13.** $\dfrac{(5x+2)(x-1)}{x(3x-7)}$ **14.** $\dfrac{2}{3}$

15. $-\dfrac{3}{2}$ **16.** $2\pm\sqrt{6}$ **17.** 0 **18.** $f(1)$ is undefined. **19.** $-2, 5$ **20.** $w=\dfrac{3}{49}t^2$ **21.** $y=\dfrac{40}{\sqrt{x}}$ **22. a)** $C(n)=200n+10,000$

b) $B(n)=\dfrac{200n+10,000}{n}$ **c)** $P(n)=\dfrac{350n+10,000}{n}$ **d)** 450; If the bike manufacturer makes and sells 100 bikes in a month, they

should price the bikes at \$450 to make a profit of \$150 per bike. **23. a)** $T(a)=\dfrac{400}{a+70}+\dfrac{920}{a+75}$ **b)** 16.83; When the student drives

5 mph above the speed limits, the trip takes 16.8 hours. **c)** $-71.58, 4.23$; When the trip takes 17 hours, the student is driving 4.2 mph

above the speed limits. **24. a)** $g(L)=\dfrac{3200}{L^2}$ **b)** 88.89 hertz **c)** 4 cm **d)** Decreasing; The longer the length of the prongs, the lower

the frequency.

Chapter 9

Exercise Set 9.1

1. $x^{2/5}$ **3.** $x^{3/4}$ **5.** $x^{1/2}$ **7.** $\sqrt[7]{(2x+9)^3}$ **9.** $(3x+2)^{4/7}$ **11.** $\sqrt{7x+4}$ **13.** 7 **15.** $5\sqrt{2}$

17. x **19.** x^4 **21.** $6x^3$ **23.** $x\sqrt{5}$ **25.** $x^4\sqrt[4]{x}$ **27.** $2x^2\sqrt{6x}$ **29.** $4xy^4\sqrt{5x}$ **31.** $10xy^2\sqrt{2xy}$ **33.** $(2x+5)^4$

35. $(6x+3)^2\sqrt{6x+3}$ **37.** 3 **39.** x **41.** $2x$ **43.** $2x$ **45.** $3x^3$ **47.** $x^2\sqrt[6]{x^5}$ **49.** $5x^5\sqrt[3]{x^2}$ **51.** $2x^8y\sqrt[5]{2y^2}$ **53.** $6xy$

55. $3x+6$ **57.** $(4x+7)^4$ **59.** $(x+7)^6$ **61.** $(2x+9)^5\sqrt[6]{2x+9}$ **63.** $\sqrt[4]{x^3}$ **65.** $\sqrt[3]{x^2}$ **67.** $\sqrt[6]{(2x+7)^5}$ **69.** $x^2\sqrt[3]{x}$ **71.** $\sqrt{3}$

73. $\sqrt[5]{4x^4}$ **75. a)** $d=\dfrac{\sqrt{6h}}{2}$ **b)** 46.64 miles **c)** 212.13 miles

81. a) i. 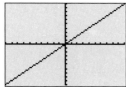 **ii.** 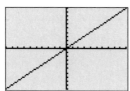 **iii.**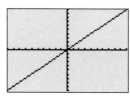

b) The graphs coincide with the graph of $y=x$.

c) i.  **ii.**  **iii.**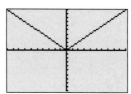

d) The graphs coincide with the graph of $y=|x|$

Exercise Set 9.2

1. $9\sqrt{x}$ **3.** $-2.5\sqrt{x}$ **5.** $-4\sqrt{3x}+10\sqrt{5x}$ **7.** $-3\sqrt{x}+5\sqrt[3]{x}$ **9.** $3\sqrt[3]{x-1}-2\sqrt{x-1}$

11. $8\sqrt{x}+6\sqrt[3]{x}$ **13.** $2.6\sqrt[4]{x}$ **15.** $25-4\sqrt{x}$ **17.** $-10\sqrt{x}-20$ **19.** 14 **21.** $7\sqrt{x}$ **23.** $12\sqrt{5x}$ **25.** $-2x\sqrt{x}$ **27.** $7x$

29. $7x\sqrt{3x}$ **31.** $x\sqrt[3]{x^2}$ **33.** $-x^2\sqrt[4]{x^3}$ **35.** $6x$ **37.** $-8x\sqrt{15}$ **39.** $2x\sqrt{14}+14x$ **41.** $-10x+38\sqrt{x}-24$ **43.** $2x-7\sqrt{x}-4$

45. $-x+1$ **47.** $x-25$ **49.** $36x+60\sqrt{x}+25$ **51.** $49x-14\sqrt{x}+1$ **53.** $16x+40\sqrt{x}+25$ **55.** $x+2\sqrt{x}+1$ **57.** $\sqrt[10]{x^7}$

59. $x\sqrt[5]{x^2}$ **61.** $-5\sqrt[4]{2x^3}+20\sqrt{x}$ **63.** $\sqrt[3]{x^2}+2\sqrt[3]{x}+1$ **65.** $\sqrt[3]{x^2}-4\sqrt[3]{x}+4$ **67.** $\sqrt{x}+2\sqrt[12]{x^7}+\sqrt[3]{x^2}$

69. $6\sqrt[6]{x^5}+2\sqrt{x}-18\sqrt[3]{x}-6$ **71.** $9\sqrt{x}-25$ **73. a)** 2.49; It takes 2.49 seconds for an object to fall 100 feet. **b)** 9.60 seconds

c) f is an increasing function. It takes more time to fall greater distances. **75.** $\sqrt[6]{x}$ **77.** $5\sqrt{x}+5$ and $4\sqrt{x}+2$; Answers may vary.

79. a) $\sqrt[12]{x^7}$ **b)** $\sqrt[kn]{x^{k+n}}$ **c)** $\sqrt[12]{x^7}$ **d)** $\sqrt[35]{x^{12}}$

Exercise Set 9.3

1. $\dfrac{2\sqrt{3}}{3}$ **3.** $\dfrac{8\sqrt{x}}{x}$ **5.** $\dfrac{3\sqrt{5x}}{5x}$ **7.** $\dfrac{\sqrt{5x}}{2x}$ **9.** $\dfrac{2\sqrt{2x}}{3x}$ **11.** $\dfrac{2\sqrt{x}}{x}$ **13.** $\dfrac{\sqrt{14}}{2}$ **15.** $\dfrac{\sqrt{2x}}{x}$ **17.** $\dfrac{\sqrt{3x}}{3}$

19. $\dfrac{3\sqrt{x-4}}{x-4}$ **21.** $\dfrac{2\sqrt[3]{25}}{5}$ **23.** $\dfrac{5\sqrt[3]{4}}{4}$ **25.** $\dfrac{4\sqrt[3]{x^2}}{5x}$ **27.** $\dfrac{3\sqrt[3]{4x}}{x}$ **29.** $\dfrac{7\sqrt[4]{4x}}{2x}$ **31.** $\dfrac{\sqrt[6]{x^5}}{x}$ **33.** $\dfrac{\sqrt[5]{2x^2}}{x}$ **35.** $\dfrac{\sqrt{6x}}{3x}$ **37.** $\dfrac{\sqrt[5]{24x}}{2x}$

39. $\dfrac{5-\sqrt{3}}{22}$ **41.** $\dfrac{8-2\sqrt{7}}{9}$ **43.** $\dfrac{\sqrt{x}+7}{x-49}$ **45.** $\dfrac{x+\sqrt{x}}{x-1}$ **47.** $\dfrac{12x+15\sqrt{x}}{16x-25}$ **49.** $\dfrac{6x+14\sqrt{x}}{9x-49}$ **51.** $\dfrac{x-10\sqrt{x}+25}{x-25}$

53. $\dfrac{x+7\sqrt{x}+12}{-x+16}$ **55.** $\dfrac{6x+17\sqrt{x}+5}{9x-1}$ **57.** $\dfrac{18x+6\sqrt{7x}+3\sqrt{5x}+\sqrt{35}}{9x-7}$ **59.** Student 1 did the work correctly. **63.** $\dfrac{x}{3\sqrt{x}}$

65. $\dfrac{1}{\sqrt{x+2}+\sqrt{x}}$ **67.** $\dfrac{2x-7\sqrt{x}+3}{4x-1}$ **69.** $3\sqrt{10}$

Exercise Set 9.4

1.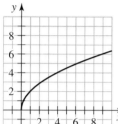

3.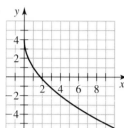

5.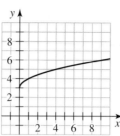

7.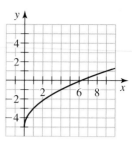

9.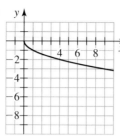

11.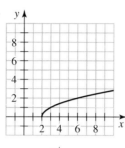

13.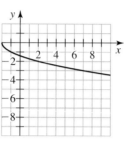

15.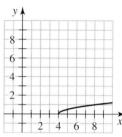

17.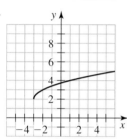

19.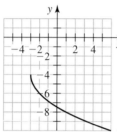

21.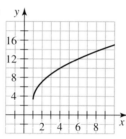

23.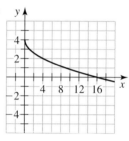

25.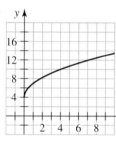

27. Domain: $x \geq 0$
Range: $y \leq 0$

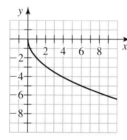

29. Domain: $x \geq 0$
Range: $y \geq 2$

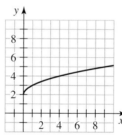

31. Domain: $x \geq -5$;
Range: $y \leq 4$

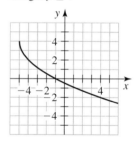

33. (a) $a < 0, h = 0$, and $k > 0$; (b) $a > 0, h < 0$, and $k < 0$; (c) $a > 0, h < 0$, and $k > 0$; (d) $a < 0, h > 0$, and $k = 0$

37. If $a < 0$, f has a maximum point at (h, k). If $a > 0$, then f has a minimum point at (h, k).　**39.** 11　**41.** $21\sqrt{c} - 3$

43. $h(x) = 2\sqrt{x} - 2$　**45.** $h(x) = -8$　**47.** $h(x) = \dfrac{x + 8\sqrt{x} + 15}{x - 25}$　**49.** $h(x) = 9\sqrt{x} - 8$　**51.** $h(x) = -\sqrt{x} + 10$

53. $h(x) = \dfrac{20x - 41\sqrt{x} + 9}{16x - 1}$　**55.** $h(x) = 4\sqrt{x}$　**57.** $h(x) = 6\sqrt{5}$　**59.** $h(x) = \dfrac{4x - 12\sqrt{5x} + 45}{4x - 45}$　**61.** $h(x) = 2\sqrt{x + 1}$

63. $h(x) = 4$　**65.** $h(x) = \dfrac{-4\sqrt{x + 1} + x + 5}{x - 3}$　**67. a)** S　**b)** 297　**c)** 299　**69.** 0　**71.** 2.4　**73.** -6　**75.** 3

77. The graph is a parabola.; No;
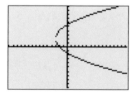

Exercise Set 9.5

1. 25　**3.** No real number solutions　**5.** 4　**7.** 4　**9.** $\dfrac{4}{7}$　**11.** 5　**13.** No real number solutions

15. $\dfrac{97}{6}$　**17.** 5　**19.** $-\dfrac{1}{4}$　**21.** 2.31　**23.** 11　**25.** 3　**27.** No real number solutions　**29.** 8　**31.** 4　**33.** 4　**35.** $3 + 2\sqrt{2}$　**37.** No

real number solutions　**39.** 4　**41.** No real number solutions　**43.** 1　**45.** 121　**47.** 1　**49.** $\dfrac{3 \pm 3\sqrt{5}}{2}$　**51.** $-4\sqrt{x} + 5$　**53.** 9　**55.** 0

57. $x + 4\sqrt{x} + 3$　**59.** $(7, 0)$　**61.** $(-7, 0)$　**63.** $(5, 0)$　**65.** There are no x-intercepts.　**67.** 4　**69.** There is no such value of x.

71. a) Although test score averages have improved since 1982, they are less than the average in 1970.   **b)** 2023   **c)** 5164; Model breakdown has occurred.   **73.** No   **75.** (4, 2)

Exercise Set 9.6
1. $y = \sqrt{x} + 3$ **3.** $y = 1.33\sqrt{x} + 2$ **5.** $y = 1.73\sqrt{x} + 2$ **7.** $y = \sqrt{x} + 1$
9. $y = 3.15\sqrt{x} - 0.45$ **11.** $y = -2.43\sqrt{x} + 9.44$ **13.** Increase b

15. a) $f(t) = 5.45\sqrt{t} + 10$; Answers may vary;

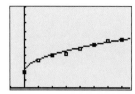

b) 37.79; In 2008, 37.8% of single-parent fathers will have never married.   **c)** 28.31; In 2010, 39% of single-parent fathers will have never married.
d) (0, 10); In 1982, 10% of single-parent fathers had never married.

17. a) $S(h) = 0.27\sqrt{h}$; Answers may vary.   **b)** ii. S;   iii. Q;   iv. S;   v. The approximate form is $T = 0.25\sqrt{h}$, which is "close" to the model $S(h) = 0.27\sqrt{h}$.   **c)** 123.46 feet   **d)** 9.55 seconds

19. a) $S(t) = 41\sqrt{t}$; Answers may vary.   **b) i)**

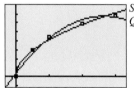

c) 82%   **d)** 6 years

ii) The function S is increasing for $t \geq 0$.

21. a) $f(n) = 31.92\sqrt{n} + 9.16$;

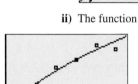

b) 93.61; 93.6% of 7th births happened in spite of contraception.
c) 8.10; All 8th births happened in spite of contraception.
d) The higher the birth order, the higher the percent of births that happened in spite of contraception.

23. a) $f(n) = 72.15\sqrt{n} - 16.74$ **b)** f is an increasing function; This suggests that the greater the number of people living with the child, the less likely the child tries to take control.

Chapter 9 Review
1. $\sqrt[7]{x^3}$ **2.** $x^{1/2}$ **3.** $\sqrt[9]{(2x+1)^2}$ **4.** $(3x+4)^{7/5}$ **5.** $4\sqrt{x}$ **6.** $2x^3\sqrt{2}$ **7.** $x^3\sqrt{3x}$
8. $\sqrt[4]{x^3}$ **9.** $2x^3\sqrt[3]{3x}$ **10.** $(6x+11)^5\sqrt[5]{(6x+11)^2}$ **11.** 0   **12.** $12\sqrt[3]{x} + 2\sqrt{x}$ **13.** $28\sqrt{x} - 7\sqrt[3]{x}$ **14.** $3x - 21\sqrt{x}$
15. $8x - 2\sqrt{x} - 3$ **16.** $24x + 76\sqrt{x} + 32$ **17.** $x - 1$ **18.** $9x - 16$ **19.** $4\sqrt[3]{x^2} + 20\sqrt[3]{x} + 25$ **20.** $25x - 20\sqrt{x} + 4$ **21.** $\sqrt[20]{x^9}$
22. $\sqrt[18]{x}$ **23.** $\sqrt[12]{x}$ **24.** $\dfrac{x\sqrt{2}}{2}$ **25.** $\dfrac{\sqrt{3x}}{x}$ **26.** $\dfrac{5\sqrt[3]{x^2}}{x}$ **27.** $\dfrac{\sqrt[5]{63x^3}}{3x}$ **28.** $\dfrac{15 - 5\sqrt{x}}{9 - x}$ **29.** $\dfrac{4 + 6\sqrt{x}}{4 - 9x}$ **30.** $\dfrac{10x - 23\sqrt{x} + 12}{4x - 9}$
31. $h(x) = -\sqrt{x} + 7$ **32.** $7\sqrt{x} + 3$ **33.** $h(x) = -12x - 14\sqrt{x} + 10$ **34.** $\dfrac{12x + 26\sqrt{x} + 10}{4 - 16x}$ **35.** 9   **36.** 4   **37.** 5   **38.** 2, 4
39. 9   **40.** Empty set solution   **41.** 13   **42.** 7

43. **44.** **45.** **46.**

47. (4, 0)   **48.** (7, 0)   **49.** Decrease a and increase b.   **50.** $y = 2.47\sqrt{x} + 2$ **51.** $y = 2.5\sqrt{x} + 3$ **52.** $y = -\dfrac{4}{3}\sqrt{x} + 7$
53. $y = 3.15\sqrt{x} + 0.55$ **54.** $y = -5.95\sqrt{x} + 17.31$ **55. a)** $f(t) = 6.4\sqrt{t} + 16$; Answers may vary.   **b)** (0, 16); In 1998, 16 million households had Sony PlayStations.   **c)** 9.77; In 2008, 36 million households will have Sony PlayStations.   **d)** 30.3 million households

Chapter 9 Test
1. $4x^4\sqrt{2x}$ **2.** $4x^7\sqrt[3]{x}$ **3.** $(2x+8)^6\sqrt[4]{(2x+8)^3}$ **4.** $\dfrac{2\sqrt[15]{x^2}}{3}$ **5.** $\dfrac{2x + 5\sqrt{x} + 3}{4x - 9}$
6. $-18\sqrt{3x} - 3\sqrt[4]{x}$ **7.** $18x - 15\sqrt{x}$ **8.** $-20x + 2\sqrt{x} + 6$ **9.** $16 - 9x$ **10.** $16\sqrt[5]{x^2} - 24\sqrt[5]{x} + 9$ **12.** $2\sqrt{x} + 11$
13. $h(x) = 3 - 8\sqrt{x}$ **14.** $-15x + 23\sqrt{x} + 28$ **15.** $h(x) = \dfrac{15x - 47\sqrt{x} + 28}{-25x + 16}$ **16.** -6 **17.** 3

18.

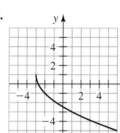

19. a) $a < 0$ and $k > 0$, or $a > 0$ and $k < 0$ **b)** $\left(\dfrac{k^2 + a^2h}{a^2}, 0\right)$

20. 25   **21.** 17   **22.** $\dfrac{144}{25}$ **23.** $(4, 0)$

24. Increase a and decrease b.   **25.** $y = 2.43\sqrt{x} + 0.56$

26. a)

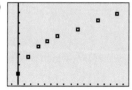

$f(t) = 2.90\sqrt{t} + 20.5$; Answers may vary.

b) 45.1 inches

c) 29 months

d) $(0, 20.50)$; The median height of boys at birth is 20.5 inches.

Chapter 10

Exercise Set 10.1

1. Not arithmetic   **3.** Arithmetic; $d = -2$ **5.** Arithmetic; $d = 8$ **7.** Not arithmetic
9. $a_n = 6n - 1$ **11.** $a_n = -11n + 7$ **13.** $a_n = -6n + 106$ **15.** $a_n = 2n - 1$ **17.** 113   **19.** -196 **21.** 156.1   **23.** -433
25. 107   **27.** 87   **29.** 313   **31.** 3571   **33.** 255   **35.** 500   **37.** Yes   **39.** No   **41.** No   **43.** $f(x) = 9x - 1$
45. a) $a_n = 800n + 26{,}700$ **b)** \$44,300   **c)** 30th year   **47. a)** $a_n = \dfrac{1}{6}n + 35$ **b)** 35.17, 35.33, 35.50, 35.67; These values represent
the number of hours the instructor would work if she had 1, 2, 3, or 4 students, respectively.   **c)** 56.67 hours per week   **d)** 150 students
49. a) $a_n = 1.8n - 50$ **b)** 170 people   **c)** \$247   **d)**

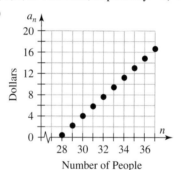

51. a) $a_n = 0.23n + 0.14$ **b)** \$3.13   **c)** No   **d)** \$18.54

Exercise Set 10.2

1. Geometric; $r = 7$ **3.** Arithmetic; $d = -7$ **5.** Neither arithmetic nor geometric
7. Geometric; $r = \dfrac{1}{5}$ **9.** $a_n = 3(2)^{n-1}$ **11.** $a_n = 800\left(\dfrac{1}{4}\right)^{n-1}$ **13.** $a_n = 100\left(\dfrac{1}{2}\right)^{n-1}$ **15.** $a_n = 4^{n-1}$ **17.** $a_n = 5n + 9$
19. 4.66×10^{23} **21.** 1.19×10^{-6} **23.** 33,554,432   **25.** 332.5   **27.** 10   **29.** 12   **31.** 122   **33.** 19   **35.** 16   **37.** Geometric
sequence   **39.** Arithmetic   **41.** $f(x) = 8(3)^{x-1}$ **43.** No   **45. a)** $a_n = 27{,}000(1.04)^{n-1}$ **b)** \$38,429.42   **c)** 17th year
47. a) 2, 4, 8, 16, 32   **b)** $a_n = 2^n$ **c)** 256 ancestors   **d)** 34.36 billion ancestors   **49. a)** $a_n = 5(3)^{n-1}$ **b)** 405 students
c) 295,245 students; Yes

Exercise Set 10.3

1. 20,205   **3.** 30,294   **5.** $-38{,}232$ **7.** 21,978   **9.** $-10{,}807$ **11.** 468   **13.** 77,875
15. 36,288   **17.** 151,468   **19.** 10,850   **21.** 5187   **23.** 0   **25.** 19,436   **27.** 50,005,000   **29.** S_n is positive.   **31.** S_n is positive.
33. Yes   **35. a)** \$58,200   **b)** \$1,213,800   **37.** Company A; \$8000   **39. a)** 3250 seats   **b)** \$50,000
c) Between \$43,500 and \$44,750, inclusive

41. a)

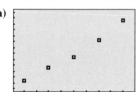

b) $f(t) = 12.62t + 174.40$

c) 174.4; In 1990, sales were \$174.4 billion.

d) 414.18; In 2009, sales will be \$414.18 billion.

e) 5885.8; Total sales from 1990 to 2009 will be \$5.89 trillion.

f) \$211.89 billion; Answers may vary.

Exercise Set 10.4

1. 671,088,635 **3.** 52,379.91 **5.** 64.84 **7.** 22.88 **9.** 610,351,562 **11.** 857.14
13. 8.84 **15.** 89,478,485 **17.** 103.9945088 **19.** 19,995.12 **21.** 100 **23.** 10,443 **25.** 36 **27.** Positive **29.** Arithmetic
31. $699,784.85 **33.** Company A; $252,572.15 **35.** 2046 ancestors **37. a)** 11th round; The entrepreneur could receive as much as
approximately $6.14 billion. **b)** There will be ten full rounds and part of an 11th round. **c)** 9 people; The entrepreneur will receive
approximately $6.14 billion. The 8 people besides the entrepreneur will receive an average of $2.98 billion. **39. a)** $f(t) = 85.89(0.47)^t$
b) 85.89; In 1980, 85.89 million cartridges were sold. **c)** 160.31 million cartridges **d)** 158 million cartridges
e) 162.06 million cartridges

Chapter 10 Review

1. Geometric sequence **2.** Arithmetic series **3.** Arithmetic sequence **4.** Geometric series
5. Geometric series **6.** None of these **7.** $a_n = 2(3)^{n-1}$ **8.** $a_n = 3n + 22$ **9.** $a_n = -5n + 14$ **10.** $a_n = 200\left(\dfrac{1}{2}\right)^{n-1}$
11. $a_n = 2.7n + 0.5$ **12.** $a_n = 800(0.7)^{n-1}$ **13.** 4.22×10^{14} **14.** 0.01171875 **15.** -204 **16.** 225.9 **17.** 505 **18.** 77 **19.** 24
20. -204 **21.** -3182 **22.** 671,173.0723 **23.** 3,221,225,469 **24.** 120,540 **25.** $-179,679$ **26.** 797,161 **27.** Geometric series
28. Arithmetic sequence **29. a)** $71,772.52; $70,000 **b)** $1,166,085.43; $1,300,000 **30. a)** $f(t) = 0.32t + 5.42$
b) 11,500 deaths **c)** 169,200 deaths **d)** Overestimates

Chapter 10 Test

1. Geometric sequence **2.** None of these **3.** Geometric series **4.** Arithmetic series
5. $a_n = -6n + 37$ **6.** $a_n = 6(4)^{n-1}$ **7.** 262 **8.** 0.000000093 **9.** 455 **10.** 9 **11.** 40.50 **12.** 4,294,967,292 **13.** -462
14. 1,124,800 **15.** 2,098,620 **16.** None of these **17.** Negative **18. a)** $f(t) = 6.69t - 41.67$ **b)** 5.16, 11.85, 18.54, 25.23, 31.92;
These values estimate the numbers of fatalities in 1987, 1988, 1989, 1990, and 1991, respectively. **c)** 1513 fatalities
19. a) $a_n = 32(1.03)^{n-1}$ **b)** The 9th year **c)** $65,049.41 **d)** $1,166,696.46

Cumulative Review for Chapters 1–10

1. $\pm\dfrac{7}{2}$ **2.** 2.9755 **3.** 1.6013 **4.** 22 **5.** $-\dfrac{5}{2}, \dfrac{1}{3}$
6. $\dfrac{1 \pm \sqrt{69}}{2}$ **7.** 3.9927 **8.** 2.0492 **9.** 2, 10 **10.** 3 **11.** $-1, \dfrac{5}{2}$ **12.** $\dfrac{2 \pm \sqrt{2}}{3}$ **13.** $-\dfrac{43}{18}$ **14.** 3 **15.** 2 **16.** $\dfrac{5 \pm \sqrt{13}}{6}$
17. $(2, -1)$ **18.** $(-2, 3)$ **19.** $x \le 7$; $(-\infty, 7]$; **20.** $\dfrac{2916}{b^{18}c^8}$ **21.** $\dfrac{4c}{5b^{1/4}}$ **22.** $\dfrac{4b^{24}c^{18}}{9}$ **23.** $\dfrac{(a-b)(a+b)}{ab}$
24. 0 **25.** $2x^3\sqrt{3x}$ **26.** $(5x-7)^5 \sqrt[4]{5x-7}$ **27.** $\dfrac{\sqrt[3]{4x^2}}{x}$ **28.** $\dfrac{15x - 22\sqrt{x} + 8}{9x - 4}$ **29.** $\ln(1125x^{35})$ **30.** $\log_b\left(\dfrac{x^{10}}{2}\right)$
31. $9x^2 - 30x + 25$ **32.** $6x + 13\sqrt{x} - 28$ **33.** $-2x^6 - 2x^4 - 10x^3 - 10x$ **34.** $4x^3 - 9x^2 - 10x + 3$ **35.** $\dfrac{(x+1)(x+3)}{5(x-4)}$
36. $\dfrac{2x^2 - 3x + 10}{(x-5)^2(x-2)}$ **37.** $\dfrac{7x}{2(x-4)(x+2)}$ **38.** $\dfrac{2x^2 + 9x + 15}{(x-3)(x+1)(x+3)(x+9)}$ **39.** $\dfrac{3}{(x-8)(x+2)}$ **40.** $f(x) = -3x^2 - 18x - 34$
41. $(x-2)(2x-5)(2x+5)$ **42.** $2x(x-5)(x+3)$ **43.** $2(x+2)(3x-5)$ **44.** $(10x-1)(10x+1)$ **45.** 3 **46.** 0, 2 **47.** 1
48. No such value of x **49.** $(0, 3)$ **50.** $(-1, 0), (3, 0)$ **51.** $y = -x^2 + 2x + 3$

52.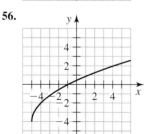

53.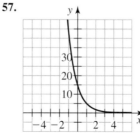

54.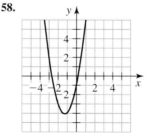

55.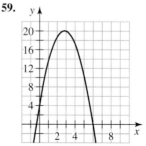

56.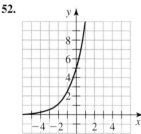

57.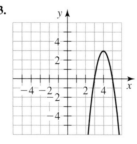

58.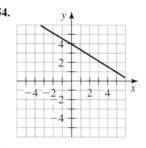

59.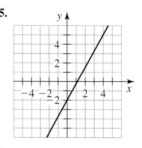

60. $y = -\dfrac{7}{5}x - \dfrac{11}{5}$ or $y = -1.4x - 2.2$ **61.** $y = 752.08(0.50)^x$ **62.** $y = 2x^2 - 5x + 3$ **63.** $y = 11.59\sqrt{x} - 11.39$

64. a) $f(x) = 2x + 2$, $g(x) = 2(2)^x$, $h(x) = 2x^2 + 2$; Answers may vary for h.   **b)**

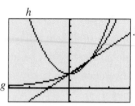

65. 4   **66.** $\dfrac{1}{2}$ **67.** -4 **68.** 1.8928   **69.** $g^{-1}(x) = 2^x$ **70.** $f^{-1}(x) = -\dfrac{1}{4}x - \dfrac{7}{4}$ **71.** All real numbers except -5 and 7.

72. $a_n = 96\left(\dfrac{1}{2}\right)^{n-1}$ **73.** 39,366   **74.** 65   **75.** 196,605   **76.** 5597   **77. a)** $B(t) = 1.24t + 6.82$

b) $R(t) = 0.97t^2 + 0.29t + 177.04$ **c)** $P(t) = \dfrac{124t + 682}{0.97t^2 + 0.29t + 177.04}$ **e)** 2013   **78. a)** $L(t) = 16.2t + 690$

b) $E(t) = 690(1.019)^t$ **c)** $L(70) = 1824$, $E(70) = 2576.57$; India's population will be 1.824 billion in 2050, according to the linear model. India's population will be 2.577 billion in 2050, according to the exponential model.   **d)** 752.57; The exponential model's prediction for 2050 exceeds the linear model's prediction for that year by 753 million people.

79. a) $f(t) = -0.045t^2 + 2.67t + 9.97$ **b)** 47.6%   **c)** 2005, 2012

80. a) $f(t) = 0.0058t^2 - 0.396t + 10.02$ **b)** 4.54; In 2009, there will be 4.5 thousand men studying to be priests in a graduate program.
   c) $t = 16.82$ or $t = 51.46$; There were 5 thousand men studying to be priests in a graduate program in 1977 and there will be that many in 2011.   **d)** The function f is decreasing for $t < 34.14$ and f is increasing for $t > 34.14$. The number of men studying to be priests in a graduate program was declining up untill 1994 and it has been increasing since 1994 and will continue to do so.

Chapter 11

Exercise Set 11.1

1. ± 7 **3.** Empty set solution   **5.** $\pm\dfrac{18}{5}$ **7.** Empty set solution   **9.** $-7, 3$ **11.** 5   **13.** $-\dfrac{10}{3}, 4$

15. $-2, 7$ **17.** Empty set solution   **19.** $-9, -1$ **21.** 3   **23.** 3, 4   **25.** $1, \dfrac{17}{3}$ **27.** ± 2.70 **29.** $-5.32, -0.20$

31. $-4 < x < 4$, or $(-4, 4)$ **33.** $x \le -3$ or $x \ge 3$, or $(-\infty, -3] \cup [3, \infty)$ **35.** Empty set solution

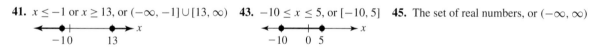

37. $x < -4$ or $x > 4$, or $(-\infty, -4) \cup (4, \infty)$ **39.** $x \le -2$ or $x \ge 2$, or $(-\infty, -2] \cup [2, \infty)$

41. $x \le -1$ or $x \ge 13$, or $(-\infty, -1] \cup [13, \infty)$ **43.** $-10 \le x \le 5$, or $[-10, 5]$ **45.** The set of real numbers, or $(-\infty, \infty)$

47. $x \le -5$ or $x \ge 11$, or $(-\infty, -5] \cup [11, \infty)$ **49.** $x \le 1$ or $x \ge 4$, or $(-\infty, 1] \cup [4, \infty)$ **51.** Empty set solution

53. $-\dfrac{39}{8} \le x \le -\dfrac{21}{8}$, or $\left[-\dfrac{39}{8}, -\dfrac{21}{8}\right]$ **55.** $x < -6.72$ or $x > 6.72$, or $(-\infty, -6.72) \cup (6.72, \infty)$

57. $-0.30 \le x \le 1.04$, or $[-0.30, 1.04]$ **59.** False

61. a) $-6, 5$ **b)** $-6 < x < 5$, or $(-6, 5)$ **c)** $x < -6$ or $x > 5$, or $(-\infty, -6) \cup (5, \infty)$ **63.** $\dfrac{-ac \pm (k - d)}{ab}$ **65.** No

Exercise Set 11.1 Quiz

1. ± 5 **2.** $-2, 5$ **3.** $-3, -\dfrac{1}{3}$ **4.** $\dfrac{4}{3}, \dfrac{1}{3}$ **5.** $\dfrac{5}{2}$ **6.** Empty set solution

7. $-6 \le x \le 6$, or $[-6, 6]$ **8.** $x \le -4$ or $x \ge 4$, or $(-\infty, -4] \cup [4, \infty)$ **9.** $x < -1$ or $x > 5$, or $(-\infty, -1) \cup (5, \infty)$

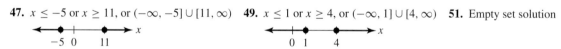

10. $-2 < x < \dfrac{4}{3}$, or $\left(-2, \dfrac{4}{3}\right)$ **11.** $-8 \le x \le -2$, or $[-8, -2]$ **12.** $-\dfrac{4}{3} \le x \le \dfrac{8}{3}$, or $\left[-\dfrac{4}{3}, \dfrac{8}{3}\right]$

13. Empty set solution   **14.** The set of real numbers, or $(-\infty, \infty)$ **15.** False   **16.** 2

Exercise Set 11.2

1.
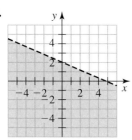

3.
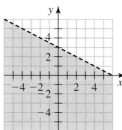

5.
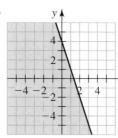

7.
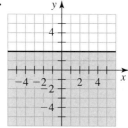

9.
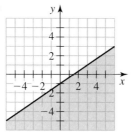

11.
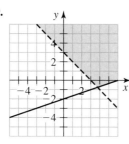

13.
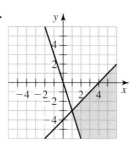

15.
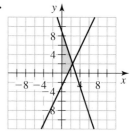

17.
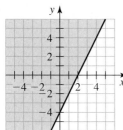

19.
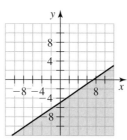

21.
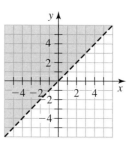

23.
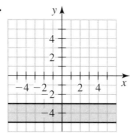

25.
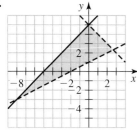

27.
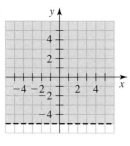

29.
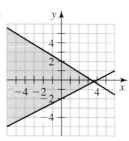

31.
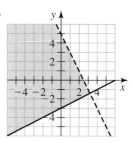

37.
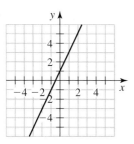

Exercise Set 11.2 Quiz

1.
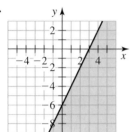

2.
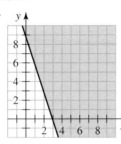

3.
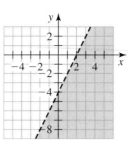

4.
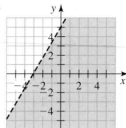

5.
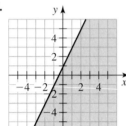

6.
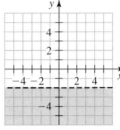

7.
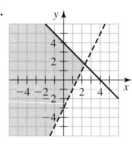

8.
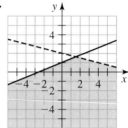

9.
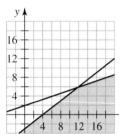

10.
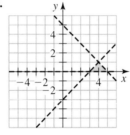

Exercise Set 11.3

1. $(x + 3)(x^2 - 3x + 9)$ **3.** $(x - 4)(x^2 + 4x + 16)$ **5.** $(x - 10)(x^2 + 10x + 100)$

7. $(x + 5)(x^2 - 5x + 25)$ **9.** $(x + 1)(x^2 - x + 1)$ **11.** $(4x - 5)(16x^2 + 20x + 25)$ **13.** $(3x + 4)(9x^2 - 12x + 16)$

15. $(10x - 1)(100x^2 + 10x + 1)$ **17.** $(2x - 3)(4x^2 + 6x + 9)$ **19.** $(5x + 2)(25x^2 - 10x + 4)$ **21.** $4(x - 2)(x^2 + 2x + 4)$

23. $7(x - 10)(x^2 + 10x + 100)$ **25.** $10(x + 4)(x^2 - 4x + 16)$ **27.** $x(x + 3)(x^2 - 3x + 9)$ **29.** $3x(x - 2)(x^2 + 2x + 4)$

31. $4x(2x - 1)(4x^2 + 2x + 1)$ **33.** $2x(x - 5)(x + 5)$ **35.** $2(x - 2)(x^2 + 2x + 4)$ **37.** $(x - 6)^2$ **39.** $(x + 1)^2(x^2 - x + 1)$

41. $(2x + 5)(3x - 2)$ **43.** $2x(x - 8)(x - 2)$ **45.** $(x + 4)(2x - 1)(4x^2 + 2x + 1)$ **47.** $x(x - 6)(x + 5)$

49. $-(4x - 5)(16x^2 + 20x + 25)$ **51.** No   **53.** No

Exercise Set 11.3 Quiz

1. $(x - 6)(x^2 + 6x + 36)$ **2.** $(x + 6)(x^2 - 6x + 36)$ **3.** $(4x + 1)(16x^2 - 4x + 1)$

4. $(3x - 5)(9x^2 + 15x + 25)$ **5.** $2x(2x + 3)(4x^2 - 6x + 9)$ **6.** $3x(5x - 2)(25x^2 + 10x + 4)$ **7.** $-5x(x - 1)(x^2 + x + 1)$

8. $-7(x - 2)(x^2 + 2x + 4)$ **9.** $(x - 2)(x + 2)(x^2 + 2x + 4)$ **10.** $(x - 5)(x + 1)(x^2 - x + 1)$

Exercise Set 11.4

1. $6i$ **3.** $-5i$ **5.** $i\sqrt{13}$ **7.** $2i\sqrt{5}$ **9.** $\dfrac{1}{2} - \dfrac{\sqrt{6}}{4}i$ **11.** $2 + \dfrac{\sqrt{2}}{2}i$ **13.** $6i$

15. $3 - 6i$ **17.** -10 **19.** $-\sqrt{15}$ **21.** $7 + 3i$ **23.** $4 - 18i$ **25.** $-2 + 14i$ **27.** -18 **29.** -50 **31.** $10 + 15i$ **33.** $-1 - 6i$

35. $-14 + 23i$ **37.** $27 - 24i$ **39.** 41 **41.** 2 **43.** $-5 + 12i$ **45.** $15 - 8i$ **47.** $\dfrac{6}{29} - \dfrac{15}{29}i$ **49.** $\dfrac{14}{15} - \dfrac{7}{15}i$ **51.** $\dfrac{11}{50} - \dfrac{23}{50}i$

53. $-\dfrac{7}{25} + \dfrac{24}{25}i$ **55.** $\dfrac{3}{5} - \dfrac{1}{5}i$ **57.** $-1 \pm i\sqrt{2}$ **59.** $1 \pm 2i$ **61.** $-\dfrac{1}{4} \pm \dfrac{\sqrt{7}}{4}i$ **63.** $\dfrac{2}{5} \pm \dfrac{1}{5}i$ **65.** $\dfrac{1}{4} \pm \dfrac{\sqrt{11}}{4}i$ **67.** $-1 \pm i$

69. $\dfrac{2}{3} \pm \dfrac{2\sqrt{2}}{3}i$ **71.** $\pm 3i$ **73.** $\pm \dfrac{\sqrt{15}}{3}i$ **75.** $-\dfrac{3}{5} \pm \dfrac{\sqrt{7}}{5}i$ **77.** $\dfrac{1}{3} \pm \dfrac{2}{3}i$ **79.** Student #2 did the work correctly.

Exercise Set 11.4 Quiz

1. $1 - 24i$ **2.** $-\sqrt{14}$ **3.** $9 - 6i$ **4.** 12 **5.** $38 - 16i$ **6.** $7 - 24i$ **7.** 89

8. $\dfrac{8}{17} + \dfrac{2}{17}i$ **9.** $\dfrac{7}{41} + \dfrac{22}{41}i$ **10.** $-i$ **11.** $1 \pm i\sqrt{2}$ **12.** $\dfrac{1}{6} \pm \dfrac{\sqrt{47}}{6}i$ **13.** $-4 \pm 2i$ **14.** $\dfrac{5}{2} \pm i$ **16.** False

Exercise Set 11.5

1. a) $f(x) = -0.0523x + 810$ **b)**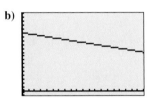

f is decreasing. The more money
that is invested in the CD (and the
less money invested in the mutual
fund), the lower the interest earned.

c) CD: $7839.39, mutual fund: $2160.61

3. a) $f(x) = -0.0695x + 850.5$ **b)** 815.75; If $500 is invested in the CD (and $8500 is invested in the mutual fund), the total interest will be $815.75.   **c)** 5043.17; If $5043.17 is invested in the CD (and $3956.83 is invested in the mutual fund), the total interest will be $500.   **d)** 155.5; If $10,000 is invested in the CD, the total interest will be $155.50. Model breakdown has occurred because only $9,000 is being invested.

5. a) $f(x) = -0.101x + 928$ **b)** The total interest will be between $120 and $675.50, inclusive. **c)** $5227.72 should be invested in the CD and $2772.28 should be invested in the mutual fund.

7. a) $f(x) = -0.0615x + 540$ **b)** (0, 540); If all of the $6000 is invested in the mutual fund, the total interest will be $540.
c) (8780.49, 0); If $8780.49 is invested in the CD, the revenue will be zero dollars. Model breakdown has occurred because only $6000 is being invested.   **d)** −0.0615; If one more dollar is invested in the CD (and one less dollar is invested in the mutual fund), the total interest will decrease by 6.15 cents.

9. a) $f(x) = -25x + 1,500,000$ **b)**

f is decreasing. The more tickets sold
for $50 (and the fewer tickets sold
for $75), the lower the total revenue.

c) 1,100,000; If 16,000 $50 tickets are sold
(and 4000 $75 tickets are sold), the total
revenue will be $1,100,000.

d) 17,000 $50 tickets, 3000 $75 tickets

11. a) $f(x) = -25x + 840,000$ **b)**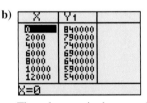

The values are in the second column of the table.
If the number of $45 tickets sold is 0, 2000, 4000,
6000, 8000, 10,000, or 12,000, the total revenue
will be $840,000, $790,000, $740,000, $690,000,
$640,000, $590,000, or $540,000, respectively.

c) The total revenue will be between
$540,000 and $840,000, inclusive.

d) 9500 $45 tickets, 2500 $75 tickets

13. a) $f(x) = 134x + 1936$ **b)** 134; If the prices of all tickets are increased by one dollar, the total revenue increases by $134.
c) Coach: $91, first class: $333   **15.** 1538 part-time students, 4615 full-time students   **17. a)** $f(p) = -2.25p^2 + 3162.5p$ **b)** (0, 0),
(1405.56, 0); If the price is $0 or $1405.56, the revenue is $0.   **c)** $702.78; 1,111,267.36   **19. a)** $n = -136.2p + 5861.64$
b) $f(p) = -136.2p^2 + 5861.64p$ **c)** (0, 0), (43.04); If the price is $0 or $43.04, the revenue will be $0.   **d)** $21.52; $63,066.86
21. a) $n = -2.16p + 2493.6$ **b)** n **c)** $f(p) = -2.16p^2 + 2493.6p$ **d)** $105.97 and $1048.47   **e)** $577.22; $719,680.67

Exercise Set 11.5 Quiz   **1. a)** $f(x) = -0.098x + 1016$ **b)** 967; If $500 is invested in the CD (and $7500 in the
mutual fund), the total interest will be $967.   **c)** 5265.31; If $5265.31 is invested in the CD (and $2734.69 in the mutual fund), the total
interest will be $500.

2. a) $f(x) = -20x + 800,000$ **b)** −20; If one more $30 ticket is sold (and one less $50 ticket is sold), the revenue will decrease
by $20.   **c)** 15,000 $30 tickets, 1000 $50 tickets

3. a) $R(p) = -120p^2 + 65,000p$ **b)** (0, 0), (541.67, 0); If the price is $0 or $541.67, the revenue will be $0.   **c)** $270.83;
$8,802,083.33

Exercise Set 11.6

1. $c = 13$ **3.** $c = \sqrt{41}$ **5.** $b = \sqrt{55}$ **7.** $a = 2\sqrt{6}$ **9.** $c = \sqrt{7}$ **11.** $c = 7\sqrt{10}$

13. 4.47 **15.** 7.62 **17.** 5 **19.** 8.01 **21.** 13.28 **23.** $x^2 + y^2 = 25$ **25.** $x^2 + y^2 = 44.89$ **27.** $(x - 5)^2 + (y - 3)^2 = 4$

29. $(x + 2)^2 + (y - 1)^2 = 16$ **31.** $(x + 7)^2 + (y + 3)^2 = 3$

33. $C(0, 0)$ and $r = 5$ **35.** $C(0, 0)$ and $r = 2\sqrt{2}$ **37.** $C(3, 5)$ and $r = 4$ **39.** $C(-6, 1)$ and $r = \sqrt{7}$

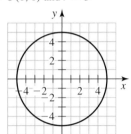 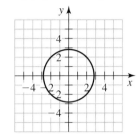 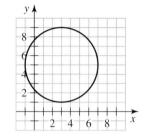 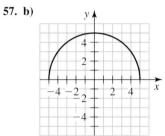

41. $C(-3, -2)$ and $r = 1$ **43.** 20.62 feet   **57. b)**   **59. c)** $3\sqrt{2}$ **d)** $\dfrac{5\sqrt{2}}{2}$

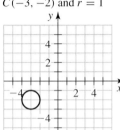

45. 2273.44 miles

47. $(x - 1)^2 + (y + 2)^2 = 9$

49. $(x - 3)^2 + (y - 2)^2 = 20$

55. No

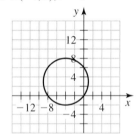

Exercise Set 11.6 Quiz

1. $4\sqrt{3}$ **2.** 10.25 inches **3.** $\sqrt{41}$ **4.** 7 **5.** $(x + 3)^2 + (y - 2)^2 = 36$

6. $x^2 + y^2 = 7.84$ **7.** $C(0, 0), r = 2\sqrt{3}$ **8.** $C(-4, 3), r = 5$ **9.** $(x - 2)^2 + (y + 1)^2 = 68$

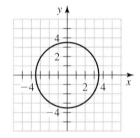 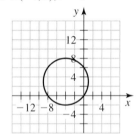

Exercise Set 11.7

1. 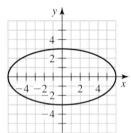   **3.** 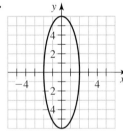   **5.** 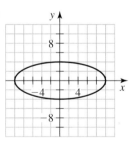   **7.**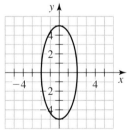

9. 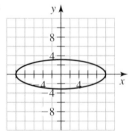   **11.** 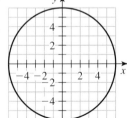   **13.** 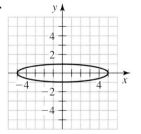   **15.**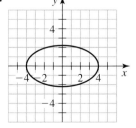

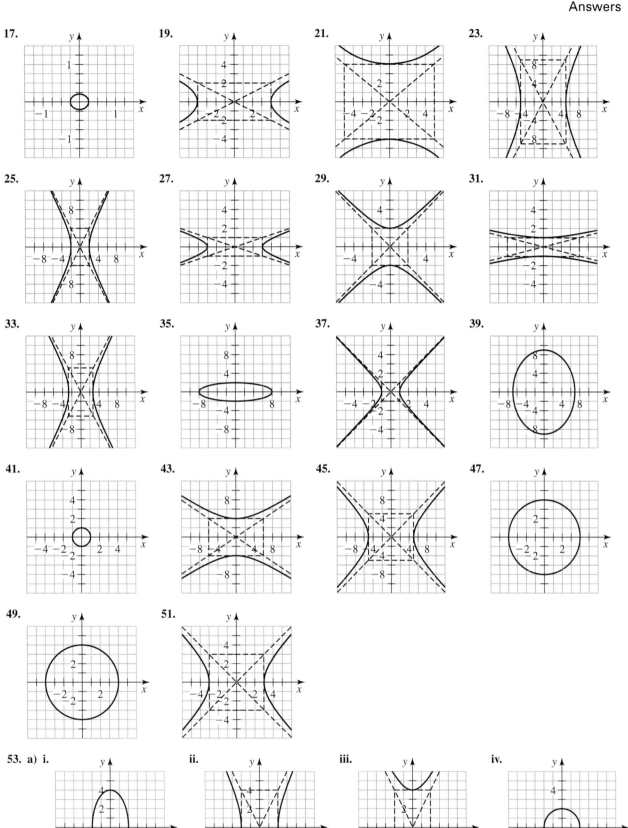

55. No   **57. b)**

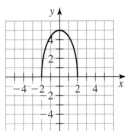

61. It is a circle and an ellipse.

Exercise Set 11.7 Quiz

1.

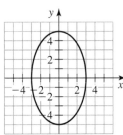

2.

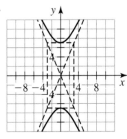

3.

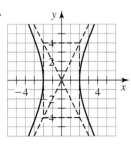

4.

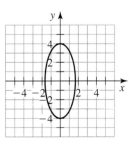

5.

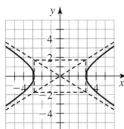

6.

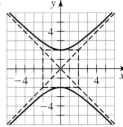

7.

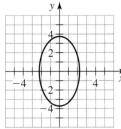

8.

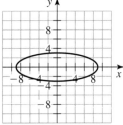

9. No

Exercise Set 11.8

1. $(-5, 0), (5, 0)$ **3.** $(-2, 5), (1, 2)$ **5.** $(-2, 2), (2, 2)$ **7.** Empty set solution
9. $(-4, 3), (3, -4)$ **11.** $(-3, -5), (0, 4), (3, -5)$ **13.** $(-3, 0), (3, 0)$ **15.** $(-1, 0), (0, 3)$ **17.** No solutions with real number
coordinates   **19.** $(1, -2)$ **21.** $(-\sqrt{3}, 1), (\sqrt{3}, 1)$ **23.** $(-0.74, -2.02), (-0.74, 2.02), (0.74, -2.02), (0.74, 2.02)$
25. $(-2.25, -3.31), (-2.25, 3.31), (2.25, -3.31), (2.25, 3.31)$ **27.** $(-3, -2), (-3, 2), (3, -2), (3, 2)$ **29.** $(-3, 22), (5, 14)$
31. $(3, 2), (2, 0)$ **33.** $(-5, 0), (5, 0)$ **37.** $c = 5, d = -2$

Exercise Set 11.8 Quiz

1. $(-3, 0), (3, 0)$ **2.** $(1, -1), (-3, 7)$ **3.** $(1, 4)$ **4.** Empty set solution   **5.** $(-4, 0)$

Appendix A

Section A.1

1.

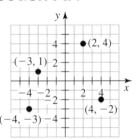

Section A.2

1. Real number   **2.** Not a real number   **3.** Not a real number   **4.** Real number   **5.** Real number
6. Real number   **7.** Real number   **8.** Real number

Section A.3

1. 3   **2.** 4.69   **3.** 0   **4.** π

Section A.4   **1. a)** -21 **b)** 36 **c)** -30 **d)** 16 **e)** 4 **f)** -9 **g)** -4 **h)** 3 **i)** 3 **j)** -3 **2. a)** -8 **b)** 6 **c)** -6 **d)** -4 **e)** -9 **f)** -2 **g)** -10 **h)** -5 **3. a)** -4 **b)** -10 **c)** 8 **d)** -5 **e)** -5 **f)** -6 **g)** 2 **h)** -4

Section A.5   **1.** Equation   **2.** Expression   **3.** Expression   **4.** Equation

Section A.6   **1.** $2x + 8$ **2.** $3x - 15$ **3.** $-5x - 30$ **4.** $-3x + 24$ **5.** $12x - 8$ **6.** $10x + 35$ **7.** $4x - 12$ **8.** $-7x + 28$ **9.** $2.8x + 11.48$ **10.** $-5.2x - 20.28$

Section A.7   **1.** $7x$ **2.** $2x$ **3.** $10x^2 + 6x$ **4.** $3x^2 + 11x + 2$ **5.** $7x^2 - 3x - 4$ **6.** $2x^2$ **7.** $6x - 1$ **8.** $-4x^2 - 6x + 14$ **9.** $4.1x^2 + 2.3x + 5.5$ **10.** $1.5x + 34.77$

Section A.8   **1.** 13 **2.** 12 **3.** 4 **4.** 1 **5.** 12 **6.** 4 **7.** 6 **8.** 1 **9.** 36 **10.** 40 **11.** -9 **12.** 9 **13.** -8 **14.** 7 **15.** 6 **16.** -17 **17.** 20.64 **18.** 12.960625

Section A.9   **1.** 3 **2.** 12 **3.** -2 **4.** -1 **5.** $\dfrac{5}{2}$ **6.** 10 **7.** 8 **8.** 1.6 **9.** -13.9739 **10.** -1.62593 **11.** $\dfrac{1}{4}$ **12.** 7

Section A.10   **1.** $y = -2x + 8$ **2.** $y = 3x - 18$ **3.** $w = -\dfrac{2}{5}t + 2$ **4.** $w = \dfrac{1}{3}t + \dfrac{7}{3}$ **5.** $d = 0.4p + 3.24$ **6.** $d = -0.625p + 6.375$ **7.** $y = 2x + 3$ **8.** $x = \dfrac{1}{2}y - \dfrac{3}{2}$ **9.** $b = -\dfrac{1}{4}a + \dfrac{5}{2}$ **10.** $a = \dfrac{3}{8}b$ **11.** $y = \dfrac{2}{3}x - \dfrac{5}{6}$ **12.** $y = -\dfrac{9}{8}x - \dfrac{3}{8}$

Section A.11   **1.** Equivalent expressions   **2.** Equivalent equations   **3.** Neither   **4.** Neither   **5.** Neither   **6.** Equivalent expressions   **7.** Equivalent equations   **8.** Neither   **9.** Neither   **10.** Neither

Index